Lecture Notes in Mechanical Engineering

Series Editors

Fakher Chaari, *National School of Engineers, University of Sfax, Sfax, Tunisia*
Francesco Gherardini , *Dipartimento di Ingegneria "Enzo Ferrari", Università di Modena e Reggio Emilia, Modena, Italy*
Vitalii Ivanov , *Department of Manufacturing Engineering, Machines and Tools, Sumy State University, Sumy, Ukraine*
Mohamed Haddar, *National School of Engineers of Sfax (ENIS), Sfax, Tunisia*

Editorial Board Members

Francisco Cavas-Martínez , *Departamento de Estructuras, Construcción y Expresión Gráfica Universidad Politécnica de Cartagena, Cartagena, Spain*
Francesca di Mare , *Institute of Energy Technology, Ruhr-Universität Bochum, Bochum, Germany*
Young W. Kwon, *Department of Manufacturing Engineering and Aerospace Engineering, Graduate School of Engineering and Applied Science, Monterey, USA*
Tullio A. M. Tolio, *Department of Mechanical Engineering, Politecnico di Milano, Milano, Italy*
Justyna Trojanowska, *Poznan University of Technology, Poznan, Poland*
Robert Schmitt, *RWTH Aachen University, Aachen, Germany*
Jinyang Xu, *School of Mechanical Engineering, Shanghai Jiao Tong University, Shanghai, China*

Lecture Notes in Mechanical Engineering (LNME) publishes the latest developments in Mechanical Engineering—quickly, informally and with high quality. Original research or contributions reported in proceedings and post-proceedings represents the core of LNME. Volumes published in LNME embrace all aspects, subfields and new challenges of mechanical engineering.

To submit a proposal or request further information, please contact the Springer Editor of your location:

Europe, USA, Africa: Leontina Di Cecco at Leontina.dicecco@springer.com
China: Ella Zhang at ella.zhang@cn.springernature.com
India, Rest of Asia, Australia, New Zealand: Swati Meherishi at swati.meherishi@springer.com

Topics in the series include:

- Engineering Design
- Machinery and Machine Elements
- Mechanical Structures and Stress Analysis
- Automotive Engineering
- Engine Technology
- Aerospace Technology and Astronautics
- Nanotechnology and Microengineering
- Control, Robotics, Mechatronics
- MEMS
- Theoretical and Applied Mechanics
- Dynamical Systems, Control
- Fluid Mechanics
- Engineering Thermodynamics, Heat and Mass Transfer
- Manufacturing Engineering and Smart Manufacturing
- Precision Engineering, Instrumentation, Measurement
- Materials Engineering
- Tribology and Surface Technology

Indexed by SCOPUS, EI Compendex, and INSPEC.

All books published in the series are evaluated by Web of Science for the Conference Proceedings Citation Index (CPCI).

To submit a proposal for a monograph, please check our Springer Tracts in Mechanical Engineering at https://link.springer.com/bookseries/11693.

Wenwei Wu · Jun Ding
Editors

Proceedings of the 22nd International Ship and Offshore Structures Congress (Volume 2)

Specialist Committee Reports

 Springer

Editors
Wenwei Wu
China Ship Scientific Research Center
Wuxi, Jiangsu, China

Jun Ding
China Ship Scientific Research Center
Wuxi, Jiangsu, China

ISSN 2195-4356   ISSN 2195-4364 (electronic)
Lecture Notes in Mechanical Engineering
ISBN 978-981-95-2785-4   ISBN 978-981-95-2786-1 (eBook)
https://doi.org/10.1007/978-981-95-2786-1

© The Editor(s) (if applicable) and The Author(s) 2026. This book is an open access publication.

Open Access This book is licensed under the terms of the Creative Commons Attribution-NonCommercial-NoDerivatives 4.0 International License (http://creativecommons.org/licenses/by-nc-nd/4.0/), which permits any noncommercial use, sharing, distribution and reproduction in any medium or format, as long as you give appropriate credit to the original author(s) and the source, provide a link to the Creative Commons license and indicate if you modified the licensed material. You do not have permission under this license to share adapted material derived from this book or parts of it.

The images or other third party material in this book are included in the book's Creative Commons license, unless indicated otherwise in a credit line to the material. If material is not included in the book's Creative Commons license and your intended use is not permitted by statutory regulation or exceeds the permitted use, you will need to obtain permission directly from the copyright holder.

This work is subject to copyright. All commercial rights are reserved by the author(s), whether the whole or part of the material is concerned, specifically the rights of translation, reprinting, reuse of illustrations, recitation, broadcasting, reproduction on microfilms or in any other physical way, and transmission or information storage and retrieval, electronic adaptation, computer software, or by similar or dissimilar methodology now known or hereafter developed. Regarding these commercial rights a non-exclusive license has been granted to the publisher.

The use of general descriptive names, registered names, trademarks, service marks, etc. in this publication does not imply, even in the absence of a specific statement, that such names are exempt from the relevant protective laws and regulations and therefore free for general use.

The publisher, the authors and the editors are safe to assume that the advice and information in this book are believed to be true and accurate at the date of publication. Neither the publisher nor the authors or the editors give a warranty, expressed or implied, with respect to the material contained herein or for any errors or omissions that may have been made. The publisher remains neutral with regard to jurisdictional claims in published maps and institutional affiliations.

This Springer imprint is published by the registered company Springer Nature Singapore Pte Ltd.
The registered company address is: 152 Beach Road, #21-01/04 Gateway East, Singapore 189721, Singapore

If disposing of this product, please recycle the paper.

Preface

The first volume contains the eight Technical Committee reports, and the second volume contains the reports of the eight Specialist Committees, presented and discussed at the 22nd International Ship and Offshore Structures Congress (ISSC 2025) in Wuxi (China), on September 22–26, 2025. The Official Discussers' reports and all floor discussions, including the replies by the committees, will be published after the Congress in electronic form.

The Standing Committee of the 22nd International Ship and Offshore Structures Congress comprises:

Chairman

Wenwei Wu — China

Co-chairman

Josko Parunov — Croatia
Yordan Garbatov — Portugal
Alex Babanin — Australia
Kim Branner — Denmark
Kazuhiro Iijima — Japan
Henk den Besten — Netherlands
Xiaozhi Wang — US
Bernt Leira — Norway
Myung-Hyun Kim — Republic of Korea
Patrick Kaeding — Germany
Sime Malenica — France
Andrea Ivaldi — Italy
Feargal Brennan — UK
Jan Czaban — Canada
Murilo Augusto Vaz — Brazil

On behalf of the Standing Committee, we would like to thank the sponsors of ISSC 2025.
Thanks to our Sponsors for their financial contribution.

September 2025

Wenwei Wu
Chairman

Jun Ding
Secretary

Contents

Committee V.1: Limit States During Accidental and Damage Conditions 1
   Kristjan Tabri, Michael Rye Andersen, Gaetano De Luca,
   Do Kyun Kim, Rafet Emek Kurt, Herve Le Sourne, Thomas Lindemann,
   Lucas Marquez, Tom Mitchell Ferguson, Jakub Montewka,
   Bianca de Carvalho Pinheiro, Bruce W. T. Quinton, Smiljko Rudan,
   Carey Walters, Yasuhira Yamada, Zhaolong Yu, and Ling Zhu

Committee V.2: Experimental Methods 92
   Paul Lara, Ilson Paranhos Pasqualino, Bin Liu, Pero Prebeg,
   Mikko Suominen, Sören Ehlers, Cesare Mario Rizzo,
   Hideaki Murayama, Apostolos Grammatikopoulos, Sigmund K. Ås,
   and Seo Jung Kwan

Committee V.3: Subsea Technology 170
   Chen An, Tauhid Rahman, Celso Kazuyuki Morooka, Carlos Montoya,
   Yoshihiro Konno, Svein Sævik, Chunsik Shim, Cihan Bayindir,
   Kourosh Parsa, and Yi Xia

Committee V.4: Offshore Renewable Energy 254
   Charles Rawson, Nagi Abdussamie, Asger Bech Abrahamsen,
   Ivan Catipovic, Thomas Choisnet, Toshiki Chujo, Jose Gaspar,
   Luca Greco, Ole A. Hermundstad, Han-Koo Jeong, Shen Li,
   Wengang Mao, Rachel Nicholls-Lee, Gabriele Notaro, Elif Ogus,
   Freeman Ralph, and Shali Sun

Committee V.5: Special Vessels 359
   E. Begovic, J. Ali-Lavroff, D. Boote, A. Egorov, M. Rodriguez,
   A. Salcedo, J. Sawamura, H. Seyffert, J. B. Souppez, L. Toshkov,
   N. Vladimir, and F. Wang

Committee V.6: Ocean Space Utilization 442
   Chao Tian, Harry Bingham, Ciro Busiello, Ingo Drummen,
   Nuno Fonseca, Zhiqiang Hu, Debabrata Karmakar, Ekaterina Kim,
   Sarat Mohapatra, Motohiko Murai, and Robert Sielski

Committee V.7: Structural Assessment During Operations 533
   J. M. Underwood, A. Barbato, A. Bekker, M. Braun, M. A. Eder,
   P. Hess, C. Jochum, D. Morato, V. Nilva, N. Osawa, P. Pahlavan,
   D. Sarsoza Burgos, Â. P. Teixeira, I. Thompson, and Y. Wang

Committee V.8: Uncertainty Modelling in Waves and Wave Responses 626
Carlos Guedes Soares, Elzbieta Bitner-Gregersen,
Apostolos Papanikolaou, Josko Parunov, Wei Qiu, Tomoki Takami,
Solomon Yim, Xueliang Wang, Takuji Waseda, and Huidong Zhang

Author Index 699

Committee V.1: Limit States During Accidental and Damage Conditions

Kristjan Tabri[1(✉)], Michael Rye Andersen[2], Gaetano De Luca[3], Do Kyun Kim[5], Rafet Emek Kurt[6], Herve Le Sourne[7], Thomas Lindemann[8], Lucas Marquez[9], Tom Mitchell Ferguson[4], Jakub Montewka[10], Bianca de Carvalho Pinheiro[11], Bruce W. T. Quinton[12], Smiljko Rudan[13], Carey Walters[14], Yasuhira Yamada[15], Zhaolong Yu[16], and Ling Zhu[17]

[1] Tallinn, Estonia
kristjan.tabri@taltech.ee
[2] Technical University of Denmark (DTU), Lyngby, Denmark
[3] Istituto Nazionale di Geofisica e Vulcanologia, Rome, Italy
[4] Australian Maritime College (AMC), Launceston, Australia
[5] Seoul National University, Seoul, Republic of Korea
[6] University of Strathclyde, Glasgow, United Kingdom
[7] ICAM School of Engineering, Nantes, France
[8] University of Rostock, Rostock, Germany
[9] Liège, Belgium
[10] Gdansk University of Technology, Gdansk, Poland
[11] Federal University of Rio de Janeiro (UFRJ), Rio de Janeiro, Brazil
[12] Memorial University of Newfoundland, St. John's, Canada
[13] University of Zagreb, Zagreb, Croatia
[14] Delft, Netherlands
[15] Tokyo, Japan
[16] Trondheim, Norway
[17] Wuhan, China

Committee Mandate. Concern for limit states of ships and offshore structures and their structural components under accidental and damage conditions during operations. Types of events considered shall include collision, grounding, dropped objects, explosion, fire, and operational conditions beyond the design envelope. Hazard identification, accidental loads and nonlinear structural consequences including resilience and resulted risks shall be considered. Special attention is to be given to methods and assessment considering structural resilience, operability and survivability. Unmanned vessels and offshore wind farms shall be considered.

Keywords: Accidental Limit States · Structural Resilience · Operability · Survivability · Collision · Grounding · Dropped Objects · Explosion · Fire · Hazard Identification · Risk Assessment · Accidental Loads · Nonlinear Structural Consequences

1 Introduction

Limit States during Accidental and Damage Conditions indicates the unexpected conditions the ship or marine structures experience beyond its typical operational state. These conditions are typically associated with loads exceeding the operational loads the ships are designed to withstand without any or significant permanent deformations or damages. Events leading to accidental, or damage conditions can include collision, grounding, dropped objects, explosion, fire, or any other operational conditions beyond the design envelope. Potential hazards are to be identified in the design stage and the associated risks, loads, structural response, and consequences to be evaluated. Without the ability to foresee and predefine all the possible extraordinary conditions, the ships still must be designed with the resilience and redundancy to provide the ability to survive in these conditions.

The report covers a range of topics related to accidental limit states, including collision, grounding, and fire, as well as discussing new areas such as human errors, resilience, and emerging technologies like offshore wind turbines, underwater infrastructure, and maritime autonomous surface ships (MASS). Additionally, the concept of structural resilience is explored in relation to ship and structure operation in conditions beyond their design parameters. The time range of publications for newer topics is broader compared to conventional topics, to offer a more comprehensive overview.

The report primarily addresses limit states in the event of accidents and damage, with efforts made to avoid overlap with other committees. This includes a brief discussion of residual strength after accidents to prevent redundancy with committee III.1 Ultimate Strength. Additionally, emerging technologies are briefly introduced and primarily evaluated in terms of their potential impact on accidental and damage conditions, to avoid overlap with V.4 Offshore Renewable Energy.

Section 2 focuses on the prevention and planning rules to minimize the occurrence and consequences of unexpected conditions. Hazard identification and risk analysis are discussed together with the role of human errors and safety learning from past incidents and accidents. Assessment methods from legislative perspective are briefly described both for offshore and ship structures. Consequence and mitigation assessment is described in Sect. 3, where experimental, theoretical and numerical methods are covered. The damages to underwater pipelines have received special attention in light of recent events involving critical infrastructure. Section 4 focuses on the resilience, operability and survivability in accidental and damaged conditions, and in conditions where structures are operating beyond the initial design scope. Section 5 discusses the new and emerging technologies such as uncrewed vessels, offshore infrastructure, and alternative fuels from the context of accidental and damaged conditions. Each main chapter concludes with recommendations on its respective scope. Summary and conclusions are provided in Sect. 6.

Report is concluded with a benchmark study carried out by some of the committee members and deals with fluid-structure interaction assessment of slamming loads yielding to plastic deformations of the structure. Experimental plate drop-tests are used to assess the effect of numerical simulation practices.

2 Prevention and Planning

Planning cannot fully prevent accidents to occur, but the consequences of extreme event scenarios can be reduced by designing resilient structures. Rules and standards provide the tools and guidance for the design of resilient structures constructed to minimize/mitigate the effects of hazards including the planning needed to adapt to longer-term changes in use or the environment. In the face of the diverse range of possible hazards for which a structure can be exposed to, a resilient structure is designed with the abilities to:

- prepare for extreme event scenarios,
- resist the effects of extreme events.
- respond to these effects,
- minimize consequences/survive/recover from such effects in a timely and efficient and most important safe manner.

Effective preparatory risk management is essential for identifying potential hazards and threats. It can facilitate development of appropriate abilities before an extreme event, which aid an adequate response to events and help with successful recovery from them. It also allows for adaption to evolving requirements and the effects of a changing environment. Finally, in-service integrity management play an important role.

This chapter gives an overview of the principles, rules, and standards applicable to design for accidental events for ships and offshore structures and identifies known shortcomings and problems with the current approaches. Where applicable, guidance and recommendations are also provided. Design standards provide good explanations of the principles underlying the design for accidental events and are referenced in this chapter. Accidental events are addressed differently in the design of offshore facilities, ships, and other assets like fixed and floating wind farms.

An important way of improving safety is to learn from accidents which has occurred. An example is the Estonia disaster, which initiated that eight European countries decided to require stricter requirement for of damage stability for ro-ro passenger ships than those prescribed by SOLAS 90. These new standards were introduced in the context of the Stockholm Agreement (Maritime Administrations of North West Europe, 1996) to which the above eight countries became parties. General, techniques for learning from accidents and "near misses" are therefore also discussed in the following as these may lead to improvement of the codes and standards.

2.1 Hazard and Risk Analysis

For all operators of ships and offshore facilities, safety should be the primary concern. The full lifecycle is considered for an offshore asset, and most national regulators require that all hazards and corresponding risks associated with the design, fabrication, installation and operation shall be minimized by adhering to governing design codes, and recognized standards. The codes and standards provide the tools for structured evaluations of hazards and corresponding risks, and as stated in DNV-OS-A101 (2023), this is with the purpose of establishing an acceptable level of safety, whilst promoting safety improvements through experience and available technology. Following the latest revision of the codes, the operators will thereby apply latest knowledge when assessing their structures.

For offshore structures, a standard limit state designs will normally consider service, fatigue and ultimate limit states, but also an accidental limit state which is based on generic Design Accidental Loads (DALs). The aim of the generic loads is to include additional safety for the accidental limit state considered acceptable by code. In DNV-OS-A101 (2019), the DALs are defined so that they affect safety functions which have an individual (per load type) frequency of occurrence in the order of 10^{-4} per year corresponding to an overall frequency of $5 \cdot 10^{-4}$ per year as the impairment frequency limit. If the "standard" DALs defined by the code are too strict for practical use, more refined analyses such as computational flow dynamics (CFD) analyses for fire and explosion loads, and probabilistic risk analyses may be carried out to set the DALs. For complex or non-standard applications, the use of generic DAL will normally have to be supported by a more comprehensive safety assessments which involve hazard assessment. Hazard assessments shall, as a minimum, focus on hazards that could directly, or indirectly, result in: loss of life, major fire or explosion, loss of structural integrity or control, the need for escape or evacuation, environmental impact and the hazard assessments can typically be split into Hazard; identification, reduction, evaluation, risk analysis comparing with safety target and criteria and verifying that dimensioning loads and assumption are included into the final evaluation of the structure.

Formalized approaches for hazard analysis can be found in several codes also including API RP 14J (2001), which presents the recommended practice for hazards analysis for Offshore Production Facilities. Applying the terminology of API RP 14J, hazards analysis is a systematic procedure for identifying, evaluating and controlling potential hazards in a facility. A hazards analysis program should be applied to all phases of the life of a facility from project inception through abandonment to assess potential hazards during design, construction and operation. An important note in API RP 14J is the minimum and acceptable hazards analysis for offshore production facilities is a check for compliance with standard practice (first step - hazard control). For this purpose, a checklist is commonly used to verify compliance and to identify areas that require further evaluation. The second step is predictive hazard evaluation, which is required for processes without sufficient previous experience or that present an unusually high risk this step may include "What if", HAZOP studies, Failure Mode and Effect Analysis and Fault Tree Analysis are all discusses in API RP 14J.

NORSOK Z-013 (2024), i.e., guidance on "risk and emergency preparedness assessment" has undergone a major update in 2024. The updated standard now proposes and describes methods for risk analysis for typical major accident hazards with the aim of obtaining more standardisation. For several of the typical major accident hazards, both simplified and detailed methods are described. The focus for the simplified methods is on providing sufficient decision support for safe design and not on detailed quantitative analysis of risk.

The above focusses on offshore structures, but the principles of outlined approaches could in general terms also be applied for ship. More specific discussion of risk and safe passage for ships is discussed in the following sections.

2.2 Human Factors

Despite its reactive nature, learning from accidents and incidents remains the main source of safety improvement in the maritime domain. Human factors (HF) significantly influence maritime safety, contributing to about 80% of accidents and incidents. To address these issues, structured HF taxonomies like the HFACS, STAMP, and FRAM are employed to facilitate deeper learning from past incidents. This review uses the Preferred Reporting Items for Systematic Reviews and Meta-Analyses (PRISMA) framework to identify relevant studies in this field. A keyword search in the ScienceDirect database yielded 35 peer-reviewed articles published between 2020 and 2024 that employ human factors taxonomies focused on maritime accidents or incidents.

2.2.1 Human Factors Analysis and Classification System (HFACS)

Developed by Shappell and Wiegmann (2000), HFACS is widely used to investigate maritime accidents like groundings and collisions. (Maternová and Materna, 2023) analysed reports from the Maritime Accident Investigation Branch (MAIB) and the Transportation Safety Board (TSB) and concluded that incidents often involve multiple causal factors. Variations of HFACS, such as HFACS-MA, HFACS-PV, and others, have been developed to improve performance in specific contexts (Kaptan et al., 2021a). Yildiz et al. (2021) applied HFACS-PV to analyse grounding accidents of passenger vessels, noting its flexibility and effectiveness, which enables it to be combined with quantitative analysis. Bayesian networks are frequently combined with HFACS for risk quantification. Several studies (Adumene et al., 2022; Bayazit and Kaptan, 2024; Wang et al., 2024) investigated human and organizational factors influencing specific incident types, effectively handling incomplete and sparse data which is commonly found in the maritime industry. Kaptan et al. (2021b) and He et al. (2022) also used Bayesian approaches to evaluate errors and relationships associated with navigational devices and unsafe acts.

HFACS has also been integrated with human reliability assessment (HRA) methods. Qiao et al. (2024) combined HFACS with the Human Error Assessment and Reduction Technique (HEART), adapting error-producing conditions to a maritime context. Kandemir and Celik (2021) used HFACS-MMO to determine EPCs, extending the Ship Operations Human Reliability Analysis (SOHRA). Aydin et al. (2022) integrated HFACS-PV with the Success Likelihood Index Method (SLIM) to assess pilot transfer operations. Furthermore, some research studies focused on exploring interrelationships between causal factors. Ma et al. used rule mining and Bayesian networks to evaluate relationships within HFACS levels (Ma, Chen, et al., 2024a), (Ma et al., 2024b). Zhang et al. (2019) and (Ma et al., 2024b) extracted risk factors and created complex networks. Bicen and Celik (2022) clustered causal factors using the analytical network process (ANP), while Ma et al. (2022) employed the decision-making trial and evaluation laboratory (DEMATEL) to rank human factors. Lan et al. (2022) used association rule mining to uncover relationships between human factors and accident characteristics.

Integration of HFACS with fault tree analysis (FTA) is also common. Sarıalioğlu et al. (2020) combined HFACS-PV with fuzzy FTA to analyse human factors in engine-room fire incidents. Qiao et al. (2020) extended this approach by integrating fuzzy FTA with artificial neural networks (ANN). Review papers highlight HFACS's widespread use due to its clarity in identifying main human factors (Shi et al., 2021). Wu et al. (2022) attribute its popularity to simplicity and foundation in Reason's model, suggesting that deep learning algorithms could enhance human error analysis. Cao et al. (2023) conducted a bibliometric analysis, concluding that HFACS represents the largest knowledge area in maritime accident research.

2.2.2 Functional Resonance Analysis Method (FRAM)

FRAM describes accidents from the perspective of functions and their aspects, useful for understanding procedural variations. Nasur et al. (2025) combined FRAM with HFACS-MA to model emergency unmooring of manned and autonomous ships. Salihoglu and Bal Beşikçi (2021) used FRAM for oil spill accident analysis, finding it more reliable than conventional methods but noting its subjectivity and recommending it to be combined with other methods to ensure reliability. Peng et al. (2023) applied FRAM to analyse human-automation interaction in offshore operations.

2.2.3 The System Theoretic Accident Model and Processes (STAMP)

STAMP, developed by Leveson (2004), is suited for complex socio-technical systems. Using the theoretical base of STAMP, CAST provides a holistic view of events. Ceylan et al. (2021) used CAST to analyse ship allision accidents, emphasising that accidents are not simple cause-effect chains. Ceylan et al. (2022) employed STAMP to evaluate marine diesel engine incidents, integrating Fuzzy Failure Mode and Effects Analysis (FMEA) to quantify risks.

2.2.4 Other Taxonomies

Taxonomies like the MAIB Human Factors Taxonomy and the IMO taxonomy are crucial due to extensive data on industry issues. Coraddu et al. (2020) integrated machine learning with a database of accidents generated by MAIB's taxonomy. They aimed to determine the influencing factors and predict accident types. Results revealed that inadequate procedures and deviations from standard operating Procedures (SOPs) were among the top causes of accidents, demonstrating the importance of the organisations' role in safety. The IMO taxonomy was used by Fedi et al. (2024) to analyse 20 years of incidents along the North-East Passage, categorising causes but highlighting issues with under-reporting and inadequate data in the GISIS database.

The Swiss Cheese Model by Reason (1990) visualises organisational defences, with accidents occurring when failures align. Chen et al. (2020) expanded on this by integrating it with the SHEL framework (Software, Hardware, Environment, Liveware) to provide a detailed understanding of system interactions. There are also new taxonomies developed by researchers. The Safety Human Incident and Error Learning Database (SHIELD) is one of them, which was created during the EU-funded SAFEMODE project, aiming for systematic reporting to prevent incidents through policy and design changes. SHIELD has shown the potential to uncover more causal factors than traditional approaches (Stroeve et al., 2023) and has been utilised by numerous stakeholders, including the European Maritime Safety Agency. Other studies have also created their own classification systems, similar to SHIELD. Fan et al. (2020a) created a taxonomy for Maritime Autonomous Surface Ships (MASS), identifying human, ship, environment, and technology-related factors influencing incidents, though acknowledging the need for updates and fuller consideration of organisational factors.

To conclude, the review shows an exponential increase in the number of research papers published on human factors in maritime topics. This review, therefore, focused on HF taxonomies and their application in marine accident learning. It was observed that different methods demonstrate different strengths. HFACS and its derivatives are the most popular accident classification tools utilized in maritime accident investigation studies, focusing on risk assessments, statistical analysis, and establishing causal relationships etc.

STAMP-CAST is another approach well-suited for examining complex socio-technical systems. Its emphasis on system design and operations makes it a valuable tool for analyzing the interplay of various factors within complex maritime systems. FRAMs key strength lies in its capacity to model complex interactions within maritime operations and became popular in studying complex human system interactions, especially autonomous or remote operated vessels with highly digitalized systems.

The maritime industry is undergoing a transformative shift driven by decarbonization, autonomy, artificial intelligence, and digitalization, presenting new challenges in how we learn from accidents. The need for standardized taxonomies has become more critical in light of these emerging complexities. Without such standardization, inconsistencies in classifying human factors errors risk undermining efforts to establish effective, industry-wide best practices. Therefore, it is expected that HF will further increase its popularity in maritime in the upcoming years.

2.3 Safety Learning for Incidents and Accidents

In this section different techniques of safety learning are discussed. Various indicators are used to measure the safety of maritime transport systems, which can be categorised into two broad groups: *reactive* and *proactive* (Kretschmann, 2020). The former focus on a type of measurement called "after the loss", which refers to past situations, such as accident or injury rates, incidents, monetary losses, which are therefore known as negative indicators (Kretschmann, 2020). Despite the simplicity and straightforward nature of data collection, the main disadvantage is that the occurrence of the assumed indicators (e.g. an accident) is rare and usually occurs with a time delay in terms of the causes or reason for it. This in turn makes it extremely difficult to react quickly or implement corrective measures. Collecting statistically meaningful data is tedious and can lead to the use of outdated and irrelevant data for the conditions under which the system is operating (Zhou X et al., 2021b), which is useless from a risk and safety management perspective.

The proactive methods for assessing the safety of socio-technical systems focus on measuring so-called positive or leading indicators, which are used to assess the compliance of the system with its current specification (Kretschmann, 2020). Leading indicators are precursors to accidents, defined as "conditions, events or actions that precede an undesirable event and have some value in predicting the occurrence of the event, be it an accident, incident, near miss or undesirable safety condition" (Govindaraj,

2008). They are not actual measurements of future states, but an interpretation of past and current states in terms of what might happen in the future. Leading indicators are associated with proactive activities to identify and assess hazards and to minimize and control risks.

As leading indicators are an interpretation of the future, they are inherently subject to irreducible uncertainty which, if too great, can be misleading rather than helpful. Therefore, the leading indicator must be carefully selected to remain reliable and valid. Such a selection process can be facilitated by the well-known SMART framework (Hollnagel, 2018), where a safety indicator must be specific, measurable, achievable, relevant and time-bound. If an indicator is specific, it must be narrow and describe exactly what needs to be measured, in other words, the indicator must be valid (Podgórski, 2015). Measurable means that the indicator can be measured in the same way regardless of who uses it. Accessible means that the data collection process remains simple and inexpensive. Relevant means that the indicator must be linked to the corresponding result. Finally, time-bound means that a time frame is linked to the indicator, e.g. the frequency with which it is collected or measured.

However, there is no industry or academic standard for defining or measuring leading indicators for maritime transport. One of these indicators, the so-called near miss, is often assessed in a conventional way using different types of proximity indicators (Wróbel et al., 2023). The most common proximity indicators, which are also used in everyday navigation and therefore easily interpreted by end users, are the distance to the closest point of approach (CPA) and the time to closest point of approach (TCPA), see Chin and Debnath (2009) and Varas et al. (2017). Alternatively, the concept of the ship domain is applied in Szłapczyński & Szłapczyńska, (2017), Zhang and Meng (2019), Pietrzykowski and Wielgosz (2021), Im and Luong (2019), Rawson and Brito (2021) and Zheng et al. (2020). The overarching goal of these indicators is to delineate a zone (spatially, temporally or both) in which a navigator feels comfortable and which they want to keep away from other objects to ensure safe passage of their own vessel. To this end, a navigator performs the necessary manoeuvres according to best practise and company regulations (Pietrzykowski et al., 2020). However, the zone itself does not quantify safety.

Zone-based methods are also used in a probabilistic way, e.g. by fuzzy systems based on expert judgement and big data analysis (Zhou and Zheng (2018); Rong et al. (2021)). However, these types of models are often based on preferences rather than evidence (Mehdi et al. (2017)), so they mislead potential end users and ultimately lead to an underestimation of accident susceptibility, see Huang and van Gelder (2020).

Qualitative, probabilistic methods can measure the risk of an accident at sea by determining the potentially accident-prone locations and the associated frequency of these accidents, as well as the proximity indicators mentioned above. These methods are generally useful for strategic spatial risk assessment for waterways (Rawson and Brito, 2021) but do not adequately account for the factors known to determine a navigator's performance in an encounter situation, such as the complexity and density of nearby traffic, proximity to shallow waters or environmental conditions (Corporate Risk Associates, (2011); Ramos et al. (2020); Fan et al. (2020a)). For an overview of the methods, see for example Goerlandt and Montewka (2015) and Du et al., (2020).

Reducing the issue of safe passage to the inviolability of the domain does not seem to fully analyse the safety factors that play a role in the holistic determination of the risk of a collision between ships. When assessing the risk of a collision between vessels, several interdependent factors must be considered, including ship-ship encounter geometry, environmental conditions and manoeuvrability, all of which can affect human performance. In the event of a collision, the perceived risk can vary depending on individual factors such as experience or licence class, as D.-H. Kim (2020a) shows. According to D. H. Kim (2020b), after analysing hundreds of collision cases, the bearing and distance to the target vessel at which the risk starts to increase is significant from a vessel operator's perspective. Sotiralis et al. (2016) conducted a study on human performance and its impact on quantitative risk analysis. The aim of the study was to calculate the probability of collision accidents and identify the main factors contributing to such incidents. This approach provided valuable insights into the role of human performance in ship collisions. These factors need to be considered in order to develop a complete picture of the encounter and provide end users with an appropriate level of situational awareness as shown by D.-H. Kim (2020a).

From the above, it can be concluded that the literature lacks an intuitive, objective, evidence-based method for assessing the level of safety for a particular ship-to-ship encounter at sea. The main driving factor for accident susceptibility is the performance of a navigator, while the literature indicates that the human factor contributes to accidents at sea (Wróbel (2021)) and that the poorer the navigator's performance, the greater the likelihood of an error and thus an accident (Corporate Risk Associates, 2011).

Navigator performance depends on several factors, one of which is mental workload, which tends to increase with the complexity of the situation (van Westrenen and Baldauf (2020). The literature on situational complexity in the maritime sector remains scarce and the topic has not yet been investigated, see Du et al. (2020). Occasionally it appears in research on strategic risk assessment for waterways (Montewka et al. (2017)), tactical conflict detection (van Westrenen and Baldauf (2020)) or operational risk assessment (Cruise Ship Safety Forum (2016)). Even then, complexity is usually defined arbitrarily and used as an explanatory variable in the risk analysis without an in-depth analysis of the variable itself. As a result, the factors that determine complexity and their impact on complexity remain unexplored.

In contrast, more extensive research has been conducted on this topic in the field of aviation safety, which has led to a detailed assessment of the factors influencing encounter complexity and its impact on human performance (Dmochowski and Skorupski (2019)). To close the knowledge gap in the maritime sector, we shall combine the existing background knowledge from aviation with the knowledge of maritime experts and the existing, albeit sparse, maritime literature.

2.4 Assessment Guidance, Codes and Standards – Offshore Structures

The offshore code and standards provide well established guidance for assessment of Accidental Limit States based on generic Design Accidental Loads (DALs). Using the nomenclature of ISO 19902 (2020), the typical hazards can be divided in to:

- hazards associated with specially identified accidental events, e.g., dropped objects, vessel collision and Fire loads and explosion.

- hazards associated with abnormal environmental events, e.g., extreme wave events including Wave in Deck (WID) assessment and seismic activities.

The two types of hazards are different by nature. In principle, accidental events can in some cases be avoided by taking appropriate measures to eliminate the source of the event or by bypassing and overcoming its structural effects. In contrast with this, the possible occurrence of abnormal events cannot be influenced by taking such measures. The assessment of the accidental state (ALS) normally comprises the occurrence of an identified accidental event or of an abnormal environmental event, in combination with expected concurrent operating conditions and associated permanent and variable actions. When considering the design of pipelines, DALs related to internal overpressure and trawl/anchor impact must also be assessed following e.g., guidelines in DNV-RP-F107 (2021). High pressure/temperature wells have caused a need for guidance on how to design pipelines systems for such fields without being overconservative but still maintaining the required safety of the system. With the aim of closing this gap, the frequency dependent pressure containment (FDPC) limit state is introduced in DNV-RP-F117 (2023) where incidents occurring with a frequency of less than 10^{-2} are considered as accidental.

The codes are in general quite mature with respect to requirements for definition and assessment of generic accidental loads, see e.g. the guidelines ISO 19902 (2020), DNV-OS-A101 (2019), NORSOK N-004 (2022) and API2A-LRFD (2019) guidelines. In following focus will be on where non-conservative have been reported and areas where improvements have been made due to advances in technology.

2.4.1 Dropped Objects

Quantitative Risk Assessment (QRA) methods such as the "Ring Risk Model" described in DNV-RP-F107 (2019) are often applied for the design of pipeline systems for possible dropped objects, see (Xiang et al., 2023). The methods applied are based on simplified and practical approach for the determination of the landing point. A general determination of the landing point can be divided into three stages: air falling stage; water entry stage and underwater sinking stage where the first two stage are relatively described. The work of (Xiang et al., 2023) therefore focusses on the third stage where authors provide a comprehensive review and analysis of the deterministic and stochastic dropping processes associated with dropped objects in the field of offshore engineering. The deterministic process of dropped objects involves trajectory measurement, prediction, and optimization, while the stochastic process involves trajectory envelope and landing point predictions during free fall, and collision risk assessment. The review provides a good overview and discusses the development history and application status of several research methods involved in the study of stochastic processes, including Monte Carlo simulation and Kalman filtering.

Accurate determination of resistance is also play an important role in the assessment of dropped objects. Two common approaches are experimental testing and assessment by non-linear explicit finite element analysis (NLFEA). Digital Image Correlation (DIC) technique provides the tools for such validations and Oka et al. (2024) presents an assessment of nuclear material container drop test using explicit finite element analysis (FEA) which is validated using DIC. Several drops at various heights are tested and assessed by FEA and predictions in form of strain behaviour and peak strains were examined comparing FEA and DIC. A good comparison was observed. More discussions on importance of using validated approaches for non-linear FE analyses including presentation of a calibrated tensile strain criterion for shell elements can be found in two papers by Eriksson et al. (2024) and Eriksson et al. (2023).

2.4.2 Vessel Collision with Offshore Structures

The advancement of high-performance computers and the availability of sophisticated software makes NLFEA a powerful tool for the analysis of large displacements involving geometric and material nonlinearities including assessments Offshore Supply Vessel (OSV) colliding with fixed-type offshore platforms. There are however still several challenges and choices to made as illustrated in the study presented by (Mujeeb-Ahmed et al., 2020) Mujeeb-Ahmed et al. (2020) who presented FEA based approach. Based on their work they concluded that the computational models can ultimately be employed for quantitative risk assessment of fixed-type offshore structures collided with an OSV. In the absence of experimental test studies, the validation of their work was based various analytical and empirical expressions available for the structural behaviour of tubes subjected to the lateral collision load of a striking body.

The worldwide increasing growth of offshore wind turbine monopile foundations is a catalyst for new solutions for traditional problems. An example is a steel fender assembly with quasi-negative stiffness (QNS), which provides a system of High Stiffness for low general loads and a low stiffness for high-impact collisions with the purpose of mitigating ship collisions with offshore wind turbine (OWT) foundations proposed by (Nie et al., 2024). The dynamic behaviour of OWT monopile foundations equipped with a HGLCS fender modules was investigated. Via the utilization of Hamilton's principle, a nonlinear model of an OWT foundation incorporating the proposed innovative fender was derived and characterized. Using HGLCS fender significant decreases in maximum collision force and corresponding acceleration were found and the fender is now installed on an OWT foundation built in Guangdong Province, China.

2.4.3 Fire Loads and Explosion

CFD is a commonly applied tool for relatively accurately description the complicated interaction between obstacles-inducing turbulence and combustion during gas explosion process considering various congestion levels. It does however have the disadvantage of being computationally intensive. With the aim of developing a tool for "real-time" natural gas explosion consequence reconstruction, J. Shi et al. (2023a) have proposed a hybrid deep learning probability model to real-time predict spatial explosion overpressure of offshore platform by using sparsely observed overpressures. The validation of the model is based on benchmarking towards experiment work, sensitivity analysis of Monte Carlo sampling number N, drop probability p on model's performance is also conducted demonstrating a $R2 = 0.955$ and real-time capability with inference time of 2.9s.

2.4.4 Wave-in-Deck Events – Current Industry Practices

In recent year there has been an on-going discussion of how best to calculate the reliability of a fixed offshores structure driven where improved physical understanding of waves arising in realistic design sea-states has been the main driver. It has been realised that that many 'design wave events' will be breaking irrespective of water depth leading to the observation that some aspects of present design practice may be non-conservative. (Ma and Swan, 2023) discuss the current practice for determination of WID loads applying a large laboratory data base of WID events as basis and they find that the recommended practice, including recent updates, consistently under-predict the maximum WID loads. Included in the evaluation is their own recently developed Lagrangian Momentum Absorption (LMA) model which combines fully non-linear wave inputs and the openness/porosity of a structure yielding highly accurate predictions when compared to laboratory data base of WID events. Further discussion on prediction of WID loads including the effect of the porosity (or openness) of the topside structure and the importance of the position, orientation and size of the topside structure relative to the incident wave crest can be found in (Ma and Swan, 2023).

2.5 Assessment Guidance, Codes and Standards – Ship Structures

Guidance for assessing limit state for accidental and damages conditions of ships is provided by various rules as described below, although there is no all-encompassing standard. Few research publications are discussed as they contribute to the interpretation of more specific rules.

ABS Guide for Fire-Fighting Systems for Cargo Areas of Container Carriers identifies the required criteria for obtaining the optional notations for Container Carriers. While focusing on fire-fighting systems, in Appendix 2 a requirement is given for Container Carriers with Container Hold Flooding (CHF) notation. A hull girder strength needs to be evaluated for specified flooding conditions, and it is assumed that the hull structure shall remain fully effective in resisting the applied loading.

ABS Guide for Dropped Object Prevention on Offshore Units and Installations provides requirements for the classification of the ships by notations DOPP and DOPP+, abbreviations for the dropped objects prevention plan. The guide defines a dropped object risk assessment process in three steps: risk identification, analysis and evaluation. The accident scenarios include dropped objects in various situations (lifting, drilling, installation, maintenance), the failure of the equipment resulting in a drop of the object etc.

DNV-RP-E304 (2019) provides assessment basis for the suitability of a polyester mooring rope to remain in service, after it has been mechanically damaged by external objects. The result of the damage assessment is a revised minimum breaking strength MBS, denoted MBSDAMAGED, and is used to determine if a damaged rope is suitable for temporary or prolonged service. Three levels of damage assessment are defined, from simplest to complex, of which level 3 also includes fatigue-life assessment.

DNV-RP-D102 (2022) recommends developing a failure mode and effect analysis (FMEA) document that may be used towards class acceptance of redundant systems. The objective of the redundancy design intention is to specify the redundancy, i.e. to describe the distribution of systems and components into redundant groups. Fire and flooding redundant systems, for example, aim to prevent failure propagation due to fire and flooding events. Redundant and separated zones should be described in detail, using drawings, equipment list and other documents.

DNV-CG-0308 (2021) aims to provide for safe operation and protection of the environment in polar waters by addressing risks and hazards identified for the area and period of operation. Freezing of the ballast and freshwater tanks may lead to expansion of the fluid and falling of heavy ice blocks in the tank after discharging. Ice damage extent should be considered and assumed when demonstrating compliance. Overall, the challenge is to evaluate the severity of the ice condition ahead versus the limitation and the capability of the vessel and for that purpose a Polar operational limit assessment risk indexing system (POLARIS) is proposed. Bergström et al. (2022) applied goal-based approach complementing POLARIS methodology for selecting a suitable Polar Class for an ice-going ship. As a case study, the calculated design loads are compared to Polar Class loads and it is concluded that the ship hull should be built to PC2 i.e. one level higher than the minimum required ice class derived by POLARIS. Kujala et al. (2023) further elaborated goal-based ship design approach, emphasizing the requirement of relevant and validated design tools, namely: (a) accident statistics, (b) a framework for risk-based design of Arctic ships, (c) a simulation model of a winter navigation system, and (d) a sustainability assessment tool for Arctic shipping. A model for holistic optimization of Arctic ships is presented. Suominen et al. (2024) performed a probabilistic analysis of operational ice damage for Polar class vessels. The study suggests that the recommendations provided by the POLARIS methodology are generally reasonable although additional research is required related the unifying the design criteria of different areas, such as non-bow regions.

IACS (2023) Unified Requirements for Polar Class Ships states that the ship hull structures are typically and traditionally designed to remain elastic under design loads, i.e. no permanent hull damage occurs under normal service conditions. However, the IACS Unified Rules for Polar Class (IACS, 2023) employ a plastic design point, i.e. the normal service design ice load will induce some permanent hull deflection. It was not understood at the time that sliding ice loads induce greater structural damage than equivalent stationary loads (Quinton, 2015). Since the IACS design ice load is a glancing stationary load, it inherently does not consider the sliding load effect and therefore underpredicts permanent hull damage (Quinton, 2019).

2.6 Required Residual Strength

It is a general requirement from the offshore codes is that in the occurrence of an accidental event or abnormal environmental loads/actions, the structure can sustain damage. Such consequential damage creates a new situation, which is characterized by the after-damage design situation. After the hazardous event having occurred, the after-damage design situation, which considers the structure's further behaviour, shall also be addressed. Most codes, including API 2A (2019), proposes that the structure shall consequently be designed following a two-stage procedure as outlined below with clarifications from NORSOK N-001 (2021) included:

- design the structure for the accidental situation involving the hazard considered; damages on the structure and marine systems are acceptable as long as they do not lead to complete loss of integrity or performance. Non-linear analysis is normally accepted in this check.
- after the hazardous event has occurred, check the after-damage design situation in relation to specified environmental loads/actions. It shall be demonstrated that the structure and marine systems do not develop progressively into total collapse (e.g. free drift, capsize, sinking or extensive damage to the external environment etc). The assessment is to include post-event inspections which are conducted to evaluate the platform's structural condition following a potential overload event (storm, earthquake, mudslide, tsunami, ice) or incident (vessel impact, dropped objects, explosion, abrasion, floating debris, riser damage from anchor drag on pipeline) as stated in SIM plan API2SIM (2020).

According to Goal Based Ship Construction Standards (GBS) under "SOLAS Convention", the residual strength assessment after accidents of ship "hull girder" is required only for Tanker and Bulk carriers of 150 m and over. However, it has been discussed that the damage assumptions of the GBS rules are not too realistic.

2.7 Recommendations

The rapid integration of decarbonisation technologies, autonomous control, artificial intelligence, and digitalisation is transforming maritime operations and introducing new accident causation pathways. However, the lack of observed accidental events involving these systems reduces the potential effectiveness of traditional post-incident safety learning. To ensure robust safety assurance, the maritime sector must (i) adopt harmonised human-factors taxonomies that capture comparable data from both rare accidents and near-misses, and (ii) utilise proactive, system-theoretic techniques (such as STPA, FRAM) that expose latent hazards and unsafe control actions before loss events occur. It is also important for the maritime industry to understand the strengths and weaknesses of available and emerging methods to identify potential areas of usage.

An intuitive, objective, evidence-based method for assessing the level of safety for a particular ship-to-ship encounter at sea is currently not available in either codes/standards or literature. Inspiration for establishing such methods could be gained from the aviation safety, which has led to a detailed assessment of the factors influencing encounter complexity and its impact on human performance.

The offshore codes and standards provide well established guidance for assessment of Accidental Limit States based on generic Design Accidental Loads (DALs). However, it has been realised that that many 'design wave events' will be breaking irrespective of water depth leading to the observation that some aspects of present design practice may be non-conservative. It is recommended ensure that these findings are reflected in the codes.

With special focus on arctic shipping and the goal-based ship design approach, more effort could be put into development of validated design tools, namely: (a) accident statistics, (b) a framework for risk-based design of Arctic ships, (c) a simulation model of a winter navigation system, and (d) a sustainability assessment tool for Arctic shipping.

For the general use of goal-based ship design approach, residual strength after accidents of ship "hull girder" is required only for Tanker and Bulk carriers of 150 m and over, by Goal Based Ship Construction Standards (GBS) under "SOLAS Convention". This rule assumes damage extent of ship bottom and ship side, but this assumption does not seem to be realistic, it is therefore recommended to evaluate this requirement.

3 Consequences and Mitigation

This chapter reviews recent publications concerning experimental, analytical and numerical studies of the consequences of accidental actions. It identifies known gaps and problems in current approaches and, where appropriate, provides guidance and recommendations.

3.1 Ship Collision

Regarding ship collisions, (Liu and Guedes Soares, 2023) recently presented a review of the developments in ship collision analysis over the last 10 years and the challenges to an Accidental Limit State (ALS) design method of ships based on collisions. In this paper, the technical difficulties in finite element analysis, and especially the use of mesh-dependent material failure criteria, are discussed, as well as the accuracy of existing analytical models. In addition, the technical challenges for accurately assessing the structural damage with numerical and theoretical methods are summarised.

3.1.1 Experimental Assessment

Structural dynamic responses in ship collisions are very complex and many details of structural damage may not be accurately modelled by theoretical and numerical approaches. Experiment is indispensable for investigating the ship structural damage mechanisms and providing data for the calibration of analytical procedures and the validation of numerical results. The experimental research of ship collision is undoubtedly an effective approach, from which the damage characteristics of the structure can be observed accurately. The experiments on the ship structural components can be used to study analytically the energy absorbing mechanics and to validate numerical methods considering the non-linear structural and material characteristics. In order to assess the internal mechanics of ship structures during accidental events, ship collision experiments have been performed recently. As full-scale experimental research is difficult to conduct and very expensive, many researchers have carried out scale model "dry" or "in-tank" tests. Dry tests can be done quasi-statically or dynamically. Early research mainly relies on energy methods and the quasi-static test became a useful way in producing force-displacement diagram.

In the scale test, quasi-static loading is often applied, which is mainly because the velocity is relatively slow in actual ship collisions, and the quasi-static load is easier to control in the test. Since the extent of the damage of ship structures can be established by adding up all the contributions of the individual structural components, a series of structural elements are selected to examine their primary deformation modes and damage mechanics, for example, the static denting tests on double-hull structures by (B. Q. Chen et al., 2022). Afterwards, Zhang et al. (2023) studied the damage characteristics of a scaled stiffened panel quasi-statically punched at the mid-span by a raked bow indenter, from deformation to a large opening. Qiu et al. (2024) carried out quasi-static penetrating experiment where both the deck and side plates of the struck ship were subjected to the impact from the bow of the striking ship.

Some experimental works were performed to analyse the dynamic collision responses of different structures (F. Liu et al. (2022a); Zhang et al. (2023)) in which drop test rigs were employed (see Fig. 1). He and Guedes Soares (2024), He and Soares (2021) and Zhu et al. (2024a) studied the dynamic response of structures under repeated mass impacts using analytical, numerical and experimental approaches.

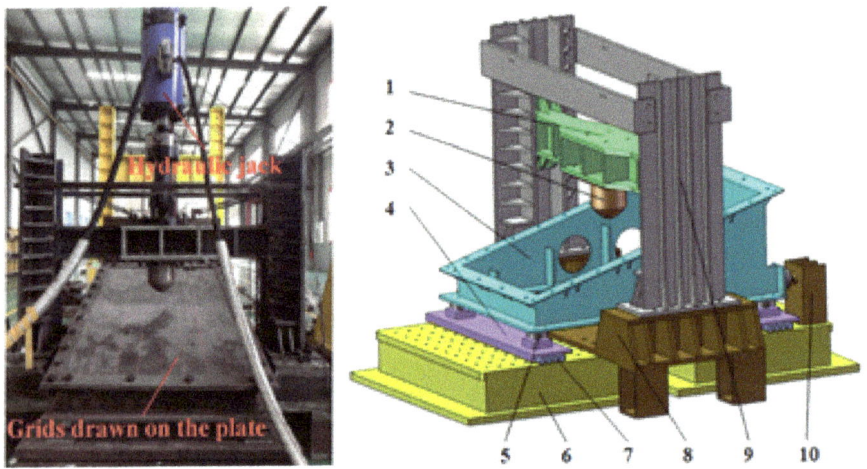

Fig. 1. Falling weight impact test setup for inclined stiffened plates (F Liu et al., 2022a)

Xu et al. (2020) used model-scale experiments in a water tank for ship-ship collision, in which both global motion and local plastic deformation were measured with the interaction of structure-fluid considered. The test set up is shown in Fig. 2. For side collision on a water tank, Zhu et al. (2020c) analysed the interaction between an elastic-plastic plate and the surrounding water in a water tank when subjected to impact. The plastic deformations for various filling levels were given for typical collision scenarios.

Cho et al. (2023) performed physical crushing tests on a cylindrical rubber fender to validate the numerical model in LS-DYNA under collision. The effects of rubber fenders on the collision energy absorption characteristics were examined in association with the structural damage of both striking OSV and struck FPSO hull structures.

3.1.2 Analytical Assessment

By considering the coupling between plates and stiffeners deformations, Zhou et al. (2023) and Zhang et al. (2024) proposed new analytical formula to predict the crushing resistance of ship decks and bulkheads in scenarios of right-angle and oblique collisions

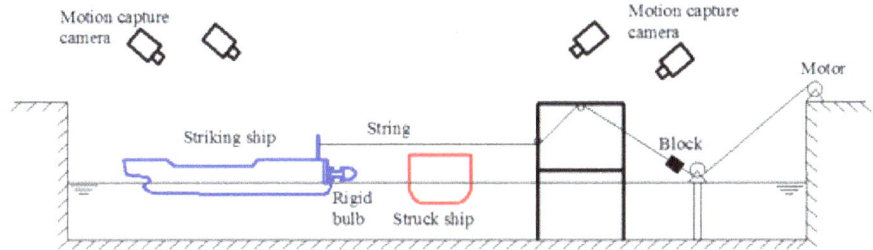

Fig. 2. In-tank ship collision model test setup (Xu et al., 2020)

by a rigid raked bow. In Cai et al. (2022a), the damage of a ship plate colliding with ice floe was assessed by combining an energy approach of ice failure based on fitted or empirical pressure-area relations with a simplified rigid-plastic solution for the plate.

Pan et al. (2022) proposed an analytical model using the Intersection Unit Method (IUM) for assessing the axial crushing force of bulbous bows by dividing them into different macro elements to derive their force-displacement relationships. Their model agreed with scaled crushing tests performed by the authors. In a similar topic, Wang et al. (2023) carried out analytical and numerical studies of ship-side structures collided by bulbous bows at oblique angles. In their work, deformation mechanisms of the involved structural components were proposed and used along with rigid-plastic analysis to obtain the side structure force-displacement relationship. An analytical model based on plastic analysis was also proposed by Liu et al. (2023) to study the crushing resistance of a cruciform structure to be inserted in double-hulled vessels subjected to collision or grounding loads. The developed model showed good agreement with both quasi-static indentation test and nonlinear FE analysis performed by the authors.

3.1.3 Numerical Assessment

Due to advances in computer technology and commercial software, dynamic numerical analysis is now commonplace and allows simulation of a real collision process, accounting for inertial effects, large deformation, contact between structural members and non-linear material behaviour including strain rate sensitivity and fracture. It is worth noting that the fracture criteria to be used for materials remain one of the main research topics in this field.

Using non-linear FE simulations, Jiang et al. (2024) investigated a novel type of ship side structure inspired by auxetic materials with negative Poisson's ratio for enhancing the crashworthiness of double-hull vessels. The innovant structure was found to have a higher energy absorption capacity than traditional ship side structures like double shell, X-core and Y-core. Kim and Sohn (2022) showed that damage metrics such as structural penetration and absorbed energy are quite sensitive to the assumed Dynamic Fracture Strain (DFS) in non-linear finite element simulations of ship collisions, providing empirical formulas for assessing these metrics based on a database of 50 simulations. In a different work, Xiong et al. (2024) reproduced numerically the ship collision accident "8–20 Shanghai" to gather further information regarding the structural response and energy absorption during the accident. They also studied different hypothetical scenarios by varying the impact position, collision angle, and structural arrangement of the collided ship.

In many ships collision analysis using FE simulations, the sea water is not explicitly modelled but its effect is considered by applying a so-called "added mass" to the vessels involved. However, realistic analyses of ship collision require the coupling of the structural response (internal mechanics) with the ship motions (external dynamics) that depend on the environment, operational conditions, and hydrodynamic forces. While such coupling implies significant computational effort, these analyses are becoming more accessible with the increased computational power that has become available over recent years. In S. J. Kim et al. (2022a), a benchmark study compares the structural dynamic responses obtained from non-linear FE and super-element models including the coupling between internal mechanics and external dynamics. Simulations of typical collision and grounding scenarios involving passenger ships confirm that considering the action of hydrodynamic forces may be necessary to correctly predict the damage sustained by the ships. Sviličić and Rudan (2023) implemented in LS-Dyna the Abkowitz model through a user-defined loading routine in order to better model a possible avoidance manoeuvre preceding a collision. The optimal hydrodynamic derivatives of Abkowitz model were determined by comparing coupled FEM-manoeuvring simulations with experimental tests and the resulting model was later confronted to partially coupled (MCOL) and fully coupled (ALE) simulations in Rudan et al. (2024), with the aim of evaluating the accuracy and efficiency of such approach.

3.1.4 Residual Strength after Collision

From the viewpoint of accidental limit state, not only collision analysis, but also the post-collision analysis can be carried out nowadays using nonlinear FEA. That is, for example, hull girder residual strength and water ingress analyses after ship collision. The residual strength assessment of ship hull girder became mandatory for tankers and bulk carriers by Goal Based Ship Construction Standards under SOLAS convention after the accident of Prestige in 2002.

The aim of the numerical simulations conducted by Do et al. (2024) was to evaluate the reduction in the residual ultimate bending strength of a container ship hull resulting from collisions at both sagging and hogging moments. The authors also formulated empirical equations to predict the ultimate residual bending strength based on numerical findings and proposed equations that exhibited good accuracy when validated against numerical and experimental data. Kuznecovs et al. (2023) proposed a framework that can be used to analyse the related consequences of ship-ship collision events using simulations and evaluations. The methodology includes nonlinear FE analyses of the collision event, a METOCEAN data analysis module, damage stability simulations, analyses of the damaged ship's ultimate strength and structural integrity, oil spill drift simulations, and finally, an evaluation of the three abovementioned consequences. Li and Kim (2022) focused on developing rapid methodologies for predicting the residual strength of ship hull girders for predefined damage scenarios. They exploited the concept of a damage index (DI) and demonstrated its application using the Intelligent Supersize Finite Element Method (ISFEM) and Modified Paik–Mansour (P–M) approach. They also discussed the Smith-type progressive collapse method codified in the Common Structural Rule (CSR) for assessing hull girder strength. (Tekgoz et al., 2020) reviewed the ultimate strength assessment of aging and damaged ship structures, emphasizing the effects of corrosion, fatigue cracking, and mechanical damage. They discussed the impact of cyclic loads on plate rigidity, re-yielding, and ultimate load capacity. (Do et al., 2022) presented numerical simulations and empirical equations to predict the ultimate residual strength of dented submarine pressure hulls. The collision scenarios considered in this study include accidents between submarine pressure hulls and attendant vessels or floating objects. The accuracy of the developed numerical method was validated by comparing its results with existing test results. Empirical equations were established based on numerical simulations. These equations are simple to use for the initial design and serviceability limit state of submarine pressure hulls. (Tabri et al., 2020) presents a time-efficient method for the residual strength assessment of a ship hull damaged in a grounding accident. The strength assessment is based on a coupled beam method that has formerly been used for intact hull ultimate strength assessment. As a case study the method is applied to a double hull tanker. Three grounding damages are defined based on IACS guidelines and with a dedicated grounding assessment tool.

3.2 Ship Grounding

Estimating the consequences of a ship grounding is slightly different from estimating the one of a collision. In case of collision, it can be assumed or estimated that both the striking and struck ships are affected. However, in case of grounding, several scenarios can be envisaged: the ship runs aground on a rock, a reef, a shoal, sand, etc. In addition, some rocks have very sharp edges, others rounded shapes, which will greatly affect the damage caused to the ship's bottom, particularly the size of the breach created in the hull. If the ship runs aground over sand, failure may not occur, but checking the ultimate strength of the damaged hull beam may be of crucial importance. Therefore, more statistical data on typical rock shapes and material properties are needed. To date, the literature providing such statistical data is limited. That is why, in many grounding analyses, the rocks are assumed to have conical or paraboloidal shape, which sometimes leads to unconservative results.

3.2.1 Experimental Assessment

For both collision and grounding analyses, researchers have been working on penetration tests to gain insight into the damage mechanism of the ship's side and bottom structures, and to validate the results of analytical or numerical approaches. Penetration forces are applied mainly by quasi-static tests and occasionally by dynamic tests. However, compared with ship collision experiments, the number of ship grounding tests is relatively small.

For the most common type of grounding scenario, bottom structures subjected to horizontal grounding forces are scratched or torn by protruding obstructions of rocks. Thus, instead of denting and penetrating as in bottom stranding, the bottom structures in raking scenarios suffer continuous scratching and tearing deformation modes that involve friction, plastic deformation and fracture. Calle et al. (2019) fabricated a 1:100

small-scale ship model with a 0.25 mm thick low-carbon steel plate and completed a grounding model test to verify the validity of the finite element method (Fig. 3). Zhou et al. (2020) carried out a set of quasi-static plate cutting tests, varying the friction between the plate and a conical rock. It was found that with the reduction of friction, the average grounding force decreases significantly, but the damage mode of the hull plate and the peak grounding force stays the same.

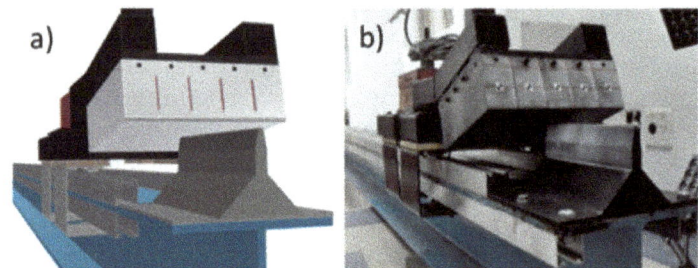

Fig. 3. Test setup for the 1:100 reduced scale dry ship grounding test (Calle et al., 2019)

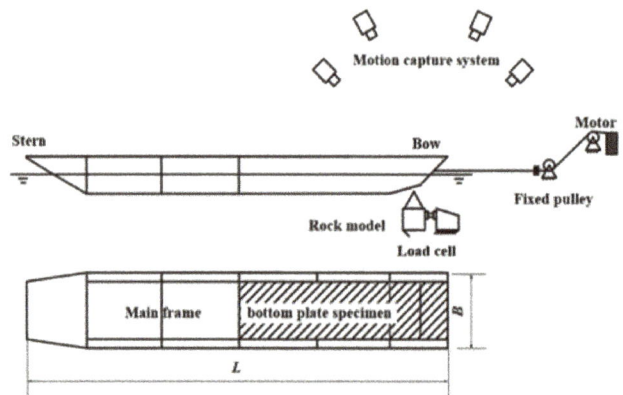

Fig. 4. In-tank ship grounding model test setup (Zhu et al., 2024b)

The above-mentioned tests are "dry" grounding tests, focusing on the internal damage mechanism of the ship grounding, without considering the influence of the surrounding water on the ship's motion in the actual grounding incidents. The importance of the interaction between ship structure and the surrounding water has been long recognized, and the simplest treatment is the introduction of an added mass in the grounding damage assessment. Zhu et al. (2024b) designed and conducted a series of 1:100 scaled in-tank ship model grounding tests over two types of sharp rocks to investigate the motion responses and structural damage modes of the ship model under various grounding initial velocities, relative heights, and rock eccentricities. An illustration of the tests is presented in Fig. 4. Zhu et al. (2024c) investigated the effects of transverse stiffeners on bottom plates in ship grounding by comparing results from grounding model tests over a sharp rock for ship bottom plating with and without transverse stiffeners.

3.2.2 Analytical and Numerical Assessments

As for ship collision, grounding analysis can be divided into external dynamics (ship motion) and internal dynamics (structural deformation and failure). In previous studies, decoupled approaches were generally used, but over the last 15 years, bi-directional coupled approaches that consider the interactions between the vessel and the surrounding water have been introduced.

Pineau et al. (2022) developed closed-form expressions based on rigid-plastic analysis to rapidly assess the damage extent in case of grounding. The rock is represented by an elliptical paraboloid, while the ship is divided into super-elements, representing the main bottom and side components. The solver was then coupled to the external dynamics program MCOL to account for the ship motion considering the loads exerted by the water. The resulting tool was used to simulate thousands of grounding scenarios considering a large cruise ship and varying the rock shape, the grounding location at ship bottom, the initial penetration and the velocity as well as the friction coefficient. The authors found that a structural reinforcement is demonstrated to be more efficient in bottom grounding than in side grounding, because in such situation, the ship is pushed away by the rock and thus undergoes significant sway, yaw and sometimes roll motions that limit the damage extent. Further, friction was shown to be of primary importance in bottom grounding, while its effect is rather limited in case of side grounding.

This two-way coupled solver was also used in Pineau and Le Sourne (2023) to assess the resisting force of outer/inner bottom plating and transverse floors when a ship undergoes combined surge and heave motions during the grounding, as well as in Conti et al. (2024) to simulate thousands of grounding events with the aim of quantifying the influence of different reinforcements on the extent of damage sustained by the vessel.

In the same way, Taimuri et al. (2022) developed a ship grounding analysis method using two-way coupled FSI method. The time domain solution is based on a simplified contact analysis model, and the plate cutting angle is calibrated based on the rock geometry and hull indentation. A ray-tracing algorithm that utilizes panels and the tip of a conical rock is implemented to assess the hull penetration. The model was successfully validated by comparing it with Ls-Dyna/MCOL simulations for a box-shaped barge of double bottom configuration and a passenger vessel. It was later used in Taimuri et al. (2023) to introduce a probabilistic analysis method for the assessment of damage extents in ship grounding. A Monte Carlo simulation was carried out to randomly generate a realistic profile that accounts for variable ship speed, conical rock geometry, rock position, and height in both deep and shallow waters. Probabilistic results were compared against existing distributions of damage extents and demonstrated an increase in the mean distribution of damage length. The authors concluded that the method allows for low-fidelity optimization of the structural arrangement of the bottom of the ship, probabilistic evaluation of loads associated with ship crashworthiness, and the assessment of operational limitations during an evasive manoeuvre.

Yamada et al. (2024) performed a series of large-scale nonlinear simulations of a VLCC running aground with a rigid rock from top of the bow. The effect of sea water was considered by applying hydrostatic pressure to the ship's hull and the fracture of the double bottom structure was investigated for several ship grounding velocities. From equilibrium of energy and momentum, the authors proposed a simplified formula to estimate the velocity beyond which rupture of cargo oil tank occurs. Resulting critical velocities were then successfully compared to the ones obtained from the numerical simulations. Zhou et al. (2024) carried out a simulation based on the Coupled Eulerian–Lagrangian (CEL) method to reproduce a grounding test case conducted in a water tank. Based on the verified simulation technology, the influences of rock shapes and friction on the grounding damage and motion responses of ships were investigated and discussed.

3.2.3 Residual Strength after Grounding

As stated above, the residual strength assessment is mandatory for tankers and bulk carriers by Goal Based Ship Construction Standards under SOLAS convention. In case of grounding accident (post-accident phase), ship owners and classification societies need to estimate risk of ship breaking before and during the ship rescue process from the grounding situation (on the rock, shoal, reef and so on). In these cases, rapid estimation of residual hull girder strength may be needed. The following methods are of great importance in the post-accident phase.

As grounding accidents are inherently unpredictable, there remains a continuous need for rapid estimation of the ship residual strength immediately following such incidents. This requires quick assessment of the location and extent of damage caused by the grounding. Over the past 12 years, various studies on grounding incidents across different ship types and sizes have been consolidated by Kim et al. (2024a). A key focus of this research is the development of a simplified method to index grounding damage. The proposed Grounding Damage Index (GDI) is calculated based on factors such as the location of the damage, the extent of penetration, and the angle of impact, utilizing the Probability Density functions suggested by the International Maritime Organization (IMO). Once the GDI is defined, it allows for the calculation of the ultimate hull girder strength for different ship types and sizes, and an empirical formula has been developed to predict this strength under various scenarios. The detailed process to develop the Residual strength versus grounding Damage index (R-D) diagram is summarized in Fig. 5. Based on the procedure above mentioned, several R-D diagrams have been developed for oil tankers, bulk carriers and container ships for the last decade. Li and Kim (2022) investigated the accuracy of the existing numerical method in developing the R-D diagram, which can be applicable to predicting the structural residual strength of grounded oil tankers.

Kim et al. (2024a) also investigated the reduction rate of section modulus in detail. They emphasized the significance of analysing variations in the section modulus, especially in the bottom part of the vessels. Their findings demonstrated a clear and consistent link between the outer bottom's section modulus and the residual ultimate longitudinal strength. The general tendency for commercial ships, residual ultimate longitudinal strength decreases as with an increasing damaged ratio of the outer bottom section modulus. They suggested that these insights could lay the groundwork for future studies focused on developing safer and more resilient ship designs to better withstand grounding accidents.

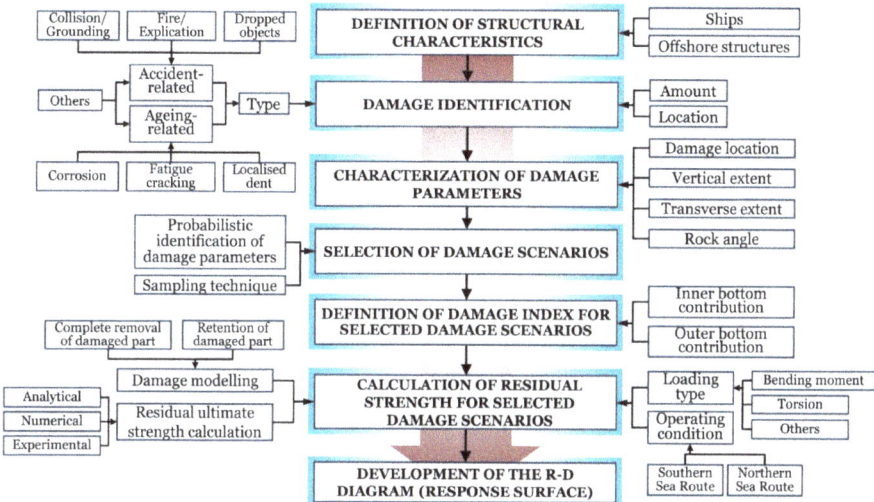

Fig. 5. General procedure for the development of R-D diagram (Kim et al. 2024a)

3.3 Explosion

Research on explosions can generally be categorized into two main areas: in-air and under-water (UNDEX) scenarios. Recent studies and developments in both areas are discussed below.

3.3.1 In-Air Explosion

In the context of in-air explosion, further distinctions can be made between onshore and offshore environments. However, the predominant focus is typically on the investigation of explosion loads, as well as the analysis and design of blast resistance structures engineered to resist these forces. Numerous studies have focused on explosion dynamics, particularly emphasizing the mandatory installation of blast walls to ensure the safety of living quarters on offshore production platforms. Offshore topsides are designed with a focus on space efficiency and weight control to optimize production and reduce costs. As a result, predicting the maximum deflection of structures, i.e. blast walls under explosive forces has become a critical area of study, with particular attention to lightweight sandwich panel systems (SPS). SPS structures are attractive due to their enhanced bending rigidity, excellent energy absorption, and cost-effectiveness.

Despite comprehensive prevention strategies, offshore operations continue to face persistent health, safety, and environmental (HSE) challenges. While explosions on offshore structures are rare, their destructive potential is significant. Blast walls act as essential passive safety barriers, protecting topside areas by absorbing and dissipating explosive energy. The saturation impulse of the beam under explosive pulse load was first developed based on the membrane force factor method, which provided a new advanced theoretical tool for studying the large plastic deflection of the beam under pulse load. The method was extended to plates structures in (Tian et al., 2019, 2020). The plastic dynamic response and saturation impulse phenomenon of metal plates were analysed by Zhu et al. (2021) based on the membrane force factor method for various pulse shape typical non-rectangular pulse loads such as linear attenuation, exponential attenuation, and linear rising exponential attenuation.

The use of lightweight SPS in blast walls not only contributes to efficient design but also offers a cost-effective solution to managing the risks posed by explosions on offshore platforms. Recently, Kim et al. (2024b) conducted in-depth reviews on mechanical response of blast wall subjected to explosive loads, including studies on advanced hexagonal honeycomb geometries, such as inclined cells, curved panels, features like indentation and perforation, graded cores, auxetic re-entrant cells, as well as panels exposed to underwater impulsive loading.

3.3.2 Underwater Explosion

He et al. (2023) carried out a near-field UNDEX test on a full-scale ship and recorded the dynamic response processes. The Coupled Eulerian–Lagrangian (CEL) numerical approach was then used to establish a model of the explosion, and the effectiveness of the method was validated by confrontation to the experiment. Based on the results, it was concluded that the near-field UNDEX could be divided into three stages: shock wave and after flow, bubble pulsation, and water jet. UNDEX experiments were also conducted by He et al. (2020), Gan et al. (2021) and Gong et al. (2023) on small-scale hull girder structures. The CEL approach was also employed by the authors to analyse the different deformation modes as well as the sensitivity of the response to the standoff distance and hull thickness. In the same way, two scaled hull girders models were built and tested by Zheng et al. (2023) to investigate the influence of superstructure on the damage characteristics of ships subjected to near-field UNDEX. The authors showed that under the same explosion condition, the presence of the superstructure increases the hull girder's resistance to the overall bending deformation and minimizes the damage of grillage at the hull bottom.

Simplified models based on analytical expressions have recently been developed to rapidly predict the UNDEX response of structures. To study the response of composite immersed cylinders, Oo et al. (2023) used double Fourier series to solve shell equations and coupled the resulting mechanical solver to plane wave and double asymptotic approximations of the fluid pressure, considering the time delay effect that occurs during the application of the incident shock wave. A simplified theoretical model was also developed by Li et al. (2021) to predict the elastic-plastic whipping response of a hull girder subjected to both the shock wave and the bubble pulses generated by an UNDEX. Finally, Sigrist et al. (2022) proposed a semi-analytical model describing the dynamic behaviour of a submerged cylindrical structure to the pressure load generated by an UNDEX. This model allowed the authors to calculate the signals and response spectra which characterize the dynamic response of submarine hulls bearing equipment.

The damage effects of shock wave, bubble pulsation, after flow, and water jet loads on a full-scale ship were numerically investigated by Huang et al. (2024). They showed that the combined effects of different loads lead to different damage modes. Shock waves and after flow loads are the main cause of local damage and hogging damage, while pressure difference and bubble adsorption are the main reasons for the hogging damage. A numerical model based on acoustic and shell elements was used by Zhou et al. (2021) to predict the response a hull girder subjected to both UNDEX and wave induced pressures. The authors concluded that the structural damage severity may be underestimated without considering the synergistic effect of the combined loads.

3.4 Dropped Objects Analysis

Plastic deformations and damages due to dropped objects in platform operations, lifting operations or anchor work are common in ships and offshore structures. Several theoretical, numerical and experimental studies have been conducted in connection with the elasto-plastic behaviour of structures under the impact of dropped objects. General ship structures and pipelines are discussed separately in the below sections.

3.4.1 Ship Structures

Markovic and Kovacevic (2019) and Kovacevic et al. (2019) investigated the influence of patch load length on the ultimate capacity of longitudinally stiffened girders. It found that by increasing the patch load length, the ultimate strength of the plate girders is significantly increased, >50% when the length of an applied patch load increased from 0 mm (concentrated load achieved by a half-round bar) to 150 mm. The nonlinear elasto-plastic response of stiffened steel plates loaded quasi-statically by a central rigid rectangular indenter is investigated both experimentally and numerically by Zhu et al. (2020b) and the experimental set-up and a stiffened plate model is shown in Fig. 6. The concept of applying the elasto-plastic method to the design of deck plates under patch loads was introduced, and a simple design formula to determine plating thickness was proposed based on an acceptable level of permanent set. The thicknesses obtained by the proposed design formula are compared with the design thickness required by BV and LR rules.

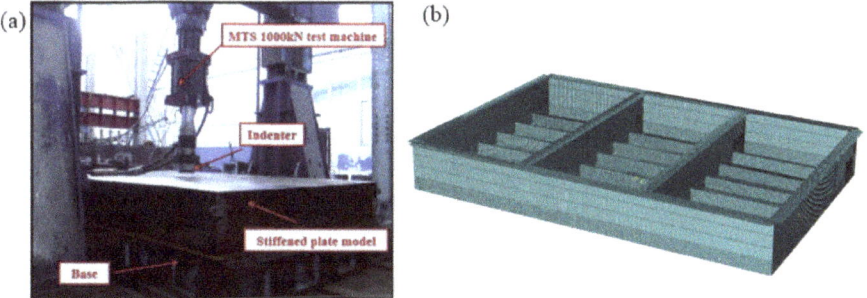

Fig. 6. Drop tests with rigid indenter: (a) experimental set-up with stiffened plate; (b) finite element model (Zhu et al., 2020b).

Ship structures may also be subjected to repeated random patch loads at different locations. Against this background, the elasto-plastic response of ship structure under repeated patch loads at different locations was studied using finite element method by

Cai et al. (2021). Xiang et al. (2023) provided a comprehensive review and analysis of the deterministic and stochastic dropping processes associated with dropped objects in the field of offshore engineering. Trajectory envelope and risk assessment of regular dropped objects were performed, and the paper concludes with a detailed description of the completed review results, current issues and future directions. Oka et al. (2024) used Digital Image Correlation (DIC) techniques and FEM analysis to evaluate the deformations of the nuclear material container during drop impact. Comparing the strain measurements from Abaqus to DIC shows similar behaviour with minor differences in the peak strain measurements.

3.4.2 Single-Walled Pipelines

Jiang et al. (2019) studied the response of single-walled pipelines to transverse impact loads by dropped objects both experimentally and numerically. A series of experiments were carried out to examine the relevant effects, including the seabed conditions (rigid or flexible) and impact energy, as shown in Fig. 7. Effect of drop height, weight and ground conditions were discussed and classified. Jiang and Dong (2022) presented a two-level quantitative risk analysis for pipeline failure caused by platform dropped objects. Numerical analysis and probabilistic approaches were coupled to quantify the effects of nonlinearity and pipe–soil interactions and the underlying uncertainties of related factors. By combining probabilistic sampling and numerical simulations, any potential contributions of dropped objects to the failure risk of pipelines could be estimated.

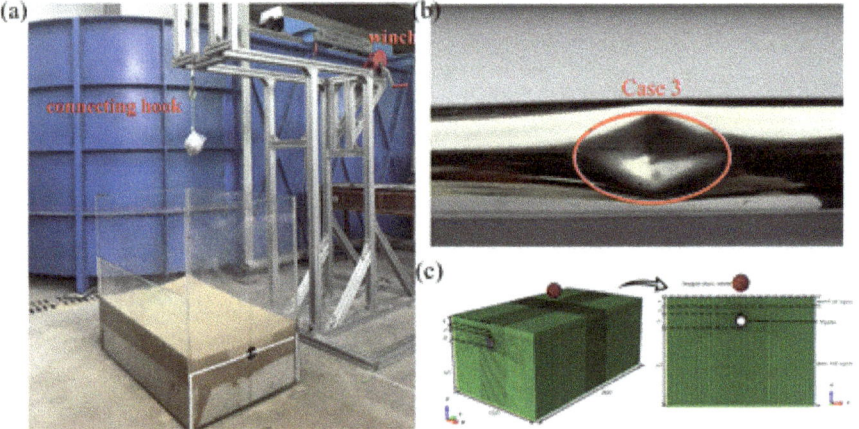

Fig. 7. (a) Experimental equipment; (b) Dent damage of pipe specimens in the experiment; (c) Configuration of the numerical model for buried pipeline (Jiang et al., 2019)

The rupture and perforation failure processes on pressurized metal pipelines under heavy transverse impact loadings were studied by Zhou and Zhang (2022) by using a cube penetrator. Based on previous authors' conclusions that the transition from rupture to perforation failure modes of pressurized pipelines mainly occurs at the impact velocity of approximately 150 m/s, impact velocities from 120 to 180 m/s were investigated, representing extreme conditions that may result in rupture or perforation failures, such as the penetration of underwater explosive projectiles or fragments, and the strike of torpedoes. They identified two typical rupture patterns (flanging and plugging) and three zones in the perforation mode (perforated, cratering, and undamaged). Different failure phenomena, such as flanging, plugging, and fragmenting, may be involved in the perforation process. When the impact speed is high, the failure mode changes from rupture to perforation affected by the shear effects (Fig. 8a). The impact limit is typically used to identify the transition between the rupture and perforation modes and the resistance of the target structure to withstand the impact loading. The rupture limit expresses the impact velocity at which the macroscopic crack is first generated within the impact zone, and the perforation limit denotes the impact velocity at which the penetrator first breaks through the pipe wall or the plugging completely falls off. The study established a modified cavity expansion model to predict the perforation limit and found that internal pressure enhances pipeline resistance against penetration. The impact limits were found to vary linearly with deflection angle, internal pressure, and diameter-to-thickness ratio. Proposed limits could be used as a guideline to supplement specifications and propose safeguards for the impact protection of submarine pipelines (Zhou and Zhang, 2022)).

Wu et al. (2023) proposed an evaluation method for dented natural gas pipelines considering a ductile damage model based on the Oyane ductile fracture criterion used in a finite element model. According to this criterion, when the damage integral value (I) equals 1, the dented pipeline fails and should be repaired or replaced. The influence of the dent depth was studied using indenters of different shapes (spherical, axial and transverse) in the FE model, which was validated with experimental results. The pipeline defect parameter D is defined as dent depth to pipe diameter ratio. The damage integral value increases with the increase of this parameter, but the increasing rate drops gradually.

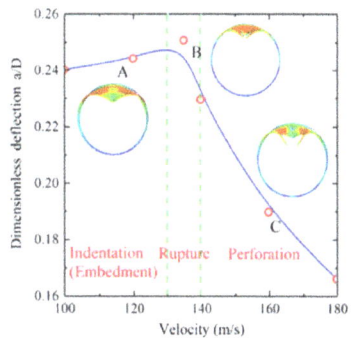
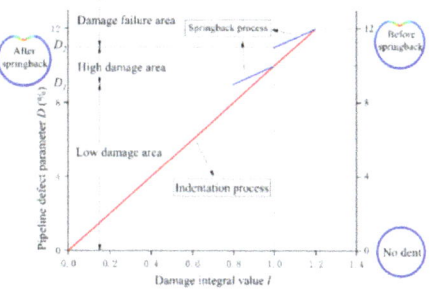

(a) Mode transition process of the impact damage zone, where a is the deflection in the transverse direction and D is the pipe outer diameter

(b) Variation of the pipeline defect parameter D (%) and pipeline thresholds (D1 and D2) with the dent damage integral value I

Fig. 8. Finite element results of the damage zone of dented pipes under different impact velocities (Zhou and Zhang, 2022); (b) Relationship between the damage integral value, the pipeline defect parameter, and the pipeline thresholds during indentation and spring back process (Wu et al., 2023)

In the formation process of unconstrained dents, the spring back due to elastic unloading with the indenter removal results in the decrease of the dent depth and final ductile damage amount. Spherical dents cause the most significant damage to the pipeline, while transverse dents cause relatively minor damage. The proposed evaluation method was defined based on the definition of monitoring and repair thresholds for the pipeline defect parameter D (Wu et al., 2023). When the damage integral value I reaches 1 before the spring back, the monitoring threshold ($D1$) is defined as the pipeline defect parameter after the dent spring back, while the pipeline defect parameter is defined as the repair threshold ($D2$) when the damage integral value I reaches 1 after the dent spring back. A threshold diagram as a function of the defect parameter was drawn and three damage areas were identified (Fig. 8b). The proposed method has the advantages of the easy acquisition of evaluation parameters (dent depth and shape, pipe diameter and material, and internal pressure), the direct engineering application, and the use of monitoring and repair thresholds as a function of dent parameters, while the ASME B31.8 and CSA Z662 codes, for instance, judge the risk of a dent based on a fix criterion of dent depth to diameter ratio of 6%, above which the damaged pipe needs to be repaired.

3.4.3 Pipe-in-Pipes and Sandwich Pipelines

Compared with single-walled transport pipelines, sandwich pipes (SPs), with higher strength-to-weight ratio, lower production cost, higher structural carrying capacity and good insulation performance, can meet the increasingly urgent needs of deep-sea transportation. The dent damage caused by dropped objects is one of the main failure forms of SPs. W. Chen et al. (2024a) investigated the damage behaviour of SPs under dropped object impact considering the effect of the seabed flexibility. Using a finite element model, the dent characteristics of SPs under different seabed flexibility were explored in a parametric study varying the mass and kinetic energy of the dropped object and the buried depth. When the kinetic energy of the falling object is low, the proportion of the final kinetic energy of the dropped object and the friction loss energy of the collision system are smaller, and the proportion of energy absorbed by the SP is larger. However, the trend is opposite when the kinetic energy of the dropped object is high, with the energy proportion dissipated by the annular layer significantly decreased, while the energy proportion of the outer pipe and the inner pipe gradually increases, resulting in a significant increase of the dent depth. The strain of each layer increases rapidly with the increase of the falling object kinetic energy. In the case of low kinetic energy, the impact velocity dominates the dent damage of the SP, while the falling object mass plays a more important role in high kinetic energy conditions. Under the same falling object kinetic energy, the impact of a falling object with a light weight and high speed is more severe and the dent depths are greater than that of a falling object with a heavy weight and low speed. With the increase of the falling object mass, for the same kinetic energy, the energy absorption ratio of each layer shows a downward trend, and the variation of dent strain caused by the increase of the falling object mass is non-linear. This is mainly attributed to the local stress concentration during the collision process that plays a more important role in affecting the level of dent strain. The dent depths of the SP layers decrease gradually from outer to inner pipe. The dent strain of the annular layer, which is a brittle light cement-based material, is largest, with the annular layer having the largest energy absorption proportion and taking the predominant failure. In the case of rigid seabed, the outer pipe takes the largest energy dissipation proportion, and the impact resistance capacity of SPs can be improved by appropriately increasing the wall thickness or the material strength of the outer pipe. Different from the rigid seabed condition, the SPs experience prominent overall bending deflection in addition to the local dent deformation in the collision process under flexible seabed condition. In this condition, the annular layer of SP absorbed most of the impact kinetic energy compared with the outer and inner pipes. The flexible seabed foundation can effectively reduce the local dent damage degree of SPs. Under the same falling object kinetic energy, the greater the buried depth is, the greater the maximum dent depth of each layer will be. The outer pipe strain increases with the increase of burial depth. The buried depth has little influence on the strain of the annular layer and inner pipe. Prediction formulas for the impact damage evaluation of dented SPs were proposed, based on dent characteristics, which can provide support for the damage evaluation of SPs under the different impact kinetic energy and mass.

Gao et al. (2022) studied the transverse impact resistance of pipe-in-pipe submarine pipelines with local dent defects formed after 15 years of service (design service life) by a FE model and experimental tests. Compared with single-layer pipes, extra interactions exist among the inner pipe, the filler (polyurethane foam (PUF) or PUF and high-density polyethylene (HDPE)) and outer pipe of pipe-in-pipes which should be considered in the impact event. The pipe-in-pipe specimens were transversally impacted on the drop hammer machine, and the same drop weight, impact velocity and impact height were adopted for all tests. The failure modes predicted by the FE models are similar to the experimental observations, and both show the mechanism of plastic hinge with a coupling deformation of local indentation and global bending. The impact force-deflection curves of pipe-in-pipes can be divided into four stages according to Gao et al. (2022): primary vibration stage (stage I), secondary vibration stage (stage II), stable stage (stage III) and unloading stage (stage IV). In the whole impact process, the total impact energy produced by the drop hammer is dissipated in three forms, i.e., the energy absorbed by global bending deformation, the energy dissipated by local indentation and the kinetic energy of the rebound of drop hammer. The energy is mainly dissipated by the outer and inner pipes. PUF and HDPE absorb very little energy in the whole impact process and have little contribution to the impact resistance of pipe-in-pipes. Although the specimens with pre-existing dents dissipated more impact energy, compared to intact specimens, the energy dissipated by the outer pipe decreases, whereas the total energy dissipated by the inner pipe, PUF and HDPE increases. Then, pipe-in-pipe specimens with pre-existing dents under transverse impact load have larger deformation, aggravated damage in the inner pipe and reduced impact resistance. As the dent width increases, the dent depth has a more pronounced effect on the residual deformation of the inner pipe. The results show that the peak impact force decreases with the increase of the suspension span, but the impact time is increased. That means longer pipe-in-pipe specimens can absorb more impact energy. When the suspension span increases, the primary failure mode changes from local indentation to global bending deformation. The local indentation deformation of pipe-in-pipes with smaller span lengths under lateral impact load will dissipate more energy, which is more detrimental to the inner pipe. The obtained results can provide data support for the life extension assessment of pipe-in-pipe submarine pipelines.

3.5 Fire

Recent book by Paik (2020) book describes principles, industry practices and methodologies for advanced safety studies within the framework of quantitative risk assessment and management and associated with the safety and resilience of structures and infrastructures with tolerance against various types of extreme conditions and accidents such as fires and explosions. Advanced computational models for characterizing structural actions and their effects in extreme and accidental conditions, which are highly nonlinear and non-Gaussian in association with multiple physical processes, multiple scales, and multiple criteria are presented together with probabilistic scenario selection practices and applications.

Recent more specific studies have focused on engine room fires. Li et al. (2022) developed a risk assessment method for ship power systems during engine room fires using the expert comprehensive evaluation (ECE) method combined with fuzzy fault tree analysis (FFTA). They analysed the main engine system and its subsystems, constructed a fuzzy fault tree, and calculated the probability of system failure. The study found that the main engine fuel system posed the highest fire risk, primarily due to fuel leakages. The method improved the accuracy of risk assessments by addressing incomplete statistics. (H. Zhang et al., 2022a) proposed a dynamic evolutionary model to quantify the domino effect of ship engine room fires. Using matrix calculation and Monte Carlo simulation, they calculated the dynamic probability of multiple accident units' domino effects. The study highlighted the importance of extinguishing fires within 2 to 4 min to prevent further spread. The results provide valuable insights for fire prevention, suppression, and protection design in ship engine rooms.

3.6 Material Description Including the Failure

Before the reporting period, considerable effort had been made in identifying realistic fracture loci. Two key drawbacks of these loci were that (1) they were often expensive enough to calibrate that they could not be used in typical industrial applications, and (2) that they were often calibrated for very small solid elements, which were far too small to be applied in realistic simulations of ship and offshore structures in accidental limit state analysis. These limitations led to specific developments in this reporting period: simplification of the calibration process, reporting of calibration parameters for materials relevant for maritime and offshore applications, and generalization to shell element simulations. These developments are addressed in more detail in the following paragraphs.

To simplify calibration, Lu et al. (2021) proposed the Maximum Shear Stress and Rice Tracey (MSSRT) criterion, which can be calibrated with only tensile test data. This is less accurate than more detailed models but makes it easier to apply, especially in an industrial setting.

Several authors have published detailed fracture calibrations on maritime steels. Park et al. (2019) performed a detailed fracture calibration on EH36 steel. Cerik and Choung

(2020) performed a detailed plasticity and fracture calibration on Grade A, AH36, and DH36 steels. X. Li et al. (2023) and R. Yu et al. (2022a) performed detailed calibrations of the failure properties of an L907A ship steel. In particular, X. Li et al. (2023b) considered temperatures in the temperature range from 20 °C to 800 °C and strain rate range from 0.001 to 5200/s. While useful for accidental limit states, these calibrations appear to be mostly aimed at ballistic response. These published calibrations offer some hope to designers, who often lack the actual material and the budget to perform a full material calibration in the design phase. By using this published data, the designers can make predictions and simulations of realistic scenarios without the material, budget and lead time that are typically required. However, there is not yet evidence in the literature that a calibration of one material is transferrable to another material, even in the same nominal grade.

To simulate fracture in shell elements, Pack and Mohr (2017) proposed a variant of the Hosford-Coulomb model that could account for both necking and fracture. Their main concept was to use a damage indicator for both necking and failure in what they referred to as the Domain of Solid-to-Shell Equivalence (DSSE). This reference has seen considerable interest in the maritime and offshore communities. For example, Cerik et al. (2019) demonstrated the model of Pack and Mohr (2017) on an S235JR steel. Likewise, Yoon and Choung (2023) applied the localized necking model of Pack and Mohr to a floating offshore wind turbine, and Cerik and Choung (2023) demonstrated the model of Pack and Mohr on a laboratory-scale structure that was intended to be representative of an offshore structure. Lu et al. (2022) updated Pack and Mohr (2017) model by scaling between necking criteria and a fracture criterion to improve mesh objectivity in shell elements. In this case, they scaled between the Bressan–Williams–Hill (BWH) necking model and the MSSRT model introduced by Lu et al. (2021).

Storheim et al. (2018) provided a critical analysis of the failure criterion advanced in classification documents such as DNV RP-C208 (2022) and DNV RP-C204 (2021). They found that the criteria dramatically underpredicted the failure behaviour of a reference structure, resulting in very low energy absorption for DNV RP-C208 and could over-predict or under-predict failure depending on element size for DNV RP-C204.

All of the foregoing discussion in this section is aimed primarily at the development of fracture in base material – away from welds or supports such as stiffeners. Collisions and blasts that are experienced on real structures often feature fracture patterns that follow either welds or supporting structures. To address this, studies like Okawa et al. (2024) have addressed the fracture behaviour of welds and their associated heat affected zones. More needs to be done to generalize these models and make them suitable for very large shell structures.

3.7 Recommendations

When evaluating failure criteria in an accidental or damaged condition, the concept of "conservative" can be poorly defined. In the case when the objective is to minimize accelerations (e.g. on crew or cargo) or to minimize damage onto a struck object, under-estimating failure could be un-conservative. Nevertheless, in this subsection, "conservative" will be taken to indicate early prediction of failure, which is often regarded as safe.

Storheim et al. (2018) have shown that DNV RP-C208 appears to be conservative, though DNV RP-C204 may not always be conservative, depending on element size. Applying necking-based methods like DSSE (Pack and Mohr, 2017) or BWH can be implemented into many commercial finite element packages, such as Abaqus or LS-DYNA, with standard features, and they can often be calibrated with only a single parameter. These offer a higher level of fidelity than codes such as DNV recommended practices without much additional effort for implementation. Caution should be applied when methods like DSSE or BWH are calibrated with only necking or localization criteria because fracture can sometimes occur at lower levels of strain than necking or localization. More sophisticated methods that scale between necking and fracture criteria to offer mesh objectivity can offer a higher level of precision, but they come at the expense of requiring considerably more calibration and implementation effort.

Existing calibrations of more sophisticated fracture criteria can be useful in the design process, but the transfer of published results to the specific materials that are used in construction is not guaranteed, even if they are in the same nominal grade. Therefore, failure calibrations based on the actual construction material are most highly recommended when safety is critical and specific scenarios are being considered.

The simulation of fracture in large-scale shell structures (such as maritime and offshore structures) will continue to develop in the near future. Mesh objectivity, especially when combined with low calibration effort and ease of implementation, remains an issue, and research is ongoing in this area. Further development in the failure of welds, especially in the presence of stiffeners and bulkheads, is required in future work.

Increasing computational capacity allows FSI simulations, where water domain is directly modelled, for example using ALE approach that is suitable when inertial effects are dominating. However, it is suggested to validate these advanced approaches with simpler model and first principles. For slamming simulations, where plastic deformations occur, numerical approaches are not necessarily precise enough. It is suggested to run wider validations and benchmark studies between experienced experts using well reported laboratory experiments.

4 Resilience, Operability, and Survivability

In the context of ships and offshore structures, resilience may broadly be defined as the ability to sustain required operations and achieve system goals under a large variety of conditions, including anticipated and unanticipated events.

Of particular concern for this committee is structural resilience, which includes consideration of structural resilience to local loads – the local strength of a structure as a function of increasing localized damage; and structural resilience to global loads – the global strength and stability of a ship structure as a function of increasing damage. Sustaining required operations and achieving system goals are the metrics by which resilience is assessed, however these are highly scenario specific, e.g. a design is generally considered successful if it is resilient when subject to operational design loads. The required operational system requirements and associated structural limit states are generally clearly defined. Resilience becomes less clear when the scope of assessment moves beyond operational design loads into accidental loads. Stated goals are less precise and confidence in structural resilience to accidental limit states is not as high. These are traditional views of structural resilience are generally well considered in ship and offshore structure design rules (see Sect. 2). This chapter explores some non-traditional ideas of resilience including hull structural resilience for beyond-design operations and resilience of unmanned vessels. The concept of resilience may also be extended to include societal expectations towards the sustainable design and operation of ships and offshore structures. Expanding understanding of hull structural resilience can lead to more efficient hull designs regarding material weight and repair efforts and may contribute to extending their useful life.

4.1 Resilience for Beyond-Design Operations

Increasingly, there is interest in defining the resilience of hull structures to loads greater than their design loads such that insight into purposefully operating vessels in conditions beyond their original design scenario may be made. Some examples of such cases are ice loads on ship hulls and slamming and green water. Additionally, insights into structural resilience against actions such as unintended docking or side-by-side operations impacts is desirable.

Recently investigations into understanding the resilience of conventional ship and offshore structures hulls in ice have been conducted. With increasing interest in the Polar regions coupled with warmer global temperatures, the assessment of resilience of conventional ships for operation in ice-infested waters is essential. Hisette et al. (2023) state that uncertainties in ice loads lead to high safety margins regarding hull structural design. They state that "… margins can either be reduced by improved understanding and modelling of the physical interaction process with ice or by integration of digital models based on a sufficiently high amount of data." Myland et al. (2021) conducted ice tank model test experiments to determine the level ice resistance of "non-typical icebreaking bow shapes" with the future goal of developing a level ice resistance prediction method. "Non-typical bow shapes" refers to hull forms with high stem and/or small waterline angles. As such bow shapes are not optimized for operation in ice, one must assume that their use in ice would be limited to relatively thin ice covers. The authors examine full-scale thicknesses of 0.52, 0.67, and 0.79 m level ice. Buckling of thinner ice covers (Hendrikse and Metrikine, 2016; Ranta and Polojärvi, 2019) is not discussed in detail but is highly important in certain cases outlined below.

The assessment of low and non-ice-class ships for operation in various ice covers has been the subject of recent work. For non-ice-class ships, it may be said that any impact with ice is beyond the design load for the ship. For low-ice-class ships, overload constitutes any impact with ice that exceeds the design ice load. Suominen et al. (2024) use empirical data recorded from ice impacts on the S.A. Agulhas II to perform a probabilistic analysis of damage to IACS Polar Class (IACS, 2023) vessels. They define a relationship between the POLARIS (IMO, 2016) Risk Index Outcome (RIO) values and the probability of ice induced hull fracture. They note that the probably of fracture for the higher polar classes (PC5-3) is generally reasonable, but that there may be reason for concern for the lower polar classes PC6-7.

Assessment of the resilience of non-ice class and low-ice class hulls to "overload" ice impacts requires a careful, scenario-based investigation that typically involves explicit time-integration finite element models of both the ship hull and ice. It is imperative to include both the ice and hull deformations, and where possible, to account for the energy associated with the external dynamics of both the ship and ice bodies. If plastic hull damage is expected or observed, it is further imperative to assess the hull structural capacity to resist moving/sliding ice loads which can incite significant losses of hull structural resistance compared to similar load cases without load sliding movement (Quinton, 2015). Daley (2015) published the first "safe speed" assessment in open literature for a notional destroyer operating in a marginal zone. He uses the Popov-Daley approach (Daley 2000) – which is the basis for the calculation of the design ice load for the IACS polar rules (IACS, 2023) – to estimate the ice loads for a glancing bow-shoulder impact, but with newly proposed updates to the ice flexural failure limit model and accounting of energy lost to hull deformations. He includes an assessment of the "moving load effect (Quinton, 2015)" and notes a significant loss of hull structural resistance compared to similar load cases without load movement. Dolny et al. (2016, 2017a, 2017b) investigated the capabilities of the Hull 3000 notional destroyer in ice and proposed an experimental plan for large non-ice class grillage tests. Quinton et al. (2017) published guidelines for the numerical assessment of hull response to moving loads on ships. Daley et al. (2017) extend the previous work of Daley (2015) to include ice loads due to head on ramming and introduces a new finite element ice model – based on an isotropic elasto-plastic material model – that develops ice loads according to process pressure-area curves. Further, he uses GEM (Daley et al., 2012) to examine the hull interaction with pack ice for various ship manoeuvres. Dolny (2018) reviewed existing approaches and presented a methodology for predicting safe speed limits for light ice-strengthened government vessels using the example of a 5000 t PC5 Patrol Vessel. He notes: the need to better define the link between ice crushing terms and actual ice properties in the Popov-Daley approach (Daley 2000); the need to consider moving/sliding ice loads in such assessments; continued development of thin-ice mechanics; and the effect of ship manoeuvres other than "forward motion". Lesar (2019) investigated structural response of the notional Hull 3000 destroyer hull structure to ice impact. For the cases examined he notes that relying on a flexural limit to limit ice loads is not advised as it does not consistently influence the prediction of impact load or resulting hull damage. Lesar (2020) extends his previous work to consider moving/sliding ice loads and includes hull rupture/tearing. He recommends explicitly modelling ship hydrodynamic motions. Sun & Huang (2020) propose criteria for scaled modelling of glancing ship-ice impacts. They observe the sliding motion of ice using a tactile sensor during model scale ship-ice impact trials. Zhu et al. (2020) study permanent deflections in ship plates resulting from dynamic ice impact laboratory experiments and explicit time-integration finite element analyses. They introduce a numerical ice model based on a soil and concrete constitutive model. Good correlation with experiments is achieved. They propose an energy absorption reduction factor (EARF) for use in design, to predict the degree of structural damage to a plate during ice impact. Bobeldijk et al. (2021) assessed the safe speed in ice for "forward motion" for a non-ice class naval surface vessel. They used the classic (i.e. rigid hull) Popov-Daley approach to calculate the ice load magnitude and applied the ice load as a moving/sliding pressure (rather than by modelling the ice with finite elements). They vary sailing speed, hull material (EH36 and Lean Duplex EN1.4062), as well as hull design. They note that the location of the hull that experiences the sliding load is dependent on sailing speed, and they note improved performance with Lean Duplex (EN1.4062). They also note that characterisation of hull structure deformation during ice impact is recommended as significant impact energy can be dissipated in hull deformation. Zhang et al. (2022) study moving/sliding ice floe impacts with a deformable grillage and validate their proposed ice material model against experiments from Quinton (2015). Andrade et al. (2023) conducted full-scale, medium energy (47.1 kJ) ice-cone impacts with a clamped grillage cut from the Canadian ex-HMCS IROQUOIS destroyer. These were the commissioning impacts in a series of similar impacts (in publication). They observed noticeable plastic hull damage and were able to account for the ice-structure deformational energy balance. Bobeldijk et al. (2024) present a direct approach for ice belt scantling design using nonlinear FEA, the Popov-Daley ice load method (Daley, 2015), and a moving/sliding ice load for design. They note an 11% decrease in hull structure weight compared with an IACS Polar Rules (IACS, 2023) design.

Assessing the resilience of low- and non-ice-class ships to ice loads requires a realistic assessment of both ice and ship hull deformation processes. Neither the ice, nor the hull structure may be considered rigid. Inclusion of all significant energy dissipation mechanisms is important. Herrnring and Ehlers (2021) developed the FEA "Mohr-Coulomb Nodal Split" (MCNS) model for ice. This method uses a Mohr-Coulomb material model in combination with tied-nodes between hexahedral (i.e. solid) finite elements to model ice of various geometries. Element erosion is not employed except to model pressure melting. Avoiding element erosion is preferred as it can lead to non-physical results for ice models that use it to simulate ice crushing and cracking (i.e. gaps due to eroded elements cannot transmit load and eroded elements violate mass and energy conservation). The tied-nodes approach (termed "nodal splitting" by the authors) enables spontaneous ice spallation to be explicitly numerically modelled, which is a significant improvement in the simulation of ice for ship-ice interaction. The MCNS method has been calibrated against laboratory ice impact experiments involving drop tower experiments of cylinders with various shaped caps that are composed of fine-grained, equiaxed, freshwater ice (Herrnring and Ehlers, 2021; Herrnring, 2023; Müller et al., 2023). Mokhtari et al. (2023) propose a viscoelastic material model with pressure- and rate-dependent progressive damage for polycrystalline ice under impact loading. They develop three numerical implementations – Lagrangian FEM, coupled FEM-SPH, and coupled FEM-ALE – and validate each against physical and ice crushing empirical data. They note good correlation with empirical data for the FEM-SPH and FEM-ALE implementations but favour the FEM-SPH model due to its ability to simulate large deformations. Progress has also been made on "contactless" ice models (i.e. where the impacting ice body is not modelled with finite elements but instead is represented as a pressure). Andrade et al. (2023) extended the 4D Pressure Method (Quinton et al., 2012) to automatically generate real time high-pressure and low-pressure zones for any given ship-ice contact area shape. High-pressure zone size and shape are based on empirical observations (Gagnon et al., 2020). This approach is coupled with the Popov-Daley method (Daley 2000) to estimate the collision energy for a given ice impact scenario. This approach has a high computational advantage over ice models that use finite elements to model ice (Andrade et al., 2022).

Of particular relevance to ship hull resilience is understanding the responses of steel plates and frames to repeated impacts. Recently, (Robbins, 2020) used Memorial University of Newfoundland's large double pendulum apparatus (Lande Andrade et al., 2023) to conduct repeated impacts of a nominally rigid spherical cap of radius 25.4 cm (10 in.) and height 5.08 cm (2 in.) with a 1.36 m × 2.03 m × 7.9 mm mild steel clamped plate. Four repeated impacts were conducted, each with ~44 kJ impact energy. Sudo shakedown (Jones, 1973) was not observed for these impacts as the experiments prematurely ceased due to the COVID-19 pandemic. These repeated impacts were used to calibrate an explicit time-integration finite element numerical model, which was then used to predict plate behaviour for various repeated impact scenarios including repeated impacts at the same location on the plate, and various categories of repeated impacts at locations slightly adjacent to the previous impact location. Repeated adjacent impacts were shown to generate considerably more damage than repeated impacts at the same location. Cai et al. (2022) conducted laboratory repeated ice impact experiments (same impact location) on 800 mm × 400 mm × 2 mm and 1200 mm × 400 mm × (2 mm and × 6 mm) clamped steel plates at impact energies of up to ~700 J and developed numerical models of the experiments that utilized eroding ice elements. They report that pseudo shakedown was achieved, however most of the impact energy was dissipated via ice deformation processes. He and Guedes Soares (2021, 2024) recently studied beams and plates (both mild steel and aluminium alloy) subject to repeated impacts using an instrumented drop tower apparatus. In the experiments, pseudo-shakedown was observed for all cases of beam and plate. The authors' numerical simulations of the plate repeated impacts represented the experiments well but differ slightly in the growth rate of accumulated permanent displacement, peak contact force, and rebound velocity. The authors note that for the design of a structure that works under repeated impacts, if the plastic deformation is allowed, the pseudo-shakedown state will be the limit state of the design, provided a fatigue failure does not occur. It is accepted that this is true for repeated impacts at the same location, however (Robbins, 2020) shows that repeated adjacent impacts cause considerably more damage than repeated impacts at the same location that experience pseudo-shakedown. This path-dependent behaviour is notionally similar to the "moving load effect" described by Quinton (2015) in that preexisting damage at a location adjacent to the present location of a load compromises the structural response at the present location of the load.

Yoon et al. (2023) studied the influence of hydrodynamic forces in collisions between an icebreaker and an idealized rigid spherical-shaped iceberg through NLFEA simulations. In their work, ship hydrodynamics were considered through a user-defined plug-in for Abaqus/Explicit named HydroQus, a subroutine based on potential flow theory developed by the authors, while the iceberg hydrodynamics were disregarded. M. Li et al. (2023a) also studied ship-iceberg collisions using a NLFEA, focusing their work on studying the structural response of the ship's containment system during the collision.

Feng et al. (2023) numerically reproduced experimental three-point flexion tests on ice beams and collisions between ice floes and stiffened panels using a combination of smoothed particles and finite elements, while assuming a von-Mises plasticity constitutive model for the ice. A good correlation between the experimental and numerical simulations was observed in both studied setups in terms of force-time histories and remaining panel deformations. Ship-ice collision numerical studies were also carried out in the works of Y. Yu et al. (2024a) and Y. H. Shi et al. (2023b), the latter coupling the ship-ice-water-air using ALE and S-ALE algorithms (Fig. 9).

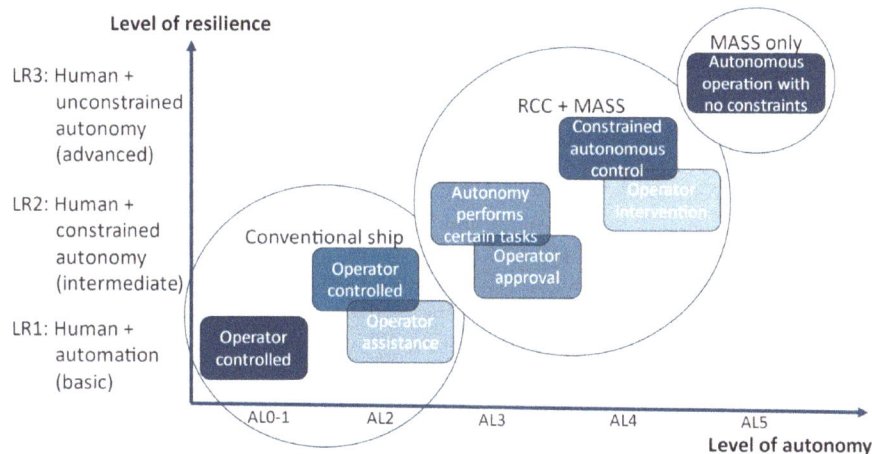

Fig. 9. Levels of resilience with respect to the level of autonomy (Fjørtoft and Mørkrid, 2021)

4.2 Resilience of Unmanned Vessels

Uncrewed vessels require additional considerations when investigating resilience. Fjørtoft and Mørkrid (2021) note that increasing the level of autonomy of a vessel also increases the level of resilience required. Considerations for the level of autonomy primarily consider the level of interaction required by the operator (controlled, assisted, approved, intervention) and the location of the operator (on-board, remote-control centre (RCC) or without real-time communication).

The three levels of resilience are described by Fjørtoft et al. (2023) as follows:

- "**LR1, Basic (AL0-1-AL2):** This resilience level includes barriers that are relevant for conventional ships where humans are in control and assisted by automation, such as extra crew on board and redundancy of critical components.
- **LR2, Intermediate (AL3-AL4):** Different barriers compared to LR1, as the crew is moved to an RCC. More focus on hand-over between automation and RCC operator and technological barriers that are feasible due to more autonomy and automation on board. Resilience-focus moves from an operational to a technological perspective, where humans are the back-up solution. Redundancy in communication is critical. Operator knowledge at the RCC must cover what today is critical crew on board.
- **LR3, Advanced (AL5):** At this level, all preventive barriers are technological and there are no operators in the loop, unless an incident happens. Hence focus on barriers must be technological and the requirements to testing and verification of software reaches a higher level. Barriers related to traffic control centres, RCCs and other

ships are crucial. Redundancy requirements will probably reach a new level. Most reactive barriers at this level will probably have to be external and linked to human intervention."

This contrasts with the degrees of autonomy used by the IMO (2021) for a scoping exercise during the development of the MASS code:

- **"Degree one:** Ship with automated processes and decision support. Seafarers are on board to operate and control shipboard systems and functions. Some operations may be automated and at times be unsupervised but ready to take control.
- **Degree two:** Remotely controlled ship with seafarers on board. The ship is controlled and operated from another location. Seafarers are available on board to take control and to operate the shipboard systems and functions.
- **Degree three:** Remotely controlled ship without seafarers on board: The ship is controlled and operated from another location. There are no seafarers on board.
- **Degree four:** Fully autonomous ship: The operating system of the ship is able to make decisions and determine actions by itself."

As an emerging area of interest there are still significant areas of research that are required to be undertaken. Chaal et al. (2023) undertook a systematic literature review of research into the risk, safety, and reliability of autonomous ships on papers published between 2011 and 2023. These included risk assessment and collision avoidance (covered in Sect. 5.1 of this report), reliability and maintenance of systems, vehicles and tracking control, artificial intelligence and risk analysis including security systems. Of note is that there is currently little literature focused on the resilience of uncrewed vessel structures.

The development of autonomous shipping networks requires resilience to be considered across the entire system to maintain successful operation. Fjørtoft et al. (2023) investigated all stages of an autonomous shipping and grouped threats into:

- human, organizational, and operational, such as workers and crew and collaboration challenges
- technological such as communication systems, remote operation and cyber-attacks, navigation and steering, and vessels port infrastructure,
- and external such as weather, tides and route closure.

While it is clear that autonomous shipping will require extensive work across the entire sector, a focus on threats particular to the vessel identified undetected damage to the vessel by water intrusion or fire, damage to control systems, loss of communication between the vessel and external or internal resources such as sensors.

A key challenge for the introduction of increased MASS is the increased risks associated with cyber-security. Schinas and Metzger (2023) identified that the increased digitalisation of shipping is an increasing threat to maritime security and highlighted the current policy challenges associated with ensuring the MASS vessels are classified as sea-worthy. They also proposed several measures to manage the risks associated with cyber-security including safety systems, training, hardware and software protection, administrative controls and physical security of computer systems.

To develop these systems a good understanding of existing and future cyberthreats is required. Mohsendokht et al. (2024) investigated cyber-attacks on the maritime industry since 2001 and collated the type of cyber threat, target, victim country, region, cyber threat origin, consequence and year. This data was utilised to generate a Bayesian network to provide further insight into the threat of cyber security to the maritime sector. However, they highlight that while cyber security is an ongoing threat to the sector there is currently a lack of a comprehensive database documenting attack.

While many of the threats target systems on board vessels, cyber-attacks can also target external information systems. identified methods for improving cyber resilience for attacks on Global Navigation Satellite Systems (GNSS), such as GPS, GLONASS and Galileo, used in autonomous navigation systems. These malfunctions can be caused either cyber-attacks or by faults. By comparing the GNSS system with radar measurements, the approach was able to identify malfunctions when within range of buoys, beacons or land.

In addition to cyber-attacks, another additional challenge for the resilience of uncrewed vessels is the unattended operation of machinery plants on the vessel. There have been several studies into understanding the maintenance and required redundancy of power and other systems. Approaches for these analyses typically use Bayesian inference to investigate these problems (Abaei et al., 2022; BahooToroody et al., 2022a, 2022b; Jovanović et al., 2024). Digital twins have also been utilised to identify faults and quickly assess their severity (Hasan et al., 2024).

While the use of MASS is increasing rapidly and key challenges in the area of cyber security and systems resilience is advancing there has been little published research into identifying the difference in structural resilience requirements for an uncrewed vessel differ from a crewed vessel.

4.3 Recommendations

The use of excessive conservatism is often prevalent when considering the large-scale interaction between ice and hull structures, which can lead to over-designed structures due to over-estimated ice loads (or equivalently underestimated hull resistance to ice loads). In beyond-design cases, such as non-ice class ships operating in ice, it is imperative to remove some conservatism by accounting for the dissipation of energy caused by structural deformations. That being said, it is unconservative to ignore the effects of sliding loads causing plastic damage. Additionally, non-trivial ship motions must be taken into consideration. For non-ice class ships operating in ice, recommended areas of study include thin ice dynamics and hull fracture due to sliding ice loads.

As the use of MASS continues to grow and advancements are made in cyber security and systems resilience, there is a lack of published research on the differences in structural resilience requirements between unmanned and manned vessels. Further exploration is needed in concepts for machinery resilience and fault prediction.

5 New and Emerging Technologies

Advancements in technology have led to increased utilization of oceans and sea areas for activities such as offshore wind energy, aquafarming, hydrogen production, and the use of new energy sources. This, along with the introduction of autonomous surface vessels, will bring new actors into the maritime industry, impacting its safety. As a result, novel methods, such as surrogate models that integrate theoretical and machine learning models, are being developed for design and simulation purposes. This article focuses on the topics of autonomous vessels, energy sources, offshore structures, and surrogate models.

5.1 Uncrewed Vessels

The integration of new technologies and increased automation on-board ships has led to the emergence of concepts for uncrewed or fully autonomous vessels, commonly referred to as Maritime Autonomous Surface Ships (MASS). Several practical examples of MASS vessels exists and dedicated companies developing such vessels are emerging (SeaMachines, SeaRobotics, MaritimeRobotics, L3Harris, OceanAlpha, Teledyne Marine, MindChip etc). Several small fully autonomous boats, typically less than 10 m long, are already conducting specialist tasks such as sea bottom bathymetry, water quality assessment and different surface level and underwater monitoring tasks. Larger pilot projects include the Norwegian container ship Yara Birkeland and Chinese 120-m-long electric container ship called Zhi Fei.

When it comes to MASS, several challenges involving safety and security due to various levels of autonomy and the changing role of humans are still to be addressed (Negenborn et al., 2023). These involve the legislative, technical, economic and social aspects. J. Liu et al. (2022b) confirmed that the MASS concept has become one of the most significant aspects to minimize human errors. In this research study the human-machine interface (HMI) based operational errors in autonomous ships are predicted to improve safety control levels. A statistical analysis regarding the impact of autonomous ships on safety at sea is performed by de Vos et al. (2021). The distribution of human casualties and lost ships over accident types, ship types and ship sizes are determined. It is concluded that the implementation of autonomy on small cargo ships with a length below 120 m will have the largest safety benefit. Furthermore, regulatory framework and guidelines are picking up. IMO aims to integrate new and advancing technologies in its regulatory framework and the aim is to adopt a non-mandatory goal-based MASS Code to take effect in 2025, which will form the basis for a mandatory goal-based MASS Code, expected to enter into force on 1 January 2028. Preliminary guidelines for using automated processes for navigation and systems maintenance have been published by several classification societies (DNV-CG-0264, 2021), (Bureau Veritas, 2019).

From technical perspective, sensor fusion, artificial intelligence and machine learning are to replace the perceptive and decision-making capacity of humans. AI-powered intelligent situation awareness and autonomous navigation algorithms must safely and efficiently adhere to the regulations, which are only designed for human interpretation without MASSs consideration (Bakdi & Vanem, 2024). One of the main challenges of the development of MASS is to ensure safe navigation and collision avoidance (CA). The International Regulations for Preventing Collisions at Sea (COLREG) provides the foundation for CA standards followed by vessels. However, with the advent of MASS, there is a need to reconcile these standards as in GOLREG standards several guidelines are linguistic or expressed in qualitative terms (e.g. "early" or "substantial") that are hard to interpret by machines. (Porathe, 2020) discussed the interaction between automatic ships and traditional manned ships in the light of prevailing regulations and new e-Navigation solutions suggested. The concept of "automation transparency" is discussed and several concrete examples are suggested.

In light of the autonomous collision avoidance problem of MASS, various solutions have been proposed by researchers, including the artificial potential field method, heuristic algorithms, reinforcement learning etc. (Chang et al., 2024) conducted an extensive literature review for publications from 2018–2022 to identify trends and weaknesses in recent studies on CA for MASS. The Conventional-Collision-Avoidance-Process (CCAP), which benchmarks manned modern ships' capacity for collision avoidance compliance under COLREG and industry requirements, was used to break down a ship's collision avoidance process into 53 CA functions under eight main categories.

Bakdi and Vanem (2024) focused on codifying COLREGs into a machine-executable system applicable to MASSs, then analysing their performance in dynamic and mixed interactions between multiple vessels in complex scenarios. Paper uses the concept of navigational complexity to support the decision-making and analysis in multi-collision-conflict scenarios. Graph models are exploited to quantify the complexity and for conflict resolution. The work is validated on a database of historical scenarios extracted from multiple data sources.

Zheng et al. (2021) proposed a decision-making method for ship collision avoidance based on improved cultural particle swarm optimization, that is used to find the optimal steering angle complying to COLREG regulations. In order to enhance the autonomous collision avoidance decision-making capability of MASS in accordance with the relevant provisions of COLREGs, Liang et al. (2024) proposed an improved heuristic NSGA-2 (Non-dominated Sorting Genetic Algorithm II) autonomous collision avoidance decision-making algorithm, which incorporates the collision hazard and the path cost of collision avoidance actions. Jadhav et al. (2023) proposed a novel artificial potential field (APF) based local planner to perform obstacle and collision avoidance for autonomous surface vessels. It is mentioned that this method works well even in scenarios imposing conflicting COLREG responsibilities on the vessel. The ability of this approach to maintain collision free navigation in narrow channels is also demonstrated (Jadhav et al., 2023). (Huang and van Gelder, 2020) proposed an improved time-varying collision risk (TCR) measure. The measurement of TCR reflects the dangerous level of the approaching ships and the difficulty of avoiding collisions. By comparing with traditional measures, e.g., Collision Risk Index (CRI), it was found that (1) the TCR can distinguish changes of risk that have identical CRI level, (2) the TCR measure offers a reasonable tool to evaluate the collision risk of entire traffic, and (3) it reflects the influence of manoeuvrability improvement on collision risk. The article reaches two conclusions: the collision risk is monotonically increasing when introducing more ships, and ignorance of ship manoeuvrability results in an underestimation of collision risk.

5.2 Offshore Wind Infrastructure

The installed capacity of Offshore Wind Turbines (OWTs) has grown exponentially in recent decades due to the ongoing climate crisis. Additionally, substantial progress has been made in developing Floating Offshore Wind Turbines (FOWTs) to exploit the stronger and more consistent winds available in areas far from the coast. As of

2023, the cumulative capacity of OWTs was around 64.3 GW worldwide, of which 188 MW belonged to FOWTs (Williams and Zhao, 2023). Collision events are considered in design standards as an Accidental Limit State because wind farms are continually exposed to ship collisions, both with service vessels operating within the farms and nearby marine traffic. Yet, the simulations of ship-OWT/FOWT collisions are challenging due to the significant interaction between aerodynamic, hydrodynamic, and structural contact forces through the impact.

X. Liu et al. (2022c) carried out a parametric study of a 5000 t Offshore Supply Vessel (OSV) striking a 3 MW OWT with a fixed jacket foundation, see Fig. 10. In their study, the ship's impact location on the jacket (i.e., braces or legs), impact angle, and impact velocities were examined and compared in terms of energy dissipations, local deformations, contact forces, and maximum tower displacements. Ladeira et al. (2023) proposed a simplified method to simulate ship collisions against tubular OWT supports that accounts for the support's local elastic and plastic denting, global elastic bending, plastic collapse, and a buckling mechanism at the support's clamped base. Nie et al. (2024) presented a numerical investigation on the dynamic behaviour of OWTs subjected to ship collisions and whose monopile foundations were equipped with different fender configurations. Ship collision loads will induce turbine responses of high order modes. Yu and Amdahl (2023) derived a Rayleigh-Ritz solution for high order natural frequencies and eigenmodes of monopile supported offshore wind turbines considering tapered towers and soil pile interactions. The model provided high accuracy for the predicted high order natural frequencies and eigenmodes of monopile OWTs by comparison of USFOS results. The importance of including the rotational inertia of the RNA was highlighted in the evaluation of high order modes. (Hammad and Yu, 2024) extended the high order modal solutions by Yu and Amdahl (2023) to derive the dynamic responses of monopile OWTs under accidental loads. For ship collisions, as a kind of hard contact loading, a contact algorithm was developed by utilizing provided nonlinear force deformation curves and numerical iterations for convergence. Ship collision loads and responses were well predicted using the proposed contact algorithm together with superposition of modal responses. The results emphasized the importance of considering turbine flexibilities in collisions with large elastic energy absorption in the struck OWT. In the works of Rong et al. (2020) and Ren et al. (2023), experimental and numerical investigations on the deformation behaviour of monopile and spar-like tubular members under lateral loads were carried out.

Zhang et al. (2021) studied the response of a 5 MW spar-type FOWT subjected to ship collisions under different operational conditions. To do so, they proposed a mathematical model to simulate the external mechanics of ship-FOWT collisions by integrating a simplified 3D dynamic model for rigid bodies with an aero-hydro-servo-elastic fully coupled simulation tool, allowing them to assess the full 6-DOF kinematics through the collision. In a later study, Zhang and Hu (2022) investigated the influence of considering the internal mechanics of both the FOWT and the ship structure in collision events using a NLFE model built in LS-DYNA in which the wind, wave, and mooring loads were included through a user-defined load subroutine. Z. Yu et al. (2022b) investigated the response to ship collision events of a 10 MW semi-submersible FOWT in both parked and operating conditions using the NLFE code USFOS (see Fig. 11), also accounting for

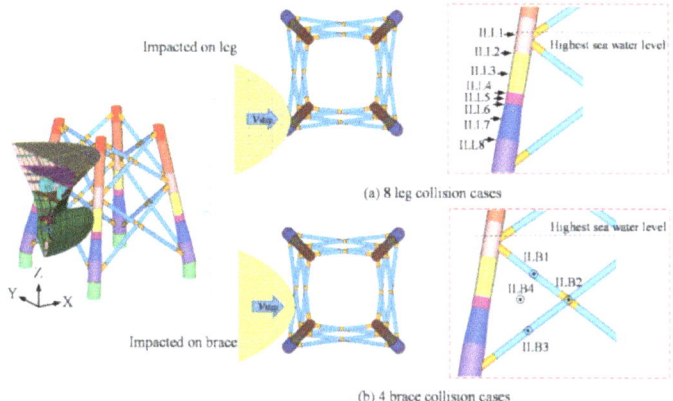

Fig. 10. Schematic of: (a) eight leg impact cases, and (b) four brace impact cases (X. Liu et al., 2022c)

aerodynamic, hydrodynamic, and mooring restoring forces. In their study, the importance of considering not only collisions in still water but also under operational conditions was remarked as the latter may lead to exceedance of the nacelle operational limits (i.e., accelerations) and increase the risk of tower buckling.

In the work of Yoon and Choung (2023), ship collisions against a spar-type FOWT were studied in the FE code Abaqus using an in-house algorithm called HydroQus, which was developed by the authors to assess the collision external dynamics including hydrodynamic effects in a similar fashion than the modified rigid-body solver MCOL proposed by (Le Sourne et al., 2001), though extended to account for 1st and 2nd order wave excitation forces. In another study, Ren et al. (2022) studied ship collisions against spar-type FOWTs in still water conditions using the FE code LS-DYNA along with the rigid-body solver MCOL to model the collision external dynamics, also including non-linear damper elements to account for viscous forces in the FOWT's surge direction. More recently, Vandegar et al. (2024) proposed a simplified model to simulate ship collisions against spar-type FOWTs. The proposed model consists of a semi-analytical approach in which the external dynamics are assessed through the rigid-body solver MCOL, while the internal mechanics of the spar FOWT are modelled using the elasto-plastic formulations developed by Ladeira et al. (2023) for standalone OWT supports. In the work of Haneda et al. (2020), experiments of ship-FOWTs collisions using scale models were performed for different impact velocities, wave conditions, and floater configurations. Lastly, in a different topic, the dynamic response of FOWTs under freak waves was studied both experimentally and numerically in the works of Zeng et al. (2023a) and Zeng et al. (2023b).

The majority of FOWT's floaters proposed or constructed to date have been made of steel. However, there is increasing interest in using concrete structures due to their economic advantages and longer lifespans as discussed in Choisnet et al. (2016) and Mathern et al. (2021). Márquez et al. (2021) studied the capabilities of different concrete constitutive laws for capturing the flexural response and shear failure of RC walls subjected to ship collisions. In their study, the authors emphasized on the non-negligible

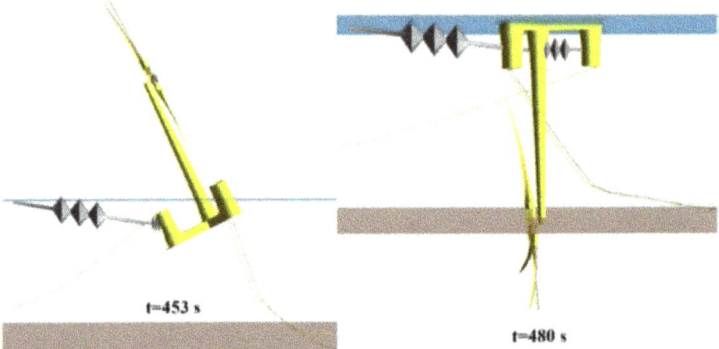

Fig. 11. Overturn of the OO-STAR floating turbine after tanker collision for the case column1-180deg-tanker side (Z. Yu et al., 2022b)

discrepancies between the studied constitutive laws when predicting force-displacement relationships and shear failure under the same collision conditions. Sha et al. (2022) investigated the response of a RC semi-submersible FOWT subjected to ship collisions using a two-step decoupled analysis framework where the local force-deformation curves were firstly gathered through the FE code LS-DYNA and later used in the assessment of the global motions of the collision using OrcaFlex. The authors also studied the effects of strain rate effects in the striking ship, which were seen to have a non-negligible influence on the impact force profiles and structural damages. In the work of Márquez et al. (2022), a simplified model to simulate ship collisions against FOWT with RC floaters was proposed. Their model accounts for the rigidity of both the ship and the RC floater, while the external dynamics are accounted for through the rigid-body solver MCOL. In their work, the simplified model predictions correlated well with NLFE simulations for different parametric variations in terms of energy balances, contact force profiles, rigid-body motions, and deflections/penetrations of both structures at the contact point.

Although advances in the modelling of ship-OWT/FOWT collisions have been made in the past years, there are still challenges to face. For instance, the development and integration of pertinent damage metrics (e.g., through-thickness crack sizes, fracture thresholds, and residual strengths) would improve the understanding of their structural vulnerability to collision events. The use of these metrics along non-deterministic frameworks is a promising area of research.

5.3 Alternative Fuels and Transportation

The International Maritime Organization (IMO) proposed the vison to reduce the greenhouse gas (GHG) emissions from international shipping. One level of ambition directing the 2023 IMO GHG strategy is to peak GHG emissions from international shipping as soon as possible and to reach net-zero GHG emissions by or around, i.e. close to, 2050 (IMO, 2023).

Bilgili (2021) mentioned, that alternative marine fuels have to be studied for its useability and whether they can replace conventional fuels. A comparison of alternative fuels for shipping in terms of lifecycle energy and cost is proposed by Law et al. (2021). A study analysing the engineering considerations of alternative fuels storage on board of large-scale international vessels with a particular focus on ammonia, hydrogen and methanol is proposed by McKinlay et al. (2021). Wang et al. (2022) reviewed low and zero carbon fuel technologies available for ships including liquefied natural gas, liquefied petroleum gas, methanol, biodiesel, hydrogen and ammonia. Strantzali et al. (2023) developed a comprehensive multicriteria evaluation approach for alternative marine fuels comprising economic, technical and environmental as well as social aspects. Bergström et al. (2023) developed a simulation-based approach to evaluate various decarbonization strategies for a selected fleet of merchant ships. Exemplarily, investigated the integration of a tank storage solution for alternative fuels on a RoRo ship. Different candidates like Liquefied Natural Gas (LNG), Methanol (MeOH), Ammonia (NH3), Liquefied Hydrogen (LH2) but also Heavy Fuel Oil (HFO) and Marine Diesel Oil (MDO) are investigated. Exemplarily, a comparison of the different fuels based on environmental impact, risks and hazards are given in Table 1. A score of 1 (good) to 5 (bad) is attributed to each fuel (Aider et al., 2023). The configuration and application of low-carbon fuels green marine power systems in diverse ship types is presented by Wang et al. (2024a). Further perspectives and challenges of large-scale shipping of low-carbon fuels are investigated by Abraham et al. (2024). The shipping industry is already very active implementing several actions to decarbonize ship operations, but more efforts are necessary to continue (Guedes Soares, 2023). Based on the proposed literature study, liquefied natural gas, methanol, ammonia and hydrogen can be treated as feasible alternatives to fossil fuels for ships. Here, hazards and resulting risks of the application and transport of alternative maritime fuels are discussed. Fuel cell and battery technologies also have to be mentioned as alternatives to fossil fuels in the maritime sector.

Table 1. Comparison of fuels based on environmental criteria from 1 (good) to 5 (bad), (Aider et al., 2023)

Criteria	HFO	MDO	LNG	NH_3	MeOH	LH_2
GHG reduction potential	5	4	2	2	1	1
Air pollutants	5	4	3	3	2	1
Aquatic ecotoxicity	5	3	1	3	1	1
Human toxicity	3	3	5	4	3	4
Flammability	1	4	5	4	4	5
Explosion risk	1	1	5	2	1	5

5.3.1 Liquefied Natural Gas

Liquefied natural gas (LNG) is recognized as one of the most efficient solutions for reducing GHG emissions in the shipping industry (Abdelmalek and Guedes Soares, 2023). LNG carriers burn the Boil-off gas (BOG) generated during the voyage via steam turbines. In the maritime industry not only LNG carriers but also bunkering vessels, LNG-fueled vessels and floating LNG units are operating. For transport and storage purpose of natural gas the liquefaction process enables a 600 times volume reduction. Natural gas is liquefied at atmospheric pressure at $-162\,°C$ with a density of 443.5 kg/m^3 and each kilogram carries 13.9 kWh of energy (Aider et al., 2023).

Abdelmalek and Guedes Soares (2023) reviewed risk analysis studies in the maritime LNG sector classified for methods, tools, objectives, ship types, ship operation and human errors. The resulting hazards from the accidental non-ignited LNG leaks (environment, human body, brittle fracture, rapid phase transition) as well as ignited LNG leaks are pointed out. Due to cryogenic and flammable nature of LNG, major attention has to be given on fire protection to prevent accidents. Therefore, Kim et al. (2023) developed a risk-based fire protection design of an LNG fuelled ship. The quantitative risk assessment (QRA) is used to optimize the fire protection design of a product carrier. Hazard analysis by leakage and diffusion in LNG ships during emergency transfer operations on costal water is performed by Zhu et al. (2022). This study shows that the diffusion of LNG vapor cloud is significantly influenced by wind speed and hull restrictions. The LNG cryogenic hazard zone exists in the vicinity of the leakage points. Further guidance to authorities to formulate disaster prevention, mitigation regulations and emergency response measures are proposed.

Further investigations on accidental release in the bunkering of LNG focusing on phenomenological aspects and safety zone are performed by Carboni et al. (2022). Furthermore, the safety of LNG-fuelled ships has to be assessed in terms of potential structural damages owing to collision accidents and resulting LNG spills (S. K. Kim et al., 2022b). Collision-accidental limit state-based safety studies are performed by S. K. Kim et al. (2022b) exemplarily for an LNG-fuelled container vessel. In this study nonlinear finite element method is used at the most unfavourable scenario of collisions between a striking ship's bow and struck ship's LNG fuel tank. It is found that the inner side hull structure of the struck ship can be damaged in ship-ship collisions and the current industrial guidelines for LNG fuel tank designs are required to amend to apply for LNG-fuelled ships. Further numerical investigations on the structural assessment ship during bunkering LNG and marine operation under collision accidents are performed by Sohn and Jung (2022). The interaction between ship motions and sloshing due to partially filled tanks may affect the seakeeping properties of an LNG carrier with partially filled tanks. Exemplarily, Lyu et al. (2021) performed numerical investigations on the ship motion-sloshing interaction using flied method.

5.3.2 Methanol

Methanol is a flammable and highly volatile compound but a liquid at standard temperature and pressure. It also can be treated as alternative fuel in the shipping industry. The commercial fleet of methanol fueled ships was counted to 10 units in 2019 and its global production reached 100 million tons in 2020 (Aider et al. (2023)). Methanol still has not been deployed to the same scale as other alternative fuels as LNG with only 16 vessels deployed and 23 on order (Ramsay et al., 2023). Methanol is classified as a low flash point fuel that its onboard utilization is regulated by the international code of safety for ships using gases or low flashpoint fuels (IGF Code). Therefore, safety concerns regarding methanol storage due to its lower flash point, invisible flame and its life cycle assessment compared to other fuels should be conducted Parris et al. (2024). Methanol is miscible with water at any rate and even small amounts of methanol are poisonous to living organisms (Yaman et al., 2024).

Bertagna et al. (2023) investigated the impact of fuel switch to methanol on the design of a cruise ship. The greater fire safety compared to common patrols is reported. It is also mentioned that methanol is toxic for humans by ingestion, skin absorption and vapours inhalation but even after prolonged exposure, a fast and reliable treatment is available to ensure full recovery. Aider et al. (2023) investigated the integration of a tank storage solution for different alternative fuels on an existing RoRo ship. Based on technical and economic aspects, methanol is one of the most promising alternative fuels for operational usage.

5.3.3 Ammonia

Ammonia is a carbon-free molecule and will not produce CO_2 during the combustion process. Therefore, it is one of the most promising alternative fuels to deliver the desired GHG reductions for the maritime industry and to reach IMO's goals. Furthermore, ammonia is characterized by toxicity, flammability, corrosiveness and a distinctive smell but it can be handled similar as conventional gas fuels like propane. Ammonia can take a liquid state at room temperature by pressurization at 10 bar and at atmospheric pressure the liquidation process requires a cryogenic condition of -33 °C. Ejder and Arslanoğlu (2022) anticipated that the risk of exposure of ship personnel to ammonia during bunkering operations will increase that personnel should wear protective equipment. A review of safety assessments of the ammonia bunkering process in the maritime sector is proposed by Duong et al. (2023).

A regulatory gap analysis for risk assessment of ammonia-fuelled ships is performed by Jang et al. (2023). In this study major hazards namely toxicity, chemical corrosion, fire and explosion as well as their potential impact on humans, environment and ship in the event of ammonia leakage are discussed critically. This paper examines and compares safety regulations, rules, standards and guidelines to ammonia-fuelled ships.

5.3.4 Hydrogen

Hydrogen is the most abundant element in the universe, but pure hydrogen gas is not directly available and results from the decomposition or reformation of other products. Hydrogen gas is not toxic, odourless, colourless and tasteless but very flammable. Explosion risk becomes emanant in an environment where hydrogen concentration ranges are given between 18.3% and 59% (Aider et al., 2023). Hydrogen can be compressed and stored in gas cylinders handling pressures between 350 bar and 700 bar. Details about the different hydrogen storage methods and challenges for its maritime applications are proposed by Van Hoecke et al. (2021).

Safety considerations of hydrogen applications in shipping are proposed by Depken et al. (2022). In this study a novel method is developed to compare the safety of fuel alternatives, and it might be helpful to enable a wider but still safe use of hydrogen in shipping. Lampe et al. (2023) proposed a simulation-based approach for the calculation of design loads as well as tank specifications for liquid hydrogen fuel tanks. Status and prospects in technical standards of hydrogen-powered ships for advancing maritime zero-carbon transformation are proposed by Wang et al. (2024d). This paper aims to provide valuable guidance and direction for the future demonstration, application and promotion of hydrogen fuel cell ship technology. Further details about hydrogen towards a marine green power system architecture are proposed by Wang et al. (2024c). Due to fire risks by using and transporting hydrogen on ships, Lim et al. (2024) performed numerical analyses on extinguishing of sprinklers in a hydrogen pool fire. In this study, optimal sprinkler parameters are evaluated when fire broke out after the liquid hydrogen leaked from hydrogen storage tank on the ship's deck using hydrogen fuel cell and formed a pool. The effect of leakage location and ventilation condition on hydrogen leakage during shipping of fuel cell vehicles is investigated by Gao et al. (2024). A review of shipping cost projections for hydrogen-based energy carriers is proposed by Schuler et al. (2024).

5.3.5 Fuel Cells

Fuel cells are representing one of the possible technologies to reduce the emissions related to the present. Fuel cells technologies are already emerging that electricity is produced from the chemical energy stored in the fuel (Aider et al., 2023).

Analyses and evaluation of fuel cell technologies for sustainable ship power with respect to energy efficiency and environmental impact are proposed by Wang et al. (2024b). When hydrogen fuel and fuel cells are used on ships, there exists a possibility of low-flash fuel leakage, leading to the risk of fire and explosion (Chen and Guan, 2021). Exemplarily, Chen and Guan (2021) developed a safety design and engineering solution of a fuel cell power ship with reduced risk level in the range of a non-hazardous zone. A feasible design and optimization of an integrated power system of solid oxide fuel cell and marine low-speed dual-fuel engine are proposed by Qu et al. (2023). Exemplarily, Chen and Guan (2021) developed a safety design and engineering solution of a fuel cell power ship with reduced risk level in the range of a non-hazardous zone. A feasible design and optimization of an integrated power system of solid oxide fuel cell and marine low-speed dual-fuel engine are proposed by Qu et al. (2023). Vairo et al. (2023) developed a machine learning model for early detection of hazardous system deviations of solid oxide fuel cells used for shipping applications. It is also mentioned that the early detection of hydrogen leakages is a very important safety factor to avoid a potential accident scenario.

Wang (2023) proposed a dynamic collision avoidance strategy of a hydrogen fuel cell unmanned ship based on an improved fusion dynamic window method. In this study simulation and experiment are verified that the proposed algorithm can effectively avoid dynamic obstacle ships and the avoidance behaviour meets the rules and constraints. Xu et al. (2023b) determined the safety of hydrogen fuel cell powered ships during navigation. Jet flame accidents and flash accidents are pointed out to be the two main accident scenarios that may occur for hydrogen fuel cell powered ships. Zhang and Jiang (2023) applied the formal safety assessment method to analyse the fire risk of hydrogen storage in a hydrogen fuel cell powered ship system. It is mentioned that optimizing the ventilation system in hydrogen fuel cell powered ships can mitigate hydrogen diffusion during leakage incidents, thereby facilitating fire spread control more rapidly and diminishing accident risks. Human induced errors are also important factors, which can be reduced by comprehensive training encompassing operational procedures, safety awareness and emergency handling (Zhang and Jiang, 2023).

5.3.6 Batteries

Battery energy storage systems play a crucial role in electrifying the maritime industry which helps to pave the way towards achieving net-zero GHG emissions (He et al., 2024). Life-cycle assessments of power batteries for all-electric vessels for short-sea navigation are performed by Perčić et al. (2022). This study includes RoRo passenger ships highlighting the electrification by a Lithium-ion battery as the most appropriate alternative according to environmental and economic indicators. However, previous battery fire accidents show that battery suppliers and regulations should have a very robust safety framework, not only single cell and module level but also on vessel level (He et al., 2024). A review of risk analysis for marine transport and power applications of lithium-ion batteries is proposed by Yin et al. (2024). In this study, the safety and risk management of lithium-ion batteries shipping and battery compartment is discussed. For collisions of ships, it is mentioned that deformed battery packs may tear the battery separators, resulting in a short circuit and leak of flammable electrolyte, resulting in fire. Wei et al. (2021) proposed a fire monitoring system for power batteries on ships.

5.4 Fixed and Floating Bridges

Ship-bridge collisions might cause damages to structural components of the ship and the bridge or even failure of a whole structure. Y. Chen et al. (2024) performed dynamic response analyses of a pontoon interception system under ship collision to protect the bridge. The proposed pontoon interception system is able to stop ships of different weight and speed to prevent ship-bridge collision. The study also shows that the tension force of the pontoon interception system is not very sensitive with the kinetic energy of the striking ship due to the fluid around the pontoon and the friction between the system anchor and the riverbed. The influence of the fluid on the impact force of a ship-bridge collision is investigated by Y Chen et al. (2024). It is summarized that the fluid has a significant influence on the impact force, speed, energy dissipation and structural damage of ship-bridge collision. A simplified method to consider the important hydrodynamic effect in oblique ship-bridge collision is proposed by Fan et al. (2023), which is superior to the constant added mass method in terms of the hydrodynamic effect, but the computational costs are reduced significantly compared to fluid-structure interaction simulations. Ye et al. (2023) performed fluid-structure interaction analyses of oblique ship-bridge collisions obtaining different results compared to the constant added mass method. An analytical formula is improved to estimate absorbed energy during collision.

Nian et al. (2023) performed studies on the crashworthiness of nature-inspired functionally graded lattice metamaterials for bridge pier protection against ship collision. Wu et al. (2024) used the FEM to performed fluid-structure interaction analysis of vessel-bridge-steel floating fender collision. This study shows that the surrounding water has a positive impact for energy dissipation but neglecting the surrounding water in the constant added mass method leads to substantial deviations in the fender performance. A novel steel box-soft body combination for bridge protection against ship collision is evaluated by Yan et al. (2023). Nonlinear FE were performed to demonstrate that the anti-collision facility reduces the ship impact force as well as the bow damage.

Experimental studies on the behaviour of precast segmental piers under ship impact loadings are performed by Yao et al. (2022). The tests indicate that segmental piers are more flexible than monolithic piers. A rational design procedure for bridge piers and pylons against ship collision impacts is proposed by Pedersen et al. (2020). A set of risk acceptance criteria are developed followed by a mathematical concept to determine the probability of critical situations for ships being close to bridges. Furthermore, the probability of ship collision accidents caused by human errors as well as technical errors are considered. Based on comprehensive numerical results, an empirical relation to predict the maximum bow impact force as function of ship impact velocity, ship loading and ship size is derived, which is suited for design against bow collisions. FE analyses and experimental studies on the assessment of the axial crushing force of bow structures are performed by Pan et al. (2022) respectively Pan et al. (2023). Simplified formulae are developed for bulbous bow and modifications for existing analytical formulae proposed rake bow.

Liu and Guo (2023) proposed studies on the risk of ship collision in the bridge life cycle based on the synergetic theory. Zhao et al. (2023) performed risk analyses ship-bridge collision based on the automatic identification system data and the FE method. Xu et al. (2023a) used a machine learning-based framework to predict the impact force in ship-bridge pier collision. In this study, a FE model is used to simulate barge collisions with a double-column pier and machine learning algorithms are combined with the fast Fourier transformation to predict the impact force accurately.

Cao et al. (2023b) investigated the dynamic performance of a triple-column bridge pier under barge collision using finite element method. In this study, a performance index based on the internal energy power slope is developed to prescribe the damage states of the impacted column. Two protective column jackets are investigated for enhancing the structural anti-impact capacity, but experimental tests are still required to validate the numerical results. T. L. Chen et al. (2022a) developed analytical impact force models for bridges under barge collision. Nonlinear finite element analyses are validated against experimental data on reduced-scale barge bow models and the flexible impactor lateral collision tests on the reinforced concrete (RC) column. A simplified model to determine the impact force time-history for bridge pier-barge collision is proposed by Song et al. (2022c). The dynamic behaviour of double-column RC bridge subjected to a barge impact is analysed by T. L. Chen et al. (2022b). Results delivered by finite element analyses (LS-DYNA) are validated against experimental data for a scaled barge bow with various impact velocities on the three-column RC bridge pier specimens. This study shows that oblique barge impact amplifies the overall collapse risk of a bridge and that the oblique angle has an increasing effect on the induced damage level. The influence of local scour effect around the bridge pile foundation on the dynamic response of bridges under barge collision is investigated by Wang et al. (2024e). The study shows that horizontal displacements of the bride pier increase with an increased scour dept but the shear force and bending moment profile is primarily attributed to the barge collision location. It is also mentioned that the influence of scour is limited on the barge-pier collision force as the interaction force level primarily depends on the relative strength between the striking barge bow and the struck pier as well as the collision energy. Guo et al. (2022) proposed experimental and numerical investigations of scoured bridges with protective bonded steel plates against barge impact. It is observed that steel plates bonded on the bridge pier and transverse beam can efficiently protect the bridge from initial damage or potential failure. The most common causes of bridge failure are hydraulic failure due to scour of the bridge foundation and collision of the vessels (Irhayyim et al., 2022). Both hazards are typically treated independently but they can co-occur. Therefore, Irhayyim et al. (2022) performed research on the assessment of scoured bridges subjected to ship impact. Parametric studies (FEA) show that the pile displacements increase for increased scour and ship impact locations close to the pile cap. Wang et al. (2023b) performed nonlinear finite element analyses to investigate the dynamic response of a scoured bridge under ship collision. This study also confirms that maximum horizontal displacement measured at the top of both bridge piers and group piles increase with an increasing scour depth. The scour has e minor influence on the impact forces of the ship bow.

The above discussion mainly focused on the fixed bridges. Floating bridges are large infrastructures used to cross wide and deep fjords, assuming a long-crested and homogeneous wave field in framework of floating bridge designs Cui et al. (2022). Due to fjord topography wave fields might be short-crested and inhomogeneous, that Cui et al. (2022) used a generic method for assessment of inhomogeneous wave load effects of very long floating bridges. The method is applied exemplarily to an end-anchored floating bridge concept for crossing the Bjørnafjord in Norway, showing that the inhomogeneous wave field can cause a smaller standard deviation in the strong axis and weak axis bending moments but a significantly larger standard deviation in the axial force component along the bridge girder. Further effects of inhomogeneous wave modelling on extreme response are investigated for the same very long end-anchored floating bridge by Cui et al. (2023). The knowledge of extreme response based on inhomogeneous wave load modelling is also applicable for ultimate strength check to ensure the safety of the bridge. Fenerci et al. (2022) analysed the hydrodynamic interaction of the floating bridge pontoons and its effect on the dynamic response of the end-anchored floating bridge crossing the Bjørnafjord. It is recommended that the hydrodynamic interaction of the pontoons should be considered in the design stage of the bridge due to increased design stresses. Luan et al. (2024) investigated the shear lag effects on the global bending moments for the same long-span pontoon bridge under self-weight, traffic and environmental loads. Practical hints are given when caution is necessary by using an elementary beam model to estimate global bending moments about the weak axis of the bridge girder. The response of the end-anchored floating bridge crossing the Bjørnafjord under combined ship collision as well as wind and wave loads is analysed by Wang et al. (2023a). This study reveals the importance of multi-hazard assessment for structural analyses of floating bridges, and it also can be used as a basis for more thorough evaluation of floating bridge response under other hazards. Z. Yu et al. (2024b) investigated the ultimate and the residual strength of columns in an initial design of the Bjørnafjord floating bridge before and after ship collision with an initial kinetic energy of 125 MJ and 150 MJ respectively corresponding to a 10,000-year event as estimated from risk assessment. The finite element method, PULS analysis and DNV rules deliver similar ultimate strength for the present structure, but the results based on DNV rules are more conservative. A significant reduction in bending and torsional strength of the bridge column is observed after collision due to the presence of fracture. It is also mentioned that the results are sensitive to the adopted fracture criterion. Eidem and Sha (2024) analysed the ship impact from a general cargo vessel on the Bergsøysund floating bridge in Norway. The possible long-term consequences of the ship impact on the steel truss are given by fatigue and nonvisible micro cracks that close inspections are recommended. Experimental tests on floating bridges are extremely rare that Rodrigues et al. (2022) performed model tests of a hydro-elastic truncated floating bridge. A truncated segment of a full straight bridge at a fjord crossing supported by floating pontoons and mooring clusters. The experimental conditions comprise combinations of regular and irregular waves, current and wind.

5.5 Other Offshore Infrastructure

5.5.1 Solar Plants

Power generation from solar photovoltaics (PV) is indispensable to accomplish a complete sustainable environment and to meet the United Nation's sustainable development goal (Ghosh, 2023). Not only land-based PV (LPV) plants but also water-based PV (WPV) might be required to reduce the costs and to reach the corresponding goals. The group of WPV is composed of floating PV (FPV), underwater PV, offshore PV and canal top PV related to specific challenges in planning, installation and maintenance to prevent damages and to protect the environment. A comprehensive review of water-based PVs is proposed by Ghosh (2023). In this paper, three critical ambient factors such as wind load, albedo and ambient temperature are discussed with respect to the performance of WPV systems. Floating solar photovoltaics are related to higher energy yield and efficiency compared to conventional land-based solar photovoltaic systems (Ramanan et al., 2024). Further concerns of FPVs including degradation, corrosion, friction, mooring and anchoring systems as well as stress on moving parts, which may lead to catastrophic failure during floods and climate change are mentioned by Ramanan et al. (2024). Exemplarily, Song et al. (2022a) investigated the dynamic response of multiconnected floating solar panel systems vertical cylinders under normal operating and extreme conditions. The effect of various inlet angle of wind and wave loads on FPVs taking stress distributions into account is investigated by Choi et al. (2023). Investigations on fully nonlinear dynamics of FPV platforms with twin hull by tubular floaters in ocean waves are proposed by Abbasnia et al. (2022). Fluid-structural analyses of modular floating solar farms under wave motions are performed by Sree et al. (2022). Layout optimization of offshore wind farms considering spatially inhomogeneous wave loads are given by Zilong and Xiao Wei (2022). On overview of possibilities of solar floating photovoltaic systems in the offshore industry is given by Vo et al. (2021). Further challenges and the need of research with respect to projecting FPV power plants at any location, temperature and solar radiation data, maximum wind speed, snow load, water current, cyclone and typhoon risks are mentioned by Cuce et al. (2022). Risk assessment and policy recommendations for a floating solar photovoltaic system are proposed by Dellosa et al. (2024). In this study, recommendations for the mitigation of environmental, technical, regulatory, economic and social based risks are given.

5.5.2 Offshore Aquaculture

Offshore aquaculture is an important industry in Norway and many other countries around the world and will continue to grow in the future. Traditional offshore aquaculture structures are generally located nearshore along the coasts with floating collar-net systems. As the space available along the coasts is getting limited for further expansion and there is a growing concern on potential pollutions with increasing numbers of fish farms, the trend in the aquaculture industry is to move the fish farms from sheltered waters to more exposed seas, where the environmental conditions are more extreme. This requires new designs of offshore aquaculture structures and systems. Many innovative designs and concepts were developed, e.g. the semi-submersible Ocean Farm 1 platform and the ship-like Havfarm 1 platform that are already in operation in Norwegian waters. The aquaculture industry faces several challenges, e.g. sea ices, fish welfare, etc. From the perspectives of accidental limit states, fish escape due to breakage of fish nets is a major challenge, which will cause significant economic loss, pollution of wild stocks and environmental issues. Yang et al. (2022) presented data and scenarios based on the investigation of 745 fish escape accidents and near accidents reported to the Norwegian Directorate of Fisheries from January 2006 to August 2019. They derived the direct and underlying causes and developed generic scenarios for fish escape that can be used for improved reporting of accidents. Fish nets made of different materials and knotting techniques may exhibit very different behaviours and failure modes, which is not well understood. It is not very clear that the failure mode of rope is due to fatigue, creep, wear or simply tensile failure. The Norwegian Standard NS 9415 (Norway, 2021) however requires the force level to be below certain strength limits. Some mechanical strength tests of fish nets were presented by Liu et al. (2024) and Slagstad (2024). More investigation will be needed for failure of fish nets, especially for offshore applications.

5.5.3 Hydrogen Production Facilities

Green hydrogen production might by very important in global endeavours to reduce greenhouse gas emissions and to expedite the transition to a low-carbon future (Kumar et al., 2024). A risk-based multi-criteria decision-making framework for offshore green hydrogen system developments is proposed by Kumar et al. (2024). Fixed or floating offshore wind farms as well as floating solar PV farms can be used as renewable energy systems and already existing respectively new jacket platforms for the electrolysis-based hydrogen production. In this study, risks of hybrid renewable powered offshore green hydrogen systems are proposed. Alternatives for transport, storage in port and bunkering systems for offshore energy to green hydrogen are proposed by Saborit et al. (2023). Experimental and numerical studies on the offshore adaptability of floating hydrogen liquefaction production storage and offloading unit are performed by Sun et al. (2024).

5.5.4 Artificial Island

Artificial island might be useful to handle an increasing burden on land resources due to the urbanization. Ma et al. (2020) proposed a systematic approach regarding the safety management process in sea reclamation engineering, including the safety system design, analysis, control and assessment using the Sanya Sea Reclamation Airport (Hainan Providence, China) as case study. A huge amount of large size steel cylinders has to be transported and safely installed in a hazardous environment. Song et al. (2022b) mentioned that large cylindrical structures are used for artificial island construction and crucial in mega sea-crossing infrastructure projects such as the Hong Kong-Zhuhai-Macao Bridge project. In this study the failure mode and mechanism of large cylinder structures are investigated for artificial islands on soft clay to ensure its safety.

On the other hand, floating artificial islands also can be used as alternative solution for urban expansion in waterfront areas around the world. Flikkema et al. (2021) defined a floating island as "an artificially created floater, or set of connected floaters, moored to the seabed of which the topside can be used for activities similar to activities on land". In this paper, a technical solution for floating islands which have identified barriers for multi-use in rules and regulations are described. Drummen and Olbert (2021) proposed a conceptual design of a modular floating multi-purpose island. It is mentioned that the distance between neighbouring modules needs to include a large safety margin, that joint applications of multiple floaters are unfeasible. Tamis et al. (2021) developed an approach for the environmental impact assessment for floating island applications. This study provides a first indication of the likely threats to the ecosystem and its components.

5.6 Metamodels – Surrogate Models

Although a promising field of research, the use of metamodels (i.e., surrogate models) in marine accidental events has been limited in the past years. Metamodels are particularly valuable in analyses requiring numerous computations, especially when simplified models lack accuracy or are unavailable. Potential applications of surrogate modelling in such events include parametric explorations, uncertainty quantification, reliability analyses, and risk assessment.

Fan et al. (2020b) built a surrogate model using response surface modelling to predict the residual compression strength of the columns of a bridge subjected to barge collisions. The surrogate model was then used along Monte Carlo simulations to assess the vulnerability of the bridge by performing a fragility analysis for different exposure times, including this way the effects of corrosion-induced deterioration. Das et al. (2022) studied and compared different metamodeling techniques, consisting of deep neural networks, polynomial regression, and gradient-boosting regression trees, for building models able to predict damage extents and oil outflow in tanker collision accidents. In their study, they found the deep neural network to be the best compromise between accuracy and computational speed. Mauro et al. (2023) investigated both the accuracy and calculation time of surrogate models based on multiple linear regressions, neural networks, and decision trees to assess the breach dimensions after passenger ship collisions. In their work, a collision database consisting of 4400 damages built with the super-element code SHARP was used for training the models, whilst the best fit was obtained with the forest trees at the expense of being computationally slower. Xu et al. (2023a) proposed a framework for predicting the impact force time-histories of ship-bridge collisions using machine learning techniques. The database used for training the models was created using a high-fidelity FE model, while the impact force duration, frequency points of the impact force, and impact force time-history were built employing a Kriging model, a Backward Propagation Neural Network, and the Inverse Fast Fourier Transform, respectively. Morán et al. (2023) investigated the capabilities of an active learning approach for surrogate modelling to effectively generate the training samples in reliability analyses involving multiple limit state functions. The proposed approach was subsequently applied in a numerical experiment to estimate the probability of repair and failure events for an offshore wind substructure subjected to ship collisions. M. Zhang et al. (2022c) proposed a machine learning method for the evaluation of ship grounding risk in real operational conditions. The method is validated for a Ro-Pax vessel operating in the Gulf of Finland over a 2.5-year ice-free period based on big data records. La Ferlita et al. (2022) developed a deep neural network (DNN) to rapidly predict the residual hull girder strength of damaged ships in vertical bending. The DNN is composed of multiple fully connected layers with a Rectified Linear Unit (ReLU) which has been applied to more than 6000 samples and validated using a leave-one-out technique. The residual hull girder strength is predicted well for different cross sections under various damage scenarios due to grounding. The corresponding ultimate bending moment is compared to simplified Smith method for sagging and hogging conditions with good agreement. Ship casualties such as groundings and collisions require fast and reliable analysis methods to predict the residual hull girder strength and to ensure that a damaged ship can be towed safely. La Ferlita et al. (2021) developed an advanced salvage method for damaged ships to ensure a short-term decision-making process for a safe ocean towage to the repair yard.

5.7 Recommendations

As the maritime industry transitions toward more sustainable, automated, and technologically advanced operations, the integration of new and emerging technologies presents both opportunities and challenges.

For Maritime Autonomous Surface Ships (MASS), the report emphasizes the need for robust integration of collision avoidance protocols, particularly through machine-readable interpretations of COLREGs. Designers are encouraged to build resilience into vessel systems, ensuring redundancy in propulsion, navigation, and communication, while also addressing cybersecurity vulnerabilities through secure-by-design principles and real-time threat monitoring. The use of digital twins and probabilistic models is recommended to support predictive diagnostics and enhance operational reliability.

In offshore wind infrastructure, particularly floating offshore wind turbines (FOWTs), structural resilience under accidental loads such as ship collisions and extreme environmental conditions is critical. Coupled aero-hydro-servo-elastic simulations are advised to capture the complex interactions between environmental forces and structural responses.

The adoption of alternative fuels—including LNG, methanol, ammonia, hydrogen, and battery-electric systems—requires tailored design strategies to manage risks related to flammability, toxicity, cryogenic storage, and structural integration. Designers must consider containment, ventilation, and emergency response systems, while also evaluating the implications of fuel choice on vessel layout and performance. Fuel cells and battery systems demand advanced thermal management and early fault detection, supported by machine learning-based diagnostics.

Surrogate modeling emerges as a valuable tool for accelerating design and risk assessment. These models, trained on high-fidelity simulations, can predict structural responses and assess residual strength in real time. However, their development must be grounded in diverse, validated datasets, with clearly defined scopes of applicability and rigorous validation protocols.

6 Summary and Conclusions

As per the mandate of this committee, this report provides a comprehensive review of recent developments in the assessment of accidental and damage limit states (ALS) for ships and offshore structures. It encompasses traditional hazards such as collision, grounding, dropped objects, fire, and explosion, while also addressing emerging challenges posed by new technologies, including autonomous vessels, offshore wind infrastructure, and alternative fuels. The committee has also expanded its scope to include human factors, structural resilience, and the implications of operating beyond design conditions.

Key findings highlight the increasing complexity of maritime systems and the need for integrated approaches that combine structural, operational, and human-centric perspectives. Advances in numerical modeling, particularly in fluid-structure interaction and nonlinear finite element analysis, have significantly improved the ability to simulate and assess structural responses under extreme conditions. However, assumptions and choices made in modeling practices underscore the importance of validation against experimental data.

The report emphasizes the growing relevance of resilience, especially for unmanned and autonomous vessels, and for ships operating in polar or ice-infested waters. The need for realistic assessment of residual strength post-accident is also underscored, particularly in grounding and collision scenarios. Furthermore, the integration of alternative fuels introduces new safety considerations, requiring updated design standards and risk mitigation strategies.

In conclusion, the maritime industry is undergoing a transformative shift driven by decarbonization, digitalization, and automation. To ensure safety and sustainability, future research should prioritize: (1) harmonization of modeling practices and validation tools; (2) development of standardized human factor taxonomies; (3) resilience-based design frameworks; and (4) risk-informed integration of emerging technologies. A systems-thinking approach, supported by cross-disciplinary collaboration, will be essential to enhance the maritime safety.

7 Benchmark Study on Hydroplastic Slamming

7.1 Benchmark No. 1: Drop Tests with 1 mm Aluminium Plates

7.1.1 Experimental Tests

The benchmark study for case 1 aims to explore the applicability of the various numerical methods by simulating the drop tests of clamped flat aluminum plates. The accuracy of the numerical results is analysed by comparing the test results of slamming pressure and plastic deformations at the drop height of 1.1 m. The slamming experiments (Zhu et al., 2023) were carried out by utilizing the drop tower in Impact and Fluid-Structure Interaction Laboratory (IFSI), Wuhan University of Technology (WHUT). The drop tower is located over a water tank with a length of 35.1 m, a width of 3.8 m and a maximum water depth of 1.8 m, see Fig. 12.

Fig. 12. The drop test tower located in the water tank

The schematic of the drop tower installed over the water tank is shown in Fig. 13. Before the test, the drop model was raised to a specified height, and then the falling model impacted on to the calm water at a certain speed. As shown in Fig. 13, a guide rail system with four vertical cylindrical rods was designed to ensure the bottom plate of the drop box being parallel to the water surface before impacting the calm water. The total weight of the drop assembly was 294.6 kg.

A rectangular fixture as shown in Fig. 14(a) consisted of the two identical specimens installed on both sides. To prevent the installation of the pressure sensors from affecting the dynamic plastic response of the test plates, the specimen on the left was used to measure the strain-time history and the transverse residual deformation, while the slamming pressure history was merely measured on the right side. A strain gage and two pressure sensors were installed on the left and right test plates, respectively. To model the clamped boundary condition, 24 bolts were used to connect each test plate with the supporting structure. The overall size of the test plate specimen was 350 × 350 mm, with a valid test area of 250 × 250 mm. Fig. 14(b) shows a picture for the test specimen, where the aluminum plates of thicknesses 1.0 mm was chosen. Uniaxial tensile tests of the aluminum alloy A1060 are shown in Fig. 15.

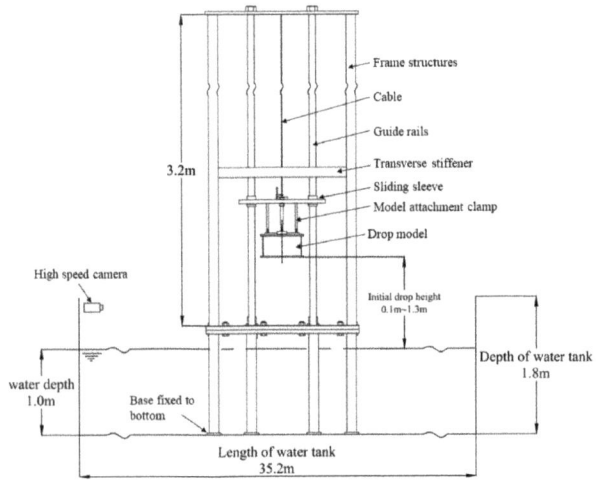

Fig. 13. Schematic of the drop test tower.

7.1.2 FSI Simulations

Seven institutions conducted numerical Fluid-Structure Interaction (FSI) simulations using a common set of input data. While each institution was free to select its own modeling approach (see Table 2), the objective was to evaluate and compare the accuracy of the simulations. The assessment focused on two primary performance indicators: slamming pressure, representing the transient hydrodynamic loads during impact, and plate deflection, reflecting the structural response of the system. This comparative study aims to highlight the influence of modeling strategies on predictive accuracy in FSI simulations.

Figures 16 and 17 show a comparison of the slamming pressure-time histories at the measuring points P1 and P2 of the flat plate, respectively. The slamming pressure curves which are outside the range are not plotted in the figures. As shown in the figures, the predictions of No.1, No.2, No.3, No.4 and No.5 are in reasonable agreement with the test results, especially for the peak values. The values of peak pressure predicted by various numerical methods at P1 and P2 are summarized in Table 3. According to the numerical simulation results, it can be found that in the FSI model with large discrepancies in the slamming pressures, the initial gap between the flat structure and the fluid was set at

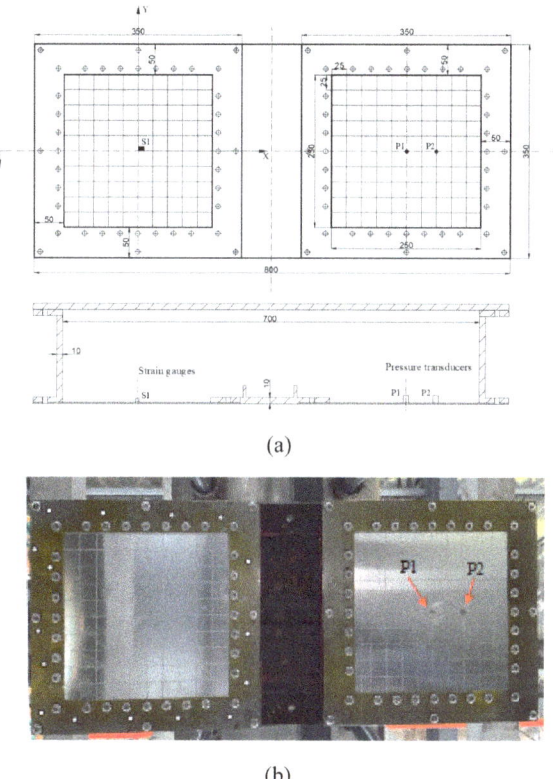

Fig. 14. Schematic and picture of the drop model including the test plate specimens

either 0 mm or 5 mm. However, for the simulation of realistic flat slamming impact, the initial gap should be sufficiently large to take into account the effect of the air cushion. When the initial gap is small, the predicted slamming pressure may increase abnormally, leading to greater plastic deformation. Figures 1 and 18 displays the variation of peak pressure and final deflection at the plate centre with respect to the initial gap. As shown in the figure, the predicted peak pressure and final deflection are the largest for the initial gap of 5 mm, and they decrease with the increase of the initial gap. It is noted that they converge at constant values when the initial gap approaches 50 mm. In addition, whether the flat structure is coupled with air in the numerical model can significantly affect the numerical predictions. If the flat structure is not coupled with air, the phenomenon of compressed air cushion will not occur, and increasing the initial gap at this point will not improve the numerical predictions.

Figure 19 shows a comparison of the transverse deformation profiles along the symmetrical axis (X-axis) based on various numerical methods. The final deflections obtained by various numerical models at the plate centre and the corresponding discrepancies are summarized in Table 4. As shown in the table, the predictions of No.2, No.3, No. 4and No. 5 are generally in reasonable agreement with the experimental results, while No. 1 estimates somewhat larger deflection.

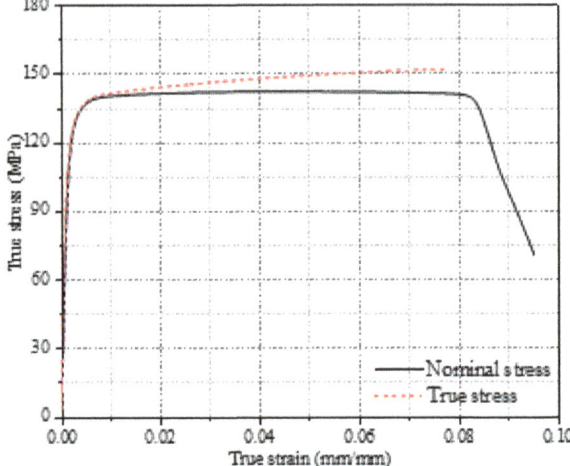

Fig. 15. Stress-strain curves obtained from uniaxial tensile tests for specimen made of aluminum alloy A1060

Table 2. Numerical FSI approaches and parameters used for the benchmark no 1

No	Method	1 initial air gap [mm]	2 yield stress [MPa]	3 Strain-rate	4 Plate/Fluid mesh size [mm/mm]	5 Penalty factor	6 Coupling with air
1	ALE	100	132	No	5/10	1.5×10^{-5}	No
2	SPH	1100	132	No	25/25	N/A	Yes
3	ALE	100	132	No	5/5	0.1	Yes
4	ALE	500	105	No	10/10	0.1	Yes
5	ALE	100	132	No	5/15	0.1	Yes
6	ALE	0	132	Yes	5/5	0.1	No
7	ALE	unknown	unknown	No	5/160	unknown	No

Table 5 presents the comparison of the various numerical simulations with the test results and the corresponding discrepancies analysis. In general, the mesh size, the penalty factor, the initial gap between the structure and the fluid and the yield strength of the material, and whether the structure is coupled with air or not are the main contributing factors for the discrepancies in the numerical results. The boundary condition was assumed the fully clamped and this may also affect the accuracy of the numerical results.

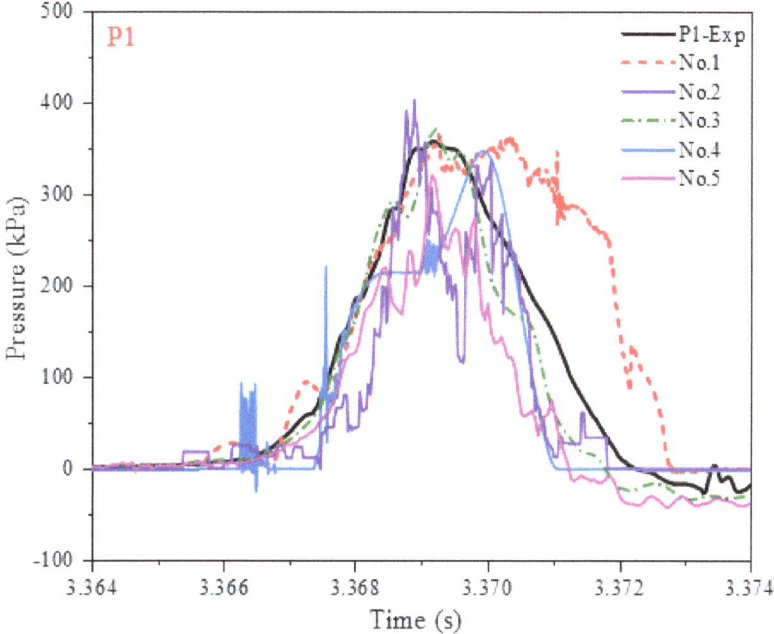

Fig. 16. Comparison of slamming pressure time histories at P1

Table 5. Error analysis of the numerical results of different FSI models.

No.	Method	Relative error E*		Permanent deflection/measured deflection at plate centre
		Pressure	Deflection	
1	ALE	✓	✗	1.36
2	SPH	✓	✓	1.16
3	ALE	✓	✓	1.11
4	ALE	✓	✓	11.1
5	ALE	✓	✓	0.93
6	ALE	✗	✗	1.50
7	ALE	✗	✗	2.31

* "✓" and "✗" indicate whether the relative error E between the numerical simulations and experiments is less than or larger than 20% respectively. Relative error E is calculated as $E = |\text{Value}(Num) - \text{Value}(Exp)|/\text{Value}(Exp) \times 100\%$.

7.2 Benchmark No. 2: Drop Tests with 0.6 mm Aluminium Plates

This benchmark 2 aims to test the capabilities of different FSI simulation techniques by reproducing the drops tests of thin aluminum plates carried out by (Abrahamsen et al., 2020). The test details are briefly described before discussion of the benchmark results.

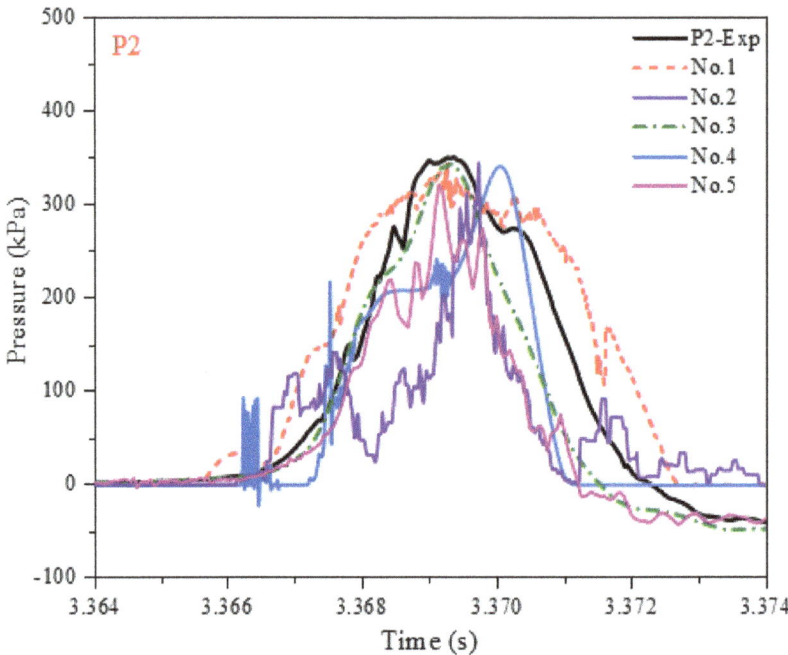

Fig. 17. Comparison of slamming pressure time histories at P2

Table 3. Peak pressures at measuring points P1 and P2

No	P1		P2	
	Peak Value [kPa]	Discrepancy [%]	Peak Value [kPa]	Discrepancy [%]
Exp	370.4	–	350.9	–
1	365.24	1.3	347.62	0.8
2	404	9.19	345	1.68
3	371	0.2	343	2.3
4	348.2	5.9	341.4	2.7
5	341.4	7.8	321.9	8.2
6	9325	2417	5645	1508
7	1594	330	837	138

7.2.1 Experimental Tests

Drop tests of aluminum plates were performed in the Ocean Basin Laboratory at SINTEF Ocean. Figure 20 shows the principal sketch of the drop test. A rotating arm was mounted on a hinge on the left side. The arm was first rotated counterclockwise and then released before the box fell freely until it hit the surface of the calm water. An open box was

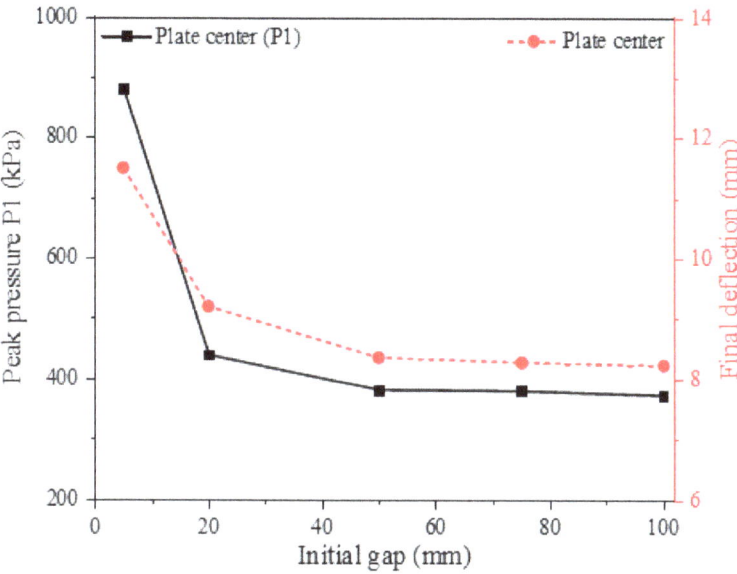

Fig. 18. The variation of numerically predicted peak pressure and final deflection at plate centre with the initial gap

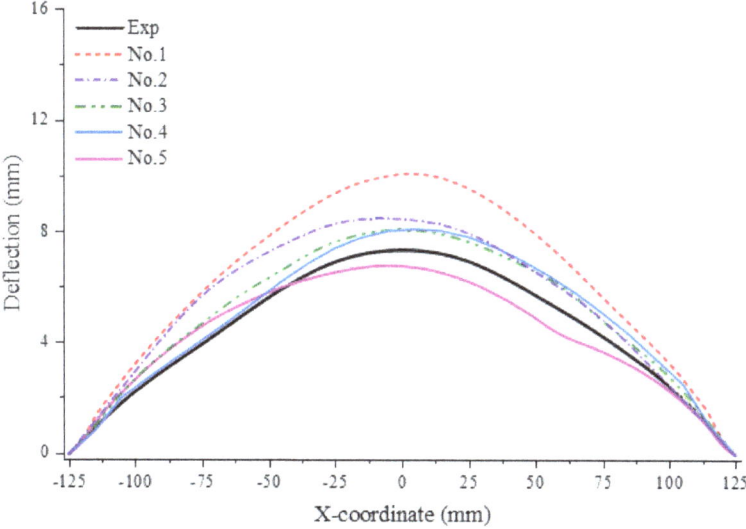

Fig. 19. Comparison of transverse deformation profiles along the symmetry axis

attached to the end of the arm; its underside was rectangular, with dimensions of 344 × 500 mm^2. The photographs in Fig. 21 show the steel box at the bottom and the steel frame that held the flexible plate. The deformable sections were clamped between the inside of a steel frame and four thick steel bars using 12 screws. The drop height is

Table 4. Final deflections at plate centre

ID	Final deflection	
	Deflection [mm]	Discrepancy [%]
Exp	7.34	–
No.1	10.03	36.6
No.2	8.56	16.4
No.3	8.13	10.7
No.4	8.13	10.8
No.5	6.81	7.2
No.6	11.1	51.2
No.7	17.02	130

adjusted with a crane and a chain connected to the impactor. Different drop heights of 118 cm, 222 cm, 443 cm and 778 cm were tested. This gives a measured velocity at the instant of water entry being 1.61 m/s, 2.21 m/s, 3.11 m/s and 4.11 m/s, respectively. The total weight of the drop test rig is 136.8 kg.

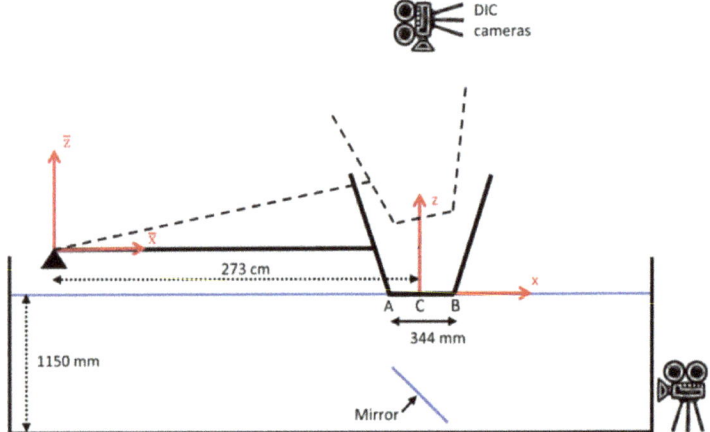

Fig. 20. A schematic illustration of the drop test set-up from Abrahamsen et al. (2020)

Fig. 21. Photographs of the steel box and the frames supporting the specimens (Abrahamsen et al. 2020)

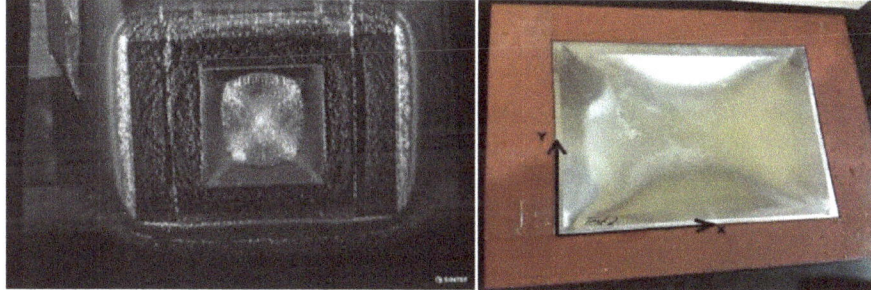

Fig. 22. *Left:* a snapshot by the high-speed camera during water entry; *right:* damaged plates (Abrahamsen et al., 2020)

High speed cameras and 3D-DIC measurements were used to track the plate deformations. The dropped box was equipped with accelerometers to monitor the motion/rotation of the rigid body. The aluminum plate has an effective plate area of 220×220 mm^2. Aluminum plates with a thickness of 0.6-mm were manufactured from low-strength, strain-hardened, and cold-rolled sheets of the commercial alloy EN AW 1050A-H111. Figure 22 shows a snapshot captured by high-speed camera during water entry and the resulting plate damage. To accurately identify the relationship between stress and strain, five material uniaxial tensile tests were carried out. Three test specimens were cut from the plate material in the direction of rolling, and two specimens were cut in the orthogonal direction. Figure 23 shows the nominal and true stress–strain curves from Abrahamsen et al. (2020).

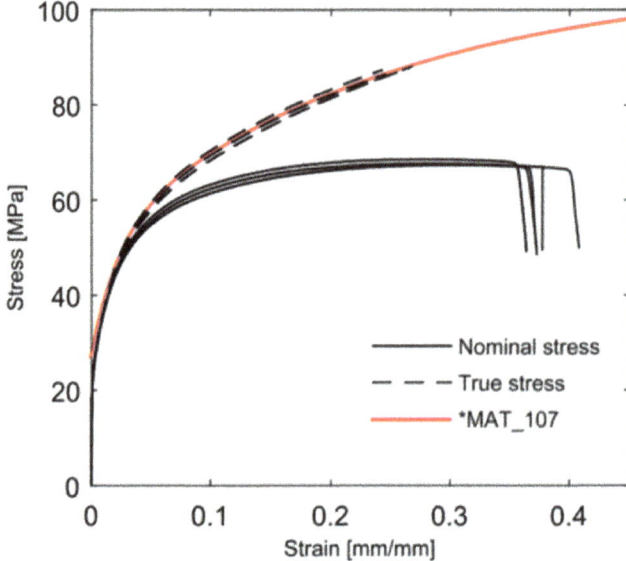

Fig. 23. Stress–strain curves from uniaxial tensile tests cut from the plate material from Abrahamsen et al. (2020)

7.2.2 FSI Simulations

Table 6 summarizes some of the key parameters for the numerical settings used in different participating groups for FSI simulations. All the simulations used two-way coupling schema. Most of the participants adopted the LS-DYNA Multi-Material Arbitrary Lagrangian Eulerian (MMALE) method and only one group used the LS-DYNA SPH method. In addition, one group reported results based on coupled hydroplastic theories by Yu et al. (2019). The model was initially developed for water impact responses for beams and was modified in (Yu, 2024) to consider membrane dominated water impact damage of square plates.

Figures 24 and 25 plot respectively the nodal deflection histories at the plate centre and the plate deflection profiles at the plate centre line $Y = 0$ reported from different groups. The reported deflections from different participants generally agree with the test results and the error is within 10% for permanent deflections. It is interesting to find that none of the participants captured the elastic rebounding as shown in the test curve.

Committee V.1: Limit States During Accidental and Damage 67

Table 6. Numerical settings for the FSI simulations of different groups

Model	Solver Release	Air gap [mm]	Average Element Size (i.e. average edge length) [mm]		Plate Mat Model	Strain-rate effects
			Plate	Fluid		
ALE1	LS-DYNA 2023 R2 Double	2.5	5	5 (at the impact zone)	*MAT_024	Cowper-Symonds
ALE2	LS-DYNA 2023 R13 Double	5	5	5 (at the impact zone)	*MAT_24	Cowper-Symonds
ALE3	LS-DYNA R13.1.0 Double	50	5	5	Johnson-Cook	No
ALE4	LS-DYNA R14.0 Double	50	5	5	*MAT_024	No
SPH1*	LS-DYNA R13.0.0 Single	696	11.5	12.5	*MAT_024	No
ALE5	LS-DYNA R14 S-ALE Double	50	5	10	*MAT_024	No
ALE6	LS-DYNA R13.1.1 Double	2	2.5	5	*MAT_024	Cowper- Symonds

*SPH1 included the pendulum motion in the numerical model

Figure 26 plots the reported pressure histories averaged over the entire plate. The nodal deflection velocities at the plate centre normalized with respect to the drop velocity are given in Fig. 27. Despite the good agreement of plate deflections, the reported pressure histories and central plate deflection velocities show some scattering. The reported peak pressures vary from around 60 KPa to approximately 140 KPa with different pressure durations disregarding the initial peaks in the acoustic phase. The pressures were not measured in the experiment due to the concern that attached pressure sensors may influence the pressure field and plate responses.

The central plate deflection velocities showed two distinct patterns. For the groups of ALE 1, 3, 5 and 6, the reported central plate deflection velocities oscillate violently during the deformation up to 6 times the drop velocity. The trend agrees with the experimental measurements with a peak deflection velocity up to 4.5 times the initial drop velocity. Groups ALE 2, 4 and SPH1 reported a different deflection velocity pattern, where the deflection velocity is relatively steady during the deformation with an average being the same as the initial drop velocity. It is not clear what causes the different deflection patterns, but this seems to be sensitive to the adopted initial air gap. The mid-plate deflection velocity predicted by the analytical hydroplastic model (Yu, 2024) agrees reasonably with an averaging of the experimental curve without oscillations since elastic effects are not included.

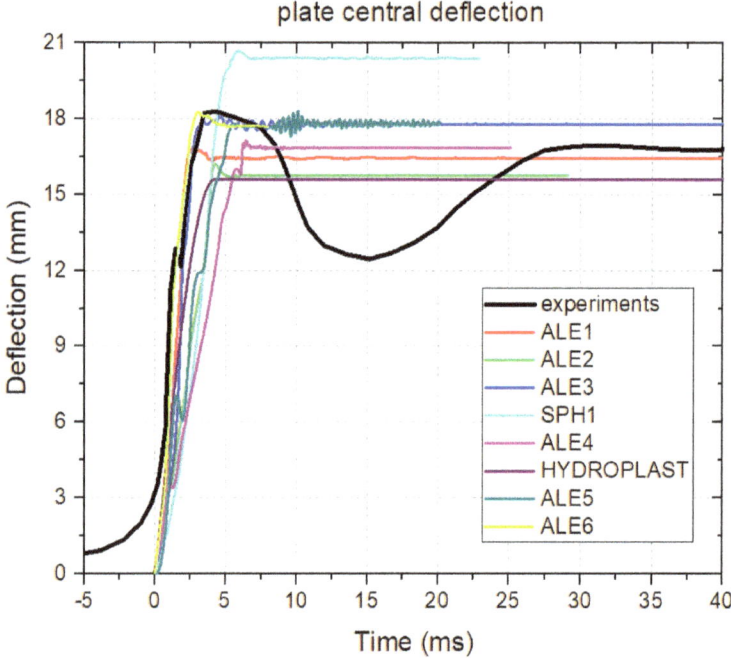

Fig. 24. Comparison of deflection histories at the plate centre

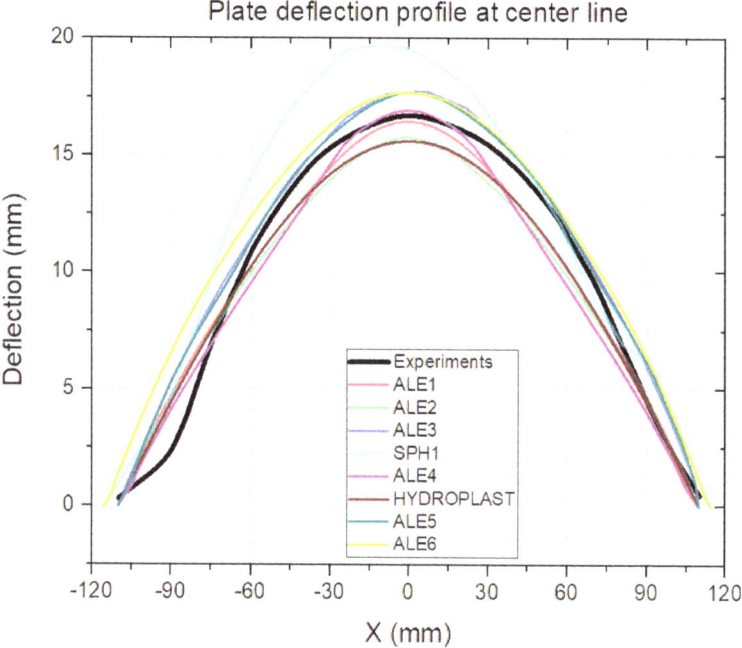

Fig. 25. Comparison of plate deflection profiles at the plate centre line $Y = 0$

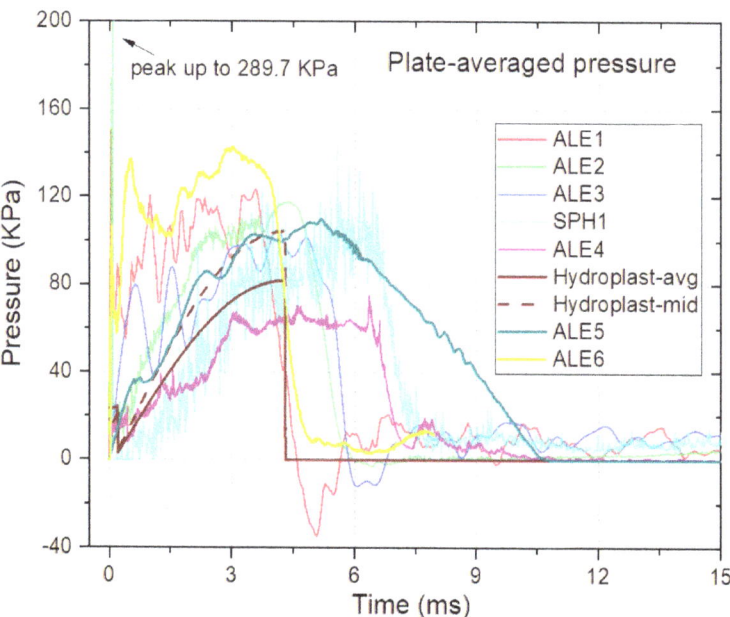

Fig. 26. Comparison of pressures averaged over the entire plate

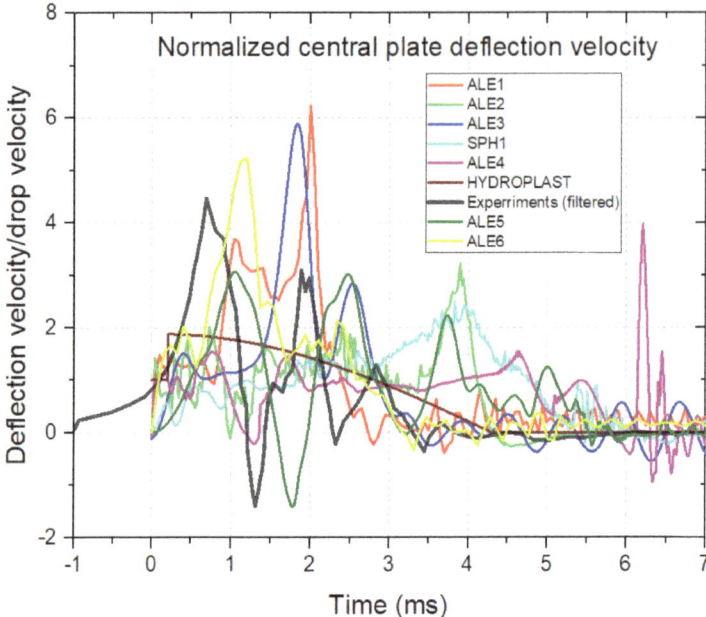

Fig. 27. Comparison of nondimensional deflection velocity histories at the plate centre normalized with respect to the drop velocity

7.3 Summary and Recommendations

Two water impact tests of thin plates were selected for the benchmark study with large permanent plastic deformations. 7 groups from worldwide participated in the benchmark with 6 using ALE and 1 group using SPH. All the information including the test results was open to all participants before the benchmark. The following observations and recommendations are reached.

There are basically two rounds of simulations in the benchmark. In the first round, participants conducted simulations independently to the best of their abilities. In the second round, participants reconducted the simulations after discussion of simulation results and modelling techniques in the group. The first-round results showed big scatter for test 1 and less for test 2. It was concluded that this was due to the use of different initial air gaps. The results showed better agreement on permanent deflections with test results by using increased air gaps in the second round. The differences of predicted pressures however remained substantial.

Overall, the ALE and SPH techniques showed strong capabilities to simulate the complicated FSI problem with large plastic damage. The results showed certain dependence on modelling techniques and numerical settings. Users are recommended to calibrate their modelling and settings to test results before conducting complicated FSI simulations.

References

Abaei, M.M., Hekkenberg, R., BahooToroody, A., Banda, O.V., van Gelder, P.: A probabilistic model to evaluate the resilience of unattended machinery plants in autonomous ships. Reliab. Eng. Syst. Saf. **219**, 108176 (2022). https://doi.org/10.1016/j.ress.2021.108176

Abbasnia, A., Karimirad, M., Friel, D., Whittaker, T.: Fully nonlinear dynamics of floating solar platform with twin hull by tubular floaters in ocean waves. Ocean Eng. **257** (2022). https://doi.org/10.1016/j.oceaneng.2022.111320

Abdelmalek, M., Guedes Soares, C.: Review of risk analysis studies in the maritime LNG sector. J. Mar. Sci. Appl. (2023). https://doi.org/10.1007/s11804-023-00376-0

Abraham, E.J., Linke, P., Al-Rawashdeh, M., Rousseau, J., Burton, G., Al-Mohannadi, D.M.: Large-scale shipping of low-carbon fuels and carbon dioxide towards decarbonized energy systems: perspectives and challenges. Int. J. Hydrog. Energy. (2024). https://doi.org/10.1016/j.ijhydene.2024.03.140

Abrahamsen, B.C., Alsos, H.S., Aune, V., Fagerholt, E., Faltinsen, O.M., Hellan, Ø.: Hydroplastic response of a square plate due to impact on calm water. Phys. Fluids. **32** (2020)

Adumene, S., Afenyo, M., Salehi, V., William, P.: An adaptive model for human factors assessment in maritime operations. Int. J. Ind. Ergon. **89**, 103293 (2022). https://doi.org/10.1016/J.ERGON.2022.103293

Aider, C.A., Roß, L., Ehlers, S., Kaeding, P., Lindemann, T.: Integration of a tank storage solution for alternative fuels on a RoRo ship. In: Advances in the Analysis and Design of Marine Structures – Proceedings of the 9th International Conference on Marine Structures, MARSTRUCT 2023, pp. 761–771. CRC Press/Balkema (2023). https://doi.org/10.1201/9781003399759-84

Andrade, S.L., Gagnon, R., Colbourne, B., Quinton, B.W.: Ice pressure distribution model: a geometry-based solution for high-pressure zone representation. Cold Reg. Sci. Technol. **210**, 103822 (2023). https://doi.org/10.1016/J.COLDREGIONS.2023.103822

API RP 2A-WSD: Recommended Practice for Planning, Designing and Constructing Fixed Offshore Platforms – Working Stress Design. API (2019)

API RP 2A-LRFD: Planning, Designing and Constructing Fixed Offshore Platforms – Load and Resistance Factor Design. API (2019)

API RP 14J: Recommended Practice for Design and Hazards Analysis for Offshore Production Facilities, 2nd edn. American Petroleum Institute, Washington, DC (2001)

Aydin, M., Uğurlu, Ö., Boran, M.: Assessment of human error contribution to maritime pilot transfer operation under HFACS-PV and SLIM approach. Ocean Eng. **266**, 112830 (2022). https://doi.org/10.1016/J.OCEANENG.2022.112830

BahooToroody, A., Abaei, M.M., Banda, O.V., Kujala, P., De Carlo, F., Abbassi, R.: Prognostic health management of repairable ship systems through different autonomy degree; From current condition to fully autonomous ship. Reliab. Eng. Syst. Saf. **221**, 108355 (2022a). https://doi.org/10.1016/j.ress.2022.108355

BahooToroody, A., Abaei, M.M., Valdez Banda, O., Montewka, J., Kujala, P.: On reliability assessment of ship machinery system in different autonomy degree; A Bayesian-based approach. Ocean Eng. **254**, 111252 (2022b). https://doi.org/10.1016/j.oceaneng.2022.111252

Bakdi, A., Vanem, E.: Complexity analysis using graph models for conflict resolution for autonomous ships in complex situations. J. Offshore Mech. Arct. Eng., 1–16 (2024). https://doi.org/10.1115/1.4066198

Bayazit, O., Kaptan, M.: Dynamic risk analysis of allision in port areas using DBN based on HFACS-PV. Ocean Eng. **298**, 117183 (2024). https://doi.org/10.1016/J.OCEANENG.2024.117183

Bergström, M., Gosala, V., Depken, J., Fitz, A., Euskirchen, F., Ehlers, S.: A simulation-based approach for evaluating merchant fleet decarbonization strategies. In: Proceedings of the International Conference on Offshore Mechanics and Arctic Engineering – OMAE (2023). https://doi.org/10.1115/OMAE2023-102401

Bergström, M., Li, F., Suominen, M., Kujala, P.: A goal-based approach for selecting a ship's polar class. Mar. Struct. **81**, 103123 (2022). https://doi.org/10.1016/j.marstruc.2021.103123

Bertagna, S., Bucci, V., Marino, A., Sulligoi, G., Vicenzutti, A.: Impact of fuel switch to methanol on the design of an all electric cruise ship. In: 2023 IEEE International Conference on Electrical Systems for Aircraft, Railway, Ship Propulsion and Road Vehicles and International Transportation Electrification Conference, ESARS-ITEC 2023. Institute of Electrical and Electronics Engineers (2023). https://doi.org/10.1109/ESARS-ITEC57127.2023.10114852

Bicen, S., Celik, M.: A hybrid approach to near-miss report investigation towards next-generation safety solutions on-board ships. Ocean Eng. **266**, 112768 (2022). https://doi.org/10.1016/J.OCEANENG.2022.112768

Bilgili, L.: Comparative assessment of alternative marine fuels in life cycle perspective. Renew. Sust. Energ. Rev. **144** (2021). https://doi.org/10.1016/j.rser.2021.110985

Bobeldijk, M., Dragt, S., Hoogeland, M., van Bergen, J.: Assessment of the technical safe limit speed of a non-ice-strengthened naval vessel with representative and alternative side shell designs in ice-infested waters. Ships Offshore Struct. **16**, 275–289 (2021). https://doi.org/10.1080/17445302.2021.1912475

Bobeldijk, M.P., Hoogeland, M.G., Nedaei, A., Groes-Petersen, P.: Ice belt weight reduction of ships operating in ice floe infested waters with the direct calculation method. In: Advances in the Collision and Grounding of Ships and Offshore Structures – Proceedings of the 9th International Conference on Collision and Grounding of Ships and Offshore Structures, ICCGS 2023, pp. 283–293. CRC Press/Balkema, Nantes (2024). https://doi.org/10.1201/9781003462170-35

Bureau Veritas: Guidance Note NI 641 DT R01 E. Guidelines for Autonomous Shipping (2019).

Cai, W., Zhu, L., Gudmestad, O.T., Guo, K.: Application of rigid-plastic theory method in ship-ice collision. Ocean Eng. **253** (2022a). https://doi.org/10.1016/j.oceaneng.2022.111237

Cai, W., Zhu, L., Qian, X.: Dynamic responses of steel plates under repeated ice impacts. Int. J. Impact Eng. **162**, 104129 (2022b). https://doi.org/10.1016/J.IJIMPENG.2021.104129

Cai, W., Zhu, L., Wang, F.: Plastic deformation of ship plate subjected to repeated patch loads at different locations – elastoplastic numerical analysis and design equation. Mar. Struct. **76** (2021). https://doi.org/10.1016/j.marstruc.2020.102901

Calle, M.A.G., Oshiro, R.E., Kõrgesaar, M., Alves, M., Kujala, P.: Combined strain rate, mesh size and calibration test influence on structural failure: miniature ship grounding test. Ocean Eng. **173**, 215–226 (2019)

Cao, Y., Wang, X., Yang, Z., Wang, J., Wang, H., Liu, Z.: Research in marine accidents: a bibliometric analysis, systematic review and future directions. Ocean Eng. **284**, 115048 (2023a). https://doi.org/10.1016/J.OCEANENG.2023.115048

Cao, Z., Wang, J., Xu, G., Shao, C., Yang, J.: Dynamic performance of triple-column bridge pier under barge collision. Ocean Eng. **271** (2023b). https://doi.org/10.1016/j.oceaneng.2023.113763

Carboni, M., Pio, G., Mocellin, P., Vianello, C., Maschio, G., Salzano, E.: Accidental release in the bunkering of LNG: phenomenological aspects and safety zone. Ocean Eng. **252** (2022). https://doi.org/10.1016/j.oceaneng.2022.111163

Cerik, B.C., Choung, J.: Fracture prediction of steel-plated structures under low-velocity impact. J. Mar. Sci. Eng. **11** (2023). https://doi.org/10.3390/jmse11040699

Cerik, B.C., Choung, J.: Ductile fracture behavior of mild and high-tensile strength shipbuilding steels. Appl. Sci. (Switz.). **10**, 1–21 (2020). https://doi.org/10.3390/app10207034

Cerik, B.C., Lee, K., Park, S.-J., Choung, J.: Simulation of ship collision and grounding damage using Hosford-Coulomb fracture model for shell elements. Ocean Eng. **173**, 415–432 (2019). https://doi.org/10.1016/j.oceaneng.2019.01.004

Ceylan, B.O., Akyuz, E., Arslan, O.: Systems-Theoretic Accident Model and Processes (STAMP) approach to analyse socio-technical systems of ship allision in narrow waters. Ocean Eng. **239**, 109804 (2021). https://doi.org/10.1016/J.OCEANENG.2021.109804

Ceylan, B.O., Akyuz, E., Arslanoğlu, Y.: Modified quantitative systems theoretic accident model and processes (STAMP) analysis: a catastrophic ship engine failure case. Ocean Eng. **253**, 111187 (2022). https://doi.org/10.1016/J.OCEANENG.2022.111187

Chaal, M., et al.: Research on risk, safety, and reliability of autonomous ships: a bibliometric review. Saf. Sci. **167**, 106256 (2023). https://doi.org/10.1016/j.ssci.2023.106256

Chang, C.H., Wijeratne, I.B., Kontovas, C., Yang, Z.: COLREG and MASS: analytical review to identify research trends and gaps in the Development of Autonomous Collision Avoidance. Ocean Eng. (2024). https://doi.org/10.1016/j.oceaneng.2024.117652

Chen, B.Q., Liu, B., Guedes Soares, C.: Experimental and numerical investigation on a double hull structure subject to collision. Ocean Eng. **256** (2022). https://doi.org/10.1016/j.oceaneng.2022.111437

Chen, D., Pei, Y., Xia, Q.: Research on human factors cause chain of ship accidents based on multidimensional association rules. Ocean Eng. **218**, 107717 (2020). https://doi.org/10.1016/J.OCEANENG.2020.107717

Chen, L., Guan, W.: Safety design and engineering solution of fuel cell powered ship in inland waterway of China. World Electr. Veh. J. **12** (2021). https://doi.org/10.3390/wevj12040202

Chen, T.L., Wu, H., Fang, Q.: Impact force models for bridge under barge collisions. Ocean Eng. **259** (2022a). https://doi.org/10.1016/j.oceaneng.2022.111856

Chen, T.L., Wu, H., Fang, Q.: Dynamic behaviors of double-column RC bridge subjected to barge impact. Ocean Eng. **264** (2022b). https://doi.org/10.1016/j.oceaneng.2022.112444

Chen, W., Wan, F., Guan, F., Liu, X., Yang, Y., Zhou, C.: The effect of seabed flexibility on the impact damage behavior of submarine sandwich pipes. Appl. Ocean Res. **142** (2024a). https://doi.org/10.1016/j.apor.2023.103838

Chen, Y., Xiao, Q., Pan, J., Xu, M.: Dynamic response analysis of pontoon interception system under ship collision for protecting bridge. In: Advances in the Collision and Grounding of Ships and Offshore Structures – Proceedings of the 9th International Conference on Collision and Grounding of Ships and Offshore Structures, ICCGS 2023, pp. 477–483. CRC Press/Balkema (2024b). https://doi.org/10.1201/9781003462170-58

Chen, Y., Xiao, Q., Pan, J., Xu, M.: Impact force analysis of ship-bridge collision considering fluid influence. In: Advances in the Collision and Grounding of Ships and Offshore Structures – Proceedings of the 9th International Conference on Collision and Grounding of Ships and Offshore Structures, ICCGS 2023, pp. 485–491. CRC Press/Balkema (2024c). https://doi.org/10.1201/9781003462170-59

Chin, H.C., Debnath, A.K.: Modeling perceived collision risk in port water navigation. Saf. Sci. **47**, 1410–1416 (2009). https://doi.org/10.1016/j.ssci.2009.04.004

Cho, H.R., Kim, H.J., Park, S.M., Park, D.K., Yun, S.H., Paik, J.K.: Effect of solid rubber fenders on the structural damage due to collisions between a ship-shaped offshore installation and an offshore supply vessel. Ships Offshore Struct. **18**, 1037–1059 (2023). https://doi.org/10.1080/17445302.2022.2103965

Choi, S.M., Park, C.D., Cho, S.H., Lim, B.J.: Effects of various inlet angle of wind and wave loads on floating photovoltaic system considering stress distributions. J. Clean. Prod. **387** (2023). https://doi.org/10.1016/j.jclepro.2023.135876

Choisnet, T., Geschier, B., Vetrano, G.: Initial comparison of concrete and steel hulls in the case of Ideol's square ring floating substructure. In: Proceedings of the WWEC 2016, pp. 1–4. WWEC, Tokyo (2016)

Conti, F., Pineau, J.P., Cardinale, M., Bertin, R., Lindroth, D.: The influence of crashworthiness on passenger ship damage stability probabilistic analysis including grounding damages. In: Advances in the Collision and Grounding of Ships and Offshore Structures – Proceedings of the 9th International Conference on Collision and Grounding of Ships and Offshore Structures, ICCGS 2023, pp. 161–172. CRC Press/Balkema (2024). https://doi.org/10.1201/9781003462170-22

Coraddu, A., Oneto, L., Navas de Maya, B., Kurt, R.: Determining the most influential human factors in maritime accidents: a data-driven approach. Ocean Eng. **211**, 107588 (2020). https://doi.org/10.1016/J.OCEANENG.2020.107588

Corporate Risk Associates: A User Manual for the Nuclear Action Reliability Assessment (NARA) Human Error Quantification Technique. Report No 2. Corporate Risk Associates, Surrey (2011)

Cruise Ship Safety Forum: Damage Stability and Survivability. Monitoring and Assessing Risk from Operation of Watertight Doors. Recommendation 312/2016 (2016)

Cuce, E., Cuce, P.M., Saboor, S., Ghosh, A., Sheikhnejad, Y.: Floating PVs in terms of power generation, environmental aspects, market potential, and challenges. Sustainability (Switzerland). (2022). https://doi.org/10.3390/su14052626

Cui, M., Cheng, Z., Moan, T.: Effects of inhomogeneous wave modeling on extreme responses of a very long floating bridge. Appl. Ocean Res. **134** (2023). https://doi.org/10.1016/j.apor.2023.103505

Cui, M., Cheng, Z., Moan, T.: A generic method for assessment of inhomogeneous wave load effects of very long floating bridges. Mar. Struct. **83** (2022). https://doi.org/10.1016/j.marstruc.2022.103186

Daley, C.: Ice Impact Capability of DRDC Notional Destroyer, Halifax (2015)

Daley, C., 2000. Background Notes to Design Ice Loads.

Daley, C., Dolny, J., Daley, K.: Safe Speed Assessment of DRDC Notional Destroyer in Ice: Phase 2 of Ice Capability Assessment. Defence Research and Development Canada, Halifax (2017)

Daley, C.G., Alawneh, S., Peters, D., Quinton, B.W.T., Colbourne, D.B.: GPU modeling of ship operations in pack ice. In: International Conference and Exhibition on Performance of Ships and Structures in Ice 2012, ICETECH 2012 (2012)

Das, T., Goerlandt, F., Tabri, K.: An optimized metamodel for predicting damage and oil outflow in tanker collision accidents. Proc. Inst. Mech. Eng. M J. Eng. Marit. Environ. **236**, 412–426 (2022). https://doi.org/10.1177/14750902211039659

de Vos, J., Hekkenberg, R.G., Valdez Banda, O.A.: The Impact of Autonomous Ships on Safety at Sea – A Statistical Analysis. Reliab. Eng. Syst. Saf. **210** (2021). https://doi.org/10.1016/j.ress.2021.107558

Dellosa, J.T., Palconit, E.V., Enano, N.H.: Risk Assessment and Policy Recommendations for a Floating Solar Photovoltaic (FSPV) System. IEEE Access. **12**, 30452–30471 (2024). https://doi.org/10.1109/ACCESS.2024.3368620

Depken, J., Dyck, A., Roß, L., Ehlers, S.: Safety considerations of hydrogen application in shipping in comparison to LNG. Energies (Basel). **15** (2022). https://doi.org/10.3390/en15093250

Dmochowski, P.A., Skorupski, J.: A method of evaluating air traffic controller time workload. pp. 363–376 (2019). https://doi.org/10.1007/978-3-030-27547-1_26

DNV-CG-0264: Class Guideline DNV-CG-0264: Autonomous and Remotely Operated Ships (2021). Retrieved from https://www.dnv.com

DNV-CG-0308: IMO Polar Code Operational Requirements. Det Norske Veritas (2021)

DNV-OS-A101: Safety Principles and Arrangements. Det Norske Veritas (2019). Retrieved from https://www.dnv.com/rules-standards/

DNV-RP-C204: Structural Design Against Accidental Loads. Det Norske Veritas (2021). Retrieved from https://www.dnv.com

DNV-RP-C208: Determination of Structural Capacity by Non-linear Finite Element Analysis Methods. Det Norske Veritas (2022). Retrieved from https://www.dnv.com

DNV-RP-D102: FMEA of Redundant Systems (2022). Retrieved from https://www.dnv.com
DNV-RP-E304: Damage Assessment of Fibre Ropes for Offshore Mooring (2019). Retrieved from http://www.dnv.com
DNV-RP-F107: Risk Assessment of Pipeline Protection (2019). Retrieved from http://www.dnv.com
DNV-RP-F117: Design of Pipelines with Pressure Protection Systems (2023). Retrieved from http://www.dnv.com
Do, Q.T., Muttaqie, T., Nhut, P.T., Vu, M.T., Khoa, N.D., Prabowo, A.R.: Residual ultimate strength assessment of submarine pressure hull under dynamic ship collision. Ocean Eng. **266** (2022). https://doi.org/10.1016/j.oceaneng.2022.112951
Do, Q.T., Xuan-Phuong, D., Tra, T.H., Van Tuyen, V., Prabowo, A.R., Hung, T.D.: Parametric study of side collision-induced denting failures on the ultimate strength of a handy-size containership under vertical bending. Ocean Eng. **309** (2024). https://doi.org/10.1016/j.oceaneng.2024.118534
Dolny, J.: SSC 473 – Methodology for Defining Technical Safe Speeds for Light Ice-Strengthened Government Vessels Operating in Ice. Ship Structure Committee, Washington, DC (2018)
Dolny, J., Daley, C.G., Quinton, B.W.T., Daley, K.H.: ONR Ice Capability Assessment and Experimental Planning: Experimental Planning for Large Non-ice Class Grillage Tests. American Bureau of Shipping, Houston (2017a)
Dolny, J., Daley, C.G., Quinton, B.W.T., Daley, K.H.: ONR Ice Capability Assessment and Experimental Planning: Advanced Modeling and Re-assessment of NSWCCD HULL 3000. American Bureau of Shipping, Houston (2017b)
Dolny, J., Daley, C.G., Quinton, B.W.T., Daley, K.H.: ONR Ice Capability Assessment and Experimental Planning: DDePS_SafeCheck Technical Background and Safe Speed Assessment of NSWCCD Hull 3000. American Bureau of Shipping, Houston (2016)
Drummen, I., Olbert, G.: Conceptual design of a modular floating multi-purpose island. Front. Mar. Sci. **8** (2021). https://doi.org/10.3389/fmars.2021.615222
Du, L., Goerlandt, F., Kujala, P.: Review and analysis of methods for assessing maritime waterway risk based on non-accident critical events detected from AIS data. Reliab. Eng. Syst. Saf. (2020). https://doi.org/10.1016/j.ress.2020.106933
Duong, P.A., Ryu, B.R., Song, M.K., Van Nguyen, H., Nam, D., Kang, H.: Safety assessment of the ammonia bunkering process in the maritime sector: a review. Energies (Basel). (2023). https://doi.org/10.3390/en16104019
Eidem, M.E., Sha, Y.: Ship impact from a general cargo vessel on the Bergsøysund floating bridge in Norway. In: Advances in the Collision and Grounding of Ships and Offshore Structures – Proceedings of the 9th International Conference on Collision and Grounding of Ships and Offshore Structures, ICCGS 2023, pp. 493–499. CRC Press/Balkema (2024). https://doi.org/10.1201/9781003462170-60
Ejder, E., Arslanoğlu, Y.: Evaluation of ammonia fueled engine for a bulk carrier in marine decarbonization pathways. J. Clean. Prod. **379** (2022). https://doi.org/10.1016/j.jclepro.2022.134688
Eriksson, M., Nyberg, M., Andersen, M., Nielsen, J., Tychsen, J.: Physical testing campaign facilitating validation of welded steel structures by non-linear finite element analysis. J. Offshore Mech. Arct. Eng. **145**, 1–31 (2023)
Eriksson, M., Nyberg, M., Andersen, M., Tychsen, J., Nielsen, J.: Validated methodology for assessment of welded steel structures by nonlinear finite element analysis. Int. J. Offshore Polar Eng. **34**, 191–198 (2024). https://doi.org/10.17736/ijope.2024.ty14
Fan, C., Wróbel, K., Montewka, J., Gil, M., Wan, C., Zhang, D.: A framework to identify factors influencing navigational risk for Maritime Autonomous Surface Ships. Ocean Eng. **202**, 107188 (2020a). https://doi.org/10.1016/J.OCEANENG.2020.107188

Fan, W., Sun, Y., Yang, C., Sun, W., He, Y.: Assessing the response and fragility of concrete bridges under multi-hazard effect of vessel impact and corrosion. Eng. Struct. **225** (2020b). https://doi.org/10.1016/j.engstruct.2020.111279

Fan, W., Ye, X., Hua, X., Sha, Y., Wu, Q., Liu, B.: A simplified method to consider hydrodynamic effect in oblique vessel-bridge collisions. Appl. Ocean Res. **134** (2023). https://doi.org/10.1016/j.apor.2023.103530

Fedi, L., Faury, O., Etienne, L., Cheaitou, A., Rigot-Muller, P.: Application of the IMO taxonomy on casualty investigation: analysis of 20 years of marine accidents along the North-East Passage. Mar. Policy. **162**, 106061 (2024). https://doi.org/10.1016/J.MARPOL.2024.106061

Fenerci, A., Kvåle, K.A., Xiang, X., Øiseth, O.: Hydrodynamic interaction of floating bridge pontoons and its effect on the bridge dynamic responses. Mar. Struct. **83** (2022). https://doi.org/10.1016/j.marstruc.2022.103174

Feng, Y., Li, H., Ong, M.C., Wang, W.: Numerical investigation of collision between massive ice floe and marine structure using coupled SPH-FEM method. Ships Offshore Struct. **18**, 380–390 (2023). https://doi.org/10.1080/17445302.2022.2055932

Fjørtoft, K., Parvasi, S.P., Nesheim, D.A., Lien Wennerberg, L.A., Mørkrid, O.E., Psaraftis, H.N.: Assessing the resilience of sustainable autonomous shipping: new methodology, challenges, opportunities. Cleaner Logist. Supply Chain. **9**, 100126 (2023). https://doi.org/10.1016/j.clscn.2023.100126

Fjørtoft, K.E., Mørkrid, O.E.: Resilience in autonomous shipping. In: Proceedings of the 31st European Safety and Reliability Conference, ESREL 2021, pp. 2046–2053. Research Publishing, Singapore (2021). https://doi.org/10.3850/978-981-18-2016-8_470-cd

Flikkema, M.M.B., Lin, F.Y., van der Plank, P.P.J., Koning, J., Waals, O.: Legal issues for artificial floating islands. Front. Mar. Sci. **8** (2021). https://doi.org/10.3389/fmars.2021.619462

Gagnon, R.E., Andrade, S.L., Quinton, B.W.T., Daley, C.G., Colbourne, B.: Pressure distribution data from large double-pendulum ice impact tests. Cold Reg. Sci. Technol. **175**, 103033 (2020). https://doi.org/10.1016/J.COLDREGIONS.2020.103033

Gan, N., Liu, L.T., Yao, X.L., Wang, J.X., Wu, W.B.: Experimental and numerical investigation on the dynamic response of a simplified open floating slender structure subjected to underwater explosion bubble. Ocean Eng. **219** (2021). https://doi.org/10.1016/j.oceaneng.2020.108308

Gao, X., Shao, Y., Chen, C., Zhu, H., Li, K.: Experimental and numerical investigation on transverse impact resistance behaviour of pipe-in-pipe submarine pipelines after service time. Ocean Eng. **248** (2022). https://doi.org/10.1016/j.oceaneng.2022.110868

Gao, Y., Liu, H., Hou, Y.: Effects of leakage location and ventilation condition on hydrogen leakage during shipping of fuel cell vehicles. Int. J. Hydrog. Energy. **54**, 1532–1543 (2024). https://doi.org/10.1016/j.ijhydene.2023.12.095

Ghosh, A.: A comprehensive review of water based PV: flotavoltaics, under water, offshore & canal top. Ocean Eng. (2023). https://doi.org/10.1016/j.oceaneng.2023.115044

Goerlandt, F., Montewka, J.: Maritime transportation risk analysis: review and analysis in light of some foundational issues. Reliab. Eng. Syst. Saf. **138**, 115–134 (2015). https://doi.org/10.1016/j.ress.2015.01.025

Gong, Y., Zhang, W., Du, Z.: Damage mechanisms of a typical simplified hull girder with thinner plates subjected to near-field underwater explosions. Ocean Eng. **285** (2023). https://doi.org/10.1016/j.oceaneng.2023.115403

Govindaraj, T.: Characterizing performance in socio-technical systems: a modeling framework in the domain of nuclear power. Omega (Westport). **36**, 10–21 (2008). https://doi.org/10.1016/j.omega.2005.07.010

Guedes Soares, C.: Decarbonization of ship operations. J. Mar. Sci. Appl. (2023). https://doi.org/10.1007/s11804-023-00384-0

Guo, X., Zhang, C., Chen, Z.Q.: Experimental and numerical assessment of scoured bridges with protective bonded steel plates against vessel impact. Eng. Struct. **252** (2022). https://doi.org/10.1016/j.engstruct.2021.113628

Hammad, A., Yu, Z.: A semi-analytical approach for dynamic responses of monopile-supported offshore wind turbines subjected to accidental loads. Appl. Ocean Res. **153**, 104251 (2024)

Haneda, K., Chujo, T., Nimura, T., Fujiwara, T., Inoue, S.: Experimental study on collision behavior between a floating offshore wind turbine and a ship. J. Jpn. Soc. Nav. Archit. Ocean Eng. **32**, 65–76 (2020). https://doi.org/10.2534/jjasnaoe.32.65

Hasan, A., Asfihani, T., Osen, O., Bye, R.T.: Leveraging digital twins for fault diagnosis in autonomous ships. Ocean Eng. **292**, 116546 (2024). https://doi.org/10.1016/j.oceaneng.2023.116546

He, L., Xiaoxue, M., Weiliang, Q., Yang, L.: A methodology to assess the causation relationship of seafarers' unsafe acts for ship grounding accidents based on Bayesian SEM. Ocean Coast. Manag. **225**, 106189 (2022). https://doi.org/10.1016/J.OCECOAMAN.2022.106189

He, W., et al.: Lessons learned from the commercial exploitation of marine battery energy storage systems. J. Energy Storage. (2024). https://doi.org/10.1016/j.est.2024.111440

He, X., Guedes Soares, C.: Experimental and numerical study on the dynamic response of rectangular plates under repeated impacts. Mar. Struct. **96**, 103606 (2024). https://doi.org/10.1016/j.marstruc.2024.103606

He, X., Guedes Soares, C.: Experimental study on the dynamic behavior of beams under repeated impacts. Int. J. Impact Eng. **147**, 103724 (2021a). https://doi.org/10.1016/J.IJIMPENG.2020.103724

He, X., Guedes Soares, C.: Numerical study on the pseudo-shakedown of beams under repeated impacts. Ocean Eng. **242**, 110137 (2021b). https://doi.org/10.1016/J.OCEANENG.2021.110137

He, Z., Chen, Z., Jiang, Y., Cao, X., Zhao, T., Li, Y.: Effects of the standoff distance on hull structure damage subjected to near-field underwater explosion. Mar. Struct. **74** (2020). https://doi.org/10.1016/j.marstruc.2020.102839

He, Z., Du, Z., Zhang, L., Li, Y.: Damage mechanisms of full-scale ship under near-field underwater explosion. Thin-Walled Struct. **189** (2023). https://doi.org/10.1016/j.tws.2023.110872

Hendrikse, H., Metrikine, A.: Ice-induced vibrations and ice buckling. Cold Reg. Sci. Technol. **131**, 129–141 (2016). https://doi.org/10.1016/J.COLDREGIONS.2016.09.009

Herrnring, H., Ehlers, S.: A finite element model for compressive ice loads based on a Mohr-Coulomb material and the node splitting technique. J. Offshore Mech. Arct. Eng. **144** (2021). https://doi.org/10.1115/1.4052746

Herrnring, H.J., 2023. Experimental and Numerical Investigation of Brittle Ice Crushing Loads. Technischen Universität Hamburg, Hamburg. https://doi.org/10.15480/882.4923

Hisette, Q., Reimer, N., Schröder, C., Myland, D.: Advanced ice model test techniques and services for future scenarios of arctic logistics and infrastructure. In: Proceedings of the Annual Offshore Technology Conference (2023). https://doi.org/10.4043/32483-MS

Hollnagel, E.: Safety-II in Practice. Developing the Resilience Potentials, p. 978. Taylor & Francis Group, London (2018)

Huang, R., et al.: Investigation on the coupling damage effects of ships subjected to near-field underwater explosion loads. Mar. Struct. **98** (2024). https://doi.org/10.1016/j.marstruc.2024.103664

Huang, Y., van Gelder, P.H.A.J.M.: Collision risk measure for triggering evasive actions of maritime autonomous surface ships. Saf. Sci. **127**, 104708 (2020). https://doi.org/10.1016/j.ssci.2020.104708

IACS: UR I Series: Unified Requirements for Polar Class Ships. IACS (2023). Retrieved from https://www.iacs.org.uk/resolutions/unified-requirements/ur-i

Im, N., Luong, T.N.: Potential risk ship domain as a danger criterion for real-time ship collision risk evaluation. Ocean Eng. **194** (2019). https://doi.org/10.1016/j.oceaneng.2019.106610

IMO: International Convention for the Safety of Life at Sea (SOLAS), 1974, as amended by the 1990 Amendments. IMO, London (1990)

IMO: MSC.1/Circ.1519. Guidance on Methodologies for Assessing Operational Capabilities and Limitations in Ice. IMO, London (2016). https://www.imorules.com/GUID-0D1F00AD-98FE-49BE-9C57-820FD9E8CEC8.html

IMO: MSC.1/Circ.1638. Outcome of the Regulatory Scoping Exercise for the Use of Maritime Autonomous Surface Ships (MASS). IMO, London (2021)

IMO: MEPC 80/17/Add.1 Annex 15 – 2023 IMO Strategy on Reduction of GHG Emissions from Ships (2023). https://www.imo.org/en/OurWork/Environment/Pages/2023-IMO-Strategy-on-Reduction-of-GHG-Emissions-from-Ships.aspx

ISO 19902: Petroleum and Natural Gas Industries – Fixed Steel Offshore Structures (2020). Retrieved from http://www.iso.org

Irhayyim, A., Gunaratne, M., Fioklou, A.: Assessment of scoured bridges subjected to ship impact. Structure. (2022). https://doi.org/10.1016/j.istruc.2021.12.023

Jadhav, A.K., Pandi, A.R., Somayajula, A.: Collision avoidance for autonomous surface vessels using novel artificial potential fields. Ocean Eng. **288** (2023). https://doi.org/10.1016/j.oceaneng.2023.116011

Jang, H., et al: Regulatory gap analysis for risk assessment of ammonia-fuelled ships. Ocean Eng. (2023). https://doi.org/10.1016/j.oceaneng.2023.115751

Jiang, C., Lin, L., Chen, N.Z.: A study on auxetic-inspired side structure for enhanced crashworthiness. Mar. Struct. **93** (2024). https://doi.org/10.1016/j.marstruc.2023.103545

Jiang, F., Dong, S.: Two-level quantitative risk analysis of submarine pipelines from dropped objects considering pipe–soil interaction. Ocean Eng. **257** (2022). https://doi.org/10.1016/j.oceaneng.2022.111620

Jiang, F., Dong, S., Zhao, Y., Xie, Z., Guedes Soares, C.: Investigation on the deformation response of submarine pipelines subjected to impact loads by dropped objects. Ocean Eng. **194** (2019). https://doi.org/10.1016/j.oceaneng.2019.106638

Jones, N.: Slamming damage. J. Ship Res. **17**, 80–86 (1973). https://doi.org/10.5957/jsr.1973.17.2.80

Jovanović, I., Vladimir, N., Cajner, H., Perčić, M.: The overview of risk analysis methods and discussion on their applicability for power system of autonomous ships. TransNav Int. J. Mar. Navig. Saf. Sea Transport. **18**, 109–113 (2024)

Kandemir, C., Celik, M.: Determining the error producing conditions in marine engineering maintenance and operations through HFACS-MMO. Reliab. Eng. Syst. Saf. **206**, 107308 (2021). https://doi.org/10.1016/J.RESS.2020.107308

Kaptan, M., Sarıalioğlu, S., Uğurlu, Ö., Wang, J.: The evolution of the HFACS method used in analysis of marine accidents: a review. Int. J. Ind. Ergon. **86**, 103225 (2021a). https://doi.org/10.1016/J.ERGON.2021.103225

Kaptan, M., Uğurlu, Ö., Wang, J.: The effect of nonconformities encountered in the use of technology on the occurrence of collision, contact and grounding accidents. Reliab. Eng. Syst. Saf. **215**, 107886 (2021b). https://doi.org/10.1016/J.RESS.2021.107886

Kim, D.-H.: Human factors influencing the ship operator's perceived risk in the last moment of collision encounter. Reliab. Eng. Syst. Saf. **203**, 107060 (2020a)

Kim, D.H.: Identification of collision risk factors perceived by ship operators in a vessel encounter situation. Ocean Eng. **200**, 107060 (2020b). https://doi.org/10.1016/J.OCEANENG.2020.107060

Kim, D.K., Kim, H.B., Park, D.H., Kim, S.G., Kim, S.J., Paik, J.K.: Updating the residual strength – damage index diagrams for hull girder collapse predictions of grounded ships. Ships Offshore Struct. **1–30** (2024a). https://doi.org/10.1080/17445302.2024.2398198

Kim, D.K., Looi, C.K., Topa, A., Cho, N.K.: Prediction of mechanical response of hexagonal honeycomb SPS blast wall under explosive loading: in-depth review and empirical formula. Ocean Eng. **293** (2024b). https://doi.org/10.1016/j.oceaneng.2023.116578

Kim, H., et al: A risk-based fire protection design of a LNG fueled ship. In: Proceedings of the International Conference on Offshore Mechanics and Arctic Engineering – OMAE (2023). https://doi.org/10.1115/OMAE2023-104652

Kim, S.J., Sohn, J.M.: The effect of dynamic fracture strain on the structural response of ships in collisions. J. Mar. Sci. Eng. (2022). https://doi.org/10.3390/jmse10111674

Kim, S.J., et al.: Comparison of numerical approaches for structural response analysis of passenger ships in collisions and groundings. Mar. Struct. **81** (2022a). https://doi.org/10.1016/j.marstruc.2021.103125

Kim, S.K., Park, S.I., Paik, J.K.: Collision-accidental limit states-based safety studies for a LNG-fuelled containership. Ocean Eng. **257** (2022b). https://doi.org/10.1016/j.oceaneng.2022.111571

Kovacevic, S., Markovic, N., Sumarac, D., Salatic, R.: Influence of patch load length on plate girders. Part II: numerical research. J. Constr. Steel Res. **158**, 213–229 (2019). https://doi.org/10.1016/j.jcsr.2019.03.025

Kretschmann, L.: Leading indicators and maritime safety: predicting future risk with a machine learning approach. J Shipp. Trade. **5**, 19 (2020). https://doi.org/10.1186/s41072-020-00071-1

Kujala, P., Bergström, M., Hirdaris, S.: Goal-based ship design towards safe and sustainable shipping in ice-covered waters. Transport. Res. Proc. **72**, 3956–3963 (2023). https://doi.org/10.1016/J.TRPRO.2023.11.484

Kumar, S., Arzaghi, E., Baalisampang, T., Abaei, M.M., Garaniya, V., Abbassi, R.: A risk-based multi-criteria decision-making framework for offshore green hydrogen system developments: pathways for utilizing existing and new infrastructure. Sustainable Prod. Consumption. **46**, 655–678 (2024). https://doi.org/10.1016/j.spc.2024.03.020

Kuznecovs, A., Ringsberg, J.W., Mallaya Ullal, A., Janardhana Bangera, P., Johnson, E.: Consequence analyses of collision-damaged ships — damage stability, structural adequacy and oil spills. Ships Offshore Struct. **18**, 567–581 (2023). https://doi.org/10.1080/17445302.2022.2071014

La Ferlita, A., Di Nardo, E., Macera, M., Lindemann, T., Ciaramella, A., Koulianos, N.: A deep neural network method to predict the residual hull girder strength. In: SNAME Maritime Convention, SMC 2022 (2022). https://doi.org/10.5957/SMC-2022-074

La Ferlita, A., Rathje, H., Lindemann, T., Kaeding, P., Bronsart, R.: An advanced salvage method for damaged ship structures. In: SNAME Maritime Convention 2021, SMC 2021 (2021). https://doi.org/10.5957/SMC-2021-046

Ladeira, I., Echeverry Jaramillo, S., Le Sourne, H.: A simplified method to assess the elasto-plastic response of standalone tubular offshore wind turbine supports subjected to ship impact. Ocean Eng. **279** (2023). https://doi.org/10.1016/j.oceaneng.2023.114313

Lampe, T., Okpeke, B.E., Roß, L., Ehlers, S.: First principle design load estimation for lh2 fuel tanks by means of 0d approach. In: Proceedings of the International Conference on Offshore Mechanics and Arctic Engineering. OMAE (2023). https://doi.org/10.1115/OMAE2023-102581

Lan, H., Ma, X., Qiao, W., Ma, L.: On the causation of seafarers' unsafe acts using grounded theory and association rule. Reliab. Eng. Syst. Saf. **223**, 108498 (2022). https://doi.org/10.1016/J.RESS.2022.108498

Lande Andrade, S., Elruby, A., Oldford, D., Quinton, B.: Assessing polar class ship overload and ice impact on low-ice class vessels using a "Quasi Real Time" Popov/Daley approach. In: SNAME Maritime Convention. SNAME, Houston (2022). https://doi.org/10.5957/SMC-2022-108

Lande Andrade, S., Elruby, A.Y., Hipditch, E., Daley, C.G., Quinton, B.W.T.: Full-scale ship-structure ice impact laboratory experiments: experimental apparatus and initial results. Ships Offshore Struct. **18**, 500–514 (2023). https://doi.org/10.1080/17445302.2022.2032993

Law, L.C., Foscoli, B., Mastorakos, E., Evans, S.: A comparison of alternative fuels for shipping in terms of lifecycle energy and cost. Energies (Basel). **14** (2021). https://doi.org/10.3390/en14248502

Le Sourne, H., Donner, R., Besnier, F., Ferry, M.: External dynamics of ship-submarine collision. In: 2nd International Conference on Collision and Grounding of Ships (ICCGS), pp. 137–144 (2001)

Lesar, D.E.: Modeling and Simulation of Ice Body/Ship Structure Collision with Inelastic Structural Deformation Including Rupture and Tearing. West Bethesda (2020)

Lesar, D.E.: Modeling and Simulation of Ice Floe/Ship Structure Collision with Inelastic Structural Deformation, Ice Crushing, and Ice Flexural Fracture Enclosure. West Bethesda (2019)

Leveson, N.: A new accident model for engineering safer systems. Saf. Sci. **42**, 237–270 (2004). https://doi.org/10.1016/S0925-7535(03)00047-X

Li, C., Zhang, H., Zhang, Y., Kang, J.: Fire risk assessment of a ship's power system under the conditions of an engine room fire. J. Mar. Sci. Eng. **10** (2022). https://doi.org/10.3390/jmse10111658

Li, H., Zhang, C., Zheng, X., Mei, Z., Bai, X.: A simplified theoretical model of the whipping response of a hull girder subjected to underwater explosion considering the damping effect. Ocean Eng. **239** (2021). https://doi.org/10.1016/j.oceaneng.2021.109831

Li, M., Wan, Z., Yuan, Y., Tang, W.: A numerical method for analysing the responses of cargo containment system of an LNGC under iceberg collision. Ships Offshore Struct. 1–14 (2023a). https://doi.org/10.1080/17445302.2023.2218324

Li, S., Kim, D.K.: A comparison of numerical methods for damage index based residual ultimate limit state assessment of grounded ship hulls. Thin-Walled Struct. **172** (2022). https://doi.org/10.1016/j.tws.2021.108854

Li, X., et al.: Plastic deformation and ductile fracture of L907A ship steel at increasing strain rate and temperature. Int. J. Impact Eng. **174** (2023b). https://doi.org/10.1016/j.ijimpeng.2023.104515

Liang, Z., Li, F., Zhou, S.: An improved NSGA-II algorithm for MASS autonomous collision avoidance under COLREGs. J. Mar. Sci. Eng. **12** (2024). https://doi.org/10.3390/jmse12071224

Lim, S.J., Woo, D.H., Lee, Y.H.: Numerical analysis on extinguishing of sprinklers in a hydrogen pool fire. Int. J. Hydrog. Energy. **54**, 118–126 (2024). https://doi.org/10.1016/j.ijhydene.2023.05.178

Liu, B., Guedes Soares, C.: Recent developments in ship collision analysis and challenges to an accidental limit state design method. Ocean Eng. **270**, 113636 (2023). https://doi.org/10.1016/j.oceaneng.2023.113636

Liu, F., Hu, Y., Feng, G., Li, C.: Experimental and numerical study on the penetration of the inclined stiffened plate. Ocean Eng. **258** (2022a). https://doi.org/10.1016/j.oceaneng.2022.111792

Liu, H., Liu, K., Wang, X., Gao, Z., Wang, J.: On the resistance of cruciform structures during ship collision and grounding. J. Mar. Sci. Eng. **11** (2023). https://doi.org/10.3390/jmse11020459

Liu, H.Y., et al.: Material mechanics properties and critical analyses of fish farm netting and trusses. Ocean Eng. **292** (2024). https://doi.org/10.1016/j.oceaneng.2023.116512

Liu, J., et al.: Prediction of human–machine interface (HMI) operational errors for maritime autonomous surface ships (MASS). J. Mar. Sci. Technol. Jpn. **27**, 293–306 (2022b). https://doi.org/10.1007/s00773-021-00834-w

Liu, X., Jiang, D., Liufu, K., Fu, J., Liu, Q., Li, Q.: Numerical investigation into impact responses of an offshore wind turbine jacket foundation subjected to ship collision. Ocean Eng. **248** (2022c). https://doi.org/10.1016/j.oceaneng.2022.110825

Liu, Y., Guo, X.: Study on risk of ship collision in bridge life-cycle based on synergetic theory. Ocean Eng. **289** (2023). https://doi.org/10.1016/j.oceaneng.2023.116148

Lu, Y., Liu, K., Wang, Z., Tang, W.: Modelling of ductile fracture in ship structures subjected to quasi-static impact loads. Int. J. Impact Eng. **156** (2021). https://doi.org/10.1016/j.ijimpeng.2021.103941

Lu, Y., Liu, K., Wang, Z., Tang, W., Amdahl, J.: Development of ductile fracture modelling approach in ship impact simulations. Ocean Eng. **252** (2022). https://doi.org/10.1016/j.oceaneng.2022.111173

Luan, C., Moan, T., Kvåle, K.A., Cheng, Z.: Shear lag effects on global bending moments in a long-span pontoon bridge under self-weight, traffic and environmental loads. Mar. Struct. **93** (2024). https://doi.org/10.1016/j.marstruc.2023.103518

Lyu, W., el Moctar, O., Schellin, T.E.: Ship motion-sloshing interaction using a field method. Mar. Struct. **76** (2021). https://doi.org/10.1016/j.marstruc.2020.102923

Ma, J., Zhong, W., Zhu, X.: Safety management in sea reclamation construction: a case study of Sanya Airport, China. Adv. Civ. Eng. **2020** (2020). https://doi.org/10.1155/2020/2910612

Ma, L., Chen, L., Ma, X., Wang, T., Zhang, J.: Incorporating human and organizational failures into the formation pattern for different Arctic maritime accidents using a data-driven Bayesian network. Ocean Eng. **312**, 119125 (2024a). https://doi.org/10.1016/J.OCEANENG.2024.119125

Ma, L., Ma, X., Lan, H., Liu, Y., Deng, W.: A data-driven method for modeling human factors in maritime accidents by integrating DEMATEL and FCM based on HFACS: a case of ship collisions. Ocean Eng. **266**, 112699 (2022). https://doi.org/10.1016/J.OCEANENG.2022.112699

Ma, L., Ma, X., Wang, T., Chen, L., Lan, H.: On the development and measurement of human factors complex network for maritime accidents: a case of ship groundings. Ocean Coast. Manag. **248**, 106954 (2024b). https://doi.org/10.1016/J.OCECOAMAN.2023.106954

Ma, L., Ma, X., Wang, T., Zhao, Y., Lan, H.: A data-driven approach to determine the distinct contribution of human factors to different types of maritime accidents. Ocean Eng. **295**, 116874 (2024c). https://doi.org/10.1016/J.OCEANENG.2024.116874

Ma, L., Swan, C.: Wave-in-deck loads: an assessment of present design practice given recent improvements in the description of extreme waves and the nature of the applied loads. Ocean Eng. **285**, 115302 (2023). https://doi.org/10.1016/j.oceaneng.2023.115302

Maritime Administrations of North West Europe: Stockholm Agreement: Stability Requirements for Ro-Ro Passenger Ships. Maritime Administrations of North West Europe, Stockholm (1996)

Markovic, N., Kovacevic, S.: Influence of patch load length on plate girders. Part I: experimental research. J. Constr. Steel Res. **157**, 207–228 (2019). https://doi.org/10.1016/j.jcsr.2019.02.035

Márquez, L., Le Sourne, H., Rigo, P.: Mechanical model for the analysis of ship collisions against reinforced concrete floaters of offshore wind turbines. Ocean Eng. **261** (2022). https://doi.org/10.1016/j.oceaneng.2022.111987

Márquez, L., Rigo, P., Le Sourne, H.: Ship collision events against reinforced concrete offshore structures. In: Developments in the Analysis and Design of Marine Structures, pp. 245–253 (2021). https://doi.org/10.1201/9781003230373-29

Maternová, A., Materna, M.: Research of maritime accidents based on HFACS framework. Transport. Res. Proc. **74**, 1224–1231 (2023). https://doi.org/10.1016/J.TRPRO.2023.11.265

Mathern, A., von der Haar, C., Marx, S.: Concrete support structures for offshore wind turbines: current status, challenges, and future trends. Energies (Basel). (2021). https://doi.org/10.3390/en14071995

Mauro, F., Conti, F., Vassalos, D.: Damage surrogate models for real-time flooding risk assessment of passenger ships. Ocean Eng. **285** (2023). https://doi.org/10.1016/j.oceaneng.2023.115493

McKinlay, C.J., Turnock, S.R., Hudson, D.A.: Route to zero emission shipping: hydrogen, ammonia or methanol? Int. J. Hydrog. Energy. **46**, 28282–28297 (2021). https://doi.org/10.1016/j.ijhydene.2021.06.066

Mehdi, R., Gluch, M., Fischer, S., Baldauf, M.: A perfect warning to avoid collisions at sea? Zesz. Nauk. Akad. Morsk. Szczecin. **49**, 53–64 (2017). https://doi.org/10.17402/245

Mohsendokht, M., Li, H., Kontovas, C., Chang, C.-H., Qu, Z., Yang, Z.: Decoding dependencies among the risk factors influencing maritime cybersecurity: lessons learned from historical incidents in the past two decades. Ocean Eng. **312**, 119078 (2024). https://doi.org/10.1016/j.oceaneng.2024.119078

Mokhtari, M., Kim, E., Amdahl, J.: A non-linear viscoelastic material model with progressive damage based on microstructural evolution and phase transition in polycrystalline ice for design against ice impact. Int. J. Impact Eng. **176**, 104563 (2023). https://doi.org/10.1016/J.IJIMPENG.2023.104563

Montewka, J., Goerlandt, F., Innes-Jones, G., Owen, D., Hifi, Y., Puisa, R.: Enhancing human performance in ship operations by modifying global design factors at the design stage. Reliab. Eng. Syst. Saf. **159**, 283–300 (2017). https://doi.org/10.1016/j.ress.2016.11.009

Morán, J., Morato, P.G., Rigo, P.: Active Learning for Structural Reliability Analysis with Multiple Limit State Functions Through Variance-Enhanced PC-Kriging Surrogate Models. Trinity's Access to Research Archive, Dublin (2023)

Mujeeb-Ahmed, M.P., Ince, S.T., Paik, J.K.: Computational models for the structural crashworthiness analysis of a fixed-type offshore platform in collisions with an offshore supply vessel. Thin-Walled Struct. **154** (2020). https://doi.org/10.1016/j.tws.2020.106868

Müller, F., Böhm, A., Herrnring, H., Von Bock Und Polach, F., Ehlers, S.: Experimental and numerical analysis of ice crushing tests with different shaped ice specimens. In: Proceedings of the ASME 2023 42nd International Conference on Ocean, Offshore and Arctic Engineering (OMAE2023), pp. 1–12. ASME, Melbourne (2023)

Myland, D., Hisette, Q., Cilkaya, E., Özhan, Y.S.: Experimental investigation of aspects influencing the level ice resistance of ships with non-typical icebreaking bow shapes. In: ASME 2021 40th International Conference on Ocean, Offshore and Arctic Engineering. ASME (2021). https://doi.org/10.1115/OMAE2021-62254

Nasur, J., Bogusławski, K., Wolska, P., Gil, M., Wróbel, K.: Toward modeling emergency unmooring of manned and autonomous ships – A combined FRAM+HFACS-MA approach. Saf. Sci. **181**, 106676 (2025). https://doi.org/10.1016/J.SSCI.2024.106676

Negenborn, R., et al.: Autonomous ships are on the horizon: here's what we need to know. Nature. **615**, 30–33 (2023). https://doi.org/10.1038/d41586-023-00557-5

Nian, Y., Wan, S., Wang, X., Zhou, P., Avcar, M., Li, M.: Study on crashworthiness of nature-inspired functionally graded lattice metamaterials for bridge pier protection against ship collision. Eng. Struct. **277** (2023). https://doi.org/10.1016/j.engstruct.2022.115404

Nie, Y., Fang, H., Meng, X.: Ship collision mitigation for offshore wind turbine monopile foundations via high-general/low-collision-stiffness steel fenders. Ocean Eng. **299** (2024). https://doi.org/10.1016/j.oceaneng.2024.117272

NORSOK N-001: Integrity of Offshore Structures. Standard Norge (2021). Retrieved from http://www.standard.no

NORSOK Z-013: Risk and Emergency Preparedness Assessment. Standard Norge (2024). Retrieved from http://www.standard.no

NORSOK N-004: Design of Offshore Structures (2022). Retrieved from http://www.standard.no

Norway, S.: Floating Aquaculture Farms – Site Survey, Design, Execution and Use. Norwegian Standard, NS 9415 (2021)

Oka, J.M., Vaidya, R.U., Davis, J.T., Webber, K.W., Walzel, R.K.: Nuclear material container drop testing using finite element analysis with verification using digital Image correlation. Nucl. Eng. Des. **421** (2024). https://doi.org/10.1016/j.nucengdes.2024.113057

Okawa, T., Nakashima, K., Yamada, Y.: Study on ductile fracture behavior of welded joints of highly ductile steel for shipbuilding. In: Advances in the Collision and Grounding of Ships and Offshore Structures – Proceedings of the 9th International Conference on Collision and Grounding of Ships and Offshore Structures, ICCGS 2023, pp. 355–359 (2024). https://doi.org/10.1201/9781003462170-43

Oo, Y.P.S., Le Sourne, H., Brunellière, K.: Time delay effect in the dynamic response of submerged cylinder subjected to an underwater explosion. In: Advances in the Analysis and Design of Marine Structures – Proceedings of the 9th International Conference on Marine Structures, MARSTRUCT 2023, pp. 669–677. CRC Press/Balkema (2023). https://doi.org/10.1201/9781003399759-74

Pack, K., Mohr, D.: Combined necking & fracture model to predict ductile failure with shell finite elements. Eng. Fract. Mech. **182**, 32–51 (2017). https://doi.org/10.1016/J.ENGFRACMECH.2017.06.025

Paik, J.: Advanced Structural Safety Studies: With Extreme Conditions and Accidents (2020). https://doi.org/10.1007/978-981-13-8245-1

Pan, J., Wang, T., Huang, S.W., Xu, M.C.: Investigation of assessment method of axial crushing force of rake bow for bridge against ship collision. Ocean Eng. **269** (2023). https://doi.org/10.1016/j.oceaneng.2022.113498

Pan, J., Wang, T., Zhang, W.Z., Huang, S.W., Xu, M.C.: Study on the assessment of axial crushing force of bulbous bow for bridge against ship collision. Ocean Eng. **255**, 111411 (2022). https://doi.org/10.1016/J.OCEANENG.2022.111411

Park, S.-J., Lee, K., Cerik, B.C., Choung, J., 2019. Ductile fracture prediction of EH36 grade steel based on Hosford–Coulomb model. Ships Offshore Struct. 14, 219–230. https://doi.org/10.1080/17445302.2019.1565300

Parris, D., Spinthiropoulos, K., Ragazou, K., Giovou, A., Tsanaktsidis, C.: Methanol, a plugin marine fuel for green house gas reduction – a review. Energies (Basel). (2024). https://doi.org/10.3390/en17030605

Pedersen, P.T., Chen, J., Zhu, L.: Design of bridges against ship collisions. Mar. Struct. **74**, 102810 (2020). https://doi.org/10.1016/j.marstruc.2020.102810

Peng, C., Zhen, X., Huang, Y.: Human-automation interaction centered approach based on FRAM for systemic safety analysis of dynamic positioning operations for offshore tandem offloading. Ocean Eng. **267**, 113249 (2023). https://doi.org/10.1016/J.OCEANENG.2022.113249

Perčić, M., Frković, L., Pukšec, T., Ćosić, B., Li, O.L., Vladimir, N.: Life-cycle assessment and life-cycle cost assessment of power batteries for all-electric vessels for short-sea navigation. Energy. **251** (2022). https://doi.org/10.1016/j.energy.2022.123895

Pietrzykowski, Z., Wielgosz, M.: Effective ship domain – impact of ship size and speed. Ocean Eng. **219**, 108423 (2021). https://doi.org/10.1016/j.oceaneng.2020.108423

Pietrzykowski, Z., Wielgosz, M., Breitsprecher, M.: Navigators' behavior analysis using data mining. J. Mar. Sci. Eng. **8**, 50 (2020). https://doi.org/10.3390/jmse8010050

Pineau, J.P., Conti, F., Le Sourne, H., Looten, T.: A fast simulation tool for ship grounding damage analysis. Ocean Eng. **262** (2022). https://doi.org/10.1016/j.oceaneng.2022.112248

Pineau, J.P., Le Sourne, H.: Analytical modelling of ship bottom grounding considering combined surge and heave motions. Mar. Struct. **88** (2023). https://doi.org/10.1016/j.marstruc.2022.103364

Podgórski, D.: Measuring operational performance of OSH management system – a demonstration of AHP-based selection of leading key performance indicators. Saf. Sci. **73**, 146–166 (2015). https://doi.org/10.1016/j.ssci.2014.11.018

Porathe, T.: Safety of autonomous shipping: COLREGS and interaction between manned and unmanned ships. In: Proceedings of the 29th European Safety and Reliability Conference, ESREL 2019, pp. 4146–4153. Research Publishing Services (2020). https://doi.org/10.3850/978-981-11-2724-3_0655-cd

Qiao, W., Liu, Y., Ma, X., Liu, Y.: A methodology to evaluate human factors contributed to maritime accident by mapping fuzzy FT into ANN based on HFACS. Ocean Eng. **197**, 106892 (2020). https://doi.org/10.1016/J.OCEANENG.2019.106892

Qiao, W., Yang, J., Zhao, Y., Deng, W., Ma, X.: On the determination of the maritime-specific EPC values in reducing human factors based on maritime foundering accidents in China. Ocean Eng. **307**, 118192 (2024). https://doi.org/10.1016/J.OCEANENG.2024.118192

Qiu, X., Zhao, Y., Wang, D.: Mechanical analysis and interior stiffeners arrangement discussion of a specific ship side collision situation. Ocean Eng. **309** (2024). https://doi.org/10.1016/j.oceaneng.2024.118417

Qu, J., Feng, Y., Wu, Y., Zhu, Y., Wu, B., Xiao, Z.: Design and optimization of an integrated power system of solid oxide fuel cell and marine low-speed dual-fuel engine. J. Mar. Sci. Appl. **22**, 837–849 (2023). https://doi.org/10.1007/s11804-023-00377-z

Quinton, B.W.T.: Lateral (sliding) motion of design ice loads on IACS polar classed structures. Ships Offshore Struct., 1–11 (2019). https://doi.org/10.1080/17445302.2019.1580844

Quinton, B.W.T.: Experimental and Numerical Investigation of Moving Loads on Hull Structures. St. John's (2015)

Quinton, B.W.T., Daley, C.G., Gagnon, R.E.: Realistic moving ice loads and ship structural response. In: Proceedings of the International Offshore and Polar Engineering Conference (2012)

Quinton, B.W.T., Daley, C.G., Gagnon, R.E., Colbourne, D.B.: Guidelines for the nonlinear finite element analysis of hull response to moving loads on ships and offshore structures. Ships Offshore Struct. **12** (2017). https://doi.org/10.1080/17445302.2016.1261391

Ramanan, C.J., Lim, K.H., Kurnia, J.C., Roy, S., Bora, B.J., Medhi, B.J.: Towards sustainable power generation: recent advancements in floating photovoltaic technologies. Renew. Sust. Energ. Rev. (2024). https://doi.org/10.1016/j.rser.2024.114322

Ramos, M.A., Thieme, C.A., Utne, I.B., Mosleh, A.: Human-system concurrent task analysis for maritime autonomous surface ship operation and safety. Reliab. Eng. Syst. Saf. **195**, 106697 (2020). https://doi.org/10.1016/j.ress.2019.106697

Ramsay, W., Fridell, E., Michan, M.: Maritime energy transition: future fuels and future emissions. J. Mar. Sci. Appl. (2023). https://doi.org/10.1007/s11804-023-00369-z

Ranta, J., Polojärvi, A.: Limit mechanisms for ice loads on inclined structures: local crushing. Mar. Struct. **67**, 102633 (2019). https://doi.org/10.1016/J.MARSTRUC.2019.102633

Rawson, A., Brito, M.: A critique of the use of domain analysis for spatial collision risk assessment. Ocean Eng. **219**, 108259 (2021). https://doi.org/10.1016/j.oceaneng.2020.108259

Reason, J., 1990. Human Error. https://doi.org/10.1017/CBO9781139062367

Ren, Y., Meng, Q., Chen, C., Hua, X., Zhang, Z., Chen, Z.: Dynamic behavior and damage analysis of a spar-type floating offshore wind turbine under ship collision. Eng. Struct. **272** (2022). https://doi.org/10.1016/j.engstruct.2022.114815

Ren, Y., Yu, Z., Hua, X., Amdahl, J., Zhang, Z., Chen, Z.: Experimental and numerical investigation on the deformation behaviors of large diameter steel tubes under concentrated lateral impact loads. Int. J. Impact Eng. **180** (2023). https://doi.org/10.1016/j.ijimpeng.2023.104696

Robbins, M.L., 2020. Characterization of post-yield behavior of a warship grillage subject to repeated impacts. Memorial University of Newfoundland, St. John's.

Rodrigues, J.M., Viuff, T., Økland, O.D.: Model tests of a hydroelastic truncated floating bridge. Appl. Ocean Res. **125** (2022). https://doi.org/10.1016/j.apor.2022.103247

Rong, H., Teixeira, A.P., Guedes Soares, C.: Spatial correlation analysis of near ship collision hotspots with local maritime traffic characteristics. Reliab. Eng. Syst. Saf. **209**, 107463 (2021). https://doi.org/10.1016/J.RESS.2021.107463

Rong, Y., Yuan, H., Liu, J.X., Liu, W.G., Luo, W., Hu, Z.Q.: Experimental and numerical research on the behavior of offshore tubular member under lateral indentation. In: Developments in the

Collision and Grounding of Ships and Offshore Structures – Proceedings of the 8th International Conference on Collision and Grounding of Ships and Offshore Structures, ICCGS 2019, pp. 96–102 (2020)

Rudan, S., Sviličić, S., Prebeg, P., Ćatipović, I.: External dynamics modelling in ship collision analysis. In: Advances in the Collision and Grounding of Ships and Offshore Structures – Proceedings of the 9th International Conference on Collision and Grounding of Ships and Offshore Structures, ICCGS 2023, pp. 213–220. CRC Press/Balkema (2024). https://doi.org/10.1201/9781003462170-27

Saborit, E., et al.: Alternatives for transport, storage in port and bunkering systems for offshore energy to green hydrogen. Energies (Basel). **16** (2023). https://doi.org/10.3390/en16227467

Salihoglu, E., Bal Beşikçi, E.: The use of Functional Resonance Analysis Method (FRAM) in a maritime accident: a case study of Prestige. Ocean Eng. **219**, 108223 (2021). https://doi.org/10.1016/J.OCEANENG.2020.108223

Sarıalioğlu, S., Uğurlu, Ö., Aydın, M., Vardar, B., Wang, J.: A hybrid model for human-factor analysis of engine-room fires on ships: HFACS-PV&FFTA. Ocean Eng. **217**, 107992 (2020). https://doi.org/10.1016/J.OCEANENG.2020.107992

Schinas, O., Metzger, D.: Cyber-seaworthiness: a critical review of the literature. Mar. Policy. **151**, 105592 (2023). https://doi.org/10.1016/j.marpol.2023.105592

Schuler, J., Ardone, A., Fichtner, W.: A review of shipping cost projections for hydrogen-based energy carriers. Int. J. Hydrog. Energy. (2024). https://doi.org/10.1016/j.ijhydene.2023.10.004

Sha, Y., Zhong, Y., Wang, B.: Responses of a floating offshore wind turbine with RC floaters subjected to ship collision loads. In: Fib Symposium, pp. 1936–1945 (2022)

Shappell, S., Wiegmann, D.: The Human Factors Analysis and Classification System – HFACS (2000)

Shi, J., et al.: Real-time natural gas explosion modeling of offshore platforms by using deep learning probability approach. Ocean Eng. **276** (2023a). https://doi.org/10.1016/j.oceaneng.2023.114244

Shi, X., Zhuang, H., Xu, D.: Structured survey of human factor-related maritime accident research. Ocean Eng. **237**, 109561 (2021). https://doi.org/10.1016/J.OCEANENG.2021.109561

Shi, Y.H., Yang, D.Q., Wu, W.W.: Numerical analysis method of ship-ice collision-induced vibration of the polar transport vessel based on the full coupling of ship-ice-water-air. J. Ocean Eng. Sci. **8**, 323–335 (2023b). https://doi.org/10.1016/J.JOES.2023.04.002

Sigrist, J.F., Marie-Dit-Dinard, M., Leblond, C.: Semi-analytical models for deriving shock signals and response spectra on immersed and fluid-filled hulls subjected to shock waves. Ocean Eng. **263** (2022). https://doi.org/10.1016/j.oceaneng.2022.112346

Slagstad, M.: Simplified Analysis of Forces Acting in Pre-tensioned Aquaculture Nets. PhD Thesis, NTNU (2024)

Sohn, J.M., Jung, D.: Structural assessment of a 500-cbm liquefied natural gas bunker ship during bunkering and marine operation under collision accidents. Ships Offshore Struct. **17**, 2379–2395 (2022). https://doi.org/10.1080/17445302.2021.1996133

Song, J., Kim, J., Lee, J., Kim, S., Chung, W.: Dynamic response of multiconnected floating solar panel systems with vertical cylinders. J. Mar. Sci. Eng. **10** (2022a). https://doi.org/10.3390/jmse10020189

Song, L., Zhao, H., Li, J., Yang, Q.: Failure mode and mechanism of large cylinder structures for artificial islands on soft clay. Appl. Ocean Res. **119** (2022b). https://doi.org/10.1016/j.apor.2021.103032

Song, Y., Wang, J., Han, Q.: A simplified model of impact force time-history for bridge pier-barge collision. Ocean Eng. **264** (2022c). https://doi.org/10.1016/j.oceaneng.2022.112464

Sotiralis, P., Ventikos, N.P., Hamann, R., Golyshev, P., Teixeira, A.P.: Incorporation of human factors into ship collision risk models focusing on human centred design aspects. Reliab. Eng. Syst. Saf. **156**, 210–227 (2016). https://doi.org/10.1016/J.RESS.2016.08.007

Sree, D.K.K., et al.: Fluid-structural analysis of modular floating solar farms under wave motion. Sol. Energy. **233**, 161–181 (2022). https://doi.org/10.1016/j.solener.2022.01.017

Storheim, M., Alsos, H., Amdahl, J., 2018. Evaluation of nonlinear material behavior for offshore structures subjected to accidental actions. J. Offshore Mech. Arct. Eng..

Strantzali, E., Livanos, G.A., Aravossis, K.: A comprehensive multicriteria evaluation approach for alternative marine fuels. Energies (Basel). **16** (2023). https://doi.org/10.3390/en16227498

Stroeve, S., et al.: SHIELD human factors taxonomy and database for learning from aviation and maritime safety occurrences. Safety. **9**, 14 (2023). https://doi.org/10.3390/SAFETY9010014

Sun, C., et al.: Experimental and numerical study on the offshore adaptability of new FLH2 floating hydrogen liquefaction production storage and offloading unit. Renew. Energy. **224** (2024). https://doi.org/10.1016/j.renene.2024.120147

Sun, J., Huang, Y.: Investigations on the ship-ice impact: part 1. Experimental methodologies. Mar. Struct. **72**, 102772 (2020). https://doi.org/10.1016/J.MARSTRUC.2020.102772

Suominen, M., Kõrgesaar, M., Taylor, R., Bergström, M.: Probabilistic analysis of operational ice damage for Polar class vessels using full-scale data. Struct. Saf. **107**, 102423 (2024). https://doi.org/10.1016/j.strusafe.2023.102423

Sviličić, Š., Rudan, S.: Modelling manoeuvrability in the context of ship collision analysis using non-linear FEM. J. Mar. Sci. Eng. **11** (2023). https://doi.org/10.3390/jmse11030497

Szłapczyński, R., Szłapczyńska, J.: A method of determining and visualizing safe motion parameters of a ship navigating in restricted waters. Ocean Eng. **129**, 363–373 (2017). https://doi.org/10.1016/j.oceaneng.2016.11.044

Tabri, K., Naar, H., Kõrgesaar, M.: Ultimate strength of ship hull girder with grounding damage. Ships Offshore Struct. **15**, S161–S175 (2020). https://doi.org/10.1080/17445302.2020.1827631

Taimuri, G., Kim, S.J., Mikkola, T., Hirdaris, S.: A two-way coupled FSI model for the rapid evaluation of accidental loads following ship hard grounding. J. Fluids Struct. **112** (2022). https://doi.org/10.1016/j.jfluidstructs.2022.103589

Taimuri, G., Ruponen, P., Hirdaris, S.: A novel method for the probabilistic assessment of ship grounding damages and their impact on damage stability. Struct. Saf. **100** (2023). https://doi.org/10.1016/j.strusafe.2022.102281

Tamis, J.E., Jongbloed, R.H., Piet, G.J., Jak, R.G.: Developing an environmental impact assessment for floating island applications. Front. Mar. Sci. **8** (2021). https://doi.org/10.3389/fmars.2021.664055

Tekgoz, M., Garbatov, Y., Guedes Soares, C.: Review of ultimate strength assessment of ageing and damaged ship structures. J. Mar. Sci. Appl. (2020). https://doi.org/10.1007/s11804-020-00179-7/Published

Tian, L.R., Chen, F.L., Zhu, L., Yu, T.X.: Large deformation of square plates under pulse loading by combined saturation analysis and membrane factor methods. Int. J. Impact Eng. **140** (2020). https://doi.org/10.1016/j.ijimpeng.2020.103546

Tian, L.R., Chen, F.L., Zhu, L., Yu, T.X.: Saturated analysis of pulse-loaded beams based on membrane factor method. Int. J. Impact Eng. **131**, 17–26 (2019). https://doi.org/10.1016/j.ijimpeng.2019.04.021

Vairo, T., Cademartori, D., Clematis, D., Carpanese, M.P., Fabiano, B.: Solid oxide fuel cells for shipping: a machine learning model for early detection of hazardous system deviations. Process Saf. Environ. Prot. **172**, 184–194 (2023). https://doi.org/10.1016/j.psep.2023.02.022

Van Hoecke, L., Laffineur, L., Campe, R., Perreault, P., Verbruggen, S.W., Lenaerts, S.: Challenges in the use of hydrogen for maritime applications. Energy Environ. Sci. (2021). https://doi.org/10.1039/d0ee01545h

van Westrenen, F., Baldauf, M.: Improving conflicts detection in maritime traffic: case studies on the effect of traffic complexity on ship collisions. Proc. Inst. Mech. Eng. M J. Eng. Marit. Environ. **234**, 209–222 (2020). https://doi.org/10.1177/1475090219845975

Vandegar, G., Oo, Y.P.S., Ladeira, I., Le Sourne, H., Echeverry, S.: A simplified method to assess the elastoplastic response of standalone tubular floating offshore wind turbine supports subjected to ship impact. In: Advances in the Collision and Grounding of Ships and Offshore Structures – Proceedings of the 9th International Conference on Collision and Grounding of Ships and Offshore Structures, ICCGS 2023, pp. 447–454. CRC Press/Balkema (2024). https://doi.org/10.1201/9781003462170-54

Varas, J.M., et al.: MAXCMAS project: autonomous COLREGs compliant ship navigation. In: Proceedings of the 16th Conference on Computer Applications and Information Technology in the Maritime Industries (COMPIT) 2017, pp. 454–464 (2017)

Vo, T.T.E., Ko, H., Huh, J., Park, N.: Overview of possibilities of solar floating photovoltaic systems in the offshore industry. Energies (Basel). (2021). https://doi.org/10.3390/en14216988

Wang, H., Chen, N., Wu, B., Guedes Soares, C.: Human and organizational factors analysis of collision accidents between merchant ships and fishing vessels based on HFACS-BN model. Reliab. Eng. Syst. Saf. **249**, 110201 (2024). https://doi.org/10.1016/J.RESS.2024.110201

Wang, Y., et al.: A review of low and zero carbon fuel technologies: achieving ship carbon reduction targets. Sustain. Energy Technol. Assess. (2022). https://doi.org/10.1016/j.seta.2022.102762

Wang, Z.: Intelligent dynamic collision avoidance strategy of hydrogen fuel cell unmanned ship via improved fusion dynamic window method. IEEE Access. **11**, 69971–69988 (2023). https://doi.org/10.1109/ACCESS.2023.3293656

Wang, Z., Dong, B., Li, M., Ji, Y., Han, F.: Configuration of low-carbon fuels green marine power systems in diverse ship types and Applications. Energy Convers. Manag. (2024a). https://doi.org/10.1016/j.enconman.2024.118139

Wang, Z., Dong, B., Wang, Y., Li, M., Liu, H., Han, F.: Analysis and evaluation of fuel cell technologies for sustainable ship power: energy efficiency and environmental impact. Energy Convers. Manag. X. **21**, 100482 (2024b). https://doi.org/10.1016/j.ecmx.2023.100482

Wang, Z., Dong, B., Yin, J., Li, M., Ji, Y., Han, F.: Towards a marine green power system architecture: integrating hydrogen and ammonia as zero-carbon fuels for sustainable shipping. Int. J. Hydrog. Energy. (2024c). https://doi.org/10.1016/j.ijhydene.2023.10.207

Wang, Z., Guo, C., Wang, C., Chen, G., Xu, Y., Li, Q.: An analytical method for predicting the structural response of ship side structures by bulbous bow in oblique collision scenarios. Ships Offshore Struct. **1–14** (2023). https://doi.org/10.1080/17445302.2023.2247839

Wang, Z., Li, M., Zhao, F., Ji, Y., Han, F.: Status and prospects in technical standards of hydrogen-powered ships for advancing maritime zero-carbon transformation. Int. J. Hydrog. Energy. (2024d). https://doi.org/10.1016/j.ijhydene.2024.03.083

Wang, Z., Sha, Y., Jakobsen, J.B.: Influence of local scour on the dynamic response of bridges under barge collisions. Eng. Struct. **308**, 117941 (2024e). https://doi.org/10.1016/j.engstruct.2024.117941

Wang, Z., Sha, Y., Jakobsen, J.B.: Floating bridge response under combined ship collision, wind and wave loads. Ships Offshore Struct. (2023a). https://doi.org/10.1080/17445302.2023.2195242

Wang, Z., Sha, Y., Ong, M.C.: Numerical investigation of scoured bridge response under ship collisions. In: Proceedings of the International Conference on Offshore Mechanics and Arctic Engineering – OMAE (2023b). https://doi.org/10.1115/OMAE2023-101126

Wei, L., Zhou, Z., Wang, Z.: Fire monitoring system for power batteries on ship. In: IOP Conference Series: Earth and Environmental Science. IOP Publishing (2021). https://doi.org/10.1088/1742-6596/1802/2/022022

Williams, R., Zhao, F., 2023. Global Offshore Wind Report 2023.

Wróbel, K.: Searching for the origins of the myth: 80% human error impact on maritime safety. Reliab. Eng. Syst. Saf. **216**, 107942 (2021). https://doi.org/10.1016/J.RESS.2021.107942

Wróbel, K., Gil, M., Krata, P., Olszewski, K., Montewka, J.: On the use of leading safety indicators in maritime and their feasibility for Maritime Autonomous Surface Ships. Proc. Inst. Mech. Eng. O J. Risk. Reliab. **237**, 314–331 (2023). https://doi.org/10.1177/1748006X211027689

Wu, B., Yip, T.L., Yan, X., Guedes Soares, C.: Review of techniques and challenges of human and organizational factors analysis in maritime transportation. Reliab. Eng. Syst. Saf. **219**, 108249 (2022). https://doi.org/10.1016/J.RESS.2021.108249

Wu, Q., He, Y., Yu, Z., Wang, J., Fan, W.: Fluid-structure interaction analysis of vessel-bridge-steel floating fender collisions. Ocean Eng. **295** (2024). https://doi.org/10.1016/j.oceaneng.2024.116828

Wu, Y., Du, Z., Li, L., Tian, Z.: A new evaluation method of dented natural gas pipeline based on ductile damage. Appl. Ocean Res. **135** (2023). https://doi.org/10.1016/j.apor.2023.103533

Xiang, G., Rao, K., Xiang, X., Yu, X.: Overview and analysis on recent research and challenges of dropped objects in offshore engineering. Ocean Eng. **281** (2023). https://doi.org/10.1016/j.oceaneng.2023.114616

Xiong, Z., Ma, M., Guo, Y., Li, R.: Study of energy absorption characteristics of coastal ship side structures based on collision accidents. Ocean Eng. **307**, 118079 (2024). https://doi.org/10.1016/J.OCEANENG.2024.118079

Xu, G., Cao, Z., Wang, J., Xue, S., Tang, M.: A novel machine learning-based framework for predicting impact force in ship-bridge pier collisions. Ocean Eng. **285** (2023a). https://doi.org/10.1016/j.oceaneng.2023.115347

Xu, L., Zhu, L., Wang, X., Pedersen, P.T.: Collision experiments of ship models in water tank. In: Proceedings of the International Conference on Offshore Mechanics and Arctic Engineering – OMAE, pp. 1–5 (2020). https://doi.org/10.1115/omae2020-18741

Xu, X., Li, X., Zhou, Y., Li, K., Yang, R., Jiang, L.: Safety assessment of hydrogen fuel cell powered ships during navigation. In: 7th IEEE International Conference on Transportation Information and Safety, ICTIS 2023. Institute of Electrical and Electronics Engineers (2023b). https://doi.org/10.1109/ICTIS60134.2023.10243810

Yamada, Y., Okawa, T., Nakashima, K.: On the simplified method to estimate critical grounding velocity for the prevention of cargo tank rupture in grounding accidents. In: Advances in the Collision and Grounding of Ships and Offshore Structures – Proceedings of the 9th International Conference on Collision and Grounding of Ships and Offshore Structures, ICCGS 2023, pp. 331–336. CRC Press/Balkema (2024). https://doi.org/10.1201/9781003462170-40

Yaman, H., Yesilyurt, M.K., Raja Ahsan Shah, R.M., Soyhan, H.S.: Effects of compression ratio on thermodynamic and sustainability parameters of a diesel engine fueled with methanol/diesel fuel blends containing 1-pentanol as a co-solvent. Fuel. **357** (2024). https://doi.org/10.1016/j.fuel.2023.129929

Yan, H., Fang, H., Zhu, L., Jia, E., Dai, Z., Zhang, X.: Evaluation of a novel steel box-soft body combination for bridge protection against ship collision. Rev. Adv. Mater. Sci. **62** (2023). https://doi.org/10.1515/rams-2022-0295

Yang, X., Holmen, I.M., Utne, I.B.: Scenario analysis of fish escapes in Norwegian aquaculture for implementation of barrier management and improved learning from accidents. Mar. Policy. **143** (2022). https://doi.org/10.1016/j.marpol.2022.105208

Yao, P., et al.: Experimental study on the behavior of precast segmental piers under ship impact loading. Ocean Eng. **253** (2022). https://doi.org/10.1016/j.oceaneng.2022.111324

Ye, X., Fan, W., Sha, Y., Hua, X., Wu, Q., Ren, Y.: Fluid-structure interaction analysis of oblique ship-bridge collisions. Eng. Struct. **274** (2023). https://doi.org/10.1016/j.engstruct.2022.115129

Yildiz, S., Uğurlu, Ö., Wang, J., Loughney, S.: Application of the HFACS-PV approach for identification of human and organizational factors (HOFs) influencing marine accidents. Reliab. Eng. Syst. Saf. **208**, 107395 (2021). https://doi.org/10.1016/J.RESS.2020.107395

Yin, R., et al.: Risk analysis for marine transport and power applications of lithium-ion batteries: a review. Process Saf. Environ. Prot. **181**, 266–293 (2024). https://doi.org/10.1016/j.psep.2023.11.015

Yoon, D.H., Choung, J., 2023. Collision simulation of a floating offshore wind turbine considering ductile fracture and hydrodynamics using hydrodynamic plug-in HydroQus. J. Ocean Eng. Technol. 37, 111–121. 10.26748/KSOE.2023.004

Yoon, D.H., Jeong, S.Y., Choung, J.: Collision simulations between an icebreaker and an iceberg considering ship hydrodynamics. Ocean Eng. **279**, 114333 (2023). https://doi.org/10.1016/J.OCEANENG.2023.114333

Yu, R., et al.: Stress state sensitivity for plastic flow and ductile fracture of L907A low-alloy marine steel: from tension to shear. Mater. Sci. Eng. A. **835**, 142689 (2022a). https://doi.org/10.1016/j.msea.2022.142689

Yu, Y., Zhao, D., Li, C., Liang, Y., Wang, Z., Sun, W.: Research on ship-ice collision model and ship-bow fatigue failure mechanism based on elastic-plastic ice constitutive relation. Ocean Eng. **306**, 118005 (2024a). https://doi.org/10.1016/J.OCEANENG.2024.118005

Yu, Z.: Water Impact Damage Considering Hydro-Plastic Interactions: Extensive Experimental and Numerical Validation, and Structural Design Recommendations (2024). SSRN 4938825.

Yu, Z., Amdahl, J.: A Rayleigh-Ritz solution for high order natural frequencies and eigenmodes of monopile supported offshore wind turbines considering tapered towers and soil pile interactions. Mar. Struct. **92** (2023). https://doi.org/10.1016/j.marstruc.2023.103482

Yu, Z., Amdahl, J., Greco, M., Xu, H.: Hydro-plastic response of beams and stiffened panels subjected to extreme water slamming at small impact angles, Part I: an analytical solution. Mar. Struct. **65**, 53–74 (2019). https://doi.org/10.1016/J.MARSTRUC.2019.01.002

Yu, Z., Amdahl, J., Rypestøl, M., Cheng, Z.: Numerical modelling and dynamic response analysis of a 10 MW semi-submersible floating offshore wind turbine subjected to ship collision loads. Renew. Energy. **184**, 677–699 (2022b). https://doi.org/10.1016/j.renene.2021.12.002

Yu, Z., Wang, X., Moan, T., Amdahl, J., Sha, Y.: Ultimate and residual strength assessments of intact and collision damaged columns of the Bjørnafjorden floating bridge. In: Advances in the Collision and Grounding of Ships and Offshore Structures – Proceedings of the 9th International Conference on Collision and Grounding of Ships and Offshore Structures, ICCGS 2023, pp. 387–393. CRC Press/Balkema (2024b). https://doi.org/10.1201/9781003462170-47

Zeng, F., Zhang, N., Huang, G., Gu, Q., He, M.: Dynamic response of floating offshore wind turbines under freak waves with large crest and deep trough. Energy. **278** (2023a). https://doi.org/10.1016/j.energy.2023.127679

Zeng, F., Zhang, N., Huang, G., Gu, Q., He, M.: Experimental study on dynamic response of a floating offshore wind turbine under various freak wave profiles. Mar. Struct. **88** (2023b). https://doi.org/10.1016/j.marstruc.2022.103362

Zhang, H., Jiang, D.: Risk analysis of fire in hydrogen storage system of ship hydrogen fuel cell based on FSA. In: 7th IEEE International Conference on Transportation Information and Safety, ICTIS 2023, pp. 2480–2490. Institute of Electrical and Electronics Engineers (2023). https://doi.org/10.1109/ICTIS60134.2023.10243733

Zhang, H., Li, C., Zhao, N., Chen, B.Q., Ren, H., Kang, J.: Fire risk assessment in engine rooms considering the fire-induced domino effects. J. Mar. Sci. Eng. **10** (2022a). https://doi.org/10.3390/jmse10111685

Zhang, J., Liu, Z., Ong, M.C., Tang, W.: Numerical simulations of the sliding impact between an ice floe and a ship hull structure in ABAQUS. Eng. Struct. **273**, 115057 (2022b). https://doi.org/10.1016/J.ENGSTRUCT.2022.115057

Zhang, L., Meng, Q.: Probabilistic ship domain with applications to ship collision risk assessment. Ocean Eng. **186**, 106–130 (2019)

Zhang, M., Kujala, P., Hirdaris, S.: A machine learning method for the evaluation of ship grounding risk in real operational conditions. Reliab. Eng. Syst. Saf. **226** (2022c). https://doi.org/10.1016/j.ress.2022.108697

Zhang, M., Liao, X., Li, S., Song, S., Liu, J., Hu, Z.: Experimental and numerical investigation of the damage characteristics of a ship side plate laterally punched by a scaled raked bow indenter. Ocean Eng. **280** (2023). https://doi.org/10.1016/j.oceaneng.2023.114808

Zhang, M., Zhang, D., Goerlandt, F., Yan, X., Kujala, P.: Use of HFACS and fault tree model for collision risk factors analysis of icebreaker assistance in ice-covered waters. Saf. Sci. **111**, 128–143 (2019). https://doi.org/10.1016/J.SSCI.2018.07.002

Zhang, Y., Hu, Z.: An aero-hydro coupled method for investigating ship collision against a floating offshore wind turbine. Mar. Struct. **83** (2022). https://doi.org/10.1016/j.marstruc.2022.103177

Zhang, Y., Hu, Z., Ng, C., Jia, C., Jiang, Z.: Dynamic responses analysis of a 5 MW spar-type floating wind turbine under accidental ship-impact scenario. Mar. Struct. **75**, 102885 (2021). https://doi.org/10.1016/j.marstruc.2020.102885

Zhang, Y., Yuan, Y., Zhou, J., Tang, W.: A new analytical model to evaluate ship side collisions with large indentation considering interaction effects between structural components. Mar. Struct. **95** (2024). https://doi.org/10.1016/j.marstruc.2024.103596

Zhao, C., Cao, X., Ren, Y.: Risk analysis of bridge ship collision based on AIS data model and nonlinear finite element. Nonlinear Eng. **12** (2023). https://doi.org/10.1515/nleng-2022-0324

Zheng, K., Chen, Y., Jiang, Y., Qiao, S.: A SVM based ship collision risk assessment algorithm. Ocean Eng. **202**, 107062 (2020). https://doi.org/10.1016/j.oceaneng.2020.107062

Zheng, X., Li, H., Zhu, Y., Lv, Y., Zhang, C., Mei, Z.: The effects of superstructure form on damage characteristics of ship subjected to underwater explosion. Thin-Walled Struct. **190** (2023). https://doi.org/10.1016/j.tws.2023.110993

Zheng, Y., Zhang, X., Shang, Z., Guo, S., Du, Y.: A decision-making method for ship collision avoidance based on improved cultural particle swarm. J. Adv. Transp. (2021). https://doi.org/10.1155/2021/8898507

Zhou, D., Zheng, Z.: Dynamic fuzzy ship domain considering the factors of own ship and other ships. J. Navig. **1**–16 (2018). https://doi.org/10.1017/S0373463318000802

Zhou, H., Kong, X., Wang, Y., Zheng, C., Pei, Z., Wu, W.: Dynamic response of hull girder subjected to combined underwater explosion and wave induced load. Ocean Eng. **235** (2021a). https://doi.org/10.1016/j.oceaneng.2021.109436

Zhou, J., Zhao, Y., Zhang, Y., Zhang, L., Yuan, Y., Tang, W.: An analytical method for evaluating the structural resistance of plate structures under in-plane impact load in ship-side collision scenarios. Ocean Eng. **288** (2023). https://doi.org/10.1016/j.oceaneng.2023.115961

Zhou, Y., Zhang, S.: Perforation analysis and limit prediction of submarine pipelines subjected to extreme impact loadings. Ocean Eng. **246** (2022). https://doi.org/10.1016/j.oceaneng.2022.110651

Zhou, Z., Zhu, L., Guo, K., 2020. Plate tearing experiments over a cone-shaped rock considering friction in ship grounding. Proceedings of the International Offshore and Polar Engineering Conference 2020-Octob, 3916–3920.

Zhou, Z., Zhu, L., Li, C.B.: Ship grounding simulation with different rock shapes and its verification. Ships Offshore Struct. **20**, 316–328 (2024). https://doi.org/10.1080/17445302.2024.2336669

Zhou, X.-Y., Liu, Z.-J., Wang, F.-W., Wu, Z.-L.: A system-theoretic approach to safety and security co-analysis of autonomous ships. Ocean Eng. **222**, 108569 (2021b). https://doi.org/10.1016/j.oceaneng.2021.108569

Zhu, L., Cai, W., Chen, M., Tian, Y., Bi, L.: Experimental and numerical analyses of elastic-plastic responses of ship plates under ice floe impacts. Ocean Eng. **218**, 108174 (2020a). https://doi.org/10.1016/j.oceaneng.2020.108174

Zhu, L., Cai, W., Frieze, P.A., Shi, S.: Design method for steel deck plates under quasi-static patch loads with allowable plastic deformations. Mar. Struct. **71** (2020b). https://doi.org/10.1016/j.marstruc.2019.102702

Zhu, L., Liang, Q., Chen, M., Zhang, S.: On the fluid-structure response of elastic-plastic ship sides subjected to impact loads. Mar. Struct. **70**, 102698 (2020c). https://doi.org/10.1016/j.marstruc.2019.102698

Zhu, L., Tian, L., Chen, F., Yu, T.X.: A new equivalent method for complex-shaped pulse loading based on saturation analysis and membrane factor method. Int. J. Impact Eng. **158** (2021). https://doi.org/10.1016/j.ijimpeng.2021.104018

Zhu, L., Wang, X., Guo, K., Cai, W., Jones, N., Dai, S.: Effects of indenter geometry on pseudo-shakedown of steel plates under repeated mass impacts. Int. J. Impact Eng. **185**, 104865 (2024a). https://doi.org/10.1016/j.ijimpeng.2023.104865

Zhu, L., Zhou, Z., Pedersen, P.T.: Ship grounding model tests in a water tank: an experimental study. Mar. Struct. **93**, 103529 (2024b). https://doi.org/10.1016/j.marstruc.2023.103529

Zhu, L., Zhou, Z.H., Guo, K.L.: Grounding model test over a sharp rock for ship bottom plating with and without transverse stiffeners. In: Advances in the Collision and Grounding of Ships and Offshore Structures – Proceedings of the 9th International Conference on Collision and Grounding of Ships and Offshore Structures, ICCGS 2023, pp. 259–264 (2024c). https://doi.org/10.1201/9781003462170-32

Zhu, L., Zhu, Z., Yu, T.X., Jones, N.: An experimental study of the saturated impulse for metal plates under slamming. Int. J. Impact Eng. **178** (2023). https://doi.org/10.1016/j.ijimpeng.2023.104601

Zhu, M., Huang, L., Huang, Z., Shi, F., Xie, C.: Hazard analysis by leakage and diffusion in liquefied natural gas ships during emergency transfer operations on coastal waters. Ocean Coast. Manag. **220** (2022). https://doi.org/10.1016/j.ocecoaman.2022.106100

Zilong, T., Xiao Wei, D.: Layout optimization of offshore wind farm considering spatially inhomogeneous wave loads. Appl. Energy. **306** (2022). https://doi.org/10.1016/j.apenergy.2021.117947

Open Access This chapter is licensed under the terms of the Creative Commons Attribution-NonCommercial-NoDerivatives 4.0 International License (http://creativecommons.org/licenses/by-nc-nd/4.0/), which permits any noncommercial use, sharing, distribution and reproduction in any medium or format, as long as you give appropriate credit to the original author(s) and the source, provide a link to the Creative Commons license and indicate if you modified the licensed material. You do not have permission under this license to share adapted material derived from this chapter or parts of it.

The images or other third party material in this chapter are included in the chapter's Creative Commons license, unless indicated otherwise in a credit line to the material. If material is not included in the chapter's Creative Commons license and your intended use is not permitted by statutory regulation or exceeds the permitted use, you will need to obtain permission directly from the copyright holder.

Committee V.2: Experimental Methods

Paul Lara[1(✉)], Ilson Paranhos Pasqualino[2], Bin Liu[3], Pero Prebeg[4], Mikko Suominen[5], Sören Ehlers[6], Cesare Mario Rizzo[7], Hideaki Murayama[8], Apostolos Grammatikopoulos[9], Sigmund K. Ås[10], and Seo Jung Kwan[11]

[1] West Bethesda, USA
paul.a.lara.civ@us.navy.mil
[2] Federal University of Rio de Janeiro (UFRJ), Rio de Janeiro, Brazil
ilson@lts.coppe.ufrj.br
[3] Beijing Institute of Technology, Beijing, China
liubin8502@whut.edu.cn
[4] University of Zagreb, Zagreb, Croatia
pero.prebeg@fsb.hr
[5] Aalto University, Espoo, Finland
mikko.suominen@aalto.fi
[6] Hamburg University of Technology (TUHH), Hamburg, Germany
soren.ehlers@dlr.de
[7] University of Genoa, Genoa, Italy
cesare.rizzo@unige.it
[8] The University of Tokyo, Tokyo, Japan
murayama@giso.t.u-tokyo.ac.jp
[9] Delft University of Technology, Delft, Netherlands
a.grammatikopoulos@tudelft.nl
[10] Norwegian University of Science and Technology, Trondheim, Norway
sigmund.k.aas@ntnu.no
[11] Pusan National University, Busan, Republic of Korea
seojk@pusan.ac.kr

Committee Mandate. Concern for advances in structural model testing and full-scale experimentation and in-service monitoring and their role in the design, construction, inspection and maintenance, structural health monitoring and digital twin of ship and offshore structures. This shall include new developments in: best practice and uncertainty analysis; experimental methods and techniques; full field imaging and sensor systems; data post processing and applications for ship and offshore structures; and correlation between model, full-scale and numerical datasets.

Keywords: Experimental methodologies · scale laws · hydrodynamic behavior of flexible structures · wave-in-deck phenomena · hybrid model evaluations · vibrational analysis · fatigue assessment · large-scale impact experiments · corrosion assessments · comprehensive iced load measurements · health monitoring frameworks · digital twin models · Digital Image Correlation (DIC)

1 Introduction

Experimental methodologies are employed to assess the performance and responses of maritime vessels and offshore structures under diverse conditions. These experiments utilize sensor systems and numerical models to corroborate and substantiate design criteria, life cycle performance, accidental scenarios, and life cycle responses. This report embodies the expertise of the authors within their respective fields of research and disciplines, encapsulating the state of the art and identifying critical technologies and existing gaps in such fields.

The report encompasses sections on scaling laws, fluid-structure interaction, hybrid models, fire, friction, corrosion prognostics, large-scale subsea structures, large impact tests, full-scale ice loads, health monitoring, digital twins, and digital image correlation. It serves as a continuation of the 2021 report, incorporating a section with updates on the prior benchmark study. The report also provides a summary of findings and recommendations at the conclusion, emphasizing areas where further research is recommended.

The collaborative efforts of the diverse group of authors from various global regions are evident in the composition of this document. Overleaf served as the collaborative online platform to assemble the LaTeX content, facilitate version control, manage references, and resolve comments. Furthermore, its grammatical tool, Windscribe, was employed for editorial purposes. This tool utilizes language models and artificial intelligence capabilities to harmonize linguistic tone and standardize readability. It is crucial to highlight that the development of content was exclusively performed by the author, whereas the deployment of editorial tools was confined to grammatical and language standardization.

2 Scaling Laws

2.1 Introduction

Model-scale testing constitutes an important practice within the scientific domain. Compared to full-scale investigations, model-scale testing provides a controlled environment for studying phenomena, performance, and structural behavior, etc. Relative to other laboratory experiments, model-scale testing facilitates the examination of global phenomena and processes while integrating these with studies concentrating on local phenomena and processes. Furthermore, it presents a cost-effective approach to design analysis. This is particularly significant for industries characterized by limited production series or where prototypes serve as the definitive product, such as in shipbuilding. To effectively conduct model-scale testing and extrapolate observations to full-scale, it is imperative to comprehend the translation of these observations across varying scales, which can be accomplished via similitude methods or scaling laws.

This chapter offers a comprehensive examination of the applications and advancements of these theories. Recently, extensive reviews authored by Coutinho et al. (2016) and Casaburo et al. (2019) have been published, concentrating on similitude methods within structural engineering, while also encompassing publications external to the marine technology domain. These reviews have been significantly reflected in relevant sections of the most recent ISSC Experimental Committee reports (Ehlers S., 2022). Consequently, this review is dedicated to the exploration of studies not encompassed within those publications. The subsequent chapter presents an overview of various similitude methods, subsequently followed by recent developments and applications related to scaling laws.

2.2 Similitude Methods

Similitudes can be classified into three primary categories: geometric, kinematic, and dynamic similitude. Geometric similitude necessitates that geometric characteristics are scaled uniformly. Kinematic similarity asserts that velocities at corresponding points must be scaled by a constant factor. Dynamic similitude requires that all forces in the model be proportionally scaled by a constant factor relative to the corresponding forces in the prototype. An alternative method for classifying similitude is predicated on the fulfillment of similitude conditions (Casaburo et al., 2019), which pertains to the extent of similitude representation within the model: complete, adequate, or distorted. Complete similitude or true models satisfy all relevant conditions. Adequate similitude or first-order models fulfill the conditions related to the principal parameters. Distorted similitude or partial models do not satisfy at least one of the first-order conditions. These classifications must be considered and comprehended in the translation of observations across scales and during test preparations.

Similitude methods can be categorized into broader groups, which can be further subdivided into more specific subgroups based on the methodological approach. The literature exhibits a variety of divisions and terminologies that are somewhat contingent on the preferences of the respective authors. In general, these methods can be classified into dimensional analysis, application to governing equations, energy methods, and empirical similitude, as delineated in the ISSC Experimental Committee report (Ehlers S., 2022), consistent with the frameworks set forth by Coutinho et al. (2016) and Casaburo et al. (2019).

Dimensional analysis represents the conventional approach for establishing scaling laws (Coutinho et al., 2016), facilitating the identification of fundamental quantities that characterize the physical phenomenon or system (Ehlers S., 2022). This analysis involves defining a set of dimensionless parameters that govern the phenomenon under investigation. A physical phenomenon is expressed by an equation wherein both sides possess identical dimensions, known as dimensional homogeneity. These principles are encapsulated in Buckingham's theorem. In certain applications, this theorem is circumvented, and scaling laws are formulated by defining a scaling factor that represents the prototype/model ratio as a power law. The advantages of dimensional analysis include its simplicity and applicability even in the absence of known governing equations. However, a significant drawback lies in the necessity for an experienced analyst to select relevant parameters for the analysis. Furthermore, the resulting dimensionless terms, derived from the physical parameters when applying the theorem, may lack physical significance and uniqueness, often necessitating a trial-and-error approach (Casaburo et al., 2019).

An alternative prevalent method for deriving similitude conditions involves the direct application of similitude theory to the governing equations. This methodology entails the application of similitude theory directly to the fundamental equations of the system, encompassing boundary and initial conditions, which define the system with relevant variables and parameters. Given that analogous models are regulated by the same set of fundamental equations and conditions, similitude conditions can be deduced by establishing scale factors and juxtaposing the equations of the prototype with those of the model (Casaburo et al., 2019). This method connects the geometric, structural, excitation, and material attributes of the system with its response. The benefit of this approach lies in its ability to yield specific similitude conditions that possess physical significance. The primary limitation, however, is the necessity for the underlying equations to be known.

The energy method utilizes the principle of conservation of energy to establish similitude. Within this framework, the potential energy, exemplified as strain energy, is equivalent to the aggregate of kinetic energy and the work exerted by external forces. As the equation dictated by this principle incorporates the structural domain, applied loads, and boundary conditions, the system is analyzed in its entirety, thus negating the necessity to ascertain explicit and implicit scale factors (Casaburo et al., 2019).

The empirical similarity method entails the evaluation of two specimens. One specimen is fabricated via rapid prototyping using a simple geometric configuration, while the other is produced through the standard manufacturing process. The state transformation is ascertained by measuring the state vectors of these two specimens, in conjunction with the scaled structure derived from rapid prototyping.

2.3 Recent Developments in Ship and Offshore Structure Testing

Environment

Historically, model-scale testing in ice has concurrently employed Froude and Cauchy scaling methods. This combined scaling approach, along with the expertise of operational ice tanks, has demonstrated a strong correlation between scales concerning the performance in ice (such as resistance and maneuverability) of vessels with conventional hull designs, as noted by Riska et al. (1994) and Lau (2015). Nevertheless, this methodology results in relatively pliable model ice with increased plasticity, as discussed in A. Palmer & Dempsey (2009, R. U. F. von Bock und Polach & Molyneux (2017), von Bock und Polach et al., (2019), which may impede scale correlation when the testing parameters are adjusted. Consequently, novel scaling approaches for ice in model-scale testing have been proposed in recent studies.

To address the limitations associated with soft model ice, von Bock und Polach et al. (2021) proposed the use of Model Ice of Virtual Equivalent Thickness (MIVET). The fundamental aim of this method is to enhance the elastic modulus of ice to mitigate plasticity while preserving the ice stiffness number by reducing its thickness. The ice stiffness number serves as an equivalent to the ratio of the characteristic length to the ice thickness (F. Li et al., 2003). In order to ensure that the ice fractures under an external load comparable to that of conventionally scaled ice, the approach recommends an accurate scaling of the critical bending moment, which is determined via cantilever beam testing. To realize this, the method requires the ice's strength to be increased by the square of the factor k applied to modify the ice thickness h, denoted as MIVET = h/k (F. von Bock und Polach et al., 2021). The authors acknowledge that this technique modifies the ice's mass; however, the error introduced is deemed negligible for wave propagation within solid ice (Fox, 2001).

Challenges associated with soft model ice can considerably impede the accurate depiction of processes pertinent to structure-ice interactions. This is particularly true for investigations centered on the vibration of vertical structures where ice crushing predominantly influences the interaction. Following the recommendation by Palmer and Dempsey (2009), Hammer and Handrikse (2023) posited that ice strength remains invariant across scales. Within their methodology, the modal characteristics of structural models are adjusted through mass scaling, which corresponds to the ratio of mean brittle crushing load between model-scale and full-scale; however, the strength of the ice is not scaled (Hendrikse et al., 2022; Hammer & Hendrikse, 2023; Hammer et al., 2023). The outcomes indicate that this approach results in an enhanced representation of ice-induced vibrations compared to the conventional Froude-Cauchy scaling.

Merchant vessels operating in ice-covered maritime regions are typically engineered to function with the assistance of icebreakers. Consequently, these vessels are generally constructed to navigate through a broken ice channel, which serves as the design ice condition for merchant vessel construction in accordance with the Finnish-Swedish Ice Class Rules. According to Matala (2021), the soft model ice scaled using Froude-Cauchy scaling fails to accurately represent the natural behavior of brash ice, resulting in conservative performance estimates for vessels with hull forms optimized for open water (Matala & Suominen, 2022). This discrepancy is thought to be associated with an increased cohesion between brash ice fragments in the soft model ice compared to solid ice fragments in nature (Matala & Suominen, 2022). To address this issue, a new dimensionless parameter, termed brash ice similitude—which represents the relationship between cohesive and inertial forces—has been proposed for application in the scaling of brash ice properties to enhance the portrayal of brash ice behavior (Matala & Suominen, 2023). In practice, the new methodology suggests that the strength of brash ice fragments can be considered scale invariant, and that the cohesive forces between the ice fragments should be negligible at the model scale due to their minimal presence at the full scale. Fundamentally, this approach is consistent with the material strength-based scaling proposed by Høyland (2010) (Matala, 2021).

Structure Impact

Mazzariola and Alves (2019b) employed dimensional analysis, specifically the Buckingham theorem, to establish a scaling law for structural impact testing based on the VSM framework (comprising Initial Velocity, Yield Stress, and Structure Mass). This approach is capable of addressing thickness distortion, variations in density, and the mechanical properties of materials, such as yield stress, strain hardening, and strain-rate hardening. These distortions were mitigated through the development of a dimensionless parameter derived from the plastic bending moment. The methodology demonstrated substantial agreement between analytical and numerical analyses (Mazzariol & Alves, 2019b), and its validity was further substantiated through impact experiments involving various materials, scales, and applied loads (Mazzariol & Alves, 2019a).

Calle et al. (2020) employed an analogous methodology to assess structural impact, where a scaled-down model of a ship structure was fabricated using Additive Manufacturing techniques and subsequently subjected to collision tests. Nonetheless, it proved challenging to accurately scale the thicknesses of smaller structural components. It was observed that uniformly increasing the thickness across all structural elements did not yield an accurate representation of the model's response. Consequently, adjustments for thickness distortion in each structural element were made, taking into consideration the anticipated dominant structural collapse modes. In collision events involving large maritime structures, the predominant collapse mechanisms are associated with membrane tension and folding. This methodology was similarly employed and validated by Calle et al. (2020) through a large-scale raking experiment on a ship's bottom structure conducted by Kuroiwa et al. (1992).

Buckling

Wang et al. (2019) established a scaling law for the nonlinear buckling of stiffened orthotropic shallow spherical shells by employing an energy-based approach alongside Donnell's nonlinear shell theory. The total energy of the system was regarded as comprising both strain and potential energy components. The derivation of the strain energy took into account the constitutive equations pertinent to orthotropic shallow spherical shells. Concurrently, the potential energy resulting from external forces was derived under the assumption of external uniform pressure. Similitude factors, or scaling factors, which articulate the relationship between prototype and model parameters, were deduced by enforcing the requirement for complete geometric similitude of the skins of the two systems within the energy function. In the absence of empirical data, this scaling law was corroborated through numerical verification. Furthermore, the investigation included an analysis of partial similitude, or distorted models, concluding that a model with distortions in material properties and geometry is capable of forecasting the prototype's behavior. However, discrepancies in the Poisson's ratio may precipitate significant deviations in the behavior of the prototype.

Offshore Structure Vibrations

The interaction between ice and offshore structures presents distinct challenges, particularly for installations situated in regions characterized by seasonal ice cover. These challenges predominantly involve heightened loading and vibrational forces as a consequence of ice interaction. The phenomenon of frequency lock-in, wherein structural vibrations and movements are amplified due to ice interaction, poses significant risks for offshore infrastructures. Ziemer (2021) investigated the frequency lock-in phenomenon and ice-induced vibrations through model-scale experimentation with vertical structures embedded in ice. A recommendation was made to forgo Froude scaling since gravitational forces do not exert a substantial influence (A. Palmer et al., 2010). The crushing load scaling ratio was derived from the load equation pertinent to vertical structures, ensuring an accurate load level for rigid structures. However, the reliance of local pressure distribution on aspect ratio and ice thickness hinders complete similarity between the two geometrically scaled models. To accurately model the process associated with frequency lock-in, precise scaling of the amplification factor was deemed essential. Acknowledging that the local failure process dictating simultaneous load accumulation is independent of geometry, it was ascertained that time remains scale-independent. To maintain strain rate similarity, velocity must be scaled in accordance with geometric proportion, while the oscillator's mass must be adjusted by the square of the geometric scale. With this methodology, the crushing forces and the incidence of frequency lock-in are adequately scaled. Nonetheless, this approach necessitates large geometric scales due to stiffness considerations, whereupon the vibrations observed in model-scale evaluations may be insignificant enough to be reduced to mere noise.

A comparable scaling methodology has been employed in investigations concerning ice-induced vibrations on vertical-sided offshore structures (Hendrikse et al., 2022; Hammer & Hendrikse, 2023), and (Hammer et al., 2023). Hammer et al. (2024) proposed that the scaling of time should remain invariant, akin to the approach by Ziemer (Ziemer, 2021), while also ensuring the preservation of kinematics such that the amplitude of structural displacement remains consistent across different scales. Since the kinematics are not subjected to scaling, the properties of ice and the physical structure should exhibit uniformity across varying scales. For ice, the transition speed from ductile to brittle failure, as well as the impact of velocity, were regarded as critical for ice-induced vibrations impacting vertically-sided structures. To sustain these characteristics, the drift speed and strength of ice should remain unaltered. A more comprehensive exposition and discourse regarding the scaling has been discussed by Hammer et al. (2024). Aiming to achieve a structural response analogous to that observed in full-scale scenarios, a hardware-in-loop hybrid setup, encompassing both physical and numerical models, was implemented in model-scale testing. In the case of the physical model, the stiffness and material properties are maintained uniformly across different scales, i.e., they are not scaled. Nonetheless, mean pressure similarity, defined as the ratio between the measured model-scale and the estimated full-scale mean load, was employed to scale the structural properties (mass, damping, and stiffness) of the full-scale structure within the numerical domain during the model scale tests (Hammer et al., 2024; Hammer & Hendrikse, 2024). This scaling approach was corroborated through experimental validation, wherein the aspect ratio and shape of the structures were varied, and was qualitatively validated by replicating the structural vibrations observed in two full-scale structures, namely Molikpaq and Norströmsgrund lighthouse.

In the context of global ship vibrations, Froude scaling is employed due to the necessity for gravity wave similarity in experiments dominated by seakeeping considerations. It is demonstrable that geometric similarity with respect to the external configuration of the vessel, alongside dimensional analysis techniques for other aspects, can be effectively utilized (Bishop & Price, 1980). This approach relies on an idealized beam-like global dynamic behavior, which can be justifiably assumed in backbone models (refer to the subsequent section). To date, no comprehensive scaling theory has been established that encompasses both global and local scaling pertaining to ship structures. The typical method for assessing the effectiveness of the scaling process involves analyzing the resultant natural frequencies. While this approach may be adequate for vertical bending beam-like responses, it may rapidly lead to inaccuracies in the case of antisymmetric vibrations (Grammatikopoulos et al., 2021).

2.4 Summary and Conclusions

Model-scale testing has remained an important method for studying structures that requires understanding and development of scaling laws. Recently, the scaling laws developed for model-scale tests with structures have focused on structural failure in impact tests, vibrations, and buckling. In the case of model-scale testing in ice, the application of Froude-Cauchy scaling has been prevalent for several decades irrespective of the testing scope. Nevertheless, recent developments have shifted the emphasis towards scaling the response of ice in accordance with the specific testing scope. Consequently, various scaling methodologies tailored to distinct testing scenarios have been introduced. However, the validation of these newly proposed scaling methods is still in progress, and the current scaling approaches address only a limited range of testing scopes. Therefore, additional scaling strategies are necessary to adequately encompass the diverse scopes of different tests.

3 Fluid-Structure Interaction of Flexible Structures

3.1 Introduction

Ship and offshore structures frequently encounter conditions in which fluid-structure interactions are prevalent. Conducting experiments to assess global responses presents substantial challenges, as it necessitates the concurrent scaling of multiple aspects related to both structure and fluid (refer to the previous section). Predominantly, structural and wave-related aspects are scaled according to Froude's law, posing significant challenges; for wind turbines, Reynolds scaling may also be imperative, further increasing the complexity inherent in experimental design. This section elucidates the methodologies employed to develop flexible scaled models of ship and offshore structures for assessment within a towing tank, ocean basin, or analogous facility. The primary focus is on the scaling of the global responses of the structures. In this context, responses of individual components, such as foils, blades, sails, etc., are not considered. Consequently, the examined structural excitations are chiefly wave-induced. Aside from a singular instance presenting the response of a stiffened panel, this section exclusively addresses the global responses of ship hulls, floaters, and wind turbines. In the antecedent report by the Experimental Methods Committee, a section was dedicated to the hydrodynamics of flexible structures. This section seeks to augment that discourse by concentrating on the structural aspects of this interaction.

3.2 Types of Models

Models of ships and offshore structures intended for fluid-structure interaction experiments must possess the capability to undergo deformation in response to fluid excitation in a realistic manner. For Froude-scaled structures, this necessitates achieving identical strains at model-scale and full-scale. Irrespective of the structure type, flexible models are classified into distinct categories based on their manufacturing methods and the mode of introducing flexibility. In most cases, the external geometry, functioning as the fluid boundary, is composed of a series of rigid segments. Flexibility between these segments is provided through either a flexible backbone or a series of flexible joints. Less commonly, the structure is continuous, serving simultaneously as a source of stiffness and a fluid boundary. Illustrations of these categories for ships are presented in Fig. 1.

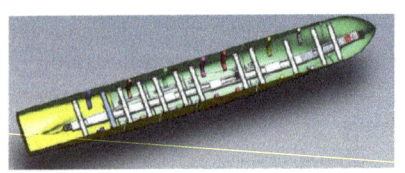

(a) Flexible backbone model (Marón & Kapsenberg, 2014)

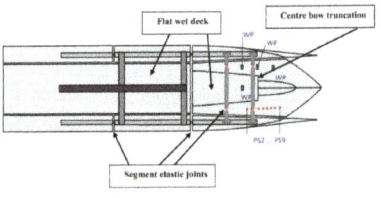

(b) Flexible joint model (Davis et al., 2017)

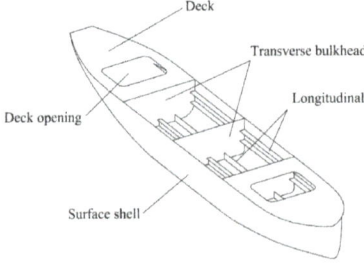

(c) Elastic model (J. Jiao et al., 2021)

Fig. 1. Schematics of the three types of flexible models

The prevalence of experimental research utilizing flexible ship models surpasses that of floating structures. As a result, the nomenclature pertaining to various model types has predominantly developed within this domain of literature. This classification will be employed throughout the current section and will be extended to encompass models of other structural forms. The categorization is partially derived from the review conducted by Grammatikopoulos (Grammatikopoulos, 2023).

3.2.1 Flexible Backbone Models

Flexible backbone models integrate a backbone system to interconnect a sequence of rigid segments, as suggested by the nomenclature. Typically composed of aluminium or steel, this backbone may possess either a uniform or non-uniform cross-sectional profile along its length. The significance of the backbone's geometry is primarily related to its structural characteristics and its influence on the initial few natural frequencies, particularly the first one. Uniform backbones are often employed due to the simplicity of production and the focus on the first natural frequency. To accurately depict the 2-node bending mode, the assembly generally consists of a minimum of four segments. Deflection within the model is quantitatively assessed through the employment of strain gauges affixed along the backbone. For antisymmetric vibrations, shear strain gauges are utilized to measure shear strains along the backbone.

There are instances of non-uniform backbones in ship structures (refer to Fig. 2), particularly notable in vessels subjected to impulse loads (Dessi et al., 2007). In such scenarios, multiple flexible symmetric modes may be activated, necessitating a non-uniform stiffness distribution to achieve the scaling of natural frequencies. An alternative version of a non-uniform backbone is employed to scale the natural frequencies associated with antisymmetric modes. The prevalent method in this context involves introducing openings on the upper side of the backbone to approximate the shear centre location and distribution of torsional stiffness, thereby simulating deck openings typically found in container ships (Marón & Kapsenberg, 2014). Nonetheless, when a U-shaped backbone fabricated from steel or aluminum adheres to the stipulated scaling rules, it often fails to precisely align with the shear center location, resulting in diminished accuracy in the coupling level between horizontal bending and torsion (Grammatikopoulos, 2021). To mitigate this issue, materials characterized by a low Young's modulus, such as acrylonitrile butadiene styrene (ABS) resin, may be incorporated into the U-shaped backbone to concurrently simulate vertical vibration stiffness, horizontal vibration stiffness, and torsional vibration stiffness (Yang et al., 2021).

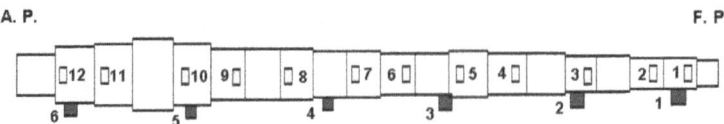

(a) Non-uniform backbone for symmetric modes (Dessi et al., 2007)

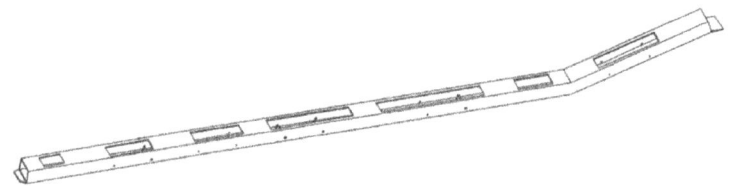

(b) Non-uniform backbone made of steel for antisymmetric modes (Marón & Kapsenberg, 2014)

Fig. 2. Schematics of non-uniform backbones

Backbone models are prevalently employed for ships due to the hull-girder's significant resemblance to beam-like dynamic behavior. Similarly, this principle applies to wind turbines, resulting in the increasing adoption of backbone models within this field. Robertson et al. conducted measurements of the global loads on various rigid wind turbine models as part of the DeepCwind project (Robertson et al., 2017). During the project's initial phase, three distinct 1/50 scale models of wind turbine concepts underwent testing. A direct consequence of the employed geometric scaling was the occurrence of unexpected behavior at low Reynolds numbers with respect to wind excitation. These rigid models were subsequently replaced by a flexible model of the same scale. Although the first natural frequency was accurately scaled, the diameter was significantly smaller than what the geometric scaling principles would suggest. As a result, the scaling of wind drag forces remained suboptimal. These experiments vividly illustrate the inherent challenges in simultaneously scaling the hydrodynamic and aeroelastic behavior of floating structures (Fig. 3).

Leroy et al. developed a 1/40 scale model of a 10 MW spar-type floating wind turbine (Leroy et al., 2022). The model's flexibility was incorporated via a flexible backbone, akin to those commonly employed in ship models. The structural geometry was represented by lightweight sections affixed to this backbone. The primary natural frequency, specifically the frequency corresponding to the 2-node bending mode, was scaled according to Froude's law. The experimental data obtained was subsequently utilized for the validation of numerical predictions (Ran et al., 2023).

Fig. 3. On the left side, a flexible wind turbine model, achieving the correct first natural frequency by featuring a smaller diameter than what geometric scaling would dictate (Robertson et al., 2017). The example of a spar configuration on the right used light sections attached to the backbone to overcome this issue (Ran et al., 2023). (Color figure online)

3.2.2 Flexible Joint Models

The second category of segmented models, referred to as flexible joint models, enjoys considerable popularity. These models offer inherent adaptability, owing to the ability to adjust the stiffness of the flexible joints without necessitating replacement. Consequently, the same set of joints can simulate diverse stiffness levels as well as uniform or distributed stiffness (M. Wu et al., 2012), thus surpassing backbone models in adaptability. Deformations are measured through sensors incorporated within the joints, such as torsional springs (Thomas et al., 2003). The primary disadvantage of these systems is the absence of a continuous structure, which limits the model's capacity to account for cross-sectional characteristics, as previously described (e.g., effective shear area, location of shear centre).

Flexible joint models have not been applied within wind turbine models due to the simple cross-sectional geometry, which makes joint flexibility unnecessary. In contrast, these models have been employed to simulate floating solar installations and other Very Large Floating Structures (VLFS). Given that VLFS are frequently modular and employ flexible joints for inter-module connections in real-world scenarios, their use in scaled models is a logical extension. Ding et al. exemplified such experimentation, by constructing and evaluating an 8-module linear VLFS under varied wave conditions (Ding et al., 2021). A critical consideration for these extremely flexible structures is their dynamic interaction with seabed morphology, particularly in regions characterized by relatively shallow water depth.

3.2.3 Elastic Models

Elastic models are predominantly applicable to ships due to the relative complexity of ship structures compared to other offshore structures. These models present significant challenges in design and production, as the structure must also function as a boundary for fluid interactions. A notable advantage of these models is the capability to perform measurements at any location on the structure, with inherent scaling of factors such as distributed stiffness and the location of the shear center. Primary challenges involve the necessity for simultaneous scaling of both external dimensions and plate thicknesses, and the limitations posed by traditional manufacturing methods in the incorporation of complex structural details. Furthermore, the structural design must ensure watertight integrity.

The aforementioned requirements considerably restrict the range of potential materials and manufacturing techniques. As a result, since the introduction of elastic models in academic literature, these models have incorporated only fundamental characteristics, such as deck openings and bulkheads (Y.-S. Wu et al., 2003), and have otherwise consisted of an external shell with transverse bulkheads. A more detailed version has been developed recently, featuring torsion boxes adjacent to the main deck, in a model constructed from acrylic sheets (Komoriyama et al., 2024).

In pursuit of a more precise structural response, certain research groups endeavor to develop models with increased structural detail, referred to as fully elastic models. These models generally incorporate most or all primary structural components. Chen et al. (2020) designed and fabricated elastic models utilizing sheets of Lexan^{TM}, a polycarbonate thermoplastic resin. Their initial model comprised a barge with three holds, featuring longitudinal, transverse, and bulkhead stiffeners. The authors identified a strain gauge stiffening effect on the local responses due to the material's high flexibility, which was considered during the calibration of the sensors. Subsequent work by the authors employed the same material and methodologies to construct a model of the S175 container ship, incorporating the cellular structure of the cross section.

Grammatikopoulos et al. (2021) proposed an innovative approach by utilizing additive manufacturing techniques to create fully elastic models. The initial demonstration featured a barge with a cellular cross section, which was similarly based on the S175. When using ABS as the printing material, discrepancies between the static and dynamic elastic moduli were identified, potentially exerting a significant impact on the dynamic responses of the final vessel, contingent upon the method of material property acquisition during design (Grammatikopoulos et al., 2020). Subsequently, an investigation was conducted involving a heavy-lifting catamaran model, comprising all primary structural elements, which was manufactured using PETG (Keser et al., 2023). Although the methodology shows promise, additional research is required to ascertain the transverse properties of additively manufactured structures, as these properties considerably influence the dynamic responses of multihull vessels.

As previously stated, the concept of elastic models is primarily applicable to ship structures. Nonetheless, in this subsection, it will be addressed in a broader context to encompass other models where the structural and fluid boundaries coincide. At the present time, this largely pertains to models of floating solar platforms. Otto et al. conducted tests on an inflatable mattress, intended to support solar panels, within a wave basin (Otto et al., 2022). In this instance, testing of the full-scale floater was feasible due to its manageable size, and it is anticipated that similar scaled tests may gain traction in the scholarly literature. Research involving membrane structures for floating solar platforms, both slack (Schreier & Jacobi, 2021) and pre-tensioned (Mukhlas et al., 2022), has begun to surface within the literature, and their prevalence is expected to increase. For slack membranes, a significant scaling challenge lies in the behavior of wrinkling; for pre-tensioned membranes, scaling primarily depends on the pre-tension load, as the membrane exhibits behavior analogous to a two-dimensional version of a string. Although these structures do not exhibit the geometric complexity challenges inherent in ship models, it is anticipated that distinctive types of issues will emerge as their utilization becomes more widespread.

In certain scenarios, only specific segments of a floating structure are modeled as flexible to facilitate the examination of localized responses. In such instances, the structural geometry is typically integrated into the scaled model, rendering the section concerning elastic models particularly pertinent. Ahani et al. conducted slamming assessments on 3D-printed panels, analogous to steel panels with stiffeners present in an offshore semi-submersible (Ahani et al., 2024). The work delineates a methodology wherein scaled models with intricate geometries, varying thickness, and stiffeners are employed in hydroelastic tests. In this study, the elastic properties of the material, influenced by strain rate, were quantified and applied in finite element simulations to ensure that the deflections upon impact corresponded with the Froude scale.

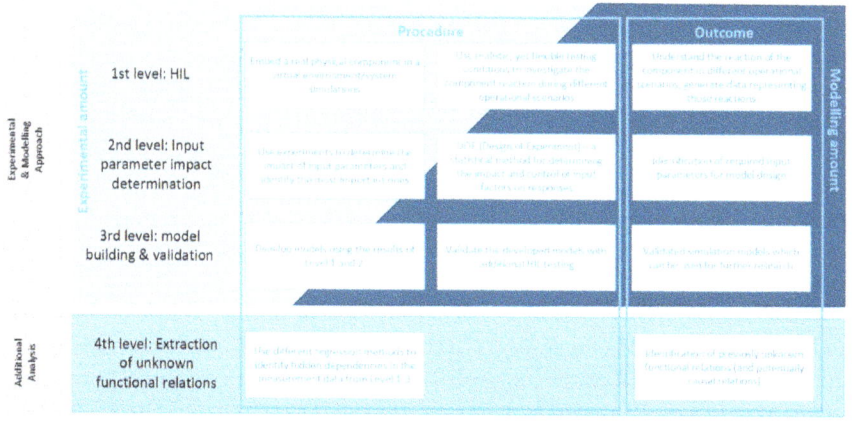

Fig. 4. Hybrid model testing levels

3.3 Summary and Conclusions

Backbone models are demonstrably the predominant choice in the literature concerning the testing of flexible structures within fluid-structure interaction (FSI) studies. This popularity is particularly pronounced when a beam-like bending response can be assumed, and the first flexible mode is dominant. Accordingly, these models are frequently employed in the analysis of monohull ships and wind turbines. In contrast, for structures that deviate from a beam-like form, encompass complex mode interactions, exhibit torsional behavior, or necessitate consideration of local responses, elastic models have emerged as a progressive alternative. These models are typically produced through additive manufacturing, foam construction, or the shaping of plastic sheets. For two-dimensional structures, such as floating membranes, the prevailing method involves employing either scaled-down replicas of the real structures (e.g., inflatable mattresses) or direct scaled models of the membranes themselves. Within this categorization, these are considered more akin to elastic models, as both the source of stiffness and the hydrodynamic boundary coincide within the same structure. Notably, the application of elastic models to wind turbines has not been documented thus far; however, an impending increase in their use is anticipated, particularly in conjunction with the rising interest in floating wind technology.

4 Hybrid Model Testing

Hybrid model testing can be systematically classified into four distinct levels, each characterized by varying degrees of complexity and content (refer to Fig. 4): First-level basic hardware-in-the-loop (HiL) testing, as depicted in Fig. 4, facilitates the emulation of both the environment and ship systems through models that deliver realistic operational scenarios for the physical component undergoing testing. The response of this component to the scenarios can be directly evaluated. "Real-time hardware-in-the-loop simulation-based testing has been recognized as an advanced method for the analysis and testing of power system phenomena and components. Realistic, yet flexible testing conditions for de-risking equipment are its salient benefits." (Kotsampopoulos et al., 2018). The accuracy of HiL test outcomes is inevitably contingent upon the precision with which the simulation models reflect the ship system and environment; as such, the component's reactions will only be as accurate as the models that portray the scenarios.

The second tier pertains to the identification of input parameters necessary for the construction of precise and effective models. Experiments are instrumental in assessing the influence and regulation of input factors on responses. It is crucial to conduct an uncertainty evaluation of these experiments to ascertain their precision (Viswanathan et al., 2022).

The tertiary stage pertains to model validation and the advancement of software-driven functionalities for the component subject to evaluation. A Hardware-in-the-Loop (HiL) testbed serves as a technological development instrument that facilitates the assessment of simulation models devised to represent the component being examined. The HiL configuration permits a direct comparison between the behavior of the simulation model and the actual component, thereby enabling comprehensive model validation.

The fourth level pertains to the derivation of previously uncharted functional relationships, including potential causal associations, from empirical data. When the established physical interactions related to the observed behavior of the component under examination are formulated through differential equations, and the unfamiliar physical interactions are encapsulated within a system of equations by data-driven or machine learning models that have been trained using the available test data, regression techniques are accessible to discern functional dependencies concealed within the measurement data, thereby facilitating the augmentation of the existing system of equations. These functional relationships have the potential to reveal cause-and-effect dynamics underlying the unanticipated behavior of the component under scrutiny or to enhance existing models in order to augment the precision with which they characterize the behavior of the component in question (Safikou & Bollas, 2023; Man & Weil, 2023).

Within the scope of the EU-funded Project HELENUS (High Efficiency Low Emission Nautical SOFC - www.helenus.eu), coordinated by the DLR Institute of Maritime Energy Systems, Hardware-in-the-Loop (HiL) testing is employed to experimentally verify the dynamic operation of a Solid Oxide Fuel Cell (SOFC) module as an integral component of a ship's comprehensive energy system. Reference W. Shi et al. (2023) provides an extensive overview of real-time hybrid model tests conducted on floating offshore wind turbines, while Ha et al. (2023) successfully developed a hybrid model test technique for evaluating the performance of a floating offshore wind turbine under conditions of asymmetrical thrust. Concerning maritime applications, reference Wei et al. (2021) introduces a hybrid model for forecasting ship roll by utilizing full-scale measurements and analytical techniques. In addition, reference Jenssen et al. (2024) details the implementation of hybrid model testing to simulate the dynamic performance of Floating Wind Turbines (FWT) within hydrodynamic laboratories, as initially outlined in Fig. 5. Specifically, two distinct 15 MW FWT models were evaluated at SINTEF Ocean to demonstrate the effectiveness of this method. The findings illustrated that hybrid model testing proficiently controlled dynamic loads, maintaining minimal tracking errors.

Considerable utilization of Hardware-in-the-Loop (HiL) systems has been observed in testing the dynamic responses of marine risers. An innovative hybrid model test technology was developed by H. Ren et al. (2024) for the investigation of vortex-induced vibrations (VIV) in a bluff body. The general framework of this hybrid control method, depicted in Fig. 6, enables the precise real-time control of key physical parameters such as mass, damping ratio, and spring stiffness in VIV model tests. Furthermore, VIV tests at very high Reynolds numbers for a bluff body were conducted using this hybrid experimental technique, as documented in Fig. 7. These tests disclosed novel VIV behaviors, including the "Soar" and "Death" stages at critical Reynolds number ranges, as reported in

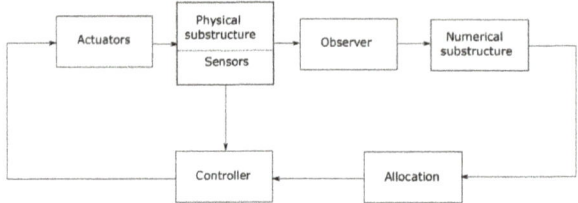

Fig. 5. Generic control loop of hybrid model testing

Fig. 8. Through these experiments, a database of hydrodynamic coefficients for VIV at high Reynolds numbers was established. This database, when used in conjunction with the non-iterative frequency-domain (Lu et al., 2018) and time-domain prediction method (Lu et al., 2019), facilitates accurate prediction of VIV for marine risers subject to high Reynolds numbers.

In summary, the integration of hybrid model testing and hardware-in-the-loop approaches persists in their application across various experimental methodologies. These approaches are utilized to mitigate risks in testing and to enhance the comprehension of intricate experimental settings. It is advocated that the continued application of these techniques be encouraged to further the integration of models with experimental data.

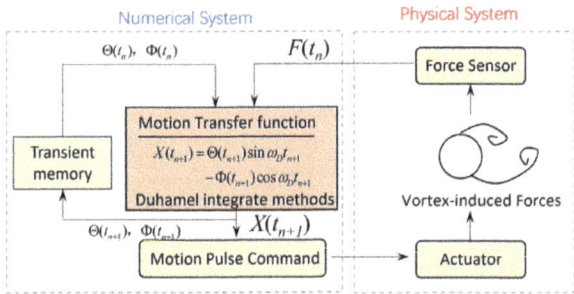

Fig. 6. A block diagram of the general framework of hybrid control methods

Fig. 7. Global diagram of the experimental apparatus, including the linear motion actuator, force sensor, and bluff cylinder model mounted under the towing carriage

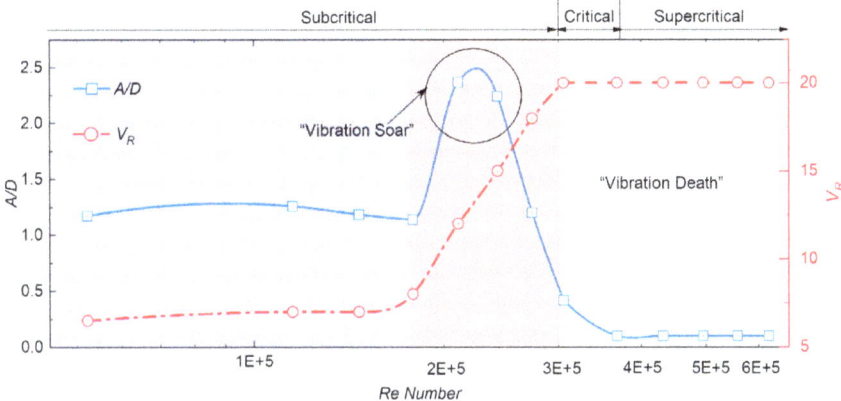

Fig. 8. Evolution of the non-dimensional maximum response amplitudes of VIV and the corresponding reduced velocity, with Re. The blue, red, and gray shaded regions denote the stable, tuning, and death stages for VIV response, respectively. (Color figure online)

5 Fire Test

This chapter gives an overview of the principles, rule, and standards applicable to design for fire accidental events for ships and offshore structures and the current state-of-the-art methods for fire testing.

5.1 Principles, Standards and Rules for Fire Test

The formulation of safety strategies aimed at mitigating fire hazards on board ships and offshore installations has traditionally been accomplished through adherence to prescriptive regulations set forth by regulatory authorities. The International Maritime Organization (IMO) has established fire safety regulations for international merchant vessels under the auspices of the International Convention for the Safety of Life at Sea (SOLAS). In July 1998, the IMO introduced the Fire Test Procedures Code (Organization & Committee, 2012), which encompasses fire test procedures for fire-resistant constructions and materials employed on ships. The FTP Code is founded upon International Standards Organization (ISO) fire tests, which encompass assessments of non-combustibility, fire resistance, flammability, flame spread, smoke production, and the toxicity of constructions and materials, as well as specific products such as textiles, upholstered furniture, and bedding. The tests, referenced tests, and analogous test methods are summarized in the table below (Table 1).

In the Oil & Gas industry, two primary fire testing scenarios are frequently employed: the hydrocarbon pool fire and the jet fire. A hydrocarbon pool fire involves the ignition of a pool of hydrocarbons in a furnace, with a crumple temperature that approximates $1100\,°C$ and a heat flux of up to $150\,kW/m^2$. The

Table 1. Fire test procedure code (IMO FTP Code, 2010) and test method

FTP Code	Type of test	Referred test method	Similar test Method
Part 1	Non-combustibility Test	ISO 1182	
Part 2	Smoke and Toxicity Test	ISO 5659-2	
Part 3	Fire Resistance Test for Fire Resistant Divisions	IMO A.754(18)	ISO 834-1 UL 1709 EN 1363-2:1999
Part 4	Fire Resistance Test for Fire Door Closing Mechanisms		
Part 5	Surface Flammability Test	IMO A.653(16) IMO A.687(17)	ISO 5658-2
Part 6	Test for Primary Deck Coverings	IMO A.653(16)	ISO 5658-2
Part 7	Flammability Tests for Curtains and Vertically Suspended Textiles and Films	IMO A.471(XII) IMO A.563(14)	ISO 6940/41 EN 1101/02
Part 8	Test for Upholstered Furniture	IMO A.652(16)	BS 5852-1 ISO 8191-1/-2 EN 1021-1/-2
Part 9	Test for Bedding Components	IMO A.688(17)	EN 597-1/-2
Part 11	Fire resistance test of load-bearing divisions of high-speed craft	-	-

test scenarios and methodologies for pool fires are generally prescribed in standards such as UL1709, BS476, IMO FTP Code 3, ISO 834-3, and ISO 1363-2. In contrast, a jet fire occurs because of the release of high-pressure hydrocarbons (gas) from a ruptured pipe or vessel. The jet fire test results in a higher temperature and increased heat flux, ranging from 250 kW/m^2 to between 350–500 kW/m^2, depending upon the test standards applied. The method for conducting a jet-fire test is defined within ISO 22899-1 and OTI95-634 standards.

It should be noted that these testing standards specifically address the evaluation of structural steel. The criteria of these tests involve measuring the time required to achieve the defined critical temperature of the steel, which is predominantly set at 400 °C. It is a standard procedure to apply the same testing conditions and critical temperature criteria to piping and other process equipment.

5.2 Fire Testing Methods

5.2.1 Fire Resistance Test

Fire resistance refers to the capability of a partition or boundary, typically a bulkhead or ceiling, to endure fire exposure, provide protection against fire, inhibit the propagation of fire to adjoining compartments, and preserve structural integrity when subjected to fire. Examples commonly used for fire resistance evaluations include pool fire tests, burner test, and furnace test with fire curves (see Fig. 9). Taking the furnace test as a specific instance, the structural specimens are evaluated in the form of panels, with the side within the furnace exposed to a temperature profile that adheres to one of the fire curves. Fire resistance is quantified by the duration required for the cooler surface of the panels to reach a temperature of 140 °C. The configuration of the selected fire curve is designed to reflect the characteristics of the particular fire hazard, acknowledging that there is significant variability among fires.

The fire resistance requirements for steel structures in oil and gas installations, specifically those involved in the transportation and production of hydrocarbons, such as tankers and offshore platforms, are examined. It is noteworthy

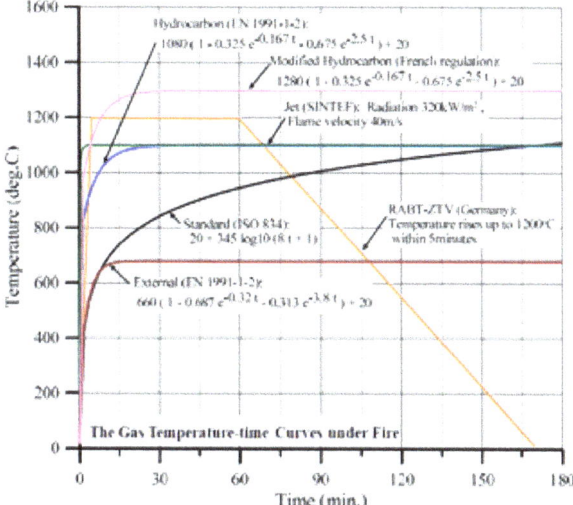

Fig. 9. Time-dependent temperatures curves

that the fire resistance values under hydrocarbon fire conditions, as contrasted with standard fire conditions, are specifically detailed for stationary offshore platforms (Gravit & Shabunina, 2022).

Fire doors are critical safety components designed to impede or halt the progression of fire and must be subjected to preventive testing and classification. Each door undergoes a specified time-temperature heating protocol based on its fire door classification, with both mechanical and thermal constraints employed to evaluate its functional performance. The design of conventional fire doors leverages well-established methodologies aimed at optimizing their thermal performance to satisfy fire resistance testing criteria. Nonetheless, thermal gradients can induce considerable deformations, as the door often bends away from its supporting frame due to uneven temperature distribution. This deformation can lead to the propagation of flames and smoke, thereby resulting in the failure of fire resistance tests (Kyaw Oo D'Amore et al., 2020).

Composite materials are increasingly utilized in various fire protection applications aboard ships and offshore installations. Figure 10 illustrates the fire resistance properties of several composite material types that have been evaluated for potential application in offshore installations. Despite the inherent combustibility of the composite's organic components, the materials analyzed exhibited notable fire-resistant properties. The fire behavior of thick composite laminates can be effectively modeled (Gibson et al., 1995; Dodds et al., 2000), facilitating the prediction of materials' integrity and fire performance given specific hot face temperature or heat flux levels. The predominant factor affecting the fire integrity of thick composite laminates is the endothermic nature of the resin degradation process, which acts to retard heat transfer through the laminate.

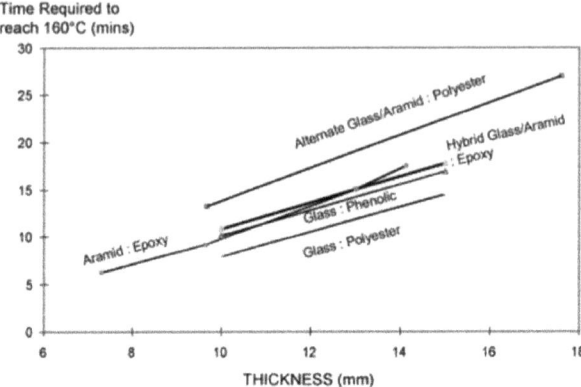

Fig. 10. Fire resistance values, measured as a function of thickness, for a range of different laminates, subject to the hydrocarbon fire curveÂă

A substantial amount of experimental research offers insights into fire integrity and performance metrics crucial for the design of fire-resistant ship compartment structures, primarily through standard fire resistance tests. Given that numerous companies often aim to attain specific classifications for their products, these tests are conducted with a designated fire load exposure and terminated when the predetermined objective is achieved. From a scientific perspective, such results are inadequate to comprehensively depict the performance of a product because (i) the continuation of the test might precipitate failure (such as collapse) shortly thereafter or after an extended period, and a similar uncertainty applies to the fire load. (ii) A minor increase in load could lead to failure before the required fire resistance duration is met. Consequently, it is always advisable to conduct fire resistance tests to the point of failure (e.g., structural failure due to collapse, significant deflection rate, or loss of integrity) to extract the maximum possible information from the test (Jones & Brischke, 2017).

5.2.2 Fire Test for Passive Fire Protection Systems

Passive Fire Protection (PFP) predominantly serves to maintain the structural integrity of steel frameworks, pipelines, including their valves and flanges, as well as other equipment that may contain hydrocarbons. The primary function of PFP is to facilitate a controlled shutdown and the safe evacuation of personnel during a fire incident. The evaluation of PFP should be performed in certified testing facilities, and these test results are utilized as a data set for designing PFP according to specific project requirements. Various thicknesses of PFP materials are assessed, and the duration required to reach the critical temperature is meticulously recorded. It is a widely acknowledged practice to interpolate, that is, to compute a theoretical line between two established points, from certified test results for optimizing the PFP layer thickness to fulfill the stipulated requirements (Paik et al., 2021a; Paik et al., 2021b). However, the extrapolation of results is generally not accepted.

5.2.3 Fire Test for Valve System

Fire protection is an essential aspect of safety design for industrial valves, particularly in environments where the occurrence of fire incidents is probable. Valves employed within the oil and gas, refining, chemical, and petrochemical sectors are required to ensure a dependable and secure shut-off during a fire. The concept underpinning fire testing is that a fire-safe valve must continue to function under pressurized conditions even after exposure to combustion at a predetermined high temperature for a specified duration, with post-burn leakage maintained within defined limits.

The standard methodology for the fire testing of valves, including fire-safe valves, entails the complete and uniform encapsulation of a water-filled, pressurized, closed valve in high-temperature flames ranging from approximately 750 °C to 1000 °C for a period of 30 min. During complete immersion of the valve in fire, which subjects the seat and sealing areas to elevated temperatures, the intensity of heat is closely monitored utilizing thermocouples and calorimeter cubes. Throughout this process, both the external and internal leakages from the valve are measured. Subsequently, upon cooling of the valve post-fire test, the pressure containment capability of the same valve seats, shell, and seals is evaluated.

5.2.4 Reaction to Fire Test

The concept of reaction to fire pertains to the extent to which a specific material contributes to a fire. Contrastingly, fire resistance pertains to a system's capability to withstand fire penetration and inhibit temperature escalation between the exposed and protected sides under a fully developed fire scenario. As the terminology implies, a material's reaction to fire refers to its behavior when subject to flame exposure. The assessment methodology evaluates the role of building materials in a fire, particularly during its nascent stages. Materials and products are classified into seven distinct Euro classes based on their fire reaction characteristics. In accordance with the European Standard EN 13501-1, products are categorized into one of seven principal classes, as detailed below, based on their combustibility level, and may also be assigned additional classes related to the volume of smoke emitted or the quantity of burning droplets or particles generated. Details regarding reaction to fire test methodologies are elucidated in Annex 1 of the IMO FTP Code. It is noteworthy that testing of a product or material may not always be necessary. Annex 2 of the FTP Code stipulates the conditions under which a material or product may be installed without undergoing testing or receiving approval. Furthermore, the provisions of the FTP Code exclude applications involving green or renewable energy in maritime contexts. Consequently, in alignment with the global trend towards eco-friendly shipping, the formulation of pertinent application guidelines will be essential when integrating green and renewable energy technologies into cargo and fuel shipping practices.

5.3 Recent Developments in Fire Experimental Methods

Fire resistance tests are formulated to assess the performance of building components concerning their load-bearing or fire-separating properties, commonly known as their fire resistance, in accordance with their regulated application in building structures. Given that fire resistance is predominantly necessitated in static structures such as buildings, which are subject to regulation by local jurisdictions and often by local building practices, a plethora of test specifications exist, and international consensus remains limited. Throughout Europe, the implementation of standardized testing is progressing via the Construction Products Directive.

In contrast to terrestrial structures, the transportation sector, encompassing aerospace, maritime, and offshore industries, represents an area where testing methodologies elucidate the principal requirements for fire resistance. Steel components within a ship's hull and the structures comprised of cargo tanks, decks, and bulkheads delineating industrial spaces are engineered according to specified fire resistance classifications, contingent upon parameters such as fire resistance limits and temperature exposure regimes, per the IMO FTP Code. Identical fire resistance classifications are prescribed for offshore platforms (ABS, 2021). Previous investigations into fire resistance have aimed to simulate experimental data to ascertain the fire resistance limits of various structural components and systems in ships of diverse classes. These investigations address several key issues, including the calculation of thermal insulation parameters, the prognostication of fire resistance limits (Zong et al., 2023), the computation of structural behavior under thermal loads (Seo et al., 2017), and the refinement of calculated coefficients of thermal conductivity and heat capacity pertinent to application within fire temperature ranges.

In the marine, offshore, and aerospace sectors, where the transition from metallic materials to polymer composites is underway, the employment of combustible composites in lieu of steel has the potential to significantly diminish weight. Conversely, this substitution may lead to an augmented risk of fire. Consequently, the development of novel testing methodologies is imperative to simultaneously address structural integrity and flammability. Evegren et al. executed fire tests to evaluate and ensure the safety of employing lightweight FRP composites within ship structures (Evegren et al., 2016). A non-load-bearing sandwich panel bulkhead concept, originally from the building sector, was identified to possess prospective utility for deckhouse structures (see Fig. 11). Two sets of tests were conducted with an emphasis on the fire performance of the exterior combustible FRP surfaces. Five series of tests were performed to assess the fire resistance, evaluating the structural fire integrity of various FRP composite structures. Three viable protective measures were identified to be suitable for safeguarding external FRP surfaces: a drencher system (3 mm/min), a fire-protective coating (LEO), and a certified balcony sprinkler. The fire resistance tests indicated that an insulated FRP composite bulkhead concept could be conservatively appraised by testing the bulkhead under the highest design load. Load-bearing double FRP composite bulkheads can provide adequate fire resistance in scenarios where traditional insulation is infeasible (e.g., for exterior surfaces).

Fig. 11. Fire resistance test of load bearing FRP composite bulkhead with design loads

The composites industry is experiencing an upsurge driven by heightened global environmental awareness. Innovative composite materials are being developed to mitigate their adverse environmental effects by employing cleaner manufacturing processes and, where feasible, substituting synthetic materials with more sustainable bio-based alternatives. Within this framework, natural fiber composites (NFC) are put forward as promising candidates to either replace or reduce the use of synthetic fibers for reinforcing polymers in various industrial sectors, such as the marine sector, where composite utilization has been extensively researched in recent years. A significant proportion of the research on the fire performance of natural fiber composites (NFCs) focuses on the response to fire parameters instead of fire resistance (Naughton et al., 2014). Moreover, the analytical techniques and prevailing theories pertinent to the residual mechanical properties of fiber reinforced polymer (FRP) composites exposed to fire are not applicable to NFCs.

A material or structural member attains a fire resistance classification through compliance with performance criteria established in a comprehensive fire resistance test. Such a test assesses the fire resistance capabilities of a building component within a configuration and scale comparable to practical applications.

The comprehensive fire test presents a challenge for the adoption of innovative materials, such as natural fibre composites (NFCs), due to its financial implications and the prescriptive nature of its pass/fail outcomes. Presently, no standardized tests exist specifically for NFCs. Nevertheless, contemporary UK standards acknowledge the incorporation of natural fibres as reinforcements within polymer composites, and much of the experimental testing of NFCs adheres to standards designed for conventional fibre-reinforced polymers (FRPs). A comprehensive test would be requisite to ascertain that an NFC adheres to building regulations and fire safety codes. Given that the fire resistance and reaction characteristics of NFCs remain largely unexplored, comprehensive testing remains the sole indicator of fire performance (Fan et al., 2017).

5.4 Summary and Recommendations

The field of fire testing is undergoing continuous evolution as standards undergo periodic revisions and updates. Nonetheless, the IMO FTP Code currently lacks applicability to green or renewable energy initiatives within the maritime domain. Composites are increasingly being employed in a range of fire protection applications on maritime vessels and offshore installations. Consequently, comprehensive full-scale testing and/or methodologies are requisite in order to develop and substantiate the compliance of novel composites, particularly those based on bio-based materials, with established marine standards and fire safety requirements.

6 Friction Tests

Friction constitutes a fundamental element in the interaction between solid surfaces, with objectives ranging from guaranteeing adhesion in components—such as bolts and rivets—to minimizing resistance in parts in motion. The discipline dedicated to the study of friction, referred to as tribology, concentrates on comprehending and regulating these interactions. Tribology encompasses the characterization and modification of surface properties, contact mechanics, and lubrication. Through the optimization of these elements, engineers can achieve specific design objectives, thereby ensuring durability, efficiency, and safety across diverse applications.

A review of the recent literature in marine and offshore publications reveals that a substantial portion of studies addressing contact mechanics depend predominantly on simplified models utilizing tabulated values for static and kinetic friction. Although such models establish a basis for fundamental calculations, they frequently fail to account for the intricate behaviors of materials in real-world scenarios. Variations in surface roughness, material degradation, and environmental factors, including moisture, temperature, and pressure, can considerably influence frictional properties. This issue is particularly pertinent to the interactions between sea ice and floating structures, where ice mechanics are considerably modified by environmental conditions.

The accelerated diminishment of polar sea ice attributed to climate change is resulting in the emergence of new maritime routes, notably the Northern Sea Route and the Northwest Passage. As these passages become increasingly accessible, a comprehensive understanding and mitigation of ice friction are imperative to ensure navigational safety. The implementation of friction-reducing coatings will enhance the operational efficiency of vessels traversing these nascent trade corridors. However, the environmental implications, particularly the release of micro-plastics and anti-fouling chemicals, warrant significant consideration due to the region's ecologically vulnerable wildlife (Qi, Li, Zhao, Zhang, & Zhou, 2024).

The prediction of friction between the hull and ice for vessels classified for ice conditions is crucial for calculating power requirements and optimizing ice-breaking performance. In experiments involving models tested in icy conditions, it is generally challenging to distinguish between forces attributable to friction and those arising from ice fracture and submersion. One study (Hisette & Myland, 2022) elaborates on a dynamic friction table setup, wherein model ice can be translated across various surfaces. Notable effects related to grain structure and speed were observed. During the initial 5–20 cycles, friction is maximized when the ice remains rough, with its granular surface layer exposed. Continued cycling reveals larger columnar grains, thereby resulting in a friction coefficient that stabilizes at less than half of the initial value. Subsequent tests will be conducted on abraded coatings.

Offshore wind turbines and wave energy converters employ submerged cables that facilitate the connection between energy-generating units and onshore power grids, where interlayer friction exerts a significant influence on torsional and bending stiffness, as well as on fatigue life (Qin et al., 2024). The prevailing industry practices, which utilize the Coulomb friction model incorporating an average of static and dynamic friction coefficients, are deemed conservative and fail to consider recent findings indicating that the interlayer friction coefficient undergoes temporal changes (Y. Yin et al., 2019). A friction and wear test apparatus, depicted in Fig. 12, was developed by Y. Yin et al. (2021), demonstrating that the evolution of the interlayer friction coefficient within an umbilical is correlated with the wear depth in the anti-wear nylon fiber tape.

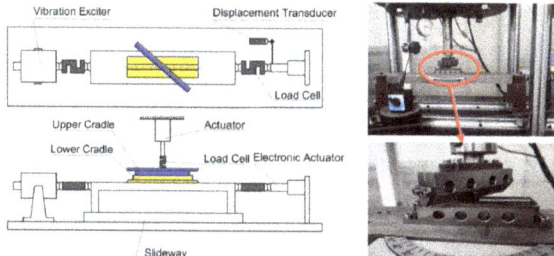

Fig. 12. Interlayer friction and wear test rig for umbilicals (Y. Yin et al., 2021).

In summary, investigations focusing on contact interactions frequently employ simplified friction models, utilizing tabulated values in lieu of conducting empirical friction tests under pertinent conditions. Validating these friction models through empirical testing and simulation becomes challenging, particularly when the measured responses encompass multiple physical domains. This challenge is notably evident in the context of sea ice interactions, as differentiating forces attributable to friction from those arising from ice fracture and submersion proves complex. Efforts are ongoing to characterize hull-ice friction concerning the crystallographic structure of ice and the properties of hull coatings. Similarly, the internal friction present within power cables and umbilicals has been examined through the use of specially designed test rigs. Moreover, comprehension of the external friction response of such structures is critical for accurately estimating installation loads under deep-water conditions. Friction within these structures significantly influences dynamic behavior and fatigue life. Therefore, friction testing is essential for enhancing the durability, efficiency, and safety of marine structures and equipment, and further research on tests conducted under relevant conditions is advocated.

7 Corrosion Prognostics Health Management

Corrosion represents a primary factor in the structural failure of ships and offshore structures, making expensive preventive maintenance a standard preventive measure, as documented in (Lin & Dong, 2023; Vieira et al., 2022). Consequently, corrosion prognostics within the realm of structural health monitoring emerges as a natural evolution and research domain aimed at enhancing structural reliability and reducing expenses. The intricacy of corrosion phenomena, notably pitting corrosion, presents considerable challenges in terms of characterization and prediction (Nugroho et al., 2021), even when utilizing laser scanners for surface capture. Laser scanning and digital image correlation demand a traceable surface, commonly necessitating the chemical cleaning of the surface to eliminate any loose subsurface corrosion deposits. Several probabilistic prediction models have been developed to address the complexity of corrosion wastage (Kim et al., 2022; Woloszyk & Garbatov, 2024). Reliable monitoring of corrosion degradation is crucial for producing meaningful predictions. Advanced monitoring systems now incorporate wave-based methodologies (Zima et al., 2022) and fibre Bragg grating-based sensing systems (Tan et al., 2017). For all approaches, the optimal placement of sensors is vital (Silionis & Anyfantis, 2024). The unpredictable nature of loading further complicates prediction efforts (Katsoudas et al., 2023a), necessitating provisions for instantaneous strength prediction (Y. Liu & Ren, 2023). Additionally, the concurrent monitoring of the impact of corrosion on the fatigue strength of welds, alongside efforts to mitigate coating failures, as summarized in Andresen-Paulsen et al. (2023), is essential.

To investigate the impact of seawater on the structural integrity of steel structures, fatigue experiments can be conducted within a seawater environment (Woitzik et al., 2023). Given the prevalence of welded structures, extending the study to include fillet and butt-welded specimens is a logical progression (Shojai et al., 2023). Digital Image Correlation (DIC) systems enable the measurement of geometry and surface parameters, such as pitting corrosion, which can subsequently be employed in numerical simulations to predict damage onset. Future research will involve comparing these findings with the fatigue life of larger specimens subjected to corrosion in artificial seawater (see Fig. 13). A critical area for analysis is determining whether the behavior of contemporary corrosion protection coatings, particularly around welds, is accurately represented by current advanced corrosion progression models. To further this analysis, employing a combination of 3D laser scanning and DIC to capture local geometries and strain configurations is recommended.

Fig. 13. Corroded specimen

8 Large-Scale Subsea Structures Test

8.1 Introduction

Comprehensive testing of subsea structures has been conducted since the 1970s, following the seminal publication by A. C. Palmer and Martin (1975) on buckle propagation in subsea pipelines. Due to the substantial costs associated with these tests, many experimental investigations utilize small-scale samples. However, to attain results that more accurately reflect the complexities of these structures, large-scale testing is occasionally necessary. This chapter will present recent studies focusing on large-scale testing of subsea equipment, emphasizing concerns related to structural integrity in components such as rigid or flexible pipes, cables, templates, and bases.

8.2 Flexible and Rigid Risers

Multiphase flow within pipes is prevalent in various engineering applications, particularly in offshore deep-water oil and gas transport. Flow-induced vibrations within the pipe can result in mechanical failure, potentially leading to the uncontrolled release of transported fluids. In subsea applications, flexible J-risers are commonly utilized to convey produced fluids from the seafloor to the host platform. Despite the significant risks associated with subsea hydrocarbon leaks, there is a notable paucity of investigations into how flow-induced vibrations in large-scale, pressurized flexible J-risers may compromise system integrity. Pickles et al. (2024) conducted an experimental study examining the response of a composite riser with a tensile armor helical structure subjected to various two-phase, water-nitrogen flows at a pressure of 10.8 barg and ambient temperature. High-speed cameras were employed to examine the flow structure at either end of the flexible riser, while synchronized surface-mounted strain gauges and accelerometers were used to analyze the pipe's response. Time-averaged data were captured to evaluate the general behavior of the pipe, while statistical analysis of fluctuations elucidated the pipe's movement. This phenomenon was primarily observed under multiphase flow conditions, specifically when the gas flow rate was increased at a constant water flow rate or under conditions of high gas flow rate. The strain gauges recorded increased average strain under these conditions, accompanied by a visually detectable whipping motion. The authors recommended future coupled analyses of fluid-structure interaction to assess the riser's structural integrity.

A combined hydraulic and power umbilical cable designed for deep-water applications, incorporating a 66 kV configuration, was subjected to an innovative testing approach as discussed in Y. Zhou et al. (2024). This approach integrated mechanical tests as outlined in API 17E and the type test consistent with IEC 63026 standards for high-voltage cables, aimed at validating the reliability of the umbilical system, including its repair joint and end termination. In addition to the standard electrical tests for power cables, the testing program encompassed a series of evaluations, including combined tensile and bending tests, end terminations strength assessments, torque trials, and fatigue as well as water penetration tests conducted under an external pressure of 15 MPa, as illustrated in Fig. 14. The proposed testing methodology serves as a potential benchmark for the qualification and assessment of such structures, particularly in the context of advancements in subsea electrification.

8.3 Rigid Pipes

Subsea rigid M-shaped jumpers provide dependable and adaptive connections to production systems by employing deflection to diminish the effects of internal flow and allow for installation tolerances. During their operational life, the gas-liquid mixed multiphase flow traversing these jumpers induces flow-induced vibration (FIV), thereby reducing their fatigue lifespan. G. Li et al. (2023) conducted an investigation into the air-water mixed two-phase flow within a rigid

Fig. 14. End termination strength test of 66 kV power and hydraulic combined umbilical.

M-shaped jumper at varying mixture velocities and water volume fractions, with the objective of analyzing the factors influencing flow patterns and determining the most vulnerable areas. Experimental procedures were undertaken to ascertain the flow patterns in each segment and the historical pressure variations. Air and tap water served as the gas-liquid two-phase media within the experimental setup. Under conditions of high-speed two-phase flow, significant equivalent stress was observed at the fixed supports at both ends of the M-shaped jumper. Concurrently, stress concentrations were identified at the elbows, with both the average and RMS of equivalent stress surpassing those in the horizontal or vertical segments attached on either side. The time-history stress response of each node on the developed structural model was obtained through one-way coupled fluid-structure analysis. The application of rain flow counting technology in conjunction with the S–N curve facilitates the performance of fatigue damage computations and the prediction of fatigue life as a subsequent analysis.

W. Li et al. (2022) conducted an investigation into gas-liquid flow and the consequent vibrations within a multi-plane jumper utilizing experimental methodologies (refer to Fig. 15). The investigation encompassed a detailed analysis of flow patterns at each characteristic section of a Z-shaped jumper with an internal diameter of 48 mm, encompassing diverse flow regimes such as dispersed bubbly, slug, churn, wavy, stratified, and annular flows. Displacement and pressure sensors were strategically installed adjacent to each elbow to effectively capture the vibrational and pressure responses of the jumper. The one-way transient fluid-structure analysis employed in this study is deemed adequate for predicting the dynamic response characteristics of the jumper. The time-averaged accelerations derived from the finite element method and the flow patterns forecasted by the volume of fluid method demonstrate a commendable agreement with the experimental observations. In addition to the internal two-phase flow phenomena, the jumper is also exposed to current flow, leading to external flow-induced vibrations within a subsea environment. Consequently, further research is warranted to explore the coupled vibrational characteristics of the jumper under the simultaneous influence of both internal and external flows.

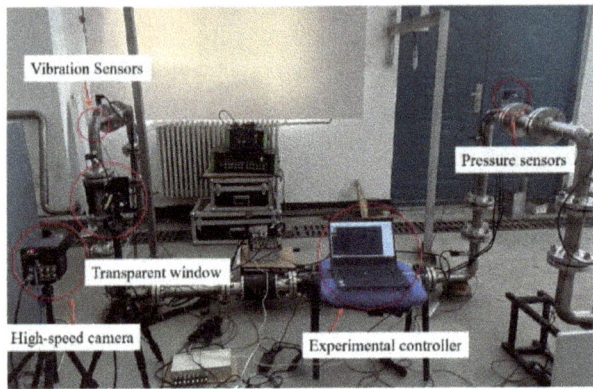

Fig. 15. Experimental setup to evaluate flow induced vibration.

The phenomenon of free spanning in subsea pipelines, resulting from localized seafloor scour, constitutes a significant operational concern. B. Zhou et al. (2024) conducted experimental analyses under clearwater scour conditions, examining the effects of impermeable, porous, and flexible spoilers on both fixed and self-burial configurations. The flow fields surrounding the pipeline were characterized utilizing the particle image velocimetry (PIV) technique (refer to Fig. 16). The results from the scour tests on a fixed subsea pipeline demonstrate that the presence of spoilers can significantly expedite the erosion rate.

Buckling damage constitutes a principal safety concern for subsea pipelines. Yu et al. (2021) undertook comprehensive full-scale experiments to investigate the buckling behavior of subsea pipelines equipped with integral buckle arrestors subjected to external pressure. The study presents pressure chamber tests of four full-sized pipelines characterized by uniform thickness but varying diameters. The tested samples, depicted in Fig. 17, encompass not only collapse and propagation tests but also crossover tests, which serve to assess the efficiency of buckle arrestors. This research was conducted to validate a finite element-based methodology employing thin shell elements for the calculation of collapse, buckle propagation, and crossover pressures in subsea pipelines. Despite demonstrating significant correlation with the experiments and other calculations, the authors acknowledge that this approach may benefit from refinement to accommodate thick-walled pipes.

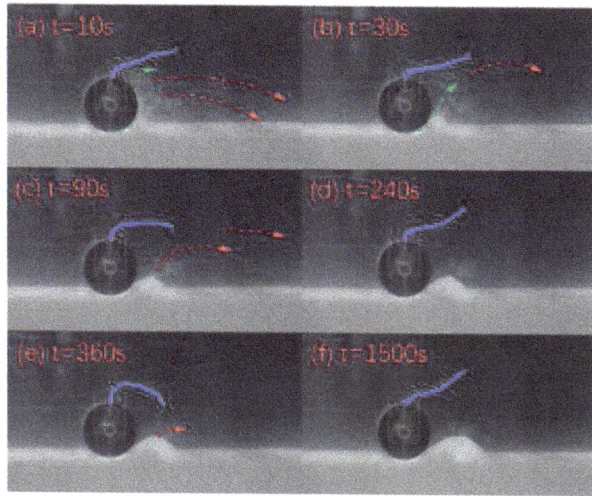

Fig. 16. Experimental snapshots on the scour process of pipeline attached with the flexible spoiler.

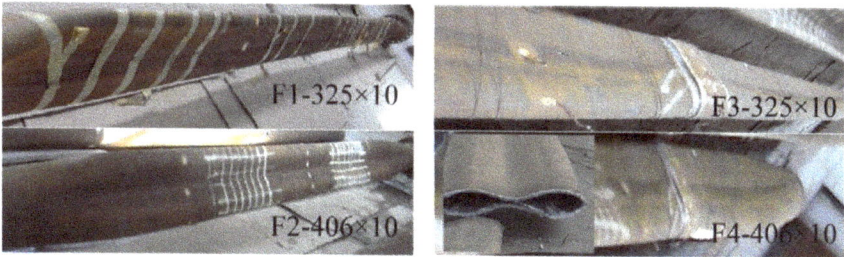

Fig. 17. Pipe morphology after tests. (a) Collapse and propagation tests, (b) Propagation and crossover tests.

8.4 Templates

During the process of elevating submerged marine modules, the most significant dynamic loads manifest within the splash zone. These forces exhibit substantial dynamism and their estimation with adequate precision proves challenging due to their irregular and complex geometrical configurations. An experimental investigation was undertaken by Zan et al. (2021) within a wave tank to examine the dynamics of a large subsea module in the splash zone under the influence of irregular wave conditions. The experiments utilized a 1:8 scale model of a subsea module, which was scrutinized at five distinct positions relative to the free surface, ranging from fully above the surface to entirely submerged. The ensuing data were subjected to analysis in both the time and frequency domains. Dynamic loads on the subsea module were examined through spectral analysis and a dynamic effect parameter. Furthermore, several time-domain numerical analyses were executed with varying peak periods to ascertain the tension in the sub-slings. The experimental findings pertaining to the relative forces and motions of the subsea module are presented.

8.5 Summary and Conclusion

In summary, numerous recent experimental investigations on large-scale subsea structures have primarily focused on fluid-structure interaction, with an aim to understand the responses concerning vibration, displacement, and stress. Although simulating two-phase fluid flow is relatively straightforward, challenges arise with the simulation of multiphase flow involving mineral oil. Additionally, the integration of internal and external fluid flows can result in costly experimental setups, which may undermine the feasibility of such experiments. This issue can be addressed by employing numerical models validated against simpler experimental setups, as these models are capable of coupling complex analytical scenarios.

9 Large-Scale Impact Tests

A fundamental aspect of research lies in the execution of experiments across varying scales, which offers profound insights into the behaviors under investigation. Experiments conducted at different scales elucidate observed behaviors, ranging from laboratory-scale models to specimens subjected to diverse scaling factors culminating in full-scale investigations. Scaling dependencies emerge when scale influences complexity and expenses, rendering some large-scale experiments more challenging and costly to perform. Nonetheless, such large-scale experiments are employed as they provide critical insights that are elusive at smaller scales. While other sections address health management systems or algorithms for monitoring large structures, this section concentrates specifically on large-scale impact testing, which is conducted infrequently.

9.1 Impact Tests

An instance of conducting a large impact test involved experiments with flexible guided anti-collision (FGAD) devices (F. Wang et al., 2023), where a series of 12 field tests were executed to assess the response of a vessel (weighing 250 t and operating at speeds of 2.5–3 m/s) upon impacting a pier at varying incidence angles (0 and 25°), with subsequent validation via numerical methods. The experiments entailed impact tests on piers equipped with these devices; the FGADs demonstrated the capability to alter the vessel's trajectory with minimal reduction in velocity, limited structural damage to the vessel over the course of 12 tests, and a reduction in forces exerted on the pier by up to 50.

A full-scale test may also transpire inadvertently, whereby unforeseen damage facilitates numerical analyses to elucidate the intrinsic behavior, as exemplified by the scenario in which wind turbine jacket foundations are impacted by ships

(X. Liu et al., 2022). In such an instance, the deformation observed on the foundation post-collision served as the foundation for numerical analyses concerning bulbous bows, wherein the evaluation of velocity and incident angles determined the positions of impact that engendered the maximal damage; localized deformation replicating the genuine damage was identified as the primary damage mode. Similarly, the impact on monopile structures that support wind turbines has been the subject of investigation, where adverse effects on the vessels have been documented. Studies derived from this data have prompted the recommendation to employ higher strength materials in the zones of impact where ruptures might occur, consequently diminishing the potential opening areas by up to 50.

Analogous to previously discussed anti-collision mechanisms, pneumatic rubber fenders serve to mitigate structural damage resulting from impacts. These fenders are implemented at piers during mooring processes and during interactions between tankers and offshore installations. Comprehensive full-scale experiments, alongside experimental evaluations, have been conducted to elucidate the intricate behavior of rubber, assess vessel damage through numerical means, and ascertain the optimal quantity of fenders required to maximally absorb impact energy between these two structures (Park et al., 2023). Additionally, solid fenders were subjected to full-scale experimental investigation to numerically assess the impact damage occurring between an offshore installation and a supply vessel, demonstrating a potential reduction in the applied energy on the impacted structure by up to 30.21.

The collision of structures is a significant concern, necessitating the execution of risk assessments, uncertainty analyses, and various other evaluations to prevent such occurrences (Sukma et al., 2022). These assessments are conducted concurrently with the advancements in technologies that support collision avoidance modeling and testing (Zhu et al., 2023).

The assessment of large-scale impact phenomena can be executed through underwater explosion (UNDEX) testing, as exemplified by the shock test conducted on a full-scale barge in conjunction with numerical simulations. The barge experiences irreversible structural damage as a result of the whipping effect induced by the shock and the initial collision, as well as from the generation of a subsequent jet impact due to the explosion. Although the test involves a barge rather than a full-sized vessel, it provides a fundamental insight into mechanisms that may elucidate the effects on larger vessels (Zhang et al., 2023).

Dedicated shock tests were carried out to calibrate numerical models by (Mannacio, Barbato, et al., 2022) and (Mannacio, Di Marzo, et al., 2022) on large scale specimens of composite laminates conceived on purpose. While the numerical model of the specimens was successfully validated using obtained experimental results, full scale shock trials of a composite hull block of a minehunter was not available for calibrating the hull model. Therefore, experimental data of a test carried out decades ago was used to validate the complete hull finite element model (Mannacio, Di Marzo, Gaiotti, Rizzo, & Venturini, 2023).

9.2 Full Ship Shock Trials: FSST

Naval vessels are susceptible to the risk of non-contact underwater explosions, a probability that arises from their operational environments. This consideration has compelled contemporary naval forces to adopt a comprehensive shock hardening testing protocol. While barge shock testing has been referenced as a method for evaluating smaller systems and components, it presents a challenge when attempting to conduct shock hardening assessments at the level of structural members or components of a vessel. This is necessary to accurately verify the behavior and error characteristics of related members and components within the entire ship system. Moreover, the scientific foundation underpinning the component testing methodologies is insufficient, and these procedures do not necessarily align with either the temporal progression of a shock impulse affecting a ship or the specific response of a component at its position on the vessel.

Full Ship Shock Trials (FSST) are undertaken to assess the resilience of hull structural integrity as well as the operational efficacy of systems and subsystems under shock conditions. Such trials are infrequent; the most recent FSST, conducted in 2021, served as a hardening evaluation of the United States Navy's latest nuclear aircraft carrier, the Gerald Ford, providing critical insights into the naval architecture and engineering systems of this platform (Fig. 18).

Fig. 18. FSST Test for Aircraft carrier USS Gerald R. Ford (2021)

The primary function of Full Ship Shock Trials (FSST) is to mitigate the risk of critical equipment failures that cannot be identified solely through component testing, to assess design and construction, and to validate shock hardening criteria. Nonetheless, FSST faces the challenge of being expensive and requiring extensive time for planning. Conversely, numerical modeling and simulation (M&S) offer insights into the intricate details of the structural and fluid models, dynamic characteristics of the ship hull, and its internal components. Within the context of the U.S. Navy, M&S is either replacing or being utilized alongside full ship trials test evaluations. Considering the swift advancements in M&S technology from a cost-effective perspective, various aspects of research and development are being pursued in applying M&S and FSST testing (Mair et al. 1997, Didoszak et al. 2004, Office of the Director 2022).

Currently, the mechanisms underlying vibration-damping in naval vessels subjected to underwater shock remain inadequately understood. Consequently, ship shock trials continue to be the most effective method for investigating issues related to ship vibration-damping. A strategy for modeling damping in naval ship systems was explored for the transient time-domain analysis of ship shock utilizing FSST (Fan et al., 2017). Madhusudhana et al. collected underwater acoustic data from FSST recordings to validate models of underwater acoustic propagation (Madhusudhana et al., 2023). A predictive approach for ship shock M&S has been developed, with predicted results compared to data from ship shock tests conducted during sea trials. These results are incorporated into an advanced M&S tool with significant modeling parameters replacing FSST (Grządziela, 2011).

Technology development is being carried out along with policy-based supplementation in related elements such as measurement-related monitoring technologies pertinent to the evaluation of FSST tests. Importantly, considerable efforts are being directed towards reducing costs associated with the application of FSST. Moreover, M&S and FSST are employed in a complementary manner for testing and evaluation, with concurrent advancement in technology development to ensure responsiveness to evolving threats from weapon systems.

It is necessary to recognize the specificity of these live ship shock tests being applied preferentially to specific nations and naval vessels. Nonetheless, it is anticipated that future maritime and offshore infrastructure will serve in the transportation and production of hazardous materials which have yet to be identified, or in the provision of energy and environmental advantages. The marine environment may be categorized into diverse accident scenarios, with extreme and accident-prone environments posing a more significant threat load than anthropogenic challenges. In light of FSST in naval ships, it is essential to examine measures for ensuring the integrity and safety of ships and marine structures within the commercial domain.

9.3 Summary and Conclusion

In summary, for this section, the execution of large-scale impact tests at full scale is constrained by their complexity and associated costs. In certain instances, unanticipated occurrences necessitate further inquiries through the integration of numerical simulation and laboratory-scale experimentation to replicate aspects of the impact conditions. The preparation of large-scale tests may extend over several years, encompassing numerical simulations, substantial financial outlay, meticulous instrumentation, platform planning, and the thorough execution of an intricate test event to capture the intended behavior.

10 Full-Scale Ice Load Measurements

10.1 Introduction

This chapter addresses methodologies for assessing ice-induced loads on ship hull structures at full scale. These methodologies are categorized into direct and indirect measurement techniques. The direct measurement methods entail the deployment of external devices affixed to, or integrated within, the structure to ascertain load parameters. Conversely, the indirect approach adopts inverse engineering principles, deducing external loadings from observed structural responses. While it is recognized that ice exerts considerable forces on ship propulsion systems and propellers, necessitating consideration in their design, the current review deliberately excludes measurement techniques pertaining to these components. In coherence with the committee's emphasis on Experimental Methods, this chapter concentrates on applicable methodologies. Any results and observations concerning measured loads have been delegated to respective loading committees. Furthermore, it is acknowledged that previous full-scale campaigns have been documented by antecedent committees, such as Committee V.6 Arctic Technology (2015). Accordingly, the purpose is not to compile an exhaustive list of measurement campaigns, but rather to highlight select instances where the methodologies have been effectively employed.

10.2 Direct Measurements

Various methodologies have been employed to directly quantify the loads induced by ice. Vuorio et al. (1979) devised a pressure gauge system that characterizes ice pressure by measuring the internal surface pressure of the hull within the confines of a rigidly structured area mounted internally. Glen and Blount (1984) implemented hull-mounted pressure gauges to measure pressure across the gauge area. Hoffmann (1985) utilized load sensors embedded in a specially constructed load panel affixed to the hull. The accuracy of the measurements attained through these systems is commendable. However, the limitation of these methods is their localized measurement scope, and the systems developed by Vuorio et al. (1979) necessitate extensive installation efforts. The load panel integrating load sensors within a specially constructed section of the outer hull (Hoffmann, 1985) demands even more comprehensive work and structural alterations. Additionally, a lack of published results from the load panel suggests its limited success.

In order to elucidate the effects of ice-induced external pressure, Riska et al. (1990) employed a PVDF film and a window on the outer hull to quantify local pressures and observe localized processes. Due to the interaction's extraordinarily high pressure and significant mechanical wear, the external gauge exhibited a brief operational lifespan, although the window facilitated qualitative observations. Gagnon (2008) and Gagnon et al. (2020) have engineered optics-based methodologies capable of enduring ice-induced pressures. These methods utilize acrylic beams and derive pressure measurements through analyzing beam deformation under load, categorizing them as indirect approaches. This technique has demonstrated the capability to deliver high-resolution pressure distributions at points of contact. However, a major limitation arises from the necessity of installing the system onto a ship, necessitating structural modifications. Moreover, by altering the structural stiffness, it remains uncertain how representatively the measured pressures reflect the actual structural conditions of the ship.

Recently, Gagnon et al. (2024) introduced a prototype of a relatively thin pressure sensor panel that employs an innovative application of fiber optics. In this configuration, the fibers feature specially engineered small indentations that enable light collection from the lateral direction, subsequently transmitting it along the fiber's length. These fibers reside between two acrylic plates, one of which is coated with a thin layer of Mylar film. When a pressure strip is applied to the suitably illuminated Mylar film, optical contact is established with the acrylic sheet, modifying the internal reflection of the side lighting, and consequently illuminating the white Mylar film within the contacted area. This configuration permits the estimation of both the area and magnitude of pressure. The outlined method has been validated under laboratory conditions but remains untested in full-scale implementations.

10.3 Indirect Measurements

Although direct methods have been employed, indirect approaches have predominated in the determination of local ice-induced loads. These methods ascertain the load acting on the structure by evaluating the structural response. Traditionally, this response has been gauged using strain gauges affixed to a frame or stiffener, although the potential of alternative methods has also been explored. The external load can be inferred from the load-strain relationship, typically determined through Finite Element Analysis (Ralph et al., 2003; Suominen, 2018). In these instances, the load-strain relationship is utilized to deduce the structural response, premised on assumed locations, shapes, and pressure distributions of the load, see e.g. Riska et al., (1983) and Ralph et al., (2003), which are generally considered constant during the measurement process. In scenarios where the instrumentation comprises multiple frames or stiffeners, the ice-induced load at a particular location can be deduced from the response of the entire instrumented structure by employing the response coefficient matrix. This matrix is derived using the unit load principle, whereby various locations are individually loaded, and the response at different instrumented locations is assessed.

The determination of hull responses is frequently derived from the measurement of shear strain differentiation between two positions on a frame, as exemplified by studies on MT Igrim (Korri & Varsta, 1979), IB Sisu (Kujala, 1989a), CANMAR Kigoriak (Ghoneim & Keinonen, 1983), MS Arcturus (Riska, 1982), MS Kemira (Kujala, 1989b), IB Oden (St. John et al., 1994), CCGS Louis S. St. Laurent (Ritch, 2008), MT Uikku (Kotisalo & Kujala, 1999), PM Teshio (Uto et al., 2006), USCGC Healy (Hänninen et al., 2001), CCGS Terry Fox (Ritch et al., 2008), PV Soya (Takimoto et al., 2006), KV Svalbard (Leira et al., 2009), PSRV S.A. Agulhas II (Suominen et al., 2013), and RV Polarstern (Kubiczek et al., 2022). This methodological approach is premised on the notion that ice-induced loads interact with the frame as shear forces. Consequently, the ice-induced loads between distinct locations, such as the upper and lower sections of transverse frames, can be ascertained by assessing the shear strain variation at these points. Occasionally, the focus has been on the compressive strain on the frame orthogonal to the shell, as observed in USCGC Polar Sea (St. John et al., 1984) and RV Nathanial B. Palmer (St. John & Minnick, 1995). This method targets the assessment of localized loads impacting the structure, given the rapid diminishment of strain normal to the hull plating beyond the contact area. Therefore, this technique facilitates the identification of loads with a direct impact at the strain gauge site. Furthermore, attempts to establish local loading through the measurement of bending strain at the frame's flange have been attempted; however, these measurements have presented interpretational challenges.

The magnitude, location, and dimensions of load patches induced by ice exhibit rapid variation in full-scale measurements (Kujala et al., 1994). The uncertainty inherent in these measurements is attributed to the presumed loading conditions. In scenarios involving load measurements on transverse frames, shear strain difference demonstrates reduced sensitivity to variations in loading conditions compared to compressive strain measurements normal to the hull; this reduced sensitivity pertains to alterations in the load patch height and its location on the frame, provided that the loading impacts occur between the gauges (Suominen, 2018). However, the shear strain difference-based methodology is susceptible to variations in the horizontal dimensions of the load, which are contingent upon the structural configuration. This vulnerability can be mitigated by extending instrumentation to adjacent frames (Suominen et al., 2017). Notwithstanding these uncertainties, measurements of ice-induced loads on transverse frames based on shear strain differences have been shown to produce magnitudes that are generally representative of the true loads involved (Suominen, 2018; Böhm et al., 2021). When an instrumented ship or structure is framed longitudinally, the variability in ice-induced load height and location exacerbates uncertainty, as the loading could affect either the spaces between frames or the location of sensors. In the former scenario, the structural response disparity is notably influenced by the local structural configuration. In the latter scenario, proximity to the load location may introduce a significant strain gradient, thereby impeding precise load determination.

More sophisticated inverse methodologies that do not presuppose known load locations or contact areas have been employed in historical full-scale assessments, as referenced in (Adams et al., 2019). Nevertheless, the variability in load position and contact area renders the inverse problem ill-posed, meaning disparate input values may produce identical outputs. Specifically, varying configurations of ice-induced loads can yield identical structural responses. To regulate the solution domain, such methods necessitate regularization to constrain the solutions (Adams et al., 2019; Y. H. Liu et al., 2016). Tikhonov regularization has been favored by Liu et al. (2016) and Adams et al. (2019) as the inverse technique to address the ill-posed nature of the problem and determine the ice-induced load on a ship's hull using full-scale measurement data. However, comparative analyses with controlled calibration pulls remain unpublished, rendering these methodologies principally verified through qualitative means.

In recent years, the rapid advancement of computational technology has facilitated the application of machine learning methodologies to the inversion of ice loads. To ascertain full-scale ice loads, Wu et al. (2021) executed an ice load measurement aboard the "Xue Long" during an Arctic expedition in August 2017. The least squares support vector machine algorithm was employed within the ice load identification model, and an experimental application was conducted to assess the feasibility of this model. Notwithstanding initial efforts and innovations in the domain of ice load inversion via machine learning, the current research is largely situated in the phase of method validation. These approaches predominantly operate within the scope of supervised learning, thereby requiring pre-existing stress and load datasets for model training. The training datasets are frequently virtual or derived from numerical simulations, which introduces uncertainties in assessing the error of inversion outcomes when applied to real-world ship data in practical engineering scenarios.

Efforts to assess ice-induced loads extending beyond instrumented regions have also been pursued. Kong et al. (2021) utilized the least square support vector machine approach to ascertain ice-induced loads acting outside the instrumented zone, based on the compressive strains measured on frames beyond the load influence area aboard RV Xue Long 2. Wang et al. (2023) employed a radial basis function neural network to the same dataset, thereby enhancing the outcomes. In addition, Wang et al. (2023) carried out model-scale tests to further investigate the method. On the whole, the numerical analyses and laboratory experiments were deemed to faithfully represent reality. Nonetheless, it should be noted that these far field measurements demonstrate susceptibility to noise. This sensitivity can be mitigated by introducing noise to the training data, although the noise level might become considerable when compared to the actual signal, particularly as strains outside the loading are negligible in full scale (Fig. 19).

10.4 Summary and Recommendations

The contact area and pressure distribution within the contact area of ice-induced loads exhibit significant spatial and temporal variability. Despite the uncertainties associated with these phenomena, measurement techniques based on hull response via shear strain gauges on transverse framing have demonstrated an ability to quantify the order of magnitude of ice-induced loads. In contrast, measurements for longitudinally framed vessels have proven more challenging, necessitating further refinement. Recently, more sophisticated inverse methodologies and machine learning techniques have been developed and implemented to enhance measurement accuracy and extract additional information, such as contact area dimensions. These advancements have been validated under laboratory conditions and qualitatively assessed in full-scale scenarios. Nonetheless, comprehensive validation and uncertainty analysis in full-scale conditions remain outstanding.

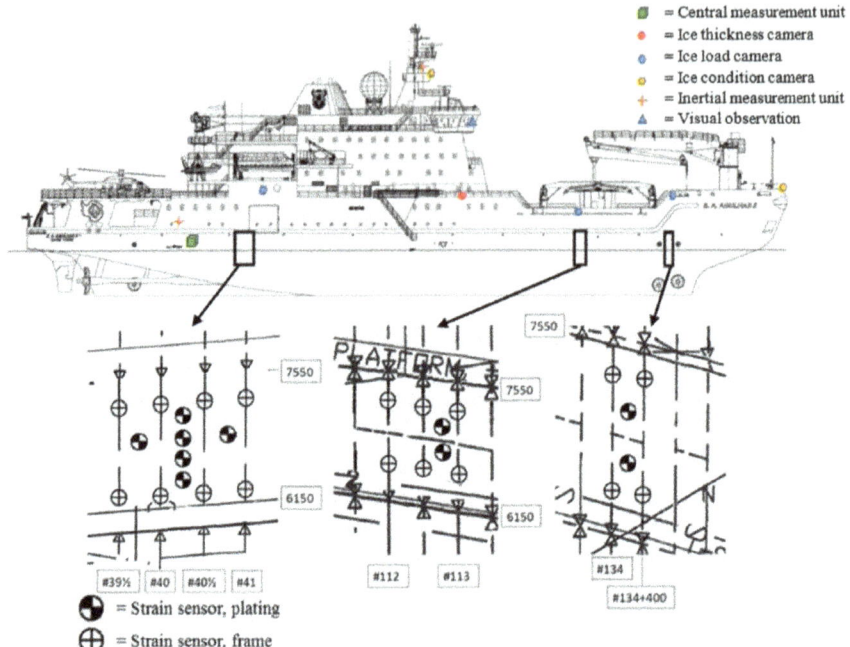

Fig. 19. Instrumentation of S.A. Agulhas II. The ship is instrumented with shear strain gauges at the starboard side on nine frames, including two at the bow, three at the bow shoulder and four at the stern shoulder. The measured shear strain can be converted to forces according to Suominen et al. (2015)

11 Health Monitoring and Digital Twin

This section delivers an exhaustive overview of two interrelated domains: Health Monitoring and Digital Twin Models, both of which are indispensable for the advanced structural analysis and operational efficiency of marine and offshore structures. These areas converge in their objective to harness real-time data to enhance the safety, reliability, and efficiency of maritime structures. Health monitoring methodologies concentrate on detecting and analyzing physical damage, whereas digital twin models expand upon this by offering a comprehensive virtual representation that not only monitors current conditions but also forecasts future performance, thereby providing a proactive approach to marine structure management. In the Health Monitoring subsection, we examine the latest approaches for overseeing the condition of ships and offshore structures, specifically the implementation of Acoustic Emission (AE) and strain sensors for real-time damage detection. Additionally, the integration of machine learning and other data-driven techniques in refining the accuracy and sensitivity of these monitoring systems is discussed, underscoring the crucial role of real-time data in preserving structural integrity.

The subsection on Digital Twin Models examines the implementation of digital twin technology within the maritime sector, wherein empirical data is harmonized with virtual simulations to optimize the lifecycle management of vessels and offshore installations. Digital twins offer essential insights into the operational efficiency and structural integrity of these assets, closely aligning with the objectives of health monitoring systems.

11.1 Health Monitoring

This subsection elucidates the advancements in Structural Health Monitoring (SHM) techniques and systems applicable to marine and offshore structures. A variety of health monitoring strategies have been devised for ship structures. Acoustic Emission (AE) constitutes an SHM method predicated on transient stress waves emanating as a result of damage initiation and progression in materials. Saccone and Pahlavan (2024) examined the effect of stiffeners situated between the damage source and AE sensors on the propagation of AE waves for ship structure surveillance through experimental and simulation approaches, aiming to enhance the sensitivity for fatigue crack detection by considering the presence of stiffeners. Monitoring strain induced by deformation is also advantageous, as excessive deformation may result in damage. Machine learning and additional data-driven processing methodologies have been developed and integrated into SHM. Katsoudas et al. (2023) utilized statistical pattern recognition and machine learning techniques to detect corrosion-induced thickness loss via strain sensors. Aravanis et al. (2023) proposed two distinct methodologies for damage detection in marine stiffened panels based on strain response data. The initial approach discriminates between healthy and damaged states employing a detection theory-based binary classifier, whereas the alternate method correlates strain to out-of-plane deflection utilizing a probabilistic regression model. These methodologies were assessed using simulation, and results demonstrated that they furnish uncertainty-informed forecasts of out-of-plane deflection levels within ship hull structures. In addition to hull structures, research concerning the SHM of rudders and propellers has been conducted. Jang et al. (2024) proposed sensor placements for effective fatigue failure monitoring of ship rudders by assessing the mode shapes derived from CFD and FEM analyses using the structural model, accounting for the structural complexity of ship rudders. Furthermore, Hamada et al. (2023) evaluated the vibration deformation of recreational boat propellers using piezoelectric line sensors integrated into the blades and determined that the amplitudes of the measured signals from damaged blades exceeded those from intact blades.

Fiber Bragg Grating (FBG) sensors possess several advantageous characteristics, including immunity to electromagnetic interference, resistance to chemical corrosion, compact size, and lightweight nature, making them highly effective for monitoring physical and chemical parameters in challenging environments such as marine settings. The foundational principles and theoretical underpinnings of their operation have been elaborated by Min et al. (2021), who also examined primary applications such as temperature, pressure, salinity, pH, heavy metal, and structural health monitoring, revealing the considerable potential of optical fiber distributed sensing technology for SHM. L. Ren et al. (2006) implemented FBG sensors for health monitoring of the offshore oil production platform CB271 situated in the Bohai Sea, East China, detailing the sensor installation process during platform construction and conducting model validation in a laboratory using a seismic simulation shaking table under various loading conditions. Additionally, L. Wu et al. (2018) evaluated the performance of FBG sensor packaging under conditions characteristic of the marine environment, including intense sunlight, heavy rainfall, and saline water, to ensure the sensors' repeatability and durability for long-term application (Fig. 20).

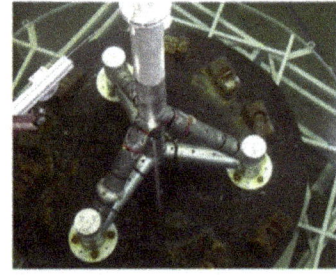

Fig. 20. The wind turbine model with FBG sensors fixed to the partially submerged (Min et al., 2021)

A variety of measurement and data processing methodologies for Structural Health Monitoring (SHM) of offshore structures, such as wind turbines and jacket platforms, have been introduced. Maetz et al. (2023) proposed a microwave SHM approach to detect damage in grouted connections between the monopile and the transition piece of wind turbines, demonstrating that the proposed methodology facilitates a preliminary localization of damage within these connections, based on experiments conducted with a scaled laboratory model. G. Wang (2024) presented a methodology for long-term offshore SHM utilizing a stand-alone Global Navigation Satellite System (GNSS), devoid of reference stations, to predict future structural submergence. This approach also assessed seafloor subsidence rates in the Gulf of Mexico's oil field area using data collected by a GNSS antenna mounted on a fixed platform. P. Jiao et al. (2024) developed a real-time SHM system for marine and offshore structures aimed at damage monitoring, identification, and early warning of underwater concrete structures through vision-based image processing conducted by highly controllable underwater robots. In support of this, P. Jiao et al. (2024) developed the YOLO-Underwater model, based on the YOLOv5 algorithm, to detect concrete damages underwater, including cracks, corrosion, and exposed reinforcement. Weil et al. (2022) implemented an unsupervised novelty detection pipeline, which combines an autoencoder with the Mahalanobis distance, to analyze SHM data from offshore wind turbines. Vieira et al. (2023) assessed the impact of implementing an SHM system for the support structures of bottom-fixed offshore wind turbines on overall energy production and the operational lifespan of wind farms using simulations based on economic models. The study demonstrated quantitative economic benefits, such as extended inspection intervals and potential prolongation of farm operation. Ye et al. (2022) designed an SHM system for an offshore platform under construction in the East China Sea and examined the characteristics and performance of the proposed system through numerical simulations across various scenarios.

The implementation of shape-sensing technology using the inverse Finite Element Method (iFEM) for the structural health monitoring of marine and offshore structures is presently under investigation. iFEM has been successfully utilized by Ghasemzadeh et al. (2022) to ascertain the locations of corrosion damage and pits within corroded offshore components and marine structures. Similarly, Miyashita et al. (2022) employed iFEM in the comprehensive modeling of a 6600 TEU container ship, successfully demonstrating that the bending and torsional deformations of the hull in waves can be accurately reconstructed from the strain data derived via fluid analysis and FEM. Furthermore, Miyashita et al. (2022) validated the feasibility of conducting real-time operational analyses with iFEM through the adoption of parallel computation techniques, thereby enhancing computational efficiency. Nevertheless, it must be noted that a significant portion of prior research remains simulation-based, with experimental assessments predominantly confined to plates and stiffened plates. On the contrary, (Riccioli, Huijer, Grasso, Rizzo, & Pahlavan, 2023) moved from the experimental side in the development of a novel sensor system intended for composite structures health monitoring: in that case, the numerical simulation is the mean for outlining and exploiting the measurements.

11.2 Digital Twin Models

In recent years, the marine industry has witnessed a marked increase in the implementation of sophisticated technologies, such as the Internet of Things (IoT), deep learning, cloud computing, and artificial intelligence (AI). These technologies, which have already transformed traditional sectors, are now poised to bring fundamental changes to the marine industry. Within this context, Digital Twin (DT) stands out as a notably promising instrument. DT constitutes a virtual representation of a physical object or system that operates as a digital mapping platform for pivotal and interconnected entities. It has found extensive application across various domains, including product design, manufacturing, mechanical analysis, construction, and engineering. By converting tangible physical data into virtual models, DT enables simulations, analysis, data accumulation, mining, and electronic application, thereby augmenting the performance, efficiency, and reliability of the physical system.

In the marine industry, digital twin technology plays a crucial role in delivering pertinent feedback throughout the entire lifecycle of marine products, encompassing design, production, operation, and maintenance. This feedback furnishes decision-makers or decision-making systems with essential insights, facilitating the optimization of product performance, cost reduction, and safety enhancement. Consequently, the application of digital twin technology serves as a practical benchmark for the intelligent and digital transformation of the marine industry. Given its potential for enhancing the efficiency, quality, and safety of marine industry products, research in this area is notably active and receives considerable support through grants specifically allocated for this purpose.

While the deployment of Digital Twin (DT) technologies in the maritime industry manifests promising applications, substantial research challenges persist. The comprehensive management of the life cycle of maritime assets via DT represents a prolonged challenge necessitating further scholarly inquiry and empirical case studies. Critical challenges to be addressed encompass:

- Achieving high-fidelity representation of physical entities through virtual models.
- Navigating the inherent complexity and uncertainties associated with ship systems and the dynamic marine environment.
- Overcoming limitations imposed by the current state of digitalization aboard vessels and the availability of shore-based communication infrastructure.
- Addressing the absence of standardization within the maritime industry, which hampers integration of disparate DT systems.
- Mitigating cybersecurity risks, as increased interconnectivity of ships and maritime systems renders them more susceptible to cyber-attacks.
- Upskilling and training of the maritime workforce in new technologies to enable the operation and maintenance of DT systems.
- Addressing the significant initial investments required for DT implementation, which may discourage some enterprises from adopting this technology.

Although digital twin (DT) technology in the maritime industry is still in its early stages, it possesses significant potential to advance the sustainable use of marine resources and contribute to environmental conservation. Ongoing research and development are imperative to address existing challenges and fully harness the advantages of DT within the maritime sector. The community's interest in the application of this technology is evident from the substantial number of review papers. For instance, Pang et al. (2021) provides an overview of the current state-of-the-art in digital twin technology, a crucial component of the Industry 4.0 digitalization process. Their research discusses the development of a novel framework that integrates the digital twin and digital thread to enhance data management, fostering innovation, improving production processes and performance, and ensuring continuity and traceability of information. The digital twin/thread framework incorporates behavior simulation and physical control elements, which depend on the connectivity between the twin and thread for effective information flow and exchange to drive innovation. The framework encompasses specifications related to organizational architecture layout, security, user access, databases, and hardware and software requirements. It is anticipated that the framework will be applicable to optimizing operational processes and information traceability in the physical realm, particularly in an Industry Shipyard 4.0. Lv et al. (2023) provides a review of the state-of-the-art DT applications across various segments of the maritime industry, including shipbuilding, offshore oil and gas, marine fisheries, and marine energy. The analysis indicates that DT significantly supports full lifecycle management within shipbuilding, including the product design phase, manufacturing, operations, and maintenance. Additionally, this work examines the challenges and opportunities associated with DT implementation in the maritime sector, aiming to offer a reference point for the development of intelligent systems and guide the sustainable use of marine resources in the future. Mauro and Kana (2023) presents a systematic review of DT applications in the maritime industry, focusing on the ship life cycle. It underscores that, unlike other industries, the shipping sector often misuses the term "Digital Twin," mistaking it for basic virtual models that lack real-time data exchange. The review identifies gaps in current research, particularly in the design and decommissioning phases of the ship life cycle, where appropriate DT models remain underdeveloped. The paper also highlights that the ship industry is 2–3 years behind other sectors in adopting DT technologies, notably in manufacturing. A key contribution of the paper is the development of a systematic methodology for evaluating DT research, categorizing 58 relevant studies from 215 identified publications. The study recommends that future research should concentrate on developing DT-based procedures for ship design and retrofitting, with initiatives such as the "Digital Twin for Green Shipping" (DTGS) project offering potential solutions to address these research gaps (Fig. 21).

The focus of the Digital Twin (DT) application in the report by the Experimental Methods Committee centers on the integration of measurement and sensor equipment, alongside the processing of the collected data, Fujikubo et al. (2024) presents the outcomes of the Digital Twin for Ship Structures (DTSS) project, which was conducted through a collaborative effort between industry and academia in Japan. The primary objective of the DTSS is to offer a more comprehensive understanding and visualization of the real-time structural performance of ships during operation, thereby facilitating more optimal, data-driven ship design, construction, and operation. Key achievements of the project include the integration of monitoring systems with numerical simulations via data assimilation methods, specifically the wave spectrum method, Kalman filter method, and iFEM. These methods were validated through measurements at both model-scale and full-scale ship levels. The DTSS system proficiently captures stress responses across the entire ship structure when encountering waves, thereby enabling more precise predictions of short-term extreme responses and long-term fatigue damage. Compared to conventional methods, DTSS offers sig-

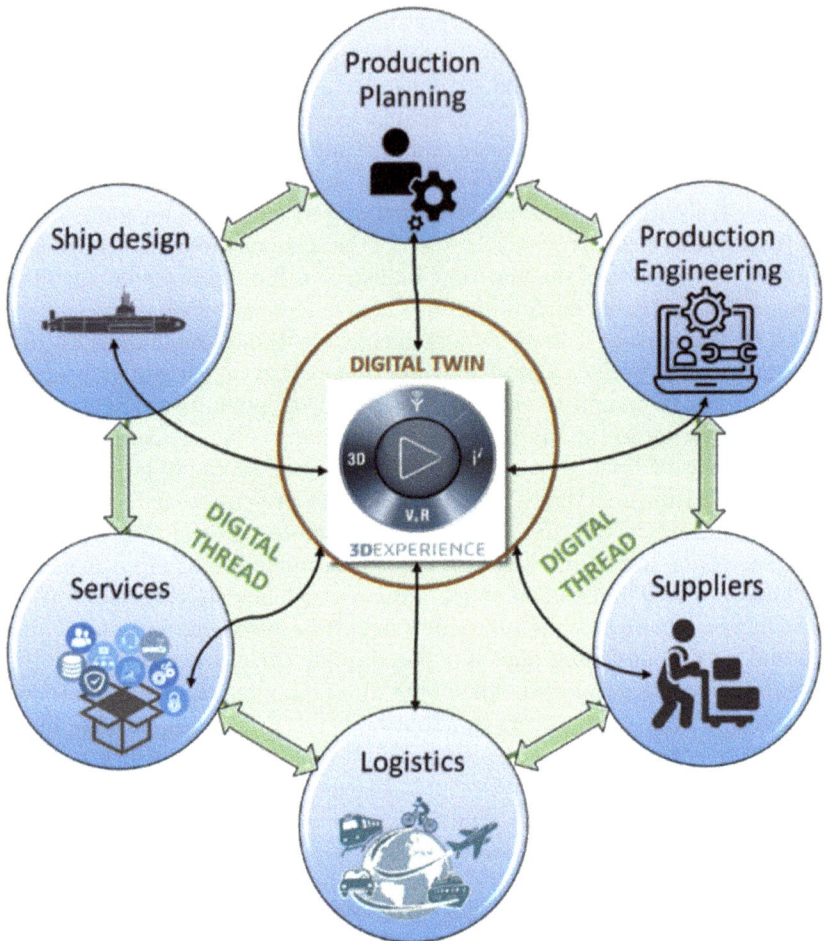

Fig. 21. Digital Twin and thread implementation scheme for a shipyard (Pang et al., 2021)

nificantly enhanced response predictions by utilizing real-time data from encountered wave conditions. The DTSS project introduced an open platform, i-SAS, which enables the application of DT technology for navigation support, maintenance planning, regulatory improvements, and product value enhancement. Furthermore, future research requirements were identified, including extending DTSS to address nonlinear responses and localized stress for fatigue assessments, as well as reducing uncertainties inherent within the DTSS system itself. The results of this project confirmed the technical feasibility of DTSS through comprehensive numerical simulations and real-world experiments, demonstrating its capability to mitigate uncertainties in load and strength estimations, thus contributing to safer and more efficient ship operations. The project underscores the broad potential for DTSS in predicting diverse ship responses, making it applicable beyond ships to marine structures such as floating wind turbines. Future endeavors aim to extend DTSS capabilities, further integrate it into the maritime industry, and explore cost-effective implementation strategies.

Hasan et al. (2024) delineates a rigorous methodology for fault diagnosis in autonomous maritime vessels, capitalizing on the capabilities of digital twin (DT) technology. The principal feature of this investigation lies in the incorporation of the Adaptive Extended Kalman Filter (AEKF) algorithm within the DT framework. This integration successfully illustrates the algorithm's ability to estimate fault parameters, particularly within the propulsion systems of ships. Digital twins function as virtual analogs of physical systems, with the AEKF algorithm performing a critical function by supplying accurate estimations of fault magnitudes. The proposed methodology received validation through extensive numerical simulations, which demonstrated the algorithm's proficiency in delivering accurate and reliable fault diagnoses. These findings highlight the algorithm's effectiveness and capability in detecting and estimating system anomalies. Additional validation was obtained via real-world experiments, substantiating the practical applicability of the proposed approach. The experiments underscored the capacity of digital twins to facilitate real-time health monitoring of ships, thereby allowing timely identification and remediation of faults. Oka et al. (2025) research endorses the 2D-AIS method, which employs two-dimensional wave spectra to more precisely estimate long-term hull stress as compared to conventional approaches. By integrating actual wave data from hindcast sources, the 2D-AIS method enhances the accuracy of predictions, reducing errors from approximately 35.

11.3 Summary and Conclusions on Health Monitoring and Digital Twin

This section delineates the critical technologies of Health Monitoring (HM) and Digital Twin (DT), both intended to augment the safety, efficiency, and resilience of marine structures through the employment of real-time data. Health monitoring techniques concentrate on damage detection by employing sensors and data-driven methodologies, whereas digital twins furnish virtual models to forecast future performance and refine lifecycle management.

Notable advancements have been realized in both domains. Health Monitoring has experienced the integration of Acoustic Emission, strain sensors, and machine learning to identify structural damages. Concurrently, DT technology has progressed through the utilization of empirical data and sophisticated algorithms, such as Kalman filters, to enhance structural assessments and operational decisions.

Nevertheless, several gaps persist. In the domain of health monitoring, a significant portion of research depends on simulations with limited empirical data, and the integration of real-time monitoring for intricate structures such as rudders and propellers remains a formidable challenge. Concerning digital twins, principal challenges encompass high-fidelity modeling, constraints in digitalization, cybersecurity issues, and substantial implementation costs. Furthermore, the maritime industry is comparatively slower in the adoption of these technologies when juxtaposed with other sectors.

12 Applications of Digital Image Correlation (DIC)

12.1 Introduction

Digital Image Correlation (DIC) represents a relatively novel measurement technique predicated upon the acquisition of extensive datasets through digital imaging. In recent years, DIC has been introduced to the market, with the advent of open-source software being developed by researchers, subsequently made accessible online and applied in selected cases. Commercial software and equipment were also developed in parallel by a few start-up firms. These methodologies signify a paradigmatic shift in the experimental analysis of structures; traditionally, displacement and strain measurements are conducted at a limited number of specific points with commendable accuracy. Conversely, emerging digital methods facilitate the acquisition of an extensive pattern of displacement and strains distribution over a surface. This, however, comes at the expense of accuracy, albeit temporarily with anticipated improvements on the horizon, and of some difficulties in the practical use of the measurement system, which is still rather complicate to deploy in the field for the time being. Significantly, such innovative methodologies enable more meaningful comparisons with numerical analyses like Finite Element Analysis (FEA), consequently fostering a robust experimental-numerical integrated approach in structural design and analysis of ships and offshore structures, possibly integrating also environmental action analysis. As a matter of facts, image processing based techniques are also applied in the assessment of hydrodynamic phenomena, both in laboratories and in the field. The 2021 Committee provided an introduction to DIC through a literature review of this emergent experimental technique and executed a benchmark study aimed at evaluating its uncertainties, with an emphasis on human factors in the application of DIC alongside other measurement techniques. The 2025 Committee undertook a review of recent applications of DIC, focusing on a detailed examination of its utilization within the domain of ships and offshore structures. Purposefully, other fields of application were excluded from this review, unless explicit connections or indications relevant to the marine industry were discerned. The fundamental objective is to offer motivating exemplars and specialized application practices, potentially contributing to the exploitation of DIC's capabilities in maritime and offshore structural contexts. Indeed, the advent of novel materials and a goal-based design strategy necessitate enhanced experimental validation of increasingly complex and elaborate numerical simulations, which are increasingly integral to project development and mandated by regulatory requirements.

12.2 Potential of DIC

In this section, the literature review is organized to underscore various aspects of the potential applications of DIC in the context of ships and offshore structures. Although the literature contains numerous studies detailing DIC applications, there is a scarcity concerning our particular areas of interest. Consequently, the objective of this review is to delineate the benefits and limitations within the ship and offshore structure sectors. It is acknowledged that DIC holds significant potential; however, the breadth of its applications remains limited due to certain anticipated challenges to be resolved in the future. An exemplary application of DIC is observed in the case of hybrid joints: for instance, to determine the optimal geometry of a joint, a structural adhesive cohesive zone model Finite Element Analysis (FEA) simulation was executed to minimize Von Mises stresses, subsequent to its validation through DIC measurements obtained from a destructive laboratory test in the adhesive-bonded area. The failure modes of fracture and debonding were throughly examined, which conventional techniques struggle to capture with adequate detail. Comprehensive descriptions are provided by (de Vicente et al., 2022). The modeling and analysis of thin-walled aluminum/steel explosion welded transition joints for shipbuilding purposes have been documented in (Boroński et al., 2020), where a successful integration of Finite Element Analysis (FEA) and DIC results has been achieved, as also noted in (Corigliano et al., 2018), which includes comparisons with infrared thermography measurements as well. Furthermore, a three-dimensional Digital Image Correlation (3D-DIC) method was utilized on a composite underwater structure specimen to evaluate its pressure-bearing capability, employing both conventional strain gauges and numerical simulations. Experimental and numerical evidence indicates that strain gauges exhibit limited accuracy and produce erratic measurements due to the substantial size discrepancy between the strain gauges and the small structural components, coupled with a restricted number of measurement points. In contrast, the 3D-DIC system features a minimal measurement unit that is smaller than the size of these small structural components, enabling precise stress-strain evaluation over an expansive measurement area, and provides a detailed illustration of the strain surface distribution. Additionally, a protective shield was engineered for the cameras to secure optimal functionality within the experimental setting, thereby presenting the feasibility of employing DIC underwater, albeit presently confined to laboratory conditions (Luo et al., 2023). The investigation of fluid–structure interaction effects was satisfactorily conducted using DIC, demonstrating its efficacy in measuring structural deformations associated with unsteady cavitating flows around both flexible and rigid NACA hydrofoils composed of polyvinyl chloride (PVC), brass, and aluminum, as reported by (Yuxing Lin & Schellin, 2022). Reference is made to the limited comparable applications available in the open literature, directing interested readers to this publication for its comprehensive literature review. An analogous application, which boasts the advantage of measuring the response without affecting model properties or altering the flow field, unlike conventional experimental methods and in situ measurements, is elucidated in

(Tödter et al., 2021). This represents a significant benefit of optical measurement techniques not attainable with traditional sensors, which are generally intrusive. The stereo digital image correlation (stereo-DIC) technique demonstrates significant potential in the comprehensive three-dimensional (3D) deformation measurement of marine propeller blades, utilizing stroboscopic lighting to characterize 3D dynamic deformation underwater in a windowed cavitation tunnel. Although enhanced calibration techniques were necessary, the feasibility of monitoring the full-field 3D dynamic response and structural health of underwater rotating structures has now been established (Su et al., 2022). Savio (2015) and Savio et al. (2024) satisfactorily employed stereo DIC to measure the deflection of resin propeller blades solving the problem caused by windows distortion. To evaluate the applicability of DIC for vortex-induced vibration (VIV) investigations, results were compared with those obtained from conventional methods utilizing triaxial accelerometers, possibly intrusive as strain sensors. For small motion amplitudes, the speckle pattern required refinement, while it was adequate for medium and large amplitudes. The influence of the DIC processing techniques was found to be negligibly small under the tested conditions, although different speckle patterns displayed varying accuracy. The repeatability of DIC measurements could not be assessed cost-effectively due to the excessive acquisition duration at high frame rates, as the memory capacity of the cameras posed a limiting factor (Tödter et al., 2021). The role of curvature in the response of air-backed composite plates within the design of naval systems has been successfully studied by Ulbricht, Han, & Porfiri (2024), who used both DIC and planar particle image velocimetry (PIV) to allow for the study of the flow physics and structural response. An impulsive loading was generated from the side of the plate in contact with the fluid, mirroring the loading conditions associated with underwater explosions. An application of DIC for analyzing rapid dynamic events is documented in Gargano et al. (2022), focusing on the response of sandwich structures to explosive blasts in naval ships. The reverse side of the specimen sandwich panels was coated with a speckle pattern comprising dots to enable DIC measurement of out-of-plane displacements and surface strains induced by explosive forces. The dots measured approximately 2 mm in diameter and were spaced roughly 1 mm apart, establishing the spatial resolution for the DIC assessments. The Aramis DIC system was effectively utilized with dual high-speed Photron SA5 cameras operating at a frame rate of

$$7000\,\mathrm{s}^{-1}$$

and offset at an angle of
$$22.5°$$
from the panel surface.

The assessment of service conditions for ships and offshore structures presents a significant avenue for investigation. For instance, corrosion assessments have been conducted and documented, as seen in Qvale et al. (2021) and Shojai et al. (2023), with respect to fatigue failures. Typically, corrosion and fatigue are evaluated independently within standard testing and measurement practices.

However, DIC facilitates the identification of interactions, particularly in differentiating between crack initiation and propagation sites at pre-existing weld notches and those arising from corrosion. It was observed that both, individual pits and uniform corrosion at the weld toe, contribute to fatigue, aligning with expectations; yet, surprisingly, residual stresses induced by clean blasting were partially mitigated by corrosion, thus influencing fatigue strength. More recently, Hu, Hua, Liu, Wang, & Wu, (2025) investigated the multiple pit corrosion interactions and their effect on fatigue crack initiation by methodically establishing a sensitive crack initiation detection scheme by combining high-frequency testing and DIC. DIC finds particularly notable applications in the analysis of highly flexible structures, such as composite foils, and fabrics, such as sails, as well as in fluid-structure interaction problems more broadly. In these contexts, conventional deformation measurement techniques may be intrusive and fail to yield reliable outcomes. Banks et al. (2015) evaluates the use of DIC in these scenarios, focusing on the examination of a curved daggerboard of a racing catamaran with complex geometry within a wind tunnel setting. Additionally, they furnish a comprehensive review of analogous applications. Of particular interest is their discussion on the generation of speckle patterns, whether manually or through a stochastic digital image creation software, which may influence measurement outcomes. Furthermore, they address the issue of image blurring due to vibrations at elevated wind speeds, advocating for the encapsulation of cameras within purpose-designed fairings, which only marginally mitigates the problem, reducing the DIC system error by 57 An interesting application of DIC together with traditional strain gauges is described by G.-J. Shi, Ji, Xu, Wang, & Xu, (2024): the ultimate strength of a scaled GFRP hull girder with hat stiffeners and foams under bending load was experimentally assessed. The strain variation is recorded by the digital image correlation (DIC) system on deck upper surface while critical locations are locally monitored by strain gauges and displacement sensors. Videos allowed recording deformed shapes as well as strain field to validate FEA. The ISSC 2025 Technical Committee III.2 on Ultimate Strength has conducted a benchmark study focusing on the buckling and ultimate strength of a transversely loaded thin plated stiffened structure. A full-scale experiment conducted at the Marine Structures Testing Lab of the University of Genoa served as the experimental target for this benchmark, (Barsotti et al., 2025). Data on applied load versus displacement was provided, alongside 3D scans of the panel's geometry prior to testing for the assessment of initial deformations, and measurements from approximately 40 strain gauge channels and potentiometer displacement gauges were also available. Interested readers are referred to the pertinent ISSC report and associated literature for detailed information. Data from collapse tests were also compiled, complemented by high-definition videos of the shell plating captured from various angles suitable for 3D reconstruction. Acquired DIC data were not utilized in the aforementioned benchmark. Presented here are some images illustrating preliminary analyses of such DIC data currently under review and available for future benchmarks. In addition to the displacements of the central zone of the panel, which is the target area of the tested specimen (being the panel ends specifically engineered to impose appropriate boundary conditions), it was feasible to obtain the temporal evolution of the strain tensor maps on the plating. Traditional gauging techniques, which offer only point measurements, cannot provide such information (see Fig. 22). Consequently, strain gradients can be estimated within the 2D domain, facilitating a comprehensive insight into the buckling behavior of elementary plate panels enclosed by stiffeners. It is pertinent to mention that, as of now, current buckling evaluations recently consolidated in IACS Rec. URS-35 (IACS, 2024) endorse local elastic plate buckling, given that a stiffened plate has reserve strength exceeding the elastic buckling of elementary plate panels. The benchmark conducted by the aforementioned ISSC Technical Committee III.2 on Ultimate Strength identified conservative results in CSR formulations, underscoring the critical importance of validating nonlinear finite element analyses for the shipbuilding industry, especially considering that CSR guidelines apply to at least 90.

12.3 Updates on the Benchmark of ISSC 2021 Committee

In an effort to supplement the ISSC 2021 benchmark, a Committee member who did not participate in the prior study was enlisted. The identical composite specimens utilized in the previous benchmark were supplied to this new participant, with instructions to metiÅŻculously adhere to the benchmark guidelines and to document the measurement techniques and their practical application. This consideration arises from the understanding, based on prior experiences, that the application of DIC techniques necessitates a distinct perspective and specific skills for successful implementation. Consequently, the ensuing description serves as a guidance document to underscore the challenges and application strategies as discussed among the current committee members. The primary objective of the 2021 benchmark was, and remains, to estimate the first natural frequency of small cantilever beams composed of fiberglass sandwich laminate. These specimens were originally cut from a sandwich panel for the prior benchmark investigation. The characteristics and properties of the specimens have been detailed in the preceding Experimental Methods Committee of the ISSC 2021 report (Ehlers S., 2022) as follows:

- The sandwich core is a 10 mm thick PVC ($75\,\text{kg/m}^3$)
- E PVC 2.60E+09 [N/m^2], PVC 0.32 [-]
- The skins' stacking sequence includes one biax layer ($\pm 45°$; $600\,\text{g/m}^2$) and one twill layer ($0°/90°$; $200\,\text{g/m}^2$) on each skin.
- Nominal dimensions of the cut sample are (length/width/thickness): 560/30/12 mm
- Measured dimensions: $563 \times 30.7 \times 11.8$ mm/70 g (Sample A)
- Material properties:
  - E glass 7.00E+10 [N/m^2], glass 0.25 [-], G glass 3.00E+10 [N/m^2]
  - E epoxy 3.00E+09 [N/m^2], epoxy 0.35 [-], G epoxy 1.50E+09 [N/m^2]

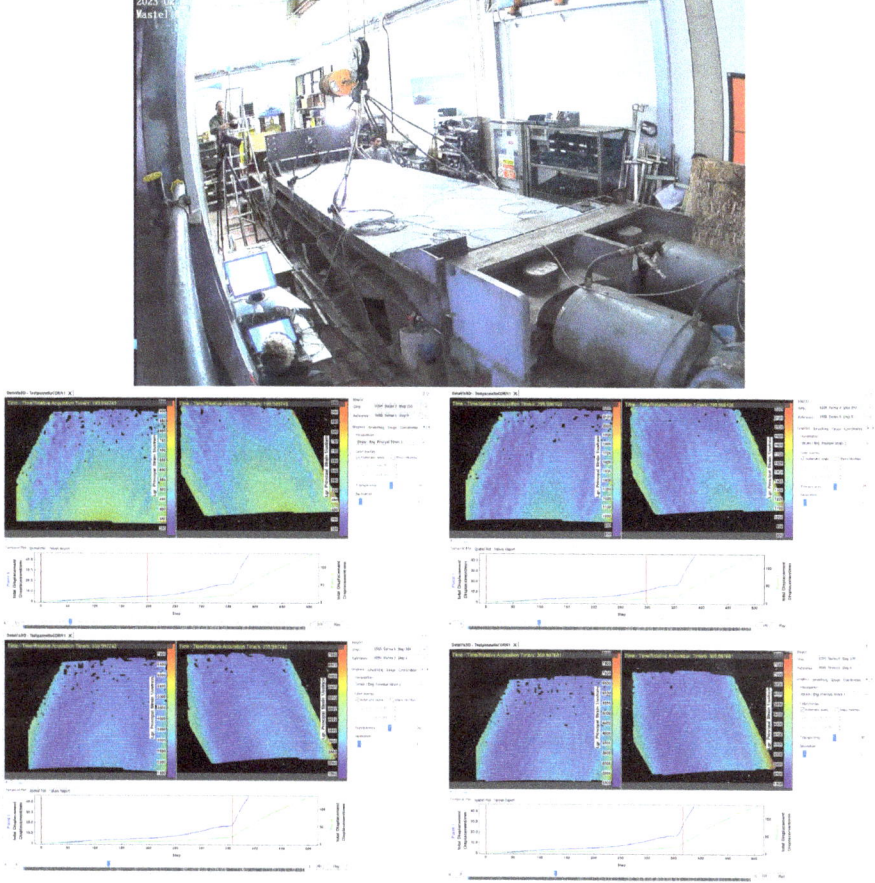

Fig. 22. examples of DIC results obtained from the experimental target of the benchmark test of the ISSC TC III.2 on Ultimate Strength, strain pattern of the plating at certain points in time

Figure 23 illustrates the specimen subjected to testing and its measurement conducted using a basic measuring tape. It is worth noting that nowadays even such trivial measurements are more and more carried out and reported in everyday practice using the aid of digital pictures. To ensure a secure and rigid clamping arrangement, the specimen was secured using a hydraulic wedge grip MTS 647.10A, maintaining a 500 mm length of unsupported span. To mitigate potential damage to the specimen due to excessive clamping force, aluminum spacers with a thickness of 11.55 mm were fabricated, thereby averting the risk of deformation under high pressure. The DIC system utilized for measurement comprised two cameras and two illumination sources.

- Cameras: Basler boost boA5328 - 100 cm
- Lenses: Schneider Kreuznach, JADE 2.8/35 C (focal length 35 mm)

- Lights: Blue-X-Focus v3

The data acquisition was performed using the Vic-Snap commercial software, and the subsequent analysis was conducted within the Vic-3D environment.

A speckle pattern was applied to the tip of the specimen. The cameras and lighting sources were mounted on a tripod. The cameras were configured at an approximately 20-degree stereo angle, such that the sample was centrally positioned within the camera views, maintaining a distance of 55 cm from the cameras to the sample, as illustrated in Fig. 23. Following the alignment of the cameras, the lighting sources were oriented and adjusted accordingly. The focus of the cameras was manually adjusted directly from the lenses and subsequently locked after achieving the sharpest image at an aperture of F/4. To facilitate an increased sampling frequency, the image was cropped to the area of interest, specifically an approximately 40×40 mm, plane at the location of the speckle (refer to Fig. 24). The frame rate was then configured to 500 Hz, with the exposure time set at 86 ms. Subsequent to these configurations, calibration was performed using a 14 mm-10-dot calibration frame, adhering to the standard calibration procedure in which the calibration table is rotated and translated within the field of view of the cameras, approximately at the distance of the sample. Polarizing lenses were affixed to the cameras prior to measurements to mitigate local overexposure during the data acquisition process.

Calibration images and measurements were acquired utilizing Vic-snap software. The sample underwent excitation through horizontal force applied via an Allen key at its lowest corner, followed by its release to facilitate free vibration. Measurements spanning approximately 20 s were recorded. Subsequent to recording the vibrations, the gathered data underwent analysis using the Frequency Analyzer tool integrated into Vic-3D software. A calibration routine was executed within the program to generate a calibrated database. To minimize projection error, the 'Auto Correct Calibration' function was employed, successfully reducing the error to 0.05 mm. During data analysis, initial displacements were excluded from consideration. To determine the first eigenmode, horizontal displacements were calculated from the images, both as an average over the defined area of interest and from a specific point, as illustrated in Fig. 24. Figure 25 displays a three-second excerpt from the measured time history, wherein the phase of average displacement over the area and displacement at the point closely corresponds. Consequently, the Fast Fourier Transform (FFT) was applied exclusively to the average area using the tool. The FFT results indicated the highest amplitude, identifying the first natural frequency as 28 ± 0.2 Hz, which is approximately 10% higher but consistent with the DIC results reported in the 2021 ISSC benchmark Fig. 25.

The preceding detailed description elucidates that multiple instrumentation parameters require appropriate configuration by the user, while numerous other parameters remain unreported due to their integration within the proprietary software employed in this instance. Nevertheless, the measurements align with those documented in the prior term of the Committee.

Committee V.2: Experimental Methods 147

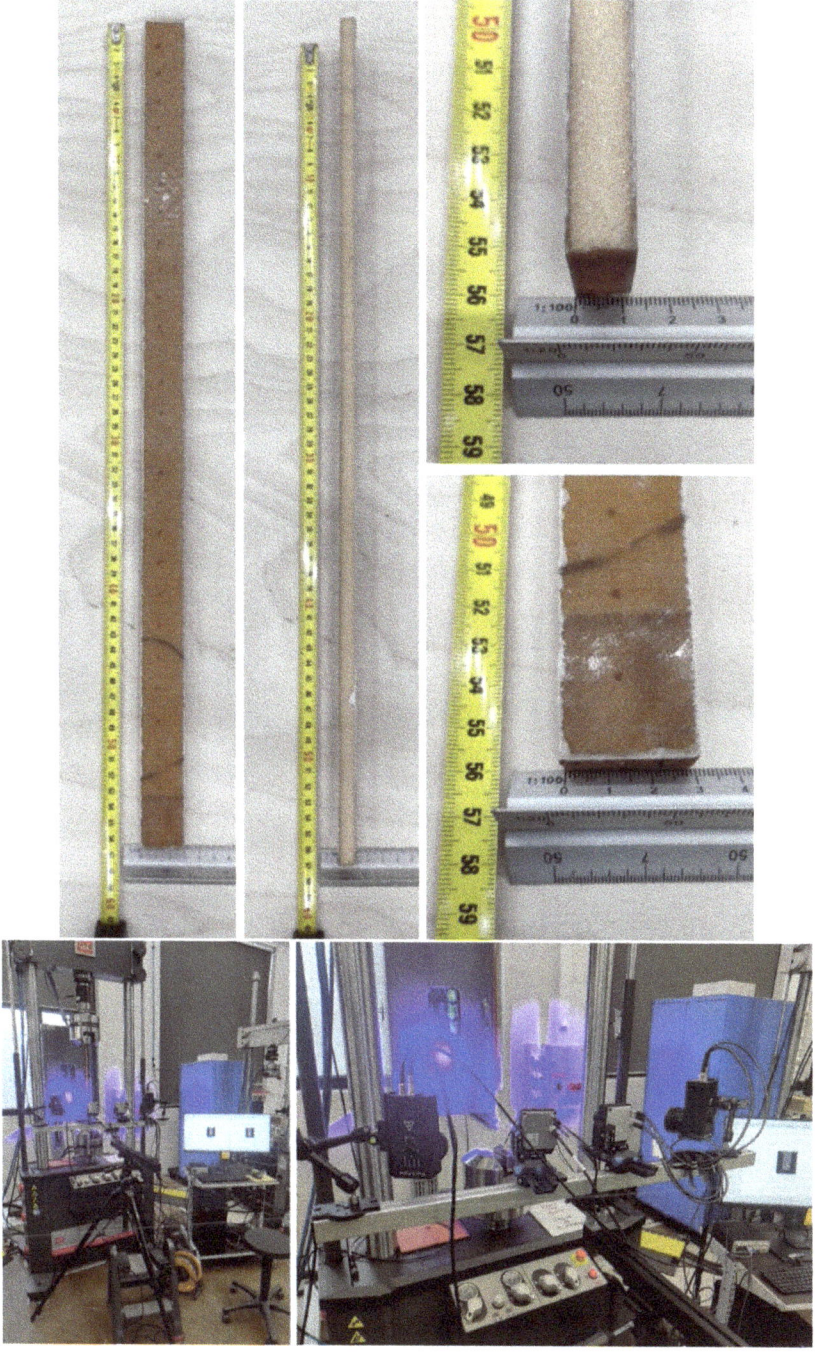

Fig. 23. Specimen and testing setup

Fig. 24. Screenshot from the build-in Frequency Analysis tool. The colored area over the speckle shows the area of interest, and the white circle with markers a secondary area of interest.

12.4 DIC Application Issues

Applications of DIC have been acknowledged in scientific literature for several years, with the 2021 ISSC Committee introducing a benchmark to probe this emerging experimental technique. The current Committee is focused on assessing DIC applications pertinent to maritime vessels and offshore structures. Supplementary benchmarking tests have been conducted to enhance previously established findings. The succeeding discussion delineates the obstacles encountered in applying DIC, particularly considering specific and in situ applications. Primarily, during the benchmarking process, researchers faced significant challenges in producing a dependable speckled pattern for the measurement of extensive structural areas, in contrast to specimen measurements. For the purposes of the benchmark, the measurement area approximated $0.1 \, \text{m}^2$, where conventional techniques for speckle pattern generation, such as the indiscriminate application via a spray can, proved inadequate at this scale. Some participants succeeded in pattern creation through random splattering, yet subsequently encountered complications in accurately calibrating the DIC system. This was ostensibly due

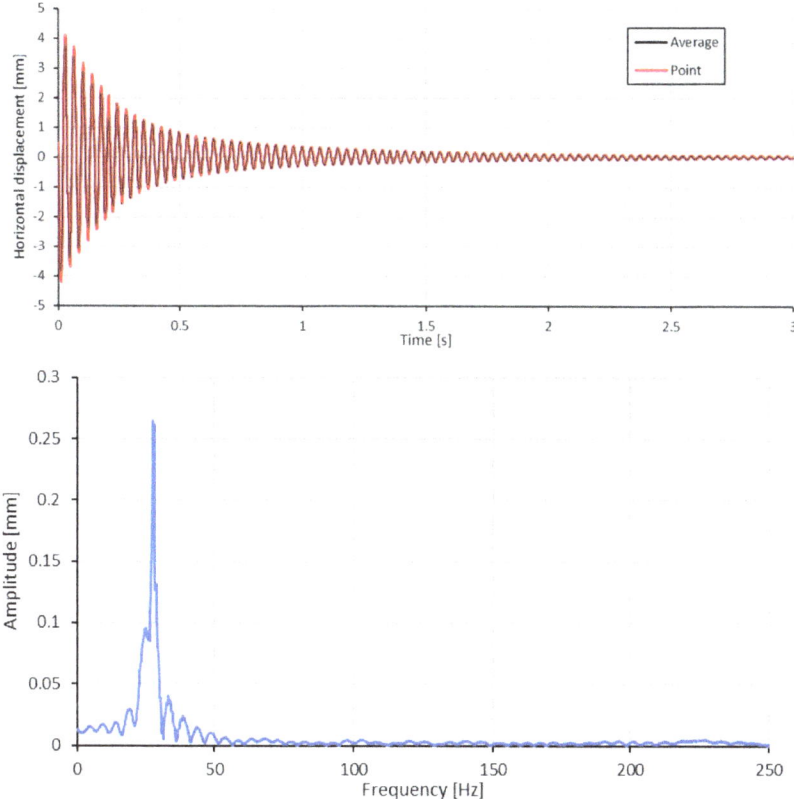

Fig. 25. Measured time history of horizontal displacement and corresponding FFT

to utilizing a calibration pattern of substantial size compared to the dots produced on the specimen's surface. Several alternative methods were proposed and experimentally evaluated: • The application of a printed speckled pattern on the surface of the structure did not result in significant measurements, suggesting suboptimal adhesion. • The employment of a plastic "stamp" failed to generate distinctly defined spots, although this method had previously proven effective for softer, rubber-like structures. The fabrication of a rubber stamp remains a potential solution under consideration. • The manual creation of a quasi-random speckled pattern yielded comparatively improved results (see Fig. 26), yet the signal-to-noise ratio in conditions involving vibrations was found to be inadequately low.

A significant issue identified in the benchmark investigation as well as in other referenced experimentation was the loss of camera focus. The structure in general exhibits both rigid body motions and vibrations: when the magnitude of these motions increases, there is a greater likelihood that portions of the surface would be outside the camera's field of view and the remaining visible surface would be at an incorrect distance for the lenses to accurately capture the

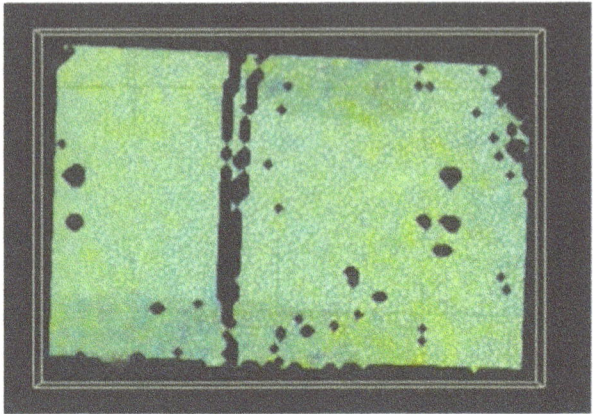

Fig. 26. the manually generated (using a marker) speckled pattern was captured by the cameras in the areas without cables and holes. Nevertheless, vibration measurements were not possible due to low signal-to-noise ratio

responses. Overall, individuals utilizing DIC techniques encountered challenges in preparing the surface for measurement, even within a laboratory setting. Such challenges are exacerbated during large or full scale tests, as demonstrated in the previously referenced panel collapse test (Barsotti et al., 2025). Figure 27 presents images from a large scale test also conducted at the Marine Structures Testing Lab of the University of Genoa on an actual propeller blade, which underscore these issues, highlighting that they are further compounded in situ.

In addition to the challenges associated with pattern preparation and system calibration, users frequently encounter difficulties related to the positioning of the system and the management of cabling components (see again Fig. 27). The abundance and fragility of power and data cables/connectors further complicate operations in environments that are not conducive to delicate equipment. While utilizing custom-developed software can enhance comprehension and offer flexible application options, employing a commercial hardware and software measurement setup provides the benefit of producing reliable results within a reasonable timeframe. This approach mitigates the need for adjusting numerous parameters and instrumentation variables, which are often complex to fine-tune accurately. Nonetheless, it is important to note that even relatively straightforward tests conducted in controlled laboratory environments with a well developed commercial system and software pose significant measurement challenges. Attention is specifically directed to the speckle pattern illustrated in Fig. 24, which differs notably from those generated by random applications of spray cans.

Admittedly, Digital Image Correlation (DIC) is still in its nascent stages, with a novel experimental data paradigm emerging, transitioning from one-dimensional point measurement to two-dimensional surface measurements as previously mentioned. Consequently, when test data are employed for the val-

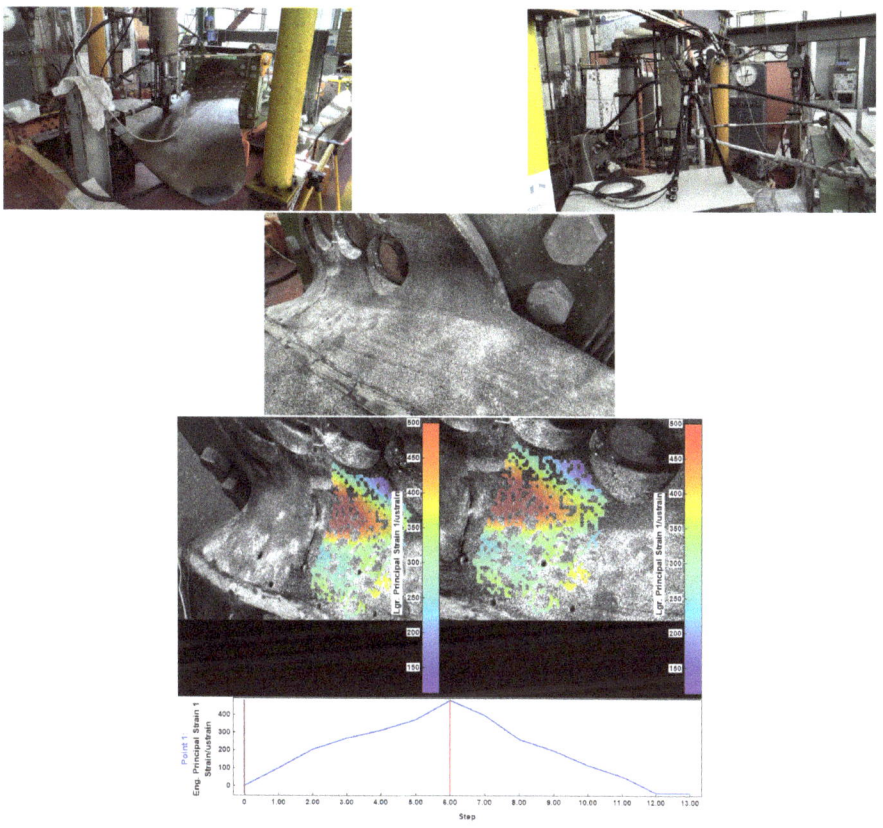

Fig. 27. issues in setting up the DIC measurements in a large scale test in the Marine Structures Testing Lab of the University of Genoa, Italy: "holes" in the strain measurements onto the blade surface are evident

idation of numerical models, as is frequently the case, contemporary literature underscores the advantageous integration of these two methodologies. Notably, commercial DIC software now provides the capability to interface with results from key Finite Element Analysis (FEA) packages. As an additional perspective, a concise account of a DIC challenge documented in the open literature further demonstrates the dynamism within this field (Reu et al., 2017). This initiative evaluates the precision and dependability of various DIC software and methodologies. Participants engage in the analysis of standardized images of a test object subjected to predetermined deformations, utilizing either commercial or open-source software. The challenge evaluates the performance of these tools based on criteria of accuracy, efficiency, and the capability to process complex datasets. This undertaking not only sets a benchmark for existing DIC software but also cultivates a collaborative environment, promoting the exchange of techniques and insights. Open-source tools are frequently emphasized for their adaptability, which facilitates the swift integration of community-driven innovations. The DIC Challenge 2.0 (https://idics.org/challenge/), a progression from the initial DIC Challenge cited in 2021 report of this committee (Ehlers S., 2022), sought to advance the comprehension and evaluation of two-dimensional DIC algorithms. The primary goal was to improve the quantification of the spatial resolution of these algorithms. An essential component of this challenge was the development of novel images specifically tailored for assessing and enhancing the execution of 2D-DIC. The findings elucidated the trade-off between displacement and strain signals noise (or measurement noise) and spatial resolution across a spectrum of DIC algorithms. The results revealed that, while the performance of 2D algorithms generally conformed with theoretical predictions, particularly concerning displacement measurement, the determination of strain spatial resolution exhibited significant variability among the different algorithms. A notable conclusion drawn from the study is that, with appropriate adjustment of solution parameters, the performance of each of the 10 codes can be more closely aligned. This suggests that the quality of outcomes for 2D-DIC is more significantly influenced by the choice of analysis software settings rather than the code's implementation itself, at least among those who participated in this challenge. Consequently, adequate training is essential for the effective adoption of DIC as a primary measurement technique in laboratory and in situ. While obtaining results through DIC is a relatively straightforward process, ensuring the accuracy and reliability of those results presents greater challenges. Moreover, it is crucial to note that, in many experimental contexts, the quality of results is predominantly affected by experimental conditions rather than by deficiencies in the DIC software itself. Commercial DIC software solutions such as Correlated Solutions' VIC-2D/3D, GOM Correlate, LaVision's DaVis, and Dantec's Istra4D provide an array of comprehensive features, including intuitive user interfaces, extensive support, and advanced processing capabilities, though their substantial cost may be a barrier for some users. This potential impediment can be mitigated by open-source alternatives available on platforms like GitHub. Software such as Ncorr and DICe are accessible and modifiable without cost. Ncorr, which operates within MATLAB, and DICe, suitable for both 2D and stereo-DIC applications, offer viable platforms for conducting DIC analysis, though they often necessitate increased user input and customization. Open source DIC software hosted at GitHub were sorted by number of (discord) stars:

- Digital Image Correlation Engine (DICe)
- MultiDIC
- µDIC
- Ncorr
- OpenCorr, (see also W. Yin et al. (2024))

to assess the practicality of various software options. An endeavor was undertaken to evaluate open-source software endowed with stereo-DIC capabilities. Software reliant on MATLAB was excluded from the investigation due to the relatively high cost associated with acquiring a MATLAB license, as the primary objective was to explore entirely cost-free alternatives. Contrary to initial expectations, the undertaking proved to be considerably more challenging, and consequently, the current ISSC report does not include comprehensive findings since the investigation remains ongoing. Nonetheless, a promising software adaptation is in development and is anticipated to be presented, if not sooner, at the forthcoming ISSC Congress. Open-source DIC software generally presents a steep learning curve and necessitates a high degree of expertise in both DIC techniques and programming. Although certain platforms provide 3D DIC functionalities, applying them to practical applications poses significant challenges. Difficulties frequently encountered include inadequate documentation, the absence of intuitive or effective graphical interfaces for process management, the requirement for Python or C++ programming to set up particular measurements, algorithmic inefficiencies resulting in untenable analysis durations, and a deficiency in post-processing tools. A prevalent issue is the inconsistency or cessation of developer engagement within many open-source initiatives. Consequently, post-initial release, the software may cease receiving updates, bug resolutions, or feature enhancements, thus undermining its reliability for long-term usage. In some instances, development is entirely abandoned, leaving users with obsolete or partial tools. Moreover, diminished or intermittent community contributions can result in prolonged periods during which even critical issues remain unresolved, thereby complicating efforts to address these challenges or adapt the software to meet evolving requirements (Fig. 28).

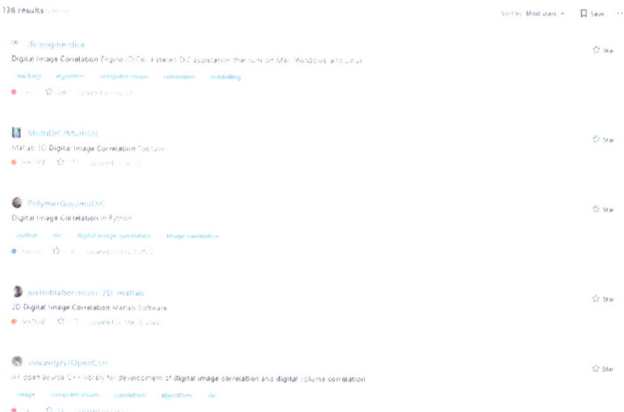

Fig. 28. Open source DIC software hosted at GitHub sorted by number of (discord) stars

12.5 Summary on DIC Applications in Ship and Offshore Structures

The literature and internet survey, along with the applications detailed in this section and the experiences documented in the 2021 ISSC Committee report, suggests that DIC is a promising experimental technique. However, its primary limitation is its applicability in challenging environments such as marine settings. Although commercial equipment permits applications with acceptable but considerable effort, leveraging open-source software, standard cameras, and other readily available market devices, to develop DIC instrumentation remains currently beyond the typical skill set of naval architects and marine engineers. The predominant challenges involve system calibration and the creation of an appropriate pattern on the specimen's surface, as well as the selection of suitable instrumentation components and their appropriate integration when not using ready-to-use equipment available from relatively few vendors. Conversely, laboratory applications provide valuable opportunities for validating numerical simulations, which conventional gauging techniques cannot achieve, offering a comprehensive view of structural deformations as evidenced by literature and direct applications by members of this ISSC Committee.

13 Summary and Conclusions

Various experimental methodologies are employed to assess the performance and responses of ships and offshore structures under diverse conditions. These methodologies utilize sensor systems and numerical simulations to verify and validate design specifications, lifecycle performance, accidental scenarios, and lifecycle responses. The report encapsulates the expertise of the authors in their specific research domains, articulating the current state of the art and identifying critical technologies and existing gaps. Summaries of each section are provided, with more comprehensive details available within each respective section.

Within the Scaling Laws section, the validation process for the newly proposed scaling method remains in progress, and the current scaling approaches address only a limited portion of the testing scopes. Therefore, additional scaling methodologies are requisite to encompass the scopes of various tests.

In the section on the Fluid Structure Interaction of Flexible Structures, it is observed that elastic models are increasingly being employed for non-beam-like structures, scenarios where multiple modes are significant, torsion, and local responses, among others. These models are fabricated utilizing additive manufacturing, foam, or shaped plastic sheets.

Within the section dedicated to Hybrid model testing, a comprehensive summary of models is presented across various categories, with particular emphasis placed on hardware-in-the-loop methodologies, which are increasingly prevalent within the domain.

Within the fire test section, it is acknowledged that fire testing methodologies are continuously advancing. Consequently, the implementation of a comprehensive full-scale testing protocol is essential to develop and confirm that novel composites, specifically bio-based materials, adhere to marine standards and fire safety prerequisites.

Within the section dedicated to friction testing, the intricacies associated with interactions involving sea ice are highlighted, particularly the challenges in differentiating forces attributed to friction from those related to ice fracture and submersion. In a similar vein, the investigation of internal friction in power cables and umbilicals is explored. The necessity of friction testing is underscored, and further research conducted under pertinent conditions is advocated.

The section on corrosion prognostics elaborates on corrosion as a principal contributor to the structural failure of maritime vessels and offshore structures. Corrosion prognostics are being integrated with structural health monitoring techniques to effectively manage life cycles and minimize costs.

Within the domain of large-scale subsea structures, numerous recent experimental investigations have been focused on the fluid-structure interaction, aiming to understand the response concerning vibration, displacement, and stress.

In the section concerning large impact tests, it is observed that the planning of large-scale tests may require several years. This includes numerical simulations, substantial financial resources, instrumentation, platform planning, as well as the meticulous execution of a complex testing event to accurately observe the desired behavior.

Within the section on full-scale ice load measurements, it is observed that the contact area and the pressure distribution within this area exhibit significant spatial and temporal variability. Measurements of ice-induced loads on ships with longitudinal framing present further complexities, necessitating additional advancement in this area.

Within the section on health monitoring and digital twins, the field of health monitoring heavily depends on simulations, which are constrained by the availability of experimental data. The integration of real-time monitoring for intricate structures, such as rudders and propellers, continues to pose significant difficulties. Digital twins are confronted with fundamental challenges including the need for high-fidelity modeling, limitations of digitalization, issues related to cybersecurity, and substantial implementation costs. Furthermore, the maritime industry is notably behind other sectors in embracing these technologies.

The section on the application of Digital Image Correlation (DIC) indicates that DIC is a promising experimental method. However, its primary limitation lies in its application challenges in harsh environments, such as marine settings. Furthermore, although commercial equipment facilitates applications with considerable but manageable effort, the potential for utilizing open-source software, standard cameras, and other widely available devices to develop DIC instrumentation exceeds the current average proficiency of naval architects and marine engineers.

Acknowledgements. The authors and committee members wish to formally acknowledge the significant contributions of Dr. Patrick Kaeding in his capacity as the committee liaison, whose exceptional efforts facilitated the development of this report. Additionally, the authors extend their appreciation to Professor Jeom-Kee Paik, who served as the official discusser and provided a comprehensive technical review of this document's contents. Finally, acknowledgment is due to Dr. Shixiao Fu, who meticulously conducted a peer review of certain sections of this document.

References

ABS: Rules for building and classing facilities on offshore installations. American Bureau of Shipping (2021)

Adams, J., Valtonen, V., Kujala, P.: Validation of the line-like nature of ice-induced loads using an inverse method. In: Proceedings of the 25th International Conference on Port and Ocean Engineering Under Arctic Conditions (2019)

Ahani, A., Greco, M., Abrahamsen, B.C.: Hybrid analysis of severe wave impact and hydroelastic effects on a rectangular vertical cylinder. Ocean Eng. (2024). https://doi.org/10.1016/j.oceaneng.2024.119846

Andresen-Paulsen, G., Braun, M., von Bock und Polach, F., Marquardt, T., Momber, A., Ehlers, S.: The integrity of corrosion protection systems of welded maritime structures under cyclic loading. In: Advances in the Analysis and Design of Marine Structures (2023)

Aravanis, G.I., Silionis, N.E., Anyfantis, K.N.: Damage detection in ship hull structures under operational variability through strain sensing. Ocean Eng. **286**, 115537 (2023). https://doi.org/10.1016/j.oceaneng.2023.115537

Banks, J., Marimon Giovannetti, L., Soubeyran, X., Wright, A., Turnock, S., Boyd, S.: Assessment of digital image correlation as a method of obtaining deformations of a structure under fluid load. J. Fluids Struct. 58, 173–187 (2015). https://www.sciencedirect.com/science/article/pii/S0889974615002066, https://doi.org/10.1016/j.jfluidstructs.2015.08.007

Barsotti, B., Battini, C., Gaiotti, M., Rizzo, C.M., Vergassola, G.: Experimental and numerical assessment of ultimate strength of a transversally loaded thin-walled deck structure. Mar. Struct. (2025). https://doi.org/10.1016/j.marstruc.2025.103793

Bishop, R., Price, W.: On the scaling of flexible ship models. J. Sound Vib. **73**(3), 345–352 (1980)

Boroński, D., Skibicki, A., Maćkowiak, P., Płaczek, D.: Modeling and analysis of thin-walled Al/steel explosion welded transition joints for shipbuilding applications. Mar. Struct. 74, 102843 (2020). https://www.sciencedirect.com/science/article/pii/S0951833920301362, https://doi.org/10.1016/j.marstruc.2020.102843

Böhm, A.M., von Bock und Polach, R.U.F., Herrnring, H., Ehlers, S.: The measurement accuracy of instrumented ship structures under local ice loads using strain gauges. Mar. Struct. 76, 102919 (2021). https://www.sciencedirect.com/science/article/pii/S0951833920302100, https://doi.org/10.1016/j.marstruc.2020.102919

Calle, M.A., Salmi, M., Mazzariol, L.M., Alves, M., Kujala, P.: Additive manufacturing of miniature marine structures for crashworthiness verification: scaling technique and experimental tests. Mar. Struct. 72, 102764 (2020). https://www.sciencedirect.com/science/article/pii/S0951833920300587, https://doi.org/10.1016/j.marstruc.2020.102764

Calle, M.A., Salmi, M., Mazzariol, L.M., Kujala, P.: Miniature reproduction of raking tests on marine structure: similarity technique and experiment. Eng. Struct. 212, 110527 (2020). https://www.sciencedirect.com/science/article/pii/S0141029619350941, https://doi.org/10.1016/j.engstruct.2020.110527

Casaburo, A., Petrone, G., Franco, F., De Rosa, S.: A review of similitude methods for structural engineering. Appl. Mech. Rev. **71**(3), 030802 (2019). https://doi.org/10.1115/1.4043787

Chen, Y., Zhang, S., Chen, W.K., Magee, A.: Design and fabrication of a fully elastic ship model. Proc. Int. Conf. Offshore Mech. Arctic Eng. OMAE 6B-2020, 1–10 (2020). https://doi.org/10.1115/omae2020-18633

Corigliano, P., Crupi, V., Guglielmino, E., Mariano Sili, A.: Full-field analysis of Al/Fe explosive welded joints for shipbuilding applications. Mar. Struct. 57, 207–218 (2018). https://www.sciencedirect.com/science/article/pii/S0951833917302290, https://doi.org/10.1016/j.marstruc.2017.10.004

Coutinho, C.P., Baptista, A.J., Dias Rodrigues, J.: Reduced scale models based on similitude theory: a review up to 2015. Eng. Struct. 119, 81–94 (2016). https://www.sciencedirect.com/science/article/pii/S0141029616301274, https://doi.org/10.1016/j.engstruct.2016.04.016

Davis, M.R., French, B.J., Thomas, G.A.: Wave slam on wave piercing catamarans in random head seas. Ocean Eng. **135**, 84–97 (2017). https://doi.org/10.1016/j.oceaneng.2017.03.007

Dessi, D., Mariani, R., Coppotelli, G.: Experimental investigation of the bending vibrations of a fast vessel. Aust. J. Mech. Eng. 4(2), 125 (2007). http://search.informit.com.au/documentSummary;dn=776398053134181;res=IELENG, https://doi.org/10.1080/14484846.2007.11464521

de Vicente, M., Silva-Campillo, A., Herreros, M.Á., Suárez-Bermejo, J.C.: Design of a structurally-welded/adhesively bonded joint between a fiber metal laminate and a steel plate for marine applications. J. Mar. Sci. Technol. **27**(2), 1002–1014 (2022). https://doi.org/10.1007/s00773-022-00885-7

Ding, J., Geng, Y.C., Xu, S.W., Yang, W.N., Xie, Z.Y.: Experimental study on responses of an 8-module VLFS considering different encounter wave conditions (2021). https://doi.org/10.1016/j.marstruc.2021.102959

Dodds, N., Gibson, A., Dewhurst, D., Davies, J.: Fire behaviour of composite laminates. Compos. A Appl. Sci. Manuf. **31**(7), 689–702 (2000)

Ehlers, S., et al.: Committee vol 6 - arctic technology. In: Proceedings of the 19th International Ship and Offshore Structures Congress, vol. 2 (2015)

Ehlers S., et al.: Report of the ISSC committee vol 2 experimental methods. In: Wang, X., Pegg, N. (eds.) Proceedings of the 21st International Ship and Offshore Structures Congress, ISSC 2022 (2022). https://doi.org/10.5957/ISSC-2022-COMMITTEE-V-2

Evegren, F., Hertzberg, T., Rahm, M.: Fire tests of FRP composite ship structures (2016)

Fan, M., Naughton, A., Bregulla, J.: Fire performance of natural fibre composites in construction. In: Fan, M., Fu, F. (eds.) Advanced High Strength Natural Fibre Composites in Construction, pp. 375–404. Woodhead Publishing (2017). https://www.sciencedirect.com/science/article/pii/B9780081004111000145, https://doi.org/10.1016/B978-0-08-100411-1.00014-5

Fox, C.: A scaling law for the flexural motion of floating ice. In: Dempsey, J.P., Shen, H.H. (eds.) IUTAM Symposium on Scaling Laws in Ice Mechanics and Ice Dynamics, pp. 135–148. Springer, Dordrecht (2001). https://doi.org/10.1007/978-94-015-9735-7_12

Fujikubo, M., et al.: A digital twin for ship structures-R&D project in Japan. Data-Centric Eng. 5 (2024) https://doi.org/10.1017/dce.2024.3

Gagnon, R.: Analysis of data from bergy bit impacts using a novel hull-mounted external impact panel. Cold Reg. Sci. Technol. 52(1), 50–66 (2008). https://www.sciencedirect.com/science/article/pii/S0165232X07000924 (Bergy Bits.), https://doi.org/10.1016/j.coldregions.2007.04.018

Gagnon, R., Andrade, S., Quinton, B., Daley, C., Colbourne, B.: Pressure distribution data from large double-pendulum ice impact tests. Cold Reg. Sci. Technol. 175, 103033 (2020). https://www.sciencedirect.com/science/article/pii/S0165232X19304793, https://doi.org/10.1016/j.coldregions.2020.103033

Gagnon, R., Bugden, A., Mackay, J.: A relatively thin prototype of NRC's pressure-sensing technology. In: Proceedings of the 27th IAHR International Symposium on Ice (2024)

Gargano, A., Das, R., Mouritz, A.: Comparative experimental study into the explosive blast response of sandwich structures used in naval ships. Compos. Commun. 30, 101072 (2022). https://www.sciencedirect.com/science/article/pii/S2452213922000171, https://doi.org/10.1016/j.coco.2022.101072

Ghasemzadeh, M., Mokhtari, M., Kefal, A.: Localized corrosion damage prediction of steel plates in marine applications using quadrilateral inverse-shell elements based on iFEM. In: Sustainable Development and Innovations in Marine Technologies, pp. 123–129. CRC Press (2022). https://doi.org/10.1201/9781003358961-17

Ghoneim, G., Keinonen, A.: Full-scale impact tests of Canmar Kigoriak in thick ice. In: Proceedings of the 7th International Conference on Port and Ocean Engineering Under Arctic Conditions, vol. 3, pp. 329–346 (1983)

Gibson, A., Wu, Y., Chandler, H., Wilcox, J., Bettess, P.: Model for the thermal performance of thick composite laminates in hydrocarbon fires. Revue de l'Institute Francais du Petrole **50**(1), 69–74 (1995)

Glen, I., Blount, H.: Measurement of ice impact pressures and loads onboard CCGS Louis S. St. Laurent. In: Proceedings of the 3rd Conference on Offshore Mechanics and Arctic Engineering, vol. 3, pp. 246–252 (1984)

Grammatikopoulos, A.: The effects of geometric detail on the vibratory responses of complex ship-like thin-walled structures. Mar. Struct. 78 (2021)

Grammatikopoulos, A.: A review of physical flexible ship models used for hydroelastic experiments. Mar. Struct. 90 (2023). http://creativecommons.org/licenses/by/4.0/, https://doi.org/10.1016/j.marstruc.2023.103436

Grammatikopoulos, A., Banks, J., Temarel, P.: Prediction of the vibratory properties of ship models with realistic structural configurations produced using additive manufacturing. Mar. Struct. 73 (2020)

Grammatikopoulos, A., Banks, J., Temarel, P.: The design and commissioning of a fully elastic model of a uniform container ship. Mar. Struct. 78 (2021)

Gravit, M., Shabunina, D.: Numerical and experimental analysis of fire resistance for steel structures of ships and offshore platforms. Fire 5(1) (2022)

Grządziela, A.: Ship impact modeling of underwater explosion. J. KONES **18**, 145–152 (2011)

Ha, Y.-J., Ahn, H., Park, S., Park, J.-Y., Kim, K.-H.: Development of hybrid model test technique for performance evaluation of a 10 MW class floating offshore wind turbine considering asymmetrical thrust. Ocean Eng. 272, 113783 (2023). https://www.sciencedirect.com/science/article/pii/S0029801823001671, https://doi.org/10.1016/j.oceaneng.2023.113783

Hamada, R., et al.: Structural health monitoring of CFRP propellers by piezoelectric line sensors. In: Proceedings of the 14th International Workshop on Structural Health Monitoring, IWSHM 2023, Stanford, CA, USA (2023)

Hammer, T.C., Hendrikse, H.: Experimental study into the effect of wind-ice misalignment on the development of ice-induced vibrations of offshore wind turbines. Eng. Struct. 286, 116106 (2023). https://www.sciencedirect.com/science/article/pii/S0141029623005205, https://doi.org/10.1016/j.engstruct.2023.116106

Hammer, T.C., Hendrikse, H.: Hardware-in-the-loop experiments in model ice for analysis of ice-induced vibrations of offshore structures. Sci. Rep. 14, 18327 (2024). https://www.nature.com/articles/s41598-024-68955-x, https://doi.org/10.1038/s41598-024-68955-x

Hammer, T.C., Puolakka, O., Hendrikse, H.: Scaling ice-induced vibrations by combining replica modeling and preservation of kinematics. Cold Reg. Sci. Technol. 220, 104127 (2024). https://www.sciencedirect.com/science/article/pii/S0165232X24000089, https://doi.org/10.1016/j.coldregions.2024.104127

Hammer, T.C., Willems, T., Hendrikse, H.: Dynamic ice loads for offshore wind support structure design. Mar. Struct. 87, 103335 (2023). https://www.sciencedirect.com/science/article/pii/S095183392200171X, https://doi.org/10.1016/j.marstruc.2022.103335

Hasan, A., Asfihani, T., Osen, O., Bye, R.T.: Leveraging digital twins for fault diagnosis in autonomous ships. Ocean Eng. 292, 116546 (2024). https://www.sciencedirect.com/science/article/pii/S002980182302930X, https://doi.org/10.1016/j.oceaneng.2023.116546

Hendrikse, H., et al.: Ice basin tests for ice-induced vibrations of offshore structures in the shiver project. In: Polar and Arctic Sciences and Technology, vol. 6, p. V006T07A009 (2022). https://doi.org/10.1115/OMAE2022-78507

Hisette, Q., Myland, D.: Investigations on hull-ice frictional effects. In: 41st International Conference on Ocean, Offshore and Arctic Engineering, ASME 2022 (2022)

Hoffmann, L.: Impact forces and friction coefficients on the forebody of the German polar research vessel Polarstern. In: Proceedings of the 8th International Conference on Port and Ocean Engineering Under Arctic Conditions, vol. 3, pp. 1189–1202 (1985)

Hu, Z., Hua, L., Liu, W., Wang, X., Wu, L.: Fatigue life assessment of marine structures accounting for multiple pit corrosion interactions: coupling effects and crack initiation detection. Ocean Eng. 333, 121581 (2025). https://www.sciencedirect.com/science/article/pii/S0029801825012879, https://doi.org/10.1016/j.oceaneng.2025.121581

Hänninen, S., Lensu, M., Riska, K.: Analysis of the ice load measurements during USCGC Healy ice trials, Spring 2000 (2001)

Høyland, K.: Thermal aspects of model basin ridges. In: Proceedings of the 20th IAHR International Symposium on Ice (2010)

IACS: Requirements concerning strength of ships-UR S35 buckling strength assessment of structural elements, Technical report. International Association of Classification Societies (2024). https://iacs.org.uk/resolutions/unified-requirements/ur-s

Jang, W.S., Hong, S.Y., Song, J.H.: Structural health monitoring system development for ship rudders. In: Proceedings of IEEE 18th International Conference on Advanced Motion Control, AMC 2024, Kyoto, Japan, pp. 1–5 (2024). https://doi.org/10.1109/AMC58169.2024.10505446

Jenssen, Y., Sauder, T., Thys, M.: Performance of a cable-driven robot used for cyber-physical testing of floating wind turbines (2024)

Jiao, J., Ren, H., Guedes Soares, C.: A review of large-scale model at-sea measurements for ship hydrodynamics and structural loads. Ocean Eng. 227 (2021). https://doi.org/10.1016/J.OCEANENG.2021.108863

Jiao, P., Ye, X., Zhang, C., Li, W., Wang, H.: Vision-based real-time marine and offshore structural health monitoring system using underwater robots. Comput. Aided Civ. Infrastruct. Eng. **39**(2), 281–299 (2024). https://doi.org/10.1111/mice.12993

Jones, D., Brischke, C.: Performance of Bio-based Building Materials. Woodhead Publishing (2017)

Katsoudas, A.S., Silionis, N.E., Anyfantis, K.N.: Structural health monitoring for corrosion induced thickness loss in marine plates subjected to random loads. Ocean Eng. 273, 114037 (2023). https://www.sciencedirect.com/science/article/pii/S0029801823004213, https://doi.org/10.1016/j.oceaneng.2023.114037

Katsoudas, A.S., Silionis, N.E., Anyfantis, K.N.: Structural health monitoring for corrosion induced thickness loss in marine plates subjected to random loads. Ocean Eng. **273**, 114037 (2023). https://doi.org/10.1016/j.oceaneng.2023.114037

Keser, A., Verdult, M., Seyffert, H., Grammatikopoulos, A.: The design, production, verification, and calibration of an elastic model of a catamaran for hydroelastic experiments. In: 13th Symposium on High Speed Marine Vehicles, HSMV 2023, pp. 139–148. IOS Press (2023). https://research.tudelft.nl/en/publications/the-design-production-verification-and-calibration-of-an-elastic-, https://doi.org/10.3233/PMST230019

Kim, C., Oterkus, S., Oterkus, E., Kim, Y.: Probabilistic ship corrosion wastage model with Bayesian inference. Ocean Eng. 246, 110571 (2022). https://www.sciencedirect.com/science/article/pii/S0029801822000452, https://doi.org/10.1016/j.oceaneng.2022.110571

Komoriyama, Y., Houtani, H., Takami, T., Matsui, S., Fujimoto, W.: Tank test of flexible acrylic ship model in waves for extreme response prediction. In: Proceedings of the ASME 2024 43rd International Conference on Ocean, Offshore and Arctic Engineering, pp. 1–9 (2024)

Kong, S., Cui, H., Wu, G., Ji, S.: Full-scale identification of ice load on ship hull by least square support vector machine method. Appl. Ocean Res. 106, 102439 (2021). https://www.sciencedirect.com/science/article/pii/S0141118720309986, https://doi.org/10.1016/j.apor.2020.102439

Korri, P., Varsta, P.: On the ice trial of a 14500 DWT tanker on the Gulf of Bothnia. In: Proceedings of the 24th Norwegian Ship Technology Conference (1979)

Kotisalo, K., Kujala, P.: Ice load measurements onboard MT Uikku during the ARCDEV voyage. In: Tuhkuri, J., Riska, K. (eds.) The 15th International Conference on Port and Ocean Engineering Under Arctic Conditions, POAC'99, Helsinki, 23–27 August 1999, pp. 974–987 (1999)

Kotsampopoulos, P., et al.: A benchmark system for hardware-in-the-loop testing of distributed energy resources. IEEE Power Energy Technol. Syst. J. **5**(3), 94–103 (2018). https://doi.org/10.1109/JPETS.2018.2861559

Kubiczek, J.M., Herrnring, H., Ehlers, S.: Raw data of strain measurements on the ship structure of RV Polarstern during the MOSAiC expedition [dataset]. PANGAEA (2022). https://doi.org/10.1594/PANGAEA.942950

Kujala, P.: Results and statistical analysis of ice load measurements on board icebreaker Sisu in winters 1979 to 1985 (1989)

Kujala, P.: Results of long-term ice load measurements on board chemical tanker Kemira in the Baltic sea during the winters 1985 to 1988 (1989)

Kujala, P., Tuhkuri, J., Varsta, P.: Ship-ice contact and design loads (1994)

Kuroiwa, T.: Study on damage of ship bottom structures due to grounding. In: First Joint Conference on Marine Safety and Environment/Ship Production (1992)

Kyaw Oo D'Amore, G., Marinò, A., Kašpar, J.: Numerical modeling of fire resistance test as a tool to design lightweight marine fire doors: a preliminary study. J. Marine Sci. Eng. 8(7), 520 (2020)

Lau, M.: Model-scale/full-scale correlation of NRC-OCRE's model resistance, propulsion and maneuvering test results. In: Ian Jordaan Honoring Symposium on Ice Engineering, vol. 8, p. V008T07A016 (2015). https://doi.org/10.1115/OMAE2015-42114

Leira, B., Børsheim, L., Espeland, Ø., Amdahl, J.: Ice-load estimation for a ship hull based on continuous response monitoring. Proc. Inst. Mech. Eng. Part M J. Eng. Marit. Environ. 223(4), 529–540 (2009)

Leroy, V., Delacroix, S., Merrien, A., Bachynski-Polić, E.E., Gilloteaux, J.C.: Experimental investigation of the hydro-elastic response of a spar-type floating offshore wind turbine. Ocean Eng. (2022). https://doi.org/10.1016/j.oceaneng.2022.111430

Li, F., Yue, Q., Shkhinek, K., Kärnä, T.: A qualitative analysis of breaking length of sheet ice against conical structures. In: Proceedings of the 17th International Conference on Port and Ocean Engineering Under Arctic Conditions, vol. 1, pp. 293–304 (2003)

Li, G., Li, W., Han, F., Lin, S., Zhou, X.: Experimental investigation and one-way coupled fluid-structure interaction analysis of gas-liquid two-phase flow conveyed by a subsea rigid m-shaped jumper. Ocean Eng. **285**, 115292 (2023)

Li, W., Zhou, Q., Yin, G., Ong, M.C., Li, G., Han, F.: Experimental investigation and numerical modeling of two-phase flow development and flow-induced vibration of a multi-plane subsea jumper. J. Mar. Sci. Eng. **10**(10), 1334 (2022)

Lin, B., Dong, X.: Ship hull inspection: a survey. Ocean Eng. 289, 116281 (2023). https://www.sciencedirect.com/science/article/pii/S0029801823026653, https://doi.org/10.1016/j.oceaneng.2023.116281

Liu, X., Jiang, D., Liufu, K., Fu, J., Liu, Q., Li, Q.: Numerical investigation into impact responses of an offshore wind turbine jacket foundation subjected to ship collision. Ocean Eng. **248**, 110825 (2022)

Liu, Y., Ren, H.: Rapid acquisition method for structural strength evaluation stresses of the ship digital twin model. Ocean Eng. 285, 115323 (2023). https://www.sciencedirect.com/science/article/pii/S0029801823017079, https://doi.org/10.1016/j.oceaneng.2023.115323

Liu, Y.H., Suominen, M., Kujala, P.: Research of ice-induced load on a ship hull based on an inverse method. J. Ship Mech. **20**(12), 1604–1618 (2016)

Lu, Z., Fu, S., Zhang, M., Ren, H.: An efficient time-domain prediction model for vortex-induced vibration of flexible risers under unsteady flows. Mar. Struct. **64**, 492–519 (2019)

Lu, Z., Fu, S., Zhang, M., Ren, H., Song, L.: A modal space based direct method for vortex-induced vibration prediction of flexible risers. Ocean Eng. **152**, 191–202 (2018)

Luo, F., et al.: Investigation of the 3D-DIC testing method for composite shell in a deep-water high-pressure environment. Thin-Walled Struct. 190, 110962 (2023). https://www.sciencedirect.com/science/article/pii/S0263823123004408, https://doi.org/10.1016/j.tws.2023.110962

Lv, Z., Lv, H., Fridenfalk, M.: Digital twins in the marine industry. Electronics 12(9) (2023). https://www.mdpi.com/2079-9292/12/9/2025, https://doi.org/10.3390/electronics12092025

Madhusudhana, S., Klinck, H., Seger, K.D., Verlinden, C.M.A., Heaney, K.D.: A passive acoustic approach to impact assessment of large underwater explosions on marine fauna-planning, analyses, and lessons learned. In: Oceans 2023, Limerick, pp. 1–7 (2023). https://doi.org/10.1109/OCEANSLimerick52467.2023.10244589

Maetz, T., et al.: Microwave structural health monitoring of the grouted connection of a monopile-based offshore wind turbine: fatigue testing using a scaled laboratory demonstrator. Struct. Control. Health Monit. **2023**, 1–18 (2023). https://doi.org/10.1155/2023/1981892

Man, M., Weil, M.: Remote monitoring method for human body electrostatic potential based on symbolic regression machine learning. Measure. Sci. Technol. 34 (2023). https://doi.org/10.1088/1361-6501/acc3b6

Mannacio, F., Barbato, A., Marzo, F.D., Gaiotti, M., Rizzo, C., Venturini, M.: Shock effects of underwater explosion on naval ship foundations: validation of numerical models by dedicated tests. Ocean Eng. 253, 111290 (2022). https://www.sciencedirect.com/science/article/pii/S0029801822006849, https://doi.org/10.1016/j.oceaneng.2022.111290

Mannacio, F., Di Marzo, F., Gaiotti, M., Guzzo, M., Rizzo, C., Venturini, M.: Characterization of underwater shock transient effects on naval e-glass biaxial fiberglass laminates: an experimental and numerical method. Appl. Ocean Res. 128, 103356 (2022). https://www.sciencedirect.com/science/article/pii/S0141118722002875, https://doi.org/10.1016/j.apor.2022.103356

Mannacio, F., Di Marzo, F., Gaiotti, M., Rizzo, C., Venturini, M.: An open benchmark to assess the effects of underwater explosions on steel panels using the volume of fluid approach. Ocean Eng. 288, 116093 (2023). https://www.sciencedirect.com/science/article/pii/S0029801823024770, https://doi.org/10.1016/j.oceaneng.2023.116093

Marón, A., Kapsenberg, G.K.: Design of a ship model for hydro-elastic experiments in waves. Int. J. Nav. Archit. Ocean Eng. 6(4), 1130–1147 (2014). http://www.degruyter.com/view/j/ijnaoe.2014.6.issue-4/ijnaoe-2013-0235/ijnaoe-2013-0235.xml, https://doi.org/10.2478/ijnaoe-2013-0235

Matala, R.: Investigation of model-scale brash ice properties. Ocean Eng. 225, 108539 (2021). https://www.sciencedirect.com/science/article/pii/S0029801820314475, https://doi.org/10.1016/j.oceaneng.2020.108539

Matala, R., Suominen, M.: Investigation of vessel resistance in model scale brash ice channels and comparison to full scale tests. Cold Reg. Sci. Technol. 201, 103617 (2022). https://www.sciencedirect.com/science/article/pii/S0165232X22001367, https://doi.org/10.1016/j.coldregions.2022.103617

Matala, R., Suominen, M.: Scaling principles for model testing in old brash ice channel. Cold Reg. Sci. Technol. 210, 103857 (2023). https://www.sciencedirect.com/science/article/pii/S0165232X23000873, https://doi.org/10.1016/j.coldregions.2023.103857

Mauro, F., Kana, A.: Digital twin for ship life-cycle: a critical systematic review. Ocean Eng. 269, 113479 (2023). https://www.sciencedirect.com/science/article/pii/S0029801822027627, https://doi.org/10.1016/j.oceaneng.2022.113479

Mazzariol, L.M., Alves, M.: Experimental verification of similarity laws for impacted structures made of different materials. Int. J. Impact Eng. 133, 103364 (2019). https://www.sciencedirect.com/science/article/pii/S0734743X19304129, https://doi.org/10.1016/j.ijimpeng.2019.103364

Mazzariol, L.M., Alves, M.: Similarity laws of structures under impact load: geometric and material distortion. Int. J. Mech. Sci. 157–158, 633–647 (2019). https://www.sciencedirect.com/science/article/pii/S0020740319304138, https://doi.org/10.1016/j.ijmecsci.2019.05.011

Min, R., Liu, Z., Pereira, L., Yang, C., Sui, Q., Marques, C.: Optical fiber sensing for marine environment and marine structural health monitoring: a review. Opt. Laser Technol. **140**, 107082 (2021)

Miyashita, T., et al.: Deformation estimation of container ship in waves by inverse finite element method. In: Proceedings of 15th International Symposium on Practical Design of Ships and Other Floating Structures, PRADS 2022, Dubrovnik, Croatia (2022)

Mukhlas, M., Kristiansen, T., Lader, P., Kristiansen, D.: Experimental and numerical investigation of a vertical pre-tensioned membrane sheet in regular waves. In: 9th International Conference on Hydroelasticity in Marine Technology, pp. 7823–7830 (2022)

Naughton, A., Fan, M., Bregulla, J.: Fire resistance characterisation of hemp fibre reinforced polyester composites for use in the construction industry. Compos. B Eng. **60**, 546–554 (2014)

Nugroho, F.A., Braun, M., Ehlers, S.: Probability analysis of pit distribution on corroded ballast tank. Ocean Eng. 228, 108958 (2021). https://www.sciencedirect.com/science/article/pii/S0029801821003930, https://doi.org/10.1016/j.oceaneng.2021.108958

Oka, M., Komoriyama, Y., Ma, C.: Accuracy verification of the 2D spectral AIS method by the hull monitoring data. Mar. Struct. 99, 103704 (2025). https://www.sciencedirect.com/science/article/pii/S0951833924001321, https://doi.org/10.1016/j.marstruc.2024.103704

International Maritime Organization, International Maritime Organization Maritime Safety Committee: FTP code: international code for application of fire test procedures. International Maritime Organization (2010)

Otto, W., Bunnik, T., Kaydihan, L.: Hydro-elastic behaviour of an inflatable mattress in waves. In: 9th International Conference on Hydroelasticity in Marine Technology, pp. 289–300 (2022)

Paik, J.K., et al.: Full-scale fire testing to collapse of steel stiffened plate structures under lateral patch loading (part 1) - without passive fire protection. Ships Offshore Struct. **16**(3), 227–242 (2021). https://doi.org/10.1080/17445302.2020.1764705

Paik, J.K., et al.: Full-scale fire testing to collapse of steel stiffened plate structures under lateral patch loading (part 2) - with passive fire protection. Ships Offshore Struct. **16**(3), 243–254 (2021). https://doi.org/10.1080/17445302.2020.1764706

Palmer, A., Dempsey, J.: Model tests in ice. In: Proceedings of the 20th International Conference on Port and Ocean Engineering Under Arctic Conditions, p. 10 (2009)

Palmer, A., Qianjin, Y., Fengwei, G.: Ice-induced vibrations and scaling. Cold Reg. Sci. Technol. 60(3), 189–192 (2010). https://www.sciencedirect.com/science/article/pii/S0165232X09002080, https://doi.org/10.1016/j.coldregions.2009.11.005

Palmer, A.C., Martin, J.: Buckle propagation in submarine pipelines. Nature **254**(5495), 46–48 (1975)

Pang, T.Y., et al.: Developing a digital twin and digital thread framework for an 'Industry 4.0' shipyard. Appl. Sci. 11(3), 1097 (2021)

Park, S.M., Kim, H.J., Cho, H.R., Kong, K.H., Park, D.K., Paik, J.K.: Effect of pneumatic rubber fenders on the prevention of structural damage during collisions between a ship-shaped offshore installation and a shuttle tanker working side-by-side. Ships Offshore Struct. **18**(4), 596–608 (2023)

Pickles, D., Hunt, G., Elliott, A., Cammarano, A., Falcone, G.: An experimental investigation into the effect two-phase flow induced vibrations have on a j-shaped flexible pipe. J. Fluids Struct. **125**, 104057 (2024)

Qi, X., Li, Z., Zhao, C., Zhang, Q., Zhou, Y.: Environmental impacts of arctic shipping activities: a review. Ocean Coast. Manage. 247, 106936 (2024). https://www.sciencedirect.com/science/article/pii/S0964569123004611, https://doi.org/10.1016/j.ocecoaman.2023.106936

Qin, X., Zhang, M., Fu, S., Li, H., Hou, J., Xu, Y.: Review on researches and main influencing factors on mechanical properties of offshore wind power cables. J. Ocean Eng. Sci. (2024). https://www.sciencedirect.com/science/article/pii/S2468013324000329, https://doi.org/10.1016/j.joes.2024.06.001

Qvale, P., Zarandi, E.P., Ås, S.K., Skallerud, B.H.: Digital image correlation for continuous mapping of fatigue crack initiation sites on corroded surface from offshore mooring chain. Int. J. Fatigue 151, 106350 (2021). https://www.

sciencedirect.com/science/article/pii/S0142112321002103, https://doi.org/10.1016/j.ijfatigue.2021.106350

Ralph, F., Ritch, R., Daley, C., Browne, R.: Use of finite element methods to determine iceberg impact pressure based on internal strain gauge measurements. In: Proceedings of the 17th International Conference on Port and Ocean Engineering Under Arctic Conditions, June 2003 (2003)

Ran, X., Leroy, V., Bachynski-Polić, E.E.: Hydroelastic response of a flexible spar floating wind turbine: numerical modelling and validation (2023). http://creativecommons.org/licenses/by/4.0/, https://doi.org/10.1016/j.oceaneng.2023.115635

Ren, H., Fu, S., Zhang, M., Xu, Y., Ren, H.: Developing a virtual physical system for vortex-induced vibration studies of a bluff body. J. Ocean Eng, Sci (2024)

Ren, L., Li, H.-N., Zhou, J., Li, D.-S., Sun, L.: Health monitoring system for offshore platform with fiber Bragg grating sensors. Opt. Eng. **45**(8), 084401–084401 (2006)

Reu, P.L., et al.: DIC challenge: developing images and guidelines for evaluating accuracy and resolution of 2D analyses. Exp. Mech. **58**(7), 1067–1099 (2017). https://doi.org/10.1007/s11340-017-0349-0

Riccioli, F., Huijer, A., Grasso, N., Rizzo, C.M., Pahlavan, L.: Development of a retrofit layer with an embedded array of piezoelectric sensors for transient pressure measurement in maritime applications. Mar. Struct. 89, 103395 (2023). https://www.sciencedirect.com/science/article/pii/S095183392300028X, https://doi.org/10.1016/j.marstruc.2023.103395

Riska, K.: Ice trials of 8000/12000 tdw ro-ro vessel interim report, data gathering system (WorkingPaper No. LAI-331A/82). Technical Research Centre of Finland (1982)

Riska, K., Jalonen, R., Veitch, B., Nortala-Hoikkanen, A., Wilkman, G.: Assessment of ice model testing techniques. In: Proceedings of the 5th International Conference on Ships and Marine Structures in Cold Regions, pp. F1–F22 (1994)

Riska, K., Kujala, P., Vuorio, J.: Ice load and pressure measurements. On Board I.B. Sisu. In: Proceedings of the 7th International Conference on Port and Ocean Engineering Under Arctic Conditions, vol. 2, pp. 1055–1069 (1983)

Riska, K., Rantala, H., Joensuu, A.: Full scale observations on ship-ice contact results from tests series. Onboard IB Sampo, Winter 1989 (WorkingPaper No. M-97). Helsinki University of Technology (1990)

Ritch, R.: First year hull-ice interaction loads measured on the Louis S. St-Laurent during the 1995 Gulf of St. Lawrence trials, Day 2, Monday, 21 July 2008 (2008). https://doi.org/10.5957/ICETECH-2008-122

Ritch, R., Frederking, R., Johnston, M., Browne, R., Ralph, F.: Local ice pressures measured on a strain gauge panel during the CCGS terry fox bergy bit impact study. Cold Reg. Sci. Technol. 52(1), 29–49 (2008). https://www.sciencedirect.com/science/article/pii/S0165232X07000912 (Bergy Bits.), https://doi.org/10.1016/j.coldregions.2007.04.017

Robertson, A.N., et al.: OC5 project phase II: validation of global loads of the DeepCwind floating semisubmersible wind turbine. Energy Procedia **137**, 38–57 (2017). https://doi.org/10.1016/J.EGYPRO.2017.10.333

Saccone, C., Pahlavan, L.: Influence of stiffeners on acoustic emission monitoring of ship structures. In: Proceedings of the 11th European Workshop on Structural Health Monitoring, EWSHM 2024, Potsdam, Germany (2024). e-Journal of Nondestructive Testing

Safikou, E., Bollas, G.M.: Symbolic regression for fault prognosis and remaining useful life estimation. In: 2023 American Control Conference (ACC), pp. 4715–4720 (2023). https://doi.org/10.23919/ACC55779.2023.10156572

Savio, L.: Measurements of the deflection of a flexible propeller blade by means of stereo imaging. In: Fourth International Symposium on Marine Propulsors, SMP'15, Austin, Texas, USA, June 2015 (2015)

Savio, L., Jenssen, Y., Henry, P., Franzosi, G.: Experimental investigation on the flow and deformation fields of an elastic propeller. In: Eighth International Symposium on Marine Propulsors, SMP'24, Berlin, Germany, March 2024 (2024). https://doi.org/10.15480/882.9328

Schreier, S., Jacobi, G.: Experimental investigation of wave interaction with a thin floating sheet. Int. J. Offshore Polar Eng. **31**(04), 435–444 (2021)

Seo, J.K., Lee, S.E., Park, J.S.: A method for determining fire accidental loads and its application to thermal response analysis for optimal design of offshore thin-walled structures. Fire Saf. J. **92**, 107–121 (2017)

Shi, G.-J., Ji, Y.-H., Xu, J.-B., Wang, D.-Y., Xu, Z.-T.: Experimental study of structural failure and ultimate strength of GFRP girder with hat stiffeners and foams under bending load. Mar. Struct. 96, 103607 (2024). https://www.sciencedirect.com/science/article/pii/S0951833924000352, https://doi.org/10.1016/j.marstruc.2024.103607

Shi, W., et al.: Real-time hybrid model tests of floating offshore wind turbines: status, challenges, and future trends. Appl. Ocean Res. 141, 103796 (2023). https://www.sciencedirect.com/science/article/pii/S0141118723003371, https://doi.org/10.1016/j.apor.2023.103796

Shojai, S., et al.: Assessment of corrosion fatigue in welded joints using 3D surface scans, digital image correlation, hardness measurements, and residual stress analysis. Int. J. Fatigue 176, 107866 (2023). https://www.sciencedirect.com/science/article/pii/S0142112323003675, https://doi.org/10.1016/j.ijfatigue.2023.107866

Silionis, N.E., Anyfantis, K.N.: Optimal sensor placement for corrosion induced thickness loss monitoring in ship structures. Mar. Struct. 93, 103524 (2024). https://www.sciencedirect.com/science/article/pii/S0951833923001570, https://doi.org/10.1016/j.marstruc.2023.103524

St. John, J., Daley, C., Blount, H.: Ice loads and ship response to ice (1984)

St. John, J., Minnick, P.: Ice load impact study on the national science foundation's research vessel Nathaniel B. Palmer (1995)

St. John, J., Sheinberg, R., Ritch, R., Minnick, P.: Ice impact load measurements abroad the oden during the International Arctic Ocean Expedition 1991 (1994)

Su, Z., Pan, J., Zhang, S., Wu, S., Yu, Q., Zhang, D.: Characterizing dynamic deformation of marine propeller blades with stroboscopic stereo digital image correlation. Mech. Syst. Sig. Process. 162, 108072 (2022). https://www.sciencedirect.com/science/article/pii/S0888327021004593, https://doi.org/10.1016/j.ymssp.2021.108072

Sukma, R.A., Handani, D.W., Nugroho, T.F., Tyasayumranani, W.: Risk assessment of ship collision on FSO Abherka and oil spill modelling due to structural damage. IOP Conf. Ser. Earth Environ. Sci. **1081**, 012029 (2022)

Suominen, M.: Uncertainty and variation in measured ice-induced loads on a ship hull (Unpublished doctoral dissertation). Aalto University, Finland (2018)

Suominen, M., et al.: Full-scale measurements on board PSRV S.A. Agulhas II in the Baltic Sea. In: Port and Ocean Engineering Under Arctic Conditions, Espoo, Finland, 9–13 June 2013 (VK: T20404) (2013)

Suominen, M., Kujala, P., Romanoff, J., Remes, H.: The effect of the extension of the instrumentation on the measured ice-induced load on a ship hull. Ocean Eng. 144, 327–339 (2017). https://www.sciencedirect.com/science/article/pii/S0029801817305759, https://doi.org/10.1016/j.oceaneng.2017.09.056

Suominen, M., Romanoff, J., Remes, H., Kujala, P.: The determination of ice-induced loads on the ship hull from shear strain measurements. In: Analysis and Design of Marine Structures V, pp. 375–383 (2015)

Takimoto, T., Uto, S., Oka, S., Murakami, C., Izumiyama, K.: Measurement of ice load exerted on the hull of icebreaker Soya in the Southern Sea of Okhotsk (2006)

Tan, C.H., Mahamd Adikan, F.R., Shee, Y.G., Yap, B.K.: Non-destructive fiber Bragg grating based sensing system: early corrosion detection for structural health monitoring. Sensors Actuators A Phys. 268, 61–67 (2017). https://www.sciencedirect.com/science/article/pii/S0924424717310385, https://doi.org/10.1016/j.sna.2017.10.048

Thomas, G.A., Davis, M.R., Holloway, D.S., Roberts, T.: The whipping vibration of large high speed catamarans. Int. J. Marit. Eng. (2003). http://eprints.utas.edu.au/2293/1/Gilesthomas.pdf

Tödter, S., el Sheshtawy, H., Neugebauer, J., el Moctar, O., Schellin, T.E.: Deformation measurement of a monopile subject to vortex- induced vibration using digital image correlation. Ocean Eng. 221, 108548 (2021). https://www.sciencedirect.com/science/article/pii/S0029801820314566, https://doi.org/10.1016/j.oceaneng.2020.108548

Ulbricht, N., Han, N., Porfiri, M.: On the role of curvature in the response of air-backed composites to hydrodynamic loading: an experimental study. Compos. Struct. 344, 118328 (2024). https://www.sciencedirect.com/science/article/pii/S0263822324004562, https://doi.org/10.1016/j.compstruct.2024.118328

Uto, S., Oka, S., Murakami, C.: Ice load exerted on the hull of icebreaker PM Teshio in the South Sea of Okhotsk, vol. 2 (2006)

Vieira, M., Henriques, E., Snyder, B., Reis, L.: Insights on the impact of structural health monitoring systems on the operation and maintenance of offshore wind support structures. Struct. Saf. 94, 102154 (2022). https://www.sciencedirect.com/science/article/pii/S0167473021000771, https://doi.org/10.1016/j.strusafe.2021.102154

Vieira, M., Snyder, B., Henriques, E., White, C., Reis, L.: Economic viability of implementing structural health monitoring systems on the support structures of bottom fixed offshore wind. Energies 16(13) (2023). https://doi.org/10.3390/en16134885

Viswanathan, V.K., et al.: Hybrid optimization and modelling of CI engine performance and emission characteristics of novel hybrid biodiesel blends. Renew. Energy 198, 549–567 (2022). https://www.sciencedirect.com/science/article/pii/S0960148122011740, https://doi.org/10.1016/j.renene.2022.08.008

von Bock und Polach, R.U.F., Ettema, R., Gralher, S., Kellner, L., Stender, M.: The non-linear behavior of aqueous model ice in downward flexure. Cold Reg. Sci. Technol. 165, 102775 (2019). https://www.sciencedirect.com/science/article/pii/S0165232X1830555X, https://doi.org/10.1016/j.coldregions.2019.05.001

von Bock und Polach, F., Klein, M., Hartmann, M.: A new model ice for wave-ice interaction. Water 13(23) (2021). https://www.mdpi.com/2073-4441/13/23/3397, https://doi.org/10.3390/w13233397

von Bock und Polach, R.U.F., Molyneux, D.: Model ice: a review of its capacity and identification of knowledge gaps. In: Polar and Arctic Sciences and Technology; Petroleum Technology, vol. 8, p. V008T07A017 (2017). https://doi.org/10.1115/OMAE2017-61808

Vuorio, J., Riska, K., Varsta, P.: Long term measurements of ice pressure and ice-induced stresses on the icebreaker sisu in winter 1978 (1979)

Wang, F., et al.: Flexible guided anti-collision device for bridge pier protection against ship collision: numerical simulation and ship collision field test. Ocean Eng. **271**, 113696 (2023)

Wang, G.: A methodology for long-term offshore structural health monitoring using stand-alone GNSS: case study in the Gulf of Mexico. Struct. Health Monit. **23**(1), 463–478 (2024). https://doi.org/10.1177/14759217231169934

Wang, J., Chen, X., Sun, K., Ji, S.: Far-field identification of ice loads on ship structures by radial basis function neural network. Ocean Eng. 282, 115072 (2023). https://www.sciencedirect.com/science/article/pii/S0029801823014567, https://doi.org/10.1016/j.oceaneng.2023.115072

Wang, J., Li, Z.L., Yu, W.: Structural similitude for the geometric nonlinear buckling of stiffened orthotropic shallow spherical shells by energy approach. Thin-Walled Struct. 138, 430–457 (2019). https://www.sciencedirect.com/science/article/pii/S0263823117310066, https://doi.org/10.1016/j.tws.2018.02.006

Wei, Y., Chen, Z., Zhao, C., Tu, Y., Chen, X., Yang, R.: A BiLSTM hybrid model for ship roll multi-step forecasting based on decomposition and hyperparameter optimization. Ocean Eng. 242, 110138 (2021). https://www.sciencedirect.com/science/article/pii/S0029801821014591, https://doi.org/10.1016/j.oceaneng.2021.110138

Weil, M., Weijtjens, W., Devriendt, C.: Autoencoder and Mahalanobis distance for novelty detection in structural health monitoring data of an offshore wind turbine. J. Phys: Conf. Ser. **2265**(3), 032076 (2022). https://doi.org/10.1088/1742-6596/2265/3/032076

Woitzik, C., Braun, M., von Bock und Polach, F., Ehlers, S., Shojai, S., Schaumann, P.: Fatigue assessment of offshore wind turbine support structures subjected to seawater (2023)

Woloszyk, K., Garbatov, Y.: A probabilistic-driven framework for enhanced corrosion estimation of ship structural components. Reliab. Eng. Syst. Saf. 242, 109721 (2024). https://www.sciencedirect.com/science/article/pii/S095183202300635X, https://doi.org/10.1016/j.ress.2023.109721

Wu, G., Kong, S., Tang, W., Lei, R., Ji, S.: Statistical analysis of ice loads on ship hull measured during arctic navigations. Ocean Eng. **223**, 108642 (2021)

Wu, L., Maheshwari, M., Yang, Y., Xiao, W.: Selection and characterization of packaged FBG sensors for offshore applications. Sensors **18**(11), 3963 (2018)

Wu, M., Lehn, E., Moan, T.: Design of a segmented model for ship seakeeping tests with hydroelastic effects. In: Proceedings of the 6th International Conference on Hydroelasticity in Marine Technology, Tokyo, Japan, pp. 135–144 (2012)

Wu, Y.-S., Chen, R.-Z., Lin, J.-R.: Experimental technique of hydroelastic ship model. In: Proceedings of the Third International Conference on Hydroelasticity, Oxford, UK, September, pp. 15–17 (2003)

Yang, P., Feng, Q., Chen, H., Wen, L.: Combined backbone application on numerical simulations and a model experiment of a 20,000 TEU container ship. Ocean Eng. **223**, 108662 (2021)

Ye, H., Jiang, C., Zu, F., Li, S.: Design of a structural health monitoring system and performance evaluation for a jacket offshore platform in East China Sea. Appl. Sci. 12(23) (2022). https://doi.org/10.3390/app122312021

Yin, W., et al.: Initializing and accelerating stereo-DIC computation using semi-global matching with geometric constraints. Opt. Lasers Eng. 172, 107879 (2024). https://www.sciencedirect.com/science/article/pii/S0143816623004086, https://doi.org/10.1016/j.optlaseng.2023.107879

Yin, Y., Lu, Q., Wu, S., Yan, J., Yue, Q., Chen, J.: Experimental study on friction of steel wires of dynamic umbilical for fatigue life analysis. In: International Conference on Offshore Mechanics and Arctic Engineering, vol. 58806, p. V05AT04A031 (2019)

Yin, Y., Lu, Q., Wu, S., Yang, Z., Yan, J., Yue, Q.: Experimental study on the interlayer friction and wear mechanism between armor wires of umbilicals. Mar. Struct. 80, 103102 (2021). https://www.sciencedirect.com/science/article/pii/S0951833921001556, https://doi.org/10.1016/j.marstruc.2021.103102

Yu, Y., et al.: Buckling analysis of subsea pipeline with integral buckle arrestor using vector form intrinsic finite thin shell element. Thin Walled Struct. **164**, 107533 (2021)

Lin, Y., Kadivar, E., el Moctar, O., Neugebauer, J., Schellin, T.E.: Experimental investigation on the effect of fluid-structure interaction on unsteady cavitating flows around flexible and stiff hydrofoils. Phys. Fluids 34 (2022). https://doi.org/10.1063/5.0099776

Zan, Y., Guo, R., Yuan, L., Ma, Q., Zhou, A., Wu, Z.: Experimental study of a suspended subsea module at different positions in the splash zone. Mar. Struct. **77**, 102935 (2021)

Zhang, X., He, Z., Du, Z., Wang, J., Jiang, Y., Li, Y.: Multi-peak phenomenon of largescale hull structural damage under near-field underwater explosion. Ocean Eng. **283**, 114898 (2023)

Zhou, B., Wang, J., Cui, X., Hu, C., Liu, H., Wang, X.: Scouring development and self-burial of a subsea pipeline attached with porous/flexible spoilers. Ocean Eng. **295**, 116763 (2024)

Zhou, Y., Wang, X., Chen, K., Shi, L., Yuan, Y., Zhang, D.: An innovative testing method to qualify deep-water 66kv power and hydraulic combined umbilical. In: International Conference on Offshore Mechanics and Arctic Engineering, vol. 87806, p. V003T04A021 (2024)

Zhu, Z., Wu, P., Liu, Y., Wei, Y., Yin, Y.: A novel route-plan-guided artificial potential field method for ship collision avoidance: Modeling, integration and test. Ocean Eng. **288**, 116088 (2023)

Ziemer, G.: Ice-induced vibrations of vertical structures (Unpublished doctoral dissertation). Technische Universität Hamburg (2021)

Zima, B., Woloszyk, K., Garbatov, Y.: Corrosion degradation monitoring of ship stiffened plates using guidedwave phase velocity and constrained convex optimization method. Ocean Eng. 253, 111318 (2022). https://www.sciencedirect.com/science/article/pii/S0029801822007107, https://doi.org/10.1016/j.oceaneng.2022.111318

Zong, S., Liu, K., Qiu, W., Gao, Z., Wang, J.: Numerical and experimental analysis of fire resistance for bulkhead and deck structures of ships and offshore installations. J. Mar. Sci. Eng. **11**(6), 1200 (2023)

Mair, H.U., Reese, R.M., Hartsough, K., Naval Business Center: Simulated ship shock tests/trials? (1997). www.ida.org/LFTEsimulation/documents/SSSTT.htm

Didoszak, J.M., Shin, Y.S., Lewis, D.H.: Shock trial simulation for naval ships. In: Proceedings of ASNE Day (2004)

Office of the Director: Operational Test and Evaluation: FY 2021 Annual Report. U.S. Department of Defense, Washington, DC (2022)

Open Access This chapter is licensed under the terms of the Creative Commons Attribution-NonCommercial-NoDerivatives 4.0 International License (http://creativecommons.org/licenses/by-nc-nd/4.0/), which permits any noncommercial use, sharing, distribution and reproduction in any medium or format, as long as you give appropriate credit to the original author(s) and the source, provide a link to the Creative Commons license and indicate if you modified the licensed material. You do not have permission under this license to share adapted material derived from this chapter or parts of it.

The images or other third party material in this chapter are included in the chapter's Creative Commons license, unless indicated otherwise in a credit line to the material. If material is not included in the chapter's Creative Commons license and your intended use is not permitted by statutory regulation or exceeds the permitted use, you will need to obtain permission directly from the copyright holder.

Committee V.3: Subsea Technology

Chen An[1(✉)], Tauhid Rahman[2], Celso Kazuyuki Morooka[3], Carlos Montoya[4], Yoshihiro Konno[5], Svein Sævik[6], Chunsik Shim[7], Cihan Bayindir[8], Kourosh Parsa[9], and Yi Xia[10]

[1] China Ship Scientific Research Center (CSSRC), Wuxi, China
anchen@cup.edu.cn
[2] DNV, Sydney, Australia
[3] University of Campinas (UNICAMP), Campinas, Brazil
[4] C-CORE, St. John's, Canada
[5] China Ship Scientific Research Center (CSSRC), Wuxi, China
[6] University of Tokyo, Tokyo, Japan
[7] Mokpo National University, Muan, Republic of Korea
[8] Istanbul Technical University, Istanbul, Turkey
[9] London, United Kingdom
[10] Houston, USA

Committee Mandate. As global oil and gas resources increasingly shift to deep-sea and remote offshore areas, subsea production equipment faces progressively harsher environmental conditions. Challenges such as immense pressure, fatigue, and seawater corrosion have rendered traditional extraction methods inadequate to meet the growing production demands. To address these severe challenges, new technological innovations have become the driving force behind the advancement of subsea production systems.

Keywords: Subsea Production System · Subsea Processing · Flow Assurance · Fabrication · Testing for Qualification · Deepwater Installation · Subsea Operations · Inspection · Maintenance · Decommissioning · Hydrates · Pipelines · Risers and Umbilicals · Reliability and Safety

1 Introduction

As global oil and gas resources increasingly shift to deep-sea and remote offshore areas, subsea production equipment faces progressively harsher environmental conditions. Challenges such as immense pressure, fatigue, and seawater corrosion have rendered traditional extraction methods inadequate to meet the growing production demands. To address these severe challenges, new technological innovations have become the driving force behind the advancement of subsea production systems.

In recent years, significant progress has been made in the design, materials, and technologies of subsea production systems. For example, the use of advanced high-strength materials, anti-corrosion coatings, and intelligent monitoring systems has significantly improved the system's durability, stability, and disaster resilience; more refined optimization of equipment has ensured that subsea systems can maintain structural integrity and reliability during long-term operations under the complex conditions of deep water and ultra-deep water environments; the integration of automation and digital technologies into subsea equipment has enabled real-time monitoring and remote control, significantly enhancing fault prediction capabilities and operational efficiency, thus improving the overall stability and safety of the systems. The comprehensive application of these technologies allows subsea production systems to better adapt to extreme environments, ensuring the smooth operation of deep water and ultra-deep water oil and gas extraction. This report will provide an overview of recent developments in these areas, discuss and summarize these innovative technologies aimed at improving the structural safety, stability and long-term operational reliability of subsea production systems.

In order to provide a comprehensive understanding, this report will be divided into nine chapters. Section 1 provides basic terms, definitions, and background information to prepare readers for the complexities of offshore platform lifecycle management. Building on this foundation, Sect. 2 presents a thorough analysis of pipeline design standards, structural assessment methodologies, and the emerging role of digital tools in enhancing design flexibility and resilience of subsea systems. Section 3 delves into hazard and risk assessments, focusing on high-risk scenarios such as collisions and fires, and emphasizes recent innovations in risk management practices for offshore structures. Subsequently, Sect. 4 reviews advanced analytical, experimental, and numerical modeling techniques employed to predict failure modes in critical subsea components, such as pipelines and risers. Section 5 shifts attention to the latest research on flexible riser systems, pipeline corrosion, and the integration of artificial intelligence-driven monitoring tools that facilitate the extension of subsea asset lifespans. In parallel, Sect. 6 explores the growing importance of structural integrity management, examining the application of automated inspection technologies, robotics, and AI-based predictive maintenance strategies. Furthermore, Sect. 7 critically examines reliability and safety protocols for subsea systems, focusing particularly on enhancing the resilience of underwater connectors, blowout preventers, and control systems through the application of sophisticated risk assessment methodologies. Building on these considerations, Sect. 8 delves into life extension strategies, decommissioning procedures, and the multifaceted technical, economic, and environmental challenges inherent in retiring or repurposing offshore platforms. Finally, Sect. 9 synthesizes key advancements in subsea technology for oil and gas development, offering a forward-looking perspective that emphasizes pipeline design, riser reliability, digital management solutions, AI applications, and sustainable decommissioning strategies.

In summary, this report provides a comprehensive and in-depth exploration of the critical advancements and methodologies in subsea production systems and offshore platform lifecycle management. By analyzing key aspects such as pipeline design, structural integrity, hazard management, and the role of digital technologies, this report highlights the continuous innovations driving the industry towards safer, more efficient, and environmentally responsible practices. Looking to the future, these advancements will undoubtedly enhance the long-term sustainability of subsea operations, addressing the evolving challenges in offshore oil and gas development and paving the way for more resilient and optimized subsea systems.

2 Subsea Systems

2.1 Field Architecture

Field architecture refers to the strategic and systematic arrangement or design of the components, systems, and technologies employed in the development of an oil and gas field. It involves the integration of flow assurance and multiphase flow expertise with technological knowledge to create a robust and efficient structure that ensures the effective and safe transfer of oil and gas from the reservoir to exportable products. The field architecture is specifically engineered to handle various operational scenarios, taking into consideration factors such as flow assurance, multiphase flow dynamics, and overall cost-effectiveness.

Field architecture serves as the blueprint that organizes and structures the elements involved in field development. Both concepts aim to optimize operations economically and operationally, considering technological advancements. For a comprehensive explanation about field development and its correlation with field architecture, guidance provided by is used.

The initial phase of field development involves a comprehensive assessment of various field architecture concepts, with a primary focus on two key aspects.

Firstly, an economic evaluation is conducted, encompassing capital expenditures, such as drilling costs, equipment costs, and installation expenses, as well as operating expenditures, covering maintenance, intervention, and flow assurance measures. To ensure the realism of this phase, input and guidance from industrial partners are crucial for ensuring the realism of this phase. Integrated asset models, often referred to as digital fields, are developed for the selected concepts. These models employ realistic parameters and are implemented using commercial software when feasible. This approach enables the running of simulations throughout the entire asset life, offering a comprehensive understanding of the operational dynamics.

In the subsequent phase, the focus shifts to addressing flow assurance issues. Operational challenges and technological bottlenecks associated with existing technologies are scrutinized. Simultaneously, exploration and analysis of new enabling technologies take place. This second activity aims to enhance the robustness of the field development plan by identifying and incorporating advancements that can mitigate challenges and optimize performance.

Most recent advances in field architecture comprise integration of machine learning in field developments, expansion of digital oil fields, which consists in the digital transformation of the oil E&P industry, and in this framework an intensification in software development and data analytics.

In the context of "field architecture," a "subsea system" can be defined as a crucial sub-asset that constitutes the integrated network of equipment, structures, and technologies positioned on or below the seafloor.

2.2 General Description of Subsea System

The subsea production system encompasses all the equipment and structures situated on, below, or within the seafloor, as mentioned in topic 0. This system plays a pivotal role in the overall design and functionality of the oil and gas field, facilitating the extraction, control, and transportation of fluids from subsea wells to exportable products. The special environment subsea demands some unique aspects related to the inaccessibility of the installation and its operation and servicing, yielding distinctive challenges that require specialized approaches.

Subsea production systems have been evolving since the world's first subsea completion in Gulf of Mexico in 1961 in a water depth of about 17 m. Currently, the world's deepest installed subsea system is situated in the Tobago field, around 200 miles south of Freeport, Texas in the Gulf of Mexico, with the wellhead positioned in a water depth of 2934 m.

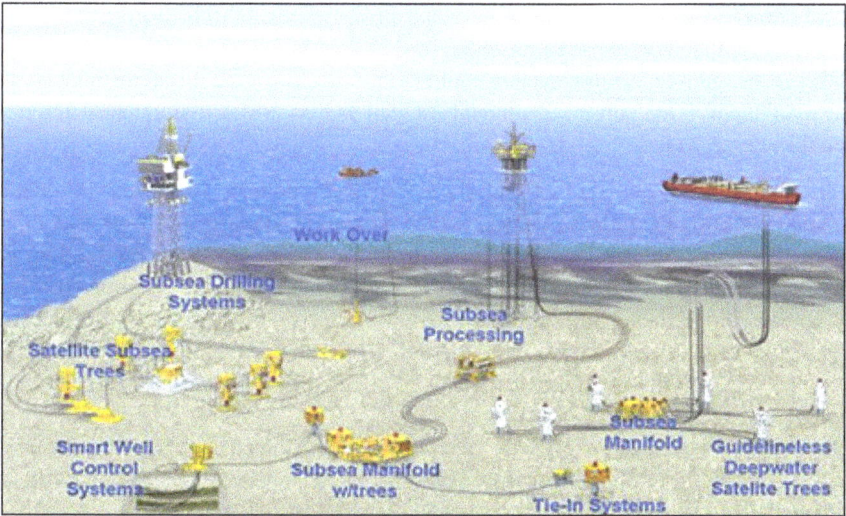

Fig. 1. Different types of subsea production systems

Figure 1 shows types of subsea production systems, which includes components strategically organized to optimize the efficiency, safety, and cost-effectiveness of field development.

According to sources, numerous subsea production systems are available, with the most fundamental being the subsea satellite—a single wellhead equipped with a subsea tree. This configuration oversees and regulates the production of an individual subsea well. Nevertheless, typically, a subsea production or injection system consists of one or more of the following components:

- A wellhead with interconnected casing strings;
- A subsea tree incorporating pressure- and flow-control valves;

- A template serving as a structural foundation for supporting and positioning diverse equipment;
- A manifold system for the directed gathering and distribution of multiple fluid streams;
- Subsea processing equipment, encompassing fluid separation devices and/or pumps, compressors, and associated electrical power distribution equipment;
- A production control and monitoring system for remotely overseeing and controlling various subsea equipment, potentially including multi-phase flowmeters, sand detection meters, and leak detection devices;
- A chemical injection system;
- An umbilical containing electrical power and signal cables, along with conduits for hydraulic control fluid and various chemicals to be injected subsea into the produced fluid streams;
- One or more flowlines for transporting produced and/or injected fluids between the subsea completions and the seabed location of the host facility;
- One or more risers for conveying produced and/or injected fluids to/from the various flowlines situated on the seafloor to the host processing facilities;
- Well entry and intervention system equipment, utilized for the initial installation and abandonment of the subsea equipment, as well as for various maintenance activities on the subsea wells.

Several authors point that commonly affected components in Subsea Production Systems (SPSs) include choke valves, cables, flanges, fasteners, and Process Control Systems (PCSs): Choke valves may suffer from erosion due to sand production. For cables, electrical insulation needs to be maintained in the saltwater conditions in which they operate, since losing insulation will result in electrical short circuits. Failures in flanges or fasteners may be caused by corrosion, overload, or fatigue. Hydraulic components involved in PCSs, such as valves, can be affected by frequent operation and may leak.

A comprehensive resume of latest technological advances is stated for processing plants, wellheads, subsea high-pressure vessel for deep waters, pumps and compressors for deep waters and electrifying subsea production, namely from topic 2.3, 2.4, 2.5, 2.6, and 2.7.

2.3 Edging towards a Subsea Processing Plant

Subsea processing involves various procedures aiming at minimizing the expenses and complexities associated with offshore field development. Key subsea processing methods include removing and re-injecting or disposing of water, boosting well fluids in single-phase and multi-phase, separating sand and solids, separating, and boosting gas/liquid, as well as treating and compressing gas.

The concept of subsea processing was initially developed to address the difficulties posed by extremely deepwater environments. Over time, it has evolved into a practical solution for fields situated in challenging conditions as well, mitigating risks associated with processing equipment on the water's surface.

The conventional method for extracting offshore reserves involves utilizing a fixed or floating production facility, where all necessary equipment, such as pumps, separators, water handling, compressors, processing, and storage is situated above the water's surface. However, in deep or remote waters, the cost of surface facilities is high, and available space is limited, posing significant challenges to production in these locations. Besides, beyond lowering development and production expenses in comparison to traditional platforms, modularized subsea technology can enhance recovery factors by alleviating backpressures on wells, achieved through methods like multiphase pumping or subsea separation, which increases the importance of further developments towards subsea equipment, such as processing plants.

The progression in traditional subsea processing began with pumping and evolved to include both separation and pumping. In this process, gas and liquids were separated subsea, with liquids pumped and gas naturally flowing to the surface to reduce back pressures and enhance efficiency, termed primary separation. This advanced into secondary separation, involving the cleaning of produced water before reinjection into a reservoir. As the trend continued, gas compression became a valuable addition to subsea processing, compressing gas on the seafloor to further decrease back pressure on the reservoir and prevent slugging. In the upcoming years, substantial growth is projected in the areas of subsea processing, long-distance tie-back, and electric control systems.

Subsea processing involves a series of technologies with different technological maturity levels, which may provide the following functions: subsea gas/liquid and gas/oil/water separation, subsea boosting, wet and dry gas compression, gas dehydration, dense gas reinjection, seawater injection, sulphate removal unit from seawater for injection, produced water reinjection or disposal at subsea, CO_2 dense phase separation and reinjection into reservoirs.

SWIT is a modular system that can be adapted to all water requirements and is based on Seabox as the initial conditioning step where microbes are killed and particles are settled out, it is possible to achieve a flexible, cost-effective, and robust oil field development by moving water treatment and injection facilities to the seabed, as close to the wellhead and reservoir as possible.

North Sea Transition Authority states, that currently operators are concentrating on the expansion of subsea facilities, emphasizing the optimization of available deck space. This involves incorporating subsea processing facilities, such as subsea gas compression and subsea water treatment. Additionally, efforts include implementing multiphase pumping to extend tieback distance and incorporating wellhead seawater injection with a seawater pump at the wellhead. All this technology is already considered as deployable (Technology Readiness Level or TRL 8-9), still according to North Sea Transition Authority.

As emerging technology (TRL 5-7), ongoing assessments of water injection treatment for existing subsea well stock to increase water injection capability and overcome platform slot constraints, by enabling treatment of raw seawater directly on the seabed for injection into oil wells for pressure support, improved sweep efficiency, and increased oil recovery.

2.4 Wellheads

Mounted at the well's opening, the wellhead features equipment designed to regulate and monitor the extraction of hydrocarbons from the underground formation. This essential component not only prevents the leakage of oil or natural gas from the well but also safeguards against blowouts caused by high-pressure formations.

In the subsea environment, the wellhead plays a crucial role. It guides the wellhead and tree system's orientation for the tree-to-manifold connection. Additionally, it interfaces with and supports the Christmas tree system and Blowout Preventer (BOP). The wellhead withstands all loads from drilling, completion, and production, considering thermal expansion, especially in the horizontal tree setup where the BOP is latched on top of the Christmas tree. It also ensures alignment, concentricity, and verticality of the low-pressure conductor housing and high-pressure wellhead housing. Prioritizing reliability, the wellhead design should be field-proven and installation should be minimally sensitive to water depth and sea conditions.

Estimation and migration of the fatigue life of wellheads is one key challenge for offshore drilling operations. Besides, running subsea systems in deeper waters and harsh high-pressure, high-temperature (HPHT) conditions bring more risks and challenges for the structural integrity and fatigue life performance of the equipment. Monitoring downhole temperature and pressure in ultra-high-pressure gas wells poses one of the greatest challenges in ensuring the integrity of the wellhead seal as well.

Latest advances on wellhead technologies concern reduction of costs, such as lower overall wellhead installation costs due to reduced rig trips, assembly of seawater pump at wellhead to eliminate need for injection water supply system to subsea wells (already set as TRL-9), new technologies designed to eliminate the need for additional lockdown devices, employing advanced seal assemblies and enabling streamlining of drilling procedures, use of machine learning, artificial intelligence and data analytics to provide digital twins, versatility to allow same wellhead case for different applications, and the development of conductor-less subsea wellheads to reduce costs and CO_2 emissions.

2.5 Subsea High-Pressure Vessel for Deep Waters

Subsea pressure vessels are essential components in the oil and gas industry, particularly within subsea production systems. They serve multiple purposes, including fluid storage to maintain pressure during the transportation of extracted oil and gas, storing hydraulic fluids for powering subsea operations, facilitating chemical injection for tasks like corrosion inhibition, housing instrumentation and control systems, accommodating subsea processing equipment such as pumps and compressors, and serving as integral components for wellhead control systems. Designed to withstand the harsh conditions of deep-sea environments, subsea pressure vessels contribute significantly to the efficient and reliable functioning of subsea oil and gas operations by providing storage, power, and containment solutions for various critical processes.

Apart from this, designing filters presents a challenge as they effectively ensure water purity in surface installations. The complexity arises when attempting to adapt this technology for remote seabed applications. A key criterion for filter selection revolves around the filter's performance, considering factors such as how efficiently this surface can be contained within a collapse-resistant pressure vessel.

Considering the ambient pressure of the seabed, a basic assessment of the collapse load on a cylindrical vessel indicates that a vessel with a 3-ft diameter would require a wall thickness of 1½ to 2 in. (3.8 to 5 cm). Manufacturers of boilers do produce vessels with significantly larger diameters and wall thicknesses, hence the primary concern lies in the cost rather than the availability of such vessels. Besides, consideration must be given to the size and weight of components. Installation related challenges are the most cited ones as far as concerning subsea pressure vessels, and collapse due to high external pressure is indicated as the most important failure mode. In this scenario, it is important to highlight as well, that fatigue-fracture failure is pointed out as the most common failure mode in HPHT pressure vessels.

First 20,000-psi subsea production system implements more than 50 seals, which needed to be rated to 20,000-psi. These seals ranged from a size as small as 0.5-in diameter to as large as 18 in. Some of these seals were made from exotic elastomers to handle the extreme pressure when combined with elevated temperatures and still be expected to seal in a gas environment.

Current ways on how the market is coping with challenges are the development of pressure vessels made of composite materials and titanium. Development of sealings also represent the newest technological advances within pressure vessels for deep waters.

2.6 Pumps and Compressors for Deep Waters

Crucial for optimizing oil and gas extraction from offshore fields, subsea pumps and compressors play a primary role. These devices elevate both pressure and hydrocarbon flow from reservoirs to surface facilities, diminishing reliance on topside processing and contributing to enhanced recovery factors.

Besides, it offers an improved business case through cost reduction and increased production, ensuring safety with remote operation, and maintaining a lower carbon footprint compared to alternative solutions. Positioning the compressor on the seabed eliminates the significant expenses associated with constructing and operating an offshore platform. By bringing the compressor closer to the wellhead, production is enhanced due to reduced pressure drop in the downstream pipeline, resulting in lower power requirements compared to alternative setups. Subsea compression stands out as an optimal solution for deep waters and challenging climates.

The electrical power needed for pumps, booster stations, or compressors is sourced from either an offshore platform or an onshore facility. To adapt to pressure drop variations during production, it becomes crucial to regulate the frequency of the power supply, facilitating the adjustment of pump or compressor speeds. The majority, if not all, motors utilized in subsea processing necessitate a variable speed drive (VSD) due to uncertainties in reservoir data, particularly over the lifespan of the field.

Main issues concerning the use of subsea multiphase pump equipment include blockage in pump intake, elevated reservoir pressure, fractured pump shaft, tubing damage caused by high water cuts, fines production, heightened salinity, and corrosion.

ABB's H. Lendenmann tested the new 8 MW multiphase compressor WGC6000. This the first time a large medium-voltage drive and a multiphase compressor were operated at 8-MW shaft power for an extended time while the entire equipment was submerged in seawater and operating at realistic process conditions. All modules are designed to operate down to depths of 3,000 m.

The collaboration between PETRONAS and FASTsubsea develop and qualify the "world's first all-electric and topside-less subsea multiphase pump technology". Compared to conventional subsea pumps, the uniqueness of this technology lies in its significantly reduced requirement for topside space, as it does not need variable speed drives or barrier fluid hydraulic power units.

As recent technological advances, following can be accounted: Improvement in power and automation, increasing of compression, reduction of number of moving parts and supporting systems towards magnetic bearings, use of lower voltage power for respective operation, cooling systems by process gas, multiphase subsea pumping through centralized subsea pumping (TRL 9), and reliability of gas compression facilities (TRL 8).

2.7 Electrifying Subsea Production

In subsea production systems (SPS), electrification offers a dependable and scalable channel that eliminates the limitations, high expansion costs, and failure modes associated with intricate hydraulic or multiplexed electrohydraulic control systems. Furthermore, opting for a fully electric approach in subsea and completion technologies fundamentally alters the technical and economic viability of certain reserves, particularly in complex reservoirs, since placing power converters and equipment in proximity to loads on the seabed yields cost-effective power supply with comparable reliability. Just to illustrate, eliminating hydraulics from the manifold reduces CAPEX, electrifying the choke valve enhances production control, and employing closed-loop self-contained valves in gas lift or carbon-capture utilization and storage applications lowers OPEX.

Electrification not only enhances operational efficiency but also contributes significantly to sustainability. By replacing fossil fuel-based power generation with electricity, particularly from renewable sources, subsea systems reduce the carbon footprint of oil and gas production. For instance, integrating offshore wind or solar energy into subsea operations creates a more environmentally friendly system, reducing reliance on conventional power sources that emit greenhouse gases. The shift to electric systems improves energy efficiency, as electric drives are typically more efficient than hydraulic systems, which helps optimize resource usage and reduce environmental impact. Additionally, fully electric subsea systems enable real-time monitoring and adjustments, improving production rates while minimizing waste and energy consumption. This transition is pivotal in making subsea production more sustainable while simultaneously increasing the cost-effectiveness and reliability of offshore operations.

World's first, in 2015, the all-electric subsea gas compression system was delivered and installed in the Åsgard Field offshore Norway. According to Rajashekara et al. (2017), the challenges in installing and operating subsea electrical systems include designing components to withstand high pressures at depths of up to 3,000 m, addressing the conductivity and corrosiveness of seawater, ensuring proper isolation between electrical equipment and the sea, and managing maintenance difficulties due to the depth, which can lead to costly and prolonged production outages. Reliability is a key factor, with equipment designed for a lifespan exceeding 20 years, necessitating a mean time between failures (MTBF) greater than 20 years and dealing with extended mean time to repair (MTTR) for subsea system. Likewise, offshore deep-water oil and gas production systems must integrate high-capacity electric submersible pumps and compressor motors on the seabed, requiring substantial electrical power, Manach et al. (2023) emphasizes the benefits of the integrated all-electric subsea system (iAES) for the development of long-tieback natural gas fields, with the increased production of decarbonized natural gas through all subsea tiebacks to shore being a key factor in reducing carbon dioxide (CO_2) emissions.

In response to environmental challenges posed by the conventional energy supply for deepwater oil and gas production, often dependent on fossil fuel combustion-based platform generation, there is an increasing focus on innovation in sustainable and environmentally friendly power processing. In this sense two primary technologies facilitating this innovation involve transmitting electrical power either from the onshore electrical grid to subsea petroleum installations or integrating offshore renewable energy sources to create a microgrid capable of powering both platform-based and subsea loads.

In the evolution of subsea electrification, the integration of backup batteries becomes pivotal as well, ensuring innovation in achieving secure, reliable, and uninterrupted operations. Apart from this, innovations relying on onshore energy supply through subsea cables and combined cycle power generation are being currently implemented to support a 100% electrifying solution.

2.8 Summary and Recommendations for Future Work

Field architecture in oil and gas development orchestrates the arrangement of components for efficient oil and gas transfer, integrating flow assurance and multiphase flow expertise with technological knowledge. Economic evaluations, including capital and operating expenditures, are pivotal in this phase, guided by input from industrial partners. Subsequent focus on flow assurance addresses operational challenges and explores new enabling technologies, with recent advances incorporating machine learning, expanding digital oil fields, and emphasizing software development and data analytics.

The subsea production system, positioned below the seafloor, plays a key role in field development. Evolving since the world's first subsea completion in 1961, these systems face unique challenges due to their inaccessibility. Various subsea production systems contribute to efficient oil extraction, and challenges include maintaining the integrity of components like choke valves, cables, flanges, and process control systems. Recent technological advances cover processing plants, wellheads, subsea high-pressure vessels for deep waters, pumps, compressors for deep waters, and electrifying subsea production.

Subsea processing minimizes expenses and complexities in offshore field development, involving methods like water removal, boosting well fluids, and gas/liquid separation. Initially developed for deepwater environments, subsea processing has evolved into a practical solution in challenging conditions, reducing costs compared to traditional platforms. The evolution includes primary and secondary separation, gas compression, and anticipated growth in subsea processing, long-distance tie-back, and electric control systems. Operators focus on expanding subsea facilities, incorporating gas compression and water treatment, with ongoing assessments for water injection treatment.

The wellhead, essential for hydrocarbon extraction, guides the wellhead and tree system's orientation in the subsea environment. Challenges include estimating and managing wellhead fatigue life and addressing risks in deeper waters. Recent advancements focus on cost reduction, lower installation costs, machine learning applications, and conductor-less subsea wellheads, prioritizing reliability and minimal sensitivity to water depth and sea conditions.

Subsea pressure vessels serve various purposes within subsea production systems, storing fluids, and facilitating chemical injection. Challenges in designing filters for remote seabed applications are addressed by considering collapse-resistant pressure vessels. Market coping strategies include developing vessels from composite materials and titanium, along with advancements in sealings for deep-water pressure vessels.

Subsea pumps and compressors optimize oil and gas extraction, reducing reliance on topside processing and enhancing recovery factors. Challenges include blockages, elevated reservoir pressure, and tubing damage. Recent advances involve improvements in power and automation, magnetic bearings, lower voltage power usage, cooling systems using process gas, centralized subsea pumping, and reliability of gas compression facilities.

Electrification emerges as a transformative strategy in subsea production systems, offering a scalable and reliable channel. Challenges in installing and operating subsea electrical systems include designing components to withstand high pressures and addressing seawater conductivity. Innovations include the world's first all-electric subsea gas compression system, integration of backup batteries, and leveraging onshore energy supply. These advancements address environmental challenges and contribute to secure, reliable, and uninterrupted subsea operations.

As recommendations for future works, it is suggested that failure modes and latest technological advances are researched, where cutting edge technologies such as additive manufacturing could be embraced and related to up-to-date challenges.

3 Structural Design and Assessment of Subsea Systems

3.1 Design Codes for Subsea Applications

In the 2022 update of the DNV-ST-F101 standard, high-strength steel and composite materials were introduced to address the challenges of extreme deep-sea environments. High-strength steel not only enhances the compressive capacity of pipelines but also maintains structural lightness. Composite materials, such as glass fiber and carbon fiber, significantly improve the durability of pipelines in acidic environments. These materials exhibit excellent fatigue resistance during long-term service, especially in weld joints and bends of pipelines. To further enhance the protective capabilities of pipelines, the standard emphasizes the application of multi-layer protective designs, such as combining corrosion-resistant coatings with cathodic protection systems, which can greatly extend the service life of pipelines. For free-spanning pipeline designs, especially in deep-water and high-current regions, the updated standard imposes stricter limits on the maximum allowable span and deflection of the free-spanning sections to reduce pipeline vibration and fatigue damage risks. In addressing vortex-induced vibrations (VIV), the standard requires the use of numerical simulation tools during the design phase to analyze the magnitude and frequency of VIV and optimize the design of the free spans to minimize vibration. Additionally, the standard recommends optimizing support structures to reduce fatigue damage in free-spanning sections, such as employing additional supports or using sandbags or concrete blocks to secure the pipeline, thus improving its stability and durability. These updates ensure the safety and long-term reliability of pipelines in extreme marine environments.

In the 2022 update of the "API Spec 17D" standard, several improvements were made to the structural design of subsea wellheads and Christmas tree systems. Firstly, higher requirements were established for high-pressure design, especially in deep-sea high-pressure environments. Through pressure vessel design and stress analysis tools, such as finite element analysis (FEA), the updated standard ensures that pressure-bearing components remain stable under extreme conditions exceeding 15,000 psi. Secondly, the standard optimizes the material selection for support structures, recommending the use of corrosion-resistant alloys (CRAs), such as nickel alloys and titanium alloys, along with multi-layer protection systems to enhance corrosion resistance. These measures, combined with cathodic protection systems, prevent electrochemical corrosion. Additionally, the design requirements for lifting equipment load capacity have been strengthened. Lifting lugs and support points must undergo testing at 2.5 times the maximum lifting load to ensure safety. Furthermore, the standard imposes stricter requirements on welding, mandating that all critical welds must be subjected to non-destructive testing (NDT), such as ultrasonic or radiographic inspection, to prevent potential structural failures.

These updates are aimed at improving the reliability and safety of subsea production systems in extreme marine environments.

3.2 Environmental Loads for Subsea Assets

In a broad sense, subsea assets can be interpreted as an underwater equipment or device to run the exploration of ocean resources. Following this understanding, few research works of environmental loads specific for subsea assets are found. Perhaps, the most relevant could be those related to fish farming cages, particularly environmental loads and response to design of fish farming cages. Jin et al. (2021) investigated fish farms suitable for aquaculture production in offshore environments. The main purpose was to develop a reliable numerical model and investigate the motion responses of the fish farm structure in waves and current. The numerical modelling scheme was verified with a systematic model test measurements with acceptable matching. A case study was conducted for the main motions of a typical flexible gravity cage underwater structure in calm water, current, regular waves, and irregular waves with/without current. Commercial available softwares were also used and they concluded that the scheme could be generalized for other offshore fish farm structures consisting of a rigid frame structure and fishnets attached to it.

Xu et al. (2021) conducted a series of experiments in a wave flume with different types of net panels in extreme waves. For a constant steepness spectrum was adopted to generate waves and observe effects due to the net characteristics. They observed that the force on the net panel in extreme waves present great dependency of the net solidity and the Keulegan Carpenter (KC) number. They proposed a wave force model to simulate the loading process for the submerged net panels in extreme waves. Furthermore, Dong et al. (2021) developed a model of uncertainties for hydrodynamic characteristics estimated from the several existing numerical models from validated data in the published researches. Due to small sample of data size available, they used two approaches to estimate the statistical uncertainty. As a case study, they did the reliability analysis of mooring line is quantified under the ultimate limit state and fatigue limit (the highest stress level below which a material can withstand an infinite number of loading cycles without experiencing failure) state and concluded the importance to account uncertainties accordingly.

Prsic et al. (2024) conducted a systematic study of hydrodynamic coefficients for simplified subsea modules and their components totally submerged in water through forced oscillation tests at the laboratory. Nearly two-dimensional test was setup in the laboratory and the main purpose was to help the planning of subsea installations. Different amplitude and periods of forced oscillations were taken and they observed that, in general, the damping dominates the hydrodynamic force and the presence of component inside the modules increases the importance of the added mass. In addition, the estimation of the hydrodynamic coefficients by summation of the coefficients for the individual structure elements generally overestimates the damping.

3.3 Installation of Subsea Structures

Oil and gas reserves decline in shallow waters, more and more offshore operations for petroleum and energy exploration in general advance for deepwater. Subsea equipments, remote and autonomous subsea installations, data monitoring and transmission from the subsea require regulatory compliance and innovations with increased complexity and safety issues, which introduce many challenges in the design, installation and operation of subsea structures and retrieval, keeping in mind sustainability and environmental aspects.

Tommasini et al. (2021) deals with the dynamics of deepwater subsea lifting operations experiencing super-harmonic resonance, when forcing frequency is near an integer fraction of the natural frequency of the system. They use the harmonic balance method to solve the non-dimensional equation of motion of the system and compare the results with time domain integration and with an equivalent model for energy dissipation. The nonlinear quadratic drag term was expanded and coefficients obtained by the least squares method to facilitate the algebraic calculations the method. They conclude that the maximum force on the lifting cable presents super-harmonic resonance amplifications on contrary with the predictions by the traditional equivalent model for energy dissipation.

Liang et al. (2021) addressed investigated the drag coefficient of the subsea christmas tree through an experiment and numerical simulations based on the Reynolds-Averaged Navier-Stokes equation with turbulence model, and the dynamic performance of the system was observed. Finite difference method is used to get the drag force and drag coefficient and they proposed a new experimental procedure was proposed for the model experiment for drag of a subsea tree.

Lopes et al. (2024) investigated the vertical stiffness of the system during the crane installation of a subsea manifold. They observed the slackening of the installation cable according to its length observing the natural frequency of vertical motions during the lowering process. A numerical model is proposed and results are compared with experimental data available.

3.4 Safety Layout Design of Subsea Systems

Wang et al. (2021b) developed a new algorithm to optimize the layout of a subsea production system to reduce facility costs. Based on an unsupervised learning method and clustering algorithm, they used a complex iterative structure to adjust continuously the optimization that account the layout scenarios of cluster manifolds, wellhead grouping, and connection between manifolds. Actual layout of subsea production system in deepwater oilfields were considered and the obtained results were compared. They verified that the proposed approach effectively reduced the development costs.

Later on, in Wang et al. (2024), authors incorporated the seabed topography in the model by focusing on cluster manifold layout, and proposed a method based on a modified adaptive particle swarm optimization (MAPSO) algorithm coupled with A-star algorithm, clustering algorithm and minimum spanning tree algorithm to optimize subsea production facilities. They show the feasibility and practicality of the model proposed by designing the layout of a deep-water oil field, by enhancing the influence of manifold number and seabed topography in the estimated costs and their reliability.

Hong et al. (2023) proposed the mixed integer nonlinear programming (MINLP) model with a series of operational constraints and requirements for the subsea oil gathering - transportation system to a floating processing terminal, in order to minimize the total layout cost. The solution figured out the structure of pipeline network topology from the each subsea allocated well, manifolds and processing terminals, routes of pipes, and the size of the facilities. For computation of the solution, decomposition strategy was built to achieve qualified initial solution and stable iteration process. Case studies are used to show the validity, feasibility and stability of the proposed model and the solution method, among other influence of other parameters of the layout design problem to show flexibility of the analysis procedure.

Meanwhile, as a part of larger research project, Sales et al. (2023) proposed to apply the mixed-integer nonlinear (MINLP) model for the optimization of a subsea petroleum production system and to compute a global optimum design, considering constraints in production, equipment duties and cost, and reliability and maintenance aspects of the system. They applied on a on a synthetic field based on an actual field in the Barents Sea, and they successfully found the best designs while other best layouts gave general insights for subsea processing layout optimization.

3.5 Flow Assurance and Subsea Structure Behavior

A stable oil and gas flow from the reservoir maximizing production with low operational costs is aimed for offshore petroleum field production. In order to keep the flow assurance for hydrocarbons production in general, several challenges are faced and they affect design and lifetime aspects of the subsea structures, risers and pipelines. For non-hydrocarbons such as in subsea mining and CO_2 injection, similar issues are also present in the flow assurance.

Heavy weight of pipelines and risers result structural resistance problem in deepwater operations, especially in deepsea mining operations. In offshore petroleum, well drilling risers are very common to relief the riser weight by attaching buoy along the riser length, which increase hydrodynamic loads linearly with the buoy diameter. Yamamoto et al. (2023) proposed a solution to assembly riser joints with the same outer diameters although with different wall thickness. They did numerical simulations in time domain with the riser in current and irregular waves. They found for the best-optimized riser structure, an axial tension reduction of 15% compared to the reference.

Novel et al. (2023) proposed an innovative method to berth the transition of a rigid riser with a seabed flowline in ultra-deep waters of the Santos Basin. Flow condition involves High Temperature and High inner Pressure (HPHT) and oil and water alternating gas (WAG). That kind of solutions is largely used for flexible lines and umbilicals, and the first time for rigid riser connections at the seabed. They addresses the benefits and challenges faced with the use of torpedo piles for anchoring of rigid riser-flowline connections, and other issues to deliver an effective pipeline system solution.

An experimental study on a small-scale laboratory model of a submerged suspended flexible pipe was presented in Bordalo et al. (2024), and an apparatus for experiment in a catenary layout, oscillating in response to the intermittent loads caused by the internal two-phase slug flow of water and air injected with sixteen combinations of flow rates was described. Oscillations of the pipe were measured and time-histories of the pipe motions were recorded. The flow rates of the liquid and gas phases, the speed of the slugs and their frequencies were measured to evaluate the slugs' lengths and hold-ups, and to assess the fluid loads on the pipe due to gravity and curvature. They conclude that a proper scaling up of experimental results in conjunction with computer simulations could be applied for designing offshore pipe operations carrying oil and gas from petroleum reservoirs submarine wells to floating facilities in the ocean.

3.6 Summary and Recommendations for Future Work

In the realm of subsea systems design and operation, remarkable progress has been achieved on multiple fronts. Design codes like DNV-ST-F101 and API Spec 17D have constantly evolved to integrate advanced materials and stricter design requirements, bolstering the safety and reliability of pipelines, wellheads, and other components in harsh marine settings. Environmental load studies on subsea assets, despite being relatively limited, have offered valuable insights through research on fish farming cages and other submerged structures, helping to better understand the forces impacting subsea systems. Installation techniques for subsea structures have advanced to handle challenges such as super-harmonic resonance and optimize the installation of various parts. The safety layout design of subsea systems has been refined using advanced algorithms, cutting down development costs. Flow assurance and subsea structure behavior research have also tackled issues related to weight and connection problems in deepwater operations.

However, there are areas that demand further attention. Future work should center on in-depth research into the failure modes of subsea systems, including comprehensive investigations of potential failure points of different components under diverse operating conditions. It's also crucial to embrace cutting-edge technologies like additive manufacturing, which has great potential in subsea component production, and the application of

AI and IoT for enhanced real-time monitoring and predictive maintenance. Moreover, continuous efforts in developing new materials and coatings are necessary to meet the needs of deeper and more extreme subsea operations. In summary, future research on subsea systems should concentrate on failure mode analysis, the application of emerging technologies, and the development of advanced materials to overcome challenges in the offshore oil and gas industry and guarantee the safe and efficient functioning of subsea systems.

4 Pipelines

Recent trends in pipeline design and evaluation have begun to adopt probabilistic methods, which account for uncertainties such as material performance, environmental conditions, and operational changes. These methods enable more accurate assessments of pipeline reliability and safety. Machine learning algorithms are increasingly being used to predict corrosion rates, identify cracks, optimize pipeline layouts, and more. These approaches not only improve design efficiency but also help reduce manual errors and provide more intelligent decision support for pipeline operations. A key focus is the strain-based design method, which considers pipeline deformation and strain during practical operation to better predict fatigue life and overall reliability. As the industry transitions from existing natural gas pipelines to those designed for hydrogen and carbon dioxide transportation, these trends signal a move toward a more intelligent, reliable, and environmentally friendly pipeline anticorrosion field.

4.1 Thermal Buckling

After years of research, the thermal buckling (the deformation or instability that occurs in structures, such as pipelines, due to thermal stresses caused by temperature changes) of submarine pipelines is still one of the hot topics. Ning et al. (2022) proposed a new line element designed for the geometrically nonlinear large deformation analysis of the pipelines subjected to high temperature and high pressure (HTHP). The axial expansion force caused by high operating temperature and pressure is incorporated into the element formulation. Furthermore, a Gauss–Legendre integral method is employed to consider continuously distributed nonlinear SPIs in both lateral and longitudinal directions. Since the pipeline might exhibit large deflections, the analysis adopts the Updated-Lagrangian approach to establish the equilibrium conditions. A Newton–Raphson procedure is developed to execute the analysis. Detailed element formulations are provided. Six groups of examples are provided to examine the robustness of the numerical method.

In another study, Wang et al. (2022a) proposed a mathematical model based on the nonlinear von-Karman assumption and Euler-Bernoulli beam theory, Wang et al. (2022b) proposed a mathematical model to simulate upheaval buckling (a type of buckling that occurs when a subsea pipeline is subjected to thermal and axial forces, causing it to lift or move upward from the seabed) of lined subsea pipelines, which considers the difference in material properties of the liner and outer pipe, and provides a closed-form solution. The influence of Young's modulus and the thermal expansion coefficient of the liner on the post-buckling response was analyzes the effects of the outer pipe and liner were compared and the effect of the ratio of the liner's thickness to the outer pipe's thickness on the post-buckling response of the lined subsea pipeline was analyzed in detail. The results show that the liner must be included in the thermal buckling analysis. The post-buckling behavior of both the liner and outer pipe were affected significantly by the thermal expansion coefficient of the liner. The displacement amplitude, axial compressive force, maximum bending moment and maximum stress all increased with increasing ratio of the liner's thickness to the outer pipe's thickness.

Zen et al. (2022) primarily focused on a 32-in. submarine pipeline, that was buried in a trench created by the a jet trenching operation. They found that the uplift distance is a critical parameter influencing the bending behavior of the pipeline uplift. This parameter is highly uncertain and makes finite element predictions for typical upheaval flexion processes, These processes include the initial phase, full development phase, and the subsequent increase in bending amplitude as temperature loading continues to rise.

As shown in Fig. 2, which presents an illustration of a typical upheaval buckling process simulated by the FE analysis, it can be clearly observed how the pipeline behaves throughout these different phases of upheaval flexion. The figure serves as a visual aid, highlighting the key characteristics and transitions in the upheaval buckling process, further emphasizing the significance of the uplift distance in influencing the pipeline's bending behavior.

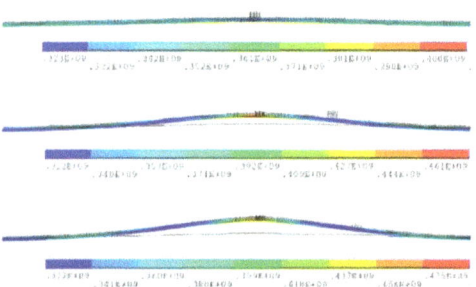

Fig. 2. Illustration of a typical upheaval buckling process simulated by the FE analysis.

Fully considering the parameters is essential to ensure a reliable design. This approach is typically understood as a conservative method, where the highest estimates are applied in the analysis. However, for the engineering applications presented here, conservative backfill failure loads (based on effective axial pressure) are predicted when a lower estimate of the pull distance is used. The final conclusion indicates that the upper extraction limit criterion is largely superior to the critical buckling load criterion and discusses the opportunity to optimize the potential cost of the stone removal requirement, focusing on the selection of the upper extraction mobilization distance, the relaxation of the design criteria, the load factor and the upper ratchet limit.

In the study, researchers observed that the buckling of submarine pipelines is often coupled with axial movement. Chen et al. (2022) derived the relationship between the length of a pipeline and lateral buckling critical temperature based on an analytical solution, considering the influence of tension forces and seabed slope angle. A theoretical analytical solution of axial walking was reviewed and summarized. The two solutions for axial walking and lateral buckling were combined, and a criterion based on critical pipeline length was proposed to estimate the possibility and priority of these two behaviors for a given pipeline under certain conditions. The influence of seabed slope and axial and lateral friction factors on the two behaviors was explored by carrying out parameter analysis, and the influence regularity of these factors on critical pipeline length and critical temperature was obtained.

Regardless of the pipeline laying method or buckling form, the interaction between the pipeline and soil cannot be overlooked when discussing the buckling and overall axial movement of pipeline, and it has been paid more and more attention by experts and scholars. Ruben et al. (2023) developed an analytical model that takes into account Engesser-Timoshenko beam theory (TBT) and considers the shear effects on pipelines to predict upheaval buckling in buried marine pipelines. Furthermore, equations that govern vertical buckling of buried pipelines considering a plastic soil with initial imperfection were considered. Analytical results were compared with finite element models of buried pipeline and other models reported in the literature, and it was observed that analytical results fall in the range of those reposted in the literature. It was also found that the incorporation shear stresses in buried marine pipelines had a minimal effect on the onset and propagation of upheaval buckling, but the soil stiffness, as can be seen from Fig. 3, had a strong influence on upheaval failure in buried marine pipelines. Wang et al. (2024) integrated a bi-linear axial pipe-soil resistance model into the mathematical framework of upheaval buckling. The mathematical model incorporated the von-Karman type of geometrical nonlinearity and the Euler-Bernoulli beam theory. The research examined typical upheaval buckling behavior and investigates the influence of axial mobilization distance and ultimate resistance on pipeline upheaval buckling behavior. The results revealed that, in contrast to rigid-plastic resistance, the inclusion of bi-linear axial pipe-soil resistance made the pipeline more susceptible to bucklin. Displacement amplitudes increased with axial mobilization distance during the post-buckling stage. Notably, a larger axial mobilization distance exerted a stronger influence on pipeline buckling. Moreover, the critical buckling temperature exhibited an almost linear negative correlation with axial mobilization distance and a positive correlation with axial ultimate resistance. Additionally, greater axial ultimate resistance magnified the impact of axial mobilization distance.

In addition, As transportation conditions evolve, a wider variety of pipelines have been applied, and prompting scholars to investigate more complex new pipeline designs. Sean et al. (2023) presented a comprehensive literature review of the state-of-the-art methods currently adopted in industry for the assessment of UHB of buried unbonded flexible pipelines. They also proposed an alternative approach that allows significant upward movement of the flexible pipe, taking advantage of the pipe's ability to accommodate such movements. A comparison of the potential failure modes, mechanisms, and consequences associated with these approaches was conducted and acceptance criteria were proposed. Parametric finite element analysis studies are performed to compare the flexible pipe behavior and download requirements using the traditional approach and the proposed alternative approach. The case studies demonstrated that permitting large deformations can result in significant savings in rock-dump volume, when compared to

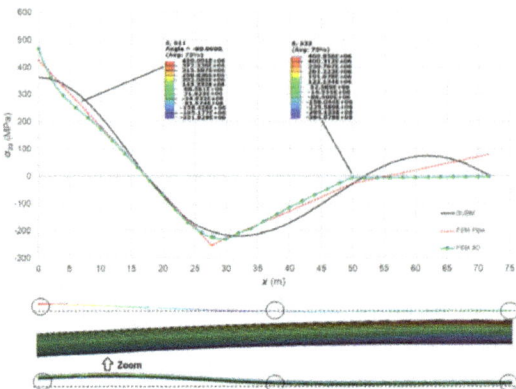

Fig. 3. Pipeline buckling behavior considering soil stiffness using FEM

the traditional UHB prevention method. Li et al. (2024) applied the energy method and thin-walled shell theory to develop analytical frameworks for assessing the thermal instability of functionally graded porous (FGP) pipelines reinforced with graphene platelets (GPL). Their findings revealed that the thermal stability of the FGP-GPL pipeline could be enhanced by adding more anchors in certain seabed conditions and laying the pipeline in an arched concave seabed.

4.2 Flexible Pipe

As mentioned above, the variety of pipelines used for oil and gas development is increasing, and with a growing range of flexible pipelines emerging in both research and engineering applications. Cao et al. (2020) proposed a study on the mechanical properties of fiberglass-reinforced flexible pipe (FGRFP) in marine applications based on the reinforced layer interface structure design, the anisotropic layer section mechanical performance test, and the structural strength analysis under the laying operation condition. Based on laboratory tests and referencing ASTM and DNV regulations, technical design indexes of flexible marine pipe that meet strength requirements have been established. Based on numerical simulation, static analysis was conducted to identify potential weak points in flexible pipelines. Especially, the feasibility of the FGRFP structural section design was further verified through dynamic analysis under the combined action of waves and currents. Finally, parameter sensitivity studies were carried out, examining the effects of wave flow direction, wave height, and period on the pipeline's performance. Figure 4 shows the FGRFP's damaged cross - section and stress distribution by FEM, supporting Cao et al.'s research on FGRFP's mechanical behavior.

Chaitanya et al. (2022) used a beam element and finite element analysis (FEA) method to simulate the behavior of buried flexible pipelines. Several vertical imperfections in the middle section were modeled to assess the potential for upheaval buckling. The model also assumes that the backfill is uniform along the pipeline route for the buried section. The bending radius of the flexible pipe is checked along the imperfect section to ensure its integrity. The modeling approach outlined in their research, coupled

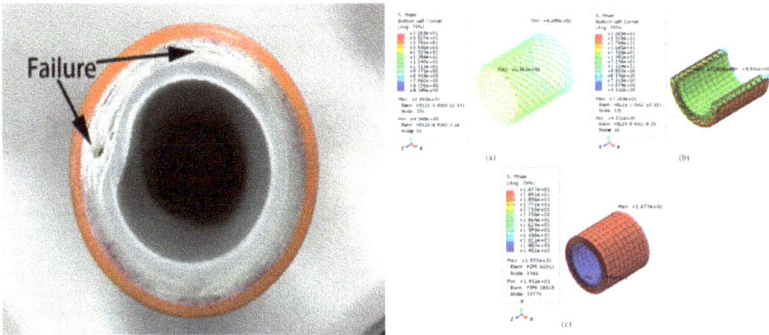

Fig. 4. Cross-section at the damage and the stress distribution of the FGRFP using FEM.

with parametric runs for seabed imperfections and soil cover depth, provides a decision-making matrix for design and installation agencies, including operators. Through this matrix, the safe working limits of the flexible pipe can be determined, or necessary mitigation measures can be identified. Utilizing this approach can help avoid costly post-installation interventions or corrective actions. An extension of this study would involve allowing significant upward movement of the flexible pipe, taking advantage of its ability to accommodate such movements. This upward movement could resemble a controlled "buckle" remaining within the limits of the pipe's structural integrity. Additionally, scenarios where the flexible pipe could be laid directly on the seabed, rather than being trenched or buried, could be identified, potentially reducing capital expenditure. Figure 5, an API 17B unbonded flexible pipe cross-section, can be a reference for Chaitanya et al.'s study on buried flexible pipelines.

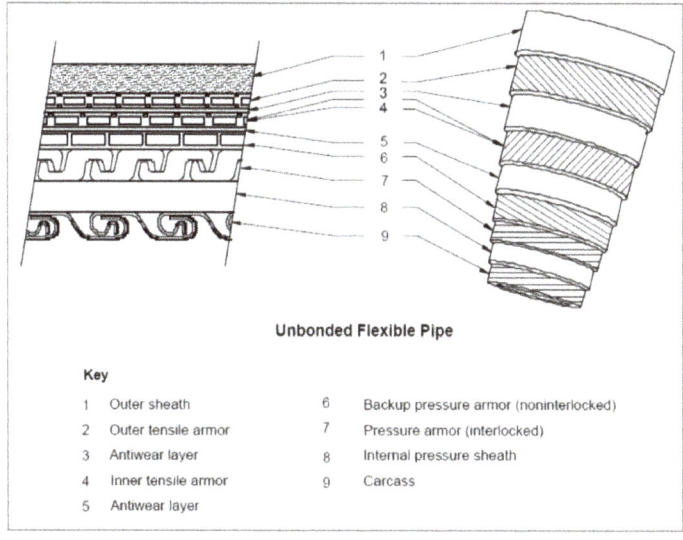

Fig. 5. Unbonded Flexible Pipe Cross-section as per API 17B

Eriksen et al. (2023) developed the design of reliable and cost-efficient cover solutions for unbonded flexible pipes. A new DNV guideline introduce the "Allowing Irreversible Displacement" design approach which permits the pipe to move within the cover, provided that the end-of-life limit states for both pipe and cover are satisfied. The research includes a description of how a model following this approach can be defined, including considerations on load case matrix and a discussion on relevant sensitivity cases to ensure conservatism in the cover design. A case study was included to show how the principles of the guideline may be applied in UHB design.

Xiao et al. (2024) investigated the structural optimization of a polyhedral composite subsea pipeline (cylinder) under pressure and thermal fields. The pipeline was confined tightly and deforms inward when it is subjected to external loadings. The interface was frictionless between the pipeline and its surrounding medium. Based on the above assumptions, thin-walled shell principles, and an admissible displacement function, the potential energy of a pipeline per unit length was obtained explicitly by simplifying the radius and bending rigidity. After taking the first derivative of the potential energy, two equilibrium equations were obtained when these equations are combined, the critical buckling pressure of the polyhedral pipeline is expressed analytically, including the effects of temperature effects. The present analytical study was compared with other numerical and experimental results, and excellent agreements are reached. Parametric studies indicate that the configuration factor decreases with an increase of thickness-to-radius ratio, the number of sides, and the increase of the temperature variation, respectively. Therefore, a polyhedral pipeline with a low thickness-to-radius ratio is recommended for engineering practices, as it can reduce material costs. Figure 6 shows the polyhedral pipeline model in a medium, visualizing Xiao et al.'s research setup.

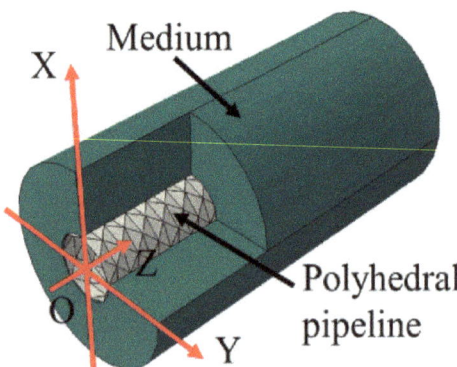

Fig. 6. The model of a polyhedral pipeline surrounded by a medium.

4.3 Pipeline Anti-Corrosion

The development of the marine industry is also directly related to the development of submarine pipelines. For many years, the research on pipeline corrosion has mainly focused on the development of anticorrosive coatings. These coatings form a protective film on the pipe's surface, preventing corrosive media from directly contacting with the pipe wall. The corrosion mechanisms in petroleum pipelines involve the cooperation of dissolved oxygen, carbon dioxide, and hydrogen sulfide, which contribute to the degradation of pipeline integrity. Anticorrosion technology in oil and gas pipelines must address issues such as top-of-line corrosion (TLC),), which occurs in multiphase flow. When water vapor condenses on the top and sides of the pipeline, it causes severe corrosion. However, at low condensation rates, corrosion is inhibited by the corrosion products that accumulate within the condensation region. Wang et al. (2023) developed an experimental method for detecting pipe steel TLC. Through macroscopic cell current mapping and local electrochemical measurements, the droplet position, droplet residence time, water condensation rate and local TLC, can be well determined. Goh et al. (2024) evaluated various acid gas treatment methods to minimize corrosion and mitigate flow protection risks in pipelines. However, in recent years, the main treatment is to choose corrosion resistant metal or non-metallic materials, metal materials, add corrosion inhibitor, coating corrosion and lining corrosion. The self-nano alloying technology of nickel plating on low-carbon steel represents a comprehensive application of advanced internal anticorrosion techniques, Combined with seamless metallurgical surfaces, this high-corrosion-resistance pipe material is considered a key development trend for the future.

4.3.1 TLC in Wet Gas Flow

Pipeline Top-of-Line Corrosion (TLC) is a globally important phenomenon in submarine pipelines. The TLC is driven by the condensation of water rich in corrosive gases. TLC occurs in multiphase flow and causes severe corrosion when water vapor condenses on the top and sides of the pipeline. Faisal M et al. (2024) discussed the corrosion mechanism of major acid corrosion of TLC, which is crucial for the pipeline design and operation of wet acid gases. When there is a significant heat exchange between the pipe and the environment, the condensate water will condense on the top of the pipe, leading to serious corrosion phenomenon. It was found that when water vapor condensation dominates, the mechanism relies on a high condensation rate over $0.25 \, g/m = 2 \, s$ to maintain corrosion because it is suppressed by the corrosion products within the condensation region. For the acidic TLC mechanism, no minimal condensation rate for TLC is inhibited. Experiments were conducted to determine the TLC risk of new sulfur gas mesh. Examining the effects of temperature, condensation rate, and of organic acids on TLC corrosion and iron sulfide (fe) scale formation. They confirmed that TLC corrosion rates in an acidic environment are mainly dependent on the ferric sulfide characteristics, which function as a function of temperature. At a gas temperature of 67 °C, the maximum weight loss rate is 0.28 mm/y. The surface is completely covered with a dense layer of iron sulfide corrosion products, which seems to provide some natural corrosion protection. However, at a gas temperature of 40 °C, the maximum corrosion rate was 1.02 mm/y. The entire surface was covered with a layer of 50-100 μm thick of iron sulfide corrosion products, which seemed to provide little corrosion protection. Based on the wire harness electrode (WBE) combined with local electrochemical measurement, Wang (2023) developed an experimental method to detect pipe steel top line corrosion (TLC). The method accurately determines droplet position, droplet residence time, water condensation rate, and local TLC are well determined by macroscopic cell current mapping and local electrochemical measurements.

4.3.2 Acidic Gas Treatment, Flow Support and Corrosion Management

Abdolreza et al. (2023) investigated corrosion and gas hydrate inhibitors when co-injected in oil and gas pipelines, developing newly synthesized dual-use inhibitors (DPI) to overcome compatibility challenges between the inhibitors, and studied the inhibitory activity of DPIs. The results shows that the inhibitor is highly adsorbed onto metals and can form a protective layer on the surface of low carbon steel. DPIs provide potential hybrid inhibition of corrosion and gas hydrate formation for flow assurance applications and reduce operating costs. Additionally, they enhance the performance of different hydrogen gas operations and improve design techniques for the removal of H_2S and CO_2. Aeman (2020) discussed the basic thermodynamic and kinetic models of the acidic gas removal process, and the absorber and stripper were improved through experimental and simulation work, and the correlation optimization studies based on chemical absorption are performed. Subpipeline flow assurance analysis considering the availability of LNG-FPSO upper systems by Kim et al. (2021) A hydrate management strategy was established, including PVCap experiment, usability analysis of LNG-FPSO upper system, pipeline hydrate risk analysis, and calculation of PVCap injection concentration. The experimental data required to determine the injection concentration of PVCap, were obtained by measuring the hydrate induction time of PVCap at supercooled temperatures of 6.1, 9.2 and 12.1 °C.

Since the erosion of smooth pipes has been widely studied, but rarely the erosion of flexible pipes, Zheng et al. (2023) conducted an erosion test study on a full-size flexible pipe with an inner diameter of 15 cm. In the straight pipe section and the 90° curved pipe section, the erosion test is conducted with the flow rate of 2.5 m/s with sand particles of different particle sizes. The bend corrosion test time is 425 h, and the straight pipe corrosion test time is 1275 h. The thickness loss was obtained by measuring the carcass cross section using a gold-y microscope. The results show that the erosion rate of the front edge of inner layer is higher in both straight tubes and bends; the corrosion rate is highest at 20–35° along the bends. The results of Zheng et al. can provide help in determining the service life of underwater flexible pipelines at risk of erosion. Yunze et al. (2023) studied the pure corrosion, erosion-corrosion and pure erosion behavior of X65 pipeline steel in acidic NaCl solutions with different pH values (1.5–5.0), and conducted an in-depth study on the erosion and corrosion interaction of pipeline steel working in acid slurry. Combined with surface characterization and electrochemical measurements of corrosion and erosion properties of steel. Through negative, polar protection, the pure erosion properties were studied. The propagation of local erosion corrosion at different pH values was detected using wire harness electrodes. It is concluded that pH = 3.5 is the critical value for the cathode reaction transition under erosion-corrosion, and in acidic fluids pH <3.5, the reduction of H+ is the main cathode reaction, leading to general erosion corrosion features interspersed with small pits initiated from the inclusions. At pH <3.5, a negative erosion - enhanced corrosion (E - C) + 3C network was found due to the material loss of the iron - containing material due to wear. In the pH 3.5 acid slurry, oxygen reduction becomes the main cathode reaction, leading to the formation of erosion and corrosion pits and craters. Elimination of the rust layer will result in a positive E-C performance at pH > 3.5. At pH 3.5, significant local corrosion occurs, where the main anode site can be seen. As oxygen reduction occurs, significant erosion pits and craters occur. More severe corrosion enhanced erosion (C-E) tends to occur in the pitting area where high macro cell current is recorded.

Using experimental and computational fluid dynamics, Rehan (2023) evaluated the erosion quantification results of the horizontal-vertical (H-V) downward long radius 90° elbow and the horizontal-horizontal-horizontal (H-H) elbow under abrasive circulation conditions using experimental and computational hydrodynamics methods. The experimental results show that the critical erosion finally ends when the angle reaches about 45°, which can be perceived in both the H-V and H-H directions. Excessive pitting and micro cutting are the main causes of bidirectional elbow corrosion wear after grinding collision.

4.4 Artificial Intelligence (AI) and Deep Learning

Imran et al. (2023) discussed state-of-the-art AI methods for the prediction and detection of ocean-related corrosion: (1) predictive maintenance methods and (2) computer vision and image processing methods. Furthermore, a brief description of the AI is described. The results of this review will bring new knowledge about the development of AI and predictive models that can help avoid unexpected failures during corrosion detection and maintenance. In addition, it will expand the understanding of computer vision and image processing methods to accurately detect corrosion in images and video.

Subrata et al. (2021) proposed an optimization method for submarine pipeline route selection using reinforcement learning (RL) algorithms. This method evaluates each possible route by combining seabed stability criteria based on the DNV GL-ST-109 standard with other constraints, such as minimizing pipeline route length, avoiding static obstacles, managing pipeline crossings, and limiting free-span segment length, with the aim of reducing mitigation costs.

Hydrate formation in submarine pipelines is one of the main reliability concerns for flow support engineers. Rapid and reliable assessment of the cooling time (CDT), the time between the closing event and possible hydrate formation in the asset, is essential for operational safety. Existing CDT prediction methods are highly dependent on the use of very complex physics-based models that require substantial computational time, hindering their use in online environments. Thus, a novel method for developing alternative models that can accurately predict CDT after an unexpected closure of a submarine pipeline was proposed by Alberto et al. The (ML) -based hybrid model consists of a support vector machine (SVM) classifier and two artificial neural networks (ANNS). The support vector machine (SVM) classifier assigns risk levels (high or low) to the measured running status of the asset, and the artificial neural network is used to predict CDT in high risk (low CDT) or low risk (high CDT) operation. According to the deep learning model, Alberto et al. (2021) proposed a hybrid optimization method based on the typical machine learning algorithm for the optimization design of the pipeline path. The proposed route optimization design approach was compared with a popular multi-objective optimization method, the genetic algorithm.

AlAzri (2023) found that the existing synthetic aperture radar (SAR) method for detecting marine pipeline leakage is limited by the complexity of the algorithm, which is difficult to process unbalanced data sets, select the best features and relatively slow detection speed. The use of deep learning method can accelerate the speed of oil detection. The convolutional neural network U-Net segmentation model for pipeline oil leak detection has shown promising automation results.

4.5 Summary and Recommendations for Future Work

Thermal buckling, flexible pipe technology, pipeline corrosion, and artificial intelligence are pivotal areas of research that hold the key to unlocking the full potential of subsea operations. As the demand for deep-water oil and gas exploration escalates, so does the complexity of the engineering challenges involved.

Thermal buckling of pipelines is a critical issue that can lead to significant disruptions in the flow of oil and gas and even catastrophic failures if not properly managed. Research in this area is essential for the stability and safety of subsea pipelines.

Similarly, the technology surrounding flexible pipes continues to evolve. These pipelines have better carrying capacity as well as lower construction costs and are essential in deep sea applications. The structural design of flexible pipelines and the study of their load-bearing characteristics are crucial to improve the durability and reliability of flexible pipelines to ensure that they can operate safely under harsh seabed conditions.

Pipeline corrosion is another area that demands focused research efforts. Corrosion not only shortens the lifespan of pipelines but also poses a significant risk to the environment and human safety. Innovative coatings, monitoring techniques, and cathodic protection systems are all active research areas aimed at mitigating the effects of corrosion and prolonging the service life of subsea infrastructure.

Lastly, the integration of artificial intelligence in subsea engineering is a frontier that promises to revolutionize the industry. AI has the potential to optimize pipeline routing, predict maintenance needs, enhance inspection processes, and improve the overall efficiency of subsea operations. By harnessing the power of machine learning and predictive analytics, the industry can move towards a more proactive and data-driven approach to asset management.

5 Risers and Umbilicals

5.1 Steel Catenary Riser

Recent studies have focused on the impact of steel catenary riser (SCR) interactions with the seabed and seawater on fatigue performance in the touchdown zone (TDZ). Janbazi et al. (2023) utilized a coupled Eulerian-Lagrangian (CEL) model to analyze the effects of seabed-seawater interactions on trench enlargement and SCR fatigue, finding that seawater entrainment and SCR oscillations can soften the seabed and enlarge the trench. Kang et al. (2023) examined the nonlinear dynamics and fatigue damage of the SCR system, considering the seafloor response and the generation of superporous pressure. Gao et al. (2023) integrated a pipe-soil interaction model to assess soil strength changes due to SCR cyclic movement. Jia et al. (2023) used OrcaFlex to develop a finite element model of an SCR, analyzing the effects of nonlinear seafloor inhomogeneity on fatigue life, and showed that uneven seabed profiles can reduce damage. These studies provide insights into SCR performance and fatigue damage in the TDZ, crucial for offshore structural integrity.

5.2 Top Tensioned Risers

Sivaprasad et al. (2023) used artificial neural networks (ANN) for data-driven analysis to predict vortex-induced vibration (VIV)-induced fatigue damage in top-tension risers (TTR). They utilized OrcaFlex and SHEAR7 to create a database that was then used to train the ANN model, which successfully provided reliable fatigue estimates with significantly reduced computational costs. Xie et al. (2021) studied the nonlinear dynamic response of TTRs with variable-density flow and VIV, using a Van der Pol oscillator model. Their computational results, compared with experimental data and CFD simulations, showed good agreement and highlighted the signficant influence of internal fluid density fluctuations on riser vibrations and fatigue damage. Duan et al. (2022) investigated the cross-flow VIV of flexible fluid transfer risers, considering the effects of vessel motion and internal flow. Their findings indicated that vessel motion significantly affects top tension and VIV characteristics, including changes in VIV amplitude, vibration frequency, and top tension frequency. Wang et al. (2021a) designed a deep-water riser VIV test system using bare fiber Bragg grating (BFBG) sensor technology and developed a numerical model based on the work-energy principle. Their study showed that the riser's vibration was dominated by the first-order vibration frequency, which increased with velocity under different top tensions, leading to a gradual reduction in the principal frequency and fatigue damage as top tensions increased. Wang et al. (2022) conducted numerical analysis and water tank tests for two key configuration modes of risers, evaluating their hydrodynamic and strength performance under various loading conditions. This research provided valuable insights into the behavior of risers under different environmental forces. Kim et al. (2022) explored the effects of 3D current on VIV using a nonlinear riser model. They compared model predictions with results from a rotating rig experiment and found that accounting for 3D current improved VIV predictions, especially at high velocities. Their study also demonstrated that using unidirectional current for predictions does not always result in the highest fatigue damage, indicating the need for more accurate modeling approaches. Wu et al. (2024) proposed a four-step approach for analyzing and predicting VIV responses. This method involves clustering measured VIV responses, identifying optimal hydrodynamic parameters, evaluating prediction accuracy, and applying classification algorithms for new cases. This approach was further explored by Andersen et al. (2024) using a Bayesian optimization framework, enhancing the prediction of VIV responses in risers.

5.3 Hybrid Riser Systems

Adegoke et al. (2023) analyzed the nonlinear dynamic response of a single hybrid riser to wave excitation, deriving the nonlinear equations of motion for the riser conveying two-phase flow were derived. Theoretical results showed that the longitudinal and transverse responses were coupled at some specific intrinsic frequencies. Korotygin et al. (2023) investigated the strength and stability performance of production hybrid composite flexible risers with composite pressure armor under harsh environmental conditions in Arctic waters. Flexible risers in various global configurations were analyzed at a water depth of 340 m to evaluate the static, dynamic, and laminar performance of their carbon fiber reinforced thermoplastic polymer composite layers. The impact of drift ice in the region was also analyzed. The results indicated that the current riser design, which includes composite layers, is insufficient to ensure the system's integrity without mitigating, the effects of ice loading. The study also provided recommendations on how to improve the service life of lightweight hybrid composite risers under arctic conditions.

5.4 Hybrid Riser Systems

The Free Standing Hybrid Riser (FSHR) system, integrating features of both rigid and flexible risers, offers an innovative solution for deepwater oil and gas development. This system includes a main riser, a buoyancy-can, and a flexible jumper. The main riser, acting as a conduit for fluids, is subjected to harsh ocean conditions and complex hydrodynamics due to wave and current excitations. It is tensioned by a buoyancy-can, typically located 50 to 250 m below the water surface, and connected to the vessel or platform via a flexible jumper. The buoyancy-can counterbalances the rigid riser's weight underwater, reducing unwanted dynamic effects from ocean waves. FSHR systems address challenges posed by deepwater environments and the limitations of traditional steel risers, such as wave motion coupling and fatigue loads.

Zhang et al. (2022) developed a structural dynamic model for FSHR, using geometric nonlinearity to simulate current loads via the Morrison equation and wake oscillator. They investigated vortex-induced motion (VIM) regimes of the buoyancy-can and phase dynamics of the riser and jumper. As presented in Fig. 7, which shows the calculating and testing results of the VIM of the buoyancy can, it provides a visual and intuitive display of the VIM characteristics of the buoyancy - can studied by Zhang et al. (2022). Their research showed that VIM significantly affects the buoyancy-can's average offset and the bending moment at the riser's top, highlighting the importance of considering VIM in FSHR dynamic analysis. Under harmonic vessel motion, the buoyancy-can's VIM exhibited periodic transformations, associated with system oscillation characteristics. Phase differences along the jumper were influenced by frequency component intensity competition, with phase trapping and locking observed under predominant fundamental frequency motion, and phase drifting and slipping occurring when fundamental and super-harmonic frequencies were similar in intensity.

Adegoke et al. (2023) developed a nonlinear dynamic model for a hybrid riser pipe, considering both internal and external flow effects. Figure 8 presents the system's schematic model, which visually depicts the setup of the hybrid riser pipe system under study by Adegoke et al. (2023). This model serves as a fundamental framework for understanding the context in which the following dynamic analysis is carried out. Their study revealed that the riser's longitudinal and transverse motions were interconnected at specific frequencies and vulnerable to resonance at certain wave frequencies, leading to complex multi-frequency excitations. Notably, harmonics resulted in bifurcated periodic solutions, and internal resonance caused energy transfer between axes, inducing axial

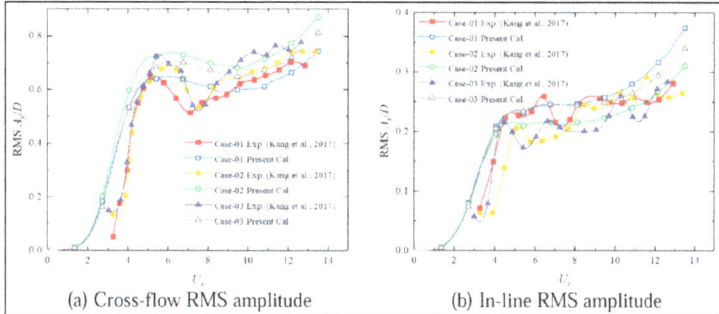

Fig. 7. Calculating and testing results of the VIM of the buoyancy can.

resonance peaks. These peaks were mitigated by the nonlinear anti-resonance effect, which helped in destabilizing the system.

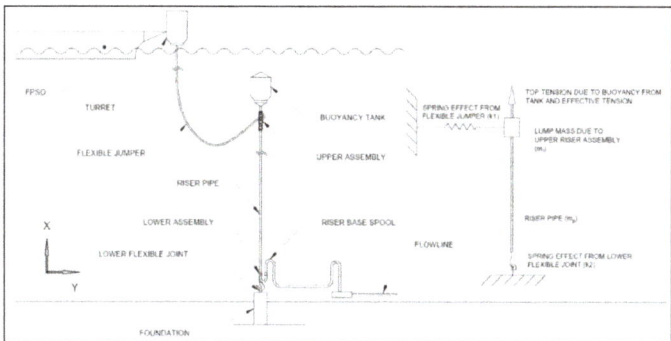

Fig. 8. System's schematic model.

5.5 Flexible Risers Configuration

5.5.1 Design Methods

Flexible risers are essential for transporting ocean resources and face complex loads due to varying oceanic conditions. These risers carry a mixture of oil, gas, and water, which, under diverse pressure and temperature conditions, create time-varying forces that can lead to significant oscillations in internal stresses and reduce the operational lifespan due to fatigue. Accurately modeling the fluid-structure interaction (FSI) between the multi-phase fluid and the riser is crucial for assessing their dynamic response.

Chen et al. (2021) studied the impact of design parameters, such as buoyancy module installation position, buoyancy ratio, and mining vehicle motion, on a saddle-shaped riser using Finite Element Method (FEM) simulations. As clearly demonstrated in Fig. 9, which shows the riser configurations, tensions and curvatures under different buoyancy ratios, it provides a visual representation of the effects of the buoyancy ratio on the riser as investigated by Chen et al. (2021). The figure vividly displays how different buoyancy ratios result in distinct riser configurations, tension levels, and curvatures, offering direct evidence for their finding that the buoyancy ratio significantly affects the riser's configuration and tension. They found that the buoyancy position and ratio significantly affect the riser's configuration and tension, while the mining vehicle's motion influences the maximum stress at the buoyancy installation's center. These insights are vital for optimizing the structural design.

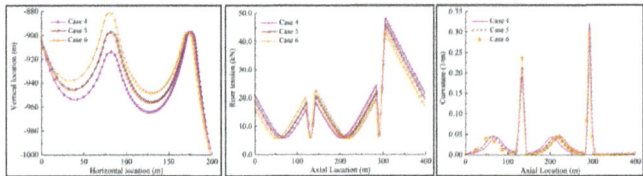

Fig. 9. Riser configurations, tensions and curvatures under different buoyancy ratios.

Vasquez et al. (2021) analyzed the 3D dynamic response of flexible catenary risers under internal slug flow. They found the slug flow, consisting of liquid slugs and gas bubbles, interacted complexly with the riser, affecting its dynamics and vibration amplitudes. The frequency of the slug flow influenced the riser's effective tension and could induce low-frequency oscillations, potentially causing resonance and structural damage. Identifying areas with low dominant frequencies is crucial for preventing structural issues. As depicted in Fig. 10, which illustrates the spatial profiles of oscillation frequencies associated with x displacement responses, the distribution of these frequencies along the riser provides valuable insights into the potential areas of concern.

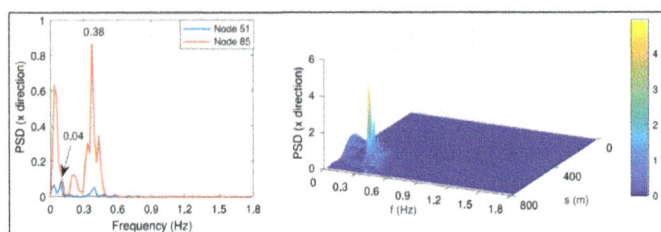

Fig. 10. Spatial profiles of oscillation frequencies associated with x displacement responses.

5.5.2 Nonlinear Dynamic Analysis

Unbonded flexible risers, known for their axial stiffness, ease of installation, and corrosion resistance, are crucial for offshore oil and gas transport. They often incorporate composite materials offering advantages in storage and transportation. These risers consist of intricately combined metallic, polymeric, and textile layers for structural diversity. Their assessment includes global and local analyses, with the global analysis focusing on dynamic response to environmental conditions and platform movements, modeled using finite elements. During installation, tensioner equipment applies compression loads to secure the riser to the seabed. However, excessive loads can cause collapse, requiring replacement. Flexible risers adapt well to seabed conditions due to their low bending stiffness but face design challenges in harsh environments due to stress variations, especially in the connection and touchdown zone, leading to concerns about fatigue damage concerns. In ultra-deep water, where traditional catenary configurations can result in significant displacements and tension variations, more complex solutions like lazy wave risers have been developed. These slender marine risers are prone to vortex-induced vibrations (VIVs), causing fatigue damage. To mitigate VIV impacts, the industry and academia are developing passive control strategies based on vortex formation mechanisms.

Sergio et al. (2023) developed a finite element modeling technique to assess the critical loads required to prevent failure in flexible risers during installation, considering the stresses and strains involved. Their model included the nonlinear plastic properties of steel layers and complex contact conditions between layers. The study found that the pressure armor layer was crucial in resisting crushing loads. This model can be adapted to predict failure in risers under installation loads. As shown in Fig. 11, which presents the relationship between applied loads and displacement of the annulus model, the results highlight the critical load levels at which significant displacements occur, further validating the model's ability to predict failure under installation conditions.

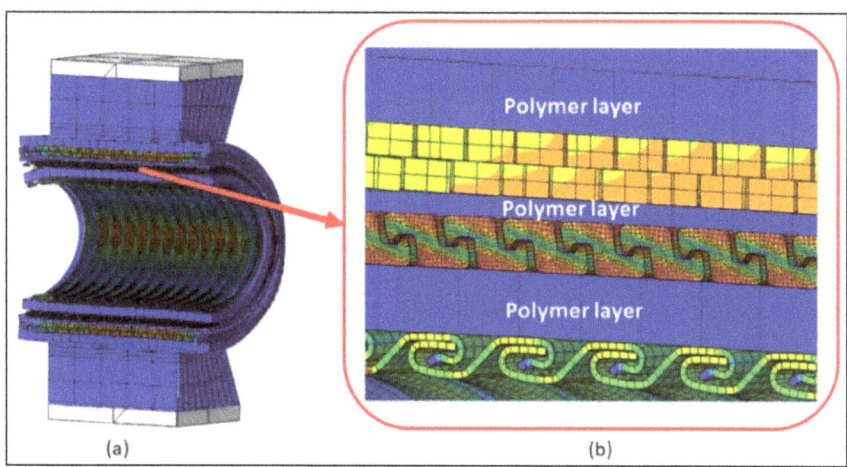

Fig. 11. Applied loads vs. displacement of annulus model.

Leonardo et al. (2022) analyzed the behavior of flexible catenary risers after rupture, emphasizing the impact of hysteretic nonlinear bending stiffness on falling riser dynamics. They found that deterministic methods were insufficient due to the unpredictable nature of dynamic buckling, with variations in initial conditions leading to different responses. A statistical analysis focused on parameters like riser slenderness and tangential drag coefficient, revealing that normal soil stiffness had a less significant impact. The study also used random wave analysis to define a safe zone around the riser, reducing collision risks with submarine elements. This analysis provided strategies for understanding post-failure behavior and mitigating risks. Figure 12 illustrates the configuration of the post-failure of a flexible catenary riser, highlighting the complex dynamics and potential trajectories of the riser after rupture.

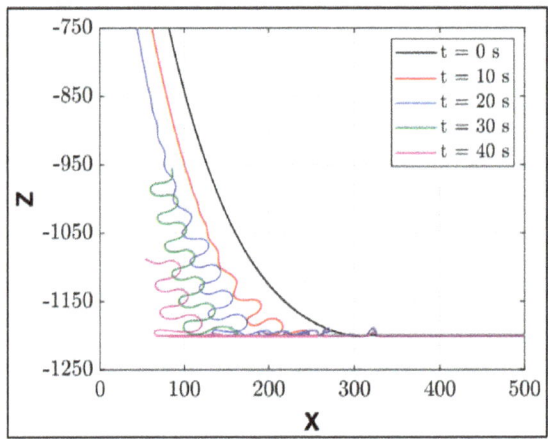

Fig. 12. Configuration of the post-failure of flexible catenary riser.

Beatriz et al. (2023) studied the post-failure behavior of flexible lazy-wave risers, focusing on the impact of dynamic buckling and the potential for damage to nearby equipment. The research highlighted the stochastic nature of the problem, with structural responses being highly sensitive to initial conditions. Key factors influencing post-failure behavior included the longitudinal drag coefficient, riser slenderness, and the use of a nonlinear moment-curvature relationship. Considering multi-directional currents was also crucial for determining safe zones around the risers. This analysis underscored the complexity of predicting riser behavior after rupture and the need for comprehensive modeling to reduce risks to surrounding infrastructure. Figure 13 illustrates the deformed configuration of the post-failure of a lazy-wave riser, a providing visual representation of the complex deformation patterns and potential trajectories that can occur after rupture.

Li et al. (2022) studied the effectiveness of serrated splitter plates in reducing flow-induced vibration (FIV) in flexible risers. Figure 14 presents the models of conventional and serrated splitter plates, which are fundamental to understanding the objects of research in Li et al.'s study. By clearly showing the distinct structures of these two types of splitter plates, the figure helps readers visualize the differences between them, which is crucial for grasping how their different geometries lead to various effects on FIV. They found that while conventional splitter plates caused a steady increase in FIV amplitude with reduced velocity, those with serrated plates initially showed stable amplitudes followed by a significant decrease and chaotic oscillations at higher velocities. Serrated splitter plates, especially with uniform serrations, were most effective at high reduced velocities in minimizing FIV.

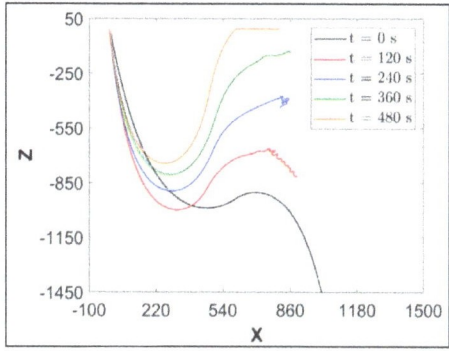

Fig. 13. Deformed configuration of the post-failure of lazy-wave riser.

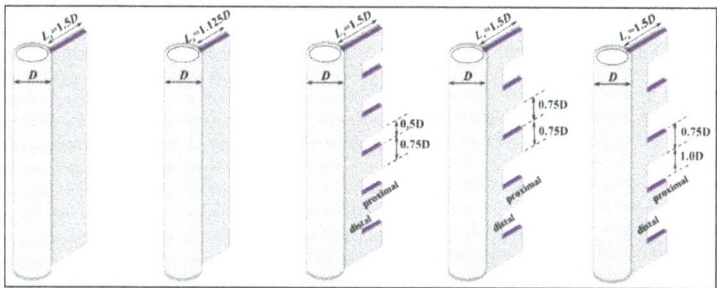

Fig. 14. Models of conventional and serrated splitter plates.

5.5.3 Full-Scale Experiments

The bend stiffener in flexible risers, often made of polyurethane, helps transition between the riser and the floating unit's bending stiffness. Despite being modeled with elastic responses, its mechanical behavior is time-dependent and can degrade over time. The helical steel wires in the tensile armor layer are crucial, especially in the bend stiffener area where they interact with neighboring layers under varying curvatures. Predicting their mechanical behavior under tension and bending is crucial for riser integrity, as these wires are prone to failure under such conditions. Accurately assessing their slippage is key to forecasting the risers' performance under complex loads.

Lu et al. (2023) performed experiments on a 6″ flexible riser and bend stiffener to study their mechanical behavior. They used a horizontal rig to measure strains at different riser rotations and conducted uniaxial tensile tests on the bend stiffener. A nonlinear finite element model (FEM) that considered contact interaction and interlayer friction was used to simulate the riser's response to combined tension and bending loads with different varying friction coefficients. The study found good qualitative agreement but significant quantitative differences between the numerical simulations and experimental data. Figure 15 shows the test rig with a 5 MN axial capacity, which was used to conduct the experiments on the flexible riser and bend stiffener, highlighting the experimental setup employed to measure strains and load responses.

Fig. 15. Test rig with 5 MN axial capacity.

Tang et al. (2022) developed a monitoring technique using 3D digital image correlation (DIC) to assess the slippage of helical wires in flexible risers under tension and bending. They tested a riser specimen using a four-point loading setup and found that the total slip distance of the wires followed a sinusoidal pattern during bending. The lateral slip component was typically smaller than the axial slip but still significant. An increase in axial tension with constant bend curvature also resulted in a greater total slip distance, highlighting the dependency of wire slip on axial load. Figure 16 presents the schematic of the four-point tension and bend loading test setup with the flexible riser assembly, illustrating the experimental configuration used to measure the slippage of helical wires.

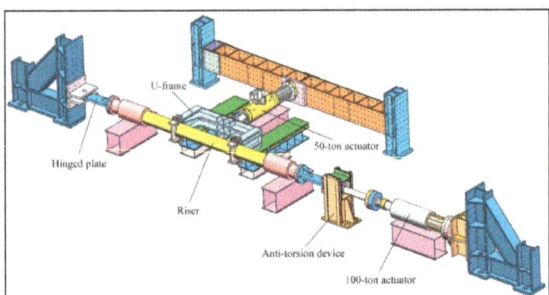

Fig. 16. Schematic of the four-point tension and bend loading test setup with flexible riser assembly.

5.6 Umbilicals

5.6.1 Design Methods

Subsea umbilicals, crucial for connecting host structures to subsea facilities, consist of electrical cables, fiber optics, steel tubes, and hoses, used for production and well control. They include power cables for energy supply and signal/communication cables for remote monitoring. Protective layers like polymeric sheaths or metallic armor wires shield the conductors from damage. Research emphasizes maintaining the bend radius and tension within safe limits, with global and local analyses ensuring durability by assessing stresses and strains in the components. Local analysis is particularly important for evaluating strain and stress in each element under global loads, indicating the umbilical's suitability for specific applications.

Finite element analysis is essential for modeling umbilicals due to the complexity of analytical solutions, particularly for 3D models that capture the effects of helical armor layers, which can pose convergence challenges due to numerous contact problems. In deepwater installations, the lazy wave configuration is favored for its ability to manage large offsets and reduce hang-off tension, with the hang-off inclination angle significantly influencing mechanical behavior. Design processes utilize capacity curves to relate tension and curvature, and subsea cable sizing is determined by power requirements and ampacity.

Material selection for umbilical tubes often favors super duplex stainless steel for its strength, while thermoplastic materials like polyethylene are used for hoses within certain temperature constraints. Electrical conductors are insulated with solid dielectric XLPE, and extruded semiconducting layers along with copper tape provide shielding for optimal electrical performance.

Armor layer choices are based on stability and cost considerations, commonly incorporating polypropylene fillers. For water-blocking sheaths, materials like lead are used for their impermeability and weight, though alternatives may be selected due to environmental concerns. Umbilical outer sheaths are designed to be moisture-resistant and include UV stabilizers to protect against sunlight damage.

Fiber-optic cables, made from glass or plastic strands, are preferred over copper cables for long-distance communication due to their lower attenuation, leading to less signal loss, and their immunity to electromagnetic interference, ensuring reliable data transmission.

Yan et al. (2023) presented an LSTM-based method for predicting umbilical cable top tension responses, using historical motion data to train the model. They found that more training data improved prediction accuracy and model stability. In a comparison with ARIMA, LSTM outperformed in forecasting cable tension, achieving precise results with a 0.3-s data sampling rate, highlighting its value for engineering applications. Figure 17 shows the predicted results by the LSTM neural network model of the time series of the top tension response of the severe hydrodynamic model, with specific time intervals (a) 8640 s–8700 s and (b) 10740 s–10800 s, demonstrating the model's effectiveness in capturing tension variations over time.

Yin et al. (2023) proposed an advanced mathematical optimization model for the cross-sectional layout of umbilicals. They combined the genetic algorithm (GA), known for its robust global search capabilities, with the generalized Lagrange multiplier (GLM), which offers rapid convergence speeds. By integrating these techniques, they aimed to solve the optimization problem efficiently. Through numerical simulations, they verified the effectiveness of the algorithm and demonstrated its ability to quickly derive the optimal cross-sectional layout. This approach provides a specific and efficient method for the rapid design of umbilicals. Figure 18 illustrates the algorithm for the optimization

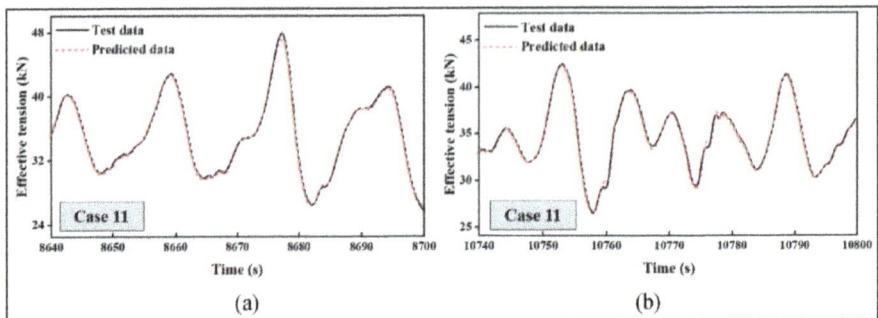

Fig. 17. The predicted results by LSTM neural network model of the time series of the top tension response of severe hydrodynamic model (a) 8640 s–8700 s; (b) 10740 s–10800 s.

process using GA and GLM, highlighting the steps and integration of these techniques to achieve the optimal cross-sectional layout of umbilicals.

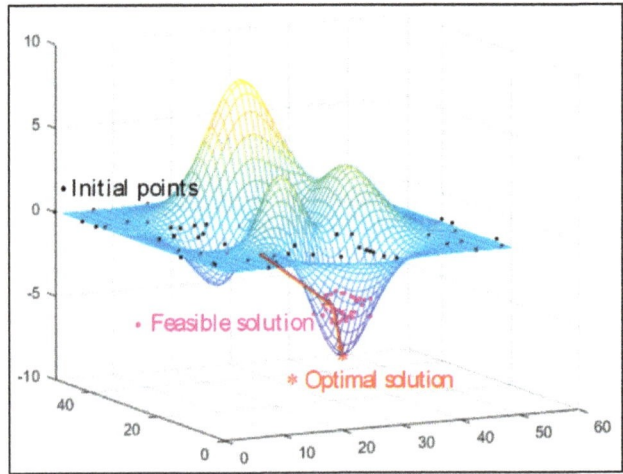

Fig. 18. The algorithm for the optimization process with GA and GLM.

5.6.2 Bending, Contact and Friction

A capacity curve is a useful tool for depicting the maximum stress levels in umbilicals under various load combinations. To accurately account for second-order effects such as axial tension, radial displacement, and torsion, a non-linear calculation is required. Analytical solutions may overestimate axial stiffness, leading to inflated allowable capacity estimates.

Euler beam theory can approximate umbilical behavior, considering the influence of contact pressure and axial strain on the neutral axis. Global curvature's impact is important, It is voften less significant than helix curvature's effect on stress. Fillers, which support and transfer loads within umbilicals, are crucial for accurately predicting load behavior prediction. Traditionally omitted in modeling, recent research suggests that their inclusion is necessary to avoid overestimating friction stress and overly conservative fatigue results.

Modeling fillers presents a trade-off between computational efficiency and accuracy. Solid elements provide a detailed model but increase computational cost, while beam elements may be less accurate. Balancing these factors is essential for effective umbilical design and analysis.

Yin et al. (2021) conducted experiments to analyze the interlayer friction and wear between armor wires in umbilicals, which are crucial for umbilical mechanical properties. They observed that the coefficient of friction (COF) increased with sliding cycles under dry conditions due to slippage between layers. The COF evolution was divided into three stages: initial slow increase, a transition with a larger increase, and a final stabilization. The introduction of nylon fiber tape between armor wires reduced the COF by 32% to 40% as shown in Fig. 19, which presents the friction model and wear scar of the armor wire, illustrating the impact of the nylon fiber tape on reducing friction and wear between the wires.

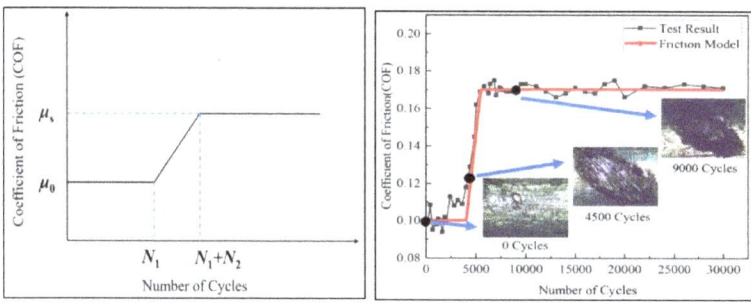

Fig. 19. Friction model and wear scar of armor wire.

Yang et al. (2021) developed a method to calculate the nonlinear tension-torsion coupling effect in umbilicals which arises from their helical structure and radial behavior influenced by polymer materials and component gaps, They employed a finite element model to analyze the umbilical's nonlinear response to tension, considering the material's stress-strain relationship. The study demonstrated that radial stiffness significantly affects the tension behavior. This methodology is useful for evaluating the tension-bending capacity and fatigue life of umbilicals, and was validated through experimental tensile tests. Figure 20 shows the comparison of test and numerical results of tension with axial strain, as well as the curve of radial stiffness with radial pressure, highlighting the accuracy of the developed method and its ability to capture the complex behavior of umbilicals under tension.

5.6.3 Corrosion and Abrasion

Umbilicals under bending or tension exhibit varying stress distributions depending on the force transmission methods. Galvanized steel, used for subsea cables and umbilicals,

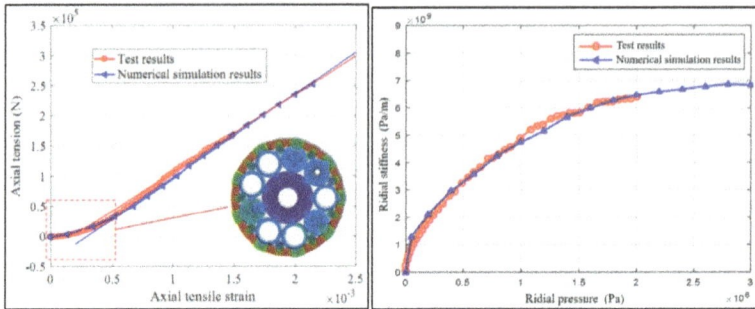

Fig. 20. Comparison of test and numerical results of tension with axial strain and the curve of radial stiffness with radial pressure.

offers high strength and fatigue resistance but may lack sufficient corrosion protection in seawater. Coating the steel armor wires with bitumen, which has complex rheological properties, can enhance corrosion resistance. Factors such as velocity, temperature, and bitumen quantity affect stress distribution in dynamic cables. Understanding the viscoelasticity of bitumen is crucial for analytical models that predict cable bending stiffness and mechanical behavior.

Reda et al. (2021b) researched the abrasion resistance of subsea cables/umbilicals, particularly the effect of lateral movement on their LLDPE outer sheaths. After 40,000 cycles corresponding to 12 km of movement, the sheath showed significant wear, especially at the touchdown point. Abrasion was observed after just 3 km of movement. The study recommends increasing the sheath thickness by 0.6 mm and using additional abrasion protection in high-wear areas to enhance cable durability. Figure 21 shows the reciprocating abrasion test rig with the umbilical sample fixed to the steel frame, illustrating the experimental setup used to simulate lateral movement and assess the wear on the LLDPE outer sheath.

Fig. 21. Reciprocating abrasion test rig showing the umbilical sample fixed to the steel frame.

5.6.4 Submarine Power Cable

Submarine power cables are crucial components in wind farm engineering, serving as vital conduits for transmitting ransmitting the electrical current generated by wind turbines or wave converters. However, these cables are subjected to combined loading in deep-sea environments, which can affect their integrity and safety during installation and operation. Dynamic power cables are complex structures, consist of various metallic components, including copper or aluminum cores for conducting electricity and steel armor layers for protection. These components are arranged in a helical configuration, with plastic sheaths providing additional insulation and fillers securing them in place. The movement of platforms, ocean swell, and current flow impose mechanical stresses on the cable, leading to potential fatigue issues, especially under cyclic bending loading conditions. To analyze the cable's response to varying sea states, a numerical beam model is employed, considering hydrodynamic loads and platform motions. The stiffness characteristics of the beam are crucial inputs for this model, as they dictate the cable's dynamic behavior. Additionally, Estimating the local stresses within the cable's components is essential for conducting fatigue analyses. There are several methods are available to obtain the overall behavior of the cable and the local stress state. These may include analytical methods, finite element analysis (FEA), or full-scale experiments, each offering insights into different aspects of the cable's mechanical performance. By combining these approaches, engineers can gain a comprehensive understanding of the cable's behavior and make informed decisions regarding its design and operation in challenging marine environments.

Menard et al. (2023) used a 3D finite element model to analyze the mechanical behavior and stress distribution in dynamic submarine power cables. They found that the friction coefficient significantly affected the cable's bending during the stick state but had less impact during slip. Residual stresses influenced the bending stiffness and stick-to-slip transition. The contact force distribution along the cable's axis depended on the contact points between armor layers, following a sinusoidal pattern. The mean contact line load was consistent across the bending cycle. Figure 22 illustrates the deflection distribution at one end of the model at a curvature of 0.2 m^{-1} around the (y_1) axis, highlighting the bending behavior and deformation patterns of the cable under the specified conditions.

Poon et al. (2023) developed a method combines global and local analyses to study the dynamic loading effects on submarine power cables, with a focus on hydro-aero-elastic interactions. The study identified factors influencing wire fretting, characterized wear on copper conductors, and modeled fretting wear-fatigue at conductor contacts. The results were consistent with existing bending fatigue data and emphasized the impact of hydrodynamic loading, wear, slip regime, and wire size on fretting fatigue life. Figure 23 outlines the four-step methodology for dynamic global-local modeling of floating offshore wind turbines and submarine power cables, illustrating the integrated approach used to analyze the complex interactions and loading effects on the system.

Chae et al. (2023) investigated the response of submarine power cables to torsion and bending during coiling, using both analytical models and finite element analysis. The research focused on the compressive stresses in the cables' armor wires and the risk of wire buckling, commonly referred to as bird-caging. It was found that shorter pitch

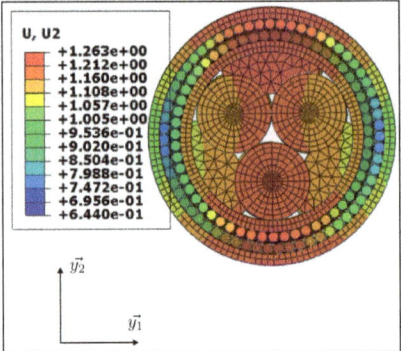

Fig. 22. Deflection distribution at one end of the model at curvature 0.2 m^{-1} around $\vec{y_1}$ axis.

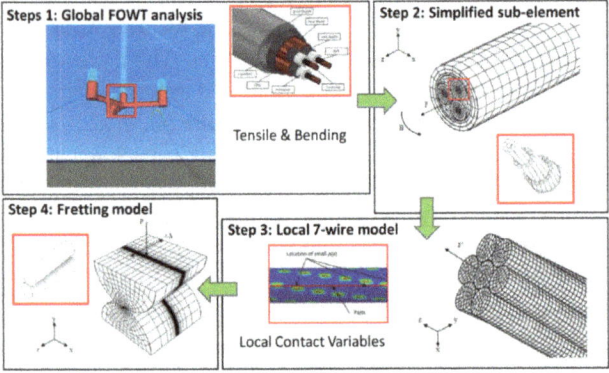

Fig. 23. Four-step methodology for dynamic global-local modelling of floating offshore wind turbine and submarine power cable.

lengths increase torsional stiffness, and the coiling process causes combined bending and twisting, leading to compressive stress on the armor wires. Torsional behavior was identified as the primary factor influencing compressive stress. The authors suggests that for cable design, torsional stiffness is more critical than bending stiffness due to its significant impact on stress during coiling. Figure 24 presents a comparison between the axial stresses predicted by numerical analyses and the radial buckling criterion, highlighting the accuracy of the models in predicting compressive stresses and the risk of wire buckling during the coiling process.

5.7 Alternative Concepts

5.7.1 Compliant Vertical Access Riser (CVAR)

The Compliant Vertical Access Riser (CVAR) is a specialized system for deep water oil and gas exploration, consisting of an upper, transitional, and lower segment with respective counterweight and buoyancy blocks. Compared to Top Tension Risers (TTR), the CVAR offers greater compliance than Top Tension Risers (TTR) and reduces fatigue

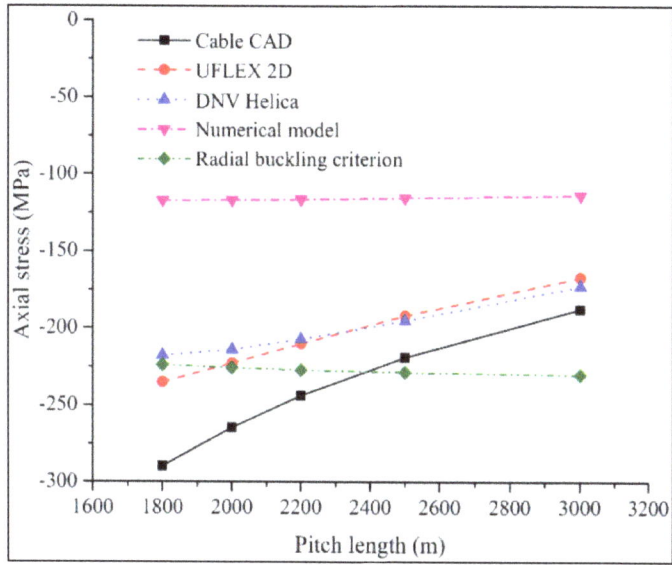

Fig. 24. Comparison between the axial stresses predicted by numerical analyses and the radial buckling criterio.

damage relative to Steel Catenary Risers (SCR) due to a single touchdown point. Unlike Steel Lazy-Wave Risers (SLWR) that require top bending stiffeners to counteract fluid forces, the CVAR's design eliminates this need, reducing costs and improving economic efficiency.

Jia et al. (2023) investigated the static performance of the Compliant Vertical Access Riser (CVAR through numerical analysis. As an innovative riser system, the CVAR offers unique configuration benefits compared to traditional marine risers. The study examined the effects of various factors, including buoyancy section configuration (such as buoyancy factor), external environmental conditions, wellhead offset distance, water depth, current velocity, and the mechanical properties of the CVAR. The parametric analysis demonstrated the suitability of the numerical method for static analysis of marine risers that lack a traditional touchdown section but feature a single touchdown point, with multiple buoyancy blocks subjected to forces of varying directions and magnitudes. Figure 25 illustrates the typical configuration of the CVAR system, highlighting its unique design features and the arrangement of buoyancy blocks that contribute to its enhanced performance compared to traditional risers.

5.8 Summary and Recommendations for Future Work

Future research on risers and umbilicals will focus on optimizing designs for reliability in deepwater, utilizing innovative materials that are resistant to high pressure and weight. The integration of artificial neural networks will be play a key role in refining component cross-sections to meet various stiffness requirements under global loading conditions. These networks have the potential to revolutionize the traditionally iterative

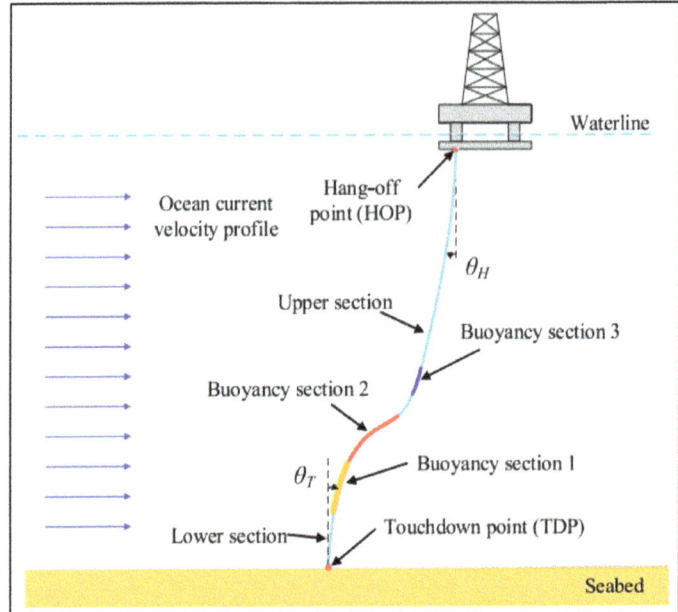

Fig. 25. Typical configuration of CVAR system.

design process by considering multiple design parameters at once, based on extensive data. Research objectives will center on overcoming deepwater technical challenges and improving computational efficiency.

6 Subsea Structural Integrity Management

6.1 Definition and Scope

Subsea Structural Integrity Management (SSIM) refers to the systematic process of securing the fit-for-purpose and integrity of subsea system over the design lifespan until decommissioning or removal. The integrity management (IM) consists of multiple aspects, such as compliance with regulatory as well as organizational requirements, regular inspection, monitoring and assessment of the system conditions, as well as protocols to apply maintenance and repairs.

Subsea Structural Integrity Management is a critical discipline within offshore industry that focuses on ensuring the safety, reliability, and longevity of subsea structures, such as pipelines, risers, and production facilities. As offshore oil and gas development expands into deep and ultra-deep waters, the management of the structural integrity of subsea and underwater facilities faces increasing challenges. Those structures operate in harsh and challenging underwater environments, where they are subjected to extreme pressures, corrosive conditions, and dynamic loads. Ensuring the integrity of subsea structures are essential in protecting the environmental and maintaining the safe production.

The concept of subsea structural integrity management has evolved significantly over the past few decades. Initially, the focus was primarily on the design and construction phases, with limited attention to long-term maintenance and monitoring. However, due to the fluctuation of oil prices and global economic events, facility owners are increasingly prioritizing the reuse and life extension of existing facilities over investing in new major constructions. With the increase of Life extension projects as many platforms approaching the end of their service life, the SSIM is becoming increasingly crucial to prevent accidents which can lead to environmental pollution and financial losses. For newer projects, the digital transformation of SSIM has become a key objective. By applying data analytics tools, facility owners can better manage the integrity of offshore assets and predict abnormalities before an incident occurs with applications of data analytics tools to allow owners to better manage the integrity of an offshore assets and predict abnormalities before onset of an event. The following paragraphs will discuss the Integrity Management methods and techniques, recent advancements and potential applications.

6.2 Integrity Management of Risers

A riser is a crucial component of a subsea production system, responsible for transporting hydrocarbon products from the subsea structure to the surface facility. To ensure the safe operation of riser systems throughout their design life, many regulatory authorities recommend managing riser operations through an integrity management program.

According to Minerals Management Service (MMS) under United States Department of the Interior (2007), Riser Integrity Management (RIM) can be defined as a continuous assessment process applied throughout design, fabrication, construction, operations, maintenance and decommissioning to assure riser systems are managed safely and to prevent major accidents. This lifecycle approach toward RIM is the responsibility of the owner and operator although guidance and minimum inspection and testing requirements are shared by regulatory authorities. RIM addresses various failure drivers such as pressure, temperature, service loads, fatigue, corrosion, fabrication, installation, and accidental damage. These can be managed through methods like inspection, cleaning, testing, monitoring, and integrity management (IM) procedures. Effective data and knowledge management of system records is essential for a successful IM program. Ideally, RIM data should be integrated into the main floating facility monitoring system, with records automatically transmitted to shore for long-term, secure storage and analysis.

As per the API RP 2RIM (2019), the integrity management process relies on gathering and collating information on the riser, periodically evaluating the data, and using the evaluation to set a strategy for subsequent inspection and monitoring. A typical IM process is shown in Fig. 1. As can be seen from the flowchart, an important link is data acquisition which can influence the riser fitness-for-service decision making loop. API RP 2RIM defines riser monitoring system into two main categories:

- Condition monitoring is concerned primarily with ensuring the conformance of the static riser arrangement to the specified design requirements and functional design conditions, such as temperature, pressure, hang-off angle, top tension, corrosion rate, fluid composition, etc.

- Riser response monitoring is primarily focused on the dynamic response of the riser. These systems are often more complex than condition monitoring systems and can involve multiple instruments placed along the riser length. The objectives of these systems are typically to capture fatigue loads, extreme loads, strain and stress, clashing, etc. For flexible pipe risers, response monitoring can also include curvatures, hang-off-angle, and displacements at key locations.

API RP 2RIM listed typical in-service parameters associated with riser system operation that can be monitored (depending on the riser system) and associated typical performance criteria such as Current speed, Wave height and period, Host FPS excursions, FPSO rotation/weather vaning, Internal corrosion, External corrosion, Fluid composition, VIV, Erosion, Temperature, Pressure, Service loads, TTR tensioner stroke range. In most cases, exceeding the performance criteria does not lead to immediate failure but can prompt further inspection or evaluation to confirm short and long-term integrity.

Song, H. et al. (2020), proposed a riser integrity management (RIM) program based on successful industry practices and the riser integrity management standard API RP 2RIM. It involves reviewing the riser design and as-built data, measuring key field data, and using a continuously calibrated digital model with riser response to monitor the riser's structural health. Key measurements include vessel motion, SCR top hang-off angles and tensions, and full water column Current velocity and direction measurements (ADCPs). A ROV inspection plan is also proposed to supplement the inspection data. A digital twin is set up to identify the trend of the riser remaining fatigue life and provide updates on the inspection and maintenance program.

In this paper, Song, H. et al. (2020) introduced how each type of measurement can be correlated with the riser stress components by empirical equations. The authors also emphasized that the responses near touch-down zone are more complex as there is no strong correlation between the top inclination angle and the maximum bending stress in the touch-down zone based on the analysis with riser model. Installing sensors near the touch-down zone is also not feasible due to durability issues. In order to address such an issue, a digital twin carrying out the real time riser analysis is suggested to monitor the stresses in the touch-down zone.

Figure 26 illustrates the recommended Integrity Management (IM) Process as per API RP 2RIM, providing a comprehensive overview of the steps involved in developing and implementing a riser integrity management plan. This figure highlights the systematic approach to ensure the structural integrity of riser systems throughout their design life and life extension phases.

The paper of Hao Song et al. systematically introduced a method of how a riser integrity management plan following the newly published standard recognized by a local authority can be developed and how it can help to ensure the structural integrity of a riser system for design life as well as the life extension phases. It also sheds light on further research opportunities in the development of calibrated digital model by joining the industry's push toward digitalization such as digital twins, machine learning, and artificial intelligence. For example, since the analysis of risers requires global performance inputs from vessel motions, the calibrated riser digital model may be interfaced with the host facility digital twin model (if available) and analyze host vessel sensor data for real time riser performance monitoring. This point is also validated by Renzi

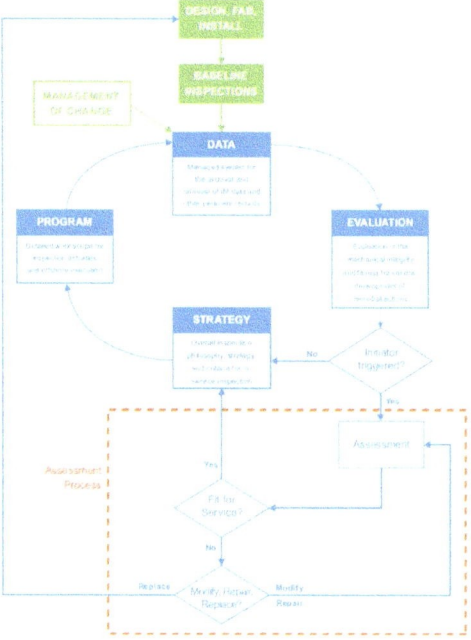

Fig. 26. Recommended IM Process as per API RP 2RIM

D. (2019) and Lee, Y. et al. (2024) that, as compared to analyse the obtained input data and perform analysis by engineers with traditional digital models which takes more efforts and time-consuming, digital twin is an up-to-date virtual representation of the system that provides functionality for performing assessments using either historical, real-time, or forecast conditions. A fully developed digital twin also incorporates a data management system, allowing for documentation and data relevant to the riser systems to be stored and accessed in a convenient and user-friendly manner for riser integrity monitoring.

Joining field recorded data with digital twin model can provide real-time conditions of riser assets and better prepare operators. Further research development may be conducted to train machine-learning models that can calculate the impact and/or failure risks by analyzing real-time data from various aspects. There has been a wide selection of machine learning models that can be used for training in a riser digital twin model. A well-trained model can be used to analyze large volumes of data from sensors and inspections to proactively identify potential issues like corrosion, damage, or anomalies in subsea pipelines and cables, allowing for early detection and preventative maintenance, ultimately improving the overall integrity of the underwater infrastructure by predicting potential failures before they occur. There have been some research efforts in the subsea structures, such as Rachman, A. et al. (2021) discussed the ML applications in the elements of a Pipeline Integrity Management (PIM) process and the aspects of ML techniques (i.e., type of input, pre-processing, learning algorithm, output and evaluation metric) that potentially can be applied in each element of PIM. Looking ahead, it is expected that similar research can be conducted by training ML models with existing riser analysis data. This approach would likely provide a comprehensive understanding of how ML can be applied to predict the performance of each element in Riser Integrity Management (RIM).

Besides the above-mentioned benefits of developing IM with advanced technology, the facility owner will also benefit by a well-maintained RIM plan and database when the facility goes into the life extension phase. There are an increasing number of platforms near service life. Nowadays, the facility owners tend to prefer to extend the service life instead of constructing all new facilities at oil fields that still have resources to exploit. In order to accurately evaluate the conditions and calculate the remaining fatigue life for a subsea structural component, designers are often faced with partial to little knowledge of the environmental history of a facility, making it difficult to quantify the existing damages. Without such data, designers may have to rely on the empirical estimation method during the original design phase which are typically more conservative than the actual loads that a facility has experienced. With a well-designed and implemented IM, the monitoring history or even the fatigue damage history may be available to life extension designer, making the evaluation of remaining fatigue life more accurate and easier and therefore prolong asset lifespan.

6.3 Integrity Management of Subsea Infrastructure

Unlike facilities on board host platform, subsea infrastructures have no direct access and have very little human interaction and intervention. The system components and design load cases of subsea infrastructures (hereafter referred to as subsea system in this section) are also very different from riser system. The focus for Subsea system IM has now moved to less regularly happening incidents whose effect sometimes will be considerably destructive. With fewer and fewer new installations being developed, the contribution from robust design and quality assured construction to future trends will be much less, whereas integrity management will become significantly more important.

As per Wood Group Guideline to Subsea Integrity Management (2020), general requirements for IM for Christmas Tree (XMT) and Wellheads are described in international standards like NORSOK D-010, NORSOK U-001 and ISO 16530-1. NORSOK D-010 standard focuses on well integrity by defining the minimum functional and performance requirements and guidelines for well design, planning and execution of well activities and operations. ISO 16530-1 document is intended to assist the petroleum and natural gas industry to effectively manage well integrity during the well life cycle by providing:

- Minimum requirements to ensure management of well integrity
- Recommendations and techniques that well operators can apply in a scalable manner based on a well's specific risk characteristics.

Similar to the RIMs process discussed earlier, several methods can be employed to implement the Subsea Integrity Management plan. For example, Wood Group (2020) guideline further states that, for successful well integrity management, the operator must consider the full life cycle risks and have approved procedures for all well workscopes. There should be a defined fit for purpose maintenance, inspection and testing programs.

There should be a defined fit-for-purpose maintenance, inspection and testing programs. In the Wood's guideline, a summary table of In-Service IM measures is tabulated the potential measures which may be applied to the various system sub-elements to mitigate specific threats / risks. These measures reflect the integrity management measures that are commonly utilized currently and can be referenced as a non-exhaustive list when preparing a subsea IM plan.

Unlike the emphasis on prediction of the riser responses in RIM, the development of the Integrity Management plan for subsea infrastructure usually starts with identifying the underlying risks for each component. A variety of process safety assessments are used by industry for risk identification and management. These include HAZID (Hazard Identification), HAZOP (Hazard & Operability Study), PHA (Process Hazard Analysis), and LOPA (level of protection analysis for safety instrumented systems). A variety of tools to manage the identified risks and potential mitigations are used by industry. These include basic "what if" approaches to more complex bow tie diagrams, fault trees, FMECA (Failure Mode Effect and Criticality Assessment), PHA (Process Hazard Analysis) and others.

Some research advancement in the Subsea IM is adoption of risk-based inspection strategies vs traditional time-based inspection strategies for subsea structures. Nasri Huang, F. J., et al. (2024) introduced how a subsea field development in Malaysia with two wells switched to a risk-based approach to optimize inspection intervals from the original time-based inspection strategy. The implementation of full diagnostic and data monitoring for the Subsea Production Control System (SPCS) and abundance of data available from the field enables the Operator to optimize the underwater inspection duration and enabled early detection of any system anomalies and degradations, hence preventing unplanned shutdown and preserving the subsea production system asset integrity.

Khan, R., et al. (2022) discussed how effective data management to drive inspection and maintenance strategy is employed in PETRONAS's Upstream Inspection and Maintenance Assurance Guideline (U-IMAGe). In order to ensure effective delivery of Subsea IM and to support and strengthen the U-IMAGe implementation, PETRONAS has embarked on a Subsea Risk Based Underwater Inspection (RBUI) strategy and methodology for its subsea assets. The Subsea RBUI involves identification of threats on a subsea asset, development of the integrity loops and risk assessments. RBUI helps to identify specific threats to a subsea asset, beginning with identifying and reviewing the hazards, threats and failure modes associated with subsea equipment and components during their life cycle. Each hazard, threat, or Likelihood of Failure (LoF) and Consequence of Failure (CoF) is then evaluated, given scores and mapped into the PETRONAS Risk matrix to obtain the Risk Ranking for each subsea asset. The Risk Ranking will then be used to determine the recommended inspection frequency and inspection methods for the subsea asset, focusing on safety, environmental and financial risks, and taking into consideration the inspection costs. By using the Subsea RBUI approach, inspection efforts can be properly prioritized and optimized by balancing risk costs (people, safety, environmental or business related) with inspection costs.

The application of digital twin technology can also significantly improve the efficiencies and accuracies in identifying and locating risks involving subsea infrastructure.

For example, Hamilton et al., 2023 proposed an innovative solution for detecting subsea pipeline leaks using visual twin technology aims to enhance the visualization and analysis of real-time data. By using game engines (such as Unity or Unreal Engine) to build high-fidelity 3D pipeline models, sensor data is integrated into the virtual environment in real time, displaying key parameters such as pipeline pressure, temperature, and flow rate. Operators can intuitively observe all operational data through the visual twin system and simulate different leakage scenarios in the virtual environment. By integrating real-time data, machine learning algorithms, and high-fidelity visualization systems, operators can more quickly and accurately identify pipeline leaks and make corresponding decisions. The future direction of this system includes further enhancing visualization and machine learning models to address more complex pipeline monitoring needs.

With the advancement in data acquisition, an AI and machine learning-based well integrity monitoring solution can be designed and implemented to enhance well safety and reduce the risk of potential catastrophic incidents (Prihandono et al. 2022). When monitoring and predicting wellbore leakage events, AI models can identify anomalies that may lead to leaks by analyzing time-series sensor data and can predict potential risks in advance. Historical data is extracted from sensors, cleaned, and resampled before being used as input for the model. The machine learning model primarily uses LSTM (long short-term memory) networks to detect abnormal patterns. LSTM is suitable for processing time-series data and can capture the non-linear relationships between complex multi-sensor data. In cases where data is limited or anomaly patterns are simple, rule-based methods can be used to detect leaks.

6.4 Subsea Infrastructure Inspection, Maintenance and Repair Techniques and Advancements

A core component of SSIM is inspection and monitoring and one of the most widely used tools for subsea inspection and monitoring is Remotely Operated Vehicles (ROVs) and Autonomous underwater vehicles (AUVs). Referring to ISO 13628-8:2002, ROVs are free-swimming underwater robots used to perform underwater observation, inspection, physical tasks such as valve operations, hydraulic functions, repairs and other general tasks within the subsea oil and gas industry. ROVs are typically connected to surface operation platforms via cables and controlled in real-time by operators. ROVs can also carry tooling packages for undertaking specific tasks such as pull-in and connection of flexible flowlines and umbilicals, and component replacement. ROV systems are typically equipped with cameras, mechanical arms, and various sensors, making them suitable for underwater works. ROV technology is advancing toward greater precision and intelligence. Traditional ROVs, which require cable connections, are complex to operate and expensive, which is why smaller ROVs are gradually replacing traditional large ROVs for shallow water inspections and light intervention tasks. Additionally, ROVs are evolving toward semi-autonomy, enhancing their operational capabilities and efficiency (Iacoponi et al., 2023). In comparison with ROVs, AUVs are underwater robots capable of executing tasks autonomously, usually without requiring cables or real-time human control, and thus can typically cover a large geographic area. AUVs are widely used for long-distance underwater tasks such as subsea pipeline inspections, environmental monitoring, and marine surveys. AUVs can carry out pre-programmed tasks autonomously, using acoustic, optical, and mechanical sensors for data collection. AUV technology has also made significant advancements in recent years, particularly in navigation control and sensing capabilities. New hybrid AUV systems combine the precision operations of ROVs with the autonomy of AUVs, allowing them to perform more complex tasks (Nevoso et al., 2024). Moreover, AUV technology is progressing toward greater automation and data-driven operations, leveraging advanced sensors and artificial intelligence to improve inspection accuracy and task execution. By reducing the need for support vessels and human operators, AUVs not only lower costs but also reduce the carbon footprint.

The latest ROVs and AUVs are usually equipped with high-definition cameras, cutting-edge sensors and even laser scanning systems. The advancements in ROVs and AUVs have significantly improved the capability of ROVs and AUVs in terms of operational scenario. For example, Medany et al., (2024) introduced an innovative pull-down camera system for inspecting subsea conductors. Equipped with a high-resolution camera and powerful LED lighting, the camera system is capable of capturing clear images of the internal structure of subsea conductors in low-light deep-sea conditions. It has flexible navigation capabilities, connected via a flexible cable, allowing it to navigate through the complex structures of subsea conductors and ensure accurate inspection of every corner. The system can also achieve real-time data transmission, with the camera able to transmit images and video data to the platform in real-time, enabling operators to analyze and make decisions during the inspection. The system is designed for internal inspections of subsea conductors, capable of identifying and detecting debris, sediments, marine growth, and structural damage, such as corrosion and mechanical damage. Traditional methods, such as ROVs or human intervention, are often inefficient and costly, whereas the pull-down camera system offers a more efficient and cost-effective alternative. The system was successfully deployed on the North Safa platform and demonstrated its accuracy and effectiveness during operation. With high-resolution imaging and real-time data analysis, operators can promptly detect potential obstructions and structural issues within conductors, thereby enhancing the safety and efficiency of offshore operations.

Besides the development in more advanced sensing and inspection techniques, the ROVs and AUVs commercial market are also advancing towards more autonomous, efficient, and cost-effective underwater solutions. Stump, J. (2021) discussed that Saipem's Hydrone-R, the first commercially deployed underwater intervention drone (UID), started system integration tests for Equinor's Njord field development in the Norwegian Sea. The Hydrone-R is a resident hybrid ROV/AUV vehicle that is proclaimed to be able to operate subsea for more than a year to a water depth of 3,000 m (9,842 ft). It can also execute light intervention tasks and autonomous inspections.

In April of 2024, Nauticus Robotics, Inc. (NASDAQ: KITT) preliminary testing results for its Aquanaut Mk2 underwater robot, as shown in Fig. 27. The company states the subsea robot exceeded expectations during its initial tests in the Gulf of Mexico, including launch and recovery, emergency procedures, and manoeuvrability. Notably, the company's 59′ autonomous surface vessel (ASV) Hydronaut, is specifically designed to transport, deploy, communicate with, recharge, and recover Aquanaut, the company's subsea robot. Enabled by Nauticus' autonomous machine intelligence software platform "toolkit", the vessels maintain a cooperative pairing, thereby powering the capabilities of the fleet to complete complex subsea tasks while significantly reducing the operational cost.

Fig. 27. Nauticus Robotics-AUV aquanaut

Furthermore, the incorporation of AI-driven functions in ROVs, such as those developed by China-based QYSEA Technology, can potentially greatly enhance the efficiency and safety of underwater inspections and operations. QYSEA has various compact ROVs tailored for non-destructive inspections. These AI functions empower operators to lock onto target objects underwater, enhance imaging clarity, enable precise measurements through augmented reality methods, and optimize images for realistic and high-quality visuals. Such advancements not only improve visual data for assessments but also enhance inspection efficiency and diver safety during maintenance operations.

Recent research trend indicates that ongoing projects are driving ROV and AUV technology towards greater capability and autonomy, increased intelligence, and reduced costs. For example, as per Ben Hayden (2023), Oceaneering is collaborating with researchers at the University of Houston and Chevron to create an autonomous robot to detect potential pipeline leaks and structural failures during subsea inspections, in response to an increasing number of severe accidents in the global oil and gas industry caused by damaged pipelines. The aim of this technology is to make the inspection process safer and more cost-effective, while also protecting subsea environments from potential disasters. Jota, J. (2024) also suggested that micro-robots may be utilized for internal inspections of pipelines or navigating complex structures with precision.

In summary, ROVs and AUVs are a critical tool used for subsea structural survey, inspection, operation. The use of ROVs/AUVs allows operators and owners to visually and quantitatively assess the conditions of subsea structures with respect of damage, corrosion. It is easy to predict that the integration of technologies like Artificial Intelligence (AI) and Machine Learning (ML) will revolutionize subsea inspections. When the sophisticated data analysis tools ML/AI models are combined with more readily available and accurate data obtained from latest ROVs/AUVs, detecting abnormalities in subsea structures becomes much more efficient compared to traditional inspection methods used in SSIM.

6.5 Summary and Recommendations for Future Work

In conclusion, subsea structural integrity management is paramount to ensuring the longevity and safety of subsea infrastructure. As subsea systems become more complex and deeper fields are explored, the role of digital tools, artificial intelligence (AI), and machine learning (ML) will continue to grow, providing operators with predictive capabilities and more accurate assessments of structural integrity management. The integration of advanced technologies, particularly digital twins, AI and ML algorithms with cutting-edge sensors and data acquisition and processing techniques, will continuously and significantly enhance subsea integrity strategies. These technologies will not only improve the precision and efficiency of assessments but also allow operators to conduct more frequent or real-time monitoring of subsea conditions, detect potential issues such as corrosion and damage at early stage, and optimize maintenance procedures.

For future research work, ongoing research into autonomous systems for inspections and interventions will help to reduce operational risks and costs, providing safer, more sustainable solutions for the subsea oil and gas industry. Additional exploration of AI and ML applications towards riser and subsea infrastructure IM is also needed, particularly for integration of real-time monitoring systems, enhancing predictive maintenance and identifying potential failure points, will play a critical role in improving riser and subsea system integrity management.

7 Structural Reliability and Safety of Subsea Systems

Subsea Production Systems (SPS) are increasingly being studied in the field of oil and gas extraction. As the global demand for energy increases, especially in deep and ultra-deep water areas, the reliability and efficiency of subsea production technologies becomes even more important. Currently, key technologies for Subsea production systems include Subsea Trees, subsea aggregation systems, pipelines and control systems. The integration and intelligent design of these components improves the efficiency and safety of the overall system. However, these systems still face some challenges in practical application. First, extreme marine environments (such as high pressure, low temperature and corrosion) place high demands on the durability of equipment. Secondly, the complex installation and maintenance process also increases the risk and cost of the operation. In addition, as oil and gas production extends into deeper and more remote areas, remote monitoring and operation of subsea systems need to be improved. In terms of reliability trends, future subsea production systems will focus more on modular design and standardized components to reduce installation and maintenance costs. At the same time, digital transformation will be a focus, using big data and artificial intelligence for predictive maintenance and improving the responsiveness of the system. In addition, environmental protection regulations are becoming increasingly stringent, and future research will pay more attention to sustainability and ecological impact, and develop environmentally friendly materials and technologies. This chapter analyses the pipes, underwater connectors, underwater blowout preventers and control systems of the submarine pipeline system, points out the risk factors of each part, discusses the reliability, and gives safety protection measures and methods.

7.1 Pipeline Reliability Analysis

The oil exploration in deep waters brings among the challenges the difficulty of inspecting subsea equipment to verify and guarantee its integrity due to the occurrence of degradation processes that depend on time. Submarine pipelines are subject to natural disasters (earthquakes, submarine landslides), environmental conditions (temperature, pressure changes), biological erosion (attachment of marine organisms), and human factors (anchor collision, construction activities), which may cause leakage or damage, resulting in environmental pollution and economic losses. Therefore, reliability analysis is particularly necessary. This analysis can identify potential failure modes, assess risk levels, and optimize design and maintenance strategies to improve the overall safety and stability of the pipeline system.

7.1.1 Submarine Pipeline Corrosion Reliability

Based on the pit depth growth model proposed by Mateus Mendes, Miguel et al. (2022) combined with the pitting evaluation of API 579-1 ASME FFS-1, the limit state equation of pitting failure mechanism was established. The corrosion failure probability of pipeline was calculated by using the advanced first and second order moments combined with ASME SA106 pipeline limit state equation (B), and the influence of increasing corrosion pit depth on the structural reliability of submarine pipeline was analyzed, Edilson et al. (2022) based on the Detection (RBI) method, proposed a new method to evaluate the uniform corrosion depth growth model for calculating the failure probability of subsea oil and gas pipelines or equipment. Considering the uniform corrosion failure mechanism referred to in ASME B31G residual life measurement standard, the failure probability was calculated by AFOSM and MCS methods based on the rupture pressure limit state equation.

- API 579-1 FFS is a result of the standardization of fitness-for-service assessment techniques for pressurized equipment that FFS is defined as the ability to demonstrate the structural integrity of an in-service component containing a flaw or damage (American Petroleum Institute, 2016). The assessment procedures can be used to evaluate four types of pitting:

  - Widely scattered pitting which occurs over a significant region of the component;
  - A local metal loss on the surface of the component in a region of widely scattered pitting;
  - Localized regions of pitting;
  - Pitting which is confined to within a local metal loss on the surface of the component.

- Review of ASME B31G standard

  This Standard is designed to assess the remaining life of the pipe, for the burst pressure, considering the degradation due to corrosion. If the equipment under study meets the application requirements listed in the standard, the determination of the remaining life of corroded pipes and its accuracy depends on the level of analysis used. According to the standard, the analysis levels vary between 0 and 3. At level 0, the assessment can be understood as a preliminary prediction of the remaining life of the tube. The loss of thickness is approved if the length of the corroded area does not exceed a limit stipulated by the standard.

  The Level 1 assessment uses information from the maximum thickness loss and axial length of the corroded area to calculate the remaining life. On the other hand, the Level 2 assessment incorporates a greater level of detail, requiring the profile of the corroded surface. Finally, the Level 3 analysis is recommended to be conducted in the evaluation of specific failures, with its own methodology which justifies the use of external loads, boundary conditions, material properties, and failure criteria not addressed by this standard.

7.1.2 Marine Pipe Leakage Failure Reliability

In recent years, the research on the reliability of submarine pipeline leakage has made remarkable progress. Increasingly, researchers are using advanced computational models and simulation techniques to more accurately predict the occurrence of leaks and their potential impacts. These models combine fluid dynamics, materials science and environmental factors to help identify key factors in the risk of leakage. In addition, the development of sensor technology enables real-time monitoring and data acquisition, improving the ability to detect leaks early. The study also focuses on post-spill emergency response strategies to reduce environmental impact and economic losses. Through these comprehensive studies, the safety and reliability of submarine pipelines have been effectively improved, laying a foundation for ensuring the sustainable use of marine resources.

DNV-RP-F107 recommends periodic risk assessments throughout the life cycle of subsea pipeline systems, Andrea et al. (2023) in order to quantify the risk to humans and the environment from an unplanned loss of CO_2 or H_2 seafloor line, modeled a seafloor plume or seafloor gas blowout based on a state-of-the-art integral model developed for seafloor release, taking into account the effects of currents, salinity, temperature, and impurities in the release stream. The goal is to provide appropriate methods for quantitative risk assessment (QRA) of the consequences of CO_2 and H_2 emissions from subsea pipelines. The safe practice of decompressing pipelines via flare systems in the hydrocarbon industry is not sufficient for pipelines in the subsea production process. Subsea production processes present many challenges due to multiphase fluid flow. If the process fails to reach its safe state, it can result in death, significant property damage, and potentially a catastrophic environment, Murugesan et al. (2023) determine the safety status of a process through the results of hazard and risk assessments under critical operating modes. A well-designed Safety Instrument function (SIF) prevents minor failures from escalating into accidents, achieving the comprehensive safety level (SIL) required for subsea High Integrity Pressure Protection Systems (HIPPS). Based on hazard assessment (HAZOP) and Risk assessment (LOPA), the SIL of the subsea HIPPS was determined by reliability calculation using the SIL of the recommended SIF, Zhang et al. (2023) evaluated and predicted the burst pressure of the pipeline based on the experimental data collected, selected the XGBoost model (a machine learning algorithm based on gradient boosting that is widely used for classification and regression tasks due to its high performance, scalability, and efficiency) with the best performance to evaluate and predict the burst pressure, and improved the accuracy of the method through the hyperparameter optimization of Bayes principle. Based on the forward feature selection method and SHAP analysis, the improved XGBoost model is explained and compared with common pipeline industry standards, and the sudden failure of subsea production pipelines is evaluated, Chen et al. (2023) used FMECA method to study the first underwater multi-function manifold in China according to IEC 60812 standard analysis method, DNVGL-RP-A203 and DNV subsea production system integrity recommendations. Failure modes of critical components are identified, key existing control measures are evaluated, and measures for design improvement and risk reduction are proposed. Based on the load resistance model in the structural reliability method, Yu et al. (2019) calculated the failure reliability index of the elbow by constructing the limit state equation of the erosion failure of the elbow and introducing parameters such as deviation factor and uncertainty, so as to predict the failure probability of the elbow in rated service state.

Several efforts are made to ensure that mechanical structures do not fail, that said, one way to prevent failures from occurring is through structural reliability. Melchers (2018) associates structural reliability with calculating and predicting the probability of a limit state violation of a structure at any time during its operational life. Ayuub and Mccuen (2016) define a limit state equation as a function of performance Z, related to structural strength R subtracted from the effects of loads L, according to Eq. 1. R and L are usually functions of basic and specific random variables for each case study.

$$Z = R - L \tag{1}$$

In this way, in terms of the performance function, where the structural strength functions and stress effects are described by probability density functions, one can define the probability of failure, P_f, of a structure that will occur when the performance function is less than zero, as represented in Eq. 2.

$$P_f = P(Z < 0) \tag{2}$$

7.2 Underwater Connector

Subsea connectors are critically important to asset integrity and environmental stewardship, delivering reliable and robust performance for extended service in exacting environments. For this reason, it is essential for connector design and testing methodology to prove subsea connection system reliability, and that can be achieved by correlating analysis and experimental data per the latest industry recommended practice, Liu et al. (2023) based on practical engineering background, established a deep-sea horizontal clamp connector fault tree model and analyzed the bottom event distribution type of its failure mechanism. In order to improve the reliability of the model, the multi-source information method is used to quantitatively solve the reliability probability of the connector by combining prior product information, expert experience and design information. The expert experience was quantified using fuzzy quantitative analysis, while the design information was estimated by developing corrosion prediction models combined with grey theory, Simon et al. (2022) propose a risk-based approach to validation of pipeline repair systems and focus on a risk-based approach that addresses potential failure modes. The methods to solve the structure, sealing and grasping functions of the connectors and the potential failure modes are discussed, and the design verification process of the submarine pipeline repair connectors is described. A type approval process was introduced, including independent design review, sample testing and quality control validation. Parag et al. (2022) addressing the lack of good methods for monitoring key components of clamp-connected systems to match computer analysis problems, A new method for designing the core system capacity of subsea flowline connectors using the elastic and plastic analysis methods of ASME Section VIII Division 2 was developed to provide reliability for the design and verification methods of subsea flowline connection systems. Recent work carried out in a controlled environment illustrates a comprehensive approach to designing and qualifying subsea flowline connection systems and establishes a better methodology which delivers superior reliability Traditionally, API 6A and 17D are the governing design codes for subsea flowline connectors. However, they do not include several key aspects of designing and qualifying a robust connection system. Recently, industry introduced a more detailed recommended practice API RP 17R to govern the design, verification, and validation of subsea flowline connectors. Parag et al. (2022) present a method to design a connector – from the metal seal gasket that is used in a topside four-bolt clamp to the final complex and proved subsea flowline connection system. Figure 28 highlights the major flow of this methodology.

Apply the elastic and plastic analysis methods of ASME VIII to design the core system capacity of subsea pipeline connectors. Improve the reliability of subsea connectivity systems based on the latest industry-recommended practices through correlation analysis and experimental data.

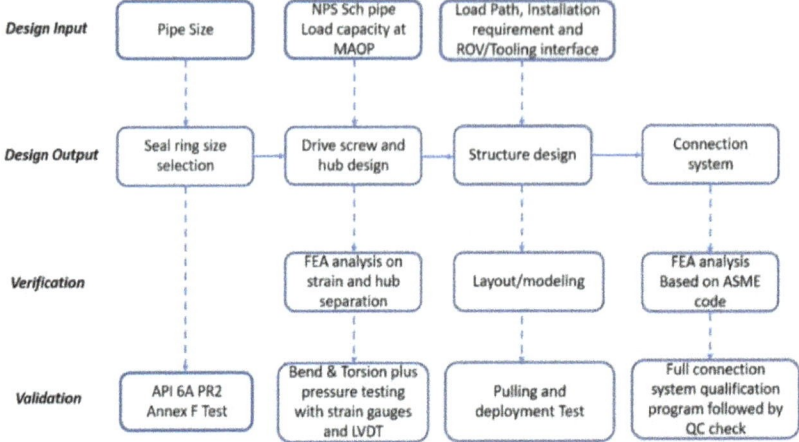

Fig. 28 Design, verification, and validation flow chart for Subsea Flowline Connection Systems

7.2.1 DNV Validation Standard for Subsea Repair Pipeline Connectors

In the oil and gas industry, subsea pipeline repair connectors are used to seal and install damaged or leaking pipelines, and it is necessary to verify this subsea pipeline repair to ensure that it is designed to meet DNV standards to reduce installation and operational risks and improve safety. DNV standards and recommended practices, such as DNV-ST-F101 for "Submarine Pipeline Systems" and DNV-RP-F113 for "Pipeline Subsea Repair" can be used for the subsea repair pipeline connector verification. DNV Standards and Verification Process Related to the Subsea Pipeline Components Currently 65% of the world's offshore pipelines are designed and installed to DNV's technical standards. One aspect of DNV's standards and verification service is that it introduces a levelled description of involvement during all phases of an asset's life. Based on the risk levels of high, medium and low, different in depth involvements might be conducted, which provide to the client an opportunity to establish efficient, cost effective, predictable and transparent verification plans. DNV standards relevant to submarine pipelines including DNV-ST-F101 and DNV-RP-F113 are introduced as follows:

- DNV-ST-F101: Submarine Pipeline Systems

    This standard has been developed and kept on updating for pipelines, which is DNV-ST-F101 for "Submarine Pipeline Systems". The objective of this standard (ST) is to provide an internationally acceptable framework for submarine pipeline systems in all lifetime phases, with a focus on structural assessment, with the aim of obtaining an appropriate and consistent level of safety. This standard provides requirements and recommendations for the concept development, design, construction, operation and abandonment of pipeline systems, with the emphasis on structural integrity. This standard has been improved over decades of work, mainly due to the contributions of Joint

    Industry Projects (JIP's) and DNV internal experts, allowing to satisfy the demands of the submarine fluids transportation industry such as increased operational complexity, more stringent government regulatory requirements, use of new

and improved technologies and materials and changing environmental conditions among others. As a result of the continuous improvement of submarine pipeline systems design, a latest version of the DNV-ST-F101 has been released in August 2021 and amended in December 2021, and the new revision incorporates recent knowledge from the Lined and Clad JIP and the research on Local Buckling performed by the European Pipeline Research Group (EPRG).
- DNV-RP-F113: Pipeline Subsea Repair

    This Recommended Practice (RP) applies for fittings used to repair and tie-in submarine pipelines. This RP provides guidance on pipeline repair methods such as pipeline hot tapping and above water tie-in. In addition, recommendations and guidance on pipeline preparedness strategies, pipeline damage assessments and testing are giving in this document. The latest edition of this RP was released in September 2019 and amended in September 2021. Besides the detailed design requirements recommended in this RP, the testing requirements are also recommended, e.g., the industry is willing to refer to the test requirements in Table 1 during the qualification testing for the repair fittings

7.2.2 Risk Assessment and FMECA Worksheet Review

DNV standards and certification processes apply a risk-based methodology. Thus, at the beginning of the design review process, Smart Flange 4 connector risk matrix, risk assessment and FMECA worksheet were created, discussed and agreed by DNV, the customer and end user subject matter experts. The Smart Flange 4 connector's failure modes, failure mechanisms and consequences were identified, and the criticality assessment was conducted. After the Smart Flange 4 connector and its eight sub-systems were assessed, the overall risks with the score of "Medium" level or higher were identified, e.g., the risk of the connector failing to seal on pipe or grip on pipe, and the housing failing to hold pressure. Thus, the critical components need verification, mainly for structure strength, gripping capacity and sealing capacity, can be focused on. An example of the applied risk matrix and the FMECA results are shown in Tables 1 and 2 below.

Table 1. The risk matrix applied for the FMECA

Category	Consequence Score				
People	No injuries	Non-recordable	First Aid Required (Recordable)	Lost time injured (Recordable)	Permanent Injury or Fatality (Recordable)
Environment	No environmental release	<1 BBL environmental release	1 -100 BBL environmental release	100 – 1,000 BBL environmental release	> 1,000 BBL environmental release
Downtime	< 1 hour	1 – 5 hours	5 – 24 hours	1 – 5 days	> 5 days
Cost	< $1,000 COPQ	$1,000 - $10,000 COPQ	$ 10,000 – $100,000 COPQ	$ 100,000 – $250,000 COPQ	> $250,000 COPQ
Technical	Negligible Effect, normal operation	Minor loss of functionality and/or redundancy	Major loss of functionality and/or redundancy	Total loss of functionality	Loss of main function and damage to interfacing and surrounding systems

Failure Rate	Descriptor	Current	1	2	3	4	5
0.25 < p	Very High: Between 2.5 and 10 failure per Mission Time, per 10 components	5	Medium	High	High	Very High	Very High
0.05 < p ≤ 0.25	High: Between 5 and 25 failure per Mission Time, per 10² components	4	Medium	Medium	High	High	Very High
0.01 < p ≤ 0.05	Medium: Between 1 and 5 failure per Mission Time, per 10² components	3	Low	Medium	Medium	High	High
0.001 < p ≤ 0.01	Low: Between 1 and 10 failure per Mission Time, per 10³ components	2	Low	Low	Medium	Medium	High
p ≤ 0.001	Very Low: Between 1 and 10 failure per Mission Time, per 10⁴ components	1	Very Low	Low	Low	Medium	Medium

Table 2. Overall risk scores for Subsea Pipeline Connector during FMECA assessment

Identifier	Sub-system/Component	Failure Mode	Overall Risk Score
00	Connector Assembly	Fail to seal and grip on pipe	Medium
01	Housing	Fail to hold pressure	Medium
02	End Cap	Fail to hold load	Medium
03	Seal Piston	Fail to create load	Medium
04	Slip	Fail to grip pipe	Medium
05	Piston Housing	Fail to hold slip in place	Medium
06	Seals	Fail to seal	Medium
07	Rings	Fail to provide proper annulus area	Medium
08	Bolts & Nuts	Fail to hold load	Medium

7.3 Reliability of Subsea Production Systems

Reliability analysis of subsea production systems is increasingly using advanced models and methods such as fault tree analysis (FTA) and Event Tree analysis (ETA) to identify and assess potential risks. Much research has focused on utilizing data-driven approaches, combined with machine learning and big data techniques, for real-time monitoring and prediction of system performance. At the same time, the introduction of standardized and modular design also enhances the maintainability and reliability of the system. In addition, experimental studies on specific environmental factors are increasing to validate and optimize existing designs. Traditional fault tree analysis is an effective tool used to evaluate system risk if the required data are sufficient. Unfortunately, the operation and maintenance data of some complex systems are difficult to obtain due to economic or technical reasons. The solution is to invite experts to evaluate some critical aspect of the performance of the system.

7.3.1 Research on Reliability of Underwater Production System

Subsea production systems are primarily used for the extraction and processing of deep-sea oil and gas through a range of equipment and technologies to safely and efficiently transport resources to the surface. The marine environment is complex and changeable, and any system failure can lead to serious safety hazards and environmental impacts. At present, many experts and scholars are committed to studying the reliability and safety of subsea production systems, using advanced modeling and simulation techniques to evaluate equipment performance and failure mechanisms in order to optimize the design and improve the overall safety of the system.

Zhao et al. (2023) proposed a system risk assessment method based on uncertain fault tree by using uncertainty measure to measure the degree of confidence in the occurrence of basic events assessed by experts, and established two general optimization models on this basis. Genetic Algorithm (GA) and Non-dominated Sorting Genetic Algorithm II (NSGA-II) were used to solve the two optimization models respectively. In addition, the proposed risk assessment method is applied to the leakage risk assessment of the subsea production system, and two general optimization models are used to optimize the leakage risk and maintenance cost of the subsea production system. The optimization results provide a theoretical basis for practitioners to ensure the safety of subsea production systems.

Chao et al. (2022) proposed a composite distribution method based on the combination of similar product distribution method and Analytic Hierarchy Process (AHP), established a typical subsea oil and gas production system model and system reliability block diagram, and comprehensively considered domestic factors, established a system reliability distribution model and constructed a judgment matrix through an expert scoring system. The system reliability is distributed and redistributed by the composite distribution method, and the system reliability index is allocated to each subsystem and equipment unit.

Wang et al. (2022) conducted a human factor reliability analysis to identify the factors that have a high impact on the probability of human factor failure of the underwater production system, built a human factor risk factor system of the underwater production system, verified the rationality of the data with reliability and validity indicators, and determined the weight coefficient of human factor risk factors with the help of structural equation model. The Bayesian network (a statistical model that represents a set of variables and their probabilistic dependencies) is constructed and the human factor probability is quantified by fuzzy theory to predict and evaluate the human factor risk of underwater production system.

Li et al. (2024) built a Bayesian risk analysis network model for the underwater production system of deep water gas field based on Bayesian network and combined with the accident tree analysis model in conventional risk analysis, and analyzed and calculated the failure probability of the gas production system of deep water underwater natural gas production system with this model, and deduced the probability of root event and intermediate event. Ranking the possibility of equipment failure saves a lot of time for staff to find the cause of failure and repair, and rescue.

7.3.2 Reliability Studies of Other Subsea Equipment Components

Subsea Tree (Christmas Tree), subsea pump, sensor and monitoring equipment, submarine cables and connectors and valves and other components as an important part of the underwater production system, play an important role, but affected by the complex conditions of the underwater production environment, prone to system failure, may lead to safety accidents, economic losses and environmental pollution. By identifying potential failures, optimizing design, and improving the operating efficiency of equipment, reliability analysis not only ensures safety and protects the environment, but also reduces operating costs and improves overall economic efficiency, thereby promoting the sustainable development and utilization of resources.

Nan et al. (2023) conducted a study on the impact of maintenance intervention on the time-varying degradation of the system under complex environment for the operation and maintenance reliability assessment of the production loop of deepwater tree, and proposed an operation and maintenance reliability analysis method based on stochastic Pe-tri network to solve the explosive problem of multiple failure modes occurring simultaneously during the reliability analysis of the production loop of deepwater tree. The complex failure modes of deepwater tree production circuits are studied, and the influence of different faults on system reliability and availability is revealed. The evaluation results of the established model are verified by OREDA failure database accuracy.

Juan et al. (2022) took the underwater control module of a Bohai oil and gas field production project as the research object, applied SimulationX modeling and simulation tool, took the abnormal function of the output port of the electromagnetic reversing valve as the top event, conducted fault tree analysis on the high-pressure hydraulic circuit, and found out the key components and main fault modes. It is suggested that the reliability levels of accumulator and solenoid directional valve should be emphasized and improved.

Subsea gate valves and actuators are indispensable equipment in subsea production systems such as subsea tree, oil production lines, and underwater manifold systems. The reliability of their seals is the key to the performance and service life, and it is also the focus of research on the localization of such valves and actuators. Wan et al. (2023), based on the analysis of the technical characteristics and applications of underwater gate valves and actuators at home and abroad, combined with the situation of the domestic development of the components, studied the key technical issue of the sealing performance of underwater gate valves and actuators. The design of seal structure, the analysis of seal performance, the research of seal reliability verification method, the process of product prototype seal test verification and the problem handling are analyzed and researched comprehensively.

The reliability of underwater blowout preventers (SBOPs) is one of the major challenges in deep-sea drilling and completion operations, accounting for a significant portion of major equipment failures and non-productive time (NPT) costs. Since the Deepwater Horizon/Mc Dondo accident in 2010, SBOP's technological advancements have focused on reliability, equipment condition monitoring, and statistical root cause analysis, with strategies to reduce NPT and improve safety through technological advancements and operational improvements as follows:

1) Maintenance and testing: After the completion of the deepwater well, the SBOP must undergo maintenance, repair if necessary, and pressure test before deploying to the next well. SBOP stress testing using an advanced digital pressure recording system.
2) Dual SBOP Strategy: Some operators work with drilling contractors in a variety of ways to prepare a second fully assembled and pressure-tested SBOP (dual SBOP), as well as deploy additional trained subsea engineers for maintenance/repair.
3) Stress testing software: Using comparative stress testing software can greatly reduce SBOP stress testing time, by eliminating human error and speeding up stress testing.
4) Leak detection and monitoring: By installing sensors and real-time test monitoring, leak detection time can be eliminated and reliability can be improved.
5) SBOP Dashboard: A dashboard that simplifies existing diagnostics and allows remote monitoring of subsea SBOP control systems, will improve communication of SBOP health and serve as a common platform that allows for standardized SBOP diagnostic data between drilling fleets, aiding in operational decision-making.
6) Redundancy and emergency systems: Ensure additional SBOP redundancy, especially when operating the Emergency Disconnect System (EDS), through the Remotely Operated Vehicle (ROV) control panel or acoustic system. Use risk analysis tools such as fault tree analysis to improve reliability and provide additional security.

7.4 Control System Reliability

The subsea production system is essential for the subsea production of oil and gas. Real-time monitoring can ensure safe production. The subsea production control system is the core of the subsea production system and the top priority to be monitored. As oil and gas production deepens, remote monitoring of subsea assets becomes increasingly important and reliable. Necessary for the reliability of subsea control module (SCM) of subsea production system is studied quantitatively and qualitatively. The Subsea Control Systems is a supporting system. As per the ISO 13628-6, the subsea controls system has the following operational requirements:

- Enable the remote operations of all the subsea control valves;
- Enable data acquisition and monitoring of the subsea production system in order to reduce the response time in case of safety or production risks;
- Enable safe control of the production in case of emergency within the response time defined as per standard and country regulations.

To improve the availability of underwater control modules (SCM), reliability studies will capture data on SCMs failure times, causes of failures, major components involved in failures, use open software to understand failure distribution, and identify potential measures that may improve asset availability.

Wei et al. (2022) adopted hazard and operability (HAZOP) analysis to carry out safety risk analysis on the test procedures and contents of a gas field development project in the South China Sea before the integrated test of the underwater control system, analyzed and studied the causes and consequences of different parameter deviations of the underwater control system in hydraulic, electric and communication aspects, and proposed corresponding safety measures.

To ensure the successful implementation of the subsea production system, Zuo et al. (2022) proposed a Bayesian network modeling method based on failure mode impact and hazard analysis (FMECA) and fault tree (FTA). BNT toolbox in MATLAB and GeNie software were used for reliability analysis of the subsea control module (SCM). The network model of underwater control module (SCM) is deduced by using the joint tree algorithm, and the weak link in SCM is identified.

Joao et al. (2023) developed embedded digital twins for subsea electric valve actuators, modeled using state-space equations based on their physical corresponding components. The digital twin model is tested using data from the physical counterpart model based on MATLAB scripts to improve the fault tolerance of electric subsea valve actuators and provide digital redundancy for key components such as position and torque sensors.

Yin et al. (2023) adopted analytic hierarchy process (AHP) to study the reliability allocation of the underwater production control emergency shutdown system, comprehensively considered the complexity of equipment units in the system, the technical level of products, fault characteristics and other factors, and reasonably allocated the given reliability index to each equipment unit of the underwater emergency shutdown system, so as to optimize the reliability design of the underwater emergency shutdown system.

Chao et al. (2023) proposed a composite fault diagnosis method driven by digital twins combined with virtual and real, aiming at the problem that multiple faults exist at the same time, resulting in mixed signals and difficulty in distinguishing the states of subsea production systems. Bernoulli equation combines loss, control and state parameters to construct a digital twin model, and constructs a fault diagnosis model based on Bayesian networks that combines virtual and real data.

Traditional reliability modeling methods that rely on historical fault data have become inadequate in view of the scarcity of fault data resulting from technological advances that have improved the reliability of subsea equipment. In this paper, Yuxin Wen et al. (2024) propose an innovative reliability modeling technique for submarine control systems, which integrates a Wiener degradation model affected by random shocks and uses Copula function to calculate the joint reliability of components and their backups. It takes into account the unique challenges of the seabed environment and the complex interactions between components under variable loads, improving the accuracy of the model. The impact of inadequate maintenance on degradation pathways is also investigated and a holistic life cycle cost model for preventive maintenance (PM) is introduced, optimized for reliability and economic considerations.

The subsea production control system is the core of the subsea production system, and it is also the most important thing to monitor. The digital twin-driven fault diagnosis is an effective method to monitor the subsea production control system. However, the combination of digital twins and troubleshooting is not comprehensive, especially when it comes to data interaction. Chao Yang et al. (2023) proposed a cross-validation enhanced digital twin-driven fault diagnosis method for minor faults in subsea production control systems. The fault diagnosis model was constructed using Bayesian networks, and the digital twin model was constructed considering control, loss and fault parameters. Individual and cumulative errors are used to measure the difference between the digital twin and the physical system.

The reliability evaluation of the subsea control module (SCM) is the key to ensure the safety and stability of subsea oil-gas production. The failure probability and reliability of SCM components depend on time and working conditions. In order to analyze the reliability of single chip computers considering different operating conditions, Tao et al. (2023) proposed a new model based on digital twins and dynamic Bayesian networks (DBN), using historical operating condition data for reliability analysis. In the proposed framework, the key operating data are obtained by sensor-based digital twin (DT) simulation and used to dynamically update the parameters in the DBN reliability analysis model. The reliability of the actual SCM electrical system is evaluated to find the most likely failure modes and the most vulnerable components in the system.

As per Birolini (2017) in his book Reliability Engineering Theory and Practice, Reliability is the characteristic of an item, expressed as probability to perform its required function under the defined conditions within a determined time interval. The study of quantitative reliability is based on the metrics that involves time to failure of a determined component, assembly, subsystem, system or the asset itself.

- Reliability, R(t): Probability of the item perform its required function until the time t;
- Failure Probability, F(t): Probability of the item fails to perform its required function until the time t;
- Average Life, or Mean Time To Failure (MTTF): Average time for a failure to occur related to the item under analysis, in case of a repairable component or system it is called Mean Time Between Failure, MTBF;
- Failure Rate, $\lambda(t)$: Instantaneous probability of the item to fail at the time t considering that item hasn't failed until that moment.

All these metrics can support decisions to be made in an effective way, however a benchmark, historical data or an industry database should be used as reference. For the oil and gas industry the Offshore and Onshore Reliability Data, or OREDA, is widely used for project design or performance evaluation. All the metrics listed can be generated since the simplest component, up to the whole asset. As shown in Fig. 29, it presents an exemplification of hierarchy from system to component level, which helps us better understand how different components are organized within a system in the context of reliability analysis.

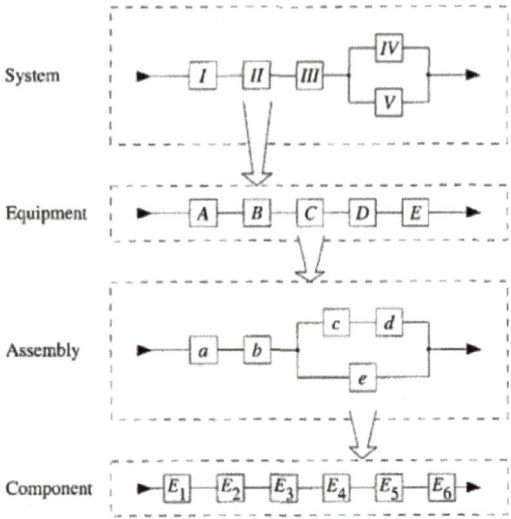

Fig. 29. Exemplification of hierarchy from system to component level.

The sequence of the items in the reliability block diagram and the hierarchy requires a deep understanding of the asset functionality and how the components interact, mainly, in a failure scenario. In order to reproduce the connection between the items that compose the asset it is applied two different types of structures: Parallel structures and Series structures. In a simplified analysis, the series structures depend on the functionality of all of its components to perform the system function. As shown in Fig. 30. On the other hand, the parallel structure can continually perform the system function even with N-1 components failed (considering N the total amount of components in the structure). As showen in Fig. 31.

Fig. 30. Series structures

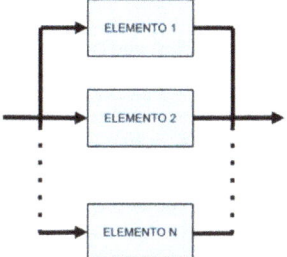

Fig. 31. Parallel structures

Weibull distribution the flexibility to fit the model into different shapes of distributions by changing the value of the 2 main parameters η (eta) and β (beta) as represented on the Fig. 32 η (eta) is also called characteristic life parameter, where 63,2% of the items will be failed up to the η time. β (beta) is called shape parameter, which gives the format of the curve depending on its value:

- $\beta < 1$ indicates an exponential decremental curve in the failure probability in time.
- $\beta = 1$ denotes the failure rate will remain constant, similar to an exponential distribution.
- $\beta > 1$ indicates an exponential incremental curve in the failure probability.

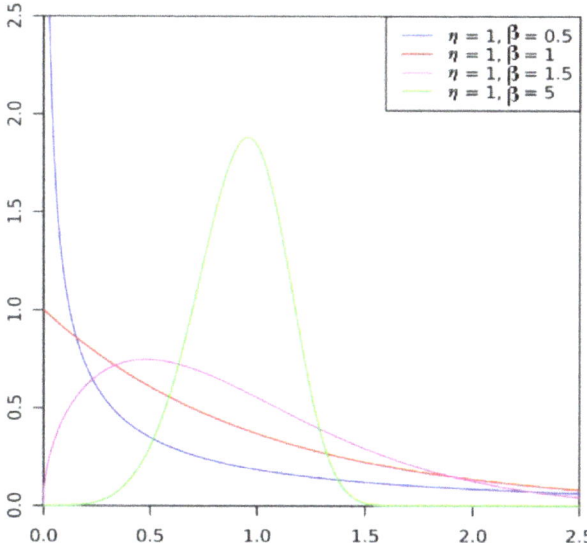

Fig. 32. Graphical representation of Weibull distribution with different parameters.

As discussed in the methodology session, this work followed the steps from 1 to 5 in order to compare and analyze the performance of the object of the study. The data collection was composed of 16 variables that would be important for this analysis and also for future evaluation in further studies, the main were:

1) SCM Serial Number
2) SCM Installation Date
3) SCM Commissioning Date
4) SCM Total amount of days without failure (Run Time)
5) Date ofFailure Reported
6) Fault Code (ISO 14224)
7) Failed Component
8) Failed Component Serial Number
9) Retrieval Date

As shown in Fig. 33, which is the data collection template for the reliability study of the SCM, it clearly presents how these variables are organized and recorded, providing a visual aid for better understanding the data collection process.

Fig. 33. Data collection template for the reliability study of the SCM.

The functional analysis of the SCM was carried out using the Failure Mode and Effect Analysis (FMEA) as a reference. A Fault Tree was built in order to illustrate and clarify the components.

The investment of time to discuss the functionality analysis and mapping of the fault tree can help to develop a RAM analysis as long as there are enough failure events to model the component level Reliability study.

In addition to this, the digitization of underwater control systems, in particular digital twin technology, plays a key role not only in designing new applications, but also in enhancing existing machines. The integration of this advanced technology enables the oil and gas industry to achieve unprecedented levels of efficiency, control and predictive maintenance, ensuring a more competitive and sustainable future. Joao Pedro Duarte da Silva and Bosch Rexroth AG/Federal University of Santa Catarina, among others, have embedded digital twins in underwater electric valve actuators, The digital twin is tested using data from the physical counterpart. In two cases, the online and recursive parameter identification is discussed. In experimental tests, the digital twin was found to expertly adjust its own parameters, mirroring the behavior of the physical twin. the embedded Digital Twin enables condition monitoring of the physical counterpart, bridging sensor gaps and compensating for sensor faults. Such proactive maintenance capabilities enhance equipment reliability significantly.

7.5 Summary and Recommendations for Future Work

This study analyzed the reliability and optimization of subsea production systems, with a focus on the Subsea Control Module (SCM). It employed advanced methodologies, such as Failure Mode and Effect Analysis (FMEA) and digital twin technology, to identify failure modes, optimize preventive maintenance, and enhance real-time fault diagnosis. Key findings included the application of Bayesian networks for risk assessment and the integration of digital twins to improve fault tolerance, bridge sensor gaps, and enable proactive maintenance. These advancements highlight the potential of combining real-time monitoring, predictive analytics, and advanced reliability models to enhance the performance and safety of subsea systems.

Future research should focus on expanding data collection and integrating real-time monitoring with advanced fault diagnosis systems, improving digital twin models through machine learning, and exploring more robust stochastic reliability models. Additionally, incorporating human reliability analysis, standardizing methodologies through cross-industry collaboration, and developing cost-effective maintenance strategies using optimization techniques are crucial. These efforts would foster greater resilience, safety, and efficiency in subsea production systems, ensuring their sustainability in increasingly complex and high-risk environments.

8 Life Extension and Decommissioning

As we all know, any device has a life cycle, and as the device gradually approaches the end of its life cycle, the ensuing challenges and changes are inevitable. Today, the world's oil and gas platforms are at this critical stage of unprecedented change.

According to Nature, many of the world's approximately 12,000 oil and gas platforms are nearing the end of their useful lives. For example, there are more than 1,500 platforms and related facilities in the North Sea with an average life span of 25 years. There are more than 1,500 platforms in the Gulf of Mexico that are more than 30 years old. In the next decade, a large number of older platforms will face the need for retirement or functional adjustment. According to Petrodata™FieldsBase, IHS Markit's proprietary database, nearly 2,800 fixed platforms and 160 floating platforms are expected to be decommissioned between 2021 and 2030, representing 33% of global fixed platforms and 43% of floating platforms, respectively. At the same time, more than 18,500 wellheads, 2,850 subsea trees, and 83,000 km of offshore pipelines and umbilical cables will be closed, revealing the enormous pressure and challenge of updating and maintaining the global energy infrastructure.

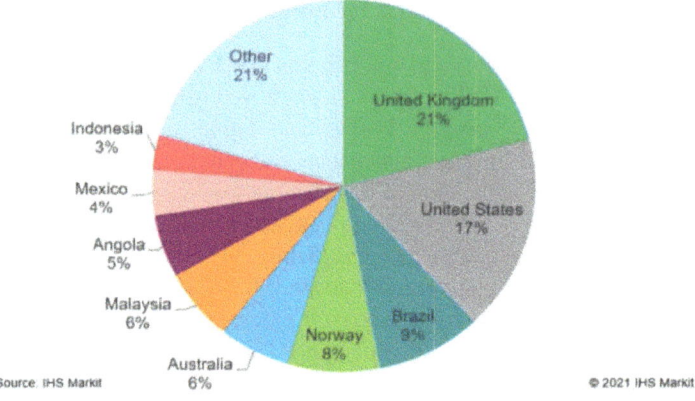

Fig. 34. Offshore oilfield decommissioning forecasts by country [IHS]

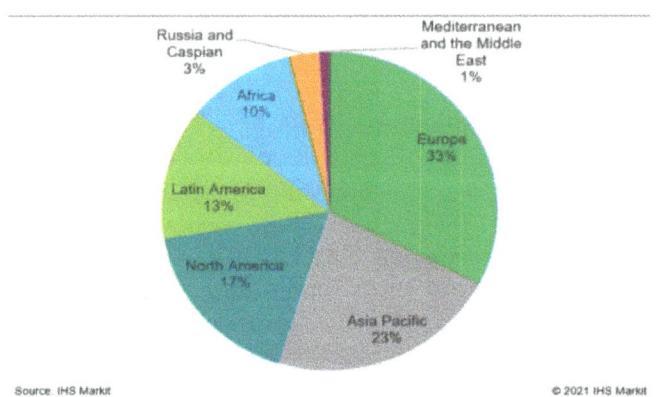

Fig. 35. Offshore oilfield decommissioning forecast by region [IHS]

As shown in the figure, Figs. 34 and 35 are pie charts of IHS Markit's offshore platform decommissioning forecasts for major countries and regions during 2021–2030. From the chart, decommissioning platforms are mainly concentrated in developed countries in Europe and the United States, as well as some oil producing countries in South America and Asia. The United Kingdom and the United States, in particular, dominate this sector and are expected to shoulder a significant amount of decommissioning spending and related projects over the next decade.

8.1 Decommissioning of Oil and Gas Platforms

Decommissioning of oil and gas platforms is often multidisciplinary and complex, requiring not only the safe cessation of production and removal of facilities, but also the restoration of environmental conditions in the operating area. The process involves not only physical dismantlement, but also extensive management, planning and legal compliance requirements, including 10 steps in project management/engineering and planning, permitting and regulatory compliance, platform preparation, well plugging and abandonment, jacket removal, rig barge mobilization and evacuation, platform removal, pipe and cable decommissioning, material disposal, and site cleanup.

8.1.1 Project Management/Engineering and Planning

The project management, engineering and planning of oil and gas platform decommissioning ensures that the project is carried out according to the established schedule, safety specifications and budget. As the decommissioning process is complex and involves multidisciplinary collaboration, the effectiveness of project management and planning directly affects the success or failure of the project. Project management mainly includes the formulation of retirement strategy, risk management, time and resource management, budget control, multi-party coordination and communication. Engineering and planning include engineering evaluation and design, platform preparation, licensing and regulatory compliance, equipment and technology selection, job site management, data acquisition and monitoring.

8.1.2 Licensing and Regulatory Compliance

Depending on the region, local regulations and licensing requirements must be met. For example, the North Sea oil and gas platform is located in the waters of the United Kingdom and Norway, so the decommissioning process must comply with the relevant laws and regulations of the United Kingdom and Norway, mainly in accordance with the OSPAR Convention, the London Dumping Convention, the International Maritime Organization (IMO) guidelines, Norway's Petroleum Act and so on.

8.1.3 Removing Submarine Devices

The removal of subsea equipment usually refers to the removal of various facilities and installations located on the seabed, such as wellhead equipment, jackets, subsea pipelines and cables. During the decommissioning process, such equipment must be handled with extreme care to avoid any adverse impact on the Marine environment. The complexity of the demolition operation depends on the water depth and the weight of the equipment. These two factors directly determine the technical resources required and the mode of operation. The water depth determines whether a remotely operated vehicle (ROV) or diver is required to perform the equipment cutting and removal operation. Remote control equipment such as ROVs must be used for remote operation. The weight of subsea equipment directly affects the size of lifting equipment and vessels that need to be used. Usually, heavier equipment requires larger lifting vessels and transport barges, which not only increases the time of equipment scheduling and mobilization, but also significantly increases the overall operating cost. The removal of subsea equipment generally includes the following items:

1) Preparing the Platform: preparation is an important early step in the decommissioning process to ensure that the platform can be safely dismantled and removed. The preparation phase of the platform covers everything from stopping production, to removing hazardous materials, to strengthening and preparing the equipment before dismantling. By properly preparing the platform, we can ensure that subsequent decommissioning can proceed smoothly while reducing potential risks to the environment, personnel and equipment.
2) Mobilization and evacuation of rig barges: A rig barge is a large floating structure specially designed for offshore platform and well operations, often equipped with heavy lifting equipment for platform removal, equipment lifting, and dismantling of subsea structures. The mobilization and evacuation phase will directly affect the overall cost and time management of the project. Mobilization refers to the process of moving a rig barge from a mooring location or other operation site to a decommissioning site. Mobilization is not only about the transport of vessels, but also about preparation, including equipment inspections, personnel arrangements and necessary work permits; Evacuation refers to the process of returning a rig barge from the decommissioning site to its base or next operating location after completion of operations.
3) Sealing and abandonment of Wells: Plugging and abandonment of a well is the process of ensuring that the well is permanently closed and poses no threat to the environment. The process involves cleaning the well hole and isolating different formations using a plugging material, such as cement, to ensure that no fluid leaks. The plugging of Wells varies slightly depending on dry and wet Wells, and dry Wells are those that have direct access to the well head from the platform. In this case, the work can be done on the platform without the need for large vessels, so the cost is relatively low: the wet well is located on the sea floor, and the plugging and abandonment work must be carried out by specialized vessels or drilling platforms. This requires more complex equipment, is more difficult to operate, and is more expensive. Especially in deep water, the decommissioning cost of a wet well can be several times that of a dry well.
4) Removal of jacket: The jacket is the support structure of a fixed oil and gas platform. It is made of steel structure and fixed to the seabed to support the upper facilities of the platform. Due to its large size, huge weight, and the complexity of the deepwater environment, the removal of the jacket requires high technical requirements and precision operations, which are usually divided into the removal and transportation of the upper facilities of the platform (integrated lifting, segmented lifting), well plugging and abandonment, and jacket lifting.
5) Platform Removal: The removal of an oil and gas platform usually involves the complete or partial removal of all of the platform's facilities from the area of operation in order to restore the seabed to its natural state. The removal process is divided into two main parts: upper facilities and support structures. The upper facilities include hundreds of tons of steel structure, oil and gas processing equipment, and personnel living quarters, which require the use of large cranes or specialized lifting equipment when dismantled. The support structure is fixed to the seabed, and its removal involves complex cutting and lifting techniques. Due to the high cost of removing oil and gas platforms, people have begun to consider whether the platform can be reused instead

of complete dismantlement. For example, Song et al. (2024) proposed that oil and gas platforms can be converted into platforms for installing renewable energy equipment such as wind turbines, thus giving them new functions.

6) Decommissioning pipes and cables: The way pipelines and cables are decommissioned depends on their length, diameter, depth, installation location and the Marine environment in which they are located. Ayeni et al. (2024) pointed out that in the world, decommissioning methods are generally divided into complete removal, partial removal and stay in place. of which complete removal of pipelines and cables is the most complete decommissioning method, especially in shallow water or Marine protected areas, which usually requires complete removal of facilities to restore the original seabed, while partial removal is often applied to deep water areas or areas where pipelines and cables are buried deeper. In this way, the pipe or cable on the seabed surface is removed, but the part buried under the seabed may remain in place. It is important to note that the remaining parts are subject to a detailed environmental assessment to ensure that there are no long-term impacts on ecosystems or navigation safety. The last option is to keep the pipelines and cables on the seabed and not remove them after proper treatment in deep-water areas or when the pipelines are buried deep, difficult to recover and have little impact on the environment. This method has the lowest cost, but there is uncertainty about the impact on the environment.

8.1.4 Material Disposal

In the decommissioning process of oil and gas platforms, material disposal is a crucial part of how to properly handle and recycle the various wastes and resources generated during the decommissioning process. These materials include platform structures, equipment, pipes, jackets, and potentially hazardous waste. Reasonable material disposal not only helps to reduce the negative impact on the environment, but also ensures that decommissioning operations comply with local and international environmental regulations, and maximize the recovery and reuse of resources, achieving a win-win situation for economic benefits and environmental protection.

At present, during the decommissioning process of oil and gas platforms, the materials removed can be divided into three main categories: recyclable materials, non-recyclable materials and hazardous waste. Disposal methods and requirements vary for each type of material, with recyclable materials, such as steel, metal, cables and pipes, sent to metal recycling centres. For example, when the Brent Delta Platform, one of four facilities at the Brent oil and gas field northeast of the Shetland Islands in the North Sea, ceased production after nearly 35 years, its superstructure (the part of the platform above the waterline) was salvaged from the sea and shipped to a specialist decommissioning company in Teesside, England, for demolition, reuse or recycling to 97%. Babadima et al. (2023) pointed out that some hazardous wastes such as residual oil and gas and asbestos will be transported to specialized hazardous waste treatment facilities for harmless treatment. recently summarized optimized sampling and analysis procedures that can accurately quantify mercury concentrations in gas-condensate fields to provide reliable support for the design of production facilities. This not only reduces risk in production, but also avoids costly facility retrofits and mercury removal costs.

It is worth noting that in the discussion of subsea equipment removal, a new development concept has emerged - to integrate the decommissioning of oil and gas platforms with the energy transition, and strive to promote a greener, low-carbon future. In Malaysia, for example, a decommissioned oil rig has been converted into a diving resort. More than 600 RIGS located in the Gulf of Mexico have been similarly converted into artificial coral for diving, recreational fishing, and other water sports. But that approach has been controversial over whether abandoned platforms leave harmful chemicals in the ocean.

8.1.5 Site Cleaning

Site cleanup is the final and crucial step in the decommissioning process to ensure that the operating area and seabed are restored to a near-natural state while complying with environmental protection and regulatory requirements. Site cleanup not only helps maintain the balance of the Marine ecosystem, but also provides the necessary security for future navigation, fishing activities and other offshore operations.

8.2 Oil and Gas Platform Life Extension and Maintenance

Extending the life of aging oil and gas platforms is a cost-effective solution compared to the high removal costs of decommissioning oil and gas platforms. Many oil and gas platforms in areas such as the North Sea have outlived their design life, but through comprehensive reliability assessments, asset integrity management, predictive maintenance and risk assessment, extending the life of these platforms is not only feasible and safe, but can also significantly reduce capex and decommissioning costs. At the same time, life extension measures are of great significance to environmental protection, taking into account economic benefits and environmental benefits.

8.2.1 Reliability Evaluation

Reliability assessment of oil and gas platforms is a key step in ensuring safe operation of platforms during their designed or extended life. The evaluation process is designed to identify potential risks, predict the likelihood of equipment failure, and optimize maintenance plans to extend the life of the platform and ensure safety. The core content of reliability assessment includes the following aspects: structural integrity analysis, equipment condition monitoring, operating environment assessment, predictive maintenance and risk assessment.

1) Structural integrity analysis: Oil and gas platforms are often exposed to harsh offshore environments such as strong winds, high waves, salt water corrosion, earthquakes, and dynamic equipment loads. Structural integrity analysis can help assess whether the platform structure can withstand these forces and maintain safe operation. In recent years, the application of digital twins has led to significant advances in the monitoring and assessment of the structural integrity of offshore platforms, especially in the aftermath of a collision. Shahir et al. (2024) combined virtual data with real-world data, enabling digital twins to be evaluated in real time as the structure of the platform changes, so as to more accurately grasp the health of the platform.

2) Equipment condition monitoring: Traditional condition monitoring methods rely on real-time measurement data, and when potential problems are first detected, there may not be enough time to take preventive measures to avoid failure. In order to improve the reliability of oil and gas platforms in this regard, the first issue is to improve the sensitivity of equipment condition monitoring, to detect potential anomalies that prevent the equipment from performing its intended function before failure occurs. At present, the main methods to improve sensitivity include assessing equipment risks through numerical simulation and risk assessment techniques, which was put forward by Li et al. (2023). Hanif et al. (2023) integrated the machine learning algorithm into state detection and utilized it to evaluate the change patterns of parameters such as temperature, pressure and flow rate collected from sensors over time. In this way, potential faults could be identified and early warnings could be given to achieve the purpose of predictive evaluation.

3) Assessment of the operating environment: The purpose of operating environment assessment is to predict and improve the long-term performance and reliability of equipment or systems by analyzing their performance in a real-world operating environment. Critical physical and operational factors in daily operations often have a significant impact on system life and failure rates. Oluka et al. (2023) put forward data-driven technical approaches. These approaches, combined with fatigue analysis, equipment aging assessment and carbon footprint reduction measures, can quantify the specific impact of the operating environment on equipment performance and further extend the service life of complex equipment such as floating offshore production facilities. Vongasemjit et al. (2023) pointed out that in the inspection and operation of extreme conditions such as offshore oil and gas wellhead platforms, semi-autonomous semi-remote robots can be adopted to avoid the requirement of personnel entering dangerous areas.

4) Predictive maintenance and risk assessment: Predictive maintenance is a data-driven technology that monitors the operating status of equipment, predicts failures in advance, and plans maintenance needs. Abili et al. (2023) stated that this technology combines advanced means such as sensors, Internet of Things, big data analytics and artificial intelligence. Sun et al. (2023) pointed out that at present, the main path to achieve predictive maintenance is to integrate production systems with digital systems, build computational models of equipment damage, and perform numerical analysis through data acquired by sensors. Risk assessment is a systematic analysis of potential risk factors for oil and gas platforms to ensure the safety and compliance of platform operations. This process includes risk identification, analysis and prioritization, covering multiple aspects of equipment, environmental and operational risks.

8.2.2 Asset Integrity Management

Asset Integrity Management is a comprehensive management framework designed to ensure the safe, reliable and efficient operation of equipment and systems throughout the life cycle of the platform. The management framework covers all stages from design, construction, operation, maintenance to decommissioning, with the core objective of avoiding major facility failures through preventive and predictive maintenance, thereby reducing the risk of accidents, reducing downtime, and extending the service life of equipment. Agala et al. (2024) claimed that an effective management framework can often increase field productivity and significantly reduce operating costs.

At present, asset integrity management is rapidly developing in the direction of digitalization and intelligence. With 3D rendering engines and engineering algorithms, data can be transmitted in a visual and data fusion manner, and engineers can obtain real-time field information and project progress with a simple mouse click. Operators can manage remote assets more efficiently with real-time access to critical information in a data lake. Wang et al. (2023) stated that through intelligent monitoring and decision support systems, managers ensure the optimal operation of the platform in complex environments, thus improving safety and sustainability.

This digital management approach not only significantly increases productivity, but also enhances flexibility and precision in addressing complex challenges, helping to ensure that oil and gas platforms continue to operate efficiently throughout their life cycle.

8.3 Summary and Recommendations for Future Work

The decommissioning of oil and gas platforms and the life cycle extension of their life cycles present apose complex and multi-faceted landscape. As a significant number of platforms worldwide approach the end of their operational lives, the challenges associated with decommissioning are becoming increasingly prominent. The process. Decommissioning involves a multitude of numerous steps, from meticulous project management and adherence to diverse regulatory frameworks to the safe removal of subsea equipment, proper disposal of materials, and comprehensive site cleanup. This requires the demanding multidisciplinary integration of various disciplines and the stakeholder coordination of multiple stakeholders to ensure to meet environmental protection and compliance with legal requirements while managing controlling costs effectively.

. In the realm of life extension and maintenance, the application of advanced technologies such as like digital twins, machine learning, and the Internet of Things has shown great IoT show potential. These technologies enable more accurate in reliability assessments, including detailed structural integrity analysis, enhanced equipment condition monitoring, comprehensive operating environment evaluations, and effective predictive maintenance and risk assessment. Asset integrity management and asset integrity management is also evolving moving towards digitalization and intelligence, providing a more efficient and sustainable framework for platform operation throughout its life cycle.

For future work efforts, several key areas warrant further attention. Firstly, continued research and development of aspects need focus. In decommissioning, continuous R&D on technologies are essential. This includes the improvement of is vital, such as improving subsea equipment removal techniques for removing subsea equipment in deep and challenging waters, aiming to reduce and exploring innovative material disposal methods to cut costs and environmental impacts. For example, the development of more advanced remotely operated underwater vehicles (ROVs) or autonomous underwater vehicles (AUVs) with enhanced cutting and lifting capabilities could facilitate the removal process. Additionally, research into innovative methods for material disposal, such as the development of more efficient recycling processes for platform structures and hazardous waste treatment technologies, is crucial to minimize the environmental footprint.

Secondly, the standardization and harmonization of Standardizing decommissioning regulations across different regions should be pursued. Given the global nature of the oil and gas industry, a unified set of standards would streamline the decommissioning process, reduce uncertainties, and facilitate promote international cooperation. This could involve the establishment of common guidelines for well plugging and abandonment, platform removal procedures, and environmental impact assessments.

In the field of life extension, further exploration of the potential of emerging technologies is needed. For instance, the should be further explored, like expanding AI application of artificial intelligence in predictive maintenance could be expanded to cover a wider range of equipment and failure modes. Additionally, the integration of and integrating blockchain technology in asset integrity management could enhance data security and transparency, providing a more reliable record of platform maintenance and operation history.

Finally, more. Moreover, comprehensive studies on the economic and environmental trade-offs in both decommissioning and life extension scenarios are required. This would help operators and policymakers make more informed decisions, balancing the costs and benefits associated with different options. By addressing these areas, the oil and gas industry can necessary to guide better manage the decision-making, enabling the oil and gas industry to handle platform end-of-life challenges of platform end-of-life stages more effectively and contribute to a more sustainable energy future.

9 Conclusions and Recommendations

This report, based on the research presented in Sects. 2, 3, 4, 5, 6, 7, and 8, delves into key areas of subsea technology in oil and gas development and offers some recommendations for future advancements. As global energy demand evolves and marine development technologies advance, subsea technology will play a crucial role in ensuring the safe and sustainable operation of offshore oil and gas platforms.

1) Subsea Pipeline Design and Corrosion Protection: With the increasing complexity of subsea oil and gas pipelines operating in extreme environments, probabilistic and strain-based design methods have matured and can effectively address pipeline failure risks. In the future, particularly for hydrogen and carbon dioxide transport systems, the focus will shift toward more advanced design tools and theoretical support to ensure safe and economical operations.
2) Reliability of Subsea Risers and Umbilicals: The fatigue failure of subsea riser systems in extreme marine environments has become more prominent, especially the soil-riser interaction in the touchdown zone (TDZ) and the fatigue caused by vortex-induced vibrations (VIV). Future research should continue to improve numerical simulations and experimental methods to better tackle the challenges of deep-water operations.

3) Structural Integrity Management and Digitalization: The development of digital and automated technologies, particularly digital twin applications, has significantly improved monitoring and maintenance efficiency for offshore platforms. By integrating sensors with automated control systems, real-time optimization of platform operations can effectively reduce unnecessary inspections and repairs, extend equipment lifespan, and lower operational costs.
4) Artificial Intelligence and Risk Assessment: The application of artificial intelligence (AI) and machine learning in the risk assessment and maintenance of subsea facilities provides new solutions for optimizing inspection and maintenance strategies. Data-driven risk assessment methods can more accurately balance the costs of inspection and maintenance with operational risks, enhancing the safety and reliability of overall systems.
5) Life Extension and Decommissioning Management: As many offshore platforms approach the end of their design life, life extension and platform decommissioning have become critical issues in the industry. Structural integrity assessments, risk evaluation, and predictive maintenance can effectively extend platform lifespans. Additionally, environmental protection and resource reuse must be integrated into decommissioning plans, and future efforts could consider repurposing decommissioned platforms for renewable energy installations or artificial ecosystems.
6) Environmental Protection and Sustainability: Environmental protection is paramount during the decommissioning of platforms and the removal of equipment. In particular, green, low-carbon technologies and methods should be employed during equipment removal and waste disposal to minimize impacts on marine ecosystems. Future research could explore more sustainable approaches to converting decommissioned platforms into eco-friendly facilities to achieve sustainability goals.

In summary, through an in-depth analysis of key areas such as subsea pipelines, riser systems, corrosion protection, and structural integrity management, this report has identified future directions for technological development. By integrating technological innovation, digital transformation, and environmental protection practices, subsea technology will play an increasingly important role in energy transition and marine resource development, supporting the sustainable supply of global energy.

References

ABB Energy Industries: INSUBSEA® Power and Automation: Expanding Capacity, Extending Lifespan and Reducing Cost for Oil and Gas Fields. ABB Energy Industries (2020)

ABB Oil, Gas and Chemicals: WHITE PAPER: Long Step-Out Systems for Subsea Pump and Compressor Applications – Economic Hydrocarbon Recovery Under Extreme Conditions. ABB Oil, Gas and Chemicals (2017)

Abidin, S.Z., Mahasan, S.: Managing offshore deepwater decommissioning: subsea and FPSO decommissioning and abandonment in West Africa. In: Proceedings of the SPE Symposium: Decommissioning and Abandonment (SPE-199214-MS), Kuala Lumpur, Malaysia, 3–4 December 2019 (2019)

Adegoke, A.S., Fashanu, A., Adewumi, O., Oyediran, A.: Modelling and nonlinear analysis of the wave-induced vibrations of a single hybrid riser conveying two-phase flow. Ocean Eng. **278**, 114305 (2023)

Adeshina, S., Adegoke, et al.: Modelling and nonlinear analysis of the wave-induced vibrations of a single hybrid riser conveying two-phase flow. Ocean Eng. **278**, 114305 (2023)

Aditama, P., Koziol, T., Dillen, M.: Development of an Artificial Intelligence-Based Well Integrity Monitoring Solution. Society of Petroleum Engineers, Abu Dhabi (2022)

Ahmed Reda, et al.: Design of subsea cables/umbilicals for in-service abrasion – Part 1: Case studies. Ocean Eng. **234**, 108895 (2021)

Ahmed, R., et al.: Design of subsea cables/umbilicals for in-service abrasion – Part 2: Mechanisms. Ocean Eng. **234**, 109098 (2021)

Al-Abbas, F.M., Alsulami, M.S., Al-Mansour, H.F., Alabdullatif, A.A., Asfoor, F.H., Al Zamanan, M.M.: Top of the line corrosion mechanism for multiphase sour gas subsea pipeline. In: AMPP 2023 Corrosion Conference & Expo, Denver, Colorado (2023)

Al-Hamati, A.A., Duan, M., An, C., Soares, C.G., Estefen, S.: Buckling properties of a subsea function chamber for oil/gas processing in deep waters. J. Mar. Sci. Appl. **19**, 642–657 (2020)

Alwi, Z.: Decommissioning of Kapal Wellbay module, its support structure an MUPO Mobile producer 1. In: Proceedings of the SPE Symposium: Decommissioning and Abandonment (SPE-193948-MS), Kuala Lumpur, Malaysia, 3–4 December 2018 (2018)

Amaechi, C.V., et al.: State-of-the-art review of composite marine risers for floating and fixed platforms in deep seas. Appl. Ocean Res. **138**, 103624 (2023)

American Petroleum Institute: API RP 17B: Recommended Practice for Flexible Pipe (5th ed.) (2014a)

American Petroleum Institute: API Spec 17J: Specification for Unbonded Flexible Pipe (4th ed., Errata 1: September 2016) (2014b)

American Petroleum Institute: API RP 17N: Recommended Practice for Subsea Production System Reliability and Technical Risk Management (Rev 2) (2017)

American Petroleum Institute: API RP 17L2: Recommended Practice for Flexible Pipe Ancillary Equipment (2nd ed.) (2020a)

American Petroleum Institute: API Spec 17L1: Specification for Flexible Pipe Ancillary Equipment (2nd ed.) (2020b)

Andersen, M.L., Sævik, S., Wu, J., Leira, B.J., Langseth, H.: Applying Bayesian optimization to predict parameters in a time-domain model for cross-flow vortex-induced vibrations. Mar. Struct. **94**, 103571 (2024)

API: API Specification 17D: Specification for subsea wellhead and tree equipment (2021)

API RP 2RIM: Integrity Management of Risers from Floating Production Systems, American Petroleum Institute (API), 1 September 2019

Bai, Y., Bai, Q.: Subsea Engineering Handbook. Gulf Professional Pub, Waltham (2012)

Bai, Y., Bai, Q.: Chapter 23 – Subsea wellheads. In: Subsea Engineering Handbook, 2nd edn. pp. 675–696. GPP – Gulf Professional Publishing (2019)

Barton, C., Albaugh, E.K., Davis, D.: 2019 Deepwater Technologies & Solutions for Concept Selection. Offshore Magazine, Houston (2019)

Basilio, L.P., et al: Subsea Processing Systems An Overview of Promising Technologies on the Subsea Factory Decarbonization Path. In OTC, Houston (2023)

Bell, M., Sriskandarajah, A., Heaulme, V., Bouillouta, F., Seinuah, A.: Remote Digital Technologies Driving Innovation – A Case Study on Subsea Operations. Society of Petroleum Engineers, Abu Dhabi (2022)

Bhowmik, S.: Machine learning-based optimization for subsea pipeline route design. In: Offshore Technology Conference, Houston, TX, USA, 16–19 August 2021 (2021)

Bhowmik, S., & Sun, T. J. (2023). Pipeline Fatigue Damage Monitoring During Hydrogen Transportation. Offshore Technology Conference, Houston, TX, USA.

Bordalo, S.N., Morooka, C.K., Trigo, C.C.O.: An experimental study of oscillations induced by two-phase slug flows of water and air on a flexible catenary shaped submerged pipe. Geoenergy Sci. Eng. **239**, 212866 (2024)

British Standards Institution: BS ISO 55000: Asset management – Overview, principles and terminology (2014)

Cai, J., Le Grognec, P.: Lateral buckling of submarine pipelines under high temperature and high pressure—a literature review. Ocean Eng. **244**, 110254 (2022)

Cai, B., Liu, Y., Liu, Z., Tian, X., Ren, C., Abulimiti, A.: Exploratory study on load and resistance factor design of pressure vessel for subsea blowout preventers. Eng. Fail. Anal. **27**, 119–129 (2013)

Cao, Y., et al.: Sensitivity analysis of key parameters on static stiffness of the power umbilical. Ocean Eng. **200**, 107055 (2020)

Chae, K., et al.: Influence of coiling behavior on axial stress in steel wires of submarine power cables: a numerical study. Ocean Eng. **288**, 116014 (2023)

Chandima Ratnayake, R.M., Markeset, T.: Technical integrity management: measuring HSE awareness using AHP in selecting a maintenance strategy. J. Qual. Maint. Eng. **16**(1), 44–63 (2010)

Chen, W., et al.: Structural configurations and dynamic performances of flexible riser with distributed buoyancy modules based on FEM simulations. Int. J. Nav. Archit. Ocean Eng. **13**, 650–658 (2021)

Chen, J., Bai, X., Vaz, M.A.: Dynamic behavior of steel catenary riser at the TDZ considering soil stiffness degeneration and trench development. Ocean Eng. **250**, 110970 (2022)

Choi, H.O., Park, H.: "Oil is the new data": energy technology innovation in digital Oil fields. Energies. **13**(21), 5547 (2020)

Cruces-Giron, A.R., Mendez Rodriguez, W.S., Correa, F.N., Jacob, B.P.: An enhanced semi-coupled methodology for the analysis and design of floating production systems. J. Offshore Mech. Arct. Eng. **143**(4), 041703 (2021)

Dai, T., et al.: Experimental and numerical studies on dynamic stress and curvature in steel tube umbilicals. Mar. Struct. **72**, 102724 (2020)

Davies, S.R., Bakke, W., Ramberg, R.M., Jensen, R.O.: Experience to Date and Future Opportunities for Subsea Processing in StatoilHydro. In Offshore Technology, Houston (2010)

Devold, H.: Oil and Gas Production Handbook. 2006–2023 ABB Oil and Gas, Oslo (2016)

Dionicio-Bravo, S., et al.: Finite element modelling and theoretical analysis of flexible risers subjected to installation/crushing loads. Ocean Eng. **272**, 113856 (2023)

DNV: DNV-RP-G101: Risk Based Inspection of Offshore Topsides Static Mechanical Equipment (2010)

DNV GL Oil & Gas: Subsea Facilities- Technology Developments, Incidents and Future Trends. DNV GL (2014)

DNVGL: DNVGL-ST-F101: Offshore Standard, Submarine Pipeline Systems (2017)

DNVGL: DNVGL-RP-F116: Integrity Management of Submarine Pipelines (2019a)

DNVGL: DNVGL-RP-F206: Riser Integrity Management (2019b)

DNVGL: DNV-ST-F101: Submarine Pipeline Systems (2022)

Dong, G., Hou, H., Xu, T.: Model uncertainty in hydrodynamic characteristics by numerical models for aquaculture plant and mooring system. Ocean Eng. **219**, 108383 (2021)

Duan, J., Zhou, J., Wang, X., You, Y.: Vortex-induced vibration of a flexible fluid-conveying riser due to vessel motion. Int. J. Mech. Sci. **223**, 107288 (2022)

Edwards, A., Zeng, P.: Future Opportunities and Challenges of Virtual Remote Operations for ROVs. Society of Petroleum Engineers, Abu Dhabi (2023)

Elosta, H., Gavouyere, T., Garnier, P.: Flexible risers lifetime extension: Riser in-service monitoring and advanced analysis techniques. In: Proceedings of the 36th International Conference on Offshore Mechanics and Arctic Engineering (Vol. 4, OMAE2017-62700). American Society of Mechanical Engineers, New York (2017)

Fang, P., et al.: Mechanical responses of submarine power cables subject to axisymmetric loadings. Ocean Eng. **239**, 109847 (2021)

Fuheng, L., et al.: Experimental investigation on flow-induced vibration control of flexible risers fitted with new configuration of splitter plates. Ocean Eng. **266**, 112597 (2022)

Gao, G., Cui, Y., Qiu, X.: Prediction of vortex-induced vibration response of deep sea top-tensioned riser in sheared flow considering parametric excitations. Pol. Marit. Res. **27**, 48–57 (2020)

Gao, Z., Wang, W., Zhou, Z., Yan, Y., Pradhan, D.L.: Dynamic response of a steel catenary riser at touch-down zone. Eng. Struct. **295**, 116839 (2023)

Gerria, A., Shokry, A., Zio, E., Montini, M.: Artificial intelligence for the online prediction of the cool-down time in a subsea pipeline after an unplanned shutdown. In: Abu Dhabi International Petroleum Exhibition & Conference, Abu Dhabi, UAE, 15–18 November 2021 (2021)

Goh, T., et al: Unlocking first deepwater sarawak sour gas molecule towards sustainable future. In: Offshore Technology Conference Asia, Kuala Lumpur, Malaysia, 27 February – 01 March 2024 (2024)

Gould, B.D.: Use of electric submersible pumping systems in offshore and subsea environments. In: OTC- Offshore Technology Conference, Houston (2011)

Gourvenec, S., White, D.J.: In situ decommissioning of subsea infrastructure. Keynote for Conference on Maritime Energy, Decommissioning of Offshore Geotechnical Structures, Hamburg, Germany, 28–29 March 2017 (2017)

Gourvernec, S.: Shaping the offshore decommissioning agenda and next-generation design of offshore infrastructure. In: Small Infrastructure and Construction, pp. 54–66 (2019)

Gower, A. S. (2023). Offshore Integrity Management and Life Extension – A Vision of the Application of Subsea Robotics in Assuring Asset Integrity. Offshore Technology Conference, Houston, TX, USA.

Hamilton, M., et al: Visual Twin for Pipeline Leak Detection. In: ADIPEC Conference, Abu Dhabi, UAE (2023)

Hansen, R.L., Rickey, W.P.: Evolution of subsea production systems: a worldwide overview. J. Pet. Technol. **47**(08), 675–680 (1995)

Hayden, B.: Oceaneering Collaborating with University of Houston and Chevron on Autonomous Robot for Pipeline Leaks and Structural Failures, Workboat Magazine, September 12, 2023 (2023)

Hegdal, T.: Subsea Water Treatment and Injection to Optimize Production and to Reduce Cost. In SPE, Baku (2023)

Hobbs, R.E.: Pipeline buckling caused by axial loads. J. Constr. Steel Res. **1**(2), 95–105 (1981)

Hong, C., Wang, Y., Estefen, S.F.: A MINLP model for the layout design of subsea oil gathering – transportation system in deep water oil field considering avoidance of subsea obstacles and pipe intersections. Ocean Eng. **277**, 114278 (2023)

Horn, A.M., Østby, E., Hyygas, B.A., Heiberg, G., Leinum, B.: Re-qualification of existing subsea pipelines for CO_2 and H_2 transport, structural integrity challenges. In: Proceedings of the ASME 2021 Asset Integrity Management – Ageing and Life Extension Conference (AIE2021-77600) (2021)

Iacoponi, S., et al.: HSURF: A New Modular Platform for Underwater Remote Semi-Autonomous Facilities Inspection. Paper presented at the ADIPEC, Abu Dhabi, UAE, October 2023

Ilangovan, K., Dindi, M., Fuglesang, A.R., Van Den Rest, B.: Qualification and Application of All Electric and Topside Less Subsea Multiphase Pump Technology in Subsea Factory Mission to Minimise the Life Cycle Cost. In IPTC (2021)

International Organization for Standardization: ISO 9001: Quality Management Systems (QMS), Requirements (2015)

International Organization for Standardization: ISO 14224: Petroleum, petrochemical and natural gas industries – Collection and exchange of reliability and maintenance data for equipment (n.d.-a)

International Organization for Standardization: ISO 20815: Petroleum, petrochemical and natural gas industries – Production assurance and reliability management (n.d.-b)

International Organization for Standardization/International Electrotechnical Commission: ISO/IEC 17020: Conformity assessment – Requirements for the operation of various types of bodies performing inspection (2012)

ISO 13628-1: Petroleum and Natural Gas Industries—Design and Operation of Subsea Production Systems – Part 1: General Requirements and Recommendations. ISO – International Organization for Standardization (2005)

ISO 13628-8:2002: Design and operation of subsea production systems—Part 8: Remotely Operated Vehicle (ROV) interfaces on subsea production systems (2002)

Jabari, R., & Cheng, T. (2020). Autonomous Evolution Robotic Systems for Underwater Surveillance and Inspection. Offshore Technology Conference, Houston, TX, USA.

Janbazi, H., Shiri, H.: Incorporation of the riser-seabed-seawater interaction effects into the trench formation and fatigue response of steel catenary risers in the touchdown zone. Ocean Eng. **289**, 116288 (2023a)

Janbazi, H., Shiri, H.: Investigation of trench effect on fatigue response of steel catenary risers using an effective stress analysis. Comput. Geotech. **160**, 105506 (2023b)

Jia, J., Gu, J., Huang, J., Gao, L., Chen, L., Wang, S.: Numerical simulation of loop formation in catenary risers on nonlinear uneven seabed. Ocean Eng. **269**, 113480 (2023a)

Jia, J., et al.: A numerical study on the installation configuration design of compliant vertical access riser (CVAR) in deepwater. Mar. Struct. **87**, 103330 (2023b)

Jiang, T., Sun, H., Wang, H., Xiao, X.: Chapter 2 – Pressure monitoring of geopressured gas wells. In: Reserves Estimation for Geopressure Reservoirs, pp. 27–52. GPP – Gulf Professional Publishing (2023)

Jin, J., et al: Numerical modelling of hydrodynamic responses of Ocean Farm 1 in waves and current and validation against model test measurements. Mar. Struct. **78**, 103017 (2021)

JIP. (2017). Handbook on Design and Operation of Flexible Pipes: Safe and Cost Effective Operation of Flexible Pipes.

Jota, J.: Below the Surface: Subsea Infrastructure Inspection Equipment and Techniques. LinkedIn, March 8, 2024 (2024)

Junior, V.J., Papaelias, M.: Design by analysis of deep-sea type III pressure vessel. Int. J. Hydrog. Energy. **46**(17), 10468–10477 (2021)

Kaculi, J.: Next Generation HPHT Subsea Wellhead Systems Design Challenges and Opportunities. In OTC, Houston (2015)

Kalyan, B., & Chitre, M. A. (2023). Concept of Operations for Collaborative Human Robot Inspection and Intervention System in Challenging Underwater Environments. Offshore Technology Conference, Houston, TX, USA.

Kane, R.D.: Commentary on standards development for selection and specification of HSC-resistant materials in subsea service with cathodic protection. In: AMPP 2022 Corrosion Conference & Expo, Denver, Colorado (2022)

Kang, Z., Li, S., Zhang, C., Qu, Y., Ma, G.: Distributions of nonlinear dynamics, stress and fatigue damage along a steel catenary riser in current affected by top-end surge. Ocean Eng. **285**, 115398 (2023)

Kaushal, D.: Digital Oil Field Future – Increasing Demands and Challenges. GreyB Services (2023) Retrieved from [https://www.greyb.com/blog/digital-oil-field-future/] (https://www.greyb.com/blog/digital-oil-field-future/). Accessed 29 Nov 2023

Khan, R., et al. (2022). Risk-Based Underwater Inspection Methodology for Subsea Integrity Management in Malaysian Operations. Offshore Technology Conference Asia, March 2022.

Khandoker, S., Landthrip, G., & Huff, P. (2008). Structural Optimization of Subsea Pressure Vessel Equipment for HPHT Application with a Design by Analysis Case Study of a Typical BOP. In OTC- Offshore Technology Conference, Houston.

Kim, S.W., Sævik, S., Wu, J., Leira, B.J.: Prediction of deepwater riser VIV with an improved time domain model including non-linear structural behavior. Ocean Eng. **236**, 109508 (2021)

Kim, S.W., Sævik, S., Wu, J., Leira, B.J.: Time domain simulation of marine riser vortex induced vibrations in three-dimensional currents. Appl. Ocean Res. **120**, 103057 (2022)

Korotygin, D., Nammi, S.K., Pancholi, K.: The effect of ice floe on the strength, stability, and fatigue of hybrid flexible risers in the Arctic Sea. J. Compos. Sci. **7**(6), 212 (2023)

Krishnamoorthy, P., Tremblay, C., & Jackson, P. (2023). Offshore Underwater Fixed Structure Integrity Assessment Using LRUT, PEC & ACFM Advanced Technologies with Minimal Marine Growth Removal. ADIPEC Conference, Abu Dhabi, UAE.

Lee, Y., et al.: Digital twin approach with minimal sensors for Riser's fatigue-damage estimation. Int. J. Nav. Archit. Ocean Eng. **16**, 100603 (2024)

Lendenmann, H., Fløisand, J.O., Vatland, S.: New Era Large Subsea Multiphase Compressor – Driven by Subsea Adjustable Speed Drive. In OTC, Houston (2021)

Leow, C.H., Ong, H.G., Khoo, C.A.: Corrosion management for offshore wet sour gas carbon steel pipeline with closed loop MEG regeneration system. In: AMPP 2023 Corrosion Conference & Expo, Denver, Colorado (2023)

Li, X., et al: Effect of hydrogen charging time on hydrogen blister and hydrogen-induced cracking of pure iron. Corros. Sci. **181**, 109200 (2021a)

Li, X., et al.: Finite element study on circular armour wire lateral buckling in umbilicals. Mar. Struct. **76**, 102895 (2021b)

Liang, X., Liu, Z., Huang, D., Wang, T., Wang, C.: Experimental and numerical investigation of the drag coefficients of subsea tree. Ocean Eng. **238**, 109701 (2021)

Liu, Q., et al.: Behavior of unbonded flexible riser with composite armor layers under coupling loads. Ocean Eng. **239**, 109907 (2021)

Lopes, F.S., Fernandes, A.C., Sales Junior, J.S., Andrade, E.M.: Nonlinear slackness anatomy during vertical installation of heavy devices in deep water. Mar. Struct. **96**, 103626 (2024)

Lu, H., et al.: Flexible riser tensile armour stress assessment in the bend stiffener region. Eng. Struct. **254**, 113849 (2022a)

Lu, H., et al.: Full-scale experimental and numerical analyses of a flexible riser under combined tension-bending loading. Mar. Struct. **86**, 103275 (2022b)

Manach, J., et al: Comparative Study Between an Integrated All-Electric Subsea Production System, Including Subsea Chemical System Injection, and a Standard Electro-Hydraulic System. In OTC, Houston (2023)

Medany et al. (2024). Inspection of Subsea Conductors Using an Innovative Drop-Down Camera. GOTECH Conference, Dubai, UAE.

Mellem, T., Hugaas, B.: A systematic approach to reduce subsea equipment failures. In: Subsea Controls and Data Acquisition 2002. Society of Underwater (2002)

Melo, J., Silva, M.: Reliability study on the life extension of subsea umbilical systems. In: Proceedings of the Offshore Technology Conference (OTC-29530-MS), Houston, Texas, USA, 6–9 May 2019 (2019)

Menard, F., et al.: A computationally efficient finite element model for the analysis of the non-linear bending behaviour of a dynamic submarine power cable. Mar. Struct. **91**, 103465 (2023)

Meneses, B.M., et al.: Post-failure behavior of lazy-wave risers. Ocean Eng. **280**, 114777 (2023)

Minerals Management Service: Deepwater Riser Design, Fatigue Life and Standards Study, 86330-20-R-RP-005, TA&R Project Number 572, Rev. 1, 22 OCT 2007 (2007)

Müller, D.T., et al: Field life extension and integrity management in Campos Basin. In: Proceedings of the Offshore Technology Conference (OTC-28750-MS), Houston, Texas, USA, 30 April–3 May 2018 (2018)

Nasri Huang, F. J., Hashim, M. F., & Omar, S. A. (2024). Integrity Management and Condition Performance Monitoring of Shallow Water Subsea Field Development in Malaysia. Offshore Technology Conference, Houston, TX, USA.

National Offshore Petroleum Safety and Environmental Management Authority (NOPSEMA): Planning for proactive decommissioning (Information Paper Document No: N-00500-IP2002 A816565). Australia (2021)

Nauticus Robotics: Testing of Aquanaut Mk2 Underwater Robot. Gulf of Mexico Initial Test Results. April 09, 2024 (2024)

Nevoso, C., Cavallini, F., Massari, G., Bernini, T., Watanabe, T., Britto, A.: Going Deeper: Inspecting Subsea Live Assets by Means of an Advanced AUV Solution. Offshore Technology Conference, Houston, TX, USA (2024)

Nikitin, N.O., Revin, I., Hvatov, A., Vychuzhanin, P., Kalyuzhnaya, A.V.: Hybrid and automated machine learning approaches for oil fields development: the case study of Volve field, North Sea. Comput. Geosci. **161**, 105061 (2022). https://doi.org/10.1016/j.cageo.2022.105061

NORSOK: NORSOK U-001: Subsea Production Systems (4th ed.) (2015)

NORSOK STANDARD U-009: Life Extension for Subsea Systems (2021)

NORSOK STANDARD Y-002: Life Extension for Transportation Systems (2021)

Norwegian Oil and Gas Association. (n.d.). Recommended Guidelines for the Management of Life Extension.

Novel, M., et al: Anchoring system of rigid riser – flowline connection with torpedo piles. In: 42nd International Conference on Ocean, Offshore and Arctic Engineering (OMAE), Melbourne, Australia (2023)

Offshore and Onshore Reliability Data (6th ed., Vols. 1–2). Topside Equipment and Subsea Equipment

Offshore Petroleum and Greenhouse Gas Storage (Greenhouse Gas Injection and Storage) Regulations 2023 (the GHG Regulations)

Offshore Petroleum and Greenhouse Gas Storage (Regulatory Levies) Act 2003

Offshore Petroleum and Greenhouse Gas Storage (Regulatory Levies) Regulations 2004

Offshore Petroleum and Greenhouse Gas Storage Act 2006 (the OPGGS Act)

Offshore Petroleum Greenhouse Gas Storage (Resources Management and Administration) Regulations 2011 (the RMA Regulations)

Petroleum Safety Authority Norway: Barriers Memorandum (2017)

Petroleum Safety Authority Norway: The management regulations. Retrieved from https://www.ptil.no/en/regulations/all-acts/ (n.d.)

Pettigrew, I.G., Barker, A., O'Brien, A., Cesan, A.: InspectTM Computed Tomography for NDT of Subsea Pipelines. In: Offshore Technology Conference, Houston, TX, USA (2021)

Pollak, J.B., et al: Project Execution of Industry First 20,000-psi Subsea Production System. In OTC, Houston (2022)

Poon, C., et al.: Fretting wear and fatigue in submarine power cable conductors for floating offshore wind energy. Tribol. Int. **186**, 108598 (2023)

Prsic, M.A., Solaas, F., Kristiansen, T.: Hydrodynamic coefficients of generic subsea modules in forced oscillation tests—importance of structural elements. J. Offshore Mech. Arctic Eng. **146**, 011401 (2024)

Rachman, A., et al.: Applications of Machine Learning in Pipeline Integrity Management: A State-of-the-Art Review. J. Energy Resour. Technol. (2021)

Rajashekara, K., Krishnamoorthy, H.S., Naik, B.S.: Electrification of subsea systems: requirements and challenges in power distribution and conversion. CPSS Trans. Power Electron. Appl. **2**(4), 259–266 (2017)

Ramos, R., Pasqualino, I., Souza, M.I.L., Nicolosi, E.R.: An application of fault tree analysis for decommissioning of subsea flexible pipeline in Brazil. In: Proceedings of the 38th International Conference on Ocean, Offshore and Arctic Engineering, Glasgow, Scotland, 9–14 June 2019 (2019)

Ray, A., Rajashekara, K.: Electrification of offshore oil and gas production: architectures and power conversion. Energies. **16**(15), 5812 (2023)

Renzi, D.F.: Digital twin model improves riser integrity management. Offshore Technology News, Nov. 25, 2019 (2019)

Ribeiro, L.G., et al.: Methodology for simulation of the post-failure behavior of flexible catenary risers. Mar. Struct. **82**, 103142 (2022)

Rokni, H.J., Shiri, H.: An alternative vessel excitation algorithm to incorporate the trench effect into the fatigue analysis of steel catenary risers in the touchdown zone. Appl. Ocean Res. **126**, 103292 (2022)

Sadeghi, A., Moe, P.T., Hilley, A.: Subsea drilling and wellhead load monitoring systems in the North Sea: a case study using a well access management system (WAMS). Int. J. Adv. Eng. Sci. Appl. **2**(1), 1–6 (2021)

Sæther, J.H.: Choke Condition and Performance Monitoring. Master's Thesis, Norwegian University of Science and Technology, Trondheim (2010)

Sales, L., Stanko, M., Jäschke, J.: Superstructure optimization of subsea processing layouts. J. Pet. Explor. Prod. Technol. **13**, 1575–1589 (2023)

Shen, Y., Birkinshaw, P., Palmer-Jones, R.: Challenges in offshore pipeline decommissioning and what we can learn from integrity management practices. In: Proceedings of the 27th International Ocean and Polar Engineering Conference, San Francisco, California, USA, 25–30 May 2017 (2017)

Shiraiwa, T., Kawate, M., Briffod, F., Kasuya, T., Enoki, M.: Evaluation of hydrogen-induced cracking in high-strength steel welded joints by acoustic emission technique. Mater. Des. **190**, 108573 (2020)

SINTEF: Ageing and life extension for offshore facilities in general and for specific systems (2010)

Sivaprasad, H., Lekkala, M.R., Latheef, M., Seo, J., Yoo, K., Jin, C.: Fatigue damage prediction of top tensioned riser subjected to vortex-induced vibrations using artificial neural networks. Ocean Eng. **268**, 113393 (2023)

Song, H., et al.: Integrity Management for Steel Catenary Risers with Design Life of 30 Years, ASME 2020 39th International Conference on Ocean, Offshore and Arctic Engineering, OMAE2020-18065, December 18, 2020 (2020)

Speight, J.G.: Chapter 6 – Production. In: Handbook of Offshore Oil and Gas Operations. Gulf Professional Publishing (2015)

Stump, J.: New subsea robotics advancing ROV, AUV technology. Offshore Journal, April 1, 2021 (2021)

Sureflex, J.I.P.: Flexible Pipe Integrity Management Guidance & Good Practice (2017)

Tang, M., et al.: Monitoring the slip of helical wires in a flexible riser under combined tension and bending. Ocean Eng. **256**, 111512 (2022)

Thorsen, M.J., Sævik, S., Larsen, C.M.: A simplified method for time domain simulation of cross-flow vortex-induced vibrations. J. Fluids Struct. **49**, 135 (2014)

Tommasini, R.B., Hill, T.L., Macdonald, J.H.G., Pavanello, R., Carvalho, L.O.: The dynamics of deepwater subsea lifting operations in super-harmonic resonance via the harmonic balance method. Mar. Struct. **80**, 103095 (2021)

Trevail, M., Lopez, C., Martins, B.: Enhancing production and operational efficiency in deepwater subsea tiebacks using artificial intelligence. In: Offshore Technology Conference, Houston, TX, USA (2024)

Tu, S., Shuai, J.: Numerical study on the buckling of pressurized pipe under eccentric axial compression. Thin-Walled Struct. **147**, 106542 (2020)

Various Authors: A recent industry understanding of upheaval buckling design study for a large-size subsea pipeline buried through a jet trenching method. In: International Society of Offshore and Polar Engineers (ISOPE) Conference, 2022 (2022a)

Various Authors: Review of non-metallic pipelines in oil & gas applications – challenges & way forward. In: International Petroleum Technology Conference, 2022 (2022b)

Vasquez, J.A.M., et al.: Three-dimensional dynamic behaviour of flexible catenary risers with an internal slug flow. J. Fluids Struct. **107**, 103409 (2021)

Wang, L., Tan, Z.: Enabling the Internet of Underwater Things (IoUT). Society of Petroleum Engineers, Abu Dhabi (2023)

Wang, Y., Wei, A., Shen, D.: Resistances to SCC and HIC for 17-4PH steel in H2S environment. Corros. Protect. **37**(2), 100–104 (2016)

Wang, C., Wang, Y., Liu, Y., Li, P., Zhang, X., Wang, F.: Experimental and numerical simulation investigation on vortex-induced vibration test system based on bare fiber Bragg grating sensor technology for vertical riser. Int. J. Nav. Archit. Ocean Eng. **13**, 223–235 (2021a)

Wang, Y., Wang, Q., Zhang, A., Qiu, W., Duan, M., Wang, Q.: A new optimization algorithm for the layout design of a subsea production system. Ocean Eng. **232**, 109072 (2021b)

Wang, J., Xu, L., Cao, J., Sheng, L., Li, C.: Numerical and experimental investigation for extreme storm-safe drilling riser. Ships Offshore Struct. **17**(7), 1462–1474 (2022a)

Wang, Y., Gao, D., Wang, J., Ning, B.: Investigation on influence of temperature and pressure on fatigue damage of subsea wellhead in deepwater drilling. J. Pet. Sci. Eng. **212**, 110328 (2022b)

Wang, Y., Wang, Q., Zhang, Y., Yue, Q., Zhang, X.: Optimization of subsea production facilities layout based on cluster manifold system considering seabed topography. Ocean Eng. **291**, 116575 (2024)

Wood Group: Guideline to Subsea Integrity Management – Wellhead to Topside ESDV, J003108-01-IM-REP-001, December 2020 (2020)

Wu, J.H., Zhen, X.W., Liu, G., Huang, Y.: Optimization design on the riser system of next generation subsea production system with the assistance of DOE and surrogate model techniques. Appl. Ocean Res. **85**, 34–44 (2019)

Wu, J., Yin, D., Lie, H., Riemer-Sørensen, S., Sævik, S., Triantafyllou, M.: Improved VIV response prediction using adaptive parameters and data clustering. J. Mar. Sci. Eng. **8**, 127 (2020)

Xie, K., et al.: Research on high quality mesh method of armored umbilical cable for deep sea equipment. Ocean Eng. **221**, 108550 (2021a)

Xie, W., Xin, W., Zhang, H.: Influence of the internal varying density flow on the vibrations and fatigue damage of a top-tensioned riser undergoing vortex-induced vibrations. Appl. Ocean Res. **117**, 102955 (2021b)

Xu, T., Dong, G., Tang, M., Liu, J., Guo, W.: Experimental analysis of hydrodynamic forces on net panel in extreme waves. Appl. Ocean Res. **107**, 102495 (2021)

Yamamoto, M., Yamamoto, J., Masanobu, S.: Study on the axial tension reduction of a dual-bore vertical riser system for deep sea mining. In: 42nd International Conference on Offshore Mechanics and Arctic Engineering (OMAE), Melbourne, Australia (2023)

Yan, J., et al.: Time series prediction based on LSTM neural network for top tension response of umbilical cables. Mar. Struct. **91**, 103448 (2023)

Yang, X., Luo, X., Chen, J., Pan, H., Li, X.: Research progress in hydrogen induced cracking of high strength aluminum alloy. Mater. Protect. **53**(8), 23–27 (2020)

Yang, Z., et al.: Integrated optimization design of a dynamic umbilical based on an approximate model. Mar. Struct. **78**, 102995 (2021a)

Yang, Z., et al.: Study on the optimization algorithm of the cross-sectional layout of an umbilical based on the layering strategy. Ocean Eng. **232**, 109120 (2021b)

Yang, Z., et al.: Study on the nonlinear mechanical behaviour of an umbilical under combined loads of tension and torsion. Ocean Eng. **238**, 109742 (2021c)

Yin, Y., et al.: Experimental study on the interlayer friction and wear mechanism between armor wires of umbilicals. Mar. Struct. **80**, 103102 (2021)

Yin, X., et al.: Study on the automatic optimization design of the cross-sectional layout of an umbilical with layers based on the GA-GLM. Mar. Struct. **88**, 103363 (2023)

Yu, C.A., Cheng, Y., Yang, G., Carballo, M.R.: Life extension of the risers used for the Hoover DDCV in Gulf of Mexico. In: Proceedings of the Offshore Technology Conference (OTC-29608-MS), Houston, Texas, USA, 6–9 May 2019 (2019)

Yuan, Y., Zheng, M., Xue, H., Duan, Z., Tang, W.: Fatigue analysis of a steel catenary riser at touchdown zone with seabed resistance and hydrodynamic forces. Ocean Eng. **244**, 110446 (2022)

Zhang, J., Guo, H., Tang, Y., Li, Y.: Effect of top tension on vortex-induced vibration of deep-sea risers. J. Mar. Sci. Eng. **8**(2), 121 (2020a)

Zhang, J., Zhang, H., Zhang, L., Liang, Z.: Buckling response analysis of buried steel pipe under multiple explosive loadings. J. Pipeline Syst. Eng. Pract. **11**(2), 04020010 (2020b)

Zhang, C., et al.: Nonlinear motion regimes and phase dynamics of a free standing hybrid riser system subjected to ocean current and vessel motion. Ocean Eng. **252**, 111197 (2022a)

Zhang, C., Lu, L., Cao, Q., Cheng, L., Tang, G.: Nonlinear motion regimes and phase dynamics of a free standing hybrid riser system subjected to ocean current and vessel motion. Ocean Eng. **252**, 111197 (2022b)

Zhang, Z., Zhang, Y., Liu, Y., Zhou, B.: Experimental study on erosion of full-scale unbonded flexible subsea pipe. In: The Thirty-third International Ocean and Polar Engineering Conference, Ottawa, Canada, June 19–23, 2023 (2023)

Zhao, S., Lan, W.: Present status and research progress of anti-corrosion technology in pipelines. Surf. Technol. **44**(11), 112–117 (2015)

Zhao, T., Yu, F.: Critical upheaval buckling forces of sandwich pipelines with variable stiffnesses of pipe material. Ocean Eng. **217**, 107547 (2020)

Open Access This chapter is licensed under the terms of the Creative Commons Attribution-NonCommercial-NoDerivatives 4.0 International License (http://creativecommons.org/licenses/by-nc-nd/4.0/), which permits any noncommercial use, sharing, distribution and reproduction in any medium or format, as long as you give appropriate credit to the original author(s) and the source, provide a link to the Creative Commons license and indicate if you modified the licensed material. You do not have permission under this license to share adapted material derived from this chapter or parts of it.

The images or other third party material in this chapter are included in the chapter's Creative Commons license, unless indicated otherwise in a credit line to the material. If material is not included in the chapter's Creative Commons license and your intended use is not permitted by statutory regulation or exceeds the permitted use, you will need to obtain permission directly from the copyright holder.

Committee V.4: Offshore Renewable Energy

Charles Rawson[1(✉)], Nagi Abdussamie[2], Asger Bech Abrahamsen[3], Ivan Catipovic[4], Thomas Choisnet[5], Toshiki Chujo[6], Jose Gaspar[7], Luca Greco[8], Ole A. Hermundstad[9], Han-Koo Jeong[10], Shen Li[11], Wengang Mao[12], Rachel Nicholls-Lee[16], Gabriele Notaro[17], Elif Ogus[13], Freeman Ralph[14], and Shali Sun[15]

[1] Waldorf, Maryland, USA
charles.e.rawson@uscg.mil
[2] Australian Maritime College (AMC), Launceston, Australia
[3] Technical University of Denmark (DTU), Roskilde, Denmark
[4] University of Zagreb, Zagreb, Croatia
[5] BW Ideol, La Ciotat, France
[6] National Maritime Research Institute (NMRI), Tokyo, Japan
[7] University of Lisbon (CENTEC), Lisbon, Portugal
[8] CNR-INM, Rome, Italy
[9] SINTEF Ocean, Trondheim, Norway
[10] Korea Institute of Ocean Science and Technology (KIOST), Busan, Republic of Korea
[11] University of Strathclyde, Glasgow, United Kingdom
[12] Chalmers University of Technology, Gothenburg, Sweden
[13] Middle East Technical University (METU), Ankara, Turkey
[14] C-CORE, St. John's, Canada
[15] Beijing, China
[16] Whiskerstay Ltd, Exeter, United Kingdom
[17] DNV, Oslo, Norway

Committee Mandate. Concern for load analysis and structural design of offshore renewable energy devices. Attention shall be given to the interaction between the load and structural response of fixed and floating installations taking due consideration of the stochastic and extreme nature of the ocean environment. Aspects related to design, prototype testing, certification, marine operations, levelized cost of energy and life cycle management shall be considered.

Keywords: Offshore wind turbine · floating wind turbine · wave energy converter · tidal energy · ocean energy · floating solar photovoltaic · hybrid systems · renewable energy

1 Introduction

This report presents a comprehensive analysis of advancements in offshore renewable energy, including the areas of offshore wind, wave, tidal, floating solar energy, and the hybridization of these technologies. While emphasizing the interaction between environmental forces and structural responses, key aspects of load analysis and structural design are the focus of this work. The committee also delves into the latest industry developments, numerical and physical modeling approaches, design standards, and operational challenges associated with fixed and floating offshore renewable energy devices.

The growing global demand for sustainable energy solutions has driven significant technological advancements in offshore wind energy, notably in floating wind turbine and hybrid systems. These technologies have seen rapid development due to increasing investments, policy incentives, and an urgent need to reduce carbon emissions. Bottom-fixed offshore wind turbines remain the predominant choice in shallow waters, while floating wind turbines offer promising solutions for deep-sea deployment, expanding the geographical feasibility of offshore wind farms. Innovations in turbine design, foundation stability, and aerodynamic performance are critical to improving efficiency and reducing maintenance costs.

Beyond wind energy, wave and tidal energy converters, as well as floating solar photovoltaic (FPV) systems, are gaining attention as viable components of the offshore renewable energy mix. Wave energy converters (WECs) harness the kinetic and potential energy of ocean waves, while tidal energy systems capitalize on predictable tidal flows to generate electricity. FPV systems, often deployed in combination with offshore wind farms, present an innovative approach to maximizing energy output within a limited marine footprint. Hybrid energy solutions integrating multiple renewable sources are becoming increasingly viable, offering enhanced energy reliability and grid stability.

This report also examines the role of digital technologies in optimizing offshore renewable energy operations. Digital twin technology, artificial intelligence (AI), and machine learning algorithms are revolutionizing predictive maintenance, structural health monitoring, and real-time performance optimization. Additionally, advanced numerical simulations and experimental testing methodologies are enhancing the accuracy of structural assessments, ensuring robust and cost-effective design solutions.

While this report primarily focuses on structural design and load analysis, it also includes discussions on marine operations, transport logistics, and total cost of energy considerations. The committee explores levelized cost of energy (LCOE) assessments, financing mechanisms, and lifecycle cost management strategies. Transport, installation, and maintenance logistics for offshore renewable energy infrastructure present unique challenges that require innovative engineering solutions and coordinated industry efforts.

2 Bottom-Fixed Offshore Wind Turbines

Bottom-fixed offshore wind turbines (BFOWTs) are mainly concentrated in shallow waters and are predominantly deployed in nearshore, shallow-water zones (<50 km from coastlines) due to cost efficiencies. Their layout follows grid-like patterns to optimize energy output and minimize infrastructure costs. Sentinel-1 SAR remote sensing and Google Earth Engine (GEE) have enabled precise mapping, revealing that 95% of global OWTs are bottom-fixed, reflecting their maturity compared to floating alternatives.

2.1 Recent Industry Developments

Wang, K. et al., (2024a) shows that bottom-fixed offshore wind turbines dominate the global offshore wind energy sector, with 12,412 units identified worldwide as of 2022.

These turbines are concentrated in Northern Hemisphere coastal regions, particularly in Europe (5,915 units) and Asia (6,490 units), while the U.S. lags significantly (7 units). Europe pioneered deployment between 2006–2018, with peak activity during 2010–2015. Europe is the traditional stronghold for offshore wind power, with most OWTs located in the nearshore areas of the North Sea and its surrounding regions. Projects often expanded incrementally, filling gaps between existing turbine arrays. The UK has the largest number of OWTs (2,737), followed by Germany (1,537), Denmark (629), the Netherlands (515) and Belgium (401). Asia, led by China (6,038 units), has experienced explosive growth since 2019, accounting for two-thirds of its installations post-2019 and its OWTs are mainly distributed in the nearshore areas of the Yellow Sea, East China Sea and South China Sea. Vietnam (419 units) and South Korea (33) are emerging players. Asia has rapid acceleration post-2019, driven by China's coastal megaprojects (e.g., Jiangsu and Guangdong provinces). Over 50% of China's OWTs were installed in 2021–2022 alone.

Due to the positive correlation between the construction and maintenance costs of wind turbines and the distance to the port or coast, most wind turbines are built in nearshore areas and distributed in a regular pattern. Therefore, the global wind turbines exhibit a regular distribution in nearshore areas, with numerous turbines aggregated to form a wind farm spatial pattern.

The status of offshore wind, as of Q2 2024, is that 73.1 GW is fully commissioned; with the majority in China (34.8 GW), followed by the UK (14.8 GW) and Germany (8.2 GW). China also has the most under construction (7.7 GW), followed by the UK (5.2 GW) and Germany (2.6 GW). Globally, a cumulative 116.2 GW is post-FID (including operational projects, under construction, and projects in the preconstruction phase).

2.2 Numerical Modeling and Analysis

While the large number of installations of BFOWT worldwide demonstrates a level of technology maturity, cost reduction and operation & maintenance (O&M) optimization of systems are still critical for their design, construction and deployment. Furthermore, the trend towards turbines characterized by higher energy harvesting capacities and greater power per mass ratios to reduce sea occupation and to compete with other sectors of the energy industry is nowadays resulting in larger and more flexible components. Consequently, these units are subject to increasingly demanding logistics concerns, extreme deployment conditions in harsh environments and, in addition, they must withstand aerodynamic, hydrodynamic, gravitational, and geotechnical loads (Jahani, et al., 2022). These aspects must be tackled into a multidisciplinary context involving the interaction among rotary-wing aerodynamics, tower/blades structural dynamics, soil-pile interaction phenomena and control strategies.

2.2.1 Aerodynamic Aspects in Simulations

With increasing maturity of OWT technology and reduced associated commercial risks, the majority of large-scale OWTs share the same features such as horizontal axis of rotation, three blades, upwind, variable-speed, and variable blade pitch. With the state-of-the-art 15 and 18 MW machines now available on the market, research effort on numerical modelling has gradually moved from wind turbine-related aspects to broader topics such as the analysis of the different types of support structures, the development of advanced control systems, new materials and installation and reliability issues.

However, comprehensive wind turbine numerical models with good levels of accuracy and low computational burden are still needed as demonstrated by results from the IEA Task 47 - TURBINIA. This work highlights the limitations of Blade Element Momentum Theory (BEMT), widely used by industry, in predicting blade aeroloads under off-design (unsteady) flow conditions, specifically when the operating conditions are characterized by significant yaw errors or wind shear/veer due to the unprecedented size of modern and next-generation wind turbine rotors. Moreover, many validation exercises on rotor aerodynamics deal only with global rotor loads, whilst it is well demonstrated that the accuracy of these predictions might hide compensating errors on blade local loads.

Among software for design and evaluation is OpenFAST (https://github.com/OpenFAST), an open-source framework for the analysis and simulation of onshore and offshore wind turbines. A real-time hybrid simulation (RTHS) framework couples experimental soil-foundation models with OpenFAST-based analytical substructures for monopile OWTs. Al-Subaihawi et al., (2024) demonstrates improved prediction of foundation nonlinearities under operational and extreme conditions, validating the framework's ability to capture coupled aero-hydro-geotechnical responses.

Further, research on rotor aerodynamics and rotor aeroelasticity is reported to be very limited (especially with respect to similar topics in the field of floating offshore wind turbine applications, see Ch. 3). For instance, Ye et al., (2023) shows a high-fidelity rotor performance analysis conducted using the Finite-Analytic unsteady Navier-Stokes code (FANS). The comparison is performed with the NTNU BT1 wind turbine tested in a wind tunnel in uniform flow conditions and with other state-of-the-art CFD codes over the rotor operating range. The paper shows that CFD predictions correctly match the experimental torque coefficient measurements trend. However, significant underestimation of CT at high TSR values is observed. Differently, local blade loads overestimation is observed on the 2 MW DANAERO rotor, thus demonstrating how the interaction between numerical modelling and experimental tests is fundamental also for the simplest steady axial flow conditions.

Melani et al., (2020) propose a method to extract rotor blades angle of attack from CFD flow fields for Darrieus turbines, comparing three post-processing techniques, 3-Point, LineAverage, Trajectory (see Fig. 1). Results show the LineAverage method achieves the highest accuracy in predicting lift/drag coefficients under stable operational conditions like Figure (TSR = 4.5). The study validates the approach using a pitching airfoil CFD model, demonstrating improved force reconstruction compared to conventional power-law profiles. Boorsma et al., (2023) validate aeroelastic simulation tools (BEM, CFD, free vortex wake) against field measurements from the DANAERO 2 MW turbine under axial, sheared, and yawed inflow. CFD and synthesized airfoil data improve agreement with measurements in sheared conditions, while BEM struggles with skewed wake effects in yawed flows. The study emphasizes the need for

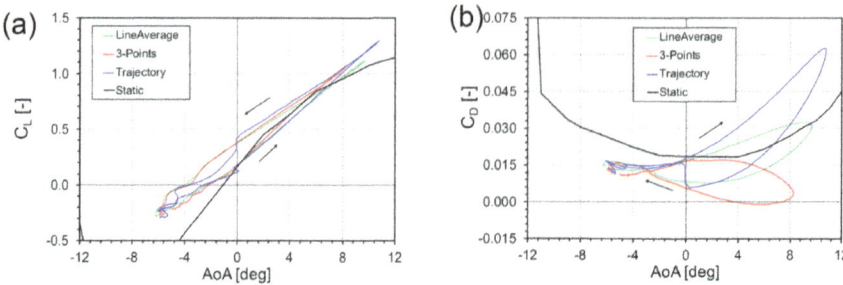

Fig. 1: Comparison of the lift (a) and drag (b) coefficient hysteresis loops obtained with different methods and static airfoil data for TSR = 4.5 (Melani et al., 2020)

turbulence model enhancements and rotational effect considerations. Using URANS and k-ω SST turbulence models, Viré et al., (2020) analyze VIVs for monopiles in supercritical flow (Re $\geq 3.6 \times 10^6$). Results reveal aerodynamic damping transitions between self-exciting and self-limiting behaviors, with lock-in frequencies validated against experiments. Castorrini et al., (2023) couples mesoscale weather models (WRF) with RANS-based CFD to predict offshore wind profiles, validated against LiDAR data. The hybrid approach improves resolution of boundary layer turbulence and wave–wind interactions, demonstrating accuracy in capturing vertical velocity and TKE profiles for large rotors.

2.2.2 Soil-Foundation Interaction (PSI)

Soil-foundation interaction (PSI) is a vital factor in ensuring structural stability, as demonstrated by Ma, et al., (2024), who analyze the dynamic responses of monopile and jacket-supported turbines using the X-SEA program and finite element analysis (FEA). Their findings emphasize that soil flexibility significantly affects natural frequencies and deflection patterns in turbine structures. By modeling lateral and axial soil resistance through nonlinear soil springs, the study underscores the necessity of PSI in OWT design guidelines to ensure stability across varying soil conditions. Building on this, Wu, et al., (2024) highlight the challenges of PSI in cold sea regions, where combined ice and aerodynamic loads cause greater displacements and reduce natural frequencies, especially in flexible foundations. Using the PISA soil modeling method, they show that accurate PSI modeling is critical for capturing the full impact of ice-induced forces, emphasizing the need for combined load analysis in cold-region OWT standards. Validation and application of integrated numerical models including all the turbine sub-components and aimed at simulating its dynamic response under the fully coupled wind/wave loads, and servo-control commands are widely reported in the literature. As an example, Ma, et al., 2024) show a time-domain analysis of the superstructure of the NREL 5 MW wind turbine by the software framework of Abaqus WT coupled with FAST. The PSI model is based on the nonlinear p-y, t-z and Q-z springs. Based on three-representative wind-wave load cases, it is found that the PSI can significantly reduce the natural frequencies of various WT components, while increasing the tower-top peak motions in normal operational conditions. For parking conditions, the PSI can largely increase the peak motion response and bending moments. This study concludes that the PSI should be considered in the design of offshore wind turbine of a monopile foundation due to its remarkable impact on its natural frequency, dynamic response, and seismic response. Sørum et al., (2022) evaluates fatigue design uncertainties in monopile offshore wind turbines using a sensitivity analysis approach. The authors identify SN curve parameters and environmental conditions as dominant factors influencing fatigue damage, while wave directionality and soil model uncertainties significantly affect dynamic response. The analysis highlights the importance of accurate environmental modeling and interdisciplinary parameter integration for reliability improvements.

Furthermore, Bergua et al., (2022) focus on the integration and verification of a new soil-structure interaction model for offshore wind design. Their work emphasizes the importance of accurately capturing soil-pile interactions to improve the reliability of OWT designs, particularly under complex loading conditions. By incorporating an elastoplastic macroelement model, this study addresses the limitations of traditional SSI methods and provides a less conservative and potentially more cost-effective design framework for OWT foundations. Bergua integrates the REDWIN macroelement model for soil-structure interaction (SSI) into OWT simulations, demonstrating reduced system loads compared to traditional methods. The macroelement's improved damping and stiffness characterization highlights the need for accurate SSI modeling in fatigue and dynamic response analysis.

Chen et al., (2023) investigate the seismic response of OWTs supported by hybrid pile-bucket foundations in liquefiable soils. Their findings demonstrate that soil-structure interactions significantly influence the stability of the turbine structures during seismic events, reinforcing the importance of detailed soil analysis in design standards. He and Ye, (2023) use the FssiCAS software and the Pastor-Zienkiewicz-Mark III (PZIII) elastoplastic soil model to analyze seismic responses of a 1.5 MW monopile OWT. Results show significant seabed liquefaction (5–6 m depth near the monopile) and tower-top horizontal oscillations (up to 2 m), but no cumulative displacement (Fig. 2). Comparisons reveal that neglecting soil plasticity or pore water effects leads to inaccurate dynamic predictions. Ngo and Kim, (2024) compare monopile, suction bucket, and jacket-supported OWTs under seismic loads using finite element modeling in ABAQUS. The monopile exhibits the largest tower-top displacement (0.73 m under El-Centro waves), while the jacket foundation experiences 17–21% higher tower stress due to structural stiffness. Fragility curves indicate monopiles are more vulnerable to displacement, whereas jackets are sensitive to stress and bending moments.

2.2.3 Reliability of Results

A fully coupled aero-servo-hydro-elastic model is developed for a 10 MW monopile OWT using nonlinear Winkler foundation-based soil-structure interaction (SSI). Xi et al., (2021) validate a bladed-style controller in FAST and demonstrates that SSI and time-varying blade-pitch angles significantly affect fatigue damage, with 90% computational time reduction compared to traditional FEA. A hydroelastic model coupling a weak-scatterer potential flow solver with FEM was developed for monopile foundations. Leroy et al., (2021) accurately predicted mudline bending moments under regular waves, matching experimental data better than Morison-based models.

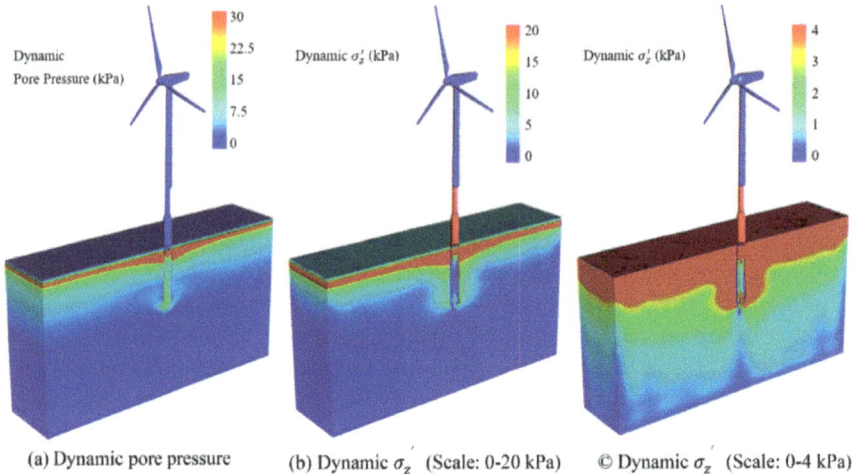

Fig. 2: Distribution of pore pressure and effective stress at t = 300 s (Noted: the initial values are excluded, and the seabed foundation is sectioned by y = 35 m). (He and Ye, 2023)

Barreto, et al., (2022) evaluates the impact of simulation length and flexible foundations on long-term extreme response extrapolation for monopile OWTs. Simulations shorter than 30 min led to significant errors (up to 40% for bending moments), while flexible foundations shifted critical wind speeds (16.5 → 18 m/s) and increased mudline bending moments by ~10% compared to rigid models

A fully coupled reliability assessment framework was developed for OWTs using surrogate models (Kriging, multivariate regression, and support vector regression) combined with dynamic simulations in OpenFAST. The study highlighted the importance of soil-structure interaction (SSI) modeling and validated the subset simulation (SS) method with Kriging as the most accurate approach for estimating reliability indices. Han et al., (2024) present a generic fully coupled framework for reliability assessment of offshore wind turbines under typical limit states. The framework revealed excessive tower vibrations as a critical failure mode under operational states and emphasized the need for SSI considerations to avoid conservative or unsafe designs. This study emphasizes the importance of soil-structure interaction (SSI) in the reliability analysis, integrating a dynamic simulation tool with surrogate models to estimate reliability indexes efficiently. The framework utilizes various methods, such as the First-Order Reliability Method (FORM) and subset simulation (SS), demonstrating that the use of equivalent coupled spring boundaries can significantly improve simulation accuracy. It advocates reliability-based calibration of partial safety factors and highlights emerging trends like risk-based inspection and digital twin integration for predictive maintenance.

2.2.4 Dynamic Response Analysis and Vibration Control

The NREL 5 MW turbine rotor is addressed by Zeng et al., (2024) where the influence of the wind field and of blade structural models on the turbine response in parked conditions and under the effect of waves is investigated. This work demonstrates that the spatial distribution of the wind speed is crucial for accurately simulating the parked turbine dynamic response where a uniform distribution of the wind field is too conservative. However, the choice of wind speed coherence function shows a weak effect. Furthermore, whilst a detailed structural model is necessary to obtain blade responses, tower structural dynamics is accurately captured with a simplified model in which the blades wind load on these are simplified as a concentrated mass and a concentrated load, respectively.

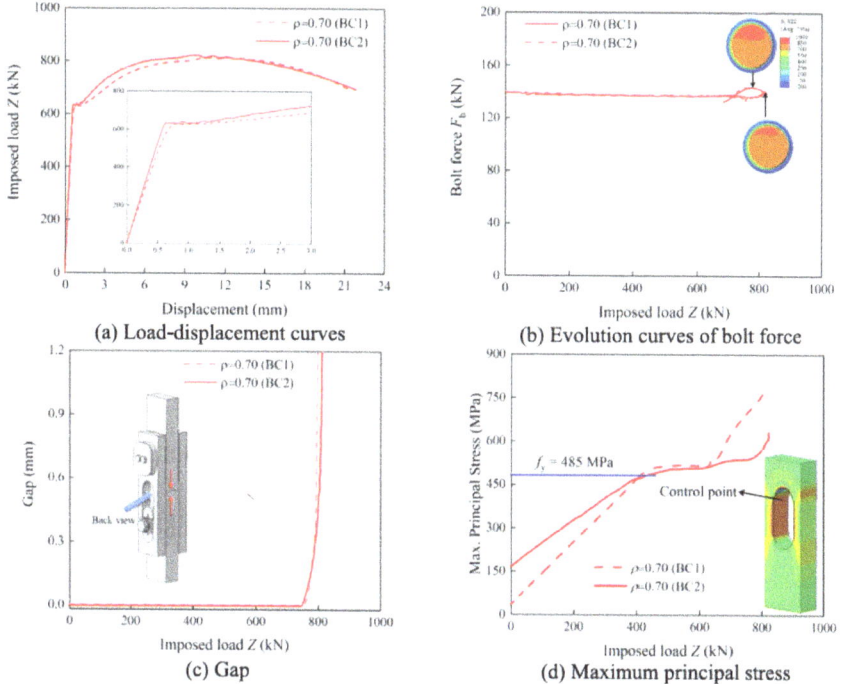

Fig. 3: Comparison between connection types (Cheng et al., 2023)

The mechanical behavior of connections within offshore wind turbine towers is also crucial. Cheng et al., (2023) investigated various connection types using finite element simulations, Fig. 3. Their findings indicate that while traditional bolted connections may present vulnerabilities due to stress concentrations, alternative designs such as the C1 wedge connection offer improved fatigue resistance and overall structural performance. This research informs installation practices, emphasizing the selection of robust connection types to enhance the longevity of OWT structures. Wang et al., (2021) compares the dynamic behavior of conventional and compact gearboxes for a 10 MW offshore wind turbine using a multi-body system (MBS) model. The compact gearbox demonstrated superior performance under torque loads, manufacturing errors, and non-torque loads, validated via decoupled global-local simulations.

An analytical solution for 1:1 internal resonance in wind turbine towers revealed coupled along-wind and cross-wind oscillations. Lenci, (2023) identified pitchfork bifurcations leading to unstable quasi-periodic responses, validated numerically for the NREL

5-MW turbine. Pezeshki et al., (2024) develops a coupled PDE model to analyze gyroscopic effects from rotor-blade rotation on OWT dynamics, including fore-aft, side-side, and yaw motions. Numerical results reveal operational natural frequency shifts and phase differences between torsional and translational responses under wave-wind-blade excitation. They also emphasize the role of gyroscopic moments in altering damping and stiffness characteristics in parked and operational conditions.

Dogru and Yilmaz, (2024) optimized a diffuser-augmented wind turbine (DAWT) using response surface methods and 3D CFD. A wide-angle GOE431 diffuser achieved a speed-up ratio of 1.59, while a six-bladed rotor increased the power coefficient by 93% (from 0.417 to 0.805). Finite blade number effects and Reynolds number sensitivity were critical limitations, with tip losses reduced to one-third of bare turbines. Machado, et al., (2024) introduces a metamaterial-based OWT design using spectral element modeling to suppress low-frequency vibrations induced by wind, waves, and blade rotation. The metastructure achieves 30% vibration amplitude reduction via tunable stopbands, outperforming traditional tuned mass dampers (TMDs). Numerical validation confirms broadband attenuation and highlights the impact of resonator mass ratios (5–30%) on bandgap formation and dynamic response.

Looking at the next generation of wind turbine rotors with increasing size, aeroelastic models considering blades' large deformation are fundamental (see, for instance Li, Z. et al., (2023b) for a FEM-based structural model). Furthermore, the investigation of blades structural integrity is addressed by Ha and Jeong, (2021) where the debonding failure at the adhesive bonded joints between spar caps and shear webs of a blade subjected to extreme bending moments (which represents one of the blade failure modes) is addressed using the 3D FEM composite analysis from the solver ANSYS. The analysis indicates peeling failure possibility of the adhesive structure close to the aft shear web for large blades with double shear webs configurations.

Given the unavailability of data related to real operating WTs, examples of detailed aeroelastic analyses focuses on "virtual" ones, such as the IEA WIND 15 MW Reference Wind Turbine (RWT). The work presented by Lapa, et al., (2023) indicates that blade torsional deformation must be considered when defining the turbine operating curves for an efficient generation of energy. Moreover, the peak shaving process is impacted when torsional deformation is not considered.

In connection with aeroelasticity, the topic of active and passive control strategies is still present in the literature with works on model predictive controllers Pustina, et al., (2022), and studies on the potential of applying viscoelastic dampers to mitigate the vibration of the tower under wind and wave loadings written by Liang, et al., (2024). In addition, Individual Pitch Controller (IPC) techniques have demonstrated their effectiveness in reducing structural loads and dynamic response of the drivetrain, achieving also the stabilization of the power output written by Xie et al., (2022).

2.2.5 Digital Twin Technology

A comprehensive overview of the state-of-the-art digital twin (DT) technology including research and industry perspectives is provided by Stadtmann et al., (2023) where critical research challenges that need to be addressed to fully realize the benefits of DTs are provided based on the industry point of view (standards, data, models and industrial acceptance). Challenges related to predictive DTs, prescriptive DTs and autonomous DTs are identified as the most impacting area of research. In addition, the complimentary role of industry and research for the success of DT technologies is underlined. An example of DT of a bottom-fixed OWT is described by Zhao, et al., (2023) where a component-based Reduced Order Modelling (ROM) is used to synthetize a DT for the modal analysis and structural response starting from a coupled FEM-CFD analysis. This model shows dramatically reduced computational costs with respect to FEM analysis and high accuracy demonstrating that it can provide almost instant predictions of the modes and responses of the turbine structure due to the wind-wave loading as well as projections of structural health conditions (Table 1). A similar strategy is proposed by Cao et al., (2023a) where the DT is fed also with physical asset data and is tailored to rapidly predict and display the distribution of the wake field, structural deformation and stress of fixed OWTs.

Table 1: Computational efficiencies of the high-fidelity FEA model and component-based ROM with and without port reduction for mode shape prediction (Zhao et al., 2023)

		High-fidelity FEA model	Comp.-based ROM	Comp.-based ROM With Port Reduction
Case 1	CPU	205.73 s	26.18 s	0.3878 s
	Time Speed-up	1.00	7.86	530.51
Case 2	CPU	201.26 s	26.61 s	0.3817 s
	Time Speed-up	1.00	7.56	527.27
Case 3	CPU	210.82 s	25.67 s	0.3979 s
	Time Speed-up	1.00	8.21	529.83
Case 4	CPU	206.19 s	25.64 s	0.3786 s
	Time Speed-up	1.00	8.04	545
Case 5	CPU	204.31 s	26.72 s	0.3866 s
	Time Speed-up	1.00	7.65	528
Case 6	CPU	214.25 s	25.62 s	0.4049 s
	Time Speed-up	1.00	8.36	529

Finally, data-driven techniques are also proposed to extend the capabilities of RANS turbulence models for wind turbines under quasi-steady conditions by using synthesized data from high-fidelity LES data. This work introduces a data-driven RANS closure model using LES data to correct turbulence anisotropy and kinetic energy production. The approach, tested on wind-tunnel-scale turbine wakes, significantly improves wake recovery and turbulence intensity predictions compared to baseline – models by Steiner, et al., (2022). An example of the application of DT-based approach to an operational (onshore) wind farm located in Yalova (Turkey) is described that a framework to retrieve and process the temporal data stream from the SCADA units of the turbines aiming at forecasting the wind speed and predicting the generated energy is proposed and successfully validated in a real-life scenario. Zhao et al., (2023) proposes a component-based reduced-order modeling (ROM) approach for real-time digital twinning of monopile-supported OWTs. The ROM framework combines static condensation and empirical port reduction, achieving a computational speedup of 650× compared to finite element analysis (FEA) with <0.2% error, enabling rapid modal and structural response predictions under wind/wave loads.

Nybø et al., (2021) compares turbulence models (Kaimal, Mann, LES, TIMESR) for a 10 MW turbine, showing TIMESR's low-frequency wind spectra increase tower and blade root moments, while Mann's negative horizontal coherence amplifies yaw moments. Nybø underscores the sensitivity of quasi-static loads to turbulence model choice, particularly in stable atmospheric conditions. Myrtvedt et al., (2020) compares Kaimal and Mann wind models for a 10 MW turbine, showing Mann's lower coherence increases yaw moments, while Kaimal amplifies low-frequency tower bending. Turbulence coherence and atmospheric stability significantly affect fatigue loads, particularly in low-frequency ranges for floating turbines. Kozmar et al., (2022) critiques IEC-standard power-law wind models, showing they underestimate dynamic loads by up to 60% compared to realistic turbulence simulations. Real-world wind conditions increase tower-top displacement variance by 550%, highlighting critical gaps in current standards.

Moving from the analysis of the single machine to the optimization of the whole wind farm, the development of efficient and accurate far wake models is still an interesting topic, especially when turbine off-design conditions such as yaw misalignment or wake steering for farm power optimization are considered. A nonlinear wake model incorporating yaw steering and vertical wind shear was developed using a Gaussian velocity deficit and a "prediction-correction" method Li, Y. et al., (2024). The model showed strong agreement with Garrad Hassan wind turbine experiments and CFD simulations, particularly capturing shear effects and wake redirection. This study validates the Dynamic Wake Meandering (DWM) model against LES for horizontal (yaw) and vertical (tilt) wake steering. Key findings include the importance of filter size in predicting wake deflection and the DWM model's limitations in capturing shear-layer-induced high-frequency wake oscillations assessment of Digital Twins for BFOWTs (Rivera-Arreba et al., 2024). Analytical far wake and mid-fidelity DWM models are addressed in the literature.

2.3 Physical Testing

Empirical research on the structural behavior of large-scale components, such as those used in offshore wind turbines, has been instrumental in validating theoretical and numerical models, especially under unique and often unpredictable load conditions. Physical testing provides insights into deformation modes, material resilience, and structural dynamics that are critical for enhancing the reliability of offshore wind systems.

One notable study by Ren et al., (2023) investigates the deformation behaviors of large-diameter steel tubes subjected to concentrated lateral impact loads. Deformation modes of steel tubes under different impact velocities, showing diamond-shaped dents under concentrated lateral loads like Fig. 4. This research is particularly relevant to

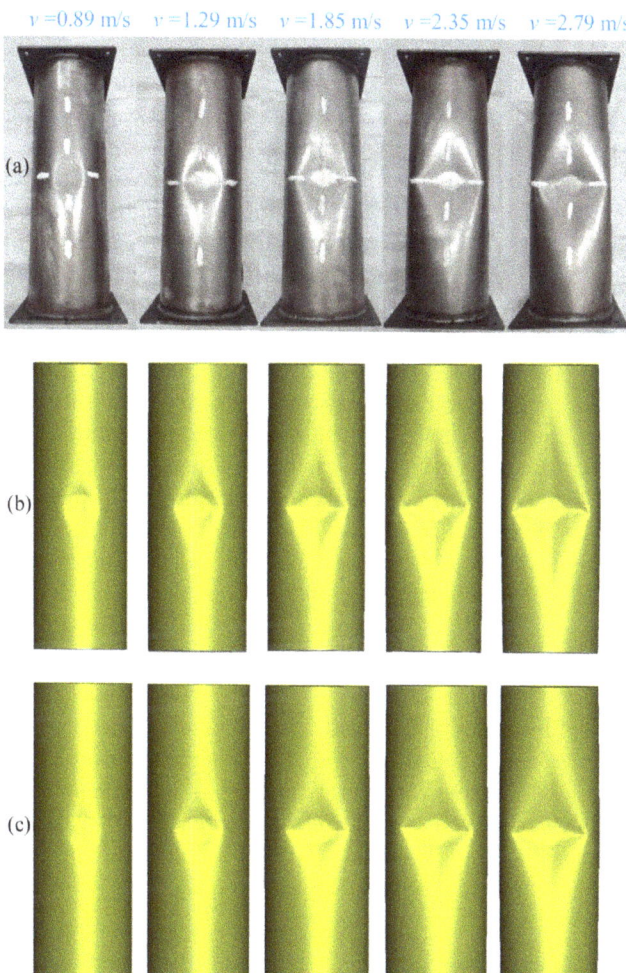

Fig. 4: Deformation modes of steel tubes under different impact velocities: (a) experiment, (b) FEM (without strain rate) and (c) FEM (C = 4000 and p = 5) (Ren et al., 2023).

offshore wind turbine (OWT) foundations, which are exposed to risks such as vessel collisions. Through scaled experiments and high-fidelity numerical simulations using the LS-DYNA software, the study evaluates the response of large-diameter tubes, which differ in deformation patterns from their smaller counterparts under similar loading. The results reveal that larger diameter tubes exhibit localized deformation, manifesting as diamond-shaped dent patterns under high-energy impacts, in contrast to the more uniform deformation typically seen in smaller diameter tubes. These findings underscore the need for tailored analytical models to predict the impact response of large-diameter tubes accurately, as existing models fall short in this regard.

The insights gained from these physical tests serve to refine design assumptions and improve the predictive accuracy of models used in OWT design. By capturing localized deformation characteristics, studies like Ren et al. contribute significantly to developing reliable safety standards and impact-resistant design strategies essential for the structural integrity of OWTs. Such empirical data is invaluable for validating and calibrating complex numerical models, ensuring that theoretical predictions align with real-world behaviors under high-impact scenarios.

It is noted that physical testing of soil resistance and foundation integrity can be challenging, both in terms of physical scale and time for observations. Centrifuge modelling uses scale models subjected to increased gravity to model full-scale nonlinear geotechnical phenomena, enabling reliable and cost-effective investigation of the response of equipment and facilities to real-world conditions; where considerable uncertainties exist in estimating the soil resistance. For bottom founded wind turbines, the performance of the piles driven into the seabed in both dense sands and hard clays can be studied over time to understand the long-term soil resistance behavior. With increased gravity, the correct stress state is achieved in the soil, allowing scaled test results to be converted to full scale field quantities. This avoids uncertainties present in 1 g scaled testing. Being more economic and efficient, a greater number of conditions can be tested to gain greater certainty in complex numerical simulations. No other practical testing is feasibility to verify numerical simulation results. Larger 5.5 m–9 m radius commercial units have the capability to model >22,000 lb and 200 g payloads spinning at 3.5 revs/s (~300 RPM).

2.4 Transport, Installation, Operation and Maintenance

While Transport and Installation (T&I) and Operations and Maintenance (O&M) challenges are often environmental, financial and safety related, there are still many areas in which structural integrity is key. With a growing body of research investigating the use of autonomous systems for T&I and O&M (Jenkins et al., 2024), and the use of digital tools for design, planning, installation and O&M (Ciuriuc et al., 2022), understanding of the industry perspective and how research can assist and improve on this is essential.

2.4.1 Transport and Installation

Jack-up Vessels are used in fixed wind turbine transport and installations. However, for more stable wind conditions and the depletion of near-shore locations, wind farms are moving farther offshore into deeper waters, challenging the current limits of offshore heavy-lift operations. Over time, the design of offshore crane vessels has converged to a semi-submersible crane-vessel (SSCV) type. Fig. 5 shows an open-source crane-vessel.

Although semi-submersibles are known for being sensitive to hanging loads, in comparison to mono hulls, they provide flexible ballasting and enough deck area to accommodate two cranes of large capacity, outreach and lifting heights. In contrast to this, in the offshore wind industry jack-up vessels have been dominating the scene due to higher stability. However, this increased stability comes at a price, which is: operational water depth limit (up to 80 m Next Generation Jack-Up, 2023), dependency on seabed conditions, vulnerability to wind direction changes and jack-up time. The effects of wave and wind loads on the structure need to be considered during transportation, especially in terms of safe transportation strategies in extreme sea conditions.

Fig. 5: Prometheus, an open-source SSCV (Domingos et al., 2024)

A comparative study by Domingos et al., (2024) assume an annual operational rate of 39% when considering only transverse waves; annual operational rate is better than sea surface operational rate (24%) due to lower transverse swing (Roll) stiffness and reduced crane acceleration. Operational rate is primarily limited by wave loading, and both wave and wind effects must be considered when evaluating the time window for installation.

2.4.2 Operations and Maintenance

The operation and maintenance (O&M) of bottom-fixed offshore wind turbines (OWTs) face challenges due to harsh marine environments and complex structural dynamics. Recent advancements focus on enhancing reliability and cost-efficiency through innovative monitoring and control technologies.

A comparative study by Wang et al., (2021) highlights the operational benefits and maintenance challenges associated with compact gearbox designs for 10 MW offshore wind turbines. By analyzing conventional and compact gearboxes through a multi-body system (MBS) approach using finite element model (FEM), the study finds that compact gearboxes can reduce fatigue damage and improve dynamic performance, primarily due to their superior load-sharing capabilities. However, the increased complexity of compact designs poses challenges in both manufacturing and maintenance, suggesting that while compact gearboxes enhance load-bearing efficiency, they require more specialized maintenance protocols, which could impact O&M costs over time.

Tuned Mass Dampers (TMDs) are widely employed to mitigate vibrations caused by combined wind-wave loads. Lu, D. et al., (2023) demonstrated that TMDs reduce dynamic responses by up to 40% under typhoon conditions, significantly improving fatigue life. Additionally, semi-active wave compensation systems (e.g., Seaqualize) enhance crane stability during maintenance operations, minimizing mechanical wear. Recent studies emphasize hybrid strategies combining digital twins, robotics, and real-time meteorological forecasting. Domingos, et al., (2024) highlighted that 2-min wave prediction intervals improve maintenance operability by 15%, while Lu, D. et al., (2023) advocated for TMDs coupled with CFD-based wake control to minimize energy losses.

Digital twins integrated with computational fluid dynamics (CFD) and machine learning enable predictive maintenance. Cao et al., (2023) developed a digital twin framework using Bayesian-regularized neural networks, achieving <4% error in stress prediction for tower structures. This approach allows rapid identification of stress concentrations (e.g., 89.44 MPa peaks at tower roots) and optimizes maintenance scheduling. Sentinel-1 SAR and Landsat SWIR time-series analyses address spatial-temporal data gaps in offshore databases. Dadmarzi and Bachynski-Polić, (2022) utilized adaptive Z-score thresholds to achieve 92% precision in turbine detection, enabling retroactive installation-year mapping (2000–2022) for lifecycle assessment. Such remote sensing tools reduce unplanned downtime by prioritizing high-risk components.

Addressing the high computational cost of fatigue assessments, Katsikogiannis et al., (2022) propose a lumping method to optimize fatigue damage predictions in monopile-based OWTs. This approach condenses extensive load cases into a manageable number, reducing computational demands by up to 96% while maintaining an accuracy of 92% to 98%. Such methods are invaluable during early design stages, offering a balance between computational efficiency and prediction accuracy. However, for detailed fatigue assessments, particularly when nonlinear dynamics are at play, fully coupled models remain essential. By significantly lowering computational time, the lumping method also supports efficient fatigue monitoring within O&M frameworks, potentially enabling more proactive maintenance practices.

Collectively, these studies underscore the importance of enhancing transport, installation, and O&M frameworks for offshore wind turbines through targeted design improvements and computational advancements. By refining load-sharing mechanisms in gearboxes and optimizing fatigue modeling approaches, the OWT industry can better address the operational challenges of larger, more powerful turbines, ensuring robust performance and extending their operational life.

2.5 Design Standards and Guidelines

The design of bottom-fixed offshore wind turbines is governed by a combination of international standards, site-specific environmental considerations, and advancements in numerical modeling. Establishing robust design standards and guidelines is critical for the safe and effective deployment of these devices, especially as turbine sizes and environmental loads increase. These standards define structural resilience, fatigue thresholds, and dynamic response characteristics under complex environmental conditions, providing a foundation for reliable and sustainable OWT systems. As highlighted in Table 2, key standards include the IEC 61400 series (e.g., IEC 61400-3 for offshore turbines), DNVGL (ST-0126, ST-0119, ST-0437), ISO19900 series, and CSA C61400; which provide comprehensive guidelines for structural integrity, load calculations, and safety factors. Liu, J. et al., (2024c) emphasize the importance of dynamic response analysis under environmental loads (e.g., waves, wind, and currents) using potential flow theory and Morison equations to simulate hydrodynamic forces.

The IEC 61400 outline numerous standard applicable to wind turbine development. Part 3-1 for fixed wind turbines and Part 3-2 for floating wind turbines deal with issues related design for the harsh marine environment (environmental loads including waves and ice), the specific types of structures (monopile, jacket, spar, semi, TLPs), foundational stability and anchoring given different seabed conditions, access for maintenance given potential storm conditions, safety of marine life and electrical systems including dynamic and static cables and design of substations and storage.

IEC 61400-50 provides standards for wind measurement like wind speed, wind direction and turbulence intensity, and provides use-case independent methodologies and requirements that will ensure consistency, accuracy and reproducibility in the measurement of the wind for the design of OWT. IEC 61400-6 includes tower and foundation design requirements. and consists of Part 50-1 for wind measurement in the application of meteorological mast, nacelle and spinner mounted instruments, Part 50-2 for wind measurement in the application of ground-mounted remote sensing technology, and Part 50-3 about use of nacelle-mounted lidars for wind measurements, respectively. The IEC 61400-50:2022 is developed from IEC 61400-12-1:2017 and IEC 61400-12-2:2013 by separating the wind measurement requirements from them. Although the IEC 614006:2020 specifies the requirements and general principles used to evaluate the tower structural integrity of onshore wind turbine including foundations, its geotechnical evaluation of soils, the flange and connection systems connected to the rotor nacelle assembly including connections to yaw bearings can be referenced in the design of OWT.

Some of these standards reference additional IEC and ISO standards and note that "all or some of their content constitutes requirements". The IEC 61400-3-1 refers to the ISO 19900 series but some differences between the two sets of standards are worth noting. For example, the IEC 61400-3-1 design cases for 'parked' scenarios reference winds and sea states with return periods of 50 years. In ISO 19900, more stringent design criteria are required for manned platforms with hydrocarbon flow and storage. For environmental loads, consideration is given to both the ULS and abnormal limit state (ALS) based on actions with associated return periods of 100 years and (in the case of L1 structures) 10,000 years, respectively.

DNVGL provides a fairly comprehensive set of standards and recommended practices related to general aspects of the design, installation and operation of offshore platforms. The DNVGL offshore wind standards DNVGL-ST-0126, DNVGL-ST-0119, and DNVGL-ST-0437 have design situations and load cases aligned fairly closely with those in the IEC standards. They cover site conditions, design scenarios, load calculation methods, installation and operations guidance for fixed and floating offshore platforms. DNV-CG-0128:2023 provides the methods and principles applicable for the assessment of buckling and ultimate strength limits (ULS) of load carrying column structural members in offshore units.

The Canadian Standards Association (CSA) is accredited to develop and maintain standards in Canada by the Standards Council of Canada. The standard CAN/CSA C61400 Wind turbines - Part 3: Design requirements for offshore wind turbines (R2021) applies specifically to offshore turbines and was adopted from the international standard IEC 61400-3:2009 with Canadian deviations.

Recent studies offer insights into various aspects of these standards, with a focus on soil-foundation interactions, hydroelastic effects, environmental load modeling, and fatigue resilience.

Hydroelastic effects play a crucial role in refining design standards. Leroy et al., (2021) develop a hydroelastic model using Weak-Scatterer potential flow theory, integrated with a Euler-Bernoulli beam structural model, specifically for monopile-supported OWTs. Their model, validated against experimental data, accurately captures dynamic responses under regular wave conditions and shows strong agreement with measured mudline bending moments. While this model has not yet proven more efficient than Morison-based methods, it provides valuable insights into fatigue and ultimate load predictions, particularly in scenarios where hydroelastic effects are significant. This supports the inclusion of hydroelastic considerations in design standards, improving the long-term resilience of OWT substructures.

Environmental conditions such as wave height, wind speed, and soil-structure interaction are critical in design phases. Standards mandate site-specific assessments using JONSWAP wave spectra and Kaimal wind models to simulate extreme conditions. For example, IEC 61400-3 requires a 50-year return period for environmental loads, while DNV guidelines specify safety factors of 1.35–1.5 for ultimate strength checks (Xia and Zou, 2023). Katsikogiannis et al., (2024) investigate the effects of different probabilistic models on extreme load predictions for monopile-based OWTs, finding that the conservative models (like the 3-parameter Weibull model) provide more reliable estimates under extreme conditions. This careful calibration of probabilistic models supports environmental load assessment standards, ensuring that OWT structures can withstand intense marine forces. Additionally, Katsikogiannis et al., (2021) propose an environmental lumping method that condenses load cases using damage-equivalent contour lines, achieving accurate fatigue predictions with a reduced computational burden. This approach balances efficiency and accuracy, offering practical tools for early-stage fatigue assessments and supporting resource-efficient design standards.

Fatigue resilience is a recurrent theme, with studies examining how various factors influence fatigue life. For example, Han et al., (2022) present a half coupling model (HCM) for fatigue assessment in jacket-type support structures, effectively separating aerodynamic and structural responses to enable accurate fatigue predictions under combined wind and wave loads. Validated against spectral methods like Dirlik's and Benasciutti-Tovo (BT), the HCM model offers a computationally feasible solution for early-stage design evaluations, making it instrumental in setting fatigue analysis standards for offshore wind applications. Sørum et al., (2022) extend this focus with a sensitivity analysis that examines the impact of SN curve parameters, wind-related factors, and soil-structure interactions across different turbine sizes. Their findings reveal that soil model choices substantially affect fatigue predictions at the monopile and tower base, highlighting the need for detailed geotechnical parameters in fatigue design standards.

Table 2: Cross-Reference Table of International Standards for Offshore Wind Turbine Structural Design Parameters

Standard Name	Key Design Parameters	Reference Standards	Unique Requirements
IEC 61400-3-1	50-year return period for extreme conditions (wind, waves)	References ISO 19900 series	Excludes hydrocarbon flow/storage criteria
	Parked scenario requirements		Focus on structural resilience
IEC 61400-3-2	Mooring system design	Additional IEC/ISO standards	Seabed adaptation requirements
	Dynamic response analysis		Marine life safety protocols
DNV-ST-0126 & 0119	Safety factors: 1.35–1.5 for ultimate strength checks	Compatible with IEC 61400 series	Includes ice load calculations
	Load case definitions		Offshore oil/gas industry alignment
DNV-ST-0437	Icing scenarios	DNV recommended practices	Specialized installation unit guidelines
	Fatigue design methodologies		
CSA C61400	Local environmental adaptations	Adopted from IEC 61400-3:2009	SCC-accredited deviations

Structural design must account for fatigue life and ultimate limit states, particularly for critical components like monopiles and transition pieces. Finite element modeling (FEM) is widely adopted to assess stress concentrations and optimize geometries, ensuring compliance with DNV's fatigue resistance criteria (Gao, L. et al., 2023a; Domingos et al., 2024). Ataei et al., (2023) highlight the need to address structural flexibility in crane systems during installation, as dynamic coupling between vessels and turbines significantly impacts load distributions.

Structural stability and resonance management are also vital, as seen in the study by Lenci, (2023) on nonlinear coupled oscillations in wind turbine towers. By focusing on along-wind and cross-wind oscillations near a 1:1 internal resonance, the study reveals that nonlinear mode coupling must be considered to avoid instability. Through an analytical model validated with NREL 5-MW reference turbine simulations, Lenci's findings highlight the need for resonance mitigation in design standards to prevent fatigue-induced failures. Moreover, new research has emerged addressing innovative solutions for vibration control in OWTs. Machado et al., (2024) explore the application of metamaterial-based vibration control for OWTs subjected to multiple hazard excitation forces, demonstrating that their approach significantly reduces vibration amplitudes compared to conventional systems. This offers promising implications for improving the operational reliability of offshore wind installations. Additionally, Pezeshki et al., (2024) investigate the gyroscopic effects of the spinning rotor-blades assembly on the dynamic response of offshore wind turbines. Their study develops an analytical solution method to derive partial differential equations governing the motions of the OWT structure, including fore-aft and side-side movements. By incorporating gyroscopic moments into the boundary conditions, the research reveals that these effects can significantly influence the operational natural frequencies of OWTs. The findings highlight the necessity of accounting for gyroscopic effects in dynamic analyses, particularly for floating offshore wind turbines, to ensure accurate predictions of structural behavior.

Seismic dynamics also warrant attention in design considerations. He and Ye, (2023) analyze the seismic response of OWTs and their interactions with seabed foundations, revealing the importance of soil-structure interactions during seismic loading. The insights from their numerical analysis inform design guidelines to enhance structural integrity in earthquake-prone areas. In a related study, Chen et al., (2023) investigate the seismic response of OWTs supported by hybrid pile-bucket foundations, demonstrating improved performance under seismic loads compared to traditional monopile foundations. Their findings emphasize the role of foundation design in ensuring stability and resilience during seismic events. Comparative seismic analysis is further explored by Ngo and Kim, (2024), who conduct a detailed assessment of the seismic performance of various foundation types (monopile, suction bucket, and pile jacket). Their dynamic analysis reveals that while the monopile foundation is more susceptible to large displacements during seismic loads, the suction bucket foundation offers better base stability, and the pile jacket foundation, while stable, experiences higher stress concentrations at the tower base.

Installation safety is governed by marine operation standards (e.g., ISO 19901-6), emphasizing risk assessments for lifting and mating processes. Numerical frameworks integrating digital twins are emerging to optimize real-time decision-making, aligning with ISO 55000 asset management principles

Finally, Sun and Fang, (2023) introduce a novel floating composite anti-collision structure designed to enhance the crashworthiness of OWTs against ship collisions. Their study shows that this structure effectively absorbs collision energy, significantly reducing damage to both the OWT and the colliding vessel, which is crucial for maintaining operational integrity.

Together, these studies form a comprehensive framework for developing robust OWT design standards. By integrating PSI considerations, hydroelastic modeling, probabilistic load assessments, fatigue resilience, and resonance control, these standards support the safe, resilient deployment of offshore wind turbines across diverse marine environments, ensuring long-term sustainability and structural integrity.

3 Floating Offshore Wind Turbines

Floating offshore wind turbines (FOWTs) represent a transformative approach in renewable energy, allowing for wind energy generation in deep-water areas where traditional fixed-bottom turbines are not feasible. This chapter discusses recent industry developments, and the current status of installed and predicted FOWT, advances in structural design through numerical and physical modelling and testing, improvements in transport and installation and operation and maintenance, and concludes with an update of the latest regulatory standards pertaining to FOWT structural qualification. Dynamic power cables are integral to the FOWT system, and have their own unique structural challenges; however, these are discussed in chapter "Uncertainty Modelling in Waves and Wave Responses".

3.1 Recent Industry Developments

By 2030, it is anticipated that 7.3 GW of floating wind projects will have started offshore installation (e.g. pre-lay of moorings or cable installation), translating to roughly 3 GW operational by 2030. By 2040, 70.9 GW of floating projects could reach the offshore installation phase, accounting for 45–50 GW of operational floating wind (4C Offshore, 2024). The UK has an ambition for 5 GW of operational floating wind by 2030, which has been reinforced by the success of Flotation Energy and Vårgrønn's Green Volt project winning in the AR6 Contracts for Difference auction in the UK in 2024 (Flotation Energy, 2024).

FOWT projects that have reached the stage of offshore installation and are either operational or where installation in underway as of Q2 2024 are shown in Fig. 6. There is a total of 235.4 MW of FOW which is currently operational, 123.6 MW under installation and 18.9 MW that have been decommissioned. The majority of this installed capacity is in Europe (301 MW), with 77 MW in APAC and < 1 MW in the Americas (4C Offshore, 2024).

Key operational projects include Hywind Tampen, Hywind Scotland, and Kincardine Offshore Wind Farm (Fig. 7).

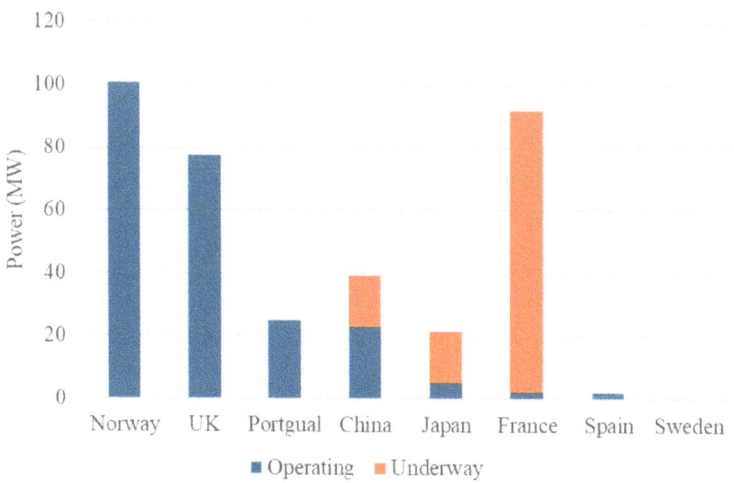

Fig. 6: Floating offshore wind projects that have reached installation globally

Edwards et al., (2024) reviewed 86 past and current, early-stage, platform designs and discuss how FOW substructures were originally influenced by floating platforms

Fig. 7: Kincardine offshore wind farm, showing three of the five 9.5 MW Vestas wind turbines on steel semi-submersible foundations designed by Principle Power (Photo Credit: Flotation Energy)

typically used in the oil and gas industry but have since deviated away from these conventional floater designs to better suit the specific needs of the technology. A previous review covers platforms that have reached at-sea deployment (Edwards et al., 2023), with the latest trend indicating these can be categorised into four types each with its own structural challenges. These four types are semi-submersible, spar, barge and tension leg platform (TLP), and industry trends towards primarily steel, but also concrete designs of each. Of the installed FOW projects globally, 160 MW utilise semi-submersibles, 155 MW spars, 37 MW barges, and 26 MW TLPs.

Floating wind levelized cost of energy (LCOE) has been stated as becoming comparable with its offshore bottom-fixed and onshore counterparts, although this depends on the extent and speed of its evolution (Maktabi and Rusu, 2024). Martinez and Iglesias (2024) mapped the costs of energy for the European Atlantic and Mediterranean and identified that the major drivers of the costs of energy are those directly tied to energy output of the farm, most notably the available wind resource. Other important parameters were the quantity of turbines, rated power, and CAEX of prominently, the turbines and substructures. Regions with the smallest LCOE (~95 EUR/MWh) were those featuring the greatest wind resource: Great Britain, Ireland, North Sea, NW Iberian Peninsula, the Gulf of Lyon and Aegean Sea. One method of reducing LCOE for floating wind, by up to 4%, has been identified as the use of shared anchors for multiple turbines (Housner and Mulas Hernando, 2024). Wake effects have been shown to be a key factor in farm layout and turbine loadings, and ultimately LCOE (Thomas et al., 2024). Sykes et al. (2024) investigated the use of Multidisciplinary Design, Analysis and Optimization (MDAO) as a method of objective assessment, and ultimately reduction, of LCOE for floating offshore wind farms in order to rank foundation designs. Overall, the LCOE related to floating wind is still very volatile, with a study by Helfer et al. (2023) indicating variance from a decrease of 50% to an increase of 100% based on sensitivity scenarios conducted.

3.2 Numerical Modelling and Analysis

3.2.1 Aerodynamics

Rotor aerodynamics is fundamental in understanding the complex behavior of floating wind turbines, particularly due to their dynamic interaction with unsteady wind and wave-induced platform motions. Despite its critical importance, recent years have seen limited research focused explicitly on this area. Most studies tend to address broader aspects of floating wind turbines, such as structural dynamics, control strategies, and mooring systems, leaving rotor aerodynamics somewhat underexplored. This gap is significant because the aerodynamic performance of the rotor directly influences energy yield, load distribution, and overall system stability. Advancing this understanding is essential to optimize designs and improve the reliability of floating wind systems in increasingly challenging offshore environments.

FOWT wake dynamics has emerged as a crucial research topic in the addressed literature, reflecting its significant role in optimizing wind farm performance. Unlike fixed-bottom turbines, FOWTs experience additional complexities in wake behavior due to platform motions induced by wind and wave forces, which can alter wake structure, direction, and recovery rates. Understanding these dynamics is vital for accurately predicting turbine interactions within a farm, minimizing wake losses, and mitigating structural loads on downstream turbines. Current studies deal with wake modeling under unsteady operating conditions including those in which the rotor works under Vortex Ring State (eventually leading to strong Blade Vortex Interactions, BVI).

In (Dong and Viré, 2022) an in-house code using the free wake vortex ring method to simulate the aerodynamic performance of the NREL 5-MW FOWT during a prescribed surge motion is proposed. Their research investigated various characteristics of the rotor as it transitions between different operational states, i.e., windmill working, vortex ring working and propeller working. The aerodynamic load changes corresponding to these states show that the vortex ring state is the most unstable of the three. Moreover, the authors propose two different criteria for the prediction of VRS (based on Wolcowitch and axial induction, respectively) which can be used for the purpose of preliminary analysis.

In Fu et al. (2023), the aerodynamic characteristics of a floating wind turbine are analyzed using *high-fidelity* CFD methods and overlapping grid technology, particularly under platform pitch motion. A detailed analysis of unsteady wake dynamics as well as blade-vortex and vortex-vortex interactions is performed (Fig. 8).

The study shows that, in case of large amplitude, the blade tip periodically enters into the wake (Fig. 8, c4) and there is strong flow mixing of the regular tip vortex and the wake, which potentially leads to faster wake recovering. Moreover, strong blade-vortex

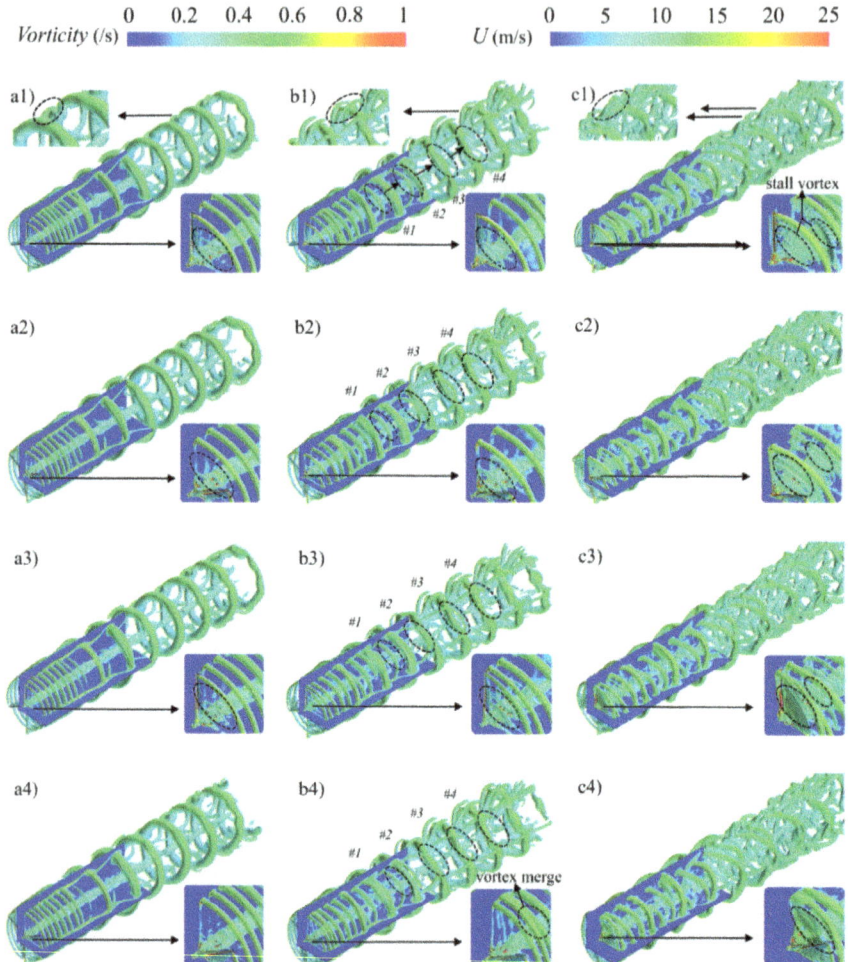

Fig. 8: Instantaneous vorticity contours from the rotor due to the pitch motion: amplitude 1° (a), 4° (b) and 10° (c). 1) $t = 0T$; 2) $t = 1/4T$; 3) $t = 1/2T$; 4) $t = 3/4T$ (T platform motion period) (Fu et al., 2023)

interactions are observed. This affects blade unsteady pressure distribution and, in turn, rotor lifetime. At the same time, more ribbon vortices appear in the wake, which further expands the influence range of the wake. The results conclude that platform motion significantly affects both the aerodynamic performance and the wake of the FOWT, especially when the pitch amplitude is large. FOWT's wake analysis is addressed also by mid-fidelity tools or simplified models. For instance, in Kleine et al. (2022) wake dynamics is investigated through numerical simulations based on linear stability theory (Fig. 9). The study introduces two simplified numerical models to capture the complex vortex behaviours under various turbine motions effectively. The findings indicate that linear theory can predict dominant flow modes when multiple motions or frequencies are involved. Additionally, the general behaviour of the wake can be understood and anticipated using relatively simplified stability models. It is concluded that the highest growth rate in vortex instabilities occurs when the motion frequency is one and a half times the turbine's rotation frequency, with lower frequencies potentially increasing fatigue or causing high amplitude motions in downstream turbines.

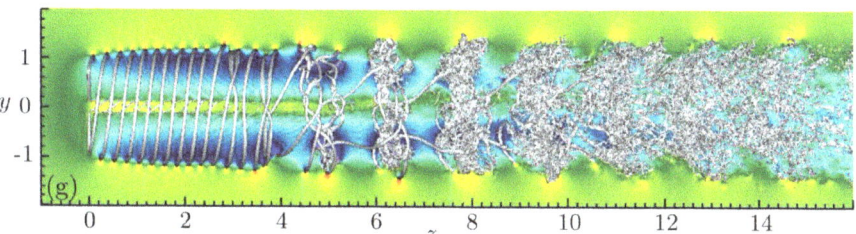

Fig. 9: Instantaneous streamwise velocity along the wake in the yz-plane and 3D iso-surfaces of vorticity magnitude for a surging rotor (Kleine et al., 2022)

Moving from the turbine level to the farm, Ramos-García et al. (2022) propose a numerical investigation on the impact of the wake of a floating IEA wind 15 MW reference wind turbine on downstream machines. Specifically, an aero-hydro-servo-elastic solver coupling free-wake vortex solver (MIRAS) and a multi-body code (HAWC2) are coupled to account for the flexibility of the different turbine components as well as to include the effect of the controller and the dynamics of floater. Moreover, a flow prediction module based on the hybrid filament-particle-mesh approach is used to describe far wake dynamics through the vorticity transport equation. The authors show that the floater motion triggers a faster breakdown of vortex structures with respect to the bottom fixed configuration, thus increasing the power production of a downstream machine. This effect is drastically reduced with the increase of upstream turbulence intensity (Fig. 10). Furthermore, the impact of surge and pitch motion of the upstream turbine on its wake induces higher blade loading on the downstream machine. Overall, the study highlights the importance of an accurate flow and wake modeling for the prediction of turbine-to-turbine interaction in a floating context.

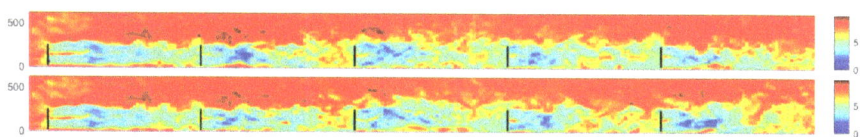

Fig. 10: Instantaneous stream-wise velocity on the extraction XZ plane for a 5-WT farm simulation of five turbines with a spacing of 15R. The fully developed wake is depicted. (Top) Bottom-fixed and (bottom) IEA-15 MW-RWT mounted on the WindCrete platform, wind speed of 15 m/s (Ramos-García et al., 2022)

FOWT farm wake modelling and prediction by a Digital Twin approach is proposed in Zhang and Zhao (2023) using a data and knowledge fusion methodology in which lidar measurements, the Navier–Stokes equations, and the turbine modeling using actuator disk method are integrated via physics-informed neural networks enabling real-time flow characterization and the possibility to retrieve unmeasured flowfield information. In the absence of real-life data, case studies of a wind farm under typical operating scenarios (i.e. a greedy case, a wake-steering case, and a partially operating case) are carried out using *high-fidelity* numerical simulations. The results show a good accuracy of the DT, which an average prediction error about 5% of its range. Nevertheless, local errors, especially downstream rotors, can be more significant, as shown in Fig. 11. Hence, further research on this topic is needed.

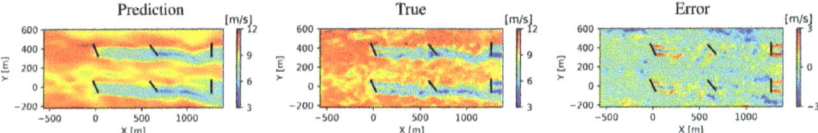

Fig. 11: DT predictions for the wake-steering case (Zhang and Zhao, 2023)

3.2.2 Hydrodynamic Loads and Structural Responses

Floater Loads and Motions For floating wind turbine support structures, assessment of mean forces and slowly varying forces due to waves and current are important in designing the mooring system, for optimizing the turbine control system and ensuring that the nacelle motions and accelerations are within acceptable limits. There is a strong coupling between the wind loads on the RNA/tower and the motions of the floater, but the hydrodynamic phenomena are similar to those for other moored floaters, like e.g. oil & gas semisubmersibles. While some barge-type floaters have been proposed, most foundations for floating offshore wind turbines (FOWT) consist of single or multiple vertical columns that penetrate the water surface, such as spars, semisubmersibles and TLPs. A short overview of the use of simplified methods for design can be found in the introduction section of Høeg and Zhang (2023). Some recent developments of simplified methods for early design stages and optimization analyses are presented later in this section. A *high-fidelity* approach to numerical loads and motion assessment is the use of field methods (CFD), solving the Navier-Stokes equations and capturing viscous effects without resolving to empirical coefficients. However, for structural integrity analysis, long time-series of the FOWT responses in various conditions are needed, and the CFD methods are too computationally demanding to be applied directly in such a process. Instead, they are often used to calculate coefficients to use with simpler methods or to calibrate these methods. They are also used to study special flow phenomena and to compare with experimental results.

Dadmarzi et al. (2022) studied the INO WINDMOOR 3-column semisubmersible using STAR-CCM+, where the VOF technique is used for free surface capturing and turbulence is solved with an improved delayed detached eddy simulation (IDDES). The structure is kept fixed and exposed to Stokes 5th order regular waves. The aim of the study was to define the drag coefficients depending on distance to free surface and

KC (Keulegan-Carpenter) number. Focus was on the single column facing the waves, but interaction effects between the columns were also investigated. Forces on the column simulated by CFD were compared with those from two Morison models; one with coefficients from CFD and one with coefficients from DNV-RP-C205. For large and steep waves, mean and first-order forces from the two Morison models deviated from those from the CFD simulations, but the model with CFD-based coefficients performed significantly better than the one with coefficients from DNV-RP-C205. Since the drag coefficient decreases rapidly with increasing KC number, and determining a representative KC number in irregular waves is challenging, the authors recommend that one should be careful not to select a too high drag coefficient. Califano et al. (2023) analyzed the same INO WINDMOOR in regular waves with the same STAR-CCM+ approach. The simulated surge and heave motions of the platform were reported to be in fair agreement with the experiments. The same floater in regular waves was also investigated with the REEF3D CFD code by Berthelsen et al. (2022), achieving good agreement with experimental surge and heave motions but less good agreement for pitch.

As part of the Reproducible CFD JIP, Wang et al., (2022) performed CFD simulations with the OC6 DeepCwind semisubmersible in 3-h irregular waves and compared with experiments. They concluded that the CFD simulations capture the low-frequency slow-drift motion well but underpredict the low-frequency pitch resonance. Wang and Chen (2022) report from an extensive validation study of CFD simulations with the URANS code ReFRESCO for the OC6 DeepCwind semisubmersible in bichromatic waves and in irregular waves. Compared to experiments, the surge, heave and pitch responses in the wave-frequency regime are well predicted. The low-frequency responses are underpredicted, but the CFD simulations agree better than the results from mid-fidelity tools. Califano et al. (2023), L. Wang et al. (2022a) and Wang and Chen (2022) all point to modelling of incident waves as one of the sources for the discrepancies between CFD and experiments. Hence, appropriate modelling of the waves is important in such validation studies.

The common engineering, or *mid-fidelity,* approach to FOWT wave load analysis is to use radiation-diffraction theory combined with Morison-type drag forces. Mean second order forces are obtained from the first order potential flow solution by including second order terms when calculating the pressure and when integrating it over the wet surface. To assess the difference-frequency and sum-frequency forces, the second order velocity potential must be calculated. These calculations give the quadratic transfer functions (QTF's) relating the difference-frequency and sum-frequency forces to the square of the amplitude of the wave components. To avoid having to calculate the full matrix of QTF's for the difference-frequency forces, Newman's approximation is often used, where only the diagonal terms are needed, and these are obtained from the first order potential. The second-order difference-frequency forces are generally small compared to the first order wave forces, but if their frequency coincides with a natural frequency, they can increase the motions significantly. The lateral motions of a column-based floater usually have low natural frequencies, and for these modes Newman's approximation generally works better than for modes with higher natural frequencies, such as heave, pitch and roll. The approximation works less well when the mode has low damping and the transfer function varies rapidly around the resonance frequency. It also becomes more uncertain when the water depth decreases. Hence, it is becoming increasingly common to calculate the full matrix of QTF's, and several commercially available potential theory solvers have this capability.

Using the full QTF's instead of Newman's approximation, reduces the underestimation of the resonant floater motions, but extensive benchmark studies within the OC5 and OC6 projects indicate that an underestimation is still present even in moderate waves (Otter et al., 2022). Reasons for the underprediction and practical remedies were discussed by Wang et al., (2022), using the OC6 DeepCwind semisubmersible as the case. They focused on ways to include viscous forces more properly, since these contribute to both excitation and damping. Most engineering tools use Morison models, and the results become sensitive to the selection of drag coefficients for the different parts of the substructure. Wang et al., (2022) included wave stretching and found that this makes the surge response more sensitive to the drag coefficients near the free surface and improved the effect of the depth-dependency. For the case studied by Wang et al., (2022) they concluded that the underpredicted surge resonant motion was primarily due to underpredicted excitation rather than overpredicted damping. Using drag coefficients that decrease with increasing depth will increase the viscous surge excitation without increasing the surge damping to the same extent.

Wang et al. (2022) also studied ways of modelling the drag forces on the heave plates at the bottom of each column. The modified model uses a reduced drag coefficient that accounts for flow separation at the plate edges while omitting effects of pressure variations that are already accounted for in the potential-flow solution. This gave better agreement with the experimental low-frequency resonant pitch motions. However, for higher frequencies, reducing the axial drag coefficient for the heave plates gives larger discrepancies for heave/pitch motions, indicating that frequency-dependent drag coefficients would be required. Hence, the reduced coefficient was only applied below a specified frequency. Instead of improving the low-frequency response predictions by modifying the drag coefficients in the Morison model, Li and Bachynski-Polić (2021) modify the QTF's based on results from CFD simulations in bi-chromatic waves. Their simulations were compared with the same OC6 DeepCwind semisubmersible experiments as used in Wang et al. (2022) and it was shown that this alternative approach also significantly improved the predictions of the low-frequency motions.

Fast, Simplified Models to Use in Optimization or Preliminary Design Phase There are many studies on optimization of the rotor nacelle assembly (RNA), but few on optimization of the support structure, as pointed out in a review by Sykes et al., (2023). Optimization requires a set of objective functions that are to be minimized under a set of constraints. Building cost is the most common objective function, but several more are discussed by Sykes et al., (2023). They also review different constraints, design variables and optimization algorithms. An analysis tool, capable of predicting the relevant FOWT responses, is needed in the process, and since many evaluations are required, the tool must be sufficiently fast. These simplified methods are also useful in the initial design process. Frequency-domain methods have often been applied (Hegseth, Bachynski and

Leira, 2021; Ferri and Marino, 2023) while some also use time-domain simulations (Faraggiana et al., 2022).

Simplifications may include the use of quasi-static models for the mooring lines, neglecting higher order potential theory forces, and using linearized Morison-type drag damping (Hegseth, Bachynski and Leira, 2021; Faraggiana et al., 2022; Ferri and Marino, 2023). In the frequency-domain methods of the latter two publications, the linearized model is established for the FOWT's steady state operating point in each weather condition. Instead of using linear hydrodynamics from a panel method, Hegseth et al., (2021) use MacCamy-Fuchs theory to calculate the wave excitation forces on a spar structure, and added mass is calculated by 2D analytic expressions. They neglect radiation damping and viscous wave excitation forces. Compared to mid-fidelity simulations, their linearized model for preliminary design and optimization predicts long-term fatigue damage and short-term extreme structural load effects within 30% agreement.

An alternative simplified method that retains the second order wave loads is presented by Carmo and Simos (2022). Instead of evaluating the loads on the instantaneous position of the body, they speed up the calculations by evaluating the second-order loads on its mean position. Carmo et al. (2023) apply the method on a four-column FOWT in irregular waves. Results are compared with experiments and with calculations using WAMIT and OpenFAST. Satisfactory agreement for long waves is obtained, especially when considering the reduction in modelling and computational cost compared to methods using second-order potential theory solvers.

Høeg and Zhang (2023) present another fast slender-body method that avoids the use of potential theory solvers. Their frequency-domain model is a combination of MacCamy-Fuchs theory and a semi-analytical Morison model, applicable to e.g. semisubmersibles with heave plates. This hybrid method uses the MacCamy-Fuchs model for horizontal wave loads on the cylinders and applies the semi-analytical Morison model for the remaining parts of the hydrodynamic loads. The authors report quite good agreement with potential flow solvers for all frequencies. The Morison coefficients must be determined by model tests or CFD. Interactions between columns are neglected.

Vortex-Induced Motions In the presence of current, or during towing, columns, such as spar-type floating wind turbines, may be subject to vortex induced motions (VIM). This phenomenon can also occur for structures with multiple columns, such as semisubmersibles. The load mechanisms exciting VIM are similar to those involved in the well-known VIV phenomenon, and Morison-type load models are used with various additional load terms. Hence, coefficients derived from physical tests and/or CFD simulations are essential input. These semi-empirical methods are frequency- or time-domain, and focus is mostly on single columns (Passano et al., 2022). With multiple columns there will be interaction effects that increase the complexity, and direct CFD simulations of the floater is a promising method for these more complex structures (F. Jiang et al., 2023a). A systematic review of recent research on VIM can be found in (Yin et al., 2022).

Floater Structural Responses For checking the integrity of the floater structure, one needs to take the pressure distribution from the water into account as well as the forces from the tower and the moorings.

Instead of mapping pressures to a finite element model, one may use the fact that the floater elements are beamlike and that stresses may be derived from cross sectional forces and moments using beam theory. Wang and Moan, (2024a, 2024b) modelled the 10 MW DTU semisubmersible floater using several rigid bodies connected by beam finite elements. Gravitational, hydrostatic, hydrodynamic, drag and inertia loads were appropriately distributed to the bodies. Coupled aero-hydro-servo-elastic time-domain simulations were carried out providing time-series of floater motions and forces/moments in the elements connecting the bodies. Assuming that the different parts of the semisubmersible floater can be modelled by Euler-Bernoulli beam theory, the stresses in the cross sections were calculated based on the forces and moments. Extreme load effects were found using the environmental contour method. The 3D contour surface spanned by the significant wave-height (Hs), the wave peak period (Tp) and the mean wind speed (Uw) is reduced to sets of 2D Hs-Tp contours by selecting a set of wind speeds (corresponding to rated, cut-off and parked conditions). Similarly, by selecting a set of critical peak periods corresponding to the peaks in the RAO's of the relevant load effects (forces/moments), 2D Hs-Uw contours are obtained. The RAO's are calculated from coupled time-domain simulations of the floater with parked turbine in regular waves covering the relevant periods and directions. A similar beam approach was used by Li et al. (2023) in the analysis of the center-column in the 4-column UMaine semisubmersible floater.

An overview of three alternative workflows for floater stress analysis is given in DNV (2024a). All three involve time-domain simulations of the complete turbine/floater/mooring system. The "Direct Load Generation" method is the classical and most general technique, where hydrodynamic pressures and Morison loads are generated in time-domain before they are mapped onto a finite-element model for structural analysis. The added mass and linear damping, and the linear and quadratic transfer functions for the wave-loads are calculated by a frequency-domain panel code. After the complete wind turbine motion simulations, a time-domain potential theory panel code is used to generate time-series of the hydrodynamic pressures based on the time-series of waves and floater motions. This method is flexible and may include local dynamic response as well as nonlinear hydrodynamic forces, but it is computationally resource demanding. A drastic reduction of the computational costs is obtained by using linear transfer functions for the radiation and diffraction pressures from the frequency-domain panel code, instead of recalculating the pressures in the time-domain panel code (Z. Gao et al., 2023b). The disadvantage of this much more efficient "Load Reconstruction" method is that nonlinear hydrodynamic (Froude Krylov) pressures cannot be directly included.

The third and most efficient method is the "Response Reconstruction" method. Instead of using a database of pressure transfer functions, a database of stress transfer functions is used. This stress transfer function database is calculated from FEA with the frequency-domain hydrodynamic pressures. In addition, stresses due to unit anchor line forces at the fairleads as well as unit forces/moments at the tower base must be included. The database may focus on only selected structural members and stress components, and it may later be extended to other members without having to rerun the time-domain simulations of the complete turbine/floater/mooring system (Bredmose et al., 2024). Variants of the method are presented by Lee et al. (2023) and Lim et al. (2023).

3.2.3 Moorings and Anchors

Most recent research uses FEM to estimate the responses of mooring lines. The FEM is usually derived by discretising some of the curved cable or beam theoretical models, (C. Zhang, Wang, et al., 2022a). The models include axial and bending stiffness along with geometric stiffness and can deal with large displacements but with the assumption of small material deformations. If large deformations are included, they are generally applied for axial deformations in cases where synthetic ropes are in question. The mass of the lines can be concentrated in the nodes (a lumped mass approach), or an appropriate mass matrix is derived using the shape functions. Hydrodynamic loads (due to waves and sea currents) that act directly on mooring lines are approximated by Morison's equation, (Guo et al., 2022). The same equation is used to estimate the added mass and drag forces.

If used for FOWT research, this kind of FEM is most often coupled with hydrodynamics of the floating support of a FOWT. It is not uncommon that the coupling includes elastic structural models of the tower and turbine blades with associated aerodynamic loads. In some research, even the turbine's control system is included to form fully coupled aero-hydro-elastic-servo-mooring codes. Zhang, C. et al., (2022a) used this approach to investigate the effects of mooring line failure on the global dynamic responses and the internal drivetrain responses of a submersible FOWT. Guo et al. (2022) studied the impacts of catenary dynamics on the global response of a large-sized FOWT under environmental loads. The dynamic stiffness of the investigated mooring system presents itself in a hysteresis loop due to the fluid and structural damping. The hysteretic behaviour becomes more evident with the increase in frequency and amplitude. Yan, X. et al. (2023b) investigated the influences of different water depths and mooring parameters of a 10 MW semi-submersible FOWT. For this purpose, the coupled model was used to validate a simplified quasi-static mooring line model. The simplified model was needed to reduce the computational time since the thirty mooring cases with different water depths and mooring parameters were observed. It was found that mooring elastic stiffness has significant influences on the mooring tension and surge response of the FOWT in shallow water.

Liang, Jiang and Merz (2023) proposed a shared mooring system that would reduce mooring and anchoring costs in a dual-spar FOWT configuration. The hydrodynamic interaction between two FOWTs was ignored in the model because of the large spacing. Compared to the single FOWT, larger horizontal platform motions and higher mooring tension were detected. An open-source code, FAST, was extended in its capabilities to understand better the potential of shared mooring for FOWT farms in Lozon and Hall (2023). The FOWTs coupling (through interconnecting mooring lines) was upgraded to the same high-resolution time step in the MoorDyn module as for an FOWT with a regular mooring system. The study revealed that the shared mooring system satisfies strength constraints without additional strengthening and offers a degree of station-keeping redundancy in the case of mooring line failure.

The coupled models are also used to investigate mooring line failures since industry standards request that dual mooring failure be examined (to ensure that an observed

FOWT will not lose without its self-sustainability), as done in Jia et al. (2023). As an alternative to the models based on the constitutive equations, a machine learning technique can be applied to detect mooring line failures. In Walker et al. (2022), a machine learning model based on kernel regularised least squares methods was used to set up two digital twins of a FOWT. The digital twins were developed to estimate the tensions in mooring lines and detect the failure of mooring lines due to extreme load and fatigue. The training data was from the Hywind Pilot Park. The first twin could predict the mooring tensions under healthy conditions of FOWT. The second twin was used to indicate the near future, of approximately 1–2 min, values of the tensions, which is enough to generate early safety-related warnings if necessary. K. Sun et al. (2023a) developed a neural network as an intelligent early detection damage model for mooring structure damage. The model was able to identify positions of mooring creep from FOWT's yaw response.

Considering the limitations of traditional methods for synthetic fibre rope analysis, a refined numerical model, Syrope, is applied in Xu et al. (2021) to avoid over-conservative mooring design. The model was developed to describe better the tension-stretch and axial stiffness characteristics of synthetic ropes, defining the static and dynamic stiffness in a single model. The model was combined with experimental data to investigate the possibility of expanding FOWT's applicability in shallow waters. The Syrope was compared to the bi-linear model in Sørum et al. (2023b). Laboratory tests were used to set up the models. Afterwards, both models were used for fatigue lifetime and extreme response predictions. The two models predict similar values of extreme tensions, while the bi-linear model predicts a longer fatigue lifetime. François and Davies (2023) developed an approach for polyamide ropes where assumptions and formulations are based on test observations. An improved representation of the non-linear load-strain response is gained using realistic load spectra rather than the usual sinusoidal loading sequences. The stochastic process was assumed to be stationary for around one to six hours. It can include a wave frequency part, a low-frequency part and possibly higher frequencies induced by turbine operation.

3.2.4 Wind Farms and Interactions

To optimize future large-scale offshore wind farms, Arabgolarcheh et al. (2023) utilized a validated actuator line CFD model to explore whether phase lag differences in the surging motions of two tandem rotors affect the load and power performance of the downstream rotor. Their findings indicate that asynchronous surging can increase root bending moment amplitudes by up to 100% in the downstream turbine. They also discovered that fatigue loads can be managed by carefully adjusting the motion phase differences between the turbines. Zhang et al. (2022) carried out a systematic analysis of the interaction of two FOWTs, utilizing the URANS method to assess their impact on the LCOE through a cost model. This work explored critical factors such as the turbines' relative rotating direction and distance, aiming to understand their effects on performance and costs. It is concluded that a tandem layout with a spacing of 9.25D offers the most practical and optimal parameter choice, providing valuable guidance and insights for future wind farm planning.

3.2.5 Applications

The design and analysis of next-generation floating wind turbines requires advanced numerical tools integrating multiple disciplines. Aero-servoelasticity, hydrodynamics, mooring and cables dynamics, controls systems are the essential disciplines to account for complex, coupled interactions between wind forces, wave dynamics, structural vibrations, and mooring tensions that floating platforms endure in harsh offshore environments. As an example, in Patryniak, et al., (2022), it is shown that the representation of the fully coupled system within the optimisation framework requires the introduction of a more complex multidisciplinary analysis workflow.

The topic of the application of widely known integrated numerical models to the analysis larger rotor is heavily represented in the literature of the addressed period. Differently, the validation of those models is much more difficult due to the lack of full-scale experimental data. The accuracy of the widely used integrated tools to address the response of large FOWTs of next generation is still uncertain; hence, literature mainly focuses on numerical assessment or benchmarks (Ramzanpoor et al., 2024) and on validation based on model scale systems tested in a wind tunnel or ocean basins. Combined wind/waves facilities are also used. For instance, Li, C. et al., (2022a) demonstrate how, respect to a concrete one, a steel floater structure with a lower centre of gravity exhibits advantages in platform pitch motion alleviation and is also subjected to lower pitch-induced tower base loads and nacelle acceleration. However, an opposite trend is found in the wave-frequency responses, leading to an insignificant difference of tower base loads and nacelle acceleration between the two structures.

An example of the analysis of an upscaled machine is described in Souza and Bachynski-Polić, (2022). In this work three spar-type 20 MW FOWTs are designed and their structural behaviour is investigated with the non-linear aero-hydro-servo-elastic software SIMA which couples a finite element aeroelastic software for structural analysis of slender marine structures (RIFLEX) and a simulator of marine operations for large bodies (SIMO). Through the integrated model, the authors show that: i) the platforms with larger restoring in pitch present less fatigue damage at the platform, but more at the tower; ii) extreme stresses are largely affected by gravitational loads, such that the designs with larger pitch at rated thrust have the highest extreme stresses at the platform and most of the tower sections; iii) load cases at the rated wind speed often govern the extreme loads, unlike previous studies with 5 MW and 10 MW machines. The same tool is used in Wang et al., (2023) where an effective and robust design method for floating wind turbine platforms, which can provide reasonable internal stress for design checks (such as intact stability, natural periods and buckling) is proposed.

A good example of analysis of the impact of modelling aspects on FOWT response is presented in Papi et al., (2023). In this work, reporting the main outcomes of the H2020 project FLOATECH, QBlade-Ocean, OpenFAST, and DeepLines Wind simulation codes are extensively benchmarked using different floater/turbine configurations, realistic environmental conditions and several DLCs. Results show a good agreement on system dynamics prediction, whilst some differences arise in the analysis of fatigue loads where, depending on the solver, under- or overestimation of lifetime damage equivalent loads is highlighted.

The most represented topic is the application of FAST (or OpenFAST) coupled with different tools for modelling subcomponents (such as the floater), to different types of analysis. For instance, Z. Wang et al. (2022d) propose an identification method applied to the tower top acceleration and root force. This tool uses a deep learning algorithm in which the FOWT response is simulated through OpenFAST to train a neural network and synthetize a Multi-Layer Perception (MLP) model. The identification method describes the coupling between tower forces and motion responses. Different input configurations are analyzed and integration strategies with Smart Health Monitoring (SHM) systems are envisaged. Fig. 12 shows the identified results of the tower top acceleration compared to the simulated values for the DTU 10 MW FOWT model.

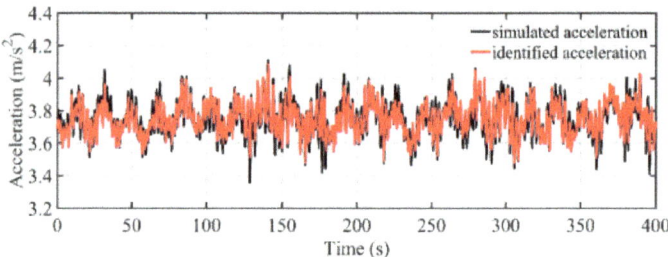

Fig. 12: Identification results of tower top acceleration compared to the true value (Z. Wang et al., 2022d)

Quantitative indicators of the MLP accuracy on acceleration and tower loads are detailed in Table 3 showing an excellent performance. This model appears to be very promising for applications in the field of SHM, which is a very important research topic with critical outcomes for the industry in terms of reliability and costs reduction. For instance, (Gorostidi, Pardo and Nava, 2023) proposes a deep learning algorithm to detect mooring line degradation and failure by monitoring the dynamic response of the Deep-CWind OC4 semi-submersible platform. Using OpenFAST, this study implements an autoencoder to predict multiple forms of damage occurring at once, with various levels of severity. The authors show that the proposed algorithm can detect mild anomalies caused by biofouling and anchor displacements, with correlation coefficients up to 98.51% and 99.16%, respectively.

Table 3: Results of evaluation metrics of the trained MLP neural network, from Wang et al. (2022). E_abs, E_max: absolute average and maximum error; R_new: curve fitting degree; MAE: mean absolute error; MSE mean square error. (Z. Wang et al., 2022d)

	E_{abs}	E_{max}	R_{new}	MAE	MSE
Acceleration	0.568%	0.812%	97.560%	0.444%	3.942%
Shear Force	0.193%	0.208%	98.837%	0.170%	2.487%
Bending Moment	0.618%	0.687%	96.374%	0.826%	4.365%

Other FAST-based applications are reported in T. Wang et al., 2024d) where a Simulink-based coupling strategy between OpenFAST and WEC-Sim is applied to the IEA 15 MW reference wind turbine and UMaine-VolturnUS-S semisubmersible platform for a reference site located at the northern North Sea. Similarly, extreme load responses of a 10 MW FOWT are studied in Xing et al., (2023) and in Gaidai et al., (2023). The latter proposes a novel reliability approach to assess multi-degree-of-freedom nonlinear system failure probability, in the case when only limited system measurements are available. Grid integration aspects using a FAST-based spar buoy FOWT model including generator, converter, and aero-hydro-servoelastic dynamics are addressed in Chen, Yang and Lou, (2024b). The FOWT model is developed using the MATLAB-Simulink cross-platform integration tool and is used to investigate the impact of wind and wave loads on wind power ramp events (WPREs). This study shows that WT power performance is highly unstable within the rated wind speed range, under WPREs within ultra short time period, resulting in failure to meet grid standards, emphasizing the external need for targeted power compensation and power signal processing. Overall, this study highlights the importance of pitch motion and wave load impact for WPRE study of FOWTs.

A few other integrated models are proposed in the literature. For instance, Luo et al., (2024) study a fully coupled floating wind turbine model using the SIMPACK multi-body dynamics environment to integrate the available numerical models of the turbine subcomponents including rotor aerodynamics (Aerodyn), control (based on a widely used FAST library), drive train dynamics, floater and moorings hydrodynamics (Hydrodyn and MAP). The integrated model is applied to investigate the effectiveness of passive 3D pendulum tuned mass dampers (3D-PTMD) and bilinear tuned mass dampers (2TMDs) in controlling the system structural vibrations. The paper shows that, if suitably designed, the TMDs can reduce the platform roll/pitch vibration frequency (mainly driven by the incoming waves). In addition, the combined use of 2TMDs and 3D-PTMD can reduce the vibration frequencies shown by the nacelle displacement. Similarly, a simplified integrated tool based on a nonlinear aeroelastic multi-body dynamic model of a FOWT is used in (Fitzgerald et al., 2023) to examine the effectiveness of TMDIs. This work shows their capability at reducing tower vibrations and improving its reliability when subjected to stochastic wind-wave loading environments. In some cases FAST submodules have been used to be integrated in higher fidelity models of FOWT components. For instance, in Festa et al., (2024) the numerical modelling of the floater, turbine and mooring system was performed using Flexcom, a commercial finite element (FE) software. Flexcom offers fully-coupled aero-hydro-servo modelling using FAST plug-ins INFLOWWIND, AERODYN and SERVODYN. The focus of the work is on FE modelling of the mooring lines to compare the impact of 4 different non-linear stiffness curves on tension reduction and platform motions. Using the IEA 15 MW reference turbine, the authors show that load reduction device stiffness curve types have little effect on out-of-plane platform motions and nacelle acceleration, but lead to an increase in surge when compared to the baseline mooring system. The increase in surge is similar regardless of the load reduction device stiffness curve shape, and is shown to be mainly driven by the length and rated tension of the device.

The analysis of FOWT-ship collision scenarios and the resulting system dynamic response are addressed in Zhang and Hu (2022) where a numerical solver based on LS-DYNA FEM solver to include wind, wave and mooring loads in the collision analysis is proposed. A simplified rotor aerodynamic model is used whereas hydrodynamic loads are based on potential-flow theory coupled with a linear wave kinematic model. In addition, a linearized model for the mooring forces is considered. Detailed FE models of the Hywind Spar-type floater and of the bulbous bow of a colliding offshore service vessel (Fig. 13, left) are built and different collision scenarios considering the influence of velocity, flexibility of tower, deformability of the ship, wind-wave loads are investigated to compute both local structural deformation and floater 6DOF rigid-body motions (Fig. 13, right). It is found that the impact velocity can greatly affect the structural response, and a high-speed impact can directly lead to tower collapse. Differently, tower flexibility influence on collision energy dissipation is more significant for bottom-fixed configurations. Finally, a rigid striking ship can be used for simplicity, anyhow the deformability can significantly influence the structural deformation and energy dissipation (local indention of FOWT is reduced more than 40% in the deformable-ship condition).

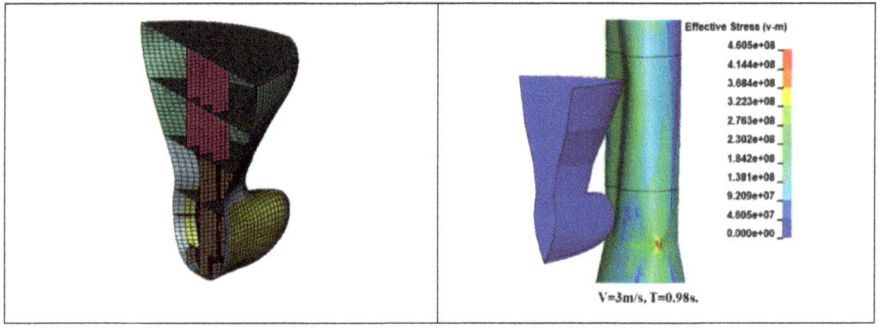

Fig. 13: Ship collision analysis: FE model of vessel bulbous bow (left) and spar deformation (right) (Zhang and Hu, 2022)

The impact of other extreme events, such as typhoons, on the FOWT response is investigated in Zhang, Z. et al., (2024d) where an integrated tools based on the commercial hydrodynamic software ANSYS-AQWA is used to analyze the dynamic responses and mooring forces of different types of platforms, including a novel fully submersible platform with vertical carbon fiber reinforced polymer tendons and circumferential catenary chains. Similarly, Kang et al. (2022) analyze the structural response of semi-submersible floating offshore wind turbine structures in waves generated in hurricane environments.

3.3 Physical Testing

With the development of new technology and components for FOW, physical testing of models and scale prototypes is essential for validation of numerical simulations. Considering the LEG 1 exclusion relating to insurance D'Andrea et al., (2023), physical testing of innovative structures at various scales is essential to improve market confidence. FOWTs are highly complex dynamic systems and therefore coupled testing under wind, wave and current loading is important for development and verification of designs. Primarily carried out in ocean basins, there are challenges with discrepancies between Froude and Reynolds scaling of the substructure and WTG respectively, accuracy of aerodynamic loads at small scale, and scaling highly dynamic components such as moorings and dynamic power cables (Fig. 14) (Holcombe et al., 2025).

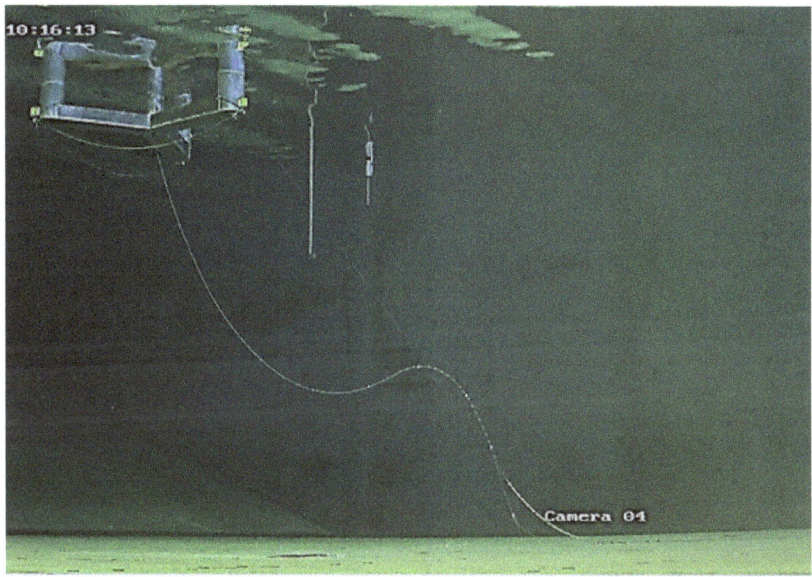

Fig. 14: 1:70 scale testing of the Volturn US platform, moorings and dynamic cable in the COAST ocean basin (Photo Credit: A. Holcombe)

While wind tunnel testing is well understood, techniques are being optimised specifically for generation of accurate aerodynamic loads (Wen et al., 2022; Schulz et al., 2024; Taruffi, Novais and Viré, 2024). Software-in-the-loop (SIL) methods are also becoming increasingly prevalent to assist in discrepancies between Froude and Reynolds scaling (Ransley et al., 2023; Bonnefoy et al., 2024; Jiang et al., 2024). WTG control strategies can also be modelled and assessed as method of load-mitigation (L. Wang et al., 2024b).

Mooring materials and components require validation and testing for long term fatigue performance (Sørum et al., 2023a), effect of breakage (Ren et al., 2024) and structural implications of shared moorings (Lopez-Olocco et al., 2023; Liang et al., 2024).

The tower plays a critical role among a FOWT system, serving as a link between the RNA and the supporting platform. It facilitates the transmission of both aerodynamic and hydrodynamic loads, and the vibrations it undergoes significantly influence the dynamic response of the nacelle and the platform. With turbine designs alternating between soft-stiff and stiff-stiff tower configurations, and floating foundation designs compensating with offset and centrally located turbines, understanding of tower loads is key for design optimisation of floating substructures.

Li, C. et al., (2022a) undertook 1:60 scale, coupled wind-wave, model tests of both a steel and concrete Y-shaped, semi-submersible platform to investigate the effects of different materials. It was found that the steel structure experienced lower tower base loads at the pitch natural frequency with implications for fatigue performance of the tower. Through an experimental study undertaken on a 12 MW semi-submersible FOWT in a wave basin, Guo et al. (2024) systematically investigated the tower load responses. It was determined that the bending moment at the tower top of a FOWT is greater than that of a fixed wind turbine, likely due to the pitch motion of the FOWT, the additional deformations of the tower due to motion, and the significantly amplified inertial load effects. The combination of pitch motion and rotor rotation induced a gyroscopic moment, resulting in the initial yawing of the FOWT.

Full scale FOWT data, from the Wind Float Atlantic wind farm, was used to assist development and validation of a new methodology that explicitly incorporates tower accelerations for the fatigue estimate of FOWTs (Pimenta et al., 2024). The technique was aimed at eliminating the need to install strain gauges on structures. The approach uses tower top accelerations to estimate the tower bending moments and fatigue life consumption, replacing more common data driven approaches based on environmental conditions and/or operation variables by analytical considerations. Feng et al. (2024a) have also undertaken trials of an indirect measurement method to acquire the thrust and pitch moment loads of the turbine without the use of strain gauges and fibre Bragg gratings with a mean relative error of less than 10%.

Measured wave data has been shown to be a potential method for complementary feedforward control of FOWTs though wave tank experiments (Hegazy et al., 2024). Freak wave loading has new impacts on FOWT structures. Experimental investigations into wave slamming effects on a 1:50 scale semi-submersible foundation of the X30 platform with a 5 MW WTG showed strong nonlinearity (Huo et al., 2023). The measured slamming pressures indicated a double-peak phenomenon where the FOWT suffered the second severe slamming after experiencing the initial freak wave slamming.

While many currently leased FOWT farms are in wave dominated regions with lower current, there is always a concern for Vortex Induced Motions (VIM) and the related structural implications. A comprehensive review by Yin et al., (2022) indicated that, while a few experimental investigations of FOWTs have been carried out, VIM model tests on integrated models (floater–mooring–subsea power cables) at small/moderate Reynolds numbers under representative sea states was an area that was lacking detail.

Structural and hydrodynamic performance validation of new FOWT designs is essential on the path to certification. Recent model tests of the WindCrete platform have indicated good performance to the design basis (Somoano et al., 2024). The OUCwind design was assessed through coupled testing in a wave basin with the data currently being used to compare and validate numerical models (Bai et al., 2024).

With regard to mooring and anchoring system validation and verification, Sect. 2.3 provides some discussion on the testing anchor/seabed interaction effects and system integrity.

3.4 Transport & Installation, Operation & Maintenance

While these items as they relate to FOWT are similar to that described in Chapter 2.4 for BFOWT, there are notable differences as well.

3.4.1 Transport & Installation

Despite the compelling environmental and economic prospects of floating wind technology, its implementation is challenging; complex installation procedures, associated high costs, and evolving regulations can hinder widespread adoption. Installation operations often utilise several vessels at one time (Fig. 15) with dynamic load regimes proving challenging for the WTG nacelle, blades, tower and FOW platform itself.

Fig. 15: One of the Kincardine floating wind turbines being towed to site illustrating the multiple vessels required for a relatively small FOWT (Photo Credit: Flotation Energy)

Hong et al. (2024) discuss the technical, operational, and economic aspects of floating offshore wind farm installation, providing a comprehensive overview of the current state-of-the-art. Critical research areas include foundation design optimisation, not least materials and structural design to optimise for fabrication and installation, and anchor design and installation considering novel anchor concepts and the anchor-seabed interaction. A novel multi-bucket foundation, aimed at improving installation and structural reliability, using integrated transportation, has been proposed by J. Li et al. (2024a). Initially numerical assessment using SESAM_ABAQUS indicated the roll and pitch of the foundation do not exceed 2.5° and the stresses in all parts of the structure during installation are below 250 MPa at 2.5 m significant wave height and 11 s peak period, which is an improvement on more conventional installation methods.

Environmental loads are key to successful installation operations, with recent research into the wind loads during transport and installation of a 10 MW turbine investigated through wind tunnel tests (Sim, 2023). The wind load coefficients at each stage of fabrication, transportation, and installation are presented for use during concept design assessment for FOWT installation.

Mooring attachment and hookup are key aspects of installation that many are endeavouring to optimise. With short weather windows and a need to be able to reach a level of integrity at which the FOWT can be left if the installation vessel needs to retreat to safety, learnings from the oil and gas industry are proving essential in the development of smaller, lighter and more cost-effective solutions for FOW. Atallah (2024) presents a case study from a recent successful mooring hook up and tensioning operation for a Floating Production Unit (FPU) where they evaluated an optimized strategy for the installation of a typical wind farm. The case study demonstrates quick connection and re-tensioning capabilities, and a need for innovation in design to increase the weather windows for mooring installation for operational success.

One area of limited research is that of quayside mooring and load requirements. With FOW platforms and WTGs increasing in size, WTG integration and subsequent mooring of the whole FOWT system in port is proving challenging, with bollards and quayside loading not rated for the larger dynamic loads present – this is an area that recent stakeholder engagement has indicated may be assisted through collaboration between industry and academia.

3.4.2 Operation & Maintenance

Chitteth Ramachandran et al. (2022) reviewed the various marine operations challenges towards the commercialization of floating wind in the context of spar-type, semi-submersible and tension leg platform (TLP) technologies. Many of the recommendations are related to cost savings and improving safety, and while OPEX is often the key driver for optimisation of O&M activities, structural health monitoring and assessment is imperative for formulation of proactive O&M strategies. Access to platforms is often required to undertake maintenance tasks and retrieve data from monitoring equipment, Fig. 16, therefore predictive and remote monitoring is an integral area for improvement.

Failure Mode and Effect Analysis methods are often used to identify key failure modes in FOW farms (McMorland et al., 2022; Y. Sun et al., 2023b; Ågotnes and Eik, 2024; Busby, Thethi and Fulton, 2024; Q. D. Feng et al., 2024a; Saetren Nornes et al., 2024). Many of these failure modes are structural in nature, with accurate prediction of structural loads and responses being a primary driver for improvement of O&M philosophies.

Robotics are being considered for integration into O&M planning and assessment of FOW structures offshore (Khalid et al., 2022, 2024). Unmanned aerial inspection of assets such as the WTG and blades can show areas where damage is occurring (K. Zhang et al., 2024c) to enabling prompt action to be taken before failure occurs.

Digital Twin (DT) technology is an area in which much progress has been made, despite the challenging computational and instrumentational requirements (B. Q. Chen et al., 2024a). DTs are based on simulated models and or Machine learning models, thereby improving predictions throughout the life of the asset (Mousavi et al., 2024).

Fig. 16: A worker accesses the base of the tower on one of the Kincardine FOWTs (Photo Credit: Flotation Energy)

Surrogate models are being considered for fatigue estimation of FOWTs (Liu et al., 2024). DTs have also been developed for assessment of specific areas of concern such as protective coating systems enabling corrosion to be minimised (Momber et al., 2022). While in their infancy, with a true digital twin requiring a phenomenal amount of computational hardware, and structural monitoring equipment, even the emerging DTs require validation through comparison with both model and dull scale trials (Branlard et al., 2024; Lotfizadeh, 2024).

Two key areas of concern that have recently been addressed through research are those of moorings and tower response and failure. The failure performance of different mooing configurations, and shared anchors, was assessed by the Offshore Renewable Energy Catapult's Floating Wind Centre of Excellence (Apollo and DOF, 2024; Weller et al., 2024). Chain catenary, semi-taut and taut mooring systems were considered comprising 3, 6 and 9 mooring lines. The key implications of redundancy provision were explored via the identification of candidate designs (based on a commercial-scale turbine), followed by ALS simulations and lifecycle analysis. It was recommended that the number of failures could be reduced by improving the reliability of mooring components, and/or reducing the system complexity buy reducing the number of components. Coraddu et al. (2024) have continued development of a DT specifically aimed at mooring line integrity and maintenance. Complementary methods of mooring integrity assessment and monitoring include; the use of autoencoders which can detect various issues including those related to marine fouling (Gorostidi et al., 2023), improved convolutional neural networks for machine learning (Sharma and Nava, 2024), and recurrent neural networks which indicated a very high rate of prediction accuracy of 98% over various scenarios (Saetren Nornes et al., 2024). Collision loads, based on the coupling of wind-wave-mooring loads, during maintenance operations were investigated through the use of Star-CCM+ and ABAQUS and assisted in informing weather windows and safe operational conditions, alongside structural design improvements (Zong et al., 2023).

The tower of a wind turbine is the primary load transfer path of the WTG loads to the floating structure and vice versa. Adaptive control methods of blade pitch, turbine yaw etc. can be used to improve corrosion-fatigue of tower fixings, such as bolts (Heng et al., 2024; J. Zhang et al., 2024b), and LIDAR assisted feedforward strategies to improve fatigue performance of both the blades and tower (Russell et al., 2024). Zhu, Z. et al., (2023b) developed a DT for wind turbine towers based on joint load-response, but including rotational effects, estimation which showed good correlation to laboratory tests. Sensor optimisation based on the effective independence method and modal assurance criteria methods has been investigated to utilise as much structural information as possible to predict progressive failures through identification of tower top accelerations and forces and moments at the tower base (Z. Wang et al., 2024e).

3.5 Design Standards and Guidelines

Further to the standards summary provided in Sect. 2.5 for fixed offshore installations, this section outlines the updated design standards and guidelines for FOWTs for the period 2022–2024, integrating recent advancements and industry best practices to ensure the safe, efficient, and reliable operation of these systems. There are several industry standards providing the framework for FOWT design, related to structures and structural performance. Table 4 summarizes the main design criteria specified in these standards:

Table 4: Design load criteria specified in different industry standards

Standard	Mooring Redundancy	Stability Structures	Damaged Stability	Type of Criteria	Materials	References
IEC	Optional, Increased safety factor	Optional	Quasi-static or dynamic-response-based	LRFD or WSD	Not specified	(IEC, 2020, 2021)
ABS	Optional, Safety factor increase 20%	Yes, in 1YRP	Quasi-static or dynamic-response-based	LRFD	Steel, concrete	(ABS, 2024)
BV	Optional, Safety factor increase 20%	Optional	Quasi-static or dynamic-response-based	WSD, LRFD optional	Steel	(BV, 2024a, 2024b)
DNV	Optional, Safety factor increase 15% to 25%	Optional	Quasi-static or dynamic-response-based	LRFD	Steel, concrete	(DNV, 2023b, 2023a, 2024e, 2024a, 2024f)
LR	Optional, Safety factor increase 50%	Yes	To IMO MODU or other	LRFD	Steel, concrete	(LR, 2024a, 2024b)
ClassNK	Mandatory, 1YRP check	Yes	Quasi-static	LRFD	Steel	(ClassNK, 2022)

Design loads for FOWTs are largely based on the fixed offshore wind turbine standard IEC 61400-3-1 (IEC, 2020), with additional considerations for floating structures (IEC, 2021). Key additional load cases include:

- Compartment damage;
- Mooring line damage;
- Stoppage and maximum operating conditions; and
- Robustness under extreme conditions (e.g., turbine still producing electricity beyond its specified operating threshold).

These additional load cases are explicitly included in IEC, DNV, and BV standards. While not specifically listed in ClassNK, LR, and ABS documents, these standards require verifications of damaged mooring and compartment conditions. There are two key design methodologies prevalent for the structural aspects of FOWTS: Loads and Resistance Factors Design (LRFD) and Working Stress Design (WSD). Table 4 indicates which of these are relevant to each society's regulations.

Loads and Resistance Factors Design (LRFD):
- Preferred for its practicality in real-world projects.
- Involves dividing material strength by a safety factor and multiplying design loads by a load factor.

Working Stress Design (WSD):
- Offered as an alternative by BV and IEC.
- Applies a global safety factor on stresses depending on material, loading condition, and failure mode.

As has been discussed in previous sections in this chapter, the two main materials used for FOW substructures deign are steel and concrete. Steel is covered across all standards and is common in the offshore industry due to its strength and durability. Concrete is only included in guidance form DNV, ABS and LR, and is seen as an emerging material for the construction of FOW substructures, primarily due to facilitation of economy of scale of manufacturing facilities and ports.

Since 2022, further advancements in FOW technology and design standards have been made. These include:

1. Updated Mooring System Guidelines:
   (a) Emphasis on redundancy and robustness to handle extreme weather conditions.
   (b) New materials and technologies for mooring lines to enhance durability and reduce maintenance.
2. Enhanced Stability and Control Systems:
   (a) Advanced dynamic response models to better predict and mitigate risks associated with floating structures.
   (b) Integration of real-time monitoring systems for continuous stability and damage assessment.
3. Sustainability and Environmental Impact:

(a) Increased focus on minimizing environmental impact during installation and operation.
(b) Guidelines for the use of eco-friendly materials and sustainable construction practices.

DNV have also recently updated the vessel standards which directly relate to T&I and O&M operations, and should be noted regarding coupled loading, collision risk and weather loading on FOWT structures (DNV, 2024d, 2024b, 2024c).

While the design standards and guidelines for floating offshore wind turbines from 2022 to 2024 reflect significant advancements in technology and industry practices. By harmonizing these standards, the industry ensures the development of safe, efficient, and reliable FOWTs. These guidelines support stakeholders, including engineers, designers, project developers, and regulatory bodies, in creating robust and sustainable wind energy solutions. Some independent research has been undertaken with suggested improvements to the standards: Wang, S. et al. (2022b) investigated wind and wave loading to floating wind turbine drive train damage, Kozmar et al., (2022) critiqued wind load assessment in offshore engineering standards and Gudmestad & Schnepf (2023) considered the critical aspects requiring inclusion in the Design Basis.

4 Wave Energy Converters

4.1 Recent Industry Developments

Wave energy convertors (WECs) have various types of power generation system and many kinds of WECs has been studied and developed. In these years, EMEC (European Marine Energy Centre), located off Scotland, PLOCAN (Oceanic Platform of the Canary Islands) located off Canary Islands and other areas were used for on sea trial projects.

Table 5 is a summary of on sea trial projects conducted in recent years (Ocean Energy Systems, 2023). There are some types of the installation as well as power generation types. The installation types are classified into a floating type, a sea bottom fixed type and a breakwater fixed type. Most of projects are still at the R & D stage which includes technological demonstration, and a few projects are in the commercial operation.

The classification of WEC type in Table 5 is according to Fig. 17 and based on EMEC and descriptions in the developers' web sites.

Small power generation WECs is the most in Table 5. Most of them supply electricity to marine equipment which includes ROV, marine observation sensors and so on. WECs which supply electricity to the land have about 200 kW power generation.

The directions of research and development are diverse, including response characteristics of floating structures, power generation performance, optimization of power generation control, mooring, coupling of multiple analysis methods, and reliability. The following section summarizes the R&D trends.

4.2 Numerical Modeling and Analysis

Wave energy converters have various types of power generation mechanics, shapes, and installation methods. Furthermore, many of the WECs have moving parts. Therefore,

Table 5: Major sea trials in recent years

Project	Foundation	Site	Power Generation	WEC type
xWave	Floating	USA	100 kW	Point absorber
SeaRay	Floating	USA	2 kW	Attenuator
PB3 PowerBuoy	Floating	USA	3 kW	Point absorber
Triton-C	Floating	USA	100 kW	Point absorber
C4 WEC	Floating	Portugal	300 kW	Point absorber
Blue X	Floating	UK	10 kW	Attenuator
Waveswing	Floating	UK	16 KW	Pressure differential
Mutriku	Breakwater	Spain	16*18.5 kW	OWC
DIKWE	Breakwater	France	800 kW	OWC
SEATURNS	Floating	France	200 kW	Rolling Mass
Exowave	Bottom	Belgium	–	Pendulum
Wavepiston	Floating	Spain	200 kW	Surge convertor
Slow Mil	Floating	Netherland	40 kW	Point absorber
ISWEC	Floating	Italy	250 kW	Gyroscopic
Wave Rudder WEC	Bottom	Japan	45 kW	Pendulum
Intelligent wave absorber	Breakwater	Japan	19.5 kW	OWC
Zhoushan	Floating	China	500 kW	Pendulum
Penghu	Floating	China	60 kW	Pendulum
Yongsoo	Bottom	Korea	250 kW	OWC
UniWave200	Bottom	Australia	200 kW	OWC

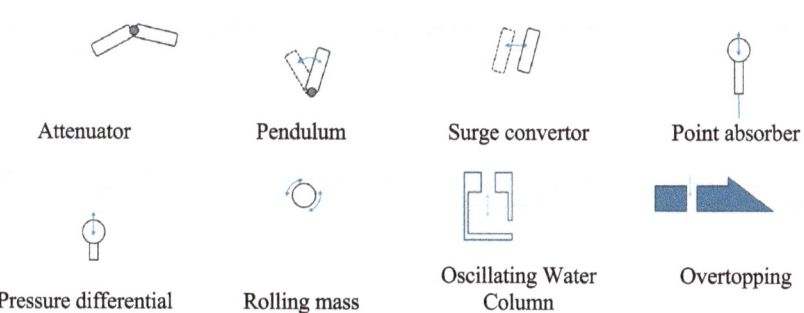

Fig. 17: Classification of WEC types (Esteban et al., 2017)

load and motion response of floaters are studied in many research works. In addition to the conventional analysis of the floating response, many studies have combined the optimization of power generation and the development of control algorithms. There is also an increasing number of studies combining multiple analysis methods.

4.2.1 Power Take Off and Control Algorithm Analysis

Parsa et al., (2022) estimated the power generation performance of the OPT PB3 Power Buoy for waves with multiple components by separating the spectra for each component. The estimation results were verified through simulation of the power simulation of the OPT PB3 Power Buoy at a point off Chile and show the effectiveness of this method for WECs with strong wave period dependence (Fig. 18).

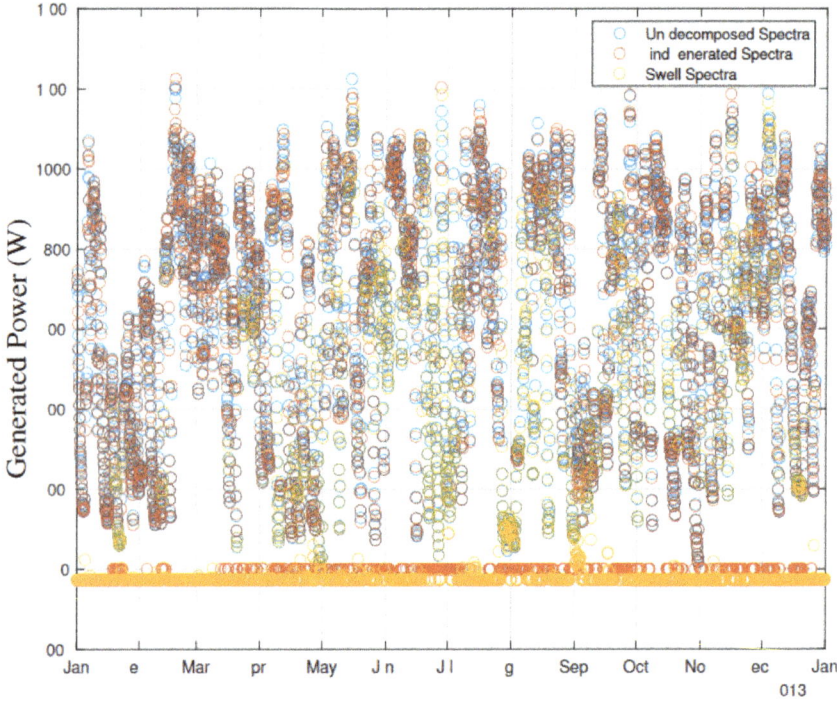

Fig. 18: Power estimates for un-decomposed and component spectra, based on simulations run on hindcast spectra for a site offshore Chile

Rahimi et al., (2022) introduced a new mechanism for a mechanical power take-off (PTO) system that can be used in a variety of wave energy converters (WECs). The mechanism uses two ball screws to convert the relative linear motion of the WEC bodies into unidirectional rotary motion. Guerrero-Fernández, J. L. et al., (2022) used WEC-Sim with a detailed Nonlinear model predictive control (NMPC) approach based on the real time iteration (RTI) scheme and revealed that RTI-NMPC clearly outperforms a simple resistive controller. Marley & Skjetne, (2023) introduced novel feedback control strategy which they called control barrier functions (CBFs). They simulated the effectiveness of CBFs against the Bolt Lifesaver point absorber WEC developed by Fred. Olsen Ltd. The control strategy mitigated the large force oscillations in power take off unit which were observed during initial on sea trials (Fig. 19).

Fig. 19: Bolt Lifesaver pictured outside Falmouth Bay, England, where she was deployed in 2012. Courtesy Fred. Olsen Ltd.

Faedo et al., (2022) introduced a nonlinear moment-based energy-maximizing control solution for WECs under non-ideal PTO behavior. The point absorber type WEC was studied and the nonlinear moment-based energy-maximizing control showed better absorbed energy than a benchmark PI controller. Umeda et al., (2024) proposed a data-driven reactive control strategy for a point absorber type WEC using Gaussian process regression and validated its efficiency through a tank test. The validation results confirmed that the proposed data-driven reactive control strategy effectively controlled the WEC and improved its performance using the Gaussian model. Moreover, performance was improved under different wave conditions compared to those used during training.

4.2.2 Hull Forms

Ghigo A., et al., (2022) proposed the optimization results for a cylindrical point absorber in the Mediterranean Sea. The optimization parameters were shape, dimensions, mass properties, ballast and draft. Huang, S., et al., (2023a) introduced the optimization method using NSGA-II and optimized results for heaving buoy type WEC in bimodal wave spectrum.

Ahmed et al., (2023) proposed a new bulbous-bottomed buoy designs for an optimal oscillating body type WEC (Fig. 20). The optimization was carried out under the condition which the radius, mass, and volume were equal. The optimized body showed about 12% enhancement of the performance (Fig. 21).

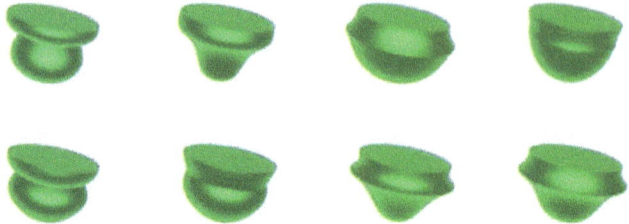

Fig. 20: The variants of the first generation (geometry definition-I)

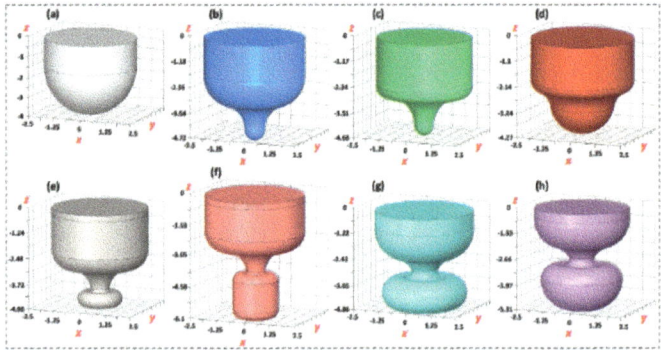

Fig. 21: The state-of-the-art buoy geometries (a) Cylindrical-hemispherical (C-HS) reference buoy (b) S-1 buoy (c) S-2 buoy (d) S-3 buoy (e) SB-1 buoy (f) SB-2 buoy (g) SB-3 buoy (h) SB-4 buoy.

4.2.3 Moorings

Yim, et al., (2022) proposed a combined nonlinear mooring-line and umbilical dynamics with bending capability (MUDB) model. The developed model was combined with WEC-Sim, and they simulated mooring lines and dynamic cable. Shahroozi, et al., (2023) proposed a neural network approach for the minimization of mooring tension of a point absorber type WEC in survival conditions. The simulation and experimental results showed that the deep neural network could reduce the mooring line tension. Gubesch et al., (2023) introduce the experimental results of a response of a floating OWC type WEC in extreme waves. A TLP (Fig. 22) and a taut mooring were the mooring setups considered in the experiment, and several differences in tension characteristics were observed between them.

4.2.4 FEM and CFD/Coupled Analysis

Huang et al., (2023b) introduced the simulation results with CFD and FEM coupled analysis for a flexible tube WEC. The nonlinear behavior of Natural Rubber is considered by using YEOH hyper-elastic model and fluid-structure interaction responses of the WEC were compared considering the impact of incident wave speed on the performance of the device. Shao et al., (2023) introduced simulation results with CFD and FEM

Fig. 22: Focused wave slammed into the TLP WEC

coupled analysis for a heave-point buoyant WEC called NoviOcean. They compared 2 commercial software packages. Yue et al., (2023) introduced the analysis results of a semi-submergible WEC which is called "Penghu". The FEM analysis was carried out with the wave load estimated by 3-dimensional potential flow theory, Fig. 23.

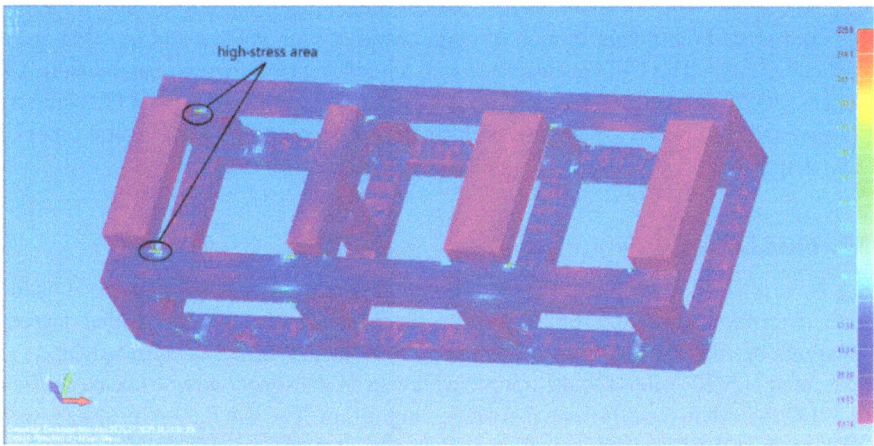

Fig. 23: Results of von Mises stress distribution of the hull.

Li, X. et al., (2022c) introduced the analysis method of coupling CFD and multi-body dynamic (MBD) for arrayed point absorber WEC (Fig. 24). The coupled simulation method was combined with CFD and MBD, and they introduced the power take off performance.

Fig. 24: Hex WEC array with sub structures

4.2.5 Reliability

Bao, X. et al., (2023) proposed a turbine fault diagnosis of the OWC WEC based on a correlation analysis of ensemble empirical modal decomposition (EEMDCA) and the fusion of multi-lead residual neural (MLRN) networks.

4.2.6 CFD

Prakash, R., et al., (2022) simulated of the motion of floats which shapes were rectangular, trapezoidal, and hemisphere with CFD solver. The results showed that the trapezoidal shape with fin found to be optimum float shape. Katsidoniotaki, E., et al., (2022) introduced the floater motion of point absorber type WEC using by CFD and a model experiment. The CFD results were well agreed with the experimental results. Li, M. et al., (2022b) introduced the optimized floater shapes of OWC type WEC using by CFD analysis. The power generation performance was compared with several types of floater shape.

4.2.7 Field Layout

Zeng, X., et al., (2023) proposed the optimized layout of the array of WECs based on the genetic algorithm (Fig. 25). From the results, the influence of introducing other degrees of freedom, besides the heave mode, on the performance of arrays was investigated.

He, et al., (2022) proposed the optimized layout of the square array of heaving buoy type WECs based on the differential evolution algorithm. The simulation results showed that the array of WECs has a large effect on the power generation performance. Abdulkadir & Abdelkhalik, (2023) introduced the results on the power generation performance with several dimensions of WECs. The studied parameters were a radius and a draft of a single float. The generic algorithm was applied to this study and the maximum 40% advantage of the power generation performance was acquired with the algorithm.

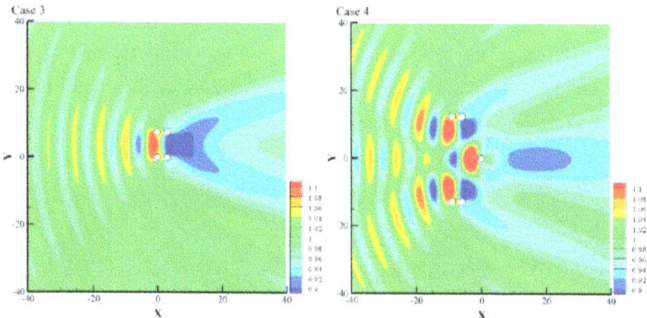

Fig. 25: Optimal layouts and the corresponding free surface amplitudes for case 3 & 4.

4.3 Physical Testing

Since there are a variety of WEC types and single installation and installation with arrays are considered, no dominant system has been found at this time. Therefore, in addition to numerical simulation studies, tank tests are also conducted. This section introduces studies of several types of tank tests.

4.3.1 Laboratory Study (Tank Tests)

Stansby et al., (2022) conducted a tank test and numerical simulations for an attenuator type WEC, M4 WEC which consists of 6 floats with 2 hinges. This model was moored to a single point buoy with elastic mooring cable. In the experiment, the floats showed occasional deck submergence (dunking) limiting relative angular motion as wave height increases, in effect providing a passive end stop, for very large waves.

Fig. 26: Aerial photo of WEC with mooring in the wave basin

Ahmed et al., (2023) conducted a tank tests for a point absorber type WEC with S shaped buoy (Fig. 26). Various types of buoys were tested through the tank tests and S shaped buoy showed a 34% greater RMS heave response than the cylindrical hemispherical buoy.

Niosi et al., (2023) introduced a tank test and numerical simulations for a Pendulum Wave Energy Converter (PEWEC) with 1:25 scale ratio carried out at the University of

Naples Federico II (Fig. 27). The experiments consist of free decay and static pull out tests to assess the inertial properties of the model and mooring system; tests in operative and extreme regular and irregular waves to fully characterize the mooring system and the device dynamics.

Fig. 27: PEWEC Mooring system 1:25 scaled model.

Meyer et al., (2023) studied a relationship between mooring types and a potential of power generation of point absorber arrays by a tank test. They conducted a tank test with 3 types of mooring types (a taut line mooring, a vertical tension leg mooring, and a conventional slack catenary mooring).

4.3.2 Field Tests

Although several sea trials have been conducted, not many studies have introduced measurement data in real sea areas. However Gato, et al., (2022) introduced the results on sea trials of biradial and Wells turbines at Mutriku WEC power plant (Fig. 28). They conducted on sea trials with 30 kW biradial turbines at 3 sites, IST variable flow test rig, the shoreline Mutriku wave power plant and the offshore IDOM MARMOK-A5 spar buoy device in EU H2020 OPERA project. The mean time averaged peak efficiency by the biradial turbine at Mutriku WEC power plant showed 37% higher results than with the Wells turbine.

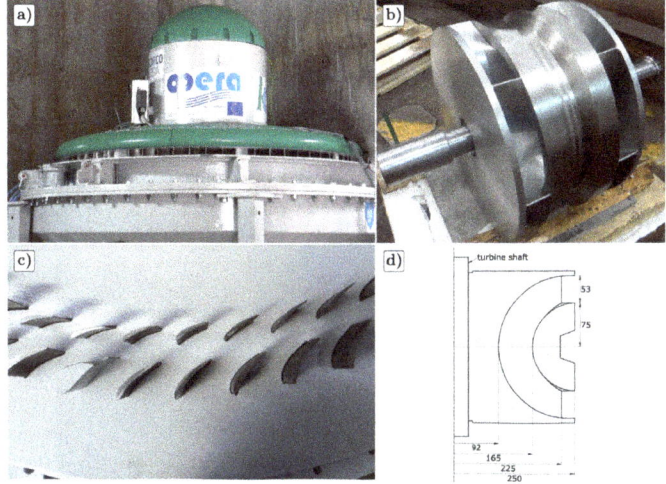

Fig. 28: a) OPERA's 30 kW biradial turbine at Mutriku's wave power plant. b) Biradial turbine rotor and c) guide vanes. d) Main dimensions of tested turbine rotor (in mm).

5 Tidal and Ocean Current Turbines

5.1 Recent Industry Developments

Tidal energy, due to its predictability, is considered vital for energy mix and security. The authors Coles et al., (2023) explored the impacts of tidal stream energy on overall energy system security using the Isle of Wight as a case study. The findings revealed that, compared to the best-performing solar and wind system, adopting tidal stream energy reduced the magnitude of maximum power shortages and surpluses by 11% and 24% respectively. Additionally, the adoption of tidal stream energy reduced the total land/sea space required by 33%. Furthermore, the energy storage capacity and charge/discharge capacity requirement for an additional inter-seasonal energy storage system to absorb curtailed energy were reduced by 21%. Assuming a 100€/MWh Levelised Cost of Energy (LCOE) by 2030, the analysis by Grattan and Jeffrey, (2023) indicated that the UK tidal stream sector could achieve commercial deployment by the early 2030s, reaching around 1GW by the mid-to-late 2030s, and ultimately deploying 6GW by 2050.

An overview of key achievements in ocean energy, including tidal stream, was summarised in the annual report issued by International Energy Agency – Ocean Energy System (Stratigaki et al., 2023). Notably, Sustainable Marine delivered the first floating in-stream tidal power to Nova Scotia's grid, harnessing the tidal currents in Canada's Bay of Fundy (Fig. 29) (Garanovic, 2022b). Altum Energy, formerly MAKO Tidal Turbines Pty Ltd., received financial support to advance their modular tidal turbines for slow-flowing tidal and river sites, with a demonstration currently underway in northwestern Australia. In the UK, Orbital Marine Power's flagship device, the O2 (Fig. 30), has been in continuous operation at EMEC since 2021, achieving a peak power of 2.5 MW. MeyGen Phase 1, operational since 2018, installed four 1.5 MW turbines and delivered over 45 GWh to the local Shetland distribution network as of October 2022. In France,

Sabella's D10-1 MW tidal turbine was redeployed in the Fromveur passage, and the company connected a small electrolyzer to experiment with green hydrogen production (Garanovic, 2022a). In Sweden, Minesto completed commissioning of the Dragon 4 (100 kW) tidal power plants in Vestmannasund, Faroe Islands. In China, the tidal stream energy demonstration project at Zhoushan installed and connected a new turbine to the grid in 2022, achieving a total installed capacity of 3.3 MW. The new turbine, Endeavour (Fig. 31), with a total weight of 325 tons, a rated power of 1.6 MW, and a designed annual power generation capacity of 2 million kWh, is expected to reduce carbon dioxide emissions by 1,994 tons annually.

Fig. 29. Sustainable Marine's floating tidal platform (Garanovic, 2022b)

Fig. 30. Orbital Marine Power's O2 turbine (Stratigaki et al., 2023)

Several review articles discuss the state-of-the-art developments in tidal energy technologies and future R&D trends. For example, Chowdhury et al., (2021) examined current trends, ecological effects, and prospects for tidal energy technology. A review of tidal current energy converters' developments in China was presented by Si et al., (2022),

Fig. 31. China's Endeavour tidal stream turbine (source: https://www.seetao.com/details/141111.html)

concluding that the development in China has diverged into large-scale turbines for grid connection and small-scale turbines for diverse applications in marine environments such as multi-energy complementary systems, in situ power supply, seawater desalination, and mariculture. The UK's Engineering and Physical Sciences Research Council funds the *Co-design to deliver scalable tidal stream energy* (CoTide) programme grant research (*Co-design to deliver Scalable Tidal Stream Energy*, no date). The CoTide program aims to develop integrated tools and design processes for tidal stream energy to reduce costs by eliminating unnecessary redundancy and improving engineering solutions' confidence. This initiative seeks to significantly contribute to achieving climate change objectives by 2030–40.

5.2 Environmental Conditions

Regarding the characterization of environmental conditions, a comparison was presented by McIlvenny et al., (2023) concerning field measurements of flow speeds in tidally energetic channels using large-scale particle image velocimetry (LSPIV) and dense optical flow techniques with the Gunnar-Farnback algorithm. Compared against underway Acoustic Doppler Current Profilers (ADCP), LSPIV was recommended due to the optical flow technique's underprediction of high velocities (>1 m/s). The authors of Thiébot et al., (2022) investigated the long-term variability of the tidal stream energy resource associated with the 18.6-year lunar cycle. Three sites in north-western Europe with strong potential for tidal array development were considered: the Alderney Race (English Channel), the Fromveur Strait (western Brittany), and the Ramsey Sound (Irish Sea). Results based on harmonic analysis and predictions of depth-averaged tidal currents indicated comparable variability in predicted annual power densities at the three locations. A numerical investigation by Spicer et al., (2023) examined the influence of a hypothetical tidal stream turbine array on barotropic tidal processes in the Piscataqua River - Great Bay estuary, potentially producing 44.7 GWh annually with 180 tidal turbines. Simulation results showed that tidal turbines can reduce tidal elevation and current magnitudes, diminishing tidal asymmetry due to reduced water storage over tidal flat regions at the estuary. Weakened ebb dominance keeps depths from the mouth to mid-reaches deep enough to minimize shallow water and frictional effects on storage mechanisms, with upstream shallow regions' frictional mechanism reduced by 10% due to the turbines. Environmental considerations such as sediment transport, water quality, and ecology are also discussed. Sea level rise (SLR) can change estuarine tidal energy. A systematic review by Khojasteh et al., (2022) on factors related to sea level rise categorized them into primary (e.g., tidal prism, range, current, asymmetry), secondary (e.g., sediment transport), and tertiary (e.g., shifts in estuarine shape/landform) factors. High uncertainty regarding SLR impacts on tidal energy resources exists due to spatial variability of estuaries. SLR may strengthen or weaken tidal ranges or currents, depending on estuarine shape and boundary conditions (e.g., levees, low-lying areas).

5.3 Tidal Turbine Loads and Response Analysis

5.3.1 Numerical Methods

An integrated floating energy system combining wind and tidal/current energy generators was investigated in Yang et al., (2023). An integrated analysis tool was developed to consider aero-hydro-servo-elastic coupling effects under wind, wave, and current loadings. It was shown that increasing the number of tidal turbines led to smaller pitch motion of the platform, and the presence of tidal turbines had an insignificant effect on fatigue damage at the tower base. Additionally, the total power of the integrated platform with three tidal turbines was expected to increase by 9.46% compared to the floating wind turbine alone. The authors of Zhang, J. et al., (2022) applied a continuous array optimization approach based on the open-source coastal ocean modeling framework Thetis to derive optimal configurations for four turbine arrays around Zhoushan Islands, Zhejiang Province, China. Different optimization scenarios were tested to investigate interactions between turbine arrays and their hydrodynamic footprint. The analysis showed that optimizing all turbine arrays simultaneously minimized array competition effects (which otherwise led to an average power decrease of 42.2%) and reduced the cost of energy. Computational fluid dynamics simulation by Mercier and Guillou, (2022) investigated spatial and temporal variations of flow characteristics at a tidal stream power site. The analysis indicated that flow characteristics are highly variable in space, especially laterally. Turbine positioning should thus be optimized locally with precision between 10 to 100 m. Longitudinal velocity can vary significantly over short periods, with average power of the flow reaching a turbine varying by a factor of 2 in less than a minute. The authors of Gao et al., (2022) employed the CFD-DPM (Discrete Phase Model) approach to study lift and drag coefficients of turbine blades under multiphase flow, considering fluid–particle interactions. The numerical analysis was validated by experimental tests of a 120-kW tidal turbine, showing small diameter particles can improve tidal current turbine power while large diameter particles can reduce power. Impact loads to tidal turbine blades from sea animals were analyzed by Gavriilidis and Huang, (2021) using Finite Element Modeling software ABAQUS, where an adult killer whale was modeled to study its impact on the blade. Various magnitudes and trajectories of animal entry into the blades were analyzed. Results indicated that equivalent plastic deformation of the blades is lower for carbon steel materials than for stainless steel. Based on 45 consecutive impacts, it was concluded that turbine blades' ultimate strength is sufficient to withstand impact loads from animals, but plastic accumulation in the blades requires further study. Increasing tidal turbine structures can lead to fatigue failure of turbine components due to load variations. The authors of Zhang, Y. et al., (2022h) applied a bidirectional fluid-structure coupling method to analyze the hydrodynamic performance and structural response of tidal-stream turbine blades. Results showed that structural responses of blades are minimally influenced by water depth but significantly affected by turbine speeds, which are critical for determining structural safety. Maximum stress responses occur near the blade root, but fatigue life remains a concern for tidal turbine blades. The structure dynamics of Horizontal Axis Tidal Turbines (HATTs) was also studied by Wang et al., (2024), analyzing inflow turbulence and unsteady force excitations on a three-blade HATT using Large Eddy Simulation. Distribution of unsteady loads on blades, the relationship between upstream and downstream flow velocities, and unsteady blade characteristics were analyzed. Under 10% turbulent intensity, HATT blades' hydrodynamic fluctuations reached 57.44%, with maximum loads located between 0.7R to 0.8R section of the blades. Middle and tip sections contributed to third-order blade passing frequency harmonics. The authors of Borg et al., (2021) developed a FEM solver for a full-scale ducted, high-solidity tidal turbine rotor made of fiber-composite, investigating structural performance under unsteady blade-resolved computational fluid dynamic analysis with configurations from aligned and yawed flows. Three internal blade designs were used for fluid-structure interaction analysis to study structural deformation, with maximum axial deflection-to-blade span ratio of 0.04 and maximum strain of 0.9%. Fatigue life assessment ensured sufficient operational time.

5.3.2 Laboratory Tests and Field Measurements

Static and fatigue tests were presented by Glennon et al., (2022) for composite tidal stream blades. Design, material selection, and manufacturing of candidate blades were carried out by SCHOTTEL Hydro (Germany) and Sustainable Marine Energy (UK): a 3 m length blade for a 6.3 m diameter rotor and a 2 m length blade for a 4.0 m diameter rotor, both resin-infused carbon fibre blades. Tests demonstrated survival of both blades in fatigue at lifetime-equivalent load, confirming structural strength and design life of the SCHOTTEL HYDRO blade. An experimental study was conducted by Zhang, J. et al., (2022) on a twin-rotor horizontal axis tidal stream turbine tested various approaching velocities and yaw angles. Results showed increased approaching velocity significantly boosted thrust and its fluctuation range. Yawed inflow reduced thrust gradually with increasing yaw angle, with thrust fluctuation range rising as yaw angle increased from 10° to 30°, then decreasing beyond 30°. Increased approaching velocity accelerated wake velocity recovery and enhanced turbulence intensity. Higher yaw angles resulted in long strip-shaped wake flow velocity distribution, reduced near-wake flow velocity deficit, and faster wake merging but negatively impacted further downstream wake recovery. Array spacing effect on wake interaction of tidal stream turbines was experimentally investigated by Zhang, Y. et al., (2023b) using a wave-current flume. Tests on single- and twin-turbine piled systems with three mid-passage distances (1.2D, 1.5D, and 2D, where D = rotor diameter) for the twin-turbine case indicated that a staggered layout with a 2D spanwise distance is preferred over an aligned layout due to better wake recovery.

5.4 Design Standards and Guidelines

Design rules and standards relevant to tidal and current turbines include the following:
- DNV-ST-0164, 2021. Tidal turbines (Det Norske Veritas, 2021b).
- DNV-SE-0163, 2021. Certification of tidal turbines and arrays (Det Norske Veritas, 2021a).
- BV Guidance note NI 603 DT R01 E, 2015. Current and tidal turbines (Bureau Veritas, 2015).
- IEC TS 62600 (International Electrotechnical Commission, 2019).

DNV-ST-0164 provides principles, technical requirements, and guidance for the design, construction, and in-service inspection of tidal turbines, covering structures, machinery, safety, controls, instrumentation, and electrical systems. DNV-SE-0163 defines specific certification requirements and respective deliverables for tidal turbines. The BV Guidance Note NI 603 DT R01 E sets requirements for fully submerged Current and Tidal Turbines (CTT) installed on the seabed, addressing support structures, turbine components (blade, hub, nacelle, duct), and electrical installations. The International Electrotechnical Commission (IEC) offers comprehensive design requirements to ensure the engineering integrity of marine energy converters (e.g., tidal energy), aiming to protect against hazards that could cause catastrophic failures in structural, mechanical, electrical, or control systems.

6 Floating Solar Photovoltaic Systems

6.1 Recent Industry Developments

The first developments of floating photovoltaic (FPV) systems started during the late 2000s, and since then, the cumulative installed capacity has doubled year on year (Friel et al., 2019). Some early FPV projects reported an improved energy yield of more than 10% over ground-mounted photovoltaic (PV) systems due to the cooling effect of water on PV panels, increasing their efficiency. In some cases, spurred by the high cost or limited availability of land, the PV industry has started to look into using water bodies for PV applications (World Bank Group, 2019). The FPV sector is still a fast-growing segment in the global PV industry, with more than 3 GWp of installed capacity by the end of 2021 (Zhao et al., 2023). Various floating technologies have emerged in the market. They can be categorised into four types: pure floats, floats with metal structures, membranes or mats that are directly installed on the water surface, and semi- or fully-closed platforms.

The deployment of FPV systems at sea is limited due to harsh environmental conditions, mainly induced by wind and waves. The vast majority of the available technology and projects in operation are located in inland freshwater bodies. It was found that the most common type of floating structures used inland is unsuitable for marine deployment (Oliveira-Pinto and Stokkermans, 2020). Furthermore, installing the FPV system in offshore environments is complicated, including lifting, towing, manoeuvring, and positioning heavy structures (Vo et al., 2021). This novel sector faces many challenges, including the unavailability of FPV-specific standards and technical guidelines, water body data, FPV plant component safety, and long-term reliability. Despite the difficulties, several projects installed at sea showed promise for offshore FPV systems worldwide.

Thus, research activities for marine FPV systems should focus on developing new concepts for FPV for the ocean instead of scaling up existing technologies. There is momentum in the sector to develop FPVs to be deployed in marine environments (Oliveira-Pinto and Stokkermans, 2020). Recent research has focused on FPV installations in coastal sites and sites with natural or artificial barriers that reduce severe wave impacts (World Bank Group, 2019). The research aims to standardise and optimise the used technology to provide commercial use of the marine segment of FPVs.

In 2021, Det Norske Veritas released the world's first Recommended Practice (RP) for floating solar energy projects, DNVGL-RP-0584, 2021 (DNV GL, 2021). The RP was released following a joint industrial project involving 24 industry participants. It applies to maritime locations close to the coast (which are reasonably protected) where significant wave heights up to 2 or 3 m are expected. These recommendations can also be used for FPV systems in protected and inland water bodies.

The analysis of the different FPV technologies revealed that pontoon-type systems will probably play a major role in the transition to the marine environment (Claus et al., 2022). These systems use joints between the floats. Truss structures are applied to this type of floating support to keep the modules safe. Also, thin-film-based designs have great potential for offshore applications. These systems minimise stress by allowing deformation and could be cheaper than pontoons at the cost of harvesting fewer resources. Integration with offshore wind, using the existing power grid, provides all types of FPVs with cheaper deployment (Ghosh, 2023).

Environmental loads, like wind and waves, are the primary loads on marine FPV systems, for which estimations are referred to the standards for relatively mature marine engineering, such as those of the oil and gas industry (Shi et al., 2023). The robust design of connections between floats is essential for the reliability of modular FPV platforms. So, different methodologies are applied to assess the structural response of FPV systems to environmental loading. However, since the dynamic response of FPV plants cannot be ignored in the marine environment, rigid solid dynamics or hydroelastic methods are applied (Claus et al., 2022). A mooring system proved to be much more critical for FPVs than traditional floating structures. With FPVs, the mooring system plays an essential part in the structural integrity of the whole system, so recent research is focused on coupled models (Catipovic et al., 2022).

Further improvements in FPV technology are needed to fully realise its potential, particularly in floating structure design, instrumentation, and monitoring systems. Addressing safety concerns, standardisation issues, and policy considerations will be crucial for widely adopting FPV systems (Ramanan et al., 2024).

6.2 Numerical Modelling and Analysis

As noted, most of the FPV installations are made of connected multiple floating bodies, i.e. made as modular structures. Incoming waves will cause structural loads on the floats, the connections between floats, and tensions in associated mooring lines. The wave loads can be determined by methods based on the potential flow theory. These methods are already well developed for the needs of the oil and gas industry, except that they have not yet been applied and tested extensively for structures with relatively large numbers of floating bodies, as current and undoubtedly future FPVs will have. One way to accommodate many floats is to use a hydroelastic approach, as presented in (Michailides et al., 2013), to analyse a modular-type floating structure with flexible connectors to determine motions and connection loads. The model was assumed linear, so the numerical results were obtained in the frequency domain. A simplified hydroelastic approach was used in (Jin et al., 2023), which presented a lightweight and high-stiffness floating platform composed of an ultrahigh performance concrete surface panel and an expanded polystyrene geofoam bottom panel as a large area bi-layered structure. The mechanical performance parameters of the upper layer are designed by adopting the representative volume element method.

The hydroelastic approach is also well suited for floating thin membranes as carriers of PV panes at sea. For example, Ma, et al., (2020) combined SPH and the linear FEM to study the structural response of the membrane. The SPH method has an advantage over the potential flow methods since it can model waves with large amplitudes. As a non-linear method, it was solved by time domain simulations. Further, due to large angular displacements, non-linear membrane response was studied in Xu and Wellens, (2022) to estimate the influence of such response on the strain-stress relation.

Some recent research is oriented on single-float support for FPVs but with the assumption that the float is rigid instead of being hydroelastic. A parametric study using a simplified approach based on the potential flow of a single floater regarding its dimensions and incoming wave characteristics is a good example (Al-Yacouby et al., 2020). A more advanced approach used for single support, including Froude-Krylov, radiation, diffraction, and viscous drag forces determined by Morison's formula, was presented in Friel et al., (2020). The response of the structure and the mooring lines was determined by FEM.

The assumption that the floats are rigid is also commonly used when dealing with multi-float support. In these studies, the connections between the floats are part of the numerical models, so more realistic, i.e. the coupled response of the floats is gained. In Catipovic et al., (2023), the ball connections between floats are considered in a frequency-domain seakeeping model. An additional stiffness matrix is presented, which introduces the influence of ball connections on multiple-float motions. The matrix can also be used to estimate the loads due to waves that occur in the connections. In Shi et al., (2023), the floating support structure is discretised into an array of floats connected by equivalent elastic beams based on Euler–Bernoulli beam theory. The corresponding structural stiffness matrix is assembled using coordinate transformations and matrix reorganisation techniques. Afterwards, the vulnerable area of the structure can be assessed using the equivalent stress based on the von Mises stress theory. A combination of rubber rings and anti-collision pads as connecting elements was proposed in Kang et al., (2023) for two triangular floats carrying PV panes. The forces in these connections were determined by time-domain simulations based on the potential flow. A similar time-domain numerical model was used in Zhang et al., (2024) to include non-linearities in the model tests and prototypes, for example, the viscous effects from waves, drag forces from currents, and the non-linearities from mooring lines and fenders. Flexible joints, whose stiffness was derived from the structural tests, were considered model constraints. The model was used for the design and verification of the recently deployed world's largest 5 MW nearshore floating PV farm in the coastal region of Singapore. The previous models include the hydrodynamic interaction between the floats, which can be computationally demanding for a large array of floats. So Zhang, D. Q. et al., (2023a) proposed a hydrodynamic interaction cut-off scheme for the multi-body potential solvers to save computational time. An optimal cut-off radius can be determined from an acceptable truncation error, so the interactions between bodies reaching outside the radius can be neglected, making the calculation less extensive.

More advanced methods for determining the wave loads were also used for multiple-float supports. The SPH method was used on a structure composed of 10 individual modules that were connected by rings in Wu et al., (2022). It was established that wave-breaking, which potential solvers can not detect, causes a significant increase in the loads, possibly making forces in the connections and tensions in mooring lines above design values. A safety analysis procedure that is composed of CFD simulations and FEM analyses with validations by experimental data was presented in Choi et al., (2023). The procedure was used to estimate the stress distribution on an FPV system for various incoming angles of the winds and waves. The Arbitrary Lagrangian-Eulerian (ALE) method is used for the fluid-structural analysis of yet again modular FPV system under wave action in Sree et al., (2022).

Wind loads make a significant part of the environmental loads of FPV structures because installed PV panels have large surfaces that are exposed to the wind. To define these loads, measurements in a wind tunnel were conducted in Chung et al., (2018), where a stand-alone PV array was put in a uniform flow with a given average air speed and turbulence intensity. The study was done for rooftop PV panels. Also, for the rooftop panels, Su, et al., (2020) applied CFD simulations to determine the wind resistance and uplift coefficients. Several turbulence models were compared with the wind tunnel experimental results, and the SST k-ω model has proven to be the most accurate. Hsu and Su, (2020) also used CFD simulations to get the wind-pressure coefficient of a PV panel under extreme conditions when the wind and wave are in the same direction. In such cases, the PV panel is subjected to a more concentrated non-uniform pressure near its edges compared to cases without waves. The SST k-ω turbulence model was used in simulations conducted by Bei et al., (2022) to resolve wind loads on a large-scale FPV system. The results show that the upstream PV modules facing the wind have an occlusion effect on the downstream PV modules. Stress analysis of the PV unit revealed that the maximum stress is mainly concentrated in the upper part of the supporting columns connecting the floating support and the PV modules. A structural response under the action of winds in combination with waves was investigated in Yang and Yu, (2021) for a single-float FPV. CFD modelled the pressure distribution on the PV panels due to wind action. The total loads were used to estimate the tensions in the associated mooring system.

An offshore PV system with 420 mooring lines was studied in Ikhennicheu et al., (2021). A quasi-static analytical analysis was applied to estimate the tensions of mooring lines so that the first-order wave loads were omitted. Waves drift loads were computed using Maruo's formula. The sea current loads were estimated based on drag and shielding coefficients. The analyses established that the mooring lines must be homogenously placed over the FPV system, i.e., non-uniform load distribution between mooring lines is a design problem. The mooring system was also studied in Song et al., (2022) for an FPV system made in a 10 × 10 assembly of floating supports that were connected by hinge joints. A catenary mooring system made of 80 steel wire ropes was observed. The lumped mass approach was used to model the mooring lines in combination with Morison's equation for hydrodynamic forces acting directly on the lines. It was found that the floaters' response was amplified in extreme wave conditions due to the resonance effect caused mainly by the stiffness of the catenary mooring. A similar number of lines in a mooring system was studied in Kanotra and Shankar, (2022). Here, the lines were made of chain and polyester, with different lengths and anchor positions to accommodate the bathymetry of the site and the water level changes. Such a large number of mooring lines was necessary to ensure that the expected load on each line was well below the structural capacity of interconnecting pins and fairleads. In Claus et al., (2023), an FPV plant was structurally assessed for marine environmental conditions. This work aimed to compare several chain sections for its mooring system. It was found that the heavier chain sections are better from the stationkeeping point of view and may prevent instantaneous chain wrenches that could result in snaps.

6.3 Physical Testing

The paper by Jiang et al., (2023b) presented an innovative floating photovoltaic (FPV) concept designed to endure harsh offshore conditions, including extreme wave heights exceeding 10 m. The FPV array employed semi-submersible floats constructed from lightweight circular materials, connected via ropes to form a soft-connected lattice structure. This design ensured modular motion with minimal wave overtopping and prevents contact between adjacent modules under extreme conditions. The study utilized a 1:60 scale model of a 2 × 3 FPV array, tested in a wave tank at the Canal de Ensayos Hidrodinámicos of the Universidad Politécnica de Madrid. The tank's dimensions were 100 m in length, 3.8 m in width, and 2.245 m in depth, equipped with a single flap wavemaker and a wave absorption beach. The experiments involved testing nine regular wave conditions (WR1 to WR9). These wave conditions represented typical wind-generated ocean waves with periods ranging from 7.8 to 12.8 s and wave-heights ranging from 1.9 to 5.1 m. The wave amplitudes were adjusted to specific wave steepness values. An additional survival condition (WR10) was included, having a higher wave height of 15.3 m to simulate extreme conditions. For irregular waves, the Pierson-Moskowitz Spectrum was employed. Three conditions (WIRR1, WIRR2, and WIRR3) represented different sea states with significant wave-heights ranging from 1.2 to 7.2 m and significant wave-periods ranging from 6.7 to 10.9 s. Moreover, Calm water free-decay tests were conducted to determine the natural frequencies and damping coefficients for the heave, roll, and pitch motions of the floats. Natural frequencies and damping coefficients for heave, pitch, and roll were determined to be 2.19 s (23%), 2.03 s (17%), and 1.78 s (14%), respectively.

Schreier and Jacobi, (2022) investigated the interaction between waves and a thin, flexible floating sheet. This study was motivated by the increasing interest in flexible floating structures for offshore applications, such as floating solar installations. The main objective was to understand the hydroelastic behavior of a floating sheet with a high length-to-height ratio in regular long-crested head waves. The experiments were conducted in the Ship Hydromechanics Lab at Delft University of Technology. The tank dimensions are 80 m long and 2.75 m wide, with a water depth of 1.00 m. The model is a closed-pore neoprene foam rubber sheet with dimensions of 4.95 m in length, 1.02 m in width, and 5 mm in thickness. The model was moored at the center of the tank using four mooring lines connected to the sidewalls. The surface deformation of the floating structure was measured using the Digital Image Correlation (DIC) technique. Regular waves were generated with a wavelength ranging from $L/\lambda = 5$ to $L/\lambda = 20$ and wave steepness in the range of 0.02 to 0.05 however, the researchers reported the results of only two specific conditions $L/\lambda = 5$ and $L/\lambda = 10$. The floating sheet mainly followed the local wave elevation, with a reduction in motion amplitude observed over the length of the structure. The wave condition $L/\lambda = 5$ showed increased hydroelastic interaction compared to $L/\lambda = 10$ which indicates stronger three-dimensional effects for shorter waves. The elevation amplitude of the model was found to be larger than the amplitude of the incoming wave, suggesting significant hydroelastic interaction. Claus et al., (2024) investigated hydrodynamic response of an innovative floating photovoltaic (FPV) system HelioSea under regular and irregular wave conditions. The HelioSea device features a pole-mounted solar platform with a double-axis tracker, supported by a tension-leg platform (TLP). HelioSea comprised a solar platform mounted on a pole with a dual-axis tracker which was supported by a tension-leg platform (TLP) to maintain stability. The experiments were conducted at the wave basin of the University of Porto, featuring a $12 \times 28 \times 1.2$ m^3 basin equipped with piston-type wave paddles. A 1:30 scale model of HelioSea was tested. Regular waves with nine different periods ranging from 0.73 to 5.11 s and three target wave heights (4–6–8 cm) were generated to establish the Response Amplitude Operators (RAOs) of the structure. Additionally, irregular wave tests using the JONSWAP spectrum were performed to evaluate the device's response under more realistic sea conditions. Both long-crested and short-crested irregular waves were tested with peak wave periods of 2.19 and 3.65 s and a significant wave height of 4 cm. Directional spreading functions were applied to generate short-crested waves. In all degrees of freedom, the HelioSea device showed a modest amplitude response for wave periods under 20 s, with surge responses reaching up to 4 m/m and yaw responses reaching up to 1 deg/m. Regular wave tests showed negligible sway, heave motions and a dominating surge motion. The anchoring mechanism significantly limited the pitch and roll movements, resulting in amplitudes below 0.5 deg/m. The yaw motions were consistently below 1 deg/m under normal wave conditions. The highest surge motion observed in the prototype scale was 1.8 m, which corresponds to a mooring line angle of 2.3°.

6.4 Design Standards and Guidelines

As noted in Sect. 6.1, the only design recommendation for floating solar projects was released in 2021, DNVGL-RP-0584 (DNV GL, 2021). Its scope is limited to significant wave heights of "up to 2–3m", which makes it inapplicable to most offshore sites. A number of demonstrators are being developed and tested (e.g. Merganser and Sun'Sete projects) yet there is still no offshore standard available. The recommended practice consequently needs to be extended to more severe environments.

The floating photo-voltaic structures expected to be installed are of considerable dimensions, potentially causing hazards to the surrounding sea-spaces and shores in case of a mechanical failure. The first demonstrators and prototypes will need to be designed following a goal- and risk-based approach to confirm that risks are taken at an acceptable level. It would be more desirable that certification schemes and standards would be available, but the maturity of these systems and the limited understanding of their behaviour still prevents it.

Indeed, most of the research so far focused on hydro-elastic modeling of the modular structures that will support the panels, and addressed specific aspects of loading and/or response (the influence of the stiffness of connections, non-linear breaking wave loads, diffraction of waves within the array, wind load/motion coupling etc.). This research helped identify the shortcomings of simulations and test models. These findings still need to be combined into complete higher fidelity models that will be required before standards can actually be issued. More recommended practices or guidance notes would be needed to guide designers in the estimation of design loads, coupling between modules, the type of dynamic simulations to be investigated, etc.

Important aspects of the structural strength of FPV systems have been so far scarcely investigated and will require further work and guidance: mooring systems and in general the long-term structural strength of these systems. A number of designs currently consider that arrays of solar panels will need to be moored by means of a large number of mooring lines, which poses the question of the safety of these systems which loads are shared between many non-linear members. In addition, limited research was made on the long term behaviour and resistance of these structures. A number of modules consider using polymer materials, which will need to be qualified in the marine environment. Elastic members connect modules; again made of polymer materials which long term behaviour (resistance, stiffness, creep, etc.) will need to be ascertained.

There is hence significant research to be undertaken before mature standards can be issued. In the interim, specific guidance notes on the testing and qualification of polymer load-bearing components, mooring system analysis as well as hydro-aero-elastic modeling would be very useful to the industry.

7 Other Offshore Renewable Energy Technologies and Hybrid Solutions

This chapter reviews the most relevant contributions regarding load analysis and structural design of floating and fixed offshore renewable energy devices. It also considers aspects related to prototype testing and the levelized cost of energy (LCoE). The review is structured into three main engineering design steps, the design requirements (e.g. LCoE), conversion methodologies (e.g. load analysis, structural design), and design specifications (e.g. structure architecture).

7.1 Ocean Thermal Energy Conversion

The Ocean Thermal Energy Conversion (OTEC) device removes and transfers seawater heat to a second fluid, which evaporates and acts on a turbine connected to an electrical generator (Fig. 32). OTECs may be located onshore and offshore (Fig. 33) and used as hybrid systems for district heating and cooling, desalination of seawater hydroponic cultivation, aquaculture, and extraction of seawater minerals. OTEC has been proposed to provide energy and extend the service life of unmanned underwater vehicles (UVV) (Jung et al., 2022).

Fig. 32. Conventional open cycle OTEC system. Adapted from (Aresti et al., 2023).

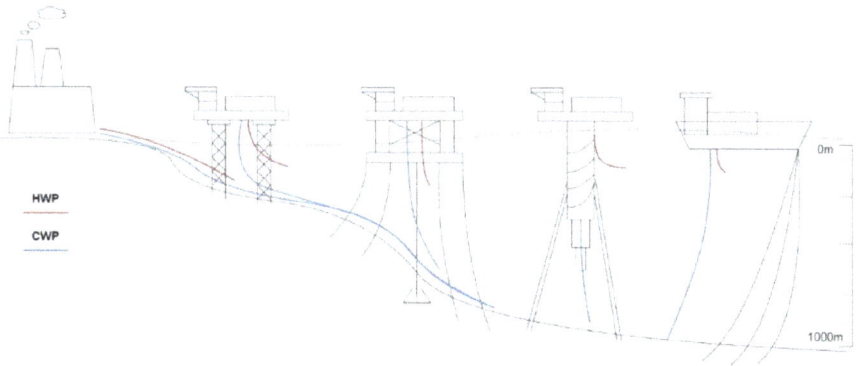

Fig. 33. OTEC positioning. Adapted from (Aresti et al., 2023).

Early development stage projects examine and propose the OTEC requirements. LCoE is the most used and important requirement. It is currently the major drawback of

OTEC devices since it is comparatively higher than conventional and other renewable energy systems (about 0.05 to 0.45 USD/kWh) (Aresti et al., 2023)

Thus, it is hard to attract research funding from investors. Hence, OTEC LCoE is used to study upscaling scenarios to make it feasible (Langer et al., 2022), and control systems are identified as a potential and important LCoE modulator since OTEC technological requirements are not focused on control as an enabling technology (Ringwood, 2022). For example, a proportional-integral controller has been designed to control the OTEC-generated net power and numerical simulations have confirmed the effectiveness of the proposed control system (Matsuda et al., 2023).

Other scenarios include hybrid solutions to improve LCoE, like integrating OTEC with solar and solar wind renewable technologies (Dezhdar et al., 2023; Hoseinzadeh et al., 2023). The OTEC system utilizes solar and wind energy to enhance its feasibility. Also, OTEC integration with Pumped Thermal Energy Storage (PTES) technology has been proposed but with a different requirement to improve, the Levelised Cost of Storage (LCOS) (Ghilardi et al., 2024)

The design specifications are related to OTEC energy harvesting performance and conversion efficiency (Langer et al., 2022, Aresti et al., 2023), LCoE (Langer et al., 2021; Langer, et al., 2022) and LCOS (Ghilardi et al., 2024).

OTEC requirements for conversion methodologies are proposed but without producing design specifications related to OTEC platform structures, since these parametric studies are performed on onshore applications, to determine OTEC energy harvesting system specifications (Dezhdar et al., 2023; Hoseinzadeh et al., 2023).

7.2 FPV Combined with Floating Wind Turbines and WECs

Floating photovoltaic (FPV) is used to exploit large water surfaces for energy generation, such as the ones in the free area between wind-floating platforms, to increase the production of energy per area of the sea bed used. (Fig. 34) (Solomin et al., 2021; Garrod et al., 2024). This technique can also be combined with WECs and the same floating structure, as illustrated in Fig. 35.

The LCoE is one of the most important indicators for assessing system performance; yet, the scientific community is divided on the best way to evaluate this metric and others (Solomin et al., 2021). Furthermore, system structural needs are not included in the list, indicating that these notions are substantially unknown and developed. Fig. 36 illustrates the only known hybrid offshore wind-solar power installation.

The harsh sea climate will drive the creation of structures capable of supporting these features, which will be done first for pilot plants (Garrod et al., 2024). Thus, it is predicted that a methodology for evaluating the structural integrity of FPV hybrid structures would be established from the single FPV mature methodologies, like the one by Claus and López, (2023) (Fig. 37)

7.3 WECs Combined with Floating Breakwater

Using WECs with a breakwater combines renewable energy production with coastal protection, against erosion and waves, resulting in more effective space usage and cost

Fig. 34. Combined floating wind and solar energy farm. Adapted from (Solomin et al., 2021).

Fig. 35. Conventional open cycle OTEC system. Adapted from (Solomin et al., 2021).

savings (e.g., construction and maintenance). WECs are located inside or in front of the breakwater, which can be either floating (Fig. 38) or stationary (Fig. 39).

WECs used in these applications are Oscillating Water Column (OWC) (Guo et al., 2021; Han and Wang, 2021; Cheng, Du, et al., 2022b; Ram et al., 2022; Mayon et al., 2023) and point absorber type (Cheng et al., 2021; Ji, et al., 2021; Cheng, Fu, et al., 2022a; Cheng, Xi, et al., 2022c; Ji and Jiao, 2021; Yang et al., 2023; Jeong and Koo, 2023; Ji and Chen, 2023; Cheng, Du, et al., 2024b; Wei et al., 2024; Cheng, S. et al., 2024; Peng et al., 2024). Also, a hybrid WEC, consisting of an OWC and a horizontal floating cylinder (HFC), is proposed by Shahabi-Nejad and Nikseresht, (2022).

Most of these studies are dedicated to studying and improving the WEC hydrodynamic performance. Very few include structural safety, and the analysis is mostly simplified. They evaluate the loadings on the system and moorings caused by the WEC

Fig. 36. Hybrid offshore wind-solar power plant. Adapted from (Garrod et al., 2024).

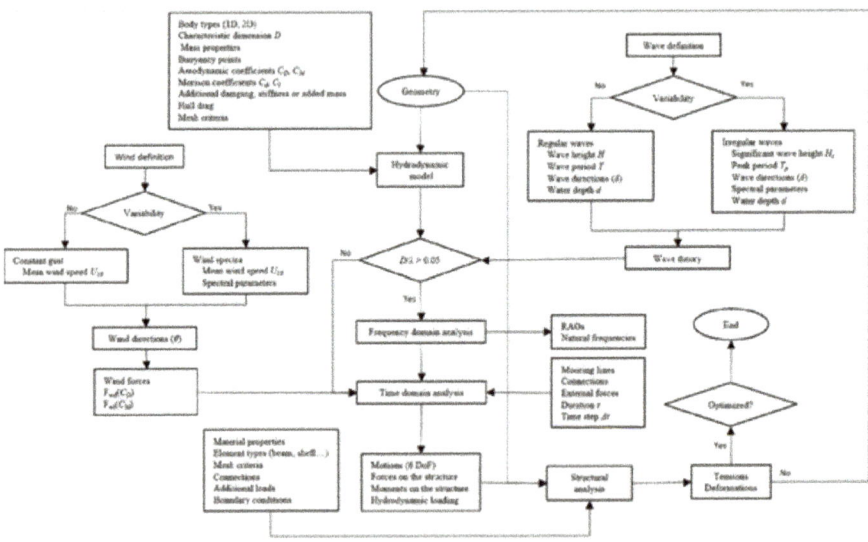

Fig. 37. Design methodology for FPV systems. Adapted from Claus and López, (2023).

motions and give indications about their impact on structural integrity without quantifying them with dedicated parameters (Guo et al., 2021; Ji and Chen, 2023; Cheng, Du, et al., 2024b). Consequently, no structural-related, requirement conversion methodologies and design specifications are presented; however, these few studies offer guidance for future research in this area.

Some of these indications include the influence of WEC Power Take-Off (PTO) damping force on the system survivability in extreme waves (Ji and Chen, 2023) and

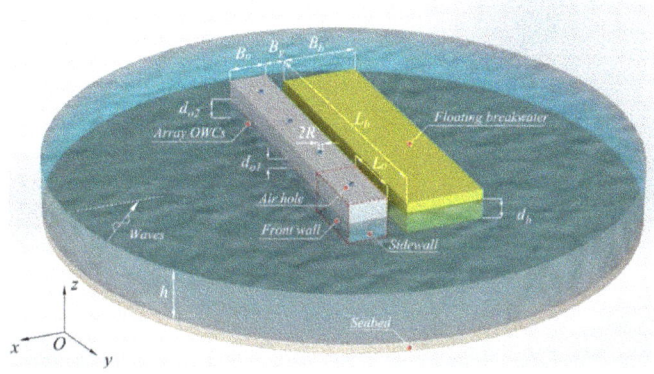

Fig. 38. OWCs in front of the floating breakwater. Adapted from (Cheng, Du, et al., 2024b).

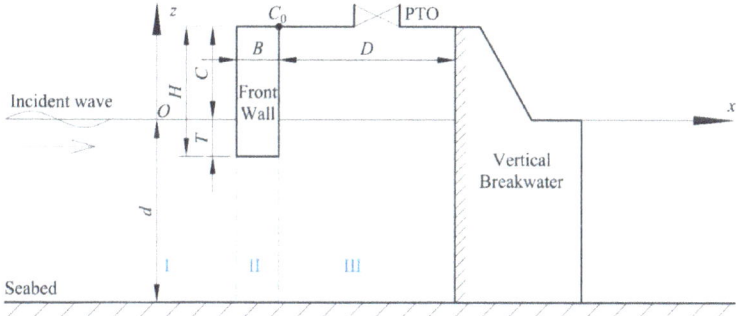

Fig. 39. OWC mounted breakwater integrated system. Adapted from (Guo et al., 2021).

the need to adjust system parameters to reach an optimal state in terms of WEC energy conversion ability and system structural safety (Guo et al., 2021).

7.4 WECs Combined with Floating Wind Turbines

Floating offshore wind turbine (FOWT) platforms are subjected to combined cyclical wind and wave interactions, rotor-induced forces, and control operations, making them susceptible to large oscillatory motion responses, undesirable resonant motions, and fatigue loads, especially in rough sea conditions systems (Uzunoglu et al., 2016).

Wind and waves cause cyclic loads on the tower and platform, resulting in significant wear damage at the tower's base and top (blades, rotor shaft, yaw bearing, gearbox, generator). This impacts the structural behavior, fatigue life, safety, and operating circumstances (Aboutalebi et al., 2023; Aboutalebi et al., 2024; Ahmad et al., 2023; Chen et al., 2022; Tian et al., 2023; Wang, Y. et al., 2022c). Thus, extreme oscillatory vibrations lead to reduced energy efficiency, greater maintenance and repair costs, and economic losses due to irreversible and inadvertent structural defects, resulting in downtime, shortened lifetime, and revenue loss (Aboutalebi et al., 2024); (Ahmad et al., 2023; M'Zoughi et al., 2023; Tian, et al., 2023).

Fig. 40. FWWPs. (a) Spar–torus WEC, (b) Semi-OWC (Oscillating Water Column), (c, d) Semi–torus WEC. Adapted from (Chen et al., 2024; Hallak, et al., 2023; Jaya Muliawan et al., 2012; Sarmiento et al., 2019; Zhu et al., 2023).

These have a detrimental impact on the levelized cost of electricity (LCoE) and undermine wind turbine upscaling to reduce LCoE, as oscillatory motions are increased with turbine upscaling (greater height and mass), resulting in higher accelerations and loads (Stansby and Li, 2023; D. Zhang et al., 2022b). These issues have a significant impact on structural design. A 5 MW HAWT (Horizontal Axis Wind turbine) blade's sectional modulus is enhanced by 50% to prevent fatigue failure caused by a 10% increase in external load at 5° pitching (Dong et al., 2022; Zhu, H. 2023; Zhu et al., 2024). Thus, the wind turbine must be strengthened, resulting in increased weight. This results in structural design needs such as increased displacement, structural reinforcements to resist larger bending moments, additional stability to counteract turning moments, and regulated motion response to minimize significant accelerations and dynamic loads on the turbine (Hallak et al., 2018).

Furthermore, this issue is more significant with a 20 MW HAWT because the rotor diameter is approximately twice the 5 MW HAWT (Kamarlouei et al., 2022). Then, suppressing platform motions for structural load reduction should be sought to extend the life of the overall structure while enhancing power output and lowering maintenance and monitoring costs (Ayub et al., 2023; da Silva et al., 2022). Thus, WECs for controlling FOWT motions are presented. The basic goal is to build these hybrid systems, also known as Floating Wind-Wave Platforms (FWWPs), to share infrastructure and, thus, increase LCoE. However, the research trend is now committed to the preliminary design of FWWPs for controlled platform motions, as rated power is delivered mostly by the wind turbine (Cao et al., 2023; Zhu et al., 2024).

Several FWWP configurations are proposed, each with unique WECs, arrays, layouts, and controls (Gaspar et al., 2021). Furthermore, many modeling and simulation approaches are proposed, as merging these technologies raises the complexity of platform dynamic analysis (M'Zoughi et al., 2023; Zhu et al., 2024).

7.4.1 Design Requirements

The design requirements can be divided into two categories: functional (FR) (features and functions) and non-functional (NFR) (architecture traits or "ilities"). In this context, FWWP feasibility is linked to economic viability and profitability, and both are related to other system characteristics such as reliability, efficiency, capacity, and maintainability.

LCoE (Yi et al., 2024) is used to assess feasibility, while the platform's structural integrity and associated metrics such as FOS (Factor of Safety) (Yi et al., 2024), ULS (Ultimate Limit State), FLS (Fatigue Limit State), ALS (Accidental Limit State) (Jaya Muliawan et al., 2012), and DEL (Damage Equivalent Load) (Chen et al., 2022) are used to assess reliability, survivability, and maintainability. The system capacity is measured using WEC CWR (Capture Width Ratio) (Shi et al., 2022). Other metrics, such as platform pitch and heave, are employed to assess the impacts of motion on structural integrity (Yi et al., 2024), but they do not provide a quantitative measure like the ULS, FLS, ALS, and DEL.

7.4.2 Conversion Methodologies

The FWWP requirements for design specification conversion methods include the integration of hydrodynamics, aerodynamics, mooring and station-keeping systems, structural dynamics, and PTO systems, among others. Thus, researchers use diverse techniques, including code and software, given the numerous disciplines involved. These combinations, known as numerical frameworks, consist of several interconnected software components that perform calculations in all required domains at each timestep. FAST@, OpenFAST@, AeroDyn@, HydroDyn@, and MoorDyn@ packages, as well as ANSYS AQWA@, Nemoh@, WAMIT@, and WEC-Sim@, are commonly utilized in these frameworks (Gaspar et al., 2024).

7.4.3 Design Specifications

The conversion methodology models are based on engineering parameters, which are WEC-measurable qualities. WEC parameters are grouped into categories, depending on its subsystems, such as PTO and PTO two-step damping and PTO spring-damping for the control system (Cao et al., 2023; Chen et al., 2022; da Silva et al., 2022; Tian et al., 2023; Zhu et al., 2024), buoy radius and draft for the absorber (Ghafari et al., 2022; ZHU1 and piston area, motor displacement, accumulator pre-charge pressure, accumulator initial gas volume and throttle valve orifice size for oil-hydraulic PTOs (Y. Wang et al., 2022).

7.4.4 FWWP Geometries

System architectures are involved in the FWWP requirement conversion approaches (e.g., geometries in Fig. 40 Approximately 77 distinct FWWP designs have been proposed over the previous 13 years (Gaspar et al., 2024). Since 2020, research on this topic has increased, with 7 studies in 2020, fully dedicated to semi-submersible hybrids, 12 in 2021, mostly dedicated to semi-submersibles and very few papers to barge and spar hybrids, 11 in 2022, dedicated to semi-submersible hybrids, and 23 in 2023, mostly dedicated to semi-submersible hybrids and very few to spar types (Gaspar et al., 2024).

Furthermore, continuous growth in these investigations is noticeable in 2024, almost reaching the same number as the previous year, with 17 research till the end of May, most dedicated to semi-submersibles and one to TLP hybrids (Gaspar et al., 2024).

Because semi-submersible FWWPs outnumber other platforms, a comparative analysis is conducted by counting research for each platform type each year and adding them up over time. Fig. 41 shows that contributions to semi-submersible FWWPs have increased exponentially since 2019, while contributions to other FWWPs have remained low and consistent (Gaspar et al., 2024).

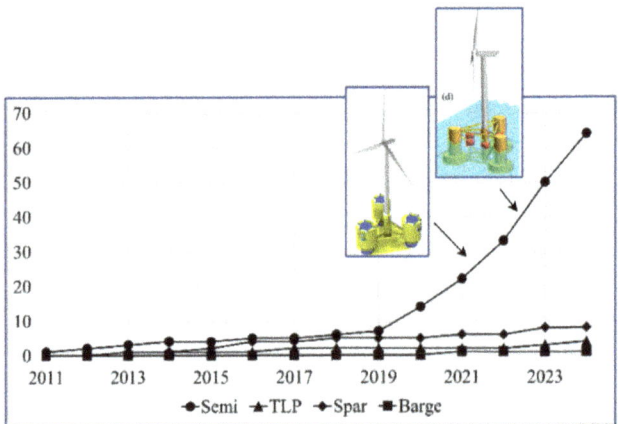

Fig. 41. Cumulative studies over the years per platform type. Adapted from (Gaspar et al., 2024).

The same approach is used for semi-submersible platform waterplane layouts, and it is revealed that only waterplane layouts 2, 3, and 4 are meaningful, as shown in Fig. 42, with contributions increasing exponentially since 2019. Layout 2 consists of one central column with a wind turbine (TCL) on top, and three exterior columns (CL). These are based on DeepCWind (OC4) (Fig. 40b–d). Layout 4 uses the same column configuration but fewer connecting pontoons (i.e., braceless platforms). Layout 3 likewise features three outer columns distributed in the same triangle pattern, but not a central one. One of these columns supports the TCL. It is based on modifications to the WindFloat and OC4 platforms. Furthermore, semi-submersible Layout 3 is catching up with Layout 4 and may overtake it by 2024.

The last analysis is performed on waterplane configurations of WEC absorbers utilized in semi-submersible FWWPs. The results are shown in Fig. 43, which demonstrates an exponential and relevant trend for type 3 absorbers (A) positioned outside the platform waterplane layout against absorbers located inside (types 1 and 5) or at the platform waterplane's boundaries (type 2). Overall, FWWP designs are moving toward more spread and less concentrated waterplanes.

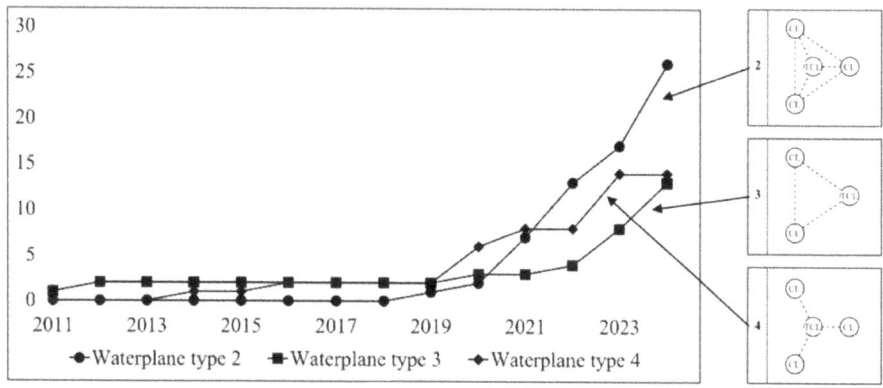

Fig. 42. Cumulative studies by SSB waterplane layout. Adapted from (Gaspar et al., 2024).

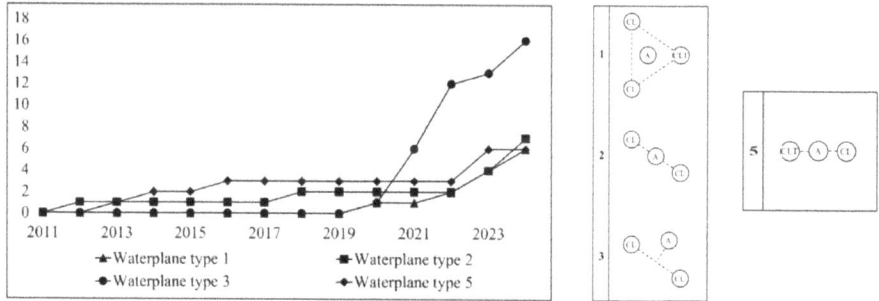

Fig. 43. Cumulative studies by SSB absorber waterplane layout. Adapted from (Gaspar et al., 2024).

8 Power Infrastructure at sea

8.1 Necessity and Application of Power Infrastructures

The main function of offshore renewables remains to feed onshore grids (International Energy Agency, 2023). This comes with the challenge of connecting to the shore a large numbers of remotely located electricity generators. This connection uses infrastructures which includes essentially cable systems that connect energy generators within an array, export substations that converts the inter-array electricity to export voltage, and subsea cables routed from the substations to shore (Feltes et al., 2012). Substations typically handle several hundreds of MW, which make these structures and the export cables critical not only for the farms but also for the grid (Boakye-Boateng, et al., 2021).

Robak and Raczkowski, (2018), Jump et al., (2021) and MacDonald, et al., (2018) list essential functions of offshore substations. These can be summarised as follows:

(i) interconnection of array to the export system,
(ii) raising voltage to its export threshold,
(iii) compensating cable capacity losses,
(iv) connecting to the export cable,

(v) allowing monitoring, maintenance and operation in general.

There is no example of normally manned substation; hence unlike most hydrocarbons production facilities, export substations are manned only when maintenance takes place. There is a single example of a floating offshore substation, installed on the Fukushima Forward site in Japan; all other substations have been installed on fixed structures, anchored to the seabed.

Previous work on cables confirmed the feasibility of high voltage dynamic cable systems, whereas certain electrical equipment suppliers (Hitachi ABB, 2021) already propose a range of equipment for floating substations. Hence the only barrier to floating substations remain their economic performance and risk profile (perceived or actual). Jump et al., (2021) compare fixed and floating substations, concluding that floating substations can be cost-competitive to fixed substations from water depth in the range of 55-60 m. shallow draft "barge-like" floaters being more competitive than semi-submersible floaters.

Offshore renewable energy being competitive in a number of places, it is also possible to use the electricity produced offshore to an energy carrier or other products needed in e.g. chemical processes in Power-to-X applications (Wulf, et al., 2020). This use of offshore renewables is expected to grow (Bossmann et al., 2018).

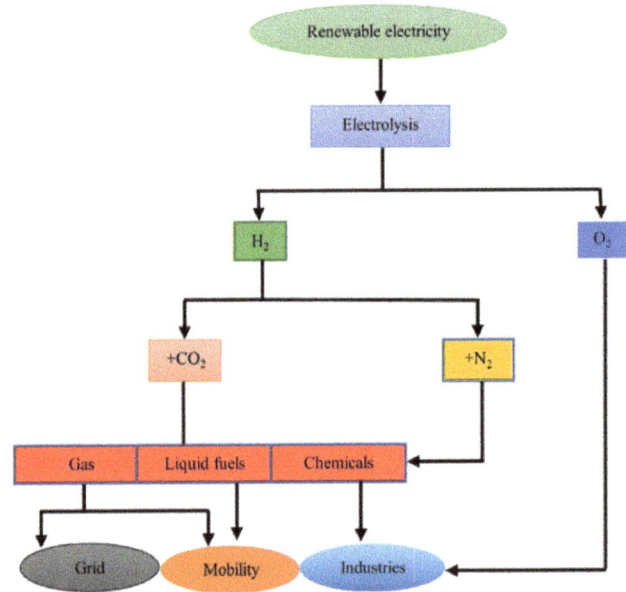

Fig. 44: Schematic of the P2X conversion pathways (Dahiru et al., 2022)

The initial feedstock of all Power-to-X projects is hydrogen that is generated by water electrolysis using renewable electricity. The flowchart in Fig. 44 from Dahiru, et al., (2022) shows a variety of pathways of Power-to-X processes. Both Singlitico, et al., (2021) and (Groenemans et al., 2022) found that hydrogen generation from offshore wind is economically viable. However, Groenmans found that the most competitive solution is to distribute electrolysers on individual wind turbines, whereas Singilitco found that centralised offshore production for a farm/group of farms is a better trade-off. Both options are however only 25% off in levelized cost of hydrogen, which is remarkable for such early-stage studies. It is hence probably worth considering both options for further development.

The first production of hydrogen from an offshore site combined Lhyfe's electrolyser and BW-Ideol's floating wind turbine. In this case, the electrolysers are located off the coast of France. Another approach followed by Dolphyn was to install the hydrogen production equipment on a barge, moor it in a sheltered site and deploy it offshore at a latter stage. No information was found on the results of these early experiments.

Little information was made available on the modifications required to operate electrolysers and all related equipment at sea. This data is proprietary to Lhyfe and Dolphyn. It can however be expected that fixed structures would need no modifications compared to onshore processes, and that centralized production on floaters would be less challenging than distributed production on floaters as the support structure would be larger on a centralized production unit.

These early experiments on hydrogen production addresses the first steps of the Power-to-X processes, but there is more to be done. In particular, (i) for direct hydrogen export; designing and qualifying hydrogen compression/export/offloading solutions (ii) for hydrogen use; designing and qualifying e.g. ammonia or methanol production platforms (iii) for designing and qualifying storage; export and offloading solutions for the produced hydrogen carrier (e.g. ammonia, methanol, etc.).

Structures currently installed and under construction deal with electricity generation, which makes electrical cable systems critical. We propose in the next subsection a state of the art of power cables design and issues. We will review in the following sections how large structures (substations or Power-to-X platforms) are (or could be) designed.

8.2 Cable Systems

Techno-economic studies help quantify the criticality of power cable systems for electricity generation; globally, around 149,300 km of cable is expected to be installed for offshore wind projects by 2035. Haq Duggal, (2024) estimates at around 150,000 km the length of cables to be installed by 2035 for offshore wind systems, leading to around 2200 cable failures between 2024–35. The Global Underwater Hub, (2023) concludes that these failures can lead to up to EUR 20.5 billion repair cost for the same period;[40% for inter-array and 60% for export cables].

Scrutiny to failure causes and claims reveal that the highest insurance settlement costs are linked to Cable Protection Systems and manufacturing faults (Marcollo and Efthimiou, 2024). This relates to the top four causes of cable failure: installation, mechanical failures, external aggression (T. Zhang, Du, et al., 2022f; T. Zhang, Li, et al., 2022g) and cable design (4C Offshore Limited, 2020; Haq Duggal, 2024). Failure rates of static subsea cables were estimated at around 0.00165–0.0213 failures/km/year (Harvey et al., 2024; Marcollo and Efthimiou, 2024), which is several orders of magnitude higher than the target failure rates [10^{-4} for the whole cable system].

Some research was made on the modelling of installation loads (Kuang et al., 2022; Li et al., 2024; Okkerstrom, 2021; Hansen, 2023) to identify critical phases, and analyse the typical loads that impact cable installation to determine the parameters affecting the process. This allowed to determine the optimal environmental conditions for cable installation through the moonpool of a vessel to minimise structural damage were determined.

Another track followed by Jordal et al., (2023) is to test cable strength. They presented a novel axial compression test rig to investigate combined bending/compression of cables. The rig captured adequately the bend curvature and axial compression behaviour of cables by testing.

Large scale mechanical testing of dynamic subsea power cables is essential for qualification, but also enables the determination of key properties and validation of numerical models (Ringsberg et al., 2023; Thies and Georgallis, 2024). These large-scale tests are the first step of a cable system analysis: material tests provide the mechanical properties of the cable, which are used as a basis to global dynamic analysis which itself feeds local strength verification (Marcollo and Efthimiou, 2024). The global models are used as part of the design process to provide a holistic picture of the hydrodynamic response of the cable and estimates of the fatigue life (Liu et al., 2024; Yan et al., 2023) and derive cable cross section loads. These global models can be complemented by local structural modelling to determine the load path through the different layers of the cable, which gives access to stiffness properties as well as the hysteretic behaviour of the cable (Skeie et al., 2023). Nicholls-Lee, et al., (2021) coupled global and local models to improve the understanding of the response of the cable and the influence of local parameters to the behaviour of the whole cable system.

The case of dynamic cables gains attention as more floating structures are being installed. Dynamic cables were found (Harvey et al., 2024) to be potentially subject to failure rates an order of magnitude higher than static cables because of their exposure to fatigue loads. Key to deriving the fatigue damage of each cable layer, various methods have been investigated to improve the understanding of the response of the complex layup of power cables to cyclic loads (Beier et al., 2023; Beier et al., 2024; Beier, 2023; Martinez-Puente et al., 2024; Sobhani, 2021; Sobhaniasl et al., 2020; Svensson, 2020).

The focus has primarily been on copper conductor analysis, which ensure the cables' main function of energy transmission (Bakken, 2020; Poon et al., 2023; Poon et al., 2024; Wan et al., 2022). Another area of focus has been on the water-tight barrier that protects higher voltage conductors from degradation by water. Cables with voltage ratings in excess of 132 kV utilise a metallic lead alloy sheath to do so (Moreno et al., 2021;.Viespoli, 2020; Viespoli et al., 2020; Viespoli et al., 2021), which is prone to work hardening under small strain cycles. The strength of these components is a concern for the industry with the advent of floating offshore substations requiring very high voltage dynamic cables (Guignier et al., 2020; Hung and Yang, 2023), or inter array cables being considered at 132 kV. A solution leading to acceptable performance was to use corrugated-copper instead of lead-based watertight sheathing (Guignier et al., 2020).

The dynamic environment also poses the threat of Vortex Induced Vibrations (VIV) on the cables (Marcollo and Efthimiou, 2024). Numerical investigations have indicated that there may be preferable orientation for the cables, depending on local metocean conditions, to minimise fatigue caused by VIV (Börü, 2021; Delizisis et al., 2022; Shim et al., 2023); however, it has been noted that more information is required for appropriate values, such as Strouhal Number, to improve modelling accuracy. Experimental investigations of a flexible dynamic power cable in lazy wave configurations found that the peak frequencies usually aligned with a Strouhal number of 0.15 for most of the cases (Elrick and Venugopal, 2023; Moideen, et al., 2024).

The industry proposes novel cable designs to overcome these issues. These novel cable structures pose new modelling challenges. Fig. 45 compares conventional cable structures and a novel design where structural members are made of aramid fibres laid with the conductors. This differs from conventional cable structures where the load bearing members are steel wires helically laid around the conductors in so-called tension armours.

Conventional dynamic subsea power cable design (Technip FMC, 2024)

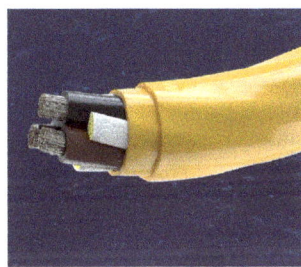
Novel dynamic subsea power cable design (NKT, 2024)

Fig. 45: Current subsea power cables designs

This novel design improves the fatigue performance of the cable by using aramid tension members and reducing the diameter of the copper wires that make up each conductor. The construction also improves the thermal dissipation from the cable which eventually reduces the risk of overheating. Research in understanding these novel material combinations will be required to allow the development of internationally-recognized standards.

With structural health being of paramount concern for dynamic subsea power cables, monitoring and fault identification is key to enable prompt action and reduce farm downtime, and hence revenue. Typical monitoring methods, such as Time Domain Reflectometry, Distributed Strain and Temperature Measurement and Partial Discharge Measurement, heavily depend on thresholds and residual trace analysis, imposing conservative limits or can be easily deteriorated by possible errors leading to an incorrect cable system assessment that does not correspond to the failure condition (Caruso, 2020). Newer monitoring system are being developed, such as SSTDR, that have the potential to detect defects and degradation in the cable and pin-point the fault location to within one meter, but can also provide real-time data streamed to shore for potential development and optimization of a digital twin of the dynamic cable system (Nicholls-Lee et al., 2022).

Digital twins have the potential to help predict failures and optimise operation and maintenance strategies, lowering farm downtime and lost revenue (Jin et al., 2022; Yin et al., 2024).

8.3 Design Standards and Guidelines

As any offshore structure, offshore substations are subject to national requirements. In addition to these regulations, classification societies, which usually act as project certifiers, have issued requirements for these. ABS has issued Requirements for Offshore Substations and Electrical Service (American Bureau of Shipping, 2023) which specifically cover fixed and floating substations. This standard calls for offshore oil and gas rules when it comes to structural strength, floating stability, mooring system design or performance assessment.

Another approach proposed by LR as part of project certification LR-GN-007 Guidance Notes for Offshore Wind Farm Project Certification (Lloyd's Register, 2024a) is to verify the structures of the substations per the same approach as the wind turbines' substructures. In this case, the primary design standard is the family of IEC 61400 standards (International Electrotechnical Commission, 2024), complemented by relevant rules for offshore units (Lloyd's Register, 2024b). Bureau Veritas approach is in the Guidance Note for the Certification of Fixed Offshore Substations for Renewable Energy Projects (Bureau Veritas, 2022), is that the structural design should comply with ISO 19900-series (International Standards Organisation, 2019), which are international oil and gas production standards.

DNV adopts a mixed approach where the safety principles, structural load cases and partial safety factors are defined in the standard for Offshore Substation ST-0145 (DNV, 2020), while the material factors, failure modes, etc. are taken from the offshore structures standards. This leads to having ABS, BV and DNV standards designing against the 100-year return period whereas LR asks to design against the 50-year return period with a different set of safety factors. Each of these classification societies consequently proposes a different design approach with its own safety factors and principles, although eventually accept to design against ISO 19900 series as an alternate. Although these standards explicitly deal with the design of mooring system as well as offshore installation, they do not cover the cable systems.

The Carbon Trust floating offshore wind joint industry project recently reviewed standards related to submarine dynamic power cables, primarily focussing on mechanical, electrical, and fatigue testing (Harvey et al., 2024). The review covered 62 publications from multiple regulatory bodies, including DNV, ISO, API, IEC, CIGRE, ABS, and BV. The standards were categorised based on three criteria:

- Type of cable (static/dynamic)
- Field of application (oil and gas/offshore renewable energy)
- Type of component (cable/ancillaries)

The review identifies CIGRE TB 862 (CIGRE, 2022) as the only directly applicable recommendation to dynamic subsea cables in the renewable energy sector; however, this covers only mechanical testing. The other standards still used are primarily related to the oil & gas industry and bring inherent levels of conservatism that may not be applicable to the floating wind industry. Cable ancillary qualification is still very focused on the API standard RP 17L2 (American Petroleum Institute, 2021).

Power-to-X platforms will include additional systems for the production and processing of hydrogen. Flexible and rigid pipe standards can effectively deal with properties of hydrogen and other products such as methanol or ammonia; only the safety of the installations need to be addressed. Owing to the nature of these fluids, it is likely that extensive explosion, collision and accidental leak load cases should be considered. Material selection will also need to be revised to prevent embrittlement or corrosion of the materials by these products (Hydrogen can embrittle materials (Findley, et al., 2022), ammonia can trigger stress corrosion cracking (National Association of Corrosion Engineers, 2021) while methanol is a solvent for a number of polymers (American Chemical Society, 2013)). In addition, current process piping standards for these products which are mostly used in an onshore environment may need to be revised to reflect the larger volumes handled in a production facility. It is likely that the first installation would be designed by a goal-based approach, and the provisions taken will gradually be reflected in rules. The subjects of the chemical compatibility of offshore construction materials as well as accident-related loads should then be studied to support the revision of standards and the first projects due to be started.

9 Life-Cycle Cost and Operational Management of Offshore Renewable Energy

Offshore renewable energy sources, such as wind farms, wave, and tidal energy, are essential to the global transition to sustainable energy systems. These technologies offer the potential for significant energy generation, contributing to energy security and reducing greenhouse gas emissions. However, the financial viability and operational management of these projects are complex, requiring a thorough understanding of life-cycle costs (LCC) and effective strategies to manage operations throughout the project's lifespan. This report explores the LCC and operational management of offshore renewable energy, focusing on offshore wind farms, wave, and tidal energy. It also provides relevant capacity and cost data to contextualize these discussions. Table 6 summarizes lifecycle cost and operational management aspects of offshore wind farms, wave energy, and tidal energy based on 2023–2024 data.

9.1 Offshore Wind Farms

As of 2023, global installed offshore wind capacity exceeds 75 GW, with substantial expansion expected, particularly in Europe, Asia, and the United States. Offshore wind farms generally have higher capacity factors (40–50%) than onshore wind farms, leading to more consistent energy generation. The costs for offshore wind have been decreasing due to technological advancements and economies of scale. According to the 2024 Global Offshore Wind report by GWEC, the average LCOE (Levelized Cost of Energy) for offshore wind in the second half of 2023 was 114 USD/MWh in the UK and 95 USD/MWh in Germany, with the most competitive projects deployed in China achieving LCOE below 50 USD/MWh. For the US, the 2023 NREL's report provided the LCOE values for the 2022 representative fixed-bottom and floating offshore wind systems (12 MW unit) are estimated at $95/MWh and $145/MWh

- **Capital Expenditure (CAPEX):** Offshore wind farms require significant upfront investment, including costs for turbines, foundations, electrical infrastructure, and installation. The CAPEX for offshore wind projects typically ranges from $2,500 to $4,000 per kW installed.
- **Operational Expenditure (OPEX):** O&M costs are a major component of LCC for offshore wind, often ranging from $80,000 to $150,000 per MW per year. These costs are influenced by the distance from shore, water depth, and the availability of maintenance infrastructure.
- **Decommissioning Costs:** Decommissioning offshore wind farms involves dismantling turbines and foundations and restoring the site. Estimated costs for decommissioning range from $200,000 to $600,000 per MW.

9.2 Wave Energy

Wave energy is an emerging technology, with global installed capacity currently less than 1 GW. The potential for wave energy is significant, particularly in regions with strong wave climates like the North Atlantic, the Pacific Northwest, and parts of Australia. The LCOE for wave energy remains high, ranging from $300 to $500/MWh, reflecting the nascent stage of the technology.

- **CAPEX:** Wave energy projects involve significant CAPEX due to the innovative nature of the technology and the harsh marine environment. CAPEX for wave energy systems can range from $5,000 to $10,000 per kW.
- **OPEX:** O&M costs for wave energy are high due to the technical challenges of maintaining devices in harsh ocean conditions. OPEX can be around $200,000 per MW per year.
- **Decommissioning Costs:** Decommissioning costs are currently difficult to estimate due to the limited number of projects that have reached end-of-life. However, these costs are expected to be high due to the need for specialized vessels and equipment.

9.3 Tidal Energy

Tidal energy, like wave energy, is at an early stage of development, with installed capacity also under 1 GW globally. However, tidal energy is more predictable than wind or wave energy due to the regularity of tidal cycles. The LCOE for tidal energy is currently between $200 and $400/MWh, with expectations of significant reductions as the technology matures.

- **CAPEX:** The CAPEX for tidal energy systems is high, often ranging from $4,000 to $8,000 per kW, due to the engineering challenges posed by underwater installations and the need for robust materials to withstand corrosive environments.
- **OPEX:** O&M costs are influenced by the accessibility of the tidal installation, with OPEX ranging from $150,000 to $250,000 per MW per year.
- **Decommissioning Costs:** Similar to wave energy, decommissioning costs for tidal energy are expected to be significant, reflecting the technical challenges of removing underwater infrastructure.

9.4 Floating PV Systems

Floating Photovoltaic (FPV) systems present a promising renewable energy solution for such applications, especially in regions with limited land availability and extensive water bodies. According to the Solar Energy Research Institute of Singapore (SERIS), as of the end of 2023, global installed FPV capacity had reached approximately 7.6 GWp, marking an increasing global adoption of this technology. Notable global projects in this field include the world's first offshore FPV farm in Leiden in 2019, the 181 MWp offshore solar project in Taiwan, and the 1 MW PV-bos project at Valencia harbor in Spain in 2024. The increase in FPV installations, along with the advancements in technology, has brought down FPV CAPEX cost from a median of 2.41 USD/Wp in 2015 to 1.05 USD/Wp in 2023

9.5 LCC Optimization Strategies

For all offshore renewable energy types, optimizing LCC involves:

- **Design for Reliability:** Integrating reliability into the design phase to minimize maintenance needs and extend operational life.
- **Predictive Maintenance:** Utilizing real-time monitoring to anticipate and address maintenance issues before they lead to costly failures.
- **Modular Design:** Adopting modular approaches to facilitate easier maintenance and upgrades, thereby reducing both OPEX and downtime.

Operational Management of Offshore Renewable Energy

Operating offshore renewable energy installations poses several challenges, including harsh environmental conditions, high maintenance costs, and logistical difficulties. Effective operational management is crucial for overcoming these challenges and ensuring the longevity and profitability of the project.

Remote Monitoring and Control

- **Real-Time Data Collection:** Advanced sensors and data acquisition systems are essential for monitoring performance, detecting anomalies, and optimizing energy output.
- **Automated Control Systems:** These systems enable remote adjustments to operational parameters in response to environmental changes, ensuring optimal performance and safety.

Maintenance Strategies

- **Preventive Maintenance:** Regularly scheduled maintenance activities are essential to prevent equipment failures and extend the life of the installation.
- **Corrective Maintenance:** Rapid response to operational issues minimizes downtime and ensures that energy production is restored quickly.
- **Condition-Based Maintenance:** Leveraging data from monitoring systems to perform maintenance only when needed, reducing unnecessary costs.

Supply Chain and Logistics

- **Logistics Planning:** Efficiently managing the supply chain and logistics, particularly for remote or hard-to-access sites, is crucial to minimizing delays and costs.
- **Inventory Management:** Maintaining a strategic inventory of spare parts and critical components ensures that repairs can be carried out promptly without extended downtimes.

Technological Innovations

- **Digital Twin Technology:** Creating digital replicas of offshore installations allows for simulation and optimization of operational strategies.
- **Autonomous Maintenance Systems:** The use of drones and autonomous underwater vehicles (AUVs) for inspection and maintenance is reducing human intervention and improving safety.
- **Integrated Energy Systems:** Integrating offshore renewable energy with other systems, such as hydrogen production or battery storage, enhances energy efficiency and provides greater flexibility in managing energy output.

The life-cycle cost and operational management of offshore renewable energy systems, including wind, wave, and tidal energy, are critical to the economic viability and long-term success of these projects. As the offshore renewable energy sector continues to grow, ongoing innovation in technology and management practices will be essential to overcoming the challenges posed by the harsh marine environment. By optimizing LCC and implementing effective operational management strategies, stakeholders can ensure that offshore renewable energy projects contribute significantly to global energy goals while maintaining financial and environmental sustainability.

Table 6: Summarizing lifecycle cost and operational management aspects of offshore wind farms, wave energy, and tidal energy based on 2023–2024 data.

Aspect	Offshore Wind Farm	Wave Energy	Tidal Energy	Floating Solar
Installed Capacity	>75 GW globally	<1 GW globally	<1 GW globally	>7 GW globally
Capacity Factor	40–50%	25–35%	25–45%	18–25%
LCOE	$50–$70/MWh	$300–$500/MWh	$200–$400/MWh	$60–$120/MWh
Capital Expenditure (CAPEX)	$2,500–$4,000 per kW	$5,000–$10,000 per kW	$4,000–$8,000 per kW	$1,000–$2,500 per kW
Operational Expenditure (OPEX)	$80,000–$150,000 per MW/year	$200,000 per MW/year	$150,000–$250,000 per MW/year	$25,000–$50,000 per MW/year
Decommissioning Costs	$200,000–$600,000 per MW	High (limited data)	High (limited data)	$50,000–$150,000/MW
Key Maintenance Strategies	Preventive, Corrective, Condition-Based	Preventive, Corrective, Condition-Based	Preventive, Corrective, Condition-Based	Preventive, Corrective, Condition-Based

(*continued*)

Table 6: (*continued*)

Aspect	Offshore Wind Farm	Wave Energy	Tidal Energy	Floating Solar
Monitoring & Control	Real-time data collection, automated control systems	Real-time data collection, automated control systems	Real-time data collection, automated control systems	Real-time monitoring, automated cleaning systems
Supply Chain & Logistics	Complex due to remote locations and large components	Complex due to specialized equipment and harsh conditions	Complex due to underwater installations	Moderately complex
Technological Innovations	Digital twins, autonomous maintenance, integrated energy systems	Autonomous maintenance systems, digital twins	Autonomous maintenance systems, digital twins	Floating array optimization, hybrid systems, anti-fouling coatings
Operational Challenges	Harsh marine conditions, accessibility, high O&M costs	Harsh marine conditions, high O&M costs, early-stage tech	Harsh marine conditions, accessibility, early-stage tech	Biofouling, water-body regulations, storm resilience
Regulatory Compliance	Established frameworks, evolving environmental regulations	Evolving regulations, less established than offshore wind	Evolving regulations, less established than offshore wind	Emerging frameworks, less established than offshore wind

10 Main Conclusions and Recommendations for Future Work

This report documents the significant progress made in offshore renewable energy technologies while highlighting key challenges that must be addressed for further advancement. Offshore wind energy, particularly bottom-fixed and floating wind turbines, continues to be the most mature sector, benefiting from ongoing innovations in turbine efficiency, structural integrity, and digital monitoring. The development of wave, tidal, and floating solar photovoltaic (FPV) systems is also gaining momentum, offering additional renewable energy sources to complement offshore wind farms.

One of the primary takeaways is the critical role of digital technologies in enhancing the performance, reliability, and economic viability of offshore renewable energy systems. Digital twins, artificial intelligence (AI), and predictive maintenance strategies are proving to be invaluable in optimizing operations, reducing unplanned downtime, and extending the lifespan of offshore installations. The integration of real-time data analytics with structural health monitoring further ensures that maintenance and operational strategies remain efficient and cost-effective.

From a structural perspective, the industry must address challenges associated with extreme environmental loading, fatigue life, and material performance. Offshore renewable structures are exposed to highly dynamic and unpredictable forces, including wind, waves, currents, and seismic activity. Innovations in structural materials, including advanced composites and corrosion-resistant alloys, are essential in ensuring the longevity and reliability of offshore installations. Additionally, improvements in foundation design, mooring systems, and hydrodynamic modeling are critical in mitigating structural fatigue and optimizing load distribution. Additional research into mechanical behavior of connections within offshore wind turbine towers is also crucial.

The ongoing development of floating renewable energy systems presents unique structural challenges, particularly in the areas of stability, dynamic response, and mooring integrity. Advanced numerical simulations, coupled with large-scale physical testing, are critical in validating design methodologies and ensuring the successful development and deployment of floating structures in deep-water environments. Hybrid solutions that integrate wave and wind energy must also be carefully engineered to balance load interactions and maximize efficiency without compromising structural integrity.

The committee believes that future investigation of digital twins integrated with computational fluid dynamics (CFD) and machine learning is key. Such integration enables predictive maintenance extending service life and wider financial viability, and further optimisation and improvements of structural health monitoring will reduce the fundamental differences between numerical models and operation reality.

Economic and logistical considerations play a pivotal role in the viability of offshore renewable energy projects. These topics extend beyond traditional marine structural design but are addressed here due to their critical impact on the practical implementation and economic feasibility of offshore renewable energy systems. Future work should explore these interdisciplinary aspects further, and their impact on the unique design features of offshore renewable energy devices.

Looking ahead, the offshore renewable energy industry must focus on expanding research and development efforts, scaling up successful pilot projects, and fostering international cooperation to accelerate the transition to a cleaner and more resilient global energy system. By leveraging technological innovations, improving economic strategies, and refining regulatory frameworks, offshore renewable energy can become a cornerstone of sustainable energy production for future generations.

References

1C Offshore: Market Overview Report: Q2 2024, Global Offshore Wind Farms. (2024)
2C Offshore Limited: Subsea Cable Insight Report: Analysis of Subsea Cable Failures. (2020)
Abdulkadir, H., Abdelkhalik, O.: Optimization of heterogeneous arrays of wave energy converters. Ocean Eng. (2023) [Preprint]
Aboutalebi, P., et al.: Control technique for hybrid floating offshore wind turbines using oscillating water columns for generated power fluctuation reduction. J. Comput. Design Eng. **10** (2023) [Preprint]
Aboutalebi, P., et al.: Hydrostatic stability and hydrodynamics of a floating wind turbine platform integrated with oscillating water columns: a design study. Renew. Energy. **221**(November 2023), 119824 (2024). https://doi.org/10.1016/j.renene.2023.119824
ABS: Guide for Building and Classing Floating Offshore Wind Turbines. (2024)
Ågotnes, T., Eik, V.: FMEA of Catenary Mooring Systems for Floating Offshore Wind Turbines-Insights from FPSOs. Bachelor's thesis in Ocean Technology, Western Norway University of Applied Sciences (2024)

Ahmad, I., et al.: Fuzzy logic control of an artificial neural network-based floating offshore wind turbine model integrated with four oscillating water columns. Ocean Eng. **269** (2023). https://doi.org/10.1016/j.oceaneng.2022.113578

Ahmed, A., et al.: Design of an S-shaped point-absorber wave energy converter with a nonlinear PTO to power the satellite-respondent buoys in the East China Sea. Ocean Eng. (2023) [Preprint]

Al-Subaihawi, S., et al.: Coupled aero-hydro-geotech real-time hybrid simulation of offshore wind turbine monopile structures. Eng. Struct. **303** (2024). https://doi.org/10.1016/j.engstruct.2024.117463

Al-Yacouby, A.M., Halim, E.R.B.A., Liew, M.S.: Hydrodynamic analysis of floating offshore solar farms subjected to regular waves. In: Lecture Notes in Mechanical Engineering, pp. 375–390. Springer Science and Business Media Deutschland GmbH (2020). https://doi.org/10.1007/978-981-15-5753-8_35

American Bureau of Shipping: Requirements for offshore substations and electrical service platforms (2023). http://www.eagle.org

American Chemical Society: Molecule of the week: methanol (2013). https://www.acs.org/molecule-of-the-week/archive/m/methanol.html

American Petroleum Institute: Recommended Practice for Ancillary Equipment for Flexible Pipes and Subsea Umbilicals, 2nd ed. (API RP 17L2) (2021)

Apollo and DOF: DC07 PROJECT gigawatt scale cable and mooring installation (2024). https://fowcoe.co.uk/.

Arabgolarcheh, A., Micallef, D., Benini, E.: The impact of platform motion phase differences on the power and load performance of tandem floating offshore wind turbines. Energy, **284** (2023). https://doi.org/10.1016/j.energy.2023.129271

Aresti, L., et al.: Reviewing the energy, environment, and economy prospects of Ocean Thermal Energy Conversion (OTEC) systems. Sustain. Energy Technol. Assess. Elsevier Ltd. (2023). https://doi.org/10.1016/j.seta.2023.103459

Ataei, B., et al.: Effects of structural flexibility on the dynamic responses of low-height lifting mechanism for offshore wind turbine installation. Mar. Struct. **89**, 103399 (2023). https://doi.org/10.1016/j.marstruc.2023.103399

Atallah, N.: Learnings from offshore oil and gas: key translational success factors in mooring hookup and tensioning methods for floating wind. In: Offshore Technology Conference 2024. Society of Petroleum Engineers (SPE), Houston (2024). https://doi.org/10.4043/35163-ms

Ayub, M.W., et al.: A review of power co-generation technologies from hybrid offshore wind and wave energy. Energies. **16**(1) (2023). https://doi.org/10.3390/en16010550

Bai, H., et al.: Theoretical and experimental study of the high-frequency nonlinear dynamic response of a 10 MW semi-submersible floating offshore wind turbine. Renew. Energy. **231** (2024). https://doi.org/10.1016/j.renene.2024.120952

Bakken, K.: Fatigue of Dynamic Power Cables Applied in Offshore Wind Farms. Norwegian University of Science and Technology (2020)

Bank Group, W.: Where Sun Meets Water Floating Solar Market Report (2019). www.worldbank.org

Bao, X., et al.: Turbine fault diagnosis of the oscillating water column wave energy converter based on multi-lead residual neural networks. Ocean Eng. (2023) [Preprint]

Barreto, D., Karimirad, M., Ortega, A.: Effects of simulation length and flexible foundation on long-term response extrapolation of a bottom-fixed offshore wind Turbine. J. Offshore Mech. Arctic Eng. **144**(3) (2022). https://doi.org/10.1115/1.4053030

Bei, Y., Yuan, B., Wu, Q.: Numerical simulation of wind load characteristics of floating photovoltaics. In: 2022 5th International Conference on Energy, Electrical and Power Engineering, CEEPE 2022, pp. 1104–1108. Institute of Electrical and Electronics Engineers Inc. (2022). https://doi.org/10.1109/CEEPE55110.2022.9783316

Beier, D.: Dynamic and Fatigue Analyses of Suspended Power Cables for Multiple Floating Offshore Wind Turbines. University of Stavanger, Stavanger (2023)

Beier, D., et al.: Fatigue analysis of inter-Array power cables between two floating offshore wind turbines including a simplified method to estimate stress factors. J. Mar. Sci. Eng. **11**(6), 1254 (2023). https://doi.org/10.3390/jmse11061254

Beier, D., et al.: Fatigue assessment of suspended inter-array power cables for floating offshore wind turbines. Eng. Struct. **308**, 118007 (2024). https://doi.org/10.1016/j.engstruct.2024.118007

Bergua, R., et al.: OC6 phase II: integration and verification of a new soil–structure interaction model for offshore wind design. Wind Energy. **25**(5), 793–810 (2022). https://doi.org/10.1002/we.2698

Berthelsen, P.A., et al.: Numerical simulation and comparison with experiments of a floating wind turbine using a direct forcing method. In: Proceedings of the 5th International Conference on Renewable Energies Offshore, Lisbon, Portugal (2022)

Boakye-Boateng, K., Ghorbani, A.A., Lashkari, A.H.: RiskISM: a risk assessment tool for substations. In: 2021 IEEE 9th International Conference on Smart City and Informatization (iSCI), pp. 23–30. IEEE (2021). https://doi.org/10.1109/iSCI53438.2021.00013

Bonnefoy, F., et al.: Multidimensional hybrid software-in-the-loop modeling approach for experimental analysis of a floating offshore wind turbine in wave tank experiments. Ocean Eng. **309** (2024). https://doi.org/10.1016/j.oceaneng.2024.118390

Boorsma, K., et al.: Progress in the validation of rotor aerodynamic codes using field data. Wind Energy Sci. **8**(2), 211–230 (2023). https://doi.org/10.5194/wes-8-211-2023

Borg, M.G., et al.: A numerical structural analysis of ducted, high-solidity, fibre-composite tidal turbine rotor configurations in real flow conditions. Ocean Eng. **233** (2021). https://doi.org/10.1016/j.oceaneng.2021.109087

Börü, M.E.: VIV Fatigue of Dynamic Power Cables Applied in Offshore Wind Turbines. NTNU (2021)

Bossmann, T., et al.: METIS Study S8 – The Role and Potential of Power-to-X in 2050 (2018).

Branlard, E., et al.: A digital twin solution for floating offshore wind turbines validated using a full-scale prototype. Wind Energy Sci. **9**(1), 1–24 (2024). https://doi.org/10.5194/wes-9-1-2024

Bredmose, H., et al.: Efficient calculation of deformation and FEM stress for substructures of floating wind turbines. In: Journal of Physics: Conference Series. Institute of Physics (2024). https://doi.org/10.1088/1742-6596/2875/1/012027.

Bureau Veritas: BV Guidance Note NI 603 DT R01 E, Current and tidal Turbines (2015)

Bureau Veritas: Certification of Fixed Offshore Substations for Renewable Energy Projects, Guidance Note NI682 DT R00 E (2022). https://marine-offshore.bureauveritas.com/bv-rules

Busby, M., Thethi, R., Fulton, T.: Floating offshore wind subsea balance of plant cost savings using a risk based assessment approach. In: Offshore Technology Conference 2024. Society of Petroleum Engineers (SPE) (2024). https://doi.org/10.4043/35399-ms

BV: Classification and Certification of Floating Offshore Wind Turbines NR572 (2024a)

BV: NI631 Certification Scheme for Marine Renewable Energy Technologies (2024b)

Califano, A., Berthelsen, P.A., Magalhaes Duque Da Fonseca, N.M.: Effect of body motion on the wave loads computed with CFD on the INO-WINDMOOR floater. In: Journal of Physics: Conference Series. Institute of Physics (2023). https://doi.org/10.1088/1742-6596/2626/1/012034

Cao, F., et al.: Progress of combined wind and wave energy harvesting devices and related coupling simulation techniques. J. Mar. Sci. Eng. MDPI. (2023a). https://doi.org/10.3390/jmse11010212

Cao, Y., et al.: Flow field distribution and structural strength performance evaluation of fixed offshore wind turbine based on digital twin technology. Ocean Eng. **288**, 116156 (2023b). https://doi.org/10.1016/j.oceaneng.2023.116156

Carmo, L.H.S., Simos, A.N.: On the complementarity of the slender-body and Newman's approximations for difference-frequency second-order wave loads on slender cylinders. Ocean Eng. **259** (2022). https://doi.org/10.1016/j.oceaneng.2022.111905

Carmo, L.H.S., Simos, A.N., de Mello, P.C.: A second-order slender-body approach for computing wave induced forces with application to a simplified semi-submersible FOWT model. Ocean Eng. **285** (2023). https://doi.org/10.1016/j.oceaneng.2023.115410

Caruso, D. Review and Critical Evaluation of Shore Based Fault Detection Systems for! Subsea Cables. Global Offshore Wind (2020) [Preprint].

Castorrini, A., et al.: Investigations on offshore wind turbine inflow modelling using numerical weather prediction coupled with local-scale computational fluid dynamics. Renew. Sust. Energ. Rev. **171**, 113008 (2023). https://doi.org/10.1016/j.rser.2022.113008

Catipovic, I., et al.: A review on marine applications of solar photovoltaic systems. In: Proceedings of the 15th International Symposium on Practical Design of Ships and Other Floating Structures – PRADS 2022 (2022)

Catipovic, I., et al.: Seakeeping assessment of a floating object with installed photovoltaic system. In: Sustainable Development and Innovations in Marine Technologies – Proceedings of the 19th International Congress of the International Maritime Association of the Mediterranean, IMAM 2022, pp. 217–228. CRC Press/Balkema (2023) https://doi.org/10.1201/9781003358961-28

Chen, B.Q., et al.: Enhancing reliability in floating offshore wind turbines through digital twin technology: a comprehensive review. Energies. Multidisciplinary Digital Publishing Institute (MDPI). (2024a). https://doi.org/10.3390/en17081964.

Chen, L., Yang, J., Lou, C.: Characterizing ramp events in floating offshore wind power through a fully coupled electrical-mechanical mathematical model. Renew. Energy. **221** (2024b). https://doi.org/10.1016/j.renene.2023.119803

Chen, M., et al.: Performance analysis of a floating wind–wave power generation platform based on the frequency domain model. J. Mar. Sci. Eng. **12**(2), 206 (2024c). https://doi.org/10.3390/jmse12020206

Chen, W., et al.: Seismic response of hybrid pile-bucket foundation supported offshore wind turbines located in liquefiable soils. Ocean Eng. **269** (2023). https://doi.org/10.1016/j.oceaneng.2022.113519

Chen, Z., et al.: Load reduction of semi-submersible floating wind turbines by integrating heaving-type wave energy converters with bang-bang control. Front Energy Res. **10** (2022). https://doi.org/10.3389/fenrg.2022.929307

Cheng, L., et al.: FE-assisted investigation for mechanical behaviour of connections in offshore wind turbine towers. Eng. Struct. **285**, 116039 (2023). https://doi.org/10.1016/j.engstruct.2023.116039

Cheng, Y., et al.: Performance characteristics and parametric analysis of a novel multi-purpose platform combining a moonpool-type floating breakwater and an array of wave energy converters. Appl. Energy. **292** (2021). https://doi.org/10.1016/j.apenergy.2021.116888

Cheng, Y., Fu, L., et al.: Experimental and numerical investigation of WEC-type floating breakwaters: a single-pontoon oscillating buoy and a dual-pontoon oscillating water column. Coast. Eng. **177** (2022a). https://doi.org/10.1016/j.coastaleng.2022.104188

Cheng, Y., Du, W., et al.: Hydrodynamic characteristics of a hybrid oscillating water column-oscillating buoy wave energy converter integrated into a π-type floating breakwater. Renew. Sustain. Energy Rev. **161** (2022b). https://doi.org/10.1016/j.rser.2022.112299

Cheng, Y., Xi, C., et al.: Wave energy extraction and hydroelastic response reduction of modular floating breakwaters as array wave energy converters integrated into a very large floating structure. Appl. Energy. **306** (2022c). https://doi.org/10.1016/j.apenergy.2021.117953

Cheng, Y., Song, F., et al.: Experimental investigation of a dual-pontoon WEC-type breakwater with a hydraulic-pneumatic complementary power take-off system. Energy. **286** (2024a). https://doi.org/10.1016/j.energy.2023.129427

Cheng, Y., Du, W., et al.: Wave energy conversion by an array of oscillating water columns deployed along a long-flexible floating breakwater. Renew. Sust. Energ. Rev. **192**, 114206 (2024b). https://doi.org/10.1016/j.rser.2023.114206

Chitteth Ramachandran, R., et al.: Floating wind turbines: marine operations challenges and opportunities. Wind Energy Sci. Copernicus Publications., 903–924 (2022). https://doi.org/10.5194/wes-7-903-2022

Choi, S.M., et al.: Effects of various inlet angle of wind and wave loads on floating photovoltaic system considering stress distributions. J. Clean. Prod. **387** (2023). https://doi.org/10.1016/j.jclepro.2023.135876

Chowdhury, M.S., et al.: Current trends and prospects of tidal energy technology. Environ. Dev. Sustain. Springer Science and Business Media B.V., 8179–8194 (2021). https://doi.org/10.1007/s10668-020-01013-4

Chung, P.-H., et al.: Wind loads on offshore floating photovoltaic panels performance rating measurement of emerging PV for outdoor solar (AM1.5G/IEC 60904 & AM0) and indoor lighting (CIE/SEMI PV80) view project wind loads on offshore floating photovoltaic panels (2018). https://www.researchgate.net/publication/339301748

CIGRE (2022) Recommendations for Mechanical Testing of Submarine Cables for Dynamic Applications. CIGRE TB 862 [Preprint]

Ciuriuc, A., et al.: Digital tools for floating offshore wind turbines (FOWT): a state of the art. Energy Rep. Elsevier Ltd., 1207–1228 (2022). https://doi.org/10.1016/j.egyr.2021.12.034

ClassNK: Guidelines for Floating Offshore Wind Turbines -Classification Survey NKRE-GL-FOWT01 (2022)

Claus, R., et al.: Design, evaluation and location of an offshore photovoltaic plant to supply energy to a port. In: Proceedings of the IAHR World Congress. International Association for Hydro-Environment Engineering and Research, pp. 2482–2491 (2022). https://doi.org/10.3850/IAHR-39WC2521716X2022354

Claus, R., et al.: Structural assessment of a pontoon-type floating photovoltaic plant for the marine environment. In: Trends in Renewable Energies Offshore – Proceedings of the 5th International Conference on Renewable Energies Offshore, RENEW 2022, pp. 709–716. CRC Press/Balkema (2023). https://doi.org/10.1201/9781003360773-79

Claus, R., et al.: Experimental proof-of-concept of HelioSea: a novel marine floating photovoltaic device. Ocean Eng. **299** (2024). https://doi.org/10.1016/j.oceaneng.2024.117184

Claus, R., López, M.: A methodology to assess the dynamic response and the structural performance of floating photovoltaic systems. Solar Energy. **262** (2023). https://doi.org/10.1016/j.solener.2023.111826

Co-design to deliver Scalable Tidal Stream Energy (n.d.). https://cotide.ac.uk/. Accessed 16 Feb 2025

Coles, D., et al.: Impacts of tidal stream power on energy system security: an Isle of Wight case study. Appl. Energy. **334** (2023). https://doi.org/10.1016/j.apenergy.2023.120686

Coraddu, A., et al.: Floating offshore wind turbine mooring line sections health status nowcasting: from supervised shallow to weakly supervised deep learning. Mech. Syst. Signal Process. **216** (2024). https://doi.org/10.1016/j.ymssp.2024.111446

Dadmarzi, F.H., et al.: Comparison of Morison forces with CFD modelling for a surface piercing column of a FOWT. In: Proceedings of the ASME 2022 International Offshore Wind Technical Conference. ASME (2022)

Dadmarzi, F.H., Bachynski-Polić, E.E.: Comparison of laboratory wave generation techniques on response of a large monopile in irregular sea. J. Phys. Conf. Ser. **2362**(1), 012011 (2022). https://doi.org/10.1088/1742-6596/2362/1/012011

Dahiru, A.R., Vuokila, A., Huuhtanen, M.: Recent development in Power-to-X: part I – a review on techno-economic analysis. J. Energy Storage. Elsevier Ltd. (2022). https://doi.org/10.1016/j.est.2022.105861

D'Andrea, F., et al.: IMIA Working Group WGP 131 (23) Floating Offshore Wind: Risk Management & Insurance (2023)

Delizisis, P., et al.: Numerical investigation of the effect of Vortex Induced Vibrations (VIV) parameters on the behavior of submarine power cables: a comparison with scaled down results. In: Proceedings of the ASME 2022 41st International Conference on Ocean, Offshore and Arctic Engineering (2022)

Det Norske Veritas: DNV-SE-0163, Certification of tidal turbines and arrays (2021a)

Det Norske Veritas: DNV-ST-0164, Tidal turbine (2021b)

Dezhdar, A., et al.: A transient model for clean electricity generation using solar energy and ocean thermal energy conversion (OTEC) – case study: Karkheh dam – Southwest Iran. Energy Nexus. **9** (2023). https://doi.org/10.1016/j.nexus.2023.100176

DNV: Offshore Substations, DNV-ST-0145. DNV (2020)

DNV: DNV-OS-C401 Fabrication and Testing of Offshore Structures (2023a)

DNV: DNV-RU-OU-0512 Floating Wind Installations (2023b)

DNV: DNV-RP-C203 Fatigue Design of Offshore Steel Structures (2024a)

DNV: DNV-RU-OU-0101 Offshore Drilling and Support Units 2024b

DNV: DNV-RU-OU-0104 Self-Elevating Units, Including Wind Turbine Installation Units and Liftboats (2024c)

DNV: DNV-RU-OU-0300 Fleet in Service (2024d).

DNV: DNV-SE-0422 Certification of Floating Wind Turbines (2024e)

DNV: DNV-ST-0437 Loads and Site Conditions for Wind Turbines (2024f)

DNV: Sesam for Floating Wind – Time Domain Workflows. DNV Sesam release notes 2024-06-11 (2024g)

DNV GL: Recommended Practice Design, Development and Operation of Floating Solar Photovoltaic Systems (2021)

Dogru, S., Yilmaz, O.: Extensive design and aerodynamic performance investigation of diffuser augmented wind turbine (DAWT) guided by generalized actuator disc theory. Renew. Sust. Energ. Rev. **192**, 114212 (2024). https://doi.org/10.1016/j.rser.2023.114212

Domingos, D.F., Wellens, P., van Wingerden, J.-W.: Frequency-domain framework for floating installation of wind-turbine towers. Ocean Eng. **297**, 116952 (2024). https://doi.org/10.1016/j.oceaneng.2024.116952

Dong, J., Viré, A.: The aerodynamics of floating offshore wind turbines in different working states during surge motion. Renew. Energy. **195**, 1125–1136 (2022). https://doi.org/10.1016/j.renene.2022.06.016

Dong, X., et al.: A state-of-the-art review … wind-wave energy converter. Prog. Energy. (2022) [Preprint]

Edwards, E.C., et al.: Evolution of floating offshore wind platforms: a review of at-sea devices. Renew. Sustain. Energy Rev. Elsevier Ltd. (2023). https://doi.org/10.1016/j.rser.2023.113416

Edwards, E.C., et al.: Trends in floating offshore wind platforms: a review of early-stage devices. Renew. Sustain. Energy Rev. Elsevier Ltd. (2024). https://doi.org/10.1016/j.rser.2023.114271

Elrick, P., Venugopal, V.: Vortex-induced vibrations of dynamic power cable for floating wind turbines. In: Proceedings of the ASME 2023 42nd International Conference on Ocean, Offshore and Arctic Engineering (2023)

Esteban, M.D., López-Gutiérrez, J.-S., Negro, V.: Classification of wave energy converters. Recent Adv. Petrochem. Sci. (2017) [Preprint]

Faedo, N., et al.: Nonlinear moment-based optimal control of wave energy converters with non-ideal power take-off systems. In: The Proceedings of 42nd International Conference on Ocean, Offshore & Arctic Engineering (2022)

Faraggiana, E., et al.: An efficient optimisation tool for floating offshore wind support structures. Energy Rep. **8**, 9104–9118 (2022). https://doi.org/10.1016/j.egyr.2022.07.036

Feltes, J., et al.: Twixt land and sea: cost-effective grid integration of offshore wind plants. IEEE Power Energy Magazine (March 2012)

Feng, Q.D., et al.: Failure analysis of floating offshore wind turbines based on a fuzzy failure mode and effect analysis model. Qual. Reliab. Eng. Int. **40**(5), 2159–2177 (2024a). https://doi.org/10.1002/qre.3505

Feng, X., et al.: Indirect load measurement method and experimental verification of floating offshore wind turbine. Ocean Eng. **3** (2024b). https://doi.org/10.1016/j.oceaneng.2024.117734

Ferri, G., Marino, E.: Site-specific optimizations of a 10 MW floating offshore wind turbine for the Mediterranean Sea. Renew. Energy. **202**, 921–941 (2023). https://doi.org/10.1016/j.renene.2022.11.116

Festa, O., Gourvenec, S., Sobey, A.: Comparative analysis of load reduction device stiffness curves for floating offshore wind moorings. Ocean Eng. **298** (2024). https://doi.org/10.1016/j.oceaneng.2024.117266

Findley, K.O., Lawrence, S.K., O'Brien, M.K.: Engineering challenges associated with hydrogen embrittlement in steels. In: Caballero, F.G. (ed.) Encyclopedia of Materials: Metals and Alloys. Elsevier, Amsterdam (2022)

Fitzgerald, B., et al.: Enhancing the reliability of floating offshore wind turbine towers subjected to misaligned wind-wave loading using tuned mass damper inerters (TMDIs). Renew. Energy. **211**, 522–538 (2023). https://doi.org/10.1016/j.renene.2023.04.097

Flotation Energy: Breakthrough success for Green Volt floating windfarm in renewable power auction. Flotation Energy News (2024). https://flotationenergy.com/breakthrough-success-for-green-volt-floating-windfarm-in-renewable-power-auction/. Accessed 5 Dec 2024

François, M., Davies, P.: An empirical model to predict elongation of polyamide mooring lines. Ocean Eng. **289**, 116154 (2023). https://doi.org/10.1016/j.oceaneng.2023.116154

Friel, D., et al.: A review of floating photovoltaic design concepts and installed variations (2019).

Friel, D., et al.: Hydrodynamic investigation of design parameters for a cylindrical type floating solar system. In: Developments in Renewable Energies Offshore Proceedings the 4th International Conference on Renewable Energies Offshore, RENEW 2020. CRC Press (2020). https://doi.org/10.1201/9781003134572

Fu, S., et al.: Study on aerodynamic performance and wake characteristics of a floating offshore wind turbine under pitch motion. Renew. Energy. **205**, 317–325 (2023). https://doi.org/10.1016/j.renene.2023.01.040

Gaidai, O., et al.: Gaidai-Xing reliability method validation for 10-MW floating wind turbines. Sci. Rep. **13**(1) (2023). https://doi.org/10.1038/s41598-023-33699-7

Gao, L., et al.: Numerical analysis of the offshore wind turbine pre-mating process using a low-height lifting system for a nonconventional installation vessel. Ocean Eng. **286**, 115555 (2023a). https://doi.org/10.1016/j.oceaneng.2023.115555

Gao, Y., et al.: Hydrodynamic analysis of tidal current turbine under water-sediment conditions. J. Mar. Sci. Eng. **10**(4) (2022). https://doi.org/10.3390/jmse10040515

Gao, Z., et al.: Time-domain floater stress analysis for a floating wind turbine. J. Ocean Eng. Sci. **8**(4), 435–445 (2023b). https://doi.org/10.1016/j.joes.2023.08.001

Garanovic, A.: Sabella formalizes partnership for tide-powered green hydrogen production. Offshore-energy.biz (2022a). https://www.offshore-energy.biz/sabella-formalizes-partnership-for-tide-powered-green-hydrogen-production/. Accessed 16 Feb 2025

Garanovic, A.: UK-based marine energy company Sustainable Marine has harnessed the enormous tidal currents in Canada's Bay of Fundy, delivering the first floating in-stream tidal power to Nova Scotia's grid. Offshore-Energy.biz (2022b). https://www.offshore-energy.biz/sustainable-marine-exports-first-floating-tidal-power-to-nova-scotia-grid/#:~:text=Sustainable%20Marine%20exports%20first%20floating%20tidal%20power%20to%20Nova%20Scotia%20grid-Business%20Developments%20%26%20Projects&text=UK%2Dbased%20marine%20energy%20company,power%20to%20Nova%20Scotia's%20grid. Accessed 16 Feb 2025

Garrod, A., et al.: An assessment of floating photovoltaic systems and energy storage methods: a comprehensive review. Results Eng. **21**, 101940 (2024). https://doi.org/10.1016/j.rineng.2024.101940

Gaspar, J.F., et al.: Review of wave energy converter design for offshore hybrid platforms. In: Innovations in Renewable Energies Offshore, pp. 795–805. CRC Press, London (2024). https://doi.org/10.1201/9781003558859-87

Gato, L.M.C., Henriques, J.C.C., Carrelhas, A.A.D.: Sea trial results of the biradial and Wells turbines at Mutriku wave power plant. Energy Convers. Manag. **268** (2022). https://doi.org/10.1016/j.enconman.2022.115936

Gavriilidis, I., Huang, Y.: Finite element analysis of tidal turbine blade subjected to impact loads from sea animals†. Energies. **14**(21) (2021). https://doi.org/10.3390/en14217208

Ghafari, H.R., Ghassemi, H., Neisi, A.: Power matrix and dynamic response of the hybrid Wavestar-DeepCwind platform under different diameters and regular wave conditions. Ocean Eng. **247**(February), 110734 (2022). https://doi.org/10.1016/j.oceaneng.2022.110734

Ghigo, A., Sirigu, S.A., Carapellese, F.: Design and optimization of a point absorber for the Mediterranean Sea. In: The Proceedings of 41st International Conference on Ocean, Offshore & Arctic Engineering (2022) [Preprint]

Ghilardi, A., et al.: Integration of ocean thermal energy conversion and pumped thermal energy storage: system design, off-design and LCOS evaluation. Appl. Therm. Eng. **236** (2024). https://doi.org/10.1016/j.applthermaleng.2023.121551

Ghosh, A.: A comprehensive review of water based PV: Flotavoltaics, under water, offshore & canal top. Ocean Eng. Elsevier Ltd. (2023). https://doi.org/10.1016/j.oceaneng.2023.115044

Glennon, C., et al.: Tidal stream to mainstream: mechanical testing of composite tidal stream blades to de-risk operational design life. J. Ocean Eng. Mar. Energy. **8**(2), 163–182 (2022). https://doi.org/10.1007/s40722-022-00223-4

Global Underwater Hub: Subsea Cable Insurance Conference & Workshop Post Event Report (2023)

Gorostidi, N., Pardo, D., Nava, V.: Diagnosis of the health status of mooring systems for floating offshore wind turbines using autoencoders. Ocean Eng. **287** (2023). https://doi.org/10.1016/j.oceaneng.2023.115862

Grattan, K., Jeffrey, H.: Delivering net zero: forecasting wave and tidal stream deployment in UK waters by 2050 the Policy and Innovation Group (P&IG) supergen offshore renewable energy hub (2023). http://www.policyandinnovationedinburgh.org/

Groenemans, H., et al.: Techno-economic analysis of off-shore wind PEM water electrolysis for H2 production (2022)

Gubesch, E., et al.: Dynamic response of a taut-moored floating oscillating water column wave energy converter in extreme waves. In: The Proceedings of 42nd International Conference on Ocean, Offshore & Arctic Engineering (2023)

Gudmestad, O.T., Schnepf, A.: Design basis considerations for the design of floating offshore wind turbines. Sustain. Mar. Struct. **5**(2), 26–34 (2023). https://doi.org/10.36956/sms.v5i2.913

Guerrero-Fernández, J.L., Tom, N.M., Rossiter, J.A.: Nonlinear model predictive control based on real-time iteration scheme for wave energy converters using WEC-Sim. In: The Proceedings of 41st International Conference on Ocean, Offshore & Arctic Engineering (2022)

Guignier, L., et al.: Design of dynamic high voltage cables for floating substation. In: Proceedings of the ASME 2020 39th International Conference on Ocean, Offshore and Arctic Engineering (2020)

Guo, B., et al.: Hydrodynamics of an oscillating water column WEC – breakwater integrated system with a pitching front-wall. Renew. Energy. **176**, 67–80 (2021). https://doi.org/10.1016/j.renene.2021.05.056

Guo, J., et al.: Tower loads characteristics of a semi-submersible floating wind turbine: an experimental study. Ocean Eng. **311** (2024). https://doi.org/10.1016/j.oceaneng.2024.118967

Guo, S., et al.: Structural responses of large-sized floating wind turbine with consideration of mooring-line dynamics based on coupled FEM simulations. Ships Offshore Struct. **17**(6), 1345–1359 (2022). https://doi.org/10.1080/17445302.2021.1918942

Ha, K., Jeong, J.-H.: Stress states investigation of adhesive bonded joint between spar cap and shear webs of a large wind turbine rotor blade. J. Mech. Sci. Technol. **35**(5), 2107–2114 (2021). https://doi.org/10.1007/s12206-021-0426-2

Hallak, T.S., et al.: Numerical and experimental analysis of a hybrid wind-wave offshore floating platform's hull. In: Proceedings of the International Conference on Offshore Mechanics and Arctic Engineering – OMAE (2018). https://doi.org/10.1115/OMAE2018-78744

Hallak, T.S., Gaspar, J.F., Guedes Soares, C.: Dynamic simulation of wave point absorbers connected to a central floating platform. In: Proceedings of the European Wave and Tidal Energy Conference, p. 15 (2023). https://doi.org/10.36688/ewtec-2023-496

Han, C., et al.: An efficient fatigue assessment model of offshore wind turbine using a half coupling analysis. Ocean Eng. **263**, 112318 (2022). https://doi.org/10.1016/j.oceaneng.2022.112318

Han, F., et al.: Generic fully coupled framework for reliability assessment of offshore wind turbines under typical limit states. Eng. Struct. **304**, 117692 (2024). https://doi.org/10.1016/j.engstruct.2024.117692

Han, M.M., Wang, C.M.: Coupled analytical-numerical approach for determining hydrodynamic responses of breakwater with multiple OWCs. Mar. Struct. **80** (2021). https://doi.org/10.1016/j.marstruc.2021.103097

Hansen, T.D.: Utsira Nord Floating Wind Farm-Optimalisation of Marine Operations Related to Inter-Array Cable Installation. University of Stavanger, Stavanger (2023)

Haq Duggal, R.: Transmission & Cables Outlook: Offshore Transmission & Cables Intelligence (2024)

Harvey, M., et al.: Floating Wind JIP Phase V Summary Report (2024)

He, K., Ye, J.: Seismic dynamics of offshore wind turbine-seabed foundation: insights from a numerical study. Renew. Energy. **205**, 200–221 (2023). https://doi.org/10.1016/j.renene.2023.01.076

He, Z., et al.: Optimization of a wave energy converter square array based on the differential evolution algorithm. Ocean Eng. (2022) [Preprint]

Hegazy, A., et al.: The potential of wave feedforward control for floating wind turbines: a wave tank experiment. Wind Energy Sci. **9**(8), 1669–1688 (2024). https://doi.org/10.5194/wes-9-1669-2024

Hegseth, J.M., Bachynski, E.E., Leira, B.J.: Effect of environmental modelling and inspection strategy on the optimal design of floating wind turbines. Reliab. Eng. Syst. Saf. **214** (2021). https://doi.org/10.1016/j.ress.2021.107706

Helfer, T., Ward, K., McNaught, D.: Review of Technical Assumptions and Generation Costs Floating Offshore Wind Levelised Cost of Energy Review. Prepared for Department for Energy Security and Net Zero, Leatherhead (2023). https://assets.publishing.service.gov.uk/media/655371f7019bd600149f1ffa/floating-offshore-wind-lcoe-report_.pdf/preview. Accessed 10 Sep 2024

Heng, J., et al.: Influence of adaptive controlling strategies of floating offshore wind turbine on corrosion fatigue deterioration of supporting towers. In: Ungureanu, V., et al. (eds.) Proceedings of the 4th International Conference "Coordinating Engineering for Sustainability and

Resilience" & Midterm Conference of CircularB "Implementation of Circular Economy in the Built Environment" (2024)

Hitachi ABB: Transformers for floating applications floating offshore substations and wind turbines (2021). https://publisher.hitachienergy.com/preview?DocumentID=9AKK107992A3261&LanguageCode=en&DocumentPartId=&Action=Launch. Accessed 1 Mar 2025

Høeg, C.E., Zhang, Z.: A semi-analytical hydrodynamic model for floating offshore wind turbines (FOWT) with application to a FOWT heave plate tuned mass damper. Ocean Eng. **287** (2023). https://doi.org/10.1016/j.oceaneng.2023.115756

Holcombe, A., et al.: Sensitivity of small-scale global models of dynamic power cables to model material properties. In: Innovations in Renewable Energies Offshore – Proceedings of the 6th International Conference on Renewable Energies Offshore, RENEW 2024, pp. 583–589. CRC Press/Balkema (2025). https://doi.org/10.1201/9781003558859-64

Hong, S., et al.: Floating offshore wind farm installation, challenges and opportunities: a comprehensive survey. Ocean Eng. Elsevier Ltd. (2024). https://doi.org/10.1016/j.oceaneng.2024.117793

Hoseinzadeh, S., et al.: Ocean thermal energy conversion (OTEC) system driven with solar-wind energy and thermoelectric based on thermo-economic analysis using multi-objective optimization technique. Energy Rep. **10**, 2982–3000 (2023). https://doi.org/10.1016/j.egyr.2023.09.131

Housner, S., Mulas Hernando, D.: Levelized cost of energy comparison of floating wind farms with and without shared anchors (2024). Available at: www.nrel.gov/publications

Hsu, S.-T., Su, K.-C.: Wind and wave effect on floating solar panel. In: 37th European Photovoltaic Solar Energy Conference and Exhibition (2020). http://photino.cwb.gov.tw/rdcweb/lib/beaufort.htm

Huang, S., Liu, W., Wang, K.: Shape optimization design of a heaving buoy of wave energy converter under bimodal spectral waves near Islands and Reefs. In: The Proceedings of 42nd International Conference on Ocean, Offshore & Arctic Engineering (2023a)

Huang, Y., et al.: Numerical analysis of flexible tube wave energy convertor using CFD-FEA method. In: The Proceedings of 42nd International Conference on Ocean, Offshore & Arctic Engineering (2023b)

Hung, Y.-C., Yang, R.-Y.: Experimental and numerical study on suitable configuration of dynamical power cable connected to floating substation in shallow water (2023). https://doi.org/10.2139/ssrn.4349135

Huo, F., et al.: Study on wave slamming characteristics of a typical floating wind turbine under freak waves. Ocean Eng. **269** (2023). https://doi.org/10.1016/j.oceaneng.2022.113464

IEC: BS EN IEC 61400-3-1:2019+A11:2020 Wind Energy Generation Systems – Design Requirements for Fixed Offshore Wind Turbines (2020)

IEC: BS EN IEC 61400-3-2. Wind Energy Generation Systems – Part 3-2. Design Requirements for Floating Offshore Wind Turbines (2021)

Ikhennicheu, M., et al.: Analytical method for loads determination on floating solar farms in three typical environments. Sol. Energy. **219**, 34–41 (2021). https://doi.org/10.1016/j.solener.2020.11.078

International Electrotechnical Commission: IEC TS 62600 (2019)

International Electrotechnical Commission: Wind Energy Generation Systems. IEC 61400 (2024) [Preprint]

International Energy Agency: Renewables 2023 – Analysis and Forecast to 2028 (2023)

International Standards Organisation: Petroleum and Natural Gas Industries – General Requirements for Offshore Structures. ISO 19900 (2019) [Preprint]

Jahani, K., Langlois, R.G., Afagh, F.F.: Structural dynamics of offshore wind turbines: a review. Ocean Eng. Elsevier Ltd. (2022). https://doi.org/10.1016/j.oceaneng.2022.111136

Jaya Muliawan, M., et al.: STC (Spar-Torus Combination): a combined spar-type floating wind turbine and large point absorber floating wave energy converter – promising and challenging (2012). http://proceedings.asmedigitalcollection.asme.org/pdfaccess.ashx?url=/data/conferences/asmep/75842/

Jenkins, B., Khalid, O., Twaddle, C.: Status: Public Floating Offshore Wind Centre of Excellence PR43: Application of Remote and Autonomous Systems (RAS) for FOW Construction, Installation, Operations and Maintenance Activities (2024)

Jeong, H.J., Koo, W.: Analysis of various algorithms for optimizing the wave energy converters associated with a sloped wall-type breakwater. Ocean Eng. **276** (2023). https://doi.org/10.1016/j.oceaneng.2023.114199

Ji, Q., Chen, G.: Hydrodynamic responses of a reversed L type floating breakwater integrated with a wave energy converter impacted by extreme waves. Ocean Eng. **272** (2023). https://doi.org/10.1016/j.oceaneng.2023.113898

Ji, Q., Xu, C., Jiao, C.: Numerical investigation on the hydrodynamic performance of a vertical pile-restrained reversed L type floating breakwater integrated with WEC. Ocean Eng. **238** (2021). https://doi.org/10.1016/j.oceaneng.2021.109635

Jia, W., et al.: Dynamic analysis of a 5 MW Barge-type FOWT with two-mooring failure of wind-wave misalignment scenarios. Ocean Eng. **285**, 115456 (2023). https://doi.org/10.1016/j.oceaneng.2023.115456

Jiang, F., et al.: Application of CFD on VIM of semi-submersible FOWT: a case study. In: Journal of Physics: Conference Series. Institute of Physics (2023a). https://doi.org/10.1088/1742-6596/2626/1/012041

Jiang, X., et al.: Development and characterisation of an AI-in-the-loop testing platform for floating wind turbines PART I: construction, validation, and benchmark testing. Ocean Eng. **297** (2024). https://doi.org/10.1016/j.oceaneng.2024.116968

Jiang, Z., et al.: Design and model test of a soft-connected lattice-structured floating solar photovoltaic concept for harsh offshore conditions. Mar. Struct. **90** (2023b). https://doi.org/10.1016/j.marstruc.2023.103426

Jin, C., et al.: Digital twin method for global motion and stress monitoring of a steel lazy wave riser. In: OCEANS 2022, Hampton Roads, pp. 1–7. IEEE (2022). https://doi.org/10.1109/OCEANS47191.2022.9977010

Jin, Z., et al.: A novel analytical model coupling hydrodynamic-structural-material scales for very large floating photovoltaic support structures. Ocean Eng. **275** (2023). https://doi.org/10.1016/j.oceaneng.2023.114113

Jordal, L., et al.: A presentation of a novel test rig for simulating combined axial compression and bending of subsea power cables and umbilicals during installation. In: International Ocean and Polar Engineering Conference (2023)

Jump, E., et al.: Offshore Substations: Fixed or Floating? Techno-Economic Analysis (2021)

Jung, H., et al.: Extracting energy from ocean thermal and salinity gradients to power unmanned underwater vehicles: state of the art, current limitations, and future outlook. Renew. Sustain. Energy Rev. Elsevier Ltd. (2022). https://doi.org/10.1016/j.rser.2022.112283

Kamarlouei, M., et al.: Experimental study of wave energy converter arrays adapted to a semi-submersible wind platform. Renew. Energy. **188**, 145–163 (2022). https://doi.org/10.1016/j.renene.2022.02.014

Kang, T.W., Kim, E.S., Yang-ik, H.: Effects of dynamic motion and structural response of a semi-submersible floating offshore wind turbine structure under waves generated in a hurricane environment. Int. J. Precis. Eng. Manuf. Green Technol. **9**(2), 537–556 (2022). https://doi.org/10.1007/s40684-021-00331-w

Kang, W., Lian, Z., Han, Y.: Design and hydrodynamic performance analysis of a two-module wave-resistant floating photovoltaic device. J. Phys. Conf. Ser. Institute of Physics. (2023). https://doi.org/10.1088/1742-6596/2565/1/012014

Kanotra, R., Shankar, R.: Floating solar photovoltaic mooring system design and analysis. In: Oceans Conference Record (IEEE). Institute of Electrical and Electronics Engineers Inc. (2022). https://doi.org/10.1109/OCEANSChennai45887.2022.9775352

Katsidoniotaki, E., et al.: Validation of a CFD model for wave energy system dynamics in extreme waves. Ocean Eng. (2022) [Preprint]

Katsikogiannis, G., et al.: Environmental lumping for efficient fatigue assessment of large-diameter monopile wind turbines. Mar. Struct. **77**, 102939 (2021). https://doi.org/10.1016/j.marstruc.2021.102939

Katsikogiannis, G., Haver, S.K., Bachynski-Polić, E.E.: Assessing some statistical and physical modelling uncertainties of extreme responses for monopile-based offshore wind turbines, using metocean contours. Appl. Ocean Res. **143** (2024). https://doi.org/10.1016/j.apor.2024.103880

Katsikogiannis, G., Hegseth, J.M., Bachynski-Polić, E.E.: Application of a lumping method for fatigue design of monopile-based wind turbines using fully coupled and simplified models. Appl. Ocean Res. **120**, 102998 (2022). https://doi.org/10.1016/j.apor.2021.102998

Khalid, O., et al.: Applications of robotics in floating offshore wind farm operations and maintenance: literature review and trends. Wind Energy. John Wiley and Sons Ltd., 1880–1899 (2022). https://doi.org/10.1002/we.2773

Khalid, O., et al.: Cost-benefit assessment framework for robotics-driven inspection of floating offshore wind farms. Wind Energy. **27**(2), 152–164 (2024). https://doi.org/10.1002/we.2881

Khojasteh, D., et al.: Sea level rise will change estuarine tidal energy: a review. Renew. Sustain. Energy Rev. Elsevier Ltd. (2022). https://doi.org/10.1016/j.rser.2021.111855

Kleine, V.G., et al.: The stability of wakes of floating wind turbines. Phys. Fluids. **34**(7) (2022). https://doi.org/10.1063/5.0092267

Kozmar, H., et al.: Wind load assessment in marine and offshore engineering standards. Ocean Eng. **252**, 110872 (2022). https://doi.org/10.1016/j.oceaneng.2022.110872

Kuang, J., et al.: Dynamic interactions of a cable-laying vessel with a submarine cable during its landing process. J. Mar. Sci. Eng. **10**(6), 774 (2022). https://doi.org/10.3390/jmse10060774

Langer, J., et al.: Plant siting and economic potential of ocean thermal energy conversion in Indonesia a novel GIS-based methodology. Energy. **224** (2021). https://doi.org/10.1016/j.energy.2021.120121

Langer, J., Infante Ferreira, C., Quist, J.: Is bigger always better? Designing economically feasible ocean thermal energy conversion systems using spatiotemporal resource data. Appl. Energy. **309** (2022a). https://doi.org/10.1016/j.apenergy.2021.118414

Langer, J., Quist, J., Blok, K.: Upscaling scenarios for ocean thermal energy conversion with technological learning in Indonesia and their global relevance. Renew. Sustain. Energy Rev. **158** (2022b). https://doi.org/10.1016/j.rser.2022.112086

Lapa, G.V.P., Gay Neto, A., Franzini, G.R.: Effects of blade torsion on IEA 15MW turbine rotor operation. Renew. Energy. **219**, 119546 (2023). https://doi.org/10.1016/j.renene.2023.119546

Lee, H., et al.: An efficient time domain structural assessment of a floating wind turbine structure. In: Proc. OMAE 2023-108155. ASME (2023)

Lenci, S.: Along-wind and cross-wind coupled nonlinear oscillations of wind turbine towers close to 1:1 internal resonance. Renew. Sustain. Energ. Rev. **187**, 113698 (2023). https://doi.org/10.1016/j.rser.2023.113698

Leroy, V., et al.: A weak-scatterer potential flow theory-based model for the hydroelastic analysis of offshore wind turbine substructures. Ocean Eng. **238** (2021). https://doi.org/10.1016/j.oceaneng.2021.109702

Li, C., et al.: Dynamics of a Y-shaped semi-submersible floating wind turbine: a comparison of concrete and steel support structures. Ships Offshore Struct. **17**(8), 1663–1683 (2022a). https://doi.org/10.1080/17445302.2021.1937801

Li, H., et al.: Effect of floater flexibility on global dynamic responses of a 15-MW semi-submersible floating wind turbine. Ocean Eng. **286** (2023a). https://doi.org/10.1016/j.oceaneng.2023.115584

Li, H., Bachynski-Polić, E.E.: Analysis of difference-frequency wave loads and quadratic transfer functions on a restrained semi-submersible floating wind turbine. Ocean Eng. **232** (2021). https://doi.org/10.1016/j.oceaneng.2021.109165

Li, J., et al.: Concept design and floating installation method study of multi-bucket foundation floating platform for offshore wind turbines. Mar. Struct. **93** (2024a). https://doi.org/10.1016/j.marstruc.2023.103541

Li, M., et al.: Hydrodynamic performance and optimization of a pneumatic type spar buoy wave energy converter. Ocean Eng. (2022b) [Preprint]

Li, X., et al.: Coupled CFD-MBD numerical modeling of a mechanically coupled WEC array. Ocean Eng. (2022c) [Preprint]

Li, Y., et al.: A nonlinear wake model of a wind turbine considering the yaw wake steering. J. Oceanol. Limnol. **42**(3), 715–727 (2024b). https://doi.org/10.1007/s00343-023-3040-6

Li, Y., et al.: Advancements and challenges in power cable laying. Energies. **17**(12), 2905 (2024c). https://doi.org/10.3390/en17122905

Li, Z., et al.: Unsteady aeroelastic performance analysis for large-scale megawatt wind turbines based on a novel aeroelastic coupling model. Renew. Energy. **218**, 119370 (2023b). https://doi.org/10.1016/j.renene.2023.119370

Liang, C., Yuan, Y., Yu, X.: Applying new high damping viscoelastic dampers to mitigate the structural vibrations of monopile-supported offshore wind turbine. Ocean Eng. **295**, 116910 (2024a). https://doi.org/10.1016/j.oceaneng.2024.116910

Liang, G., et al.: Experimental investigation of two shared mooring configurations for a dual-spar floating offshore wind farm in irregular waves. Mar. Struct. **95** (2024b). https://doi.org/10.1016/j.marstruc.2024.103579

Liang, G., Jiang, Z., Merz, K.: Dynamic analysis of a dual-spar floating offshore wind farm with shared moorings in extreme environmental conditions. Mar. Struct. **90**, 103441 (2023). https://doi.org/10.1016/j.marstruc.2023.103441

Lim, H.-J., et al.: Time domain structural analysis and digital twin application for floating offshore wind turbine. In: Proc. OMAE 2023-105074. ASME (2023)

Liu, D.P., et al.: On long-term fatigue damage estimation for a floating offshore wind turbine using a surrogate model. Renew. Energy. **225** (2024a). https://doi.org/10.1016/j.renene.2024.120238

Liu, G., et al.: Global dynamics of a hang off power cable immersed in the flow near a fixed offshore wind turbine. In: Proceedings of the ASME 2024 43nd International Conference on Ocean, Offshore and Arctic Engineering (2024b)

Liu, J., et al.: Comparative study on dynamic responses of integrated installation process of a 5-MW and a 15-MW offshore wind turbine considering a pre-installed foundation. Ocean Eng. **299**, 117399 (2024c). https://doi.org/10.1016/j.oceaneng.2024.117399

Lloyd's Register: Guidance Notes for Offshore Wind Farm Project Certification, LR-GN-007. Lloyd's Register (2024a) [Preprint]

Lloyd's Register: Rules and Regulations for the Classification of Offshore Units, LR-RU-003. Lloyd's Register (2024b) [Preprint]

Lopez-Olocco, T., et al.: Experimental comparison of a dual-spar floating wind farm with shared mooring against a single floating wind turbine under wave conditions. Eng. Struct. **292** (2023). https://doi.org/10.1016/j.engstruct.2023.116475

Lotfizadeh, O.: Life Cycle Assessment of Offshore Wind Farms-A Comparative Study of Floating Vs. Fixed Offshore Wind Turbines. University of South-Eastern Norway (2024). www.usn.no

Lozon, E., Hall, M.: Coupled loads analysis of a novel shared-mooring floating wind farm. Appl. Energy. **332**, 120513 (2023). https://doi.org/10.1016/j.apenergy.2022.120513

LR: LR-RP-003 Recommended Practice for Floating Offshore Wind Turbine Support Structures (2024a)

LR: LR-RU-003 Rules for the Classification of Offshore Units (2024b)

Lu, D., Wang, W., Li, X.: Experimental study of structural vibration control of 10-MW jacket offshore wind turbines using tuned mass damper under wind and wave loads. Ocean Eng. **288** (2023). https://doi.org/10.1016/j.oceaneng.2023.116015

Luo, M., et al.: Bidirectional vibration control for fully-coupled floating offshore wind turbines. Ocean Eng. **292** (2024). https://doi.org/10.1016/j.oceaneng.2023.116523

Ma, C., Ban, J., Zi, G.: Comparative study on the dynamic responses of monopile and jacket-supported offshore wind turbines considering the pile-soil interaction in transitional waters. Ocean Eng. **292** (2024). https://doi.org/10.1016/j.oceaneng.2023.116564

Ma, C., Iijima, K., Oka, M.: Nonlinear waves in a floating thin elastic plate, predicted by a coupled SPH and FEM simulation and by an analytical solution. Ocean Eng. **204** (2020). https://doi.org/10.1016/j.oceaneng.2020.107243

MacDonald, M., Moran, P., Kelleher, C.: Functional specification: offshore substation general requirements. EirGrid. (2018) [Preprint]

Machado, M.R., Dutkiewicz, M., Colherinhas, G.B.: Metamaterial-based vibration control for offshore wind turbines operating under multiple hazard excitation forces. Renew. Energy. **223**, 120056 (2024). https://doi.org/10.1016/j.renene.2024.120056

Maktabi, M., Rusu, E.: A review of perspectives on developing floating wind farms. Inventions. Multidisciplinary Digital Publishing Institute (MDPI). (2024). https://doi.org/10.3390/inventions9020024

Marcollo, H., Efthimiou, L.: Floating Offshore Wind Dynamic Cables: Overview of Design and Risks (2024).

Marley, M., Skjetne, R.: Mitigating force oscillations in a wave energy converter using control barrier functions. In: The Proceedings of 41st International Conference on Ocean, Offshore & Arctic Engineering (2023)

Martinez, A., Iglesias, G.: Levelized cost of energy to evaluate the economic viability of floating offshore wind in the European Atlantic and Mediterranean. e-Prime Adv. Electr. Eng. Electron. Energy. **8** (2024). https://doi.org/10.1016/j.prime.2024.100562

Martinez-Puente, E., et al.: Benchmarking of spectral methods for fatigue assessment of mooring systems and dynamic cables in offshore renewable energy technologies. Ocean Eng. **308**, 118311 (2024). https://doi.org/10.1016/j.oceaneng.2024.118311

Matsuda, Y., et al.: Net power generation control of otec plant using rankine cycle with seawater pumps by warm seawater flow rate regulation. In: IFAC-PapersOnLine, pp. 2275–2280. Elsevier B.V (2023). https://doi.org/10.1016/j.ifacol.2023.10.1193

Mayon, R., et al.: Experimental investigation on a novel and hyper-efficient oscillating water column wave energy converter coupled with a parabolic breakwater. Coast. Eng. **185** (2023). https://doi.org/10.1016/j.coastaleng.2023.104360

McIlvenny, J., et al.: 'Comparison of dense optical flow and PIV techniques for mapping surface current flow in tidal stream energy sites. Int. J. Energy Environ. Eng. **14**(3), 273–285 (2023). https://doi.org/10.1007/s40095-022-00519-z

McMorland, J., et al.: A review of operations and maintenance modelling with considerations for novel wind turbine concepts. Renew. Sustain. Energy Rev. Elsevier Ltd. (2022). https://doi.org/10.1016/j.rser.2022.112581

Melani, P.F., et al.: How to extract the angle attack on airfoils in cycloidal motion from a flow field solved with computational fluid dynamics? Development and verification of a robust computational procedure. Energy Convers. Manag. **223** (2020). https://doi.org/10.1016/j.enconman.2020.113284

Mercier, P., Guillou, S.S.: Spatial and temporal variations of the flow characteristics at a tidal stream power site: a high-resolution numerical study. Energy Convers. Manag. **269** (2022). https://doi.org/10.1016/j.enconman.2022.116123

Meyer, J., et al.: On the mooring methodology of heaving point absorber arrays. Ocean Eng. (2023) [Preprint]

Michailides, C., Loukogeorgaki, E., Angelides, D.C.: Response analysis and optimum configuration of a modular floating structure with flexible connectors. Appl. Ocean Res. **43**, 112–130 (2013). https://doi.org/10.1016/j.apor.2013.07.007

Moideen, R., Venugopal, V., Chaplin, J.: VIV of a lazy wave dynamic power cable under various currents and propagations directions – analysis of strain. In: Proceedings of the ASME 2024 43nd International Conference on Ocean, Offshore and Arctic Engineering (2024)

Momber, A.W., et al.: A digital twin concept for the prescriptive maintenance of protective coating systems on wind turbine structures. Wind Eng. **46**(3), 949–971 (2022). https://doi.org/10.1177/0309524X211060550

Moreno, E., et al.: Effect of geometrical irregularities on fatigue of lead sheathing for submarine high voltage power cable applications. Int. J. Fatigue. **151**, 106399 (2021). https://doi.org/10.1016/j.ijfatigue.2021.106399

Mousavi, Z., et al.: A digital twin-based framework for damage detection of a floating wind turbine structure under various loading conditions based on deep learning approach. Ocean Eng. **292** (2024). https://doi.org/10.1016/j.oceaneng.2023.116563

Myrtvedt, M.H., Nybø, A., Nielsen, F.G.: The dynamic response of offshore wind turbines and their sensitivity to wind field models. J. Phys. Conf. Ser. **1669**(1), 012013 (2020). https://doi.org/10.1088/1742-6596/1669/1/012013

M'Zoughi, F., et al.: Fuzzy airflow-based active structural control of integrated oscillating water columns for the enhancement of floating offshore wind turbine stabilization. Int. J. Energy Res. **2023** (2023). https://doi.org/10.1155/2023/4938451

National Association of Corrosion Engineers: Guidelines for Maintaining Integrity of Equipment in Anhydrous Ammonia Storage and Handling. NACE TR5A192 (2021) [Preprint]

Ngo, D.V., Kim, D.H.: Seismic responses of different types of offshore wind turbine support structures. Ocean Eng. **297** (2024). https://doi.org/10.1016/j.oceaneng.2024.117108

Nicholls-Lee, R., et al.: Non-destructive examination (NDE) methods for dynamic subsea cables for offshore renewable energy. Prog. Energy. **4**(4) (2022). https://doi.org/10.1088/2516-1083/ac8ccb

Nicholls-Lee, R., Thies, P.R., Johanning, L.: Coupled modelling for dynamic submarine power cables: interface sensitivity analysis of global response and local structural engineering models. (2021). http://hdl.handle.net/10871/127469

Niosi, F., et al.: Experimental validation of Orcaflex-based numerical models for the PEWEC device. Ocean Eng. (2023) [Preprint]

NKT: Dynamic cables – forming the link to the offshore industry (2024). https://www.nkt.com/products-solutions/medium-voltage/dynamic-cables

Nybø, A., Nielsen, F.G., Godvik, M.: Quasi-static response of a bottom-fixed wind turbine subject to various incident wind fields. Wind Energy. **24**(12), 1482–1500 (2021). https://doi.org/10.1002/we.2642

Ocean Energy Systems: Wave Energy Developments Highlights 2023 (2023). Available at: www.formasdopossivel.com

Okkerstrom, L.: Sensitivity Analysis of Peak Tension Loads on Subsea Power Cables During Laying Operations. Western Norway University of Applied Sciences (2021)

Oliveira-Pinto, S., Stokkermans, J.: Assessment of the potential of different floating solar technologies – overview and analysis of different case studies. Energy Convers. Manag. **211** (2020). https://doi.org/10.1016/j.enconman.2020.112747

Otter, A., et al.: A review of modelling techniques for floating offshore wind turbines. Wind Energy. John Wiley and Sons Ltd., 831–857 (2022). https://doi.org/10.1002/we.2701

Papi, F., et al.: A code-to-code comparison for floating offshore wind turbine simulation in realistic environmental conditions: quantifying the impact of modeling fidelity on different substructure concepts. Wind Energy Sci. Discuss. (2023) [Preprint].). https://doi.org/10.5194/wes-2023-107

Parsa, K., Kim, M., Williams, N.: Accurate WEC power estimation for multi-modal wave spectra. In: The Proceedings of 41st International Conference on Ocean, Offshore & Arctic Engineering (2022)

Passano, E., et al.: Simulation of VIM of an offshore floating wind turbine. In: Proc. OMAE 2022, Paper 79006. ASME (2022)

Patryniak, K., Collu, M., Coraddu, A.: Multidisciplinary design analysis and optimisation frameworks for floating offshore wind turbines: state of the art. Ocean Eng. Elsevier Ltd. (2022). https://doi.org/10.1016/j.oceaneng.2022.111002

Peng, W., et al.: Effect of varying PTO on a triple floater wave energy converter-breakwater hybrid system: an experimental study. Renew. Energy. **224** (2024). https://doi.org/10.1016/j.renene.2024.120100

Pezeshki, H., et al.: Gyroscopic effects of the spinning rotor-blades assembly on dynamic response of offshore wind turbines. J. Wind Eng. Ind. Aerodyn. **247**, 105698 (2024). https://doi.org/10.1016/j.jweia.2024.105698

Pimenta, F., et al.: Predictive model for fatigue evaluation of floating wind turbines validated with experimental data. Renew. Energy. **223** (2024). https://doi.org/10.1016/j.renene.2024.119981

Poon, C., et al.: Fretting wear and fatigue in submarine power cable conductors for floating offshore wind energy. Tribol. Int. **186**, 108598 (2023). https://doi.org/10.1016/j.triboint.2023.108598

Poon, C., et al.: Three-dimensional representative modelling for fretting wear and fatigue of submarine power cable conductors. Int. J. Fatigue. **184**, 108302 (2024). https://doi.org/10.1016/j.ijfatigue.2024.108302

Prakash, R., et al.: A numerical study on float design for wave energy converter. Ocean Eng. (2022) [Preprint]

Pustina, L., Biral, F., Serafini, J.: A novel economic nonlinear model predictive controller for power maximisation on wind turbines. Renew. Sust. Energ. Rev. **170**, 112964 (2022). https://doi.org/10.1016/j.rser.2022.112964

Rahimi, A., et al.: Dimensional optimization of a two-body wave energy converter under irregular waves for the strait of hormuz. Ocean Eng. (2022) [Preprint]

Ram, G., et al.: Hydrodynamic performance of a hybrid system of a floating oscillating water column and a breakwater. Ocean Eng. **264** (2022). https://doi.org/10.1016/j.oceaneng.2022.112463

Ramanan, C.J., et al.: Towards sustainable power generation: recent advancements in floating photovoltaic technologies. Renew. Sustain. Energy Rev. Elsevier Ltd. (2024). https://doi.org/10.1016/j.rser.2024.114322

Ramos-García, N., et al.: Investigation of the floating IEA wind 15-MW RWT using vortex methods part II: wake impact on downstream turbines under turbulent inflow. Wind Energy. **25**(8), 1434–1463 (2022). https://doi.org/10.1002/we.2738

Ramzanpoor, I., Nuernberg, M., Tao, L.: Benchmarking study of 10 MW TLB floating offshore wind turbine. J. Ocean Eng. Mar. Energy. **10**(1), 1–34 (2024). https://doi.org/10.1007/s40722-023-00295-w

Ransley, E.J., et al.: Real-time hybrid testing of a floating offshore wind turbine using a surrogate-based aerodynamic emulator. ASME Open J. Eng. **2** (2023). https://doi.org/10.1115/1.4056963

Ren, Y., et al.: Experimental and numerical investigation on the deformation behaviors of large diameter steel tubes under concentrated lateral impact loads. Int. J. Impact Eng. **180**, 104696 (2023). https://doi.org/10.1016/j.ijimpeng.2023.104696

Ren, Y., et al.: Experimental study of tendon failure analysis for a TLP floating offshore wind turbine. Appl. Energy. **358** (2024). https://doi.org/10.1016/j.apenergy.2024.122633

Ringsberg, J.W., et al.: Characterization of the mechanical properties of low stiffness marine power cables through tension, bending, torsion, and fatigue testing. J. Mar. Sci. Eng. **11**(9), 1791 (2023). https://doi.org/10.3390/jmse11091791

Ringwood, J.V.: Marine renewable energy devices and their control: an overview. In: IFAC-PapersOnLine, pp. 136–141. Elsevier B.V. (2022). https://doi.org/10.1016/j.ifacol.2022.10.421

Rivera-Arreba, I., et al.: Comparison of the dynamic wake meandering model against large eddy simulation for horizontal and vertical steering of wind turbine wakes. Renew. Energy. **221**, 119807 (2024). https://doi.org/10.1016/j.renene.2023.119807

Robak, M., Raczkowski, R.M.: Substations for offshore wind farms: a review from the perspective of the needs of the Polish wind energy sector. Bull. Pol. Acad. Sci. Tech. Sci. **66** (2018)

Russell, A.J., et al.: LIDAR-assisted feedforward individual pitch control of a 15 MW floating offshore wind turbine. Wind Energy. **27**(4), 341–362 (2024). https://doi.org/10.1002/we.2891

Saetren Nornes, I., et al.: Structural health monitoring of the mooring system for a dual-spar floating wind farm. Master's in Renewable Energy. University of Agder (2024)

Sarmiento, J., et al.: Experimental modelling of a multi-use floating platform for wave and wind energy harvesting. Ocean Eng. **173**, 761–773 (2019). https://doi.org/10.1016/j.oceaneng.2018.12.046

Schreier, S., Jacobi, G.: Measuring hydroelastic deformation of very flexible floating structures. In: Proceedings of the Second World Conference on Floating Solutions, Rotterdam, pp. 347–371 (2022). https://doi.org/10.1007/978-981-16-2256-4_21

Schulz, C.W., et al.: Wind turbine rotors in surge motion: new insights into unsteady aerodynamics of floating offshore wind turbines (FOWTs) from experiments and simulations. Wind Energy Sci. **9**(3), 665–695 (2024). https://doi.org/10.5194/wes-9-665-2024

Shahabi-Nejad, M., Nikseresht, A.H.: A comprehensive investigation of a hybrid wave energy converter including oscillating water column and horizontal floating cylinder. Energy. **243** (2022). https://doi.org/10.1016/j.energy.2021.122763

Shahroozi, Z., Göteman, M., Engström, J.: A neural network approach to minimize line forces in the survivability of the point-absorber wave energy converters. In: The Proceedings of 42nd International Conference on Ocean, Offshore & Arctic Engineering (2023)

Shao, X., et al.: FSI simulations and analyses of a non-resonant buoyant wave energy converter. In: The Proceedings of 42nd International Conference on Ocean, Offshore & Arctic Engineering (2023)

Sharma, S., Nava, V.: Condition monitoring of mooring systems for floating offshore wind turbines using convolutional neural network framework coupled with autoregressive coefficients. Ocean Eng. **302** (2024). https://doi.org/10.13039/501100011033

Shi, W., et al.: Dynamic load effects and power performance of an integrated wind–wave energy system utilizing an optimum torus wave energy converter. J. Mar. Sci. Eng. **10**(12) (2022). https://doi.org/10.3390/jmse10121985

Shi, Y., et al.: Hydroelastic analysis of offshore floating photovoltaic based on frequency-domain model. Ocean Eng. **289** (2023). https://doi.org/10.1016/j.oceaneng.2023.116213

Shim, C., et al.: A fundamental study of VIV fatigue analysis procedure for dynamic power cables subjected to severely sheared currents. J. Soc. Naval Archit. Korea. **60**(5), 375–387 (2023). https://doi.org/10.3744/SNAK.2023.60.5.375

Si, Y., et al.: State-of-the-art review and future trends of development of tidal current energy converters in China. Renew. Sust. Energ. Rev. **167**, 112720 (2022). https://doi.org/10.1016/j.rser.2022.112720

da Silva, L.S.P., et al.: Dynamics of hybrid offshore renewable energy platforms: heaving point absorbers connected to a semi-submersible floating offshore wind turbine. Renew. Energy. **199**, 1424–1439 (2022). https://doi.org/10.1016/j.renene.2022.09.014

Sim, I.-H.: Wind load estimation of a 10 MW floating offshore wind turbine during transportation and installation by wind tunnel tests. J. Wind Energy. **15** (2023)

Singlitico, A., Østergaard, J., Chatzivasileiadis, S.: Onshore, offshore or in-turbine electrolysis? Techno-economic overview of alternative integration designs for green hydrogen production into Offshore Wind Power Hubs. Renew. Sustain. Energy Trans. **1** (2021). https://doi.org/10.1016/j.rset.2021.100005

Skeie, G., et al.: Cross section analysis of dynamic cables. In: Proceedings of the Thirty-Third International Ocean and Polar Engineering Conference (2023)

Sobhani, M.: Dynamic Analysis for Fatigue Performances Assessment in Floating Offshore Wind Turbines Electrical Cables. University of Rome, Rome (2021)

Sobhaniasl, M., et al.: Fatigue life assessment for power cables in floating offshore wind turbines. Energies. **13**(12), 3096 (2020). https://doi.org/10.3390/en13123096

Solomin, E., et al.: Hybrid floating solar plant designs: a review. Energies. MDPI AG. (2021). https://doi.org/10.3390/en14102751

Somoano, M., et al.: Experimental modelling of a novel concrete-based 15-MW spar wind turbine. Ocean Eng. **309** (2024). https://doi.org/10.1016/j.oceaneng.2024.118612

Song, J., et al.: Dynamic response of multiconnected floating solar panel systems with vertical cylinders. J. Mar. Sci. Eng. **10**(2) (2022). https://doi.org/10.3390/jmse10020189

Sørum, S.H., et al.: Fatigue design sensitivities of large monopile offshore wind turbines. Wind Energy. **25**(10), 1684–1709 (2022). https://doi.org/10.1002/we.2755

Sørum, S.H., et al.: Assessment of nylon versus polyester ropes for mooring of floating wind turbines. Ocean Eng. **278** (2023a). https://doi.org/10.1016/j.oceaneng.2023.114339

Sørum, S.H., et al.: Modelling of synthetic fibre rope mooring for floating offshore wind turbines. J. Mar. Sci. Eng. **11**(1), 193 (2023b). https://doi.org/10.3390/jmse11010193

Souza, C.E.S.d., Bachynski-Polić, E.E.: Design, structural modeling, control, and performance of 20 MW spar floating wind turbines. Mar. Struct. **84** (2022). https://doi.org/10.1016/j.marstruc.2022.103182

Spicer, P., et al.: Tidal energy extraction modifies tidal asymmetry and transport in a shallow, well-mixed estuary. Front. Mar. Sci. **10** (2023). https://doi.org/10.3389/fmars.2023.1268348

Sree, D.K.K., et al.: Fluid-structural analysis of modular floating solar farms under wave motion. Sol. Energy. **233**, 161–181 (2022). https://doi.org/10.1016/j.solener.2022.01.017

Stadtmann, F., et al.: Digital twins in wind energy: emerging technologies and industry-informed future directions. IEEE Access. **11**, 110762–110795 (2023). https://doi.org/10.1109/ACCESS.2023.3321320

Stansby, P., et al.: Experimental study of mooring forces on the multi-float WEC M4 in large waves with buoy and elastic cables. Ocean Eng. (2022) [Preprint]

Stansby, P., Li, G.: A wind semi-sub platform with hinged floats for omnidirectional swell wave energy conversion (2023). https://doi.org/10.21203/rs.3.rs-3287235/v1

Steiner, J., Dwight, R.P., Viré, A.: Data-driven RANS closures for wind turbine wakes under neutral conditions. Comput. Fluids. **233**, 105213 (2022). https://doi.org/10.1016/j.compfluid.2021.105213

Su, K.C., Chung, P.H., Yang, R.Y.: Numerical simulation of wind loads on an offshore PV panel: the effect of wave angle. J. Mech. **37**, 53–62 (2020). https://doi.org/10.1093/jom/ufaa010

Sun, J., Fang, H.: The analysis of crashworthiness and dissipation mechanism of novel floating composite honeycomb structure against ship-OWT collision. Ocean Eng. **287**, 115819 (2023). https://doi.org/10.1016/j.oceaneng.2023.115819

Sun, K., et al.: Dynamic response analysis of floating wind turbine platform in local fatigue of mooring. Renew. Energy. **204**, 733–749 (2023a). https://doi.org/10.1016/j.renene.2022.12.117

Sun, Y., et al.: Failure analysis of floating offshore wind turbines with correlated failures. Reliab. Eng. Syst. Saf. **238** (2023b). https://doi.org/10.1016/j.ress.2023.109485

Svensson, G.: Fatigue Prediction Models of Dynamic Power Cables by Laboratory Testing and FE Analysis. Norwegian University of Science and Technology (2020)

Sykes, V., Collu, M., Coraddu, A.: A review and analysis of optimisation techniques applied to floating offshore wind platforms. Ocean Eng. Elsevier Ltd. (2023). https://doi.org/10.1016/j.oceaneng.2023.115247

Sykes, V., Collu, M., Coraddu, A.: TOPSIS methodology applied to floating offshore wind to rank platform designs for the Scotwind sites. Ocean Eng. **310** (2024). https://doi.org/10.1016/j.oceaneng.2024.118634

Taruffi, F., Novais, F., Viré, A.: An experimental study on the aerodynamic loads of a floating offshore wind turbine under imposed motions. Wind Energy Science. **9**(2), 343–358 (2024). https://doi.org/10.5194/wes-9-343-2024

Technip FMC: How a dynamic inter array cable system works (2024). https://www.technipfmc.com/en/campaign/the-best-solution-to-transmit-floating-offshore-wind/

Thiébot, J., et al.: On nodal modulations of tidal-stream energy resource in North-Western Europe. Appl. Ocean Res. **121** (2022). https://doi.org/10.1016/j.apor.2022.103091

Thies, P.R., Georgallis, G.: Determining submarine dynamic cable stiffness and fatigue characteristics through physical testing. In: Proceedings of the ASME 2024 43nd International Conference on Ocean, Offshore and Arctic Engineering (2024)

Thomas, B., et al.: Wake effect impact on the levelized cost of energy in large floating offshore wind farms: a case of study in the northwest of the Iberian Peninsula. Energy. **304** (2024). https://doi.org/10.1016/j.energy.2024.132159

Tian, H., Soltani, M.N., Nielsen, M.E.: Review of floating wind turbine damping technology. Ocean Eng. Elsevier Ltd. (2023a). https://doi.org/10.1016/j.oceaneng.2023.114365

Tian, W., et al.: Numerical study of hydrodynamic responses for a combined concept of semisubmersible wind turbine and different layouts of a wave energy converter. Ocean Eng. **272** (2023b). https://doi.org/10.1016/j.oceaneng.2023.113824

Umeda, J., Taniguchi, T., Katayama, T.: Experimental validation of data-driven reactive control strategy for wave energy converters: a Gaussian process regression approach. Ocean Eng. (2024) [Preprint]

Uzunoglu, E., Karmakar, D., Guedes Soares, C.: Floating offshore wind platforms, pp. 53–76. (2016). https://doi.org/10.1007/978-3-319-27972-5_4

Vicky Stratigaki, B., et al.: Annual Report: An Overview of Ocean Energy Activities in 2022 (2023)

Viespoli, L.M.: Mechanical Characterization of Lead Alloys for Subsea High Voltage Power Cable Applications. Norwegian University of Science and Technology (2020)

Viespoli, L.M., et al.: Subsea power cable sheathing: an investigation of lead fatigue performance. Procedia Struct. Integr. **28**, 344–351 (2020). https://doi.org/10.1016/j.prostr.2020.10.040

Viespoli, L.M., et al.: Tape winding angle influence on subsea cable sheathing fatigue performance. Eng. Struct. **229**, 111660 (2021). https://doi.org/10.1016/j.engstruct.2020.111660

Viré, A., et al.: Two-dimensional numerical simulations of vortex-induced vibrations for a cylinder in conditions representative of wind turbine towers. Wind Energy Sci. **5**(2), 793–806 (2020). https://doi.org/10.5194/wes-5-793-2020

Vo, T.T.E., et al.: Overview of possibilities of solar floating photovoltaic systems in the offshore industry. Energies. MDPI. (2021). https://doi.org/10.3390/en14216988

Walker, J., et al.: Digital twins of the mooring line tension for floating offshore wind turbines to improve monitoring, lifespan, and safety. J. Ocean Eng. Mar. Energy. **8**(1), 1–16 (2022). https://doi.org/10.1007/s40722-021-00213-y

Wan, D., et al.: In-situ tensile and fatigue behavior of electrical grade Cu alloy for subsea cables. Mater. Sci. Eng. A. **835**, 142654 (2022). https://doi.org/10.1016/j.msea.2022.142654

Wang, K., et al.: Remote sensing unveils the explosive growth of global offshore wind turbines. Renew. Sustain. Energy Rev. **191** (2024a). https://doi.org/10.1016/j.rser.2023.114186

Wang, L., et al.: Validation of CFD simulations of the moored DeepCwind offshore wind semisubmersible in irregular waves. Ocean Eng. **260** (2022a). https://doi.org/10.1016/j.oceaneng.2022.112028

Wang, L., et al.: Experimental investigation of advanced turbine control strategies and load-mitigation measures with a model-scale floating offshore wind turbine system. Appl. Energy. **355** (2024b). https://doi.org/10.1016/j.apenergy.2023.122343

Wang, P., et al.: Generation and distribution of turbulence-induced loads fluctuation of the horizontal axis tidal turbine blades. Phys. Fluids. **36**(1) (2024c). https://doi.org/10.1063/5.0186105

Wang, S., et al.: A comparative study on the dynamic behaviour of 10 MW conventional and compact gearboxes for offshore wind turbines. Wind Energy. **24**(7), 770–789 (2021). https://doi.org/10.1002/we.2602

Wang, S., et al.: Design, local structural stress, and global dynamic response analysis of a steel semi-submersible hull for a 10-MW floating wind turbine. Eng. Struct. **291** (2023). https://doi.org/10.1016/j.engstruct.2023.116474

Wang, S., Moan, T.: Analysis of extreme internal load effects in columns in a semi-submersible support structure for large floating wind turbines. Ocean Eng. **291** (2024a). https://doi.org/10.1016/j.oceaneng.2023.116372

Wang, S., Moan, T.: Methodology of load effect analysis and ultimate limit state design of semi-submersible hulls of floating wind turbines: with a focus on floater column design. Mar. Struct., **93** (2024b). https://doi.org/10.1016/j.marstruc.2023.103526

Wang, S., Moan, T., Jiang, Z.: Influence of variability and uncertainty of wind and waves on fatigue damage of a floating wind turbine drivetrain. Renew. Energy. **181**, 870–897 (2022b). https://doi.org/10.1016/j.renene.2021.09.090

Wang, T., et al.: A coupling framework between OpenFAST and WEC-Sim. Part I: validation and dynamic response analysis of IEA-15-MW-UMaine FOWT. Renew. Energy. **225** (2024d). https://doi.org/10.1016/j.renene.2024.120249

Wang, Y., et al.: Influence of hydraulic PTO parameters on power capture and motion response of a floating wind-wave hybrid system. J. Mar. Sci. Eng. **10**(11), 1660 (2022c). https://doi.org/10.3390/jmse10111660

Wang, Y., Chen, H.-C.: Verification and validation of CFD simulations of a FOWT semisubmersible under bichromatic and random waves. J. Offshore Mech. Arct. Eng. (2022) [Preprint].). https://doi.org/10.1115/1.4056421

Wang, Z., et al.: An identification method of floating wind turbine tower responses using deep learning technology in the monitoring system. Ocean Eng. **261** (2022d). https://doi.org/10.1016/j.oceaneng.2022.112105

Wang, Z., et al.: Monitoring system framework design for floating wind turbine using the deep learning technology and tower response identification considering sensor optimization. Ocean Eng. **299** (2024e). https://doi.org/10.1016/j.oceaneng.2024.117316

Wei, Y., et al.: Hydrodynamic analysis of a heave-hinge wave energy converter combined with a floating breakwater. Ocean Eng. **293** (2024). https://doi.org/10.1016/j.oceaneng.2023.116618

Weller, S., et al.: Failure Implications of Different Mooring Spreads and Lines (2024). www.fowcoe.co.uk

Wen, B., et al.: On the aerodynamic loading effect of a model spar-type floating wind turbine: an experimental study. Renew. Energy. **184**, 306–319 (2022). https://doi.org/10.1016/j.renene.2021.11.009

Wu, T., Zhang, C., Guo, X.: Dynamic responses of monopile offshore wind turbines in cold sea regions: ice and aerodynamic loads with soil-structure interaction. Ocean Eng. **292** (2024). https://doi.org/10.1016/j.oceaneng.2023.116536

Wu, X., et al.: Research on hydrodynamic characteristics of an offshore flexible floating photovoltaic in waves. In: 2022 3rd International Conference on Geology, Mapping and Remote Sensing, ICGMRS 2022, pp. 841–845. Institute of Electrical and Electronics Engineers Inc. (2022). https://doi.org/10.1109/ICGMRS55602.2022.9849362

Wulf, C., Zapp, P., Schreiber, A.: Review of power-to-X demonstration projects in Europe. Front. Energy Res. Frontiers Media S.A. (2020). https://doi.org/10.3389/fenrg.2020.00191.

Xi, R., et al.: Dynamic analysis of 10 MW monopile supported offshore wind turbine based on fully coupled model. Ocean Eng. **234**, 109346 (2021). https://doi.org/10.1016/j.oceaneng.2021.109346

Xia, J., Zou, G.: Operation and maintenance optimization of offshore wind farms based on digital twin: a review. Ocean Eng. **268**, 113322 (2023). https://doi.org/10.1016/j.oceaneng.2022.113322

Xie, S.Y., et al.: Modeling and analyzing dynamic response for an offshore bottom-fixed wind turbine with individual pitch control. China Ocean Eng. **36**(3), 372–383 (2022). https://doi.org/10.1007/s13344-022-0033-8

Xing, Y., et al.: Characterisation of extreme load responses of a 10-MW floating semi-submersible type wind turbine. Heliyon. **9**(2) (2023). https://doi.org/10.1016/j.heliyon.2023.e13728

Xu, K., et al.: Design and comparative analysis of alternative mooring systems for floating wind turbines in shallow water with emphasis on ultimate limit state design. Ocean Eng. **219**, 108377 (2021). https://doi.org/10.1016/j.oceaneng.2020.108377

Xu, P., Wellens, P.R.: Theoretical analysis of nonlinear fluid–structure interaction between large-scale polymer offshore floating photovoltaics and waves. Ocean Eng. **249** (2022). https://doi.org/10.1016/j.oceaneng.2022.110829

Yan, J., et al.: Optimization design method of the umbilical cable global configuration based on representative fatigue conditions. IEEE J. Ocean. Eng. **48**(1), 188–198 (2023a). https://doi.org/10.1109/JOE.2022.3200130

Yan, X., et al.: Numerical investigations on nonlinear effects of catenary mooring systems for a 10-MW FOWT in shallow water. Ocean Eng. **276**, 114207 (2023b). https://doi.org/10.1016/j.oceaneng.2023.114207

Yang, I., Tezdogan, T., Incecik, A.: Numerical investigations of a pivoted point absorber wave energy converter integrated with breakwater using CFD. Ocean Eng. **274** (2023a). https://doi.org/10.1016/j.oceaneng.2023.114025

Yang, R.Y., Yu, S.H.: A study on a floating solar energy system applied in an intertidal zone. Energies. **14**(22) (2021). https://doi.org/10.3390/en14227789

Yang, Y., et al.: Effects of tidal turbine number on the performance of a 10 MW-class semi-submersible integrated floating wind-current system. Energy. **285** (2023b). https://doi.org/10.1016/j.energy.2023.128789

Ye, M.k., et al.: CFD simulations targeting the performance of the NTNU BT1 wind turbine using overset grids. J. Hydrodyn. **35**(5), 954–962 (2023). https://doi.org/10.1007/s42241-023-0065-4.

Yi, Y., et al.: Experimental investigation into the dynamics and power coupling effects of floating semi-submersible wind turbine combined with point-absorber array and aquaculture cage. Energy. (2024) [Preprint].). https://doi.org/10.1016/j.energy.2024.131220

Yim, S.C., Chen, M., Ai, S.: A combined nonlinear mooring-line and umbilical cable dynamics model and application. In: The Proceedings of 41st International Conference on Ocean, Offshore & Arctic Engineering (2022)

Yin, D., et al.: State-of-the-art review of vortex-induced motions of floating offshore wind turbine structures. J. Mar. Sci. Eng. MDPI. (2022). https://doi.org/10.3390/jmse10081021

Yin, Y., et al.: Digital twin-driven identification of fault situation in distribution networks connected to distributed wind power. Int. J. Electr. Power Energy Syst. **155** (2024). https://doi.org/10.1016/j.ijepes.2023.109415

Yue, W., et al.: Feasibility of co-locating wave energy converters with offshore aquaculture: the pioneering case study of China's Penghu platform. Ocean Eng. (2023) [Preprint]

Zeng, S., et al.: Reliability of simplified wind fields and structure models for the response simulation of parked offshore wind turbines. Structure. **59**, 105714 (2024). https://doi.org/10.1016/j.istruc.2023.105714

Zeng, X., et al.: Hydrodynamic interactions among wave energy converter array and a hierarchical genetic algorithm for layout optimization. Ocean Eng. (2023) [Preprint]

Zhang, C., Wang, S., et al.: Effects of mooring line failure on the dynamic responses of a semisubmersible floating offshore wind turbine including gearbox dynamics analysis. Ocean Eng. **245**, 110478 (2022a). https://doi.org/10.1016/j.oceaneng.2021.110478

Zhang, C., et al.: Development of compliant modular floating photovoltaic farm for coastal conditions. Renew. Sustain. Energy Rev. **190** (2024a). https://doi.org/10.1016/j.rser.2023.114084

Zhang, D., et al.: A coupled numerical framework for hybrid floating offshore wind turbine and oscillating water column wave energy converters. Energy Convers. Manag. **267**(June), 115933 (2022b). https://doi.org/10.1016/j.enconman.2022.115933

Zhang, D.-Q., et al.: Hydrodynamic modelling of modularized floating photovoltaics arrays (2023a). http://asmedigitalcollection.asme.org/OMAE/proceedings-pdf/OMAE2023/86908/V008T09A013/7041685/v008t09a013-omae2023-102530.pdf

Zhang, J., Zhou, Y., et al.: Experimental investigation on wake and thrust characteristics of a twin-rotor horizontal axis tidal stream turbine. Renew. Energy. **195**, 701–715 (2022c). https://doi.org/10.1016/j.renene.2022.05.061

Zhang, J., Zhang, C., et al.: Interactions between tidal stream turbine arrays and their hydrodynamic impact around Zhoushan Island, China. Ocean Eng. **246** (2022d). https://doi.org/10.1016/j.oceaneng.2021.110431

Zhang, J., et al.: Coupling multi-physics models to corrosion fatigue prognosis of high-strength bolts in floating offshore wind turbine towers. Eng. Struct. **301** (2024b). https://doi.org/10.1016/j.engstruct.2023.117309

Zhang, J., Zhao, X.: Digital twin of wind farms via physics-informed deep learning. Energy Convers. Manag. **293** (2023). https://doi.org/10.1016/j.enconman.2023.117507

Zhang, K., et al.: Inspection of floating offshore wind turbines using multi-rotor unmanned aerial vehicles: literature review and trends. Sensors. Multidisciplinary Digital Publishing Institute (MDPI). (2024c). https://doi.org/10.3390/s24030911

Zhang, L., et al.: Systematic analysis of performance and cost of two floating offshore wind turbines with significant interactions. Appl. Energy. **321** (2022e). https://doi.org/10.1016/j.apenergy.2022.119341

Zhang, T., Du, A., et al.: Analysis of three-Core composite submarine cable damage due to ship anchor. IEEE Access. **10**, 93910–93920 (2022f). https://doi.org/10.1109/ACCESS.2022.3203589

Zhang, T., Li, L., et al.: Research on anchor damage and protection of three-core composite submarine cable considering impact angle. Ocean Eng. **265**, 112668 (2022g). https://doi.org/10.1016/j.oceaneng.2022.112668

Zhang, Y., Liu, Z., et al.: Fluid–structure interaction modeling of structural loads and fatigue life analysis of tidal stream turbine. Mathematics. **10**(19) (2022h). https://doi.org/10.3390/math10193674

Zhang, Y., et al.: Research of the array spacing effect on wake interaction of tidal stream turbines. Ocean Eng. **276** (2023b). https://doi.org/10.1016/j.oceaneng.2023.114227

Zhang, Y., Hu, Z.: An aero-hydro coupled method for investigating ship collision against a floating offshore wind turbine. Mar. Struct. **83** (2022). https://doi.org/10.1016/j.marstruc.2022.103177

Zhang, Z., et al.: Dynamic responses and mooring line failure analysis of the fully submersible platform for floating wind turbine under typhoon. Eng. Struct. **301** (2024d). https://doi.org/10.1016/j.engstruct.2023.117334

Zhao, S., et al.: Potential root causes for failures in floating PV systems (2023a). http://asmedigitalcollection.asme.org/OMAE/proceedings-pdf/OMAE2023/86908/V008T09A012/7041659/v008t09a012-omae2023-101868.pdf

Zhao, X., Dao, M.H., Le, Q.T.: Digital twining of an offshore wind turbine on a monopile using reduced-order modelling approach. Renew. Energy. **206**, 531–551 (2023b). https://doi.org/10.1016/j.renene.2023.02.067

Zhu, H.: Optimal semi-active control for a hybrid wind-wave Energy system on motion reduction. IEEE Trans. Sustain. Energy. **14**(1), 75–82 (2023). https://doi.org/10.1109/TSTE.2022.3202805

Zhu, K., et al.: Hydrodynamic analysis of hybrid system with wind turbine and wave energy converter. Appl. Energy. **350**, 121745 (2023a). https://doi.org/10.1016/j.apenergy.2023.121745

Zhu, K., et al.: Analytical study on dynamic performance of a hybrid system in real sea states. Energy. **290**(November 2023), 130259 (2024). https://doi.org/10.1016/j.energy.2024.130259

Zhu, Z., et al.: Digital twin technology for wind turbine towers based on joint load–response estimation: a laboratory experimental study. Appl. Energy. **352** (2023b). https://doi.org/10.1016/j.apenergy.2023.121953

Zong, S., et al.: The dynamic response of a floating wind turbine under collision load considering the coupling of wind-wave-mooring loads. J. Mar. Sci. Eng. **11**(9) (2023). https://doi.org/10.3390/jmse11091741

Open Access This chapter is licensed under the terms of the Creative Commons Attribution-NonCommercial-NoDerivatives 4.0 International License (http://creativecommons.org/licenses/by-nc-nd/4.0/), which permits any noncommercial use, sharing, distribution and reproduction in any medium or format, as long as you give appropriate credit to the original author(s) and the source, provide a link to the Creative Commons license and indicate if you modified the licensed material. You do not have permission under this license to share adapted material derived from this chapter or parts of it.

The images or other third party material in this chapter are included in the chapter's Creative Commons license, unless indicated otherwise in a credit line to the material. If material is not included in the chapter's Creative Commons license and your intended use is not permitted by statutory regulation or exceeds the permitted use, you will need to obtain permission directly from the copyright holder.

Committee V.5: Special Vessels

Ermina Begovic[1(✉)], Jason Ali-Lavroff[2], Dario Boote[3], Alexander Egorov[4], Mark Rodriguez[5], A. Salcedo[6], Junji Sawamura[7], Harleigh C. Seyffert[8], Jean Baptist Souppez[9], Lyubomir Toshkov[10], Nikola Vladimir[11], and Fuhua Wang[12]

[1] University of Naples Federico II, Naples, Italy
begovic@unina.it
[2] Australian Maritime College (AMC), Launceston, Australia
[3] University of Genoa, Genoa, Italy
[4] Marine Engineering Bureau, Odesa, Ukraine
[5] Jacksonville, USA
[6] Navantia, Madrid, Spain
[7] Osaka University, Suita, Japan
[8] Delft University of Technology, Delft, Netherlands
[9] Birmingham City University, Birmingham, United Kingdom
[10] Sofia, Bulgaria
[11] University of Zagreb, Zagreb, Croatia
[12] Shanghai, China

Committee Mandate. Concern for structural challenges of non-conventional, special surface craft, including uncertainties in established design methods and modelling techniques.

Particular attention shall be given to mega yachts, naval craft, offshore service vessels and work boats, which can be characterized by particular structural configurations and materials (wide openings, large unsupported structures, unconventionally shaped superstructures, etc.) and/or are to sustain specific loading conditions (harsh environment, severe cyclic loads or extreme operational ones).

Keywords: High Speed Vessels · High-Speed Catamarans · Yachts · Sailing Vessels · Fishing vessels · Naval Unmanned Surface Vessels (USV) · Ice Vessels · Offshore service vessels · Fishing Vessels · River – sea ships · Ballast Free Ships · Slamming · Dynamic Structural Loads · Vibration and Noise · Glazing · Comfort on Board · Fatigue

1 Introduction

This report is the continuation of the Special Craft report (Truelock et al. 2018) and Special Vessel Committee (Truelock et al. 2022). In the first report, the Committee's focus was broad and largely on highlighting "Special" craft through market segments (naval, offshore operation vessels and yachts) or as special craft with the appurtenance structure. Market analysis of naval vessels, offshore operation vessels and yachts has been reported, together with the list of recommended vessel types for future V.5 Committees. The second report further discussed those markets giving an accent to reduced emission vessels, offshore installation vessels, and ice polar vessels, and as conclusions and recommendations identified unmanned surface vessels, digital twin and offshore renewable energy vessels. Yachts have been discussed in the 2018 and 2022 reports and the benchmark was undertaken by the Committee investigating various calculation and analysis methods on window glazing as a structural material.

This report reviews recent research on the structural design of various vessel types: High-Speed Craft, Pleasure Craft, Workboats, and Ice-Polar vessels. It highlights challenges and advancements in methodologies for predicting wave loads and ship responses and design, both experimental and numerical.

For all considered special vessels, from the state of the art publications, it can be seen that the scientific community is working on the validations of advanced numerical simulations, assessment of the realistic loads, and impact of the required environmental friendliness on design. It is also noted that the industry and design offices are proposing very advanced solutions but do not publish their results. More in detail, in High Speed Craft field, can be observed an increasing number of very complex experimental campaigns with the measures of motions and wave induced loads. The application of Fluid-Structure Interaction (FSI) calculations is very relevant for this field. Can be highlighted the interest of the scientific community in the verification and validation of case studies, as well as the wider use of full-scale data and machine learning techniques for improved accuracy in slamming and fatigue prediction. For high-speed catamaran, ride control systems have been seen as promising solutions in mitigating wave loads and improving passenger comfort. Large pleasure crafts increasingly prioritize comfort, leading to research on applicable seakeeping criteria, vibration and noise reduction, which was finalized in the issuing of the new ISO 22834 Standard. Superyachts aesthetics standards impose demanding challenges related to large openings and glazing. In the small pleasure crafts field, mainly driven by offshore sailing race, innovations, particularly in composite materials and multihull configurations, influenced the design of new boats and updates of the ISO standards. The naval Unmanned Surface Vessels (USVs) and the development of structural digital twins are among the most important emerging trends in naval architecture, presenting both opportunities and challenges. The extensive research in ice load measurements in full scale and laboratory and structural challenges, such as ice-propeller interaction and fatigue damage testimonies the growing interest in human activities and shipping routes in the Arctic region. The Committee members identified and discussed challenges for offshore service vessels, ballast free ships, fishing vessels and river/sea vessels and emerging trends in both commercial and government sectors.

1.1 ISSC 2022 Special Craft Recommendations and Official Discusser Feedback

Truelock et al. (2022) concluded their work recommending that future work focus on new vessel types such as unmanned surface vessels, digital twin vessels and offshore renewable vessels. The importance of the impact of the exponential growth of ships on green shipping and the structural challenges associated with very large ships has been pointed out by both the Committee and Official Discusser. These were the starting points for the Committee members for the present report. Committee members agreed to consider only the publications from 2020 to 2024; the older references can be taken into consideration only for the vessels which have never been considered before.

1.2 Special Vessels Addressed

The committee report is a consolidation of previous reports and the most significant references that contribute to research and development in the areas of load predictions, analysis techniques and design improvements in Special Vessels. The covered vessel types and report structure are schematized in Table 1.

Table 1. Special Vessel Coverage

Special Vessel Type	Committee Mandate Focus	Special Vessels
HIGH SPEED CRAFT Bolide 80 https://www.boatinternational.com/yachts/the-register/fastest-yachts-in-the-world%2D%2D25053, accessed 15th November 2024	Specific Loading Conditions – *Slamming, Accelerations* Structural Challenges: *Hydroelasticity, Ride control* Methodologies: *EFD, CFD, FEM, FSI. ML* Number of references: *48*	• Planing Craft • High Speed Monohull • Multihulls • ACV
LARGE PLEASURE CRAFTS Benetti BNow 50 M Courtesy of Azimuth - Benetti group	Specific Loading Conditions – *not relevant* Structural Challenges – *Solar radiation - Glazing – Noise -Vibration - Comfort* New regulations Number of references: *32*	• Motor yachts larger than 24 m • Sailing yachts larger than 24 m

(*continued*)

Table 1. (*continued*)

Special Vessel Type	Committee Mandate Focus	Special Vessels
SMALL PLEASURE CRAFT Figaro 3, Beneteau https://www.beneteau.com/it/figaro/figaro-beneteau-3, accessed 15th November 2024	Specific Loading Conditions – *slamming, rig loads, torsional moment, bending of crossbeams* Structural Challenges – *strength of composite laminates – appendages design* New regulations Number of references: *56*	• Motor yachts under 24 m • Sailing yachts under 24 m
NAVAL UNMANNED VESSELS Vanguarda LUSV https://aresdifesa.it/varato-loverlord-unmanned-surface-vanguard-negli-stati-uniti/	Specific Loading Conditions – *classifications according to mission and/or size* Structural Challenges – *Not relevant* Class rules Number of references: *9 + 37 URL references for database*	• Naval Unmanned Surface Vessels
WORK BOATS DAMEN Fishing Vessel https://www.damen.com/vessels/fishing-vessels, Courtesy of DAMEN shipyard	Specific Loading Conditions – *Accident analysis, Seakeeping operability* Structural Challenges – *underwater radiated noise. Use of new materials, energy efficiency* Number of references: *16*	• Fishing vessels

(*continued*)

Table 1. (*continued*)

Special Vessel Type	Committee Mandate Focus	Special Vessels
WORK BOATS https://www.damen.com/insights-center/news/damen-presents-floating-offshore-wind-support-vessel, accessed 13th January 2025	Specific Loading Conditions –*Seakeeping operability for safe transfer* Structural Challenges – *motion-compensating equipment*, Number of references: *9*	• Offshore service vessels
WORK BOATS Courtesy of A. Egorov	Specific Loading Conditions –*Cargo loading and unloading* Structural Challenges – *buckling failure*, Number of references: *9*	• River and river/sea ships
ICE POLAR VESSELS S.A._Agulhas_II https://en.wikipedia.org/wiki/S._A._Agulhas_II, accessed 27th November 2024	Specific Loading Conditions – *Ice Loads* Structural Challenges – *fatigue, ice-hull/ ice-propeller interaction* Number of references: *71*	• Icebreakers • Research vessels • Offshore wind turbine in ice

* EFD – Experimental Fluid Dynamics
* CFD – Computational Fluid Dynamics
* FEM –Finite Elements Methods
* FSI – Fluid – Structure Interaction
* ML – Machine Learning

2 High Speed Craft

2.1 Introduction - Concepts and Challenges

High-speed marine crafts are particularly advantageous options for ferry operators for transporting passengers and cargo over shorter durations when compared to conventional craft. To achieve high Froude numbers, these vessels are designed with special features including slender hulls or a hard chine hull form for reduced resistance. High-speed displacement ships include catamarans, trimarans, and high-speed monohulls. Hull form optimization for high speed, vessel motions and structural loads must be considered since the early design stage. To this end, research on these special vessels is continuing to identify methods for reducing ship motions and shiploads to improve passenger comfort and structural design.

High speed vessels have been thoroughly addressed in the previous report and this one recalls the same classifications (fast monohulls, fast multihulls, and planing craft), adds Air Cushion Vehicles (ACV) and brings the updates from the period 2021–2024. The research in the High Speed Craft field has been mainly focused on load identifications using model experiments, full-scale analysis, and CFD analysis. Development and validation of nonlinear potential flow coupled with structural analysis is the first step forward based on the existing fast numerical methods. More sophisticated methods, based on Computational Fluid Dynamics (CFD) and Finite Elements Methods (FEM) two-way hydroelastic coupling, are gaining importance as numerical power is increasing. New techniques based on big data analysis using machine learning are foreseen as prominent methods to obtain slamming and fatigue prediction based on full-scale measurements.

Specifically for multihulls, a major structural consideration is slamming loads on the wet deck. Even though there have already been many investigations on this subject in the previous period, research has continued in this area to learn more about the slamming phenomenon, determine links between slamming pressures and different design features, and possibly implement solutions to mitigate slam events.

2.2 Loads and Structural Challenges

2.2.1 High-Speed Monohulls and Planing Craft

Due to the complexity of the interaction between water waves and high speed ships, the research in this area has been mainly focused on improving numerical and experimental techniques. To better understand nonlinear fluid–structure interaction and rarely occurring events on ships in the realistic ocean environment, advanced CFD-FEM simulations, hydroelasticity and model measurements (including large-scale models) are recognised as promising tools for both slender monohulls and high-speed planing craft.

Van Walree and Thomas (2023) reported the validation of PanShip(NL) for the Rigid Hulled Inflatable Boat (RHIB) in heavy seas against the experimental campaign, shown in Fig. 1. PanShip(NL) is a potential flow method based on 3D Green functions for diffraction and radiation forces, 3D panel methods to account for the Froude-Krylov, and restoring forces on the instantaneous submerged body. The results for the motions and accelerations show that PanShip(NL) provides adequate predictions of motions and accelerations for operability analysis purposes in low amplitude yet steep, regular waves.

Even non-linear events in head seas, such as jumping out of wave crests and acceleration peaks in steep and heavy irregular seas, are well predicted. Similar conclusions are reported in Van Walree and Sgarioto (2023) where authors compared two potential flow methods against experimental results for predicting landing craft motions.

Fig. 1. Model of landing vessel shipping water, after van Walree and Thomas (2023)

Parunov et al. (2024) presented a benchmark study on Canadian patrol frigates for motion and global wave loads, with nine participating institutions, quantifying the hydro-elastic responses. The results for heave, pitch, vertical wave bending moment (VWBM) and whipping moment (WHBM) at midship, obtained by codes based on the strip theory, 3D panel method in frequency and time domain and CFD are given in Fig. 2. As expected, uncertainties in ship motions are lower than those of rigid-body load effects. The Authors underlined that the number of codes in each group of methods is too small to group uncertainties according to the method. Much larger uncertainties are found in the whipping responses, which deserve further attention. The differences in the seakeeping codes are not so large to be the prime reason for discrepancies in the whipping prediction, therefore the conclusion is that large uncertainty in the whipping response is the consequence of different approaches to the modelling of the slamming load.

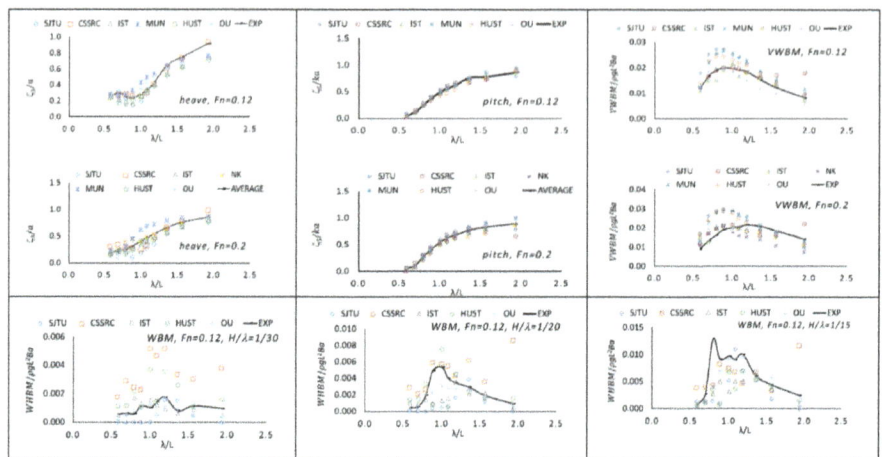

Fig. 2. Comparison of heave, Pitch, VWBM and WHBM, after Parunov et al. (2024)

Li et al. (2022a) performed an experimental study on the stern slamming of a large cruise ship using a segmented model in the head and following waves. Under moderate and large wave conditions, the greater relative velocity and impact area lead to significant stern slamming. However, the longitudinal deadrise angle is much smaller in the following waves, which reduces the slamming duration. The distribution of pressure peaks indicates that an increase in the longitudinal deadrise angle directly reduces the peak values of the slamming pressure. The combined influence of these two factors means that the whipping problem is much more serious in the following seas. Stern slamming significantly increases the magnitude of the VBM at midship. Ibrahim and Judge (2024) presented experimental results on the hull girder slamming factor of the segmented model of U.S. Coast Guard Fast Response Cutter (FRC-B) vessel. The tests in regular waves, performed at three Froude numbers, were conducted to investigate the effect of wave parameters and vessel motion on the slamming factor. The irregular wave tests were conducted to estimate slamming factors in moderate operational sea conditions (sea state 4) that could be useful for fatigue life calculations and in sea state 6 that could be used for structural design.

Jiao et al. (2021) discussed opportunities and challenges in load assessment (vertical bending moment, impact pressures, slamming and whipping) based on the published experimental campaign for large scaled models of 7 naval and 2 container ships. The Authors concluded that the large-scale model testing technique is positioned as a supplement to the classical tank tests and full-scale trial, expecting that they can be conducted prior to the design and construction of full scale ship to derisk new technology and commercial cost.

The works on wedge impacts have been numerous, dealing with curved wedge (Wen et al. (2022), deformable edge using Smooth Particle Methods (Zhao et al. 2022a), Cui et al. (2023), Zhang et al. (2022a)), hydro elastic wedge drop experiments Ren et al. (2021a), symmetric and asymmetric wedges (Xin et al. (2022)).

Hosseinzadeh et al. (2023a) reported an overview of the research on wedge impacts and a concise summary table of 19 works dealing with the experimental wedge impacts, 11 of them were performed from 2017 to 2022. In their research, the impact loads were measured on a non-prismatic aluminium wedge with a stiffened panel during free-fall water entry. Two different plates were considered on the bottom of the wedge to study the influence of flexural rigidity on hydroelastic slamming. A description of the experimental conditions, including the geometry of the wedge, material properties, and the test plan is provided. The same conditions were studied numerically using two-way CFD-FEM coupling in Hosseinzadeh et al. (2023b). Both works discuss in detail the effects of water impact velocity, deadrise angle, mass of the wedge, and bending stiffness on the slamming pressures and structural responses. Comparison of vertical acceleration, slamming pressure, and strain responses showed a slight discrepancy in the maximum value of pressure and strain predicted by numerical methods, in the Authors' opinion due to the different coupling techniques that were applied in the Fluid-Structure-Interaction (FSI) simulations.

Allaka and Groper (2020), Rosén et al. (2020) presented a validation of a mathematical model based on the Zarnick theory with incorporated previously developed improvements for added mass estimation, deadrise-dependent hydrodynamic coefficients, near-transom pressure correction, and dynamic drag. Allaka and Groper (2020) commented that from the comparison of the measured vs. predicted accelerations and motions, it was observed that implementing a full near-transom pressure correction in the initial stages of the planing mode (inception of planing) does not provide sufficiently accurate results. It was also observed that the near-transom pressure correction has to be applied as a linear/quadratic growing function of increasing Froude number. To examine the effects of the incorporated improvements in the calculation of hydromechanic forces, the authors reported that the accuracy of the new method predictions for motion and acceleration of two different crafts was in the range of 3–23% for vertical accelerations and 4–20% for pitch motion when compared with the experimental results. Shao et al. (2023) developed a new Modified Logvinovich Model for a prediction method based on a 2D+t and variable deadrise angle for the vertical force of planing craft. The authors compared the results against CFD and the original 2D+t method in calm water, showing improvements in the prediction of the trim angle and longitudinal vertical force distribution along the hull with respect to the 2D+t and very close agreement with the CFD results. Detailed validation of the numerical model from Rosén et al. (2020) has been published in Begovic et al. (2020).

Diez et al. (2022) developed tools for multidisciplinary design optimization of a deep-V grillage panel on a generic prismatic planing hull subject to slamming loads, first considering regular waves only. The authors considered fluid-structure interaction experiments, computational fluid dynamics, and computational structural dynamics, and optimized a design with a weight reduction of 35% and a safety factor of 1.72. The authors follow up in part II of the publication, Lee et al. (2024), by further considering the effect of irregular waves. Here, the authors find that their computational methods can lead to more efficient designs than the American Bureau of Shipping Rules for Building and Classing Light Warships, Patrol, and High-Speed Naval Craft (2021). From the results presented by Begovic et al. (2024) and Diez et al. (2020) it can be seen excellent agreement of both methodologies: Nonlinear strip method+FEM and 3D CFD methods+FEM, against the experimental data, and clearly there is an important advantage in terms of computational efforts when use of nonlinear strip methods, in particular if the irregular waves or extreme accelerations have to be assessed.

Marlantes and Maki (2022) presented a machine learning method for predictions of nonlinear ship motions in a range of wave conditions when trained on response data from only a single seaway. The method was applied to predict the nonlinear heave and pitch responses of a Generic Prismatic Planing Hull (GPPH) model in head seas at a single forward speed. The training and testing data were computed using the nonlinear theory of Faltinsen (2005) for the development of the method. Although the results of their study are promising, the authors underlined that there are still many limitations, such as different forward speeds and underwater hull geometries which must be addressed before the method may be useful in practice.

Dessi et al. (2023) explored Machine Learning (ML) for slamming event identification. Raw experimental data were acquired from sensors placed on a scaled flexible model of a fast monohull tested in the towing tank under different combinations of sea states and forward speeds. The authors discussed the advantages of using ML models instead of physics based or condition-based ones. The first point concerns the scalability of ML models in terms of features. Physics-based algorithms for slamming detection are based on deterministic criteria between significant variables alone. ML models take into account all the variables which the phenomenon may potentially depend on, even if this dependence is not direct. This leads to include more parameters than those strictly necessary to verify slamming criteria. On the other hand, scalability allows to predict the slamming in the presence of missing information due to sensor failure. It is interesting to note that one of the most critical pieces of information, *i.e.*, the relative motion, upon which physics-based criteria for slamming detection are based, can be recovered by other information available. The second point concerns the ability of ML models to process data in a simple and parallel way. The latter implies that the different features are computationally independent. Indeed, in physics-based identification of Ochi's slamming events, the water entry velocity must be evaluated exactly at the time of impact, that is, at a precise draft value. Thus, errors in synchronizing data or errors in draft measurement may produce rather different results. The previous considerations also indicate that ML models are more robust than physics-based ones, and more resistant to measurement noise. As a final point ML approaches can be easily reshaped to get new targets. For instance, training may be done on information about whipping levels which is not available later. In this case, the ML algorithms will predict slamming more sensitive to whipping.

2.2.2 Multi Hulls

Research on large high-speed catamarans has placed a continued focus on improving ship motions but also reducing ship structural loads and slamming through the implementation of Ride Control Systems (RCS). Incat Tasmania have incorporated RCS on their wavepiercer catamarans (Fig. 3) typically by using stern tabs mounted at the rear of the vessel and a centrally mounted T-foil located behind the center bow (Fig. 4).

Fig. 3. Incat Hull 061 98 m Wave-Piercer Catamaran (https://incat.com.au/vessel-gallery/061/, accessed 28/11/2024).

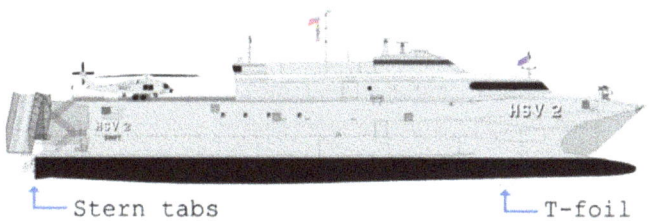

Fig. 4. Schematic of Hull 061 showing location of stern tabs and T-foil (Lau et al. 2022a).

Although recent work has been undertaken on slamming at full-scale as reported by Almallah et al. (2021), it is still necessary to quantify the effects on motions and loads to develop a better understanding for ship design. Model scale experiments and numerical analysis provide the opportunity to quantify these effects in controlled environments as opposed to the limitations encountered at full-scale due to random weather conditions and the challenges associated with sensors and data collection.

Davis (2023) performed high-speed time domain simulations of a catamaran including the use of active motion controls. He found that using (inactive) fixed T-foils can reduce slamming loads by up to 65%, and that active controls can further reduce these loads up to 79% in 3 m seas. In general, it was reported that local feedback control is the most effective way to reduce catamaran slamming.

To simulate more closely at model scale the conditions experienced at full-scale, Javanmard et al. (2023) conducted experiments in random head seas (Fig. 5). Similar outcomes were found to regular seas, thus confirming a reduction in motions, loads and Motion Sickness Incidence (MSI) (Javanmard et al. 2024) with the implementation of ride control based on an active stern tab and centrally mounted T-foil.

Fig. 5. 2.5 m hydroelastic catamaran model based on the 112 m wave-piercer catamaran shown tested in irregular head-seas with ride control using active stern tabs and centrally mounted T-foil (Javanmard et al. 2023).

It was clear from these model experiments that there were substantial gains to be achieved with the deployment of ride control systems on high-speed vessels. To expand on this further, more recent work has aimed to analyse data collected at full-scale to quantify the effects of the Ride Control System (RCS) on the full-scale catamaran vessel.

Incat Hull 061 was instrumented with strain gauges, accelerometers, wave radar and a motions inertial device to collect data during very specific and targeted wave conditions based on sea trials conducted in 2003 by the Naval Warfare Surface Centre Carderock Division. Ship data was analyzed with and without Ride Control Systems to determine the effects on motions, MSI, accelerations (loads) and passenger comfort (Lau et al. 2022a, b). From these results it was also proven at full-scale, although with greater uncertainty due to the random nature of sea conditions, that a reduction in motions and MSI was effectively achieved in head seas and bow quartering seas with the RCS deployed. These results emphasize the importance of ride control in mitigating wave loads and reducing loads on the hull girder for future design.

Further analysis of this full-scale data redirected an emphasis towards slam detection and global slam loads using various techniques. Alsallah et al. (2021) used Empirical Mode Decomposition (EMD) on vibration signals post slam events on Incat Hull 061 to identify wave impacts by decomposing the signal into many components thereby isolating the signals to identify wave impact events, as further reported by Alsallah et al. (2024). These techniques using EMD were then extended to develop new insights into the structural responses of catamarans subject to slamming. Tracing delays were found between sensors, showing the propagation of structural waves through the ship structure. These waves could then be traced back to the epicenter to identify the location of the wave slam. It was found that although the vessel was heading into head seas the slam load applied to the structure was asymmetric.

Gebrezgabir et al. (2023) developed a new method of response reconstruction using transmissibility based on wave slam load data collected on Incat catamaran Hull 061. Global wave load responses were reconstructed using transmissibility concepts based on linear response theory. It was demonstrated that longitudinal bending strains, roll, pitch and yaw rates due to slamming could be reconstructed based only on two strain gauge readings, two triaxial accelerometers at the bow and an accelerometer at the longitudinal centre of gravity. The developed techniques can reduce the number of onboard sensors needed but more importantly, extrapolate limited measurements to identify stresses at any point in the ship for quantifying design decisions on future vessels.

Fatigue remains to be an important consideration for the structural design of large high-speed catamarans to develop a greater knowledge on the parameters influencing ship longevity but also for improving structural design. Warren et al. (2022) investigated vessel stresses based on large amounts of data collected during passenger voyages on a 111 m catamaran operating in the Canary Islands. This data was used for fatigue analysis where reference was made to classification society rules in relation to fatigue design. To achieve this, long term distribution of stresses was compared to load spectra to find that the simplified methods used by classification societies were highly conservative when compared to fatigue results based on measured data (Fig. 6). A proposed combined Weibull fit method was found to increase the accuracy of fatigue analysis methods.

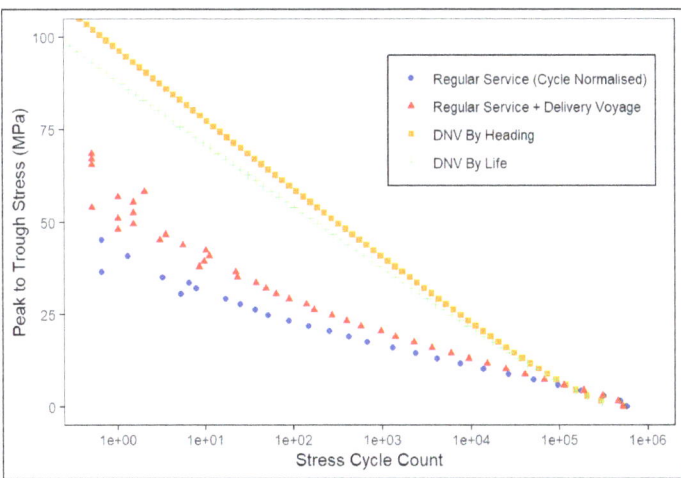

Fig. 6. 111 m wave-piercer catamaran stress spectra formulated using the DNV GL methods compared to spectra developed using in service ship data, Warren et al. (2022).

Full-scale data collected on a 111 m catamaran was further analyzed to compare results from linear elastic fracture mechanics against a linear damage hypothesis or SN curve-based methods for determining crack growth in ships (Warren et al. 2023). The results from the investigation found significant differences in the estimated life when comparing the two methods. It was speculated that the differences are due to the limitations in the linear fracture mechanics model during the crack initiation phase. Parameter changes in the fracture mechanics model also influenced the fatigue life and to obtain good accuracy, material parameters need to be known. Despite this, the linear fracture mechanics methods were shown to have a higher potential for customization when compared to linear damage hypothesis methods. This leads to benefits such as in the assessment of changes to vessel configuration, usage profiles or tracking of known flaws in the vessel.

Intelligent monitoring of catamaran vessels was further extended by Shabani et al. (2023) by introducing machine learning to classify bow entry events especially important for slam classification. The developed methods are useful in identifying slams when big data is concerned either for determining stress cycles during peak loads or to provide real time warnings due to slamming. To enable this machine learning methods were incorporated into monitoring systems (Shabani et al. (2021)). Two vessels were considered. Hull 091 a vessel operating in the Canary Islands). Machine learning models were trained using the data from both these vessels based on supervised and unsupervised models and were found to be beneficial for the classification of bow entry events based on key kinematics parameters. The events were clustered into 3 groups with respect to 6 kinematic parameters as follows: 1. Moving average bow acceleration, 2. Vertical bow acceleration above the moving average, 3. Peak frequencies obtained from wavelet analysis, 4. Peak magnitudes from the wavelet analyses, 5. Relative bow displacement and, 6. Relative bow velocity. These features were determined from strain measurements, vertical bow acceleration and bow relative motion data. Based on this ship data, a comparison was made using different algorithms including linear support vector mechanics, naïve Bayes, and decision tree for bow entry classification with results from the analysis as shown in Fig. 7. The machine learning models developed were used to group and cluster bow entry events, and by doing so this provided a basis for undergoing slam probability analyses to determine the likelihood of wave slamming that has a direct implication on structural loads and design.

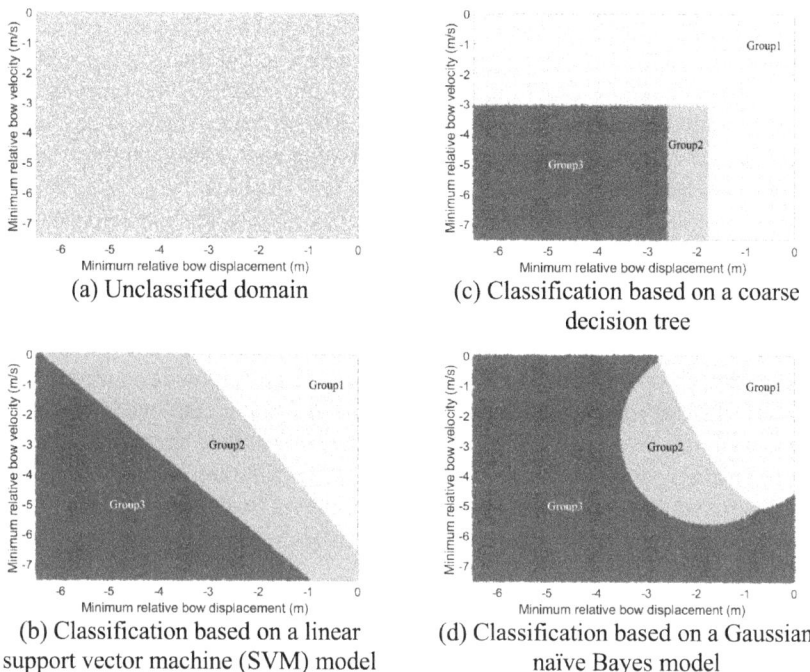

Fig. 7. Supervised machine learning models developed by Shabani et al. (2021) for classifying bow entry events using 6 key kinematic parameters clustered into 3 groups based on features from strain and acceleration data collected on Incat catamaran vessels.

2.2.3 Small Waterplane Area Twin Hull SWATH

In the recent period, there was an increasing interest for small size SWATH vessels as an option for offshore service vessels, but also there was a research by Ma et al. (2023a, b) to examine slamming loads and the possible effects of air cushion on SWATHs. Ma et al. (2023a) performed a set of drop-test experiments for a SWATH segment model to examine the air cushion effect on the resulting wetdeck slamming pressure. Their experiments showed that the slamming pressure has multi-frequency oscillations, caused by the air cushion effect, and that the air cushion effect can reduce the wetdeck slamming pressure. Further, Ma et al. (2023b) performed towing test models of a SWATH in regular waves, again confirming the pressure oscillation after the maximum slamming peak due to air cushion effect. These experiments found that the peak value of the slamming pressure at the wetdeck midpoint is larger compared to other location on the same cross section, but that slamming pressures were not strongly affected by significant wave height due to air cushion effect.

2.2.4 Trimarans

Multiple authors have used physical drop tests to better understand the effects of slamming and the expected loading. Li et al. (2021a) performed drop tests on a generic trimaran section to investigate the influence of the main hull profile on the resulting wet-deck slamming loads. The authors proposed a "weakening hull method" to modify the generic design with a lengthened main hull to hopefully minimize the flow field disturbance during hull penetration. The modified model experienced only a single slamming peak, as opposed to the original model's dual slamming due to the spray from the main hull's initial water penetration.

Duan et al. (2022) examined the effect of drop height and heel angle on slamming pressures on trimarans and offered a Gaussian distribution fit to predict the peak slamming pressure and duration time for different trimaran working conditions (considering the impact velocity and angle). Pan et al. (2023) examined slamming loads on a trimaran using an elastic model, finding that a concave trimaran shape produces more air cushion effect and that the maximum pressures are found on the side hulls and connecting bridge.

Numerical experiments and calculations are further being employed to investigate trimaran slamming. Sun et al. (2020) performed a CFD study to investigate slamming loads on trimarans and how the loads are affected by different parameters and entry types. They found that entry velocity was more dominant (over entry type) in affecting the pressure peaks and that the slamming characteristics are strongly correlated with the penetration depth.

Chen et al. (2023) examined the use of Equivalent Dynamic Coefficients (EDC), which combine static and dynamic analyses to predict the structural responses of trimarans under slamming loads. EDC are defined by applying a calculated slamming pressure to a grillage model and taking the stress from both a static and dynamic analysis; the EDC is then the ratio of the peak value from the dynamic analysis to the peak value from the static analysis. The authors found a difference of EDC values based on the added impact of the transverse bending moment, where the EDC is larger when including transverse bending modes. The authors further proposed a new method to consider material nonlinearity and an equivalent method for the transverse bending moment of the cross deck, as these two factors indirectly produced additional structural responses.

Qu et al. (2023) numerically examined the water surface evolution, slamming pressure, and structural response of a trimaran during slamming events. The authors found that relative velocity impacts the structural response but have a small impact on the resulting strain duration and pattern. Jiang et al. (2022) experimentally and numerically examined the air cushion effect for trimaran slamming via drop tests to determine the evolution of the flow field near the trimaran cross deck during the water entry moment.

2.2.5 Air Cushion Vehicles ACV

Xu et al. (2020) reviewed the progress of the studies in air cushion dynamics, skirt structure dynamics and hydrodynamics of the wave field, and analyzed relevant factors leading to the nonlinearity of ACV dynamics, such as fan characteristics, skirt materials characteristics and cushion compressibility.

The skirt, a unique flexible structure made of rubber-coated fabric material, undergoes great deformation when the ACV hovers over waves. As a consequence, analytical methods cannot hold high accuracy for the actual 3D skirt under various working conditions. In numerical simulations, high fidelity results can be achieved, however, only idealized planar finger seals or a couple of bow skirts instead of whole skirts are considered by CFD methods due to expensive calculations. Jiang and Tang (2021, 2022a) developed a numerical methodology to simulate the water entry of the flexible bag based on the Control Volume (CV) method and the Arbitrary Lagrange Eulerian (ALE) method. The reasonable accuracy of the method is validated by a water entry test of the flexible bag of an ACV. The investigations concerning the water entry of the flexible bag vertically or with pitch angles were conducted systematically by the proposed numerical method, and the results of skirt deformation, internal pressure, and maximum principal stress were discussed comprehensively. In Jiang and Tang (2022b) the same approach has been used to analyse the effects of solitary waves on the flexible bag characteristics: pressures, forces, maximum deformations, and maximum principal stresses at the joints.

Gao et al. (2021) proposed a hybrid analytic-FEM approach for the simulation of the flexible skirts of an ACV under typical working conditions, including hovering condition and skirt-water contacting condition. The method adopts the dynamic explicit algorithm to solve the large deformation of the skirts, the hydrodynamic force obtained via semi-analytical calculation is updated every time increment and then dynamic simulation of typical skirts is performed by FEM solvers. The Authors concluded that validation examples showed good agreement with the traditional analytical method and previous CFD simulation, highlighting the method's potential to model an ACV with its full skirts for simulation of six-degree-of-freedom motion in future work.

Yang et al. (2023) experimentally studied seakeeping performances of Partial Air Cushion Supported CATamaran (PACSCAT) in regular waves and discussed the effect of the waves on pressurized flow leakage and consequent change of dynamic trim and increase of accelerations. Minty et al. (2023) studied the dynamics of the air plenum of Surface Effect Ships (SES) ship analytically and experimentally performing forced oscillations of air plenum volume. The Authors discussed vibrations magnitude in the wide range of frequencies and pointed out that the measured natural frequency of the heave deck was significantly less than the theoretical natural frequency of a rigid-walled plenum, demonstrating that flexibility of the seals and water surface under SES is likely to significantly affect the natural frequency of the heave deck.

2.3 Conclusions

From the reviewed research papers dealing with the High Speed Vessels it can be underlined the increased interest in Air Cushion Vehicles and the importance of improving the methodological approach for more accurate prediction of wave loads and ship responses. The possibility of sensoring ships and the availability of big data from full scale measurements facilitate the development of machine learning techniques and it can be expected that it will be the area of researchers' interest.

3 Large Pleasure Crafts

3.1 Introduction – Concepts and Challenges

The world of yacht construction, both sailing and motor yachts, is experiencing record in Global Order Book 2023 (Montigneaux et al. (2023)), with a growth of 17.5% in orders and 6.3% increase of total gross tonnage. According to the number of projects and total

Gross Tonnage, the top builders' nations are Italy, The Netherlands, Germany, Turkey and Taiwan. The continuous growth of yachts' dimensions makes them closer to ships with similar design and construction problems, challenging designers and shipyards to develop new design approaches and construction technologies. For the same reasons, Classification Societies have dedicated more attention to develop new focused rules for design loads, structures scantling and plant layout and components.

Shenoi et al. (2009) and Boote et al. (2012) focused on sailing and motor yachts respectively; Truelock et al. (2018) addressed larger pleasure yachts, primarily in excess of 30 m, so-called "super yachts". In this report, pleasure boats are divided according to their dimensions and applicable rules to: Large pleasure crafts (over 24 m) and Small pleasure crafts, both considering motor yachts and sailing boats.

3.2 Loads and Structural Challenges in Yachts Design

Design accelerations are a driving subject for very fast vessels and often the values imposed by the Rules are too severe to keep the structural weight compatible with the required speed. This subject has been widely assessed on the occasion of the redesign of a fast patrol vessel from light alloy into composite material (Hydar et al. (2022), Souppez et al. (2020)) by comparing the acceleration values imposed by CS Rules and those obtained by experimental tank tests and direct calculations.

In Boote et al. (2023) the comparison of different composite materials: E-glass, carbon and hybrid glass/carbon fibers for hull constructions was carried out. At first, the three solutions were evaluated by laboratory tests determining the tensile, bending, shear and interlaminar ultimate values. After that, the hull structure scantling was designed for the three composite materials by RINA HSC Rules with the resulting three versions accurately weighted and then optimized by FEM evaluations. From this comparison, carbon composite showed the best results in terms of weight reduction (but not in terms of cost) for a series production where E-glass composite could assure acceptable results at lower costs.

In hulls of modern superyachts, large openings are often required for inner basins, tender garages, balconies and large side windows. These openings strongly affect the hull strength which should be accurately evaluated and compensated with additional structural elements. This subject has been widely assessed in meetings, workshops and other technical events but, as known at this moment, no specific papers are available in the literature.

The same problem affects superstructures equipped with large windows. The combined action of bending and compressive loads, together with the fact that superstructures are usually made of aluminium light alloy, can be the cause of buckling phenomena with severe consequences on structure and glass integrity. Studies carried out by Boote et al. (2017a, b) aimed to evaluate the contribution to steel hull strength of light alloy superstructures with large openings by the finite element approach. The contribution of glazing windows to the primary response to global bending moment was assessed as well.

Even if not affecting the hull strength, thermal loads could cause important effects on the deformations of the hull side plates, giving rise to unpleasant optical wave effects which can lead to harsh owner claims. The effects of solar radiation on the hull warming and the quantification on the side plate distortion were reported in Kumar et al. (2016),

Boote et al. (2017c). The study was developed by FEM structural analyses on a portion of a superyacht hull in the region of aft sides, shown in Fig. 8 where the flatness of the shell makes more visible distortions possible. Both steel and aluminium materials were considered, covered by filler layers of different kinds and thicknesses. A preliminary experimental investigation was performed on several steel and aluminium plates covered by different filler types exposed to a warming lamp to calibrate the numerical models.

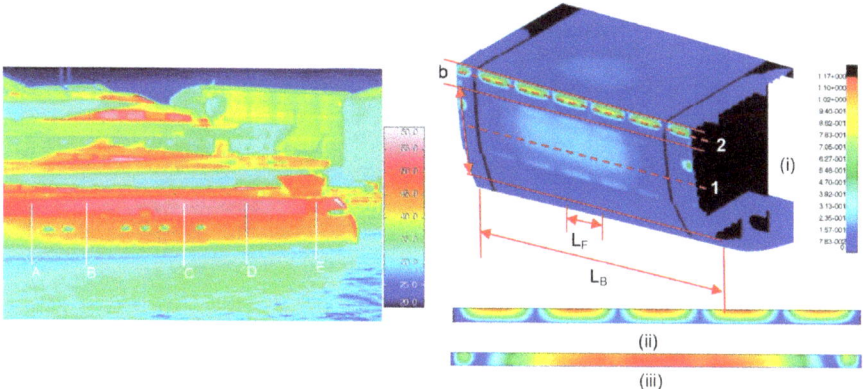

Fig. 8. Effects of solar radiation on yachts: (a) temperature measurements by thermocamera; (b) FEM simulation of plate deformation under thermal loads (Boote et al. 2017c)

3.2.1 Comfort

The comfort level on board modern superyachts is currently one the most important characteristics which make them commercially attractive, where the maximum speed previously took precedence. Comfort on board mainly means vibrations and noise control and, as they strongly depend on the hull and superstructure layout, comfort is the reference index which drives designers in structure scantling. The lack of standards and criteria for the assessment of the ship motion related to the risk of discomfort onboard large yachts was reported to be an important issue for the industry, brokers, owners and representatives. In 2022 ISO 22834 Large yachts—Quality assessment of life onboard—Stabilization and seakeeping guidelines were delivered to address the lack of a recognized and accepted procedure, criteria, and rating that can be used to compare yachts among each other and evaluate the impact of stabilization systems in the improvement of the comfort.

In Begovic et al. (2023) this new standard was applied to yachts with three different bow shapes: classic bow shape, vertical bow and bulbous bow. Even though the differences in comfort were small, the obtained ranking among the design solutions on a reference large yacht is the option nested with a bulb, contradicting the expectations of a vertical bow concept. The authors discussed some critical points of the methodology and the suitability of the proposed standard.

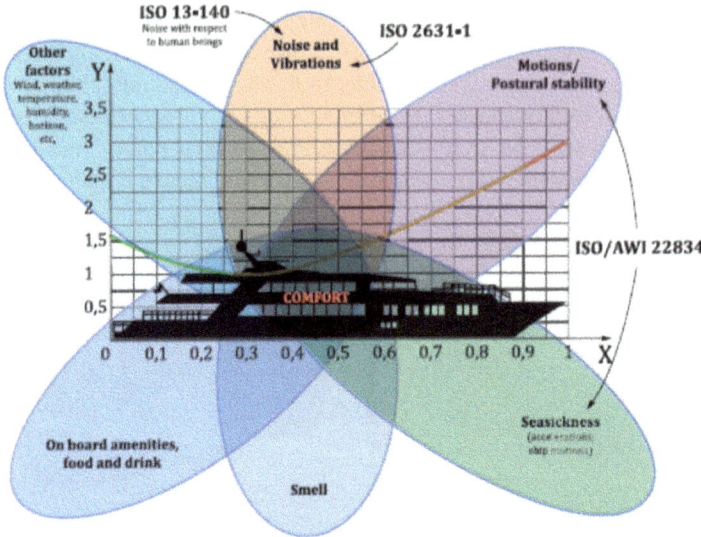

Fig. 9. List of elements contributing to comfort onboard (ISO 22834)

As seen in Fig. 9, ISO 2631 is the reference standard for the vibrations felt by humans, such as those induced by slamming as Whole Body Vibration (WBV) and Vibration Dose Value (VDV). Engelbrecht and Bekker (2023) conducted an extensive study on full-scale vertical acceleration measurements conducted near work and accommodation areas on the vessel together with the daily operational diary survey to gather human responses among passengers on polar research vessel. Results of the study reported a threshold for Vibration Dose when 50% of respondents indicated discomfort and discussed the RMS threshold. More research on this topic is needed to assist the cruise/pleasure yachts/high speed craft industry to specify criteria that will ensure comfortable vessels.

3.2.2 Vibrations

Apart from the effect of yacht motions, noise and vibrations are very important factors for the sensation of comfort on board and for these reasons the most important Classification Societies issued new rules and regulations for the evaluation of noise and vibration maximum levels. Such rules, usually named "Comfort Class Rules", contain both the general criteria for noise and vibration measurements in various yacht areas and maximum limit values which such measurements should fall into.

Vergassola et al. (2019b), Vergassola (2020) conducted research addressing the start of a measurement activity, starting from the choice and acquisition of the electronic equipment up to the first real scale measurement campaign carried out in cooperation with some Italian yacht shipyard, have been presented. A calculation procedure to preventively determine hull vibrations which can be problematic for crew and passenger comfort was set up and then applied to a case study steel yacht with light alloy superstructures. The dynamic behaviour of the same vessel was monitored during the various construction phases. Different FEM models of the reference yacht were set up by two multipurpose

FEM codes to identify the natural frequencies of local structures such as bulkheads, main deck and aluminium super-structure decks. The results of this first modal analysis were validated by real scale measurements carried out on the hull structures during construction.

The objective of this part of the investigation was to verify the reliability of numerical models with different refinement levels, starting from single parts of the hull structure, moving to more complete ones, up to the global model of the yacht. All the models were analyzed with and without outfitting. As confirmed by the analysis results, if models are properly refined, the results gathered on partial models can be as reliable as global models. This approach is useful to reduce FEM analysis time, especially when only local vibration modes are of interest (Fig. 10).

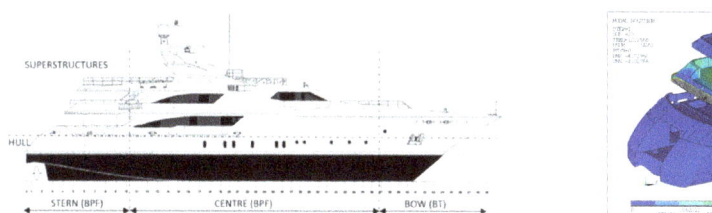

Fig. 10. Longitudinal view of the case study superyacht and FEM results of dynamic analysis (Vergassola et al. 2019b)

Lloyd et al. (2020) considered the recent advances in predicting and mitigating the noise and vibrations caused by propeller cavitation, specifically linking to concerns for yachts. The authors discuss propeller design aspects for yachts and specifically note that the empirical tip vortex method to predict broadband underwater radiated noise and hull-pressure changes due to vortex tip cavitation can be useful for design optimization.

Zambon et al. (2021) analyzed measured vibrations of super-yachts due to propellers and found that the common empirical Holden Method and similar methods suggested by classification society design standards overestimate propeller-induced hull excitations. They suggest an updated regression-statistical analysis method to consider propeller vibrations is needed for super-yachts and small cruise ships.

Fassola and Kustermann (2022) measured noise and vibration levels on a 50 m yacht and compared them against statistical energy analysis and finite element analysis, finding good agreement between prediction and measurements.

In the field of composite vessels, the modal and harmonic analyses of a 25 m motor yacht in composite materials were carried out to investigate the structure's dynamic behaviour and improve the comfort level (Hydar 2022). The modal analysis was carried out for two configurations (engines installed with rigid and elastic connections) to determine all the global and local natural frequencies and to check the possibility of the occurrence of the resonance phenomenon. Silvestri et al. (2024) reported results of sea trial measurements on a 52-metre superyacht equipped with traditional 2 five-bladed propellers, using vibration signal processing methods to identify vibrational components and shown that the proposed method can be a reliable tool for propeller cavitation monitoring.

3.2.3 Noise

Given the stricter Rules issued by Classification Societies (CS) about comfort Class, especially for yachts of more than 65 m in length, the reliable evaluation of vibration and noise levels at the earlier design stage becomes extremely important with particular attention to the difficulty of reliably predicting noise. Structure-borne noise generated by diesel engines on board yachts is strictly related to the intrinsic dynamic characteristics of both the diesel engine and the engine supporting system. To keep the structure-borne noise at low levels, the selection of the most suitable structural configuration of the engine foundation is one of the main aims of the ship designers. Prediction and optimization of the engine foundation dynamics are based on the knowledge of the diesel engine seating mobility levels. This can be achieved by FEA simulations once they have been properly calibrated on experimental data

In Vergassola et al. (2018) some numerical tools which allow prediction of noise propagation on board before sea trials to be carried out were presented. Because of the FEM technique limits in the range of high frequencies, numerical tools such as Statistical Energy Analysis (SEA) have been used to assess the noise level with accuracy at the early stages of a project, when structural changes can be made without additional time and cost. The level of uncertainties in SEA calculation is higher than for a deterministic approach like Finite Element Analysis (FEA) and this has to be taken into consideration while performing global analyses on vessels. For these reasons, experiments have to be carried out to improve the confidence of analysis to an error level up to ± 3 dB. In the paper, a procedure for the calibration of the input power due to propulsion engines has been studied in depth using a hybrid SEA+FEA model.

A study about noise propagation carried out by Statistical Energy Analysis (SEA) technique is reported in Vergassola (2020) where the propagation of noise throughout laminated glass was addressed with particular attention by a dynamic effective thickness equivalence.

The impact of Underwater Radiated Noise (URN) by ship traffic has gained increasing interest among scientists, ship designers and builders. As an example since 2017 the Port Authority of Vancouver (CA) has introduced important incentives and tax relief for those merchant ships that prove to be particularly virtuous in terms of noise emissions radiated into the water. Not many studies are available regarding pleasure crafts and large yachts. For this reason, the University of Genoa, in cooperation with an important yacht shipyard, issued an experimental campaign for the measurement of the underwater radiated noise of one large yacht (Boote et al. 2022). The noise was measured for several operative conditions and speeds ranging from zero to maximum speed.

3.2.4 Glazing

Glazed openings are particularly interesting for the V.5 Special Vessels Committee as many of the yachts have unique structural configurations allowing for very large or numerous wide glazed openings (i.e., windows, portlights, skylights) (Fig. 11).

Artefact, 2019, Nobiskrug, https://www.nobiskrug.com/fleet/artefact/, The total amount of glass on board is 750 m², photo by Francisco Martinez, accessed on 27/11/2024

Fig. 11. Artefact 2019, https://www.boatinternational.com/yachts/editorial-features/nobiskrug-superyacht-artefact, accessed on 27/11/2024, photo by Francisco Martinez

The glazing of vessel openings therefore has a very important role which brings about a variety of rules and requirements to determine the adequacy of the glazing for use in operation at sea. The Committee V.5 in the previous mandate performed a Benchmark to open the door to further research and discussion. The scantling of a large window of monolithic Tempered Toughened Safety Glass (TTG) is performed according to the class rules and results of FEM analysis by different software were compared. The conclusion of the Committee is that there is a need for a common methodology across Classification Societies and Standardisation Organisations for the calculation of specific design pressures dependent on glazing location, material and loading event. It was highlighted that each regulatory body has a different approach to the calculation of design pressure, as seen in Fig. 12. Although all utilise the same fundamental first principle formulas for flat plate bending to achieve the glazing design thickness.

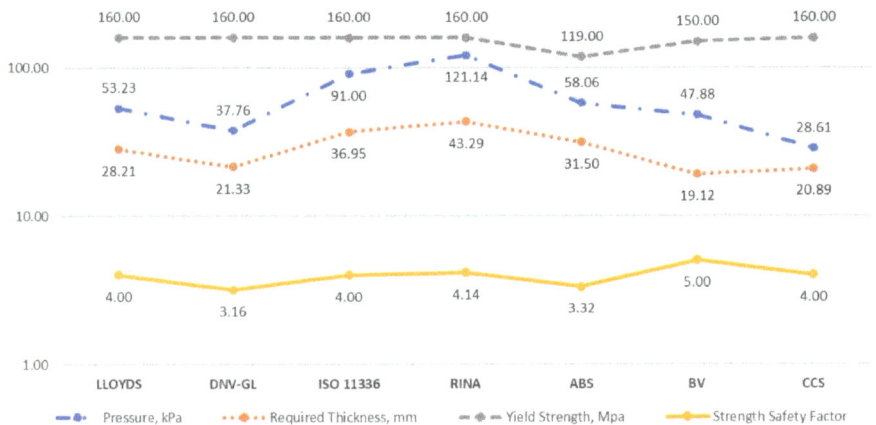

Fig. 12. Results from Class and Standard Glazing Analyses, Truelock et al. (2022)

Laminated glass is currently a common composite material in the construction of yacht windows thanks to its capabilities of safety glass and its sound insulation properties. In Boote et al. (2017c) the damping coefficient at natural frequencies of laminated glass has been obtained by different experimental modal methods and the Reverberation Time test is proposed to assess the coefficient in the whole frequency range of interest. In Vergassola and Boote (2020) a simplified method for the evaluation of a dynamic effective thickness of laminated glass plates is presented. The Authors proposed an interactive procedure to derive a dynamic equivalent monolithic glass with the same natural frequencies of the laminated one and analyzed the effect of thickness in the vibroacoustic characteristics of laminated glass. It has been highlighted that, even though the common practice in shipyards, passing from a 3 layered window to a 5 layered one, is not always the best solution, since the new natural frequencies can be resonant with the main external sources.

Wium et al. (2022) considered how to integrate glass into the yacht structural design, specifically how the glass can contribute to the structural stiffness and bear loading. The authors discussed the challenges of glass, specifically connection points/methods between the glass and other structures are examined. The authors concluded that there is a lack of knowledge on how structural glass components will perform within a ship's structure.

Moupagitsoglou (2020), Block et al. (2021) summarized the typical requirements a glass window in superyacht applications needs to fulfil, the most important design standards and testing. Furthermore, Moupagitsoglou (2020), reported an example of

FEM calculations using linear static formulas (prescribed by standards ISO 11336-1:2012 and BS MA 25:1973) and non-linear geometry analysis for two chemically strengthened panels. The author concluded that as a rule of thumb, non-linear geometry finite element analysis provides more accurate results when the yacht glass is deflected more than half of its thickness. This is usually the case for large span windows at the lower levels of the vessel due to the high design pressures. Linear static analysis using FEA or simplified formulas is adequate for smaller glass panels.

3.3 Structural Analysis and New Regulations

The rising complexity and specific needs of modern yacht design have motivated the revision and new editions of regulations by CS. Key areas of focus include:

- **Design Loads:** There have been significant revisions in structural rules to enhance safety standards, addressing the pressures on hulls, glazing practices, fire protection, and tank scantlings.
- **Technological Advancements:** The inclusion of regulations for hybrid propulsion systems and the proper management of lithium-ion batteries.
- **Innovation in Design Elements:** The regulatory updates reflect modern design practices, including new criteria for glazing, and considerations for innovative materials and construction methods.

3.3.1 Bureau Veritas (BV)

Bureau Veritas Regulations for Yachts are contained in a single collection which provides specific requirements for both commercial and private use. The first edition was issued in 2012 with some minor amendments in 2016, it was majorly revised in 2021 and subsequent amendments were issued in 2022 (Bureau Veritas 2022a). The main subjects which have been revised in the last version of their Rules reflect the present trends in yacht development.

- Particular attention is devoted to the evaluation of the pressures acting on the hull and to the scantling criteria for hull structures.
- A more specific procedure has been set up for glazing dimensioning and positioning.
- New fire protection requirements have been introduced.
- Because of the growing diffusion hybrid propulsion systems additional class notation for on-board installation of lithium batteries has been issued (reference to NR_467, Part F).
- A new section is dedicated to high density polyethylene structures with specific scantling criteria.
- Some changes have been applied to the equipment number calculation.
- Wind Propulsion System notation has been updated.
- New requirements have been introduced for composite propeller shafts

3.3.2 Lloyd's Register (LR)

For the 2024 edition of the Special Service Craft Rules (Lloyd's Register, July 2024) the following updates will be applied:

- In the structural rules updating particular attention has been addressed to the definition of the design loads for displacement vessels such as hydrodynamic and impact loads, design accelerations and wave height parameter for longitudinal strength. Other changes are planned for structure minimum thickness, for the applicability of ice class and hull openings (shell doors). Another focus point is dedicated to the structural criteria and arrangements of double bottoms
- Additional mitigation measures will be introduced to evaluate large glazing openings below the freeboard deck subject to Flag Administration acceptance.
- About tank scantling procedure a significant re-arrangement regards minimum test pressure, exemptions for very small tanks and removal of minimum overflow height for design pressure.
- Minor updates are SDA notation, sliding doors, openings in sheerstrake, azipods, fatigue limit for aluminium, bimetallic joints, virtual freeboard deck, fin stabilisers.
- New requirements will be introduced for guardrails on yachts with a load line length > 24 m.

3.3.3 Red Ensign Group (REG) (REG Yacht Code: Part A, (June 2023))

The REG rules revisions have been discussed jointly with Industry, Classification Societies, Management Companies, Associations and industry bodies in Amsterdam on May 2022 and June 2023 and will now apply to contracts signed (or keels laid if building on speculation) on or after 01 July 2024.

For what the structural aspects of motor and sailing yachts are concerned the following issues have been further developed:

- Design Criteria for shell doors, hinges and hydraulic control systems
- Alternatives to traditional sills for openings on the weather deck e.g. recessed/negative sills
- Clarification & simplification of the requirements for glazed openings, deadlights & balustrades
- Design considerations for glazed openings in way of and below the waterline
- Clarification of marking - Load Lines and Deck Lines
- Clarification of criteria for launching lifeboats, life rafts and Marine Evacuation Systems under adverse angles of heel and list
- Structural Fire Protection measures – Garages for <500GT to be protected
- Option to utilise bulkhead between ECR & engine room in lieu of the provision of a trunk
- Requirements for the storage of compressed gases and paint
- Lithium-Ion Batteries: simplification of fixed installations by harmonising with Class Rules and adoption of MGN 681 (M) Fire safety and storage of small electric powered craft on yachts
- Helidecks, Hangars & Re-fuelling Facilities: additional clarifications added
- Recreational dive facilities – Scope expanded to include wider range of diving equipment

3.3.4 Registro Navale Italiano (RINA)

The news on RINA Rules for yachting 2024 is the merge two current Rules for Pleasure and for Yachts Designed for Commercial Use in one unique «*RINA Rules for Yachting*» containing the requirements for yachts of any length (above and below 24 m) and any gross tonnage (above and below 500GT), designed both for pleasure and commercial activities carrying up to 12 passengers when in commercial use.

The 2 current set of Rules, the one for Pleasure yachts (applied to private vessel above 16 m), and the one for Yachts Designed for commercial Use (applied to Charter yachts of more than 24 m LLL) were already aligned since the last years for what it concerns the structural aspects (the major differences within the 2 set of Rules was relevant to systems and fire protection).

In particular, in the last years, the following updated have been done to Pleasure – Charter:

- The requirements for glazed bulwarks have been added. The design loads (weather and impact) for the scantling have been defined and the required test (impact test in acc. with EN 13094) for typical arrangement has been added. The use of different material has been considered and also relaxations for vessel of less than 24 m.
- Added the limitation (maximum area according to the distance to the deck of the lower edge of the window) for openable windows fitted on the hull subject to the acceptance of the Administration in case of charter service.
- Following a request of an Administration additional requirements are set for the scantling of side doors acting as platform when open. In addition to the load of the person and equipment that can be carried on the platform also the buoyancy and sea loads have to be taken into consideration for the scantling of the platform and its blocking and securing arrangement when in open position.
- Added the approach to be followed for the scantling of glazing partially submerged. Considering that these solutions have been only recently adopted in the rules it is stated that these arrangements will be considered on a case by case base (according to the dimension of the yachts, the stability criteria the yacht is subject to, the dimension of the glazing etc....)
- Revised some coefficients for the scantling of bottom plating and inner bottom stiffeners for steel and aluminium vessels. After a verification of the formulas on very large yachts (about 60–70 m) and the comparison with the required values on passenger ships made of the same material in accordance to the marine rules it has been found that a reduction of about 10% was necessary.
- Some clarifications has been set for the scantling of rudder with blades and/or stock made of composite (grp or carbon) material.
- RINA Rules for Ship (structural aspects in particular) has been recalled for yachts wishing to obtain the ice class notation (IAS, IA, IB, IC and ID).
- For the new 2024 merged edition the following amendments to the previous rules have been done:
- A new approach for the scantling of water freeing arrangement in line with the new edition 2024 of REG A. In particular the possibility of reducing the area of water freeing arrangement on upper deck of lateral external corridors has been added.
- Added the possibility of use negative sill height in line with the new edition 2024 of REG A instead of the traditional ones. The scantling (depth, length and width) of such negative sill in in line with the requirements of the most common used Large yacht Safety Codes and also the possibility of use alternative scantling subject to the satisfaction of a practical test of draining has been introduced.

- Added the possibility of use of glazed deadlights as an alternative to metallic one subject to the acceptance of the Flag Administration in case of charter service. This possibility has been added in a general way saying that the arrangement is subject to verification on a case by case base with the idea of applying the approaches in the new ISO 11336-1 as far as it is practicable.
- Small modification in the safety factors for keels of sailing yachts. The verification of grounding and pounding have been added.
- New requirements for tank testing in accordance with IACS UR S14 for "SOLAS-exempted" ships has been introduced for yachts equal or more than 500GT and new requirements for tank testing in accordance with IACS UR S14 for "non-SOLAS" ships has been introduced for yachts of not more than 500GT
- The table of minimum thickness of GRP vessels after many years of use of Part.B, Ch.4 for composite vessels has been found not more necessary and has been deleted.
- The requirements for anchors weight and chain cable length for yachts of more than 500GT have been increased and an additional class notation assignable on voluntary base has been added for yachts of any GT with anchors heavier and chain cable longer than the minimum requirements set in Part B. The notation may be assigned at different level according to the partial of full compliance with the IACS UR A1 relevant to anchoring equipment on ships.
- The minimum number of mechanical tests on GRP structures (monolithic or sandwich) for new building have been clarified in Part D.
- A dedicated section of Part A will be dedicated to the tests required and procedure for acceptance of all the hull structures and outfitting according to the different hull material for new buildings.

3.4 Conclusions

The last decade witnesses the continuous increase in the building of "superyachts", solely on powerside, sailing yachts accounts for 5.9% of Global Order Book (Boat International 2023). The design practice faces challenges to answer on key market trends like comfort and bigger spaces, the newest technologies and accessories on board, the introduction of sustainable materials, and propulsion solutions like fuel cells or hybrid propulsion. The final product is sophisticated and design practices are often very advanced to achieve the goal, but this experience is rarely shared and published within the scientific press.

4 Small Pleasure Crafts

4.1 Introduction – Concepts and Challenges

Owing to their smaller size, faster production rate and lower costs compared to ships, yachts offer an ideal design and innovation platform for proofs of concepts that could be applied to ships. For instance, the scaling of composite structures from small crafts to ships has been thoroughly investigated by the RAMSSES (Realisation and Demonstration of Advanced Material Solutions for Sustainable and Efficient Ships) (2023) and FIBERSHIP (2023) projects, demonstrating the feasibility and weight savings associated with composited, as opposed to steel, ships. Extrapolating from advances in composite ship sizes, Lowde et al. (2022) forecast the launch of the first 100 m composite ship for 2042 (see Fig. 13), showcasing the impact of advances in yachts onto ships.

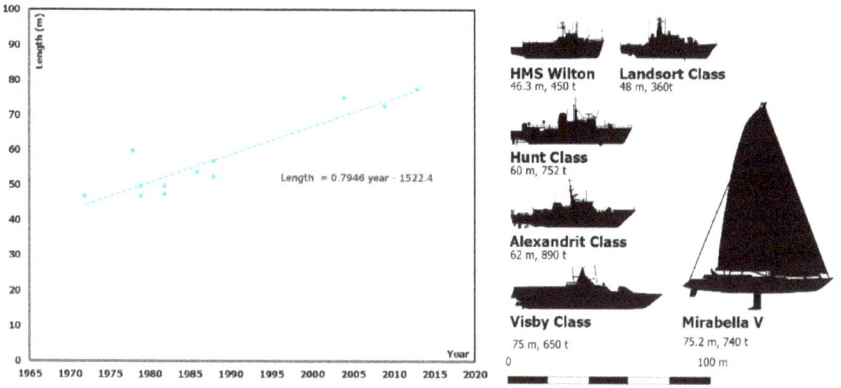

Fig. 13. Evolution in composite ship size over time (left) and scaled drawings of large compositevessels (right) (Lowde et al. 2022)

As such, recent developments in the structural design and analysis of yachts provide valuable insights into emerging trends and future developments, particularly with respect to the structural challenges associated with reduced emission vessels. Indeed, the report of the ISSC V.5 2022 (Truelock et al. (2022)) revealed there exist structural challenges to be addressed to support the development of low or reduced-emission vessels.

The main challenges identified were:

- the use of composite materials to achieve lightweight structures,
- the adoption of multihull configuration for reduced drag,
- the increasing use of hydrofoils, also intended to reduce resistance; and
- the implementation of novel propulsion methods, such as wind assisted ships, which parallels development in sailing yacht rig structure.

These four areas coincide with recent developments in small craft structures and their associated regulations, namely the ISO 12215 and its sub-parts. Additionally, the ISSC V.5 (Truelock et al. 2022) provided a benchmark on glazing given the prominence of large openings on yachts, which echoes recent developments in the ISO 12216 for small crafts and ISO 11336 for large yachts, each governing the strength and watertightness of glass windows.

4.2 Loads and Structural Challenges

4.2.1 Composite Monohull Structures and ISO 12215-5

The ISO 12215-5:2019 (ISO 2019) is concerned with the design pressures, stresses and scantlings determination for monohulls. The latest version, introduced by Souppez & Ridley (2017) and published in 2019 (ISO 2019), brought substantial changes to the structural analysis of sailing yachts (Souppez 2018a), power crafts (Souppez 2019) and commercial vessels (Souppez 2018b) below 24 m in length, prompting further research on workboats under this new regulations (Jang et al. 2019). The revision was motivated by advances in composite materials and manufacturing over the decade since the previous 2008 publication of the standard.

This prompted comparative assessments with respect to other class rules, tackling both the materials (Han et al. 2023) and loads (Souppez et al. 2020) to identify the commonalities and differences with other class rules and further experimental work to compare the mechanical properties with the default values of the ISO 12215-5. For a quasi-isotropic E-glass-epoxy laminate, Souppez and Laci (2022) revealed the conservative nature of the ISO ultimate flexural strength, the accuracy of the ultimate tensile strength, and raised concerns about the ultimate compressive strength, especially for vacuum-bagged samples, with the main cause identified as the value of the ultimate compressive breaking strain for chopped strand mat. The results are presented in Fig. 14. This provides recommendations that could inform future revisions. An additional area of further development was presented by Oh et al. (2022), who quantified the mechanical properties of glass laminate for varying fibre weight fractions. This work enables to de-rate the mechanical properties of composite laminate for fibre weight fractions beyond the ranges currently covered by the rules. Indeed, with advances in composite manufacturing techniques affecting the fibre weight fraction of the laminates and resulting mechanical properties, Han et al. (2020) demonstrated the higher accuracy of direct measurements compared to estimates and impact on calculated properties.

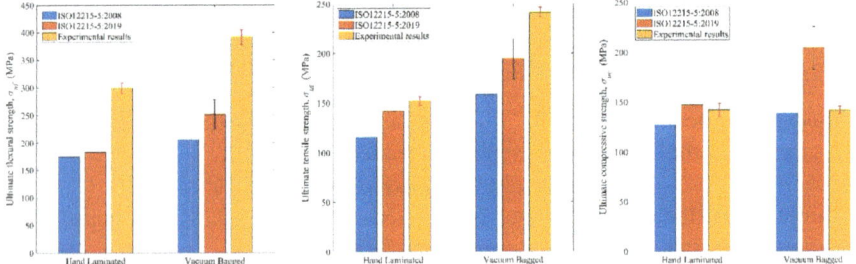

Fig. 14. Validation of the flexural (left), tensile (centre) and compressive (right) strength of quasi isotropic laminated (Souppez and Laci 2022).

Two further areas of development have attracted interest. Firstly, the use of timber as a sustainable composite material, either on its own (Souppez 2023) or combined with metal (Barry 2022), again with a focus on the comparison of the mechanical properties with that of the ISO 12215-5 (Souppez 2021). Secondly, more advanced structural designs and their optimisation, such as the concurrent multi-component optimization of stiffened-plates presented by Lorimer and Allen (2022). The advances further contribute to the design of lightweight structures, particularly when compared to steel or aluminium, thereby supporting the reduction of emissions thanks to lighter vessels. Further gains may be achieved hydrodynamically by moving to multihull configurations.

4.2.2 Multihull Configurations and ISO 12215-7

Multihull configurations include catamaran, trimarans and stabilised monohulls, all yielding significant displacement and fuel consumption compared to a monohull equivalent, as shown in the previous ISSC V.5 report (Truelock et al. 2022), thanks to a lesser hydrodynamic drag. This has, for instance, motivated work on Unmanned Robot Sailbots (URS) to maintain high speed without sacrificing stability, including that of Zhu et al. (2023), and Lin et al. (2023). Such configurations, however, are subject to global load cases. This is the purpose of ISO 12215-7:2020 (ISO 2020a), which covers the global load cases associated with multihulls. A total of six global load cases are provided, namely:

1. Diagonal load in quartering sea, ensuring suitable design torsional moment assuming the vessel is supported by two adjacent wave crests, one at the fore starboard hull (or starboard float for trimaran) and one at the aft port hull (or port float for trimaran).
2. Rig loads, specific to sailing yachts, and arising from the relevant regulations, namely the ISO 12215-10 (ISO 2020b), later discussed.
3. Asymmetric broaching loads, also specific to sailing yachts, where transverse horizontal pressure is applied at the fore end of a catamaran's hulls, or a trimaran's centre hull and leeward float.
4. Longitudinal broaching, also known as pitchpoling, where longitudinal deceleration loads are experienced as a result of the stems digging in a wave.
5. Longitudinal force, i.e. shock on the hull, representative of an impact (e.g. floating object, marine mammal), with resulting shear force and bending moment applied in the cross structure (cross beams or wet deck)
6. Bending of crossbeams, specifically those of motor vessels

Despite specific global loads for multihulls, their local scantlings remain based on ISO 12215-5. However, the slamming loads on the wetdeck of multihull vessels are of particular interest, with both numerical and experimental studies (Sun et al. (2021), Wu et al. (2021)). This is accentuated by the increasing presence of hydrofoils on multihulls to minimise emissions, be they greenhouse gas emissions thanks to lower drag (Hashimoto et al. 2021; Asgaree 2022; Firdhaus and Suastikab 2022), but also wash (Terui et al. 2023). Lastly, the inclusion of a hydrofoiling sailing catamaran class, namely the Nacra 17, since the Tokyo 2020 Olympics has triggered numerous research outputs (Graf et al. 2020; Marimon Giovannetti et al. 2022; Patterson and Binns 2022; Knudsen et al. 2023). In turn, this has triggered further interest in the structural design of appendages

4.2.3 Appendages and the ISO 12215-9

Despite the benefits of hydrofoils for both sailing (Souppez et al. 2019b; Bagué et al. 2021) and power boats (Budiarto and Firdhaus 2021; D'Amato et al. 2023), it is advances in wind assisted ship propulsion, as reported by Khan et al. (2021) and Petković et al. (2021), that have seen a novel interest for ship appendages, such as low aspect ratio keels (van der Kolk et al. 2021; Kramer and Steen 2022). Appendages, however, remain beyond the scope of ship regulations associated with wind assisted propulsion. Therefore, it is small craft regulations, namely the ISO 12215-9:2012 (ISO 2012) which covers sailing craft appendages, which can offer valuable insights.

The current 2012 version of the standard is currently undergoing revision (Lyons et al. 2024), ahead of a revised publication in 2025. The motivation for the revision is threefold. First, it follows significant regulatory advances in hull scantlings (ISO 12215-5), and the launch of new parts governing multihulls (ISO 12215-7) and rig loads and attachments

(ISO 12215-10), all covered in this chapter. Then, while investigations did not yield any concerns on the suitability of the ISO 12215-9:2012, loss of keels leading to loss of lives, such as that of the sailing vessels *Cheeki Rafiki* (Marine Accident Investigation Branch MAIB 2015), *Capella* (Federal Bureau for the Investigation of Maritime Accidents FEBIMA 2017) and *Showtime* (Lyons 2021), demonstrate the vital role of keel structures and the importance of fatigue life (Raju et al. 2010). Lastly, improvements in yacht performance over the past decade, coupled advances in the understanding of welded joints fatigue (Grimm 2016; Braun et al. 2018; Braun et al. 2021), flutter effects (Mouton and Finkelstein 2015; Mouton et al. 2018), load monitoring (Russell et al. 2016) and the behaviour of centreboard and daggerboard for dinghies (Graf et al. 2020; Guida et al. 2020; Bagué et al. 2021) further motivate a revision of the 2012 standard.

The configurations and load cases covered by ISO 12215-9 are as follows:

1. Fixed keel during a 90° knockdown.
2. Canting keel canted at 30°, including dynamics overload.
3. Vertical pounding.
4. Longitudinal impact.
5. A 90° knockdown for a capsize recoverable dinghy.
6. Dynamic loads sailing upwind.

Hydrofoils remain excluded from the scope of the Recreational Craft Directive (European Parliament 2013) and Recreation Craft Regulations (Office for Product Safety and Standards 2017) and thus may not be included in the ISO standard. However, the undeniable growth in hydrofoiling technology and configurations, depicted in Fig. 15, may lead to their eventual inclusion in small craft regulations, which would further support their wider adoption (Truelock et al. 2022).

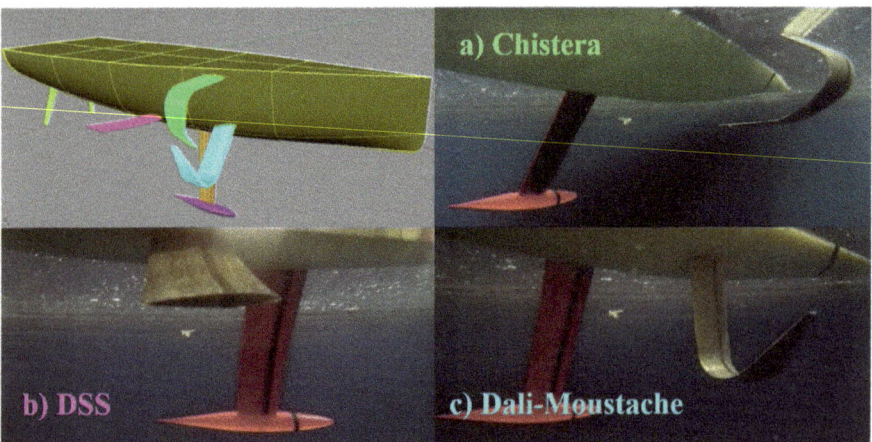

Fig. 15. Hydrofoil configuration in the benchmark of Dewavrin and Souppez (2018)

4.2.4 Rig Loads and ISO 12215-10

In parallel to the interest in keels, wind assisted ship propulsion is also concerned with the aerodynamic forces of the various systems available (Khan et al. 2021; Tillig and Ringsberg 2020; Reche-Vilanova et al. 2021; Dupuy et al. 2023). While class societies are yet to fully cover rig structures, existing regulations for yachts include the Nordic Boat Standard, reported by Larsson et al. (2022), and the guidelines for the design and construction of large modern yacht rigs by DNV-GL (2016), with most of the present research has been focused on rig loads for larger vessels.

The right way to define quantitatively the dynamic loads on a scientific base passes through the seakeeping analysis of the sailing boat under study. An interesting measurement campaign on mast acceleration values is reported in Boote et al. (2021) where two large sailing yachts were instrumented with accelerometers at the mast center of gravity during the transfer voyage from Cape Town and Genoa.

Not many papers are available about hull structure scantling of sailing yachts based on the global rig loads. One of particular interest is Ocera et al. (2017) where the authors present a simplified analytical method to evaluate the longitudinal strength of large sailing yachts taking into consideration the dock tuning load which is, for sailing yachts, of the same order of magnitude of still water and wave bending moment. The core of the method consists of the approximation of the longitudinal distribution of hull cross-section momentum of inertia by calculating cross-section geometrical properties of only 5 transversal frames, equally distributed at the stern and bow zones, at 25%, 50% and 75% of the overall length. The obtained results have then been validated by FE calculations carried out on three different case studies on which dock tuning loads, provided by the shipyard, have been applied to the numerical model.

A numerical method for the structural assessment of tall ship rigs is presented in Vergassola et al. (2023). It allows investigating complex rig layouts of large sailing vessel, composed of different types of masts (armed with both square and Lateen sails), yards, stays and spreaders. For these types of vessels, Classification Societies do not provide any direct calculation approaches, but only a mere comparison between cross sections and thicknesses with reference values as a function of the mast length. The provided approach can overcome this lack and it is based on considerations and calculations of real aerodynamic forces and the experimental evaluation of the pretension load of stays and shrouds. The herein proposed approach is very practical and can be adopted as a reference for this type of vessel, also with different mast configurations.

High performance sailing yachts are generally designed and built to a minimum weight objective often using very advanced composite materials such as pre-preg carbon fibers reinforcements not only for the hull but also for 'non structural' components, such as inner outfit and furniture. Vergassola et al. (2019b) carried out a study to verify at which level these components can be considered contributing to the hull local and global strength. Two different FE models of a 94 ft. sailing yacht, with and without 'non structural components', have been carried out in order to evaluate their contribution to the primary hull response to longitudinal bending moment and dock tuning load.

Most recently, namely since 2020, rig loads and rig attachments are also the subject of ISO 12215-10:2020 (ISO 2020b) for small crafts, building on the knowledge acquired from larger vessels. It is also anticipated these sailing yacht regulations may inform further development in regulatory structural arrangements for wind assisted ships. However, these may need to account for the greater righting moment achieved by hydrofoiling vessels (Souppez et al. 2019a), and the latest advances in sailing yacht aerodynamics that yield an increase in performance (Souppez et al. 2019b, Souppez and Viola 2024).

4.3 Conclusions

Recent advances in small pleasure crafts and their associated structural regulations have focused on composite materials, global loads for multihulls, and sailing specific areas such as appendages and rigs. These latter aspects are forecasted to develop further, owing respectively to the greater use of hydrofoils and the rise in wind assisted ships. As such, these developments contribute to the design and manufacturing of lower emission vessels, as part of a global effort to decarbonise the maritime industry.

5 Naval Vessels

Structural design methods for naval vessels including uncertainties in modeling techniques, including the blast loading, vulnerability analysis and specialised naval structures have been reported in Dow et al. (2015). Truelock et al. (2018) discussed the types of naval ships, naval rules, standards and state of the art of naval craft. The Committee evidenced that the ultimate purpose of naval craft "to deliver ordinance on target" distinguishes their design requirement and objectives from other (profit-driven) ships. In the previous report 2022, the Committee recommended focusing on digital twins and naval Unmanned Surface Vessels (USVs), which are expected to see significant development in the near future. Digital twins are another trend that will be applied across several sectors, even though most of the resources are currently focused on naval platforms.

5.1 Digital Twins

Digital Twins are virtual model of an intended or actual real-world physical product, system to process that serves as a digital counterpart of it for purposes such as simulation, integration, testing, monitoring and maintenance. (https://en.wikipedia.org/wiki/Digital_twin)

Starting from the definition of digital twins, Committee members discussed and agreed that it is a design method but still the importance and impact on the design and development of new machine learning tools, should the mention.

An example of a structural digital twin in naval vessels is the work carried out in the reference (Hageman and Thompson 2022) where real stress measurements of the "physical object" and simulation are correlated. They compared two methods: one using the ship as a wave buoy (SAWB) with motion data, which aligned well with strain measurements in high sea states but overestimated stress in mild conditions; the other using Hindcast data, which provided acceptable accuracy for fatigue accumulation in general but underpredicted it during higher sea states, leading to significant deviations. Another example is the work of Fujikubo et al. (2024) where the digital twin for ship structures aims to grasp the stress responses over the entire ship structure in waves by data assimilation that merges hull monitoring and numerical simulation. The image below shows an example of the concept of digital twin for structural monitoring (Fig. 16).

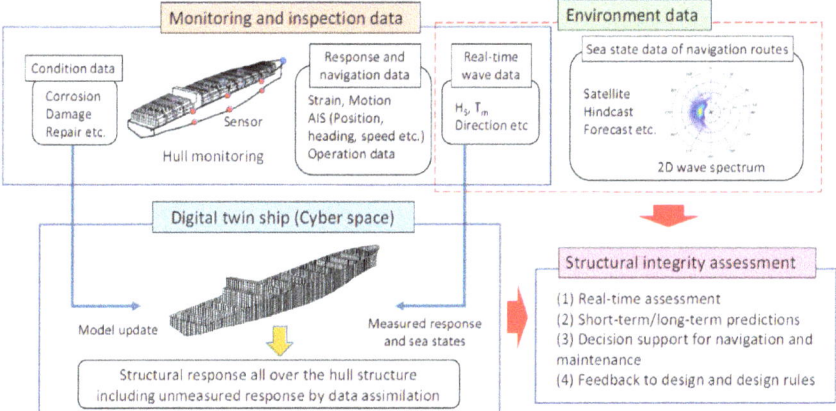

Fig. 16. Concept of digital twin system for ship structure (Fujikubo et al. 2024)

Other examples of structural digital twin are the projects ValiD I, II, III (https://www.marin.nl/en/jips/valid-participants, accessed 15th November 2024.) and FReady (https://www.marin.nl/en/jips/fready, accessed 15th November 2024), where the main goal is to optimize fleet deployment and structural integrity management through a combination of virtual and physical monitoring. The projects' objectives are to improve ship design tools, enhance hull structure monitoring, and develop forecasting techniques for fleet deployment.

The digital twins are becoming precious tools in the design and service of the ships. It is expected that the real time monitoring of ship responses: motions, accelerations, deformation and the comparison against numerical simulation, will further improve the knowledge of the phenomena and improve the definition of structural criteria used in the design such as: yielding, buckling, corrosion, fatigue, structural vibration, slamming, vulnerability, longitudinal and local structural loads, impacts loads, etc.

5.2 Naval Unmanned Surface Vessels (USV)

The first use of unmanned vessels dates back to ancient times, when they were employed as "torch ships." In modern times, they began to be used for military purposes, mainly as targets for testing in 1920. During World War II, they were also used for minesweeper.

Unmanned Surface Vehicles (USVs) represent one of the most recent and exciting developments in naval technology. These vessels, which can operate autonomously or be controlled remotely, have revolutionized the way we think about maritime exploration, ocean research, and other civilian and naval applications.

The primary use of this technology has shifted towards civilian applications, focusing on offshore and environmental domains. This transition has been marked by significant technological advancements, which in recent years have been increasingly integrated into the naval industry. Unmanned Surface Vehicles (USVs) have seen considerable improvements in autonomous navigation, high-precision sensors, and real-time communication systems. These innovations have enhanced operational efficiency and safety, enabling their deployment in various scenarios, from scientific research and environmental monitoring to complex maritime operations. The adaptation of these advanced technologies within the naval sector not only underscores the versatility and potential of USVs but also signifies a broader trend towards the modernization and digitalization of maritime assets Cubides Garzón et al. (2024), Curtolo et al. (2024), Ljulj et al. (2024).

In recent years there has been significant development of military Unmanned Surface Vehicles (USVs). According to Global Data Intelligence, the USV market is projected to grow from $894 million in 2023 to $1.5 billion in 2027, signalling potential advancements in their design (https://www.azom.com/news.aspx?newsID=61616). These USVs can be seen as automated or remotely operated vessels which can be capable of delivering explosives as a single vessel or in swarms making them difficult to counter.

All this suggests that the use of this type of vessel is promising for navies around the world. There are some advantages if we compare to traditional naval vessels:

Pros:
- Minimize risk to human lives.
- Flexibility to adjust the impact zone as it approaches the target.
- Capability to adapt to various types of weapon systems.
- Low cost compared to traditional ship price.
- Mass production in remote non-coastal locations.
- Multi-capability for simultaneous swarm attacks.
- Endurance, as it does not rely on human limitations and can operate in challenging environments.
- Automatic systems allow humans to focus on the target.

Cons:
- Susceptibility to electronic warfare.
- High vulnerability: not designed to withstand threats that a traditional military vessel could endure. Easily neutralized with an increase in specific anti-USV defenses at ports

5.2.1 Naval USV Classified by Sizing

There are different ways of classifying USVs according to their size, the following is the US NAVY's classification based on length (Navalnews 2021) as we define below:
Classification

- Large Unmanned Surface Vehicle (LUSV): Length >50 m
- Medium Unmanned Surface Vehicle (MUSV): 12 m $<$ Length < 50 m
- Small (SUSV): 7 m $<$ Length < 12 m
- Very Small (VSUSV): Length < 7 m

5.2.2 Naval USV Classified by Missions Types

Naval USV's can in turn be classified according to their mission, some of them are listed below (Navalnews (2021), Boretti (2024), Galway (2008)):

- Mine hunting:
- Mine Sweep
- Mine Neutralization
- Electronic Warfare (EW)
- Mine CounterMeasures (MCM)
- SUrfase Warfare Mission (SUW)
- Antisubmarine Surface Warfare (ASW)
- Anti Surface Warfare (ASUW)
- Logistics, Refueling
- Armed Escort
- Communication Relay
- Counter Piracy
- Intelligence, Surveillance and Reconnaissance (ISR)
- Patrol
- Defence
- Cyberwarfare
- Radar coverage

5.2.3 Naval USV Data Base

In Table 2, a list of Naval USV currently in service or under design pending completion is reported:

Table 2. List of Naval USV's currently in service or under design pending completion

Name/ID	Manufactured/designer	Manufactured Country	Sizing type	Reference Link
Seagull	Elbit systems	Israel	SUSV	https://elbitsystems.com/products/uas/unmanned-surface-vehicle/

(*continued*)

Table 2. (*continued*)

Name/ID	Manufactured/designer	Manufactured Country	Sizing type	Reference Link
Sea Hunter	RAFAEL	United States	MUSV	https://navyrecognition.com/index.php/naval-news/naval-news-archive/2023/september/13558-dsei-2023-leidos-displays-sea-hunter-usv-unmanned-surface-vehicle.html#:~:text=The%20Sea%20Hunter%20is%20an,speeds%20up%20to%2027%20knots.
Sea Baby	Security service of Ukraine (SBU)	Ukraine	VSUSV	https://defenceredefined.com.cy/ukraine-sea-baby-magura-mamai-and%CE%B25-hydra/
MAMAI	Security service of Ukraine (SBU)	Ukraine	VSUSV	https://defenceredefined.com.cy/ukraine-sea-baby-magura-mamai-and%CE%B25-hydra/
B5 Hydra	Security service of Ukraine (SBU)	Ukraine	VSUSV	https://defenceredefined.com.cy/ukraine-sea-baby-magura-mamai-and%CE%B25-hydra/
ALBATROS-T & K	ASELSAN	Turkey	SUSV	https://en.wikipedia.org/wiki/Aselsan

(*continued*)

Table 2. (*continued*)

Name/ID	Manufactured/designer	Manufactured Country	Sizing type	Reference Link
ULAQ	ARES shipyard & Meteksan defence of Turkey	Turkey	SUSV	https://www.ulaq.global/
The Protector	RAFAEL	Israel	SUSV	https://www.mindef.gov.sg/home
Inspector 90	ECA Group	France	SUSV	https://www.ecagroup.com/en/solutions/unmanned-surface-vehicle-inspector-90
Spartan Scout 7 m	US Naval Undersea Warfare Center, Radix Marine, Northrop Grumman & Raytheon	United States	VSUSV	https://www.globalsecurity.org/military/systems/ship/spartan-scout.htm
Spartan Scout 11 m	US Naval Undersea Warfare Center, Radix Marine, Northrop Grumman & Raytheon	United States	SUSV	https://www.globalsecurity.org/military/systems/ship/spartan-scout.htm
MAST	ASV Limited	United Kingdom	SUSV	http://www.navaldrones.com/MAST.html
MAST-13 (MadFox)	L3 Harris	United Kingdom	MUSV	https://www.bairdmaritime.com/work-boat-world/maritime-security-world/unmanned-systems/vessel-review-madfox-usv-designed-for-surveillance-and-force-protection-missions/
SALVO USV	DEARSAN	Turkey	MUSV	https://marinejetpower.com/references/dearsan-usv-salvo/

(*continued*)

Table 2. (*continued*)

Name/ID	Manufactured/designer	Manufactured Country	Sizing type	Reference Link
SANCAR	HAVELSAN & Yonca-Onuk shipyard	Turkey	MUSV	https://www.janes.com/defence-news/news-detail/lima-2023-turkish-navy-to-receive-first-two-usvs-during-2023
MIR	ASELSAN	Turkey	MUSV	https://wwwcdn.aselsan.com/api/file/MIR_ENG.pdf
Marlin	Sefine and TAIS shipyards & Aselsan	Turkey	MUSV	https://www.taisshipyards.com/en/usv-15-unmanned-surface-vehicle-marlin
RNMB Hebe (Catamaran)	L3 Harris	United Kingdom	MUSV	https://www.royalnavy.mod.uk/news-and-latest-activity/news/2021/june/21/210621-final-autonomous-minehunter-delivered
ARCIMS (Catamaran)		United Kingdom	SUSV	https://www.atlas-elektronik.com/solutions/mine-warfare-systems/arcims.html

(*continued*)

Table 2. (*continued*)

Name/ID	Manufactured/designer	Manufactured Country	Sizing type	Reference Link
RNMB Apollo	Thales	United Kingdom	SUSV	https://www.thalesgroup.com/en/worldwide/defence/press_release/thales-uncrewed-surface-vessel-passes-significant-milestone
Katana USV	Israel Aerospace Industries	Israel	SUSV	https://www.iai.co.il/sites/default/files/2023-03/KATANA%20D7%97%D7%96%D7%99%D7%AA%20%D7%92%D7%91.pdf
Eclipse USV	AI Seer Marine & 5G International	United States	SUSV	http://www.navaldrones.com/Eclipse.html
Textron CUSV	Textron systems	United States	SUSV	https://www.textronsystems.com/sites/default/files/_documents/003_SEA_SYSTEMS_CUSV%20Datasheet_2022_DIGITAL_SINGLE.pdf
DEVIL RAY T38	MARTAC	United States	SUSV	https://martacsystems.com/products/t38/
DEVIL RAY T24	MARTAC	United States	SUSV	https://martacsystems.com/products/t24/

(*continued*)

Table 2. (*continued*)

Name/ID	Manufactured/designer	Manufactured Country	Sizing type	Reference Link
Magura V5	Special Techno Export	Ukraine	VSUSV	https://www.navalnews.com/event-news/dsei-2023/2023/09/ukraines-magura-v5-usv-on-the-stage-at-dsei-2023/
Okham	Kayaci Defence	Turkey	SUSV	https://www.navalnews.com/event-news/dimdex-2024/2024/03/new-turkish-usv-breaks-cover-at-dimdex-2024-okhan/
Kaluga DS	Utek, Leonardo Hispania & Miltech	Spain	VSUSV	https://utek.es/kaluga_ds/
Kunai	Utek	Spain	VSUSV	https://utek.es/wp-content/uploads/2023/08/Utek_Ficha-de-producto_Kunai.pdf
OUSV 1 Ranger	Austal USA	United States	LUSV	https://www.naval-technology.com/projects/ghost-fleet-overlord-unmanned-surface-vessels-usa/
OUSV 2 Nomad	Austal USA	United States	LUSV	https://www.naval-technology.com/projects/ghost-fleet-overlord-unmanned-surface-vessels-usa/

(*continued*)

Table 2. (*continued*)

Name/ID	Manufactured/designer	Manufactured Country	Sizing type	Reference Link
OUSV 4 Mariner	Austal USA	United States	LUSV	https://www.naval-technology.com/projects/ghost-fleet-overlord-unmanned-surface-vessels-usa/
ADARO USV	SeaLandAire	United States	VSUSV	https://defence-blog.com/u-s-navy-showcases-new-type-of-unmanned-surface-vessel/
MUSCL USV	SeaLandAire	United States	VSUSV	http://www.navaldrones.com/MUSCL.html
GARC	Maritime Applied Physics Corporation	United States	VSUSV	https://defence-blog.com/us-navy-receives-new-garc-drone-boats/
Suhail USV	Performance Marine & L3Harris	Qatar	SUSV	https://breakingdefense.com/2024/03/qatari-shipbuilder-joins-forces-with-l3harris-debuts-suhail-usv-at-dimdex-2024/
B5 Hydra	Swarmly	Cyprus	VSUSV	https://www.edrmagazine.eu/swarmly-and-leonardo-join-forces-on-the-b5-hydra-armed-usv

In Fig. 17 the some of principal characteristics of USV database are summarized.

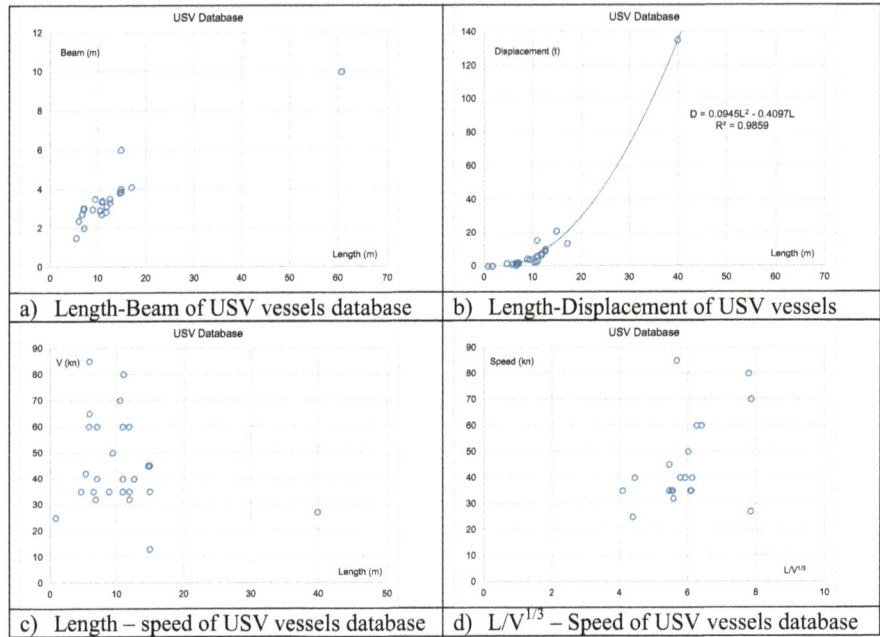

Fig. 17. Naval USV database summary

5.2.4 Classification Societies Rules for USV

Faced with this new type of ship, classification societies are developing new guidelines, standards, certificates that regulate the design, construction and operation. Below is a summary of the work of the main European and American Classification Societies

BV (Bureau Veritas 2022b) The classification society Bureau Veritas has the standard NR681 "Unmanned Surface Vessel USV", where it refers the ship's scantling to the naval standard NR483 and, in the case of requirements for the construction material to be used and analyzed:

- Steel, defined in rule note NR216
- Aluminum, defined in the rule note NR561
- Composite and/or plywood, defined in rule note NR546

LR (Lloyd's Register 2017) Lloyds Register has a Code for Unmanned Marine Systems that includes a chapter where define in a high level the: goal, functional objectives, and performance requirements.

The code defines the structure shall be designed and constructed to:

(a) *Enable the UMS to operate in all Reasonably Foreseeable Operating Conditions.*
(b) *Carry and respond to all foreseen loads in a predictable manner with a level of integrity commensurate with operational and safety requirements.*
(c) *Meet requirements for watertight, weathertight and fire integrity.*
(d) *Enable the maintenance and repair in accordance with the maintenance philosophy.*

DNV (DNV 2021) DNV has DNVGL-CG-0264 Autonomous and remotely operated ships, which does not contain any structural requirements detailed for USVs.

5.2.5 Structural Challenges Associated with the Naval USV

Reviewing the requirements of the previous classification societies and the structural publications on USVs, the technological challenge of this type of vessels is not in structural, this type of Naval USV does not present a major challenge compared to other vessels of its size, except for some naval structural requirements that should be eliminated or relaxed, such as: fatigue, residual stress analysis, vulnerability, corrosion, etc., as they should not be required in all situations.

5.3 Conclusions

USVs have made a strong appearance in recent years, with relative success in the Ukrainian-Russian war; the technology is still innovative and immature, with many questions as to whether it will replace naval warfare as we know it today. Answers to that question will require some time, as questions remain regarding the new forms of defense systems against them (Boretti, s.f.).

The use of digital twin in naval structures is a new service being evaluated and presents advantages to navigation support, maintenance support, rule improvement, and product value improvement (Fujikubo et al. 2024). In the case of naval USVs, it is a very useful tool to know the structural state and predict the structural state without crew.

Other types of unmanned military vessels include unmanned underwater vehicles (UUVs), which will also play an important role in future conflicts. Examples of UUVs range from small to extra-large, with the ORCA from the US Navy representing a significant extra-large UUVs (https://www.twz.com/orca-drone-submarine-delivered-to-navy, accessed 15/02/2024), (https://www.navsea.navy.mil/Media/News/Article-View/Article/3623016/us-navy-accepts-delivery-of-first-extra-large-unmanned-undersea-vehicle-test-as/, accessed 15/02/2024).

It is evident that navies globally will be exploring and developing various naval USV options over the coming decade.

6 Work Boats

During the last two mandates Committee analyzed the offshore operation vessels – according to their types and particular purpose (i.e. heavy lift, self-elevating vessels, subsea drilling, floatover installation vessels, walk to work vessels, etc.) and their structural challenges have been discussed. In this mandate, the Committee wanted to bring to the attention challenges related to fishing vessels and river and river/sea vessels. Design and hull forms of both types are strongly influenced by the local tradition, operating on relatively short range and often are old fleets with important structural and safety problems. Committee members observed an interest of the scientific community in these two categories and even though they are not "work boats" in the same manner as offshore operation vessels, they have been categorized here according to their specific (special) purpose. Finally according to the recommendations of the last mandate, the offshore service vessels used for floating offshore wind farm

6.1 Fishing Vessels

Recent research and developments in the fishing vessel sector are scattered among various aspects of the industry and naval architecture. The fishing industry is a prominent sector providing livelihood to billions of people across the globe and therefore the global fishing vessel market is estimated to significantly grow in near future. The topics of recent publications include risk-based maintenance methodology for fishing vessels, accident analysis, and improving hull strength using new materials and designs. New areas of research include implementing an extended emission index for energy efficiency, alternative powering options for fishing vessels, and the potential of lithium iron phosphate batteries for electrification. The integration of fully electrified fishing vessels remains as an area of future expansion as research into electric energy systems and cost reduction strategies for fishing vessel production continues to grow.

Domeh et al. (2022) presented a Risk-Based Maintenance (RBM) methodology to develop a maintenance plan for fishing vessels, to handle machinery faults, resulting in the main propulsion system (MPS) failure. The study used a new method, the "Goal-directed risk identification technique (Goal-DRIT)", to define the risk factors employed in developing the so called "MPS OOBN (Main Propulsion System Object-Oriented Bayesian network) model". The RBM methodology has been benchmarked against publicly available literature and has been shown to yield a 24.78% savings in the budgeted maintenance cost, using a fishing vessel operating in Ghana as a case study. The methodology and proposed models are recommended to the commercial fishing industry, chief engineers, and superintendents of marine vessels to aid their maintenance program design needs. Uğurlu et al. (2020) used Bayesian network, chi-square method to analyze accidents occurring between 2008 and 2018 in fishing vessels, with full lengths of 7 m and above. These networks allow to understand the occurrence of accidents in fishing vessels and to estimate the occurrence of accidents in variable conditions. It was also found that there was a significant relationship between accident category and vessel length, vessel age, loss of life and loss of vessel. Santiago Caamaño et al. (2019) employed statistical methods to consider real-time stability changes in fishing vessels. Following this Santiago Caamaño et al. (2023) proposed a combination of Empirical Mode Decomposition with the Hilbert-Huang Transform as a means of aiding a real-time stability monitoring system for fishing vessels. Rosano et al. (2023) described the development and testing of an on-board system that monitors lateral accelerations on ships, specifically a mid-sized Galician fishing vessel. The system estimates short-term extreme acceleration values and was validated using simulations and experimental data. The effectiveness of the system is verified for combinations of sea states through numerical simulations and experimental data. In all considered cases, the application of the decision support system has confirmed its capability to discriminate between safe and unsafe conditions.

Santoso and Nasution (2021) analyzed the strength of fishing vessels with composite bow by means of the finite element method in case of collision induced loads. Similarly, Windyandari et al. (2022) analyzed bow structure damage of a hybrid material combination of a coir-glass fiber composite fishing boat hull subjected to front collision load. A numerical simulation model was applied to the traditional fishing boat colliding with the fishery harbor quay wall, and the scenario was defined by varying the boat speed and the types of laminates adopted on the hull structure.

In a study conducted by Choiron (2023), a collision model for fishing vessels was developed to simulate collisions with an impactor in the form of a mooring pole during extreme weather conditions, featuring a wave height of 6 m and wind speeds 30 knots. The simulation model considered variations in ship velocity, frame spacing and weather conditions, to determine differences in structural deformation, energy absorption, and plasticity due to collisions.

Burella and Moro (2021) focused on developing noise control interventions to reduce noise-induced fatigue on small fishing vessels, rather than just addressing hearing loss risks. Using Statistical Energy Analysis (SEA), experimental measurements, and graph theory, the research analyzes noise transmission paths and the contributions of airborne and structure-borne sources on a Newfoundland and Labrador fishing vessel. The study identifies practical, cost-effective solutions to mitigate onboard noise, which could be applied to similar vessels in the region.

Helal et al. (2024) evaluated the underwater radiated noise emitted by a typical fishing vessel in Atlantic Canada to identify key noise sources and their contributions. Using hydrophones and a numerical model, the study found that over 70% of noise at lower frequencies (63 and 250 Hz) and 40% above 1 kHz was due to the diesel engine and propeller. The findings aim to inform future fishing vessel designs to reduce noise impacts on the marine ecosystem.

Recent research efforts in the field of fishing vessels have primarily focused on improving their energy efficiency and environmental friendliness. As part of this work, Koričan et al. (2023a) formulated the extended emission index (EEI) for fishing vessels, similar to the Energy Efficiency Design Index (EEDI) for merchant ships, which includes the carbon footprint in the numerator and the amount of landed fish in the denominator. However, the authors emphasize the necessity for further improvements since benefit for the society is significantly different for various fishing vessel types (i.e. purse seiners versus trawlers), while numerator values are directly dependent on the power system itself. In this sense, it is important to mention studies of Koričan et al. (2022) and Perčić et al. (2023) who analyzed alternative powering options for trawlers and purse seiners, respectively. According to these studies, electrification is recognized as a promising decarbonization option, but market-based measures are still needed to be competitive to conventional power systems with diesel engines as prime movers.

Within the scope of electrification of fishing vessels, Perčić et al. (2024) performed life-cycle performance and cost assessments of different battery chemistries. The study recognized Lithium Iron Phosphate as the most promising technology, however, the model used in the study ignores safety and specific requirements related to the marine environment covered by classification society requirements. Koričan et al. (2023b) considered the integration of fully electrified fishing vessels into isolated electricity grids of islands. It is confirmed that this holistic approach could enhance both the decarbonization of marine operations and the penetration of renewables into isolated energy systems.

In order to decrease production costs of fishing vessels, which are rather case-specific both in terms of technical and operational characteristics (compared to other vessel types)

Saral and Köse (2023) proposed a dimensionless offset table which can be utilized to ensure the standard production of Black Sea type fishing vessels with consistent quality.

6.1.1 Conclusions

The literature in the field of fishing vessels demonstrates fragmented industry challenges which involve: accident analysis, structural integrity, and energy efficiency, and therefore it seems that a more integrated and standardized approach could accelerate innovation. The push for decarbonization and improved energy efficiency is evident, with the development of metrics such as the Extended Emission Index (EEI) and research into alternative power systems. Electrification, particularly with lithium iron phosphate batteries, has emerged as a promising pathway for decarbonization, though safety and classification requirements need further exploration. Techniques like the Risk-Based Maintenance (RBM) methodology and noise control interventions demonstrate potential to optimize costs and improve safety. Case studies have validated these methods, suggesting broader applicability for the fishing industry. Studies on materials and collision modelling highlight ongoing efforts to improve structural resilience and safety under extreme operating conditions, while innovations in composite materials and numerical simulation methods offer practical solutions for vessel durability and energy absorption. The integration of fully electrified vessels, real-time monitoring systems, and extended emission indices represents good ground for further research.

6.2 Offshore Service Vessels

Europe remains the global leader in offshore wind power, having pioneered the first commercial offshore wind farm in Denmark in 1991. According to the latest data from WindEurope (https://windeurope.org/intelligence-platform/product/latest-wind-energy-data-for-europe-autumn-2024/, accessed 30th November 2024), the total installed capacity of offshore wind farms in European waters had surpassed 35 GW. The United Kingdom continues to boast the largest capacity, approximately 15 GW, followed by Germany (about 9 GW) and the Netherlands (around 4.7 GW). Some of the world's largest offshore wind farms, including the Hornsea Projects, are located off the coast of England. Looking ahead, the European offshore wind project pipeline now exceeds 100 GW, underscoring the strong momentum in this sector. In alignment with the EU's Offshore Renewable Energy Strategy, Europe has set its sights on expanding offshore wind capacity to 54 GW by 2030 and 300 GW by 2050 to support its climate and energy goals.

Offshore installations, compared to their onshore counterparts, have the advantage of being able to exploit greater wind speed, thanks to the absence of natural obstacles, and consequently to produce more power. The placement offshore of big wind farms solves also the problems of visual impact and noise because the towers are located beyond the visible horizon, more than 3 km far from the coast, and also the environmental problems related to the danger posed by the towers for birds, raptors and migratory birds in particular, and bats are much more limited.

In Hong et al. (2024) comprehensive reviews of the installation challenges and opportunities for Floating Offshore Wind Farms (FOWFs) are presented. FOWFs allow wind turbine deployment in deep waters, offering advantages like higher energy yield and reduced environmental impact compared to fixed turbines. However, the installation process is complex, involving high costs, weather-related delays, and the need for specialized vessels and equipment. The review examines different floating foundation types, including spar, semi-submersible, and tension-leg platforms (TLP), each with unique installation requirements. It highlights ongoing research efforts aimed at optimizing the installation process, reducing costs, and improving efficiency through better vessel designs and innovative installation methods. The paper also stresses the importance of strategic planning, robust logistics and maintenance.

The real negative aspect of the installation of offshore wind farms is cost. The turbine represents just one-third to one-half of the costs in offshore projects today, the rest comes from infrastructure, oversight and maintenance being this last one the activity that follows commissioning to ensure the safe and economic running of the project.

Maintenance is, therefore, a fundamental aspect on which it is necessary to work to reduce costs because it accounts for approximately one-quarter of the life-time cost of an offshore wind farm, nominally 20 years, and still, there are no standardised technical and commercial practices. In addition, because future installations will be placed farther and farther away from the coast, it is necessary to identify new logistics solutions.

Maintenance activity is the up-keep and repair of the physical plant and systems. It can be divided into preventative maintenance and corrective maintenance:

- Preventive maintenance consists of checking and replacing components on a scheduled basis in order to preserve the characteristics of the system over time.
- Corrective maintenance includes the reactive repair or replacement of failed or damaged components.

Each offshore wind project has different characteristics which determine the optimal maintenance strategy. The main factors are:

- Distance from onshore facilities;
- Average sea state;
- Number, size and reliability of turbines;
- Offshore substation design.

Among these, the distance from the mainland is the most important aspect. Currently, to access the turbines small work boats equipped for the purpose, usually catamarans (12–24 m) are used, and sometimes it is necessary to use helicopters. Workboats, usually called Wind Farm Supply Vessel (WFSV) are relatively inexpensive and can carry significant numbers of technicians, but response times and accessibility are limited by transit time and sea state. Costa et al. (2017) presented the design and development of the WFSV-PL, a 26-meter hybrid carbon composite catamaran with the main aims of reducing construction costs and time by using carbon composites instead of metal. Key features include high-speed capabilities (up to 30 knots), capacity for 12 technicians, and innovative safety and comfort measures such as shock-mitigated seats and flexible cargo decks. Structural integrity is ensured through laboratory tests on composite materials, and hydrodynamic optimization in calm water by CFD (RANSE) simulations and towing tank tests.

In the fourth edition of TNO (2020), an updated review is provided on commercially available and proven access systems for offshore wind farms. The systems are categorized into three types based on the access point to the wind turbine: (i) access to the boat landing, (ii) access to the platform on the transition piece, and (iii) access to the helicopter hoisting platform located on top of the nacelle. While the traditional method of accessing the boat landing via Crew Transfer Vessels (CTVs) remains widely used, the past decade has seen the emergence of motion-compensated gangways installed on Walk-to-Work (W2W) vessels or specialized Service Operation Vessels (SOVs). This innovation has facilitated the relocation of significant portions of the maintenance base offshore. The analysis in this report highlights that the expansion of this market aligns with the growing demand for safer and more efficient transfer solutions for technicians and equipment to wind turbines.

Kjær et al. (2024) focus on improving roll motion predictions for Wind Turbine Installation Vessels (WTIVs) through Computational Fluid Dynamics (CFD). WTIVs, crucial for offshore wind turbine installations, present unique challenges due to their wide breadth, shallow draught, and high vertical center of gravity. Traditional empirical methods often overestimate roll damping for these vessels, leading to less accurate predictions. This study uses CFD to simulate free roll decay under varying loading conditions and includes experimental model tests for validation. Results show that CFD accurately predicts roll damping, which decreases with higher centers of gravity. Adding bilge keels significantly reduces roll motions.

Ren et al. (2021b) present a study on a novel approach for floating offshore wind turbine (FOWT) installation using a catamaran vessel equipped with an active hydraulic heave compensator (AHC). The turbine's tower, nacelle, hub, and rotor are preassembled onshore and transported to the installation site. The key challenge is minimizing relative motion between the catamaran and the floating spar foundation during the mating process, particularly in harsh sea conditions. A control system is designed using singular perturbation theory to manage the AHC, which compensates for wave-induced heave motions. The goal is to ensure the smooth lowering of the turbine assembly onto the spar, reducing relative velocities and displacement. Numerical simulations, verified in MATLAB/Simulink, demonstrate that the AHC significantly improves operational success by reducing relative displacement and velocity by over 95% and 93%, respectively, in various sea states. The study concludes that the AHC-equipped catamaran enables safer and more efficient installation, expanding the weather window for FOWT deployment.

In Li et al. (2022b) the authors evaluates the suitability of SWATH vessels for walk-to-work (W2W) operations in offshore wind farms, comparing their performance to traditional monohull vessels in terms of motion behavior, fuel efficiency, and operational limits. SWATH vessels showed consistent high operability across wave headings, especially under transversal and longitudinal transfer methods. Monohull vessels performed well under head seas but had significant operability declines in beam and quartering seas due to larger sway and roll motions. Optimization of SWATH geometry and the addition of damping mechanisms could further enhance performance.

A study addresses the challenges associated with launching and recovering Remotely Operated Vehicles (ROVs) from small Offshore Service Vessels (OSVs) using a Single Point Mooring System (SPMS) has been carried out by Deng et al. (2021). The research focuses on managing wire tension, specifically snap loads, which occur when ROVs pass through the wave zone during deployment. The paper presents a numerical model validated by experimental results and provides new strategies for safer and more efficient ROV operations.

Service Operation Vessels (SOVs) are critical to the efficient maintenance of offshore wind farms, particularly those located far from shore. SOVs are equipped to carry technicians, tools, and spare parts for prolonged offshore missions, reducing downtime caused by weather or logistical delays. Maintenance operations require careful planning, as turbine components are exposed to varying weather conditions, impacting their reliability. Weather forecasts and failure probabilities are factored into planning to optimize repair kits—a selection of spare parts tailored to the expected needs of each maintenance trip. Neves-Moreira (2021) proposes a tactical model to define the optimal composition of repair kits and an operational model to validate these kits in real scenarios. These models consider constraints like vessel capacity, repair demands, and emergency resupply capabilities. An SOV's ability to stay offshore for weeks, combined with helicopter-assisted resupply, helps address unexpected failures efficiently. Simulated scenarios demonstrated the importance of adjusting repair kits based on weather forecasts, reducing downtime and maintenance costs. A decomposition approach was introduced to simplify complex planning problems, improving decision-making for large-scale wind farms. The findings offer practical insights for optimizing SOV operations and advancing offshore wind farm maintenance logistics.

Lazakis & Khan (2021) develop a novel optimization framework designed for daily route planning and scheduling of maintenance vessel activities in offshore wind farms to reduce costs, minimize fuel consumption, and maximize wind farm availability. The framework incorporates heuristic optimization and clustering techniques, considering climate data, vessel specifications, turbine failure types, and operational constraints such as weather conditions and technician requirements. By dividing tasks into sequential sessions (drop-off and pick-up), "OptiRoute", the name the authors gave to this tool, optimizes routes for both SOVs and CTVs while integrating fuel efficiency calculations. A graphical user interface (GUI) enabling real-time visualization and planning is provided as well. Case studies validate OptiRoute's effectiveness, showing reduced maintenance times, fuel costs, and improved resource allocation. The study concludes by highlighting OptiRoute's potential for practical deployment and suggests further exploration into integrating long-term maintenance planning.

The study carried out by Tusar & Sarker (2023) develops an integer programming model addressing the multi-source and multi-destination (MSMD) nature of vessel routing and technician allocation, considering vessel types, capacity, cost, and operational constraints. A case study, involving 60 turbines divided into clusters, demonstrated the model's ability to select cost-minimizing fleet configurations. Using commercial optimization tools like IBM ILOG CPLEX, the model proved effective for short-term planning. Managerial insights include prioritizing large vessels for strategic investment and integrating small vessels to optimize costs in favorable conditions. Most significant findings result to be that larger vessels improve maintenance efficiency by reducing travel and downtime costs, a mix of large vessels and smaller crew transfer vessels (CTVs) enhances cost-effectiveness, especially under favorable weather. The optimal fleet configuration depends on vessel capacities, technician demands, and operational constraints such as weather windows and maintenance schedules.

6.2.1 Conclusions

Future projects involve the installation of wind farms at increasing distances from land. To reduce transit times, it will be necessary to use offshore bases, on which the technicians will live for a period of time, usually two weeks, and will then be transported on turbines using vehicles with sea-keeping characteristics better than the current ones. The base itself may be either fixed accommodation modules, similar to those used in the oil and gas sector, jack up or offshore support vessel boats with the function of 'MotherShips' the so-called WFMS. The optimization of the fleet size, and introduction of the autonomous ships are seen as possible pathways to reduce operation and maintenance costs.

6.3 River and River-Sea Vessels

A substantial portion of river-sea vessels, which are river vessels capable of marine operations due to class restrictions, often undertakes extensive operations within marine environments. The safety of river-sea vessel operations within marine environments is facilitated by imposing restrictions based on: region, season, distance to safe harbor, and wind and wave conditions. The presence of these restrictions allows for a substantial reduction in the cost of river-sea vessel construction and is primarily achieved by lowering requirements for longitudinal and local strength, sea-keeping qualities, characteristics of equipment and gear and main engine power. Consequently, river-sea vessels feature enhanced cargo capacity at fixed draft depths at the expense of lightship weight and an increased block coefficient when compared to traditional sea vessels (Kang et al. (2024)).

However, after 5–7 years of normal operation, there is a genuine risk of fatigue damage for the most heavily loaded structural components such as longitudinal strength members, decks, and sheer strakes. Additionally, there is a possibility of water leakages in the outer shell, and inner bottom.

Typical damages to river-sea vessel hulls can be attributed to technological and constructive deficiencies that occur during the design, construction, and vessel modernization stages. These defects are aggravated by operations in exceeding limit sailing areas.

6.3.1 Structural Challenges Associated with the River and River-Sea Vessels

Analysis of the operation of these vessel types enabled the identification of factors that exert the greatest influence on risk levels throughout their entire lifetime (Nilva (2020), Pei et al. (2020), Motok et al. (2022)). Due to reduced strength standards, river and river-sea vessels possess lower strength reserves than their analogous, unrestricted counterparts. Consequently, factors contributing to overdesign for still water and wave loadings can lead to severe consequences for these vessels' hulls, due to the amplified ramifications due to these hazards.

River and river-sea vessels operate in challenging conditions characterized by shallow water and frequent lock operations during the summer and ice-covered conditions in winter. These conditions exacerbate the severity of the relevant danger by diminishing the hull's strength capacity due to the accumulation of deformation and abrasion of the shell plating. For example, a Danube river barge passes through the locks about 250 times a year during normal operations; self-propelled vessel passes the locks 300 times a year (Radojcic et al. 2022). In a typical year, with normal operational intensity, barges complete 14 voyages along the entire Danube-Main-Rhine (DMR) system, resulting in approximately 1,100 lock passages annually. Conversely, a self-propelled vessels, complete 18 voyages annually, and navigate through locks approximately 1,500 times per year. The significance of these hazards and their consequences is clear: the contact between the vessel and lock or canal walls contributes to additional scuffing of shell plating, specifically the sheer and bilge strakes. Additionally, strake stiffeners may experience deformation, with fore end members being particularly vulnerable.

Another notable characteristic is that European river vessels typically operate within shallow waterways of densely populated and developed river systems within countries with stringent environmental regulations and influential environmental organizations Economic Commission for Europe (ECE 2017–2023). As a result, grounding events occur more frequently than in other basins for river vessels within the DMR system. The hazards caused by the waterway conditions are expressed in different ways, both directly and indirectly. Direct hazards are the primary cause of emergencies, such as hull breakage, which occurs due to contact with lock and canal walls, or grounding incidents. Meanwhile, indirect hazards encompass the accumulation of damages to the bottom, bilge, or sides of the vessel, leading to a significant reduction in hull capacity. This accumulation of damage can subsequently initiate hull breakage in other situations, such as during cargo or repair operations.

The utilization of floating cranes, which are extensively used in port road transshipment complexes, can inadvertently lead to damage to the side constructions of river and river-sea vessels during boom turns, wave-induced swinging, or when navigating in close-quarters. This situation can be particularly hazardous at the onset of loading or the conclusion of unloading, as the low-depth hull of the floating crane, equipped with ample fender protection, may come into contact with the barge's unprotected side shell plating.

Comparable damages to side constructions, such as sheer and bilge strakes, can also occur when a vessel contacts the walls of locks and canals.

A relatively high share of technical faults in analyzed accidents both in Austria (where technical faults were the second most frequent cause of all accidents) and in Serbia (where technical faults, together with operational causes, were the most frequent cause of accidents) indicates that the technical reliability of vessels should be improved even if the present level of manning is maintained. The first step towards this would be certainly a thorough examination and revision of the current requirements for systems and equipment contained in technical standards for European inland cargo vessels Bačkalov et al. (2023).

An extensive analysis of repair documents, cargo operation books, and logbooks for over 140 vessels with a considerable operational history enabled the identification of typical defects and damages affecting their hulls. This analysis, in turn, facilitated the determination of the primary sources of hull damage for river and river-sea vessels. Typical hull damages for a classic river barge of the "Europe-2B" type (ECE 2020) are shown in Fig. 18.

Most hull breakages occur during cargo loading and unloading procedures, as the absence of waves results in an unchecked and potentially dangerous surge in the bending moment.

By simulating various loading and unloading scenarios for dry-cargo river barges vessels, the following configurations shown in Fig. 19 have been found to be crucial in maintaining vessel strength during cargo operations.

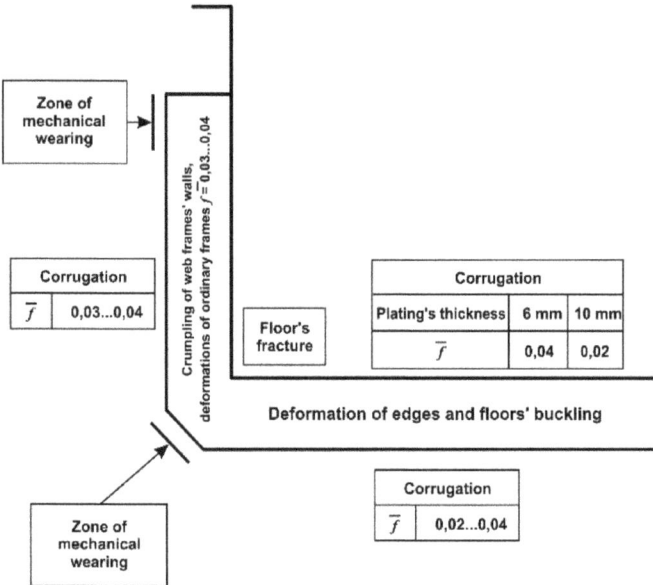

Fig. 18. Hull's damages of barge type "Europe-2B". $\overline{f}$ – relative deflection

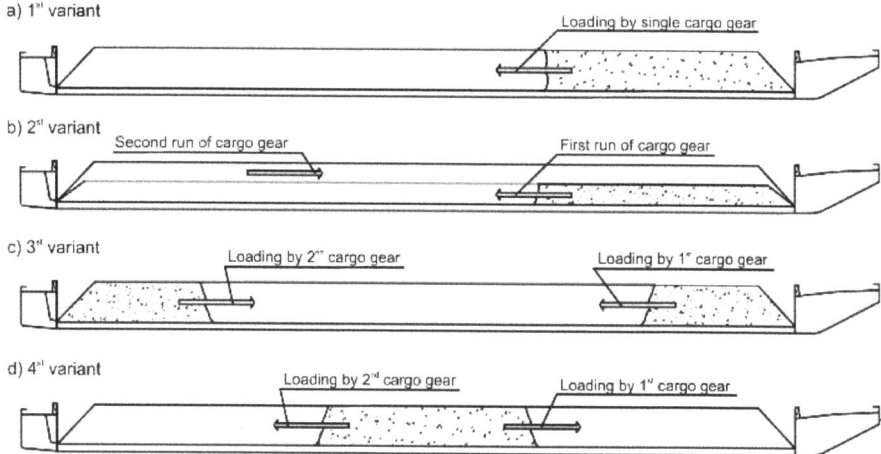

Fig. 19. Variants of cargo operations

The first and second variants are cases of "regulated" or "controlled" cargo operations. Namely such variants are typical for Loading Manuals (LM) and for typical calculations of hull's strength for the existing river and river-sea vessels. The third and fourth variants are the most dangerous cases of "uncontrolled" organization of cargo operations as the loading variant from the ends to the midship (the third variant) leads to the maximal possible hogging of the hull, and the loading variant from the midship to the ends (the fourth variant) leads to the maximum possible sagging condition. Of course, such violations may look somewhat artificial, but exactly such schemes give the maximal deviation in comparison with regulated operations (13–48%).

An overwhelming majority of river and river-sea hull breaches (particularly within the DMR system, where significant waves are absent) result from buckling failures of the compressed strake of the hull girder. This phenomenon is undoubtedly linked with the extensive utilization of transverse framing systems in European river shipbuilding.

Quantitative and qualitative enhancement of the longitudinal strength standard can be achieved through the following measures to mitigate the listed hazards:

- increasing of thickness and sizes of hull members, i.e. increasing the total of steel weight;
- modifying the load calculation method for longitudinal bending, considering additional designed wave and impact bending moments for specific vessel classes; monitoring load non-linearity and flooding; conducting longitudinal strength checks during cargo operations based on Loading Manual (LM) intermediate cases; carrying out longitudinal strength checks during trim changes for screw survey, potentially leading to enlarged member sizes;
- changing of the method of longitudinal strength calculation due to safety factor increase; checking of the strength for hull by the end of lifetime taking into account wearing and deformations that may also lead to an increase of members' sizes;
- changing of the transverse framing system of the hull girder extreme strakes to a longitudinal system;

- enhancing the buckling strength of longitudinal members within the longitudinal framing system by decreasing frame spacing and raising the cross-sectional inertia moment of these members.

Replacing the transverse framing system with a longitudinal one for the "Europe-2B" type barge resulted in a 43% increase in ultimate bending moment capacity in the sagging condition, with minimal metal consumption increase (+2.5 t). This decision enabled the attainment of equal hull strength for both hogging and sagging. This principle is crucial for river vessels, as still water bending moments during cargo operations represent the primary loading and principal hazard (Pei et al. (2020), Motok et al. (2022)).

Reliability based methods was used in analysis of the still water bending moments for more than 400 load cases of 37 prominent river-sea tankers. It is found that if some load case like 3 full after cargo tank or cargo operation procedure like one-run procedure is prohibited, the possibility of still water bending moment exceedance than the rule value will be decreased greatly (Motok et al. (2022)).

Reliability-based design optimization was also used for cargo hold structure design of a river-sea going container ship which sails in the Yangtze River and the East China E1 navigation route. The weight is reduced about 3.2%, and the value of failure probability is $10^{-7.3}$ which meets the requirement of 25 years design life. Compared to it the deterministic optimization design could reduce more weight which is about 5.1%, yet its failure probability is 0.00827 which is much lower than the 25 years design life requirement. The bilge plate, side plate, inner side plate and almost all the longitudinal scantlings are bigger by reliability-based optimization design than deterministic optimization design (Kang et al. 2024).

6.3.2 Conclusions

A systematic approach should be implemented throughout the entire vessel lifecycle encompassing classification phases, adhering to rules and requirements, designing, construction, operation, surveys, repairs, and modernization. A tailored strategy should be adopted for existing river and river-sea vessels, considering aspects such as operational conditions, transported cargoes, loading-unloading methods, waterway and port limitations based on overall vessel dimensions, and ice conditions.

New vessel construction should introduce superior quality standards for all components, including double bottom and double sides, increased hull strength reserves, thicker hull elements, more robust and dependable engines and mechanisms, modern automation and control systems, including construction and equipment redundancies

7 Ice Vessels

7.1 Introduction – Concepts and Challenges

This report is a continuation of the previous ISSC2022 report, which focused primarily on structural design, with an emphasis on the structural response based on full-scale measurements. This report expands on structural design under ice load, presenting relevant new results. The report first describes the measured structural response under ice load, which does not involve any new measurement methodology but is a crucial part of understanding the ice load mechanism in structural design. Next, the report discusses ice load evaluation, which includes ice material properties, ice load (pressure) acting on the structural component, and the structural response. These components are essential to understanding the principles of ice load. Finally, the report introduces several new topics related to ice vessels and offshore structures in ice-infested waters.

7.2 Ice Load Measurement on Vessels

Ice loading is commonly estimated using structural response data obtained from full-scale measurements in the field and model tests conducted in controlled environments. The measured structural response provides valuable information that helps us understand the transfer mechanism of ice loading during ship-ice interactions. However, there are several challenges associated with obtaining full-scale ice load measurements in the field due to the difficulties in measuring the load and the rapidly changing sea ice conditions caused by global warming. The lack of measured data is the main problem in estimating the ice loads. As an alternative to full-scale measurements, laboratory tests can be conducted to obtain data. However, these tests also present their own set of challenges. Controlling the model ice in an ice tank can be difficult, which can also create gaps in available data sets.

7.2.1 Full-Scale Measurement

Ice loads are transmitted to a ship's hull through the collision between sea ice and the vessel. When the ice pressure distribution on the hull at the collision area can be directly measured, the structural response to under the ice load can be accurately calculated. However, direct measurement of ice load using a spatial panel has difficulties in terms of spatial and temporal resolution. Therefore, a measurement system that captures the shear stress on the hull frames transferred from the ice pressure on the shell plating remains in use. Li et al. (2021a, b) analyzed full-scale data obtained from the Antarctic voyage of the Polar Supply and Research Vessels (PSRV) S.A. Agulhas II during 2018/19. Due to the lack of publicly available full-scale data covering ship performance and ice loads for various ice thicknesses, concentrations, and floe sizes, the datasets include ship navigation data, machinery data, local ice load measurements, and ice condition data, such as ice thickness, ice concentration, and ice flow size. Statistical analysis was conducted to seek suitable probability distributions that fit the measured ice loads and can be used as parent distributions for long-term estimation (Li et al. 2021b).

Ice forces converted from shear strains measured on the starboard side at the bow, bow shoulder, and stern shoulder, were utilized for the analysis. Fang Li et al. (2021a) investigated the relationship between local ice load, ship performance, and ice conditions. Suominen et al. (2021) studied the effect of manoeuvres on the characteristics and statistics of ice-induced loading at different hull areas, including the bow, midship, and stern area, using the ice trial data of S.A. Agulhas II in the Baltic Sea. Cho et al. (2021) proposed a new approach to derive an Influence Coefficient Matrix (ICM), which represents the relationship between the measured shear strains and the local ice pressure on the hull plate. The measured strain data obtained from the full-scale trial of the Korean icebreaking research vessel ARAON operating in the Ross Sea, Antarctica, in December 2019, was used for analysis. The new method used in-plane stress components (σ_x, σ_y, and τ_{xy}), taken from the hull data to derive a three-dimensional (F_x, F_y, and F_z) ice load. Oh et al. (2022), Heo et al. (2022), Wang et al. (2023a), and Kong et al. (2021) investigated the influence of missing shear strain data in the ICM. The resolution of the shear strains strongly depends on the accuracy of ICM. Due to the harsh measured conditions, the missing data will be collected occasionally during sailing in polar regions. Oh and Ha (2022) estimated the missing data of strain gauges applied to the hull of ARAON using the artificial neural network deep learning method. Heo (2022) analyzed the measured strain data using signal similarity analysis and proposed a method for reliably estimating ice load even in the case of missing data. Wang et al. (2023b) studied the effect of missing strain data on ice load identification of the ship structure using measured and simulated strain data of the Chinese icebreaker Xue Long 2 and proposed a practical solution to identify the ice load using ICM. Kong (2021) developed an ice load identification model for far-field measurements, where strain gauges were installed on the hull approximately 2 m above the waterline.

The importance of past data cannot be overstated due to the limitations of acquiring sea ice data in the field. In a recent study, Frederking (2021, 2023) reanalyzed previous ice loading data, which were measured during the icebreaking trials of the Research Vessel (R.V.) POLARSTERN in 1984. A direct measurement system was installed at two locations in the bow, where ice forces were measured using special panels. The measured ice forces of the pressure panel were then compared with strain-gauged frames on the R.V. Polarstern and the Canadian Coast Guard icebreaker Louis S. St-Laurent. The results of the comparison revealed significant differences in the nature of ice forces in terms of frequency and duration.

7.2.2 Laboratory Scale Test

Full-scale measurements can be very expensive, and identifying real-sea ice conditions, such as ice thickness, size and shape, and concentration during these measurements is challenging. Therefore, laboratory-scaled experiments have been proposed as an alternative method to understand ice-structure interaction. To date, model-scale tests of ice collision at slow speed (quasi-static) have been primarily conducted. However, due to wave-ice interaction and iceberg collision, ice collision speeds in ship-ice interactions become higher. He et al. (2022) conducted a laboratory-scale ice collision test measuring the dynamic response of a simple stiffened panel subjected to ice loading. An ice load identification model was proposed using the Green's function method and a comparison performed studying the effectiveness of the influence coefficient matrix method and the Green's function method in identifying ice loads, validating the feasibility of the Green's function method. Zhu et al. (2020) investigated the nonlinear elastic-plastic responses of plates impacted by an ice wedge striker and an idealized rigid striker, experimentally and numerically. The ice/steel striking wedge collided with the plate specimen. The plastic deformations, energy absorption of the plate, and energy dissipation of ice damage were investigated. Yu et al. (2023) reported an experimental and numerical study on the behavior of stiffened panels subjected to the impact of a wedge-shaped ice block. A new numerical ice model was proposed that represented the quasi-brittle manner and ice failure, implementing it in commercial software ABAQUS/Explicit. The numerical simulations using the proposed ice model compared well with the ice impact test.

Model scale tests of offshore structures are being conducted with increased frequency. Petry et al. (2022) performed Ice Basin experiments on mixed-mode failure of ice cones in the Aalto Ice and Wave Tank. The experiment aimed to investigate the effects of varying mechanical properties of model ice on the failure process against two types of structures: the cylinder and cone structure. Lemström (2022a, b) conducted model-scale experiments of the ice loading process against a wide, inclined structure in shallow water in the Ice Tank at Aalto. The study primarily investigated the effect of ice strength on the development of ice forces acting on the structure. Despite the advancements in model scale testing, a discrepancy between medium (laboratory) scale tests and full-scale measurements remains in the assessment of both ship structures (dynamic) and offshore structures (quasi-static). This scale problem is primarily caused by using model ice to represent the mechanical properties of sea ice.

7.2.3 Ship Monitoring under Ice Load

Ice load estimation in the field is a task that requires precise data of the ice conditions encountered. Although the estimations of ice thickness from satellite data were proposed, the spatial resolutions of satellite data are insufficient for estimating ice load on a ship's hull. Moreover, the size and features of ice forms, such as individual ice floes, level ice, pack ice, ice ridges, and multi-year ice, cannot be identified by satellite images. Shipboard monitoring can offer valuable data for ice load estimation. Chen et al. (2022) and Zhang et al. (2022b) applied the convolutional neural network method to identify areas of sea ice using sea ice images obtained from an onboard camera during the Antarctic expedition of the Xue Long vessel. Their proposed method demonstrated accurate recognition of the sea ice boundary. Similarly, Dowden et al. (2021) proposed semantic segmentation for automated detection and classification of sea ice types using camera pictures onboard an ice breaker.

In addition, Russia launched a new Ice-Resistant Self-Propelled rifting Platform (IRSPP), called "North Pole," equipped with an Ice Load Monitoring System (ILMS) (Maksimova et al. 2021). The ILMS has two functions: operational function, which ensures the safe operation of the IRSPP in ice conditions, and scientific function, which measures the ice load and the destruction of ice during ice-structure interaction.

Furthermore, Li et al. (2023) proposed using neural networks to increase accuracy of vessels operating in ice-covered waters outfitted with structural health monitoring systems. The neural networks were used to achieve optimal stress-strain prediction of ship structures. A model test was conducted, and nine strain gauges were arranged on the ship hull to monitor stress data. The results showed that the neural network could predict strain in the hull structure when operating in ice-covered waters, reducing prediction time and cost compared with conventional methods.

7.3 Evaluation of Structural Response

The determination of a structural response under ice load is crucial for evaluating ice load during ice-ship interaction, which includes ice strength, ice failure, and ice load (pressure) acting on the hull. The ice load evaluation method has been continually updated. However, research on the evaluation of structural response under ice load is not as widely published, due to the uncertainty surrounding the determination of ice properties and ice load on the hull. Recent studies on the determination of structural response are summarized below.

Structural damage to vessels operating in ice-infested waters poses a significant risk, as the loads involved are large and the interaction between deformable ice, the deformable structure, and water is complex. Experiments designed to investigate these interactions must account for hydrodynamic effects to accurately capture deformation in ship-grillage structures. To assist with this, the National Research Council of Canada, in collaboration with Defense Research and Development Canada and the US Navy, is developing a new facility called the Heavy Impact Test Theater (HITT). This facility will enable full-scale or near full-scale ice/ship-grillage impact experiments in water. In a related study, Gagnon et al. (2023) conducted a numerical simulation of a large, massive impacting object colliding with a massive hybrid ice-structure target, similar to ocean-going vessel collisions with ice masses (Gagnon et al. 2020, 2022).

Choung and Yoon (2022) and Yoon et al. (2023) conducted numerical investigations into the effects of radiation and wave excitation forces acting on the Korean icebreaker ARAON, as well as the structural damage caused by ship-iceberg impacts. Ship collision or ramming with ice floes result in energy absorption by the ship's hull, necessitating the use of suitable ice models for accurate ship-ice interactions. Yu et al. (2021) proposed a numerical solver for the coupled simulation of glacial ice impacts, which accounts for hydrodynamic-ice-structure interactions. The solver includes the BWH (Bressan-Williams-Hill) criterion for steel, a hydrodynamic pressure-dependent plasticity-based material model for constitutive modeling of ice, and liner potential flow theory for hydrodynamic loads. This solver was implemented in LS-DYNA and applied to simulate ice collisions on a semi-submersible platform column. In a separate study, Yoon et al. (2023) simulated the impact of ice blocks on a stiffened palate, utilizing the Mohr-Coulomb mathematical model (Herrnring and Ehlers 2022) for ice modeling, and was also implemented in LS-DYNA. The ice model uses a node splitting technique to preserve mass and energy during spalling and breaking of ice. The results showed that the agreement between experiments and simulations depends on the rotation angle of the ice blocks, which causes a change in the contact condition with the structure.

Mokhtari et al. (2023) conducted a numerical simulation of an aluminum plate subjected to an ice impact load, utilizing a rate and pressure-dependent elastic-plastic model for ice. The proposed ice material model was validated against physical ice-crushing tests, demonstrating a good correlation between experimental and numerical results. Wu et al. (2021) investigated the missing references dynamic response and energy absorption characteristics of aluminum honeycomb sandwich panels (AHSPs) under ice wedge impact, both numerically and experimentally. A three-dimensional nonlinear finite element model of AHSP under ice impact was established based on a concrete constitutive model of ice in the commercial package ANSYS/LS-DYNA. The numerical results of ice wedge-AHSP dynamic impact response was found to be consistent with experimental results. Banik et al. (2021) studied the low-speed impact of ice on CFRP sandwich composites, noting negligible damage with thicker face sheets. Cheemakurthy et al. (2022) investigated lightweight structural concepts in bearing impact load during ice-hull interaction using the dynamic finite element method LS-Dyna parametrically. The results indicated that the candidate structures were metal grillages and carbon-fiber reinforced plastic (CFRP) sandwich panels.

Cai et al. (2020) proposed an ice material model based on a soil and concrete material model. Using the proposed ice model, numerical simulations were conducted to study the dynamic behavior of a plate under ice impact, with results compared to a model test (Zhu et al. 2020), demonstrating the accuracy of the simulations. Cai et al. (2022a) proposed an analytical plastic damage prediction method for a ship plate impacted by an ice floe, combining the rigid-plastic theory method with the energy approach of ice crushing. The analytical results were compared with experimental (Zhu et al. 2020) and numerical (Cai et al. 2020) results. Furthermore, Cai et al. (2022b) studied the dynamic responses of steel plates subjected to repeated ice impacts based on experimental (Zhu et al. 2020) and numerical methods (Cai et al. 2020).

The Unified Polar requirements (UR) that govern the design of ice-class and polar-class vessels utilize a plastic limit state for ship structural design. However, these polar rules do not specify requirements for the connections between rule-defined shell plating and framing and the deeper supporting structure. Gosse et al. (2023) conducted a study using non-linear finite element analysis to examine connection designs on the ice strength of a vessel. The authors analyzed various connection designs between vertical stiffeners and longitudinal stringers in the ice-strengthened region of Polar Class 2 (PC 2) vessels, proposing connection designs that require less steel and welding while maintaining structural strength. Valtonen (2020) proposed a robust and straightforward assessment methodology and acceptance criteria for the practical non-linear analysis required by the IACS Polar Class Rules. The proposed method demonstrated the requirements set for non-linear analysis in the IACS PC Rules.

Shamaei et al. (2020) analyzed the relationship between ice pressure and the local design area using ship line-load data. The full-scale ice load measurements from the Kara Sea, Barents Sea, and Antarctic Ocean were considered. Pressure-area results obtained using the new definition of line load area (new definition of ice load) were compared to those obtained using the conventional local design area definition. The study found a good general agreement between the two methods. Veltheim (2022) used an inverse method to determine ice load magnitude and load patch on a ship hull. The load path and magnitude obtained using the inverse model were compared and agreed well with the full-scale data. Suominen et al. (2023) applied the Empirical Mode Decomposition (EMD) method to separate the noise from ice-induced load time histories measured onboard S.A. Agulhas II in the Baltic Sea. The full-scale data of the structural response (strain) under ice load include noise in the data. As the order of magnitude of the smallest loads and the load level related to intermediate crushing and flaking are at the same level as the noise, the identification of the smallest loads and actual changes in the load level from the measurement signal is necessary. The proposed method demonstrated its good applicability in separating the noise and the actual loading from the time history measurements.

The two main rule sets used in the marine industry are the International Association of Classification Societies (IACS) Polar Class rules and the Finnish Swedish Ice Class Rules (FSICR). Specifically, the lowest Polar Classes (PC7 and PC6) align with the two highest FSICR Ice Classes (IA super and IA). Oldford et al. (2023) examined the structural scantling differences between these ice classes (PC7 and IA) using finite element models of the structure for comparison.

During winter, the Northern Sea Route (NSR) is covered by level ice, necessitating icebreakers to assist LNG carriers in navigating the narrow ice channels. In such scenarios, ship-ice collisions typically occur on the shoulder, which is structurally weaker than the bow. Andryushin et al. (2022) analyzed sea ice forms, ice load, and hull strength during operation in the NSR for large-capacity Arctic vessels. The study presented permissible speeds, ice performance of large-capacity LNG carriers, and recommendations for ice class support vessels.

The use of simulator training for ice navigation provides an inexpensive and low-risk training method for ice-covered conditions. To effectively train individuals for ship operation in such conditions, Miller et al. (2023) developed an Ice-load algorithm that provides real-time feedback during simulator training. The ship-ice collision energy using the Popov energy method, as it was adapted for IACS Polar Class rules, was calculated using a given ice thickness, strength, and contact geometry from the simulator environment. The calculated energy was turned in to rectangular ice pressure. Using finite element analysis with the ice pressure as the input, a safe energy limit was calculated for ship structural of Polar classes (PC1-PC7). The safe collision energy limits were implemented into the simulator training to provide real-time feedback to participants about the safety of ship-ice collisions.

The interaction between ships and sea ice ridges can result in significant ice loads, which present additional challenges to the structural safety of icebreakers operating in polar regions. Sawamura et al. (2023) investigated the structural response of the Japanese icebreaker SHIRASE during its Antarctic voyage, which encountered various ice conditions including level ice, pack ice, ridge ice, and multi-year ice.

7.4 Structural Challenges Associated with Ice Vessels and Offshore Structures

This section introduces new topics, specifically structural challenges, in the Polar region. The interaction between ice and propellers results in thrust reduction and severe damage to the propeller, but this process has not been thoroughly understood due to the complexity of ice-propeller interaction. Recently, there has been a growing focus on fatigue damage in steel structures located in polar regions, particularly in offshore structures with slender bodies, such as wind turbines. This is because the vibration resulting from ice-structure interaction is especially severe in these structures. Additionally, renewable energy is a popular topic in arctic regions, with the number of studies on wind turbines in icy waters rapidly increasing.

7.4.1 Ice-Propeller Interactions

Azimuth propulsors are commonly used as the main propulsion unit on icebreakers and ice-going ships due to their high maneuverability and effective performance in ice management. These thrusters are designed to operate in ice-covered waters and must withstand impact loads from ice bodies, which can occur at high speeds when ships are traveling in channels with brash ice. Perälä et al. (2022) conducted two tests to study the impact forces and crushing phenomena associated with these impacts. The first test was a pendulum-type impact test carried out using real sea ice in the Baltic Sea, while the second test took place in the VTT towing tank and focused on the dynamics of the impact, including the movement of ice blocks and the vibration of the thruster.

The behavior of ice-classed azimuth propulsors under different ice load cases has been studied using the finite element method (Zhou et al. 2023). Ice-propeller collisions can result in extreme impulse loads, which are transferred through the propeller shaft to the sliding bearings of the stern tube and can eventually lead to the failure of the entire propeller system. Liu et al. (2023) conducted a numerical simulation of the ice-propeller milling process, using the cohesive element method and the elastoplastic softening constitutive law to simulate the crushing and deformation of sea ice. The calculated ice loads were found to be in good agreement with model test results. A sensitivity study revealed that the ice load is affected by the propeller rotation speed, advance velocity, and cutting depth to varying degrees.

In the study conducted by Xie et al. (2023), CFD-DEM coupling methods were presented to estimate the self-propulsion performance of an ice-strengthened Panamax bulk carrier in a brash ice channel. The hydrodynamic performance of a propeller was simulated by CFD, and ice load was calculated by DEM. The thrust, developed power, propulsion efficiency and ice load were calculated. The model test was carried out at HSVA and compared with numerical results. The difference of simulated and measured power is 7.38%. Gilges et al. (2023) conducted hydrodynamic simulations and multi-body simulations based on field measurements to investigate the influence of ice collision loads on the contact conditions in the bearings of the stern tube of the research vessel SA Agulhas II. They identified the operational condition thresholds, specifically the propeller torque and rotational speed that cause mixed-friction conditions in stern tube bearings during propeller-ice collisions. To monitor the health of propulsion system components in ice-covered seas, it is necessary to quantify the loading they are subjected to. However, direct measurements can be challenging in ice conditions. Therefore, it is necessary to measure elsewhere on the propulsion shaft to determine the ice-induced propeller moment.

Nickerson et al. (2022, 2023) have constructed a scale laboratory test rig to determine the ice-induced propeller moment through an inverse problem. The input loads and output responses are measured simultaneously during experiments with the scale model. Measured output responses are provided to the inverse model, and its estimation of the input load is compared to the measured loads, which are measured elsewhere on the propulsion shaft. The estimated loads are compared with the measured loads to demonstrate the accuracy of the estimations by the inverse model. To understand the interaction and failure between an ice specimen and a propeller blade-type edge, Böhm et al. (2022) conducted splitting tests of laboratory-made granular ice with a propeller-blade-like indenter at varying interaction velocities and two different blade thicknesses. The experiments indicated that the failure of the ice interacting with a propeller-blade-like indenter is a rather local process, while the failure in common compressive tests is a global process. Kistner et al. (2022) developed an FEM-based calculation procedure to identify and examine the vibratory stresses arising in the propulsion shaft of a container vessel in sea ice. The developed method provides the dynamic response of the shafting system subjected to torsional vibrations for different engine operation conditions.

7.4.2 Fatigue Damage

The current structural design rules do not sufficiently consider the effect of ice loading on fatigue life due to the lack of studies on the fatigue strength of welded joints under the combined action of wind, waves, and ice. Vessels navigating through ice-covered waters may be subject to fatigue damage due to repeated ice loading. Although researchers have proposed various approaches for the fatigue damage assessment of ships navigating in ice fields, a complete procedure is still lacking due to uncertainties in ice-ship interaction. Zhao et al. (2022b) proposed a procedure for assessing the fatigue damage of a ship's local structure caused by ice loads in level ice using numerical simulations. The study provides a numerical example for the Chinese icebreaker Xuelong 2, in which the ice-load peaks acting on the local hull due to ship-ice interaction were obtained through a long duration time-domain numerical simulation. The extreme value statistics of the line load peaks were predicted, and the fatigue stresses caused by the ice loads were estimated using beam theory. The fatigue damage was then calculated based on S-N curves and the Palmgren-Miner formula. Jeon and Kim (2022) conducted a similar fatigue damage estimation for the Korean icebreaker ARAON in level ice. They emphasized the importance of lifetime ice data collection, including ice type, ice collision frequency, ice thickness, and ice concentration, as it is crucial for the estimation of fatigue damage.

The effects of ice-induced loads and their consequences for the fatigue life of offshore structures are still not fully understood. Braun et al. (2020) and Braun (2022) have investigated the fatigue strength of the welded joints in ship and offshore structures subjected to sub-zero temperatures. The fatigue tests of butt-welded joints of normal and high-strength steel structures were conducted at room and sub-zero temperatures. They analyzed the impact of temperature on the fatigue strength. Braun et al. (2022) present the Variable-Amplitude Loading (VAL) spectrum for fixed offshore wind turbines (OWTs) and the corresponding VAL time series. The study then compared the results of fatigue tests to typical fatigue damage sums for regular and similar stress spectra. The FATICE (Fatigue Damage from Dynamic Ice Action) project, which was launched in Europe between 2018 and 2022, aims to address the challenge of fatigue damage in fixed offshore structures caused by drifting sea ice. This is a critical issue for various marine industries, including those involved in oil, gas, and offshore wind. Høyland et al. (2021) studied the fatigue damage on fixed offshore structures exposed to drifting ice in the FATICE project. For the fatigue load assessment of OWTs in sub-Arctic regions such as the Southern Baltic Sea, occurrence probabilities of ice thickness, ice-drift speed and wind-ice misalignment are required. Hornnes et al. (2022) estimated the drift ice thickness using a method based on modelled data on ice conditions from a large scale air-ice-ocean dynamics model from Copernicus, the European Union's (EU) earth observation program. They investigated the effect of the uncertainty in ice thickness and occurrence on the fatigue damage of the OWTs. Shin and Kim (2021) proposed a fatigue assessment procedure for a sloped offshore structure operating in drifting level ice. The ice breaking force induced by the ice-structure interactions was calculated using the ISO 19906 and analytical procedure. Fatigue damage was then calculated with the design S-N curve of welded joints proposed by Det Norske Veritas.

7.4.3 Offshore Wind Turbine in Sea Ice Conditions

Fixed offshore wind turbines (OWTs) are increasingly being developed for high latitude areas, where not only wind and wave loads must be considered, but also moving sea ice. To ensure that these wind turbines are safe for the environment while keeping them economically competitive, better guidelines and regulations should be developed through collaboration between European industry and academia (Høyland et al. 2021).

Many research projects related to offshore wind turbines have been launched, including the SHIVER project, which aims to address uncertainty regarding vibrations. This project is a collaboration between TU Delft, Siemens Gamesa Renewable Energy, and Aalto University. In the SHIVER project, a real-time hybrid test setup for dynamic ice-structure interaction of fixed structures has been designed for basin tests (Hendrikse et al. 2022). They proposed a classification of ice-induced vibrations that encompasses experimental observations for offshore wind turbines, based on the periodicity in the structural response at the point of ice action (Hammer et al. (2022, 2023)). An experimental campaign was carried out in the Aalto ice tank in Espoo, Finland, to investigate sea ice ridge interaction with bottom-fixed structures. The study investigated scaled ridge properties, ice growth, consolidation and failure processes, and the scaling of ridge forces with respect to cylindrical and conical structures at the water line (Heinonen et al. 2021; Salganik et al. 2021; Jiang 2021).

Fuglem et al. (2022) presented an analysis of 50-year iceberg impact parameters for four platforms off the coast of Newfoundland and Labrador, including a spar, a barge, a semisubmersible, and a Tension Leg Platform (TLP). Thijssen et al. (2022) analyzed 50-year mooring loads due to sea ice interactions for four types of floating wind platforms at various locations offshore of Newfoundland Canada. The potential advantages and disadvantages of the different designs with respect to ice actions were investigated.

Offshore platforms in the China Bohai Sea, Alaska, Canada, and Russia, as well as lighthouses and channel markers in Northern Europe, have all experienced severe Frequency Lock-In (FLI) vibration, leading to structural collapse or production shutdown. FLI vibration was examined for bottom-fixed wind turbines (Zhu et al. 2021). They developed a new ice-induced vibration (IIV) analysis model to mitigate the large structural response in FLI for monopile OWT. Ice-induced loads on offshore structures in the Baltic Sea are calculated by Tabri et al. (2022), and Heinonen and Mikkola (2023). Tabri et al. (2022) conducted the numerical assessment to evaluate the ice load history, load maxima, and vibration for a wind turbine foundation design in Estonian territorial waters near the Saaremaa of the Baltic Sea, where these locations is a potential wind farm development site. Heinonen and Mikkola (2023) simulated dynamic ice load excitation on a bottom-fixed channel marker structure (a slender monopile structure) in the Northern Bothnian Bay and analyzed the horizontal acceleration response at vertical positions in the structure.

7.5 Conclusions

Human activities in the Arctic region are increasing, primarily related to shipping routes, natural resource exploration (such as oil and gas), and the operation of ships and offshore platforms. Due to these increasing developments in the Arctic region, evaluating the safe operation and structural response will continue to be critical.

Full-scale measurements and laboratory experiments continue to be essential research topics in this area. Since the wave height is increasing due to the retreating of the sea ice area, the ice collision speed to the ship hull may increase due to the increase of wave height. Therefore, the research related to structural response under ice impact in high collision speed becomes a new, important research topic. Ice modelling including ice failure is continuously important topic for estimating structural damage with ice collision. Furthermore, the growing interest in setting up wind turbines in ice-covered waters, such as the northern Baltic Sea, has the potential to further activate research into ice-related fatigue damage.

8 Emerging Trends

8.1 "Bigger, lighter and faster"

Jagite et al. (2022) wrote: " *When technically specifying ships for the future, the following aspects are examples of what we will have even more focus on than today: bigger, lighter, and faster.*" This trend is observed in the last decades for all ships but especially for containerships, cruise ships and yachts. As reported by Rinauro et al. (2024), "*economics and logistics experts analyzed the trends of container ships growth based on economies of scale, port infrastructure, demand, and environmental tendencies, to predict the ship size limits. According to Malchow (2017), a **30000 TEU** container ship with approximately **20 m draught**, should be the ultimate limit because of the depth constraints in the Malacca Strait and the Suez Canal.* "The growing number of container losses accidents due to the parametric roll or onboard injuries due to excessive lateral accelerations and in particular, well known accidents of APL China in 1998 and Chicago express in 2008 were triggers for the IMO to start developing second generation Intact Stability Criteria and to increase awareness on safety and correct considerations of stability in waves and to related accelerations experienced by people onboard and cargo. Furthermore, large container ships are very flexible and their structural natural frequencies can fall into the range of the encounter frequencies in an ordinary sea spectrum. The classical approach to determine ship motions and wave loads based on the assumption that the ship hull acts as a rigid body is not reliable for the ultra large ships due to the mutual influence of the wave load and structure response. The methodology of predicting the ultimate strength of ULCS is continuously improving both with the most advanced numerical methods and through the validation against full scale data.

8.2 Ballast Free Vessels

The problem with the moving of invasive species from one water basin to another with the ballast water of the ships has been recognised by the end of the 20th century and in 2004 the IMO prepared regulations "Ballast Water Management Convention".

In the University of Michigan, Kotins and Parsons (2007, 2010) and Kotinis et al. (2004) proposed and patented ships with no water ballast tank. Their project excluded some compartments of the hull to reduce the displacement, there is a water flow from fore to aft and always connected to the outside water. This approach later on was developed by Godey et al. (2012) suggesting the water instead of flowing throughout the ship hull between steel structures to flow in special ducts (pipes).

Kashiro (2016) presented concept designs for Minimal Ballast water Ship (MIBS) and NO Ballast water Ship (NOBS), characterized by a midship section with an inclined bottom and reduced bow draught (greater trim compared to the conventional ship).

Since all variants of the ballast free ships have increased height of the transverse section many researchers tried to improve the arrangement of the longitudinal structures (inner bottom, inner boards, keel, bottom girders, etc.) to reduce the thicknesses and this way to minimize the increasing of the steel weight. Bending moments were found to be similar to the ones for the ballast ships and also appear during similar load cases, so the increased heights lead to increased section moduli.

Su et al. (2019) made multi-objective optimization of a cargo hold from an unmanned ballast free 30000 t oil tanker. As objective functions are used cargo hold capacity, water immersion and bending moment in a three dimensional model. As an optimization technique is used multiobjective Particle Swarm Optimization (PSO) algorithm. As a result, the optimal value and the subdivision schemes based on each object are obtained. The same ship and the same optimization objectives have been further studied by Radial Basis Function (RBF) method (Su et al. (2021)), where the Authors shown that after the optimisation the weight of the cargo hold was reduced by 1.15% and the stiffness was increased by 4.3%. In Su et al. (2022) the optimization for a compartment in the cargo hold of the same tanker is performed using the SUMP topology optimization method. As optimization constraints are taken the height of the transverse girders and the minimum width of the vertical girders as per the China Classification Society (CCS) and also structural yield. As a result the structural strain energy (compliance) is reduced by 6.8% and the stiffness is improved significantly.

The ballast free vessels are looking prospective for many types of ships and especially for those which run either loaded or ballasted, i.e. without the necessity to take some ballast in laden cases – tankers, bulk carriers, gas carriers. And even if the design of such ships seems to be a problem of the future (more or less close) it must be pointed out that Bureau Veritas, Lloyds' Register of Shipping, Class NK and probably DnV and CCS have approval in principle for similar projects. Structural analysis and uncertainty in modelling techniques is still not an issue in ballast free ships as they are in the concept stage. It has been known that the first similar ship is already built under LR Class for German company Hanseatic Ship Management −7600 m^3 LNG in Hyundai Mipo Ulsan Shipyard (Fig. 20). (https://www.lr.org/en/about-us/press-listing/press-release/building-the-worlds-first-ballast-free-lng-bunkering-vessel-with-hmd/, contacted 09/01/2024)

Fig. 20. Ballastless m/v Kairos IMO: 9819882. https://www.vesselfinder.com/it/vessels/details/9819882, accessed 30/11/2024

8.3 Life Cycle Assessment of Structural Design

While life cycle assessment (LCA) has been applied to both the end-of-life of metal ships (Rahman et al. 2016) and low-emission propulsion systems (Fernández-Ríos et al. 2022), there is an emerging interest in the application of LCA in early design stages for composite special vessels. This need was identified in the systematic review of Jacquet et al. (2024), highlighting the growth of this research field in the past decade, and particularly over the past few years. This has been applied across all special vessels covered in the scope of this chapter, namely yachts, superyachts (Del Pero et al. 2024) and ships (Han et al. 2024). The latter publication identified the advantages of sandwich composite structures, as opposed to single-skin ones, from a sustainability point of view while showcasing how material selection may be influenced by LCA in composite vessel design. However, there remain limitations associated with the underpinning LCA databases and in the application of a consistent LCA methodology to enable clear decision-making and comparisons) Jacquet et al. 2024). Consequently, this is seen as an emerging trend and it is anticipated the next ISSC V.5 committee will be able to provide a detailed account of this topic as further research emerges in the next few years.

9 Conclusions and Recommendations

This report is the continuation of previous reports Truelock et al. (2018, 2022) and provides a review of advancements, challenges, and future trends in the design and structural assessment of: high-speed crafts, pleasure boats, naval unmanned surface vessels, workboats, and ice-polar vessels. Even though a huge variety of ship types have been analyzed, some of the common features can be highlighted as the overall conclusions and recommendations for the next mandate.

The research approaches include numerical methods like CFD-FEM coupling, model and full-scale experiments, and machine learning to improve load prediction accuracy and structural design. The emerging trends from the last report remained the same: digital twins, unmanned surface vehicles, unmanned underwater vehicles, and the growth of ship size. The ballast free ships are looking prospective for ship types which run either loaded or ballasted, but at the moment, up to the knowledge of the Committee, only one of these ships has been built and it's difficult to say whether they will enter in the market.

In the field of pleasure craft and naval ships, due to the industrial competitiveness and highly confident information, there is a lack of peer reviewed papers dealing with the practical structural challenges solved by the most advanced methods. The majority of the published articles on these vessels are coming from the academy and the comparison against the real data is missing. Therefore, the closer collaboration of industry and class societies with the academy and education of new naval architects would be desirable.

As recommendations for the next Special Vessels mandate, the Committee members would highlight the following ship types and topics:

- autonomous ships and convey of ships in which only the front ship is manned, while the others follow autonomously
- unmanned surface and underwater vehicles
- wind assisted ships
- alternative fuels impact on design and structural scantling
- use of AI for scantling optimization

References

ABS American Bureau of Shipping: ABS guidance notes on ship vibrations. 2006. Updated February 2018 (2018)
Bureau Veritas: Rules for The Classification and the Certification of Yachts, NR500 DT R03 (November 2022a)
Bureau Veritas: NR 681 Unmanned Surface Vessels (USV), NR681 (July 7, 2022b). https://erules.veristar.com/dy/data/bv/pdf/681-NR_2022-07.pdf
Det Norske Veritas Germanischer Lloyds: Design and Construction of Large Modern Yacht Rigs. Det Norske Veritas Germanischer Lloyds, Baerum (2016)
DNV Det Norske Veritas: Autonomous and remotely operated ships (2021)
Lloyd's Register: Special Service Craft Rules, Southampton, (July 2024)
Lloyd's Register: Design Code for Unmanned Marine Systems (2017)
Red Ensign Group Yacht Code: Part A (June 2023)
RINA: Rules for yachting (2024)
Allaka, H., Groper, M.: Validation and verification of a planing craft motion prediction model based on experiments conducted on full-size crafts operating in real sea. J. Mar. Sci. Technol. **25**, 1199–1216 (2020). https://doi.org/10.1007/s00773-020-00709-6
Almallah, I., Ali-Lavroff, J., Holloway, D.S., Davis, M.R.: Slam load estimation for high-speed catamarans in irregular head seas by full-scale computational fluid dynamics. Ocean Eng. **234**, 1–11 (2021). https://doi.org/10.1016/j.oceaneng.2021.109160
Alsallah, A., Holloway, D., Mousavi, M., Lavroff, J.: Identification of wave impacts and separation of responses using EMD. Mech. Syst. Signal Process. **151**, 1–19 (2021). https://doi.org/10.1016/j.ymssp.2020.107385
Alsallah, A., Holloway, D., Lavroff, J.: Reducing wave impacts on high-speed catamarans through deployment of ride control: analysis of full-scale measurements. Ocean Eng. **292** (2024). https://doi.org/10.1016/j.oceaneng.2023.116581

Andryushin, A., Zuev, P., Voronin, A., Fedoseev, S., Ryabushkin, S., Kuteinikov, M.: Ensuring of the Strength of Large Arctic LNG Carriers and Oil Tankers for Year-Round Transportation in the Arctic Conditions. ISOPE2022, 1391 (2022)

Asgaree, M.: Hydrodynamic resistance reduction in catamaran assisted hydrofoils. J. Adv. Mater. Eng. (Esteghlal). **24**(1), 251–269 (2022)

Bačkalov, I., Vidic, M., Rudakovic, S.: Lessons learned from accidents on some major European inland waterways. Ocean Eng. **273** (2023). https://doi.org/10.1016/j.oceaneng.2023.113918

Bagué, A., Degroote, J., Demeester, T., Lataire, E.: Dynamic stability analysis of a hydrofoiling sailing boat using CFD. J. Sail. Technol. **6**(01), 58–72 (2021)

Banik, A., Zhang, C., Khan, M., Wilson, M., Tan, K.: Low-velocity ice impact response and damage phenomena on steel and CFRP sandwich composite. Int. J. Impact Eng. **162**, 104134 (2021)

Barry, C.D.: CNC enabled wood/metal composite construction of (relatively) high performance sailing yachts. J. Sail. Technol. **7**(01), 152–185 (2022)

Begovic, E., Della Valentina, E., Mauro, F., Nabergoj, R., Rinauro, B.: The impact of different bow shapes on large yacht comfort. J. Mar. Sci. Eng. (2023). https://doi.org/10.3390/jmse11030495

Begovic, E., et al: Experimental modelling of local structure responses for high-speed planing craft in waves. Ocean Eng. **216** (2020). https://doi.org/10.1016/j.oceaneng.2020.107986

Block, V.L., Lerner, S.D., Miller, S.R.: Glazing Safety and Resilience: New High-Performance Options, SNAME Maritime Convention 2021 27–29 October. SNAME (2021)

Böhm, A.M., Herrnring, H., Polach, F.v.B.u.: Splitting-Tests of Laboratory-Made Granular Ice With a Propeller-Like Indenter. OMAE2022-78186 (2022)

Boote, D., Hydar, V., Vergassola, G.: Re-design of a fast vessel from light alloy to composite material. International Conference on Materials, Science, Engineering & Technologies, Singapore. 7–9 September, 2023 (2023).

Boote, D., Gaggero, T., Pais, T., Rizzuto, E., Vergassola, V.: Underwater radiated noise from a large pleasure craft. In: Proceedings of the International Congress on Sound and Vibration (2022)

Boote, D., Vergassola, G., Delfino, P., Faloci, F.: Real scale measurements of yacht's mast accelerations. Ships Offshore Struct. **16**(10), 2021 (2021). https://doi.org/10.1080/17445302.2020.1816747

Boote, D., Vergassola, G., Di Matteo, V.: Strength analysis of superyacht superstructures with large openings. Int. Rev. Mech. Eng. **11**(1), 1–9 (2017a)

Boote, D., Vergassola, G., Pais, T., Kramer, M.: Finite element structural analysis of big yacht superstructures. Int. Rev. Mech. Eng. **12**(1), 1–9 (2017b)

Boote, D., Vergassola, G.M., Giannarelli, D., Ricotti, R.: Thermal load effects on side plates of superyachts. In: Marine Structures, vol. 56, pp. 39–68. Elsevier Ltd (2017c)

Boote, D., et al: Committee V.8 YACHT DESIGN. In: 18th International Ship and Offshore Structures Congress. 09–13 September 2012, Rostock, Germany (2012)

Braun, M., et al: Fatigue Strength of Fixed Offshore Structures Under Variable Amplitude Loading Due to Wind, Wave, and Ice Action. OMAE2022-78764 (2022)

Braun, M., Hensel, J., Song, S., Ehlers, S.: Fatigue strength of normal and high strength steel joints improved by weld profiling. Eng. Struct. **246**, 113030 (2021)

Braun, M., et al: Sub-zero temperature fatigue strength of butt-welded Normal and high-strength steel joints for ships and offshore structures in Arctic regions. In: Proceedings of ASME 2020 39th International Conference on Ocean, Offshore and Arctic Engineering, Fort Lauderdale, FL, USA. June 28–July 3 (2020)

Braun, M., Grimm, J.H., Hoffmeister, H., Ehlers, S., Fricke, W.: Comparison of fatigue strength of post-weld improved high strength steel joints and notched base material specimens. Ships Offshore Struct. **13**(1), 47–55 (2018)

Budiarto, U., Firdhaus, A.: Analysis of the effect of hull vane on ship resistance using CFD methods. In: IOP Conference Series: Earth and Environmental Science, vol. 649., No. 1, p. 012051. IOP Publishing (2021)

Burella, G., Moro, L.: Design solutions to mitigate high noise levels on small fishing vessels. Appl. Acoust. **172**, 107632 (2021) 16p

Cai, W., Zhu, L., Gudmestad, O.T., et al.: Application of rigid-plastic theory method in ship-ice collision. Ocean Eng. **253**, 111237 (2022a)

Cai, W., Zhu, L., Qian, X.D.: Dynamic responses of steel plates under repeated ice impacts. Int. J. Impact Eng. **162**, 104129 (2022b)

Cai, W., Zhu, L., Yu, T.X., et al.: Numerical simulations for plates under ice impact based on a concrete constitutive ice model. Int. J. Impact Eng. **143**, 103594 (2020)

Cheemakurthy, H., Barsoum, Z., Burman, M., Garme, K.: Comparison of lightweight structures in bearing impact loads during ice–hull interaction. J. Mar. Sci. Eng. **10**(6), 79 (2022)

Chen, Z., Zhao, W., Liao, X., Du, M.: A simplified methodology for dynamic responses of cross-decks of trimarans under slamming loads. Ships Offshore Struct. **18**(2), 231–239 (2023). https://doi.org/10.1080/17445302.2022.2035563

Chen, X., Ma, Y., Ji, S.: Study of Semantic Segmentation on Sea Ice Image Based on Deep Learning. POAC, 2022, Paper105 (2022)

Cho, S., Son, B., Choi, K., Jeong, S.Y., Ha, Y.S.: Enhanced Influence Coefficient Matrix for Estimation of Local Ice Loads on IBRV ARAON. POAC21-009 (2021)

Choiron M.A.: Development of fishing boat collision models in extreme weather using computer simulation (2023). https://doi.org/10.21303/2461-4262.2023.002601

Choung, J., Yoon, D.H.: Effects of Radiation and Wave Excitation Forces Acting on an Icebreaker on Ship Motions and Structural Damages during Iceberg Collision, p. 22123. IAHR ICE2022 (2022)

Costa, V., Boote, D., Pais, T., Ferrari, A., Sugalski, K.: WFSV-PL – an hybrid carbon composite wind farm supply vessel. In: Proceedings of the 27th International Offshore and Polar Engineering Conference, pp. 998–1004. San Francisco, USA (June 2017)

Cubides Garzón, D.M., Castaño Padilla, A.M., Vergara Pestaña, H.D.: Perspectives for the development of unmanned surface vehicles in Colombia: COTECMAR case. Ship Sci. Technol. **17**(34), 27–33. January 2024 – Cartagena (Colombia) (2024). https://doi.org/10.25043/19098642.247

Curtolo, P., Russo, T., Pacigìfic, G.Z., Begovic, E., Mancini, S.: Concept design on a Large Unmanned Surface Vehicle (LUSV). In: RINA Conference Autonomous Ships 2024, 20th November–21st November 2024, Copenhagen, Denmark (2024)

Cui, J., Gu, C.J., Chen, X., Li, M.Y., Masvaya, B.: Numerical study of wedge entry in still water and waves using smoothed particle hydrodynamics methods. Ocean Eng. **280**, 114776 (2023). https://doi.org/10.1016/j.oceaneng.2023.114776

D'Amato, E., Notaro, I., Piscopo, V., Scamardella, A.: Hydrodynamic design of fixed hydrofoils for planing craft. J. Mar. Sci. Eng. **11**(2), 246 (2023)

Davis, M.R.: Operation of T-foils and stern tabs to improve passenger comfort on high-speed ferries. J. Ship Res. **66**(04), 277–296 (2023). https://doi.org/10.5957/JOSR.07200047

Del Pero, F., Dattilo, C.A., Giraldi, A., Delogu, M.: LCA approach for environmental impact assessment within the maritime industry: re-design case study of yacht's superstructure. Proc. Inst. Mech. Eng. Pt. M J. Eng. Marit. Environ. **238**(1), 153–170 (2024)

Deng, Y., Ren, X., Nuernberg, M., Tao, L.: Launch and recovery of a work class ROV through wave zone in small offshore service vessel. Ocean Eng. **309**, 118541 (2024). https://doi.org/10.1016/j.oceaneng.2024.118541

Dessi, D., Sanchez-Alayo, D., Shabani, B., Ali-Lavroff, J.: Bow slamming detection and classification by machine learning approach. Ocean Eng. **287** (2023). https://doi.org/10.1016/j.oceaneng.2023.115646

Dewavrin, J., Souppez, J.B.: Experimental investigation into modern hydrofoils-assisted monohulls: how hydrodynamically efficient are they? Int. J. Small Craft Technol. **160**(B2), 111–120 (2018). https://doi.org/10.3940/rina.ijsct.2018.b2.223

Diez, M., et al: Experimental and computational fluid-structure interaction analysis and optimization of deep-V planing-hull grillage panels subject to slamming loads – part I: regular waves. Mar. Struct. **85** (2022). https://doi.org/10.1016/j.marstruc.2022.103256

Domeh, V., Obeng, F., Khan, F., Bose, N., Sanli, E.: A novel methodology to develop risk-based maintenance strategies for fishing vessels. Ocean Eng. **253**, 111281 (2022). https://doi.org/10.1016/j.oceaneng.2022.111281

Dow, R.S., et al: Committee V.5 naval vessel design. In: 19th International Ship and Offshore Structures Congress, 7–10 September 2015, Cascais, Portugal (2015)

Dowden, B., De Silva, O., Huang, W., Oldford, D.: Sea ice classification via deep neural network semantic segmentation. IEEE Sensors J. **21**(10), 11879 (2021)

Duan, W.Y., Liu, J.Y., Liao, K.P., Ma, S.: Experimental study of slamming pressure for a trimaran section with different drop heights and heel angles. Ocean Eng. **263**, 112400., ISSN 0029-8018 (2022). https://doi.org/10.1016/j.oceaneng.2022.112400

Dupuy, M., Letournel, L., Paakkari, V., Rongère, F., Sarsila, S., Vuillermoz, L.: Weather routing benefit for different wind propulsion systems. J. Sail. Technol. **8**(01), 200–217 (2023)

ECE: Inventory of Main Standards and Parameters of the E Waterway Network (Blue Book), ECE/TRANS/SC.3/144/Rev.3 (with amendments) (2017–2023)

ECE: Recommendations on Harmonized Europe-wide Technical Requirements for Inland Navigation Vessels, ECE/TRANS/SC.3/172/Rev.2, Resolution No. 61 Revision 2. (2020)

Engelbrecht, M., Bekker, A.: (2023) A discomfort threshold for impulsive whole-body vibration on a slamming-prone vessel. Appl. Ergon. **109** (2023). https://doi.org/10.1016/j.apergo.2023.103992

European Parliament: Directive 2013/53/EU of the European Parliament and of the council of 20 November 2013 on recreational craft and personal watercraft and repealing directive 94/25/EC. Off. J. Eur. Union. (2013)

Faltinsen, O.M.: Hydrodynamics of High-Speed Marine Vehicles. Cambridge University Press (2005)

Fassola, E., Kustermann, L.: Noise and vibration: comparison between prediction and measurements on yachts. In: Technology and Science for Ships of the Future. SAGE Publications Ltd (2022)

FEBIMA: Report 2017–02 on the Investigation into the Capsizing and Subsequent Rescue Mission of the Sailing Vessel CAPELLA off the Belgian Coast with the Loss of Three Lives on July 1st 2017. Federal Bureau for the Investigation of Maritime Accidents, Brussels (2017)

Fernández-Ríos, A., et al: Environmental sustainability of alternative marine propulsion technologies powered by hydrogen – a life cycle assessment approach. Sci. Total Environ. **820**, 153189 (2022)

Fibreship: Fibreship (2023). http://www.fibreship.eu/. Accessed 24 Nov 2023

Firdhaus, A., Suastikab, I.K.: Experimental and numerical study of effects of the application of hydrofoil on catamaran ship resistance. Revolution. **4**, 104–110 (2022)

Frederking, R.: Nanisivik Revisited: Ice Pressure Measurements from Winter 1985–86. POAC21-018 (2021)

Frederking R.: Ice forces on the R. V. "POLARSTERN" during 1984 labrador trials ISOPE2023 (2023)

Fuglem, M., Shayanfar, H., Liu, L., King, T., Paulin, M.: Evaluation of floating offshore wind turbine platforms with respect to iceberg impacts. IAHR Ice. **2022**, 22114 (2022)

Fujikubo, M., et al: A digital twin for ship structures – R&D project in Japan. Data-Cent. Eng. **5**, e7 (2024). https://doi.org/10.1017/dce.2024.3

Gagnon, R., Quinton, B., Mackay, J., Robbins, I., Rodriguez, M.: Design-purpose Numerical Simulations of the NRC Heavy Impact Test Theater (HITT). POAC, 2023 Paper85 (2023)

Gagnon, R., Quinton, B., Mackay, J.: A Numerical Simulation Study of Boundary – Condition Effects on the Damage of Ship Grillages Due to Ice Impacts. (R) IAHR Ice 2022, 22116 (2022)

Gagnon, R., Wang, J., Seo, D., Mackay, J.: Numerical Simulations of Naval Vessel Collisions with Ice R, pp. 339–360. IAHR 2020 (2020)

Gao, X., Xu, S., Tang, W.: Hybrid analytic-FEM approach for dynamic response analysis of air-cushion vehicle skirts. Mar. Struct. **79**, 103062 (2021). https://doi.org/10.1016/j.marstruc.2021.103062

Gebrezgabir, S., Holloway, D., Ali-Lavroff, J.: Slam and wave load response reconstruction in high speed catamarans using transmissibility on full scale sea trials. Ocean Eng. **271**, 1–12 (2023). https://doi.org/10.1016/j.oceaneng.2023.113822

Gilges, M., Saleh, A., Jain, M., Bekker, A., Lehmann, B., Jacobs, G.: Influence of Propeller-Ice Loads on the Wear Development in Stern Tube Bearings of Marine Propulsion Systems and Identification of Critical Operating Conditions. OMAE2023-102785 (2023)

Godey, A., Misra, S.C., Sha, O.P.: Development of a ballast free ship design. Int. J. Innov. Res. Dev. **1**, 10 (2012)

Gosse, J., Quinton, B., Daley, C., Kendrick, A., Bond, J.: Using Non-Linear Finite Element Analysis to Analyze the Effects of Connection Designs on the Ice Strength of a Vessel. OMAE2023-104860 (2023)

Graf, K., Freiheit, O., Schlockermann, P., Mense, J.C.: VPP-driven sail and foil trim optimization for the Olympic NACRA 17 foiling catamaran. J. Sail. Technol. **5**(01), 61–81 (2020)

Grimm, J.-H.: Extended Fatigue Recommendations for High Strength Steel Yacht Keels. Det Norske Veritas Germanischer Lloyds, Hamburg (2016)

Guida, P., Marimon Giovannetti, L., Boyd, S.: Three-dimensional variations of the NACRA 17 main foil for benchmarking shape optimizations. In: The 5th International Conference on Innovation in High Performance Sailing Yachts and Sail Assisted Propulsion, Gothenburg (2020)

Hageman, R.B., Thompson, I.: Virtual hull monitoring using hindcast and motion data to assess frigate-size. Ocean Eng. **245** (2022). https://doi.org/10.1016/j.oceaneng.2021.110338

Hammer, T.C., Owen, C.C., van den Berg, M.M., Hendrikse, H.: Classification of Ice-Induced Vibration Regimes of Offshore Wind Turbines. OMAE2022-78972 (2022)

Hammer, T.C., Willems, T., Hendrikse, H.: Dynamic ice loads for offshore wind support structure design. Mar. Struct. **87** (2023). https://doi.org/10.1016/j.marstruc.2022.103335

Han, Z., Jang, J., Souppez, J.B.R.G., Oh, D.: Environmental implications of a sandwich structure of a glass fiber-reinforced polymer ship. Ocean Eng. **298**, 117122 (2024)

Han, Z., Jang, J., Souppez, J.B.R., Seo, H.S., Oh, D.: Comparison of structural design and future trends in composite hulls: A regulatory review. Int. J. Nav. Archit. Ocean Eng. **15**, 100558 (2023)

Han, Z., Jeong, S., Noh, J., Oh, D.: Comparative study of glass fiber content measurement methods for inspecting fabrication quality of composite ship structures. Appl. Sci. **10**(15), 5130 (2020)

Hashimoto, S., et al: Hydrodynamic design and analysis of high speed catamaran with hydrofoils using CFD. In: Practical Design of Ships and Other Floating Structures: Proceedings of the 14th International Symposium, PRADS 2019, September 22–26, 2019, Yokohama, Japan-Volume I 14, pp. 508–518. Springer, Singapore (2021)

He, S., Chen, X., Ji, S.: Experimental Verification of Dynamic Ice Load Identification Method. POAC, 2022, Paper103 (2022)

Heinonen, J., Mikkola, E.: Modelling of Ice-Interaction with a Sea Channel Marker in the Bothnian Bay. POAC, 2023, Paper50 (2023)

Heinonen, J., Tikanmäki, M., MiA-kkola, E., Perälä, I., Shestov, A.: Scale-Model Ridges and Interaction with Narrow Structures, Part 3 Analysis of Ridge Keel Punch Tests. POAC21-091 (2021)

Helal, K.M., Fragasso, J., Moro, L.: Underwater noise characterization of a typical fishing vessel from Atlantic Canada. Ocean Eng. **299**, 117310 (2024) 14 p

Hendrikse, H., et al: Ice Basin Tests for Ice-Induced Vibrations of Offshore Structures in the SHIVER Project. OMAE2022-78507 (2022)

Heo, H., Choi, J., Choi, H.S., Oh, E.J., Park, S.: Dimensionality Reduction of Multivariate Sensor Data for Estimation of Ice Load on a Real Ship, p. 22110. IAHR Ice 2022 (2022)

Herrnring, H., Ehlers, S.: A finite element model for compressive ice loads based on a mohrcoulomb material and the node splitting technique. J. Offshore Mech. Arct. Eng. **144**(2) (2022)

Hong, S., McMorland, J., Zhang, H., Collu, M., Henning, H.K.: Floating offshore wind farm installation, challenges and opportunities: a comprehensive survey. Ocean Eng. **304** (2024). https://doi.org/10.1016/j.oceaneng.2024.117793

Hosseinzadeh, S., Tabri, K., Hirdaris, S., Sahk, T.: Slamming loads and responses on a non-prismatic stiffened aluminium wedge: part I. Experimental study. Ocean Eng. **279** (2023a). https://doi.org/10.1016/j.oceaneng.2023.114510

Hosseinzadeh, S., Tabri, K., Topa, A., Hirdaris, S.: Slamming loads and responses on a non-prismatic stiffened aluminium wedge: part II, numerical simulations. Ocean Eng. **279** (2023b). https://doi.org/10.1016/j.oceaneng.2023.114309

Høyland, K.V., et al: Fatigue damage from dynamic ice action – fatigue damage from dynamic ice action. The FATICE project (POAC21-026) (2021)

Hydar, V., Boote, D., Vergassola, G., Păcuraru, F.: Seakeeping analysis of a GRP fast patrol vessel. In: Proceedings of the International Conference on Offshore Mechanics and Arctic Engineering – OMAE 2022, Hamburg (Germany) (2022)

Hydar, V.: Numerical dynamic analysis of a composite pleasure craft for comfort improvement. Int. Rev. Mech. Eng. **16**(8), 379–392 (2022)

Ibrahim, A.M., Judge, C.Q.: Investigations of hull girder slamming factor for a semi-displacement vessel using model testing. Appl. Ocean Res. **150**, 104084 (2024). https://doi.org/10.1016/j.apor.2024.104084

ISO: ISO 12215-9: Small Craft – Hull Construction and Scantlings. Part 9: Sailing Craft Appendages. International Organisation for Standardization (2012)

ISO: ISO 12215-5: Small Craft – Hull Construction and Scantlings. Part 5: Design Pressures for Monohulls, Design Stresses, Scantlings Determination. International Organisation for Standardization (2019)

ISO: ISO 12215-7:2020 Small Craft – Hull Construction and Scantlings – Part 7: Determination of Loads for Multihulls and of Their Local Scantlings Using ISO 12215-5. International Organization for Standardization, Geneva (2020a)

ISO: ISO 12215-10:2020 Small Craft – Hull Construction and Scantlings – Part 10: Rig Loads and Rig Attachment in Sailing Craft. International Organization for Standardization, Geneva (2020b)

ISO: ISO 22834: Large Yachts – Quality Assessment of Life Onboard – Stabilization and Seakeeping Guidelines. International Organization for Standardization, Geneva (2022)

Jacquet, L., Le Duigou, A., Kerbrat, O.: A systematic literature review on holistic lifecycle assessments as a basis to create a standard in maritime industry. Int. J. Life Cycle Assess. **2024**, 1–23 (2024)

Jagite, G., Bigot, F., Malenica, S., Derbanne, Q., Le Sourne, H., Cartraud, P.: Dynamic ultimate strength of a ultra-large container ship subjected to realistic loading scenarios. Mar. Struct. **84**, 103197 (2022). https://doi.org/10.1016/j.marstruc.2022.103197

Jang, J.W., Han, Z., Oh, D.: Light-weight optimum design of laminate structures of a GFRP fishing vessel. J. Ocean Eng. Technol. 33(6), 495–503 (2019)

Javanmard, E., Mehr, J.A., Ali-Lavroff, J., Holloway, D.S., Davis, M.R.: An experimental investigation of the effect of ride control systems on the motions response of high-speed catamarans in irregular waves. Ocean Eng. **281**, 1–17 (2023). https://doi.org/10.1016/j.oceaneng.2023.114899

Javanmard, E., Mehr, A.J., Holloway, D.S., Davis, M.R., Ali-Lavroff, J.: Ride control system effects on motion, slam load, and passenger comfort of high-speed catamarans: an experimental study in irregular waves. In: Proceedings of the ASME 2024 43rd International Conference on Ocean, Offshore and Arctic Engineering, OMAE 2024, June 9–14, 2024 Singapore (2024). https://doi.org/10.1115/OMAE2024-126372

Jiang, Y.Y., Tang, W.Y.: Numerical study of section geometry of flexible bag of air cushion vehicle subjected to slamming loads. Ocean Eng. **227**, 108894 (2021)

Jiang, Y., Bai, J., Dong, Y., Sun, T., Sun, Z., Liu, S.: Investigations of air cushion effect on the slamming load acting on trimaran cross deck during water entry. Ocean Eng. **251**, 111161 (2022). https://doi.org/10.1016/j.oceaneng.2022.111161

Jiang, Y., Tang, W.: Numerical investigation on water entry of a three-dimensional flexible bag of an air cushion vehicle. Ocean Eng. **247**, 110653 (2022a)

Jiang, Y., Tang, W.: Numerical prediction of interaction between a flexible bag of an air cushion vehicle and solitary waves. Ocean Eng. **265**, 112583 (2022b). https://doi.org/10.1016/j.oceaneng.2022.112583

Jiang, Z.: Installation of offshore wind turbines: a technical review. Renew. Sust. Energ. Reviews **139**, 110576 (2021). https://doi.org/10.1016/j.rser.2020.110576

Jiang, Z., Heinonen, J., Tikanmäki, M., Mikkola, E., Perälä, I.: Scale-Model Ridges and Interaction with Narrow Structures, part 4 Global Loads and Failure Mechanisms. POAC21-031 (2021)

Jiao, J., Ren, H., Guedes Soares, C.: A review of large-scale model at-sea measurements for ship hydrodynamics and structural loads. Ocean Eng. **227**, 108863 (2021). https://doi.org/10.1016/j.oceaneng.2021.108863

Jeon, S., Kim, Y.: Fatigue damage estimation of icebreaker ARAON colliding with level ice. Ocean Eng. **257**, 111707 (2022). https://doi.org/10.1016/j.oceaneng.2022.111707

Kang, Y., Pei, Z., Ao, L., Wu, W.: Reliability-based design optimization of river-sea-going ship based on agent model technology. Mar. Struct. **94**, 103561 (2024)

Kashiro, R.: Innovative Ship Design with Less Ballast Water and Less Green House Gases. TSCF 2016 Shipbuilders Meeting (2016)

Khan, L., Macklin, J., Peck, B., Morton, O., Souppez, J.B.R.: A review of wind-assisted ship propulsion for sustainable commercial shipping: latest developments and future stakes. In: Wind Propulsion Conference. Royal Institution of Naval Architects (September 2021)

Kistner, G., Lal, K., Klüss, J., Kaeding, P.: Procedure for Torsional-Vibration Calculations in Ice. OMAE2022-80194 (2022)

Kjær, R.B., Shao, Y., Walther, J.H.: Experimental and CFD analysis of roll damping of a wind turbine installation vessel. Appl. Ocean Res. **143**, 103857 (2024)

Knudsen, S.S., Walther, J.H., Legarth, B.N., Shao, Y.: Towards dynamic velocity prediction of NACRA 17. J. Sail. Technol. **8**(01), 1–23 (2023)

Kong, S., Cui, H., Wu, G., Ji, S.: Full-scale identification of ice load on ship hull by least square support vector machine method. Appl. Ocean Res. **106**, 102439 (2021)

Koričan, M., Perčić, M., Vladimir, N., Alujević, N., Fan, A.: Alternative power options for improvement of the environmental friendliness of fishing trawlers. J. Mar. Sci. Eng. **10**(12), 1882, 26 p (2022)

Koričan, M., Vladimir, N., Fan, A.: Investigation of the energy efficiency of fishing vessels: case study of the fishing fleet in the Adriatic Sea. Ocean Eng. **286**(Part 2), 115734, 12 p (2023a)

Koričan, M., Frković, L., Vladimir, N.: Electrification of fishing vessels and their integration into isolated energy systems with a high share of renewables. J. Clean. Prod. **425**, 138997, 14p (2023b)

Kotinis, M., Parsons, M.G., Lamb, T., Sirviente, A.: Development and investigation of the ballast-free ship concept. Transactions SNAME. **112**, 206–240 (2004)

Kotinis, M., Parsons, M.G.: Hydrodynamic optimization testing of ballast-free ship design. Transactions SNAME. **115** (2007)

Kotinis, M., Parsons, M.G.: Hydrodynamics of the ballast-free ship. J. Sh. Prod. Des. **26**, 301–310 (2010)

Kramer, J.V., Steen, S.: Sail-induced resistance on a wind-powered cargo ship. Ocean Eng. **261**, 111688 (2022)

Kumar, V., Boote, D., Pais, T.: Development of a parametric model for analysing temperature effects of solar radiation on yachts. RINA Trans. Pt. B Int. J. Small Craft Technol. **158**., Part B1, pp. B-1,B-13, Jan-Jun 2016 (2016). https://doi.org/10.3940/rina.ijsct.2016.bl.170

Larsson, L., Eliasson, R., Orych, M.: Principles of Yacht Design. Bloomsbury Publishing, London (2022)

Lau, C.-Y., Ali-Lavroff, J., Holloway, D.S., Shabani, B., AlaviMehr, J., Thomas, G.: Influence of an active T-foil on motions and passenger comfort of a large high-speed wave-piercing catamaran based on sea trials. J. Mar. Sci. Technol. **27**, 856–872 (2022a). https://doi.org/10.1007/s00773-022-00876-8

Lau, C.-Y., Ali-Lavroff, J., Holloway, D.S., AlaviMehr, J., Thomas, G.: Influence of an active T-foil on motions and passenger comfort of a wave-piercing catamaran based on sea trials in oblique seas. J. Eng. Maritime Environ., 1–13 (2022b). https://doi.org/10.1177/14750902221111122

Lazakis, I., Khan, S.: An optimization framework for daily route planning and scheduling of maintenance vessel activities in offshore wind farms. Ocean Eng. **225** (2021). https://doi.org/10.1016/j.oceaneng.2021.108752

Lee, E.J., et al: Experimental and computational fluid-structure interaction analysis and optimization of deep-V planing-hull grillage panels subject to slamming loads – part II: irregular waves. Ocean Eng. **292** (2024). https://doi.org/10.1016/j.oceaneng.2023.116346

Lemström, I., Polojarvi, A., Tuhkuri, J., Puolakka, O.: Model-Scale Tests on the Ice-Structure Interaction Process in Shallow Water, p. 22084. IAHR Ice (2022a)

Lemström, I., Polojarvi, A., Tuhkuri, J.: Model-scale tests on ice-structure interaction in shallo water – part I: global ice loads and the ice loading process. Mar. Struct. **81**, 103106 (2022b)

Li, H., Deng, B., Zou, J., Dong, C., Liu, C., Liu, P.: Experimental free-drop test investigation into wet-deck slamming loads on a generic trimaran section considering the influence of main hull profile. Ocean Eng. **242**, 110114., ISSN 0029-8018 (2021a). https://doi.org/10.1016/j.oceaneng.2021.110114

Li, H., Zou, J., Deng, B., Liu, R., Sun, S.: Experimental study of stern slamming and global response of a large cruise ship in regular waves. Mar. Struct. **86**, 103294 (2022a). https://doi.org/10.1016/j.marstruc.2023.103563

Li, F., Ding, S., Liu, R., Wang, A.: Stress-Strain Predictive Analysis of Ship Structure in Ice Area Based on Neural Network. OMAE2023-104315 (2023)

Li F., Khawar M.B., Sandru A., Lu L., Suominen M., Kujala P. (2021b) Full-scale measurement of ship performance and ice loads in Antarctic floe ice fields POAC21-038

Li, F., Liangliang, L.L., Suominen, M., Kujala, P.: Short-term statistics of ice loads on ship bow frames in floe ice fields: full-scale measurements in the Antarctic Ocean. Mar. Struct. **80**, 103049 (2021c)

Li, H., Zou, J., Peng, Y., Zhou, X., Lu, L., Sun, S.: Numerical study of slamming and whipping loads in moderate and large regular waves for different forward speeds. Mar. Struct. **94** (2024). https://doi.org/10.1016/j.marstruc.2023.103563

Li, B., Qiao, D., Zhao, W., Hu, Z., Li, S.: Operability analysis of SWATH as a service vessel for offshore wind turbine in the southeastern coast of China. Ocean Eng. **251**, 111017 (2022). https://doi.org/10.1016/j.oceaneng.2022.111017

Lin, B., Liang, C., Zhang, L., Qian, H.: Design of a convertible autonomous sailboat. In: In 2023 IEEE International Conference on Real-Time Computing and Robotics (RCAR), pp. 566–572. IEEE (2023)

Liu, C., Zhou, L., Ding, S., Liu, R., Wang, A.: Numerical Simulation of Ice-Propeller Milling Process with Cohesive Element Method. OMAE2023-104308 (2023)

Ljulj, A., Slapnicar, V., Smiljanic, D.: Proliferation of unmanned aerial and maritime vehicles in military operations. In: Degiuli, N., et al. (eds.) Theory and Practice of Shipbuilding. IOS Press (2024). https://doi.org/10.3233/PMST240043

Lloyd, T., Foeth, E.J., Lafeber, F.H., Bosschers, J.: Progress in the prediction and mitigation of propeller cavitation noise and vibrations. In: 26th International Virtual HISWA Symposium: On Yacht Design and Yacht Construction (2020)

Lowde, M.J., Peters, H.G.A., Geraghty, R., Graham-Jones, J., Pemberton, R., Summerscales, J.: The 100 m composite ship? J. Mar. Sci. Eng. **10**(3), 408 (2022)

Lorimer, T., Allen, T.: Concurrent multi-component optimization of stiffened-plate yacht structures. J. Sail. Technol. **7**(01), 203–227 (2022)

Lyons, D.: Expert Report, Sailing Vessel Showtime, Loss of Keel, 5 January 2020. Federal Court of Australia, NSD774/2020 (2021)

Lyons, D., Bird, A., Hinterhoeller, R., Hoffmeister, H., Loiselet, K., Souppez, J.-B.R.G.: Regulatory developments in structural keel design: A revised ISO 12215-9. In: 8th High Performance Yacht Design Conference, Auckland, New Zealand (21–22 March 2024)

Ma, S., et al: Experimental study on the drop test on wet deck slamming for a SWATH segment model. Ocean Eng. **285**(part 2), 115377, ISSN 0029-8018. (2023a). https://doi.org/10.1016/j.oceaneng.2023.115377

Ma, S., Zhu, M., Liu, D., Liu, J., Wang, W.: Experimental study of wet deck slamming for a SWATH in regular waves. Ocean Eng. **288**(part 1), 115996, ISSN 0029-8018 (2023b). https://doi.org/10.1016/j.oceaneng.2023.115996

MAIB: Report on the Investigation of the Loss of the Yacht Cheeki Rafiki and its Four Crew in the Atlantic Ocean, Approximately 720 Miles East-South-East of Nova Scotia, Canada on 16 May 2014. Marine Accident and Investigation Branch, Accident Report 08-2015, Southampton (2015)

Maksimova, P.V., Chernov, A.V., Likhomanov, V.A., Svistunov, I.A.: Ice Load Monitoring System for an Ice-Resistant Self-Propelled Drifting Platform "North Pole". POAC21-042 (2021)

Marimon Giovannetti, L., Farousi, A., Ebbesson, F., Thollot, A., Shiri, A., Eslamdoost, A.: Fluid-structure interaction of a foiling craft. J. Mar. Sci. Eng. **10**(3), 372 (2022)

Malchow, U.: Growth in Container Ship Sizes to Be Stopped. Maritime Business Review (2017). ISSN: 2397-3757

Marlantes, K.E., Maki, K.J.: A neural-corrector method for prediction of the vertical motions of a high-speed craft. Ocean Eng. **262** (2022). https://doi.org/10.1016/j.oceaneng.2022.112300

Miller, L.P., Quinton, B., Soper, J., Veitch, B.: Development of Ice-Load Algorithm for Real-Time Feedback During Simulator Training. OMAE2023-101443 (2023)

Minty, P., Gohari, S., Field, B., Burvill, C.: Analytical and experimental studies on dynamic response of a SES air cushion plenum experiencing forced oscillation. Ocean Eng. **289**, 116265 (2023). https://doi.org/10.1016/j.oceaneng.2023.116265

Mokhtari, M., Kim, E., Amdahl, J.: Numerical Simulation of an Aluminium Panel Subject to Ice Impact Load Using a Rate and Pressure Dependent Elastoplastic Material Model for Ice. OMAE2023-104771 (2023)

Montigneaux, R., Macdowall, A., Mower, M.: The business of yachting. Boat Int., 50–65 (2023). https://cdn.boatinternational.com/files/2023/01/9148ea90-9003-11ed-af00-9f6e2a9c4532-GOB_2019.pdf

Motok, M., Momčilović, N., Rudaković, S.: Reliability based structural design of river–sea tankers: still water loading effects. Mar. Struct. **83**, 103202 (2022)

Moupagitsoglou, K.: Structural glass in superyacht applications: overcoming challenges, design standards and analysis methods, challenging glass 7. In: Belis, Bos, Louter (eds.) Conference on Architectural and Structural Applications of Glass. Ghent University. ISBN 978-94-6366-296-3 (September 2020). https://doi.org/10.7480/cgc.7.4609

Mouton, L., Finkelstein, A.: Exploratory study on the flutter behavior of modern yacht keels and appendages. In: 5th High Performance Yacht Design Conference, Auckland, New Zealand (2015)

Mouton, L., Leroyer, A., Deng, G., Queutey, P., Soler, T., Ward, B.: Towards unsteady approach for future flutter calculations. J. Sail. Technol. **3**(1), 1–19 (2018)

Neves-Moreira, F., Veldman, J., Teunter, R.H.: Service operation vessels for offshore wind farm maintenance: optimal stock levels. Renew. Sust. Energ. Rev. **146**, 111158 (2021)

Nickerson, B.N., Laas, J., Bekker, A.: Verification of Inverse Propeller Moment Estimation Using a Scale Laboratory Rig: Further Results and Discussion. OMAE2023-104878 (2023)

Nickerson, B.N., Laas, J., Bekker, A.: Verification of Inverse Propeller Moment Estimation Using a Scale Laboratory Rig. OMAE2022-78787 (2022)

Nilva, V.: Detailed analysis of hull damages of river and river-sea cargo vessels. In: Proceedings of Fifteenth International Conference on Marine Sciences and Technologies (Black Sea' 2020), pp. 166–171, Varna (Bulgaria) (2020)

Ocera, M., Boote, D., Vergassola, G., Faloci, F.: Simplified analytical method for the evaluation of longitudinal strength of large sailing yachts. Ocean Eng. **133**, 182–196 (2017). https://doi.org/10.1016/j.oceaneng.2016.11.064

Oldford, D., Moakler, E., Bond, J.: Ice Class IA or Ice Class PC7 for Arctic Operations. ISOPE 2023, 2009 (2023)

Office for Product Safety and Standards: Recreational Craft Regulations 2017: Great Britain. Office for Product Safety and Standards – Department for Business, Energy and Industrial Strategy, Birmingham (2017)

Oh, D., Jang, J., Jee, J.H., Kwon, Y., Im, S., Han, Z.: Effects of fabric combinations on the quality of glass fiber reinforced polymer hull structures. Int. J. Nav. Archit. Ocean Eng. **14**, 100462 (2022)

Oh, E., Ha, J.S.: Estimating Missing Hull Strain Gauge Data of the ARAON Using Artificial Intelligence. POAC, 2022, Paper76 (2022)

Pan, J., Zhang, W.Z., Sun, Z.M., Qu, X., Xu, M.C.: Experimental study on the dynamical response of elastic trimaran model under slamming load. J. Mar. Sci. Technol. (2023). https://doi.org/10.1007/s00773-023-00969-y

Parunov, J., et al: Benchmark on the prediction of whipping response of a warship model in regular waves. Mar. Struct. **94**, 103549 (2024)

Patterson, N., Binns, J.: Development of a six degree of freedom velocity prediction program for the foiling America's cup vessels. J. Sail. Technol. **7**(01), 120–151 (2022)

Pei, Z., Ma, Z., Bo, Z., Wu, W.: Research on the bending efficiency of superstructure to hull girder strength of inland passenger ship. Ocean Eng. **195**, 106762 (2020)

Perälä, I., Tikanmaki, M., Heinonen, J.: Ice Impact Loads on Azimuthing Thrusters – Small-Scale Impact Tests and Analysis, p. 22075. IAHR Ice 2022 (2022)

Perčić, M., Vladimir, N., Koričan, M., Jovanović, I., Haramina, T.: Alternative fuels for the marine sector and their applicability for purse seiners in a life-cycle framework. Appl. Sci. **13**(24), 13068, 30 p (2023)

Perčić, M., Koričan, M., Jovanović, I., Vladimir, N.: Environmental and economic assessment of batteries for marine applications: case study of all-electric fishing vessels. Batteries. **10**(1) 2024., paper no. 7, 17 p (2024)

Petković, M., Zubčić, M., Krčum, M., Pavić, I.: Wind assisted ship propulsion technologies – can they help in emissions reduction? Naše More. **68**(2), 102–109 (2021)

Petry, A., Hammer, T.C., Polojärvi, A., Hendrikse, H., Puolakka, O.: Ice Basin Experiments on Mixed-Mode Failure on Ice Cones. POAC2022, paper 29 (2022)

Qu, X., Yu, W., Wenzhe, Z., Sun, Z.: Numerical study on the dynamic response and slamming pressure of trimarans. In: Paper Presented at the 33rd International Ocean and Polar Engineering Conference, Ottawa (June 2023)

Radojcic, D., Simic, A., Momcilovic, N., Motok, V., Friedhoff, B.: Design of Contemporary Inland Waterway Vessels (The Case of the Danube River). Springer (2022)

Rahman, S.M., Handler, R.M., Mayer, A.L.: Life cycle assessment of steel in the ship recycling industry in Bangladesh. J. Clean. Prod. **135**, 963–971 (2016). https://doi.org/10.1016/j.jclepro.2016.07.014

Raju, R., Prusty, B.G., Kelly, D.W., Lyons, D., Peng, G.D.: Top hat stiffeners: a study on keel failures. Ocean Eng. **37**(13), 1180–1192 (2010)

Ramsses: Realisation and demonstration of advanced material solutions for sustainable and efficient ships (2023). https://www.ramsses-project.eu/. Accessed 24 Nov 2023

Reche-Vilanova, M., Hansen, H., Bingham, H.B.: Performance prediction program for wind-assisted cargo ships. J. Sail. Technol. **6**(01), 91–117 (2021)

Ren, Z., Javaherian, J., Gilbert, C.M.: Kinematic and inertial hydroelastic effects caused by vertical slamming of a flexible V-shaped wedge. J. Fluids Struct. **103**, 103257 (2021a). https://doi.org/10.1016/j.jfluidstructs.2021.103257

Ren, Z., Skjetne, R., Verma, A.S., Jiang, Z., Gao, Z., Henning, H.K.: Active heave compensation of floating wind turbine installation using a catamaran construction vessel. Mar. Struct. **75**, 102868 (2021b)

Rinauro, B., Begovic, E., Mauro, F., Rosano, G.: Regression analysis for container ships in the early design stage. Ocean Eng. **292** (2024). https://doi.org/10.1016/j.oceaneng.2023.116499

Rosano, G., Begovic, E., Boccadamo, G., Míguez González, M., Rinauro, B., Santiago Caamaño, L.: On-board monitoring and estimation of lateral accelerations through extreme value theory. Ocean Eng. **284**, 115177, 12 p (2023)

Rosén, A., Garme, K., Razola, M., Begovic, E.: Numerical modelling of structure responses for high-speed planing craft in waves. Ocean Eng. **217** (2020). https://doi.org/10.1016/j.oceaneng.2020.107897

Russell, S., Vanhollebeke, G., Manganelli, P.: Insights from the Load Monitoring Program for the 2014–2015 Volvo Ocean Race. SNAME Chesapeake Sailing Yacht, Annapolis (2016)

Salganik, E., Ervik, A., Heinonen, J., Høyland, K.V., Perälä, I.: Scale-Model Ridges and Interaction with Narrow Structures, Part 2: Thermodynamics of Ethanol Ice. POAC21-067 (2021)

Santiago, C.L., Galeazzi, R., Nielsen, U.D., Míguez, G.M., Díaz, C.V.: Real-time detection of transverse stability changes in fishing vessels. Ocean Eng. **189**, 106369., ISSN 0029-8018 (2019). https://doi.org/10.1016/j.oceaneng.2019.106369

Santiago Caamaño, L., Míguez González, M., Galeazzi, R., Nielsen, U.D., Díaz Casás, V.: Application of real-time estimation techniques for stability monitoring of fishing vessels. In: Spyrou, K.J., Belenky, V.L., Katayama, T., Bačkalov, I., Francescutto, A. (eds.) Contemporary Ideas on Ship Stability. Fluid Mechanics and Its Applications, vol. 134. Springer, Cham (2023). https://doi.org/10.1007/978-3-031-16329-6_21

Santoso, B.R., Nasution, P.: Analysis strength structure fishing vessel of composite sandwich plate system using finite element method. IOP Conf. Ser.: Earth Environ. Sci. **695**(1), 012021 (2021)

Saral, D., Köse, E.: A non-dimensional offset table for the Black Sea type fishing vessels. Appl. Ocean Res. **139**, 103705 (2023). https://doi.org/10.1016/j.apor.2023.103705

Sawamura, J., Yamaguchi, H., Ushio, S., Mizuno, S.: Relationship Between Ship Structural Response and Ice Conditions in Antarctic Ocean. POAC, 2023, Paper 65 (2023)

Shabani, B., Ali-Lavroff, J., Holloway, D., Penev, S., Dessi, D., Thomas, G.: Using remote monitoring and machine learning to classify slam events of wave piercing catamarans. Int. J. Mar. Eng. **163**, A15–A30 (2021). https://doi.org/10.5750/ijme.v163iA3.797

Shabani, B., Ali-Lavroff, J., Holloway, D., Penev, S., Dessi, D., Thomas, G.: Intelligent monitoring of a large catamaran ferry. Int. J. Mar. Eng. **165**, A11–A22 (2023). https://doi.org/10.5750/ijme.v165ia1.791

Shamaei, F., Bergstrøm, M., Li, F., Taylor, R., Kujala, P.: Local pressures for ships in ice: probabilistic analysis of fullscale line load data. Mar. Struct. **74**, 102822 (2020)

Shao, W., Ma, S., Duan, W., Liu, J., Zhang, Y.: Vertical force prediction of planing craft in calm water based on variable section slamming model. Ocean Eng. **287**, 115693 (2023). https://doi.org/10.1016/j.oceaneng.2023.115693

Shenoi, A., et al: Committee V.8 Sailing Yacht Design, 17th International Ship and Offshore Structures Congress, 16–21 August 2009, Seoul, Korea (2009)

Shin, Y., Kim, Y.: Fatigue Damage Estimation of a Sloped Offshore Structure in Level Ice. POAC21-071 (Conical structure) (2021)

Silvestri, P., Pais, T., Vergassola, G.: On the experimental full scale vibrational response analysis of a large pleasure yacht. Appl. Ocean Res. **150**, 104141 (2024). https://doi.org/10.1016/j.apor.2024.104141

Souppez, J.B., Ridley, J.: The Revisions of the BS EN ISO 12215. Marine Sector Showcase- Composite UK (2017)

Souppez, J.B.R.: Structural design of high performance composite sailing yachts under the new BS EN ISO 12215-5. J. Sail. Technol. **3**(1), 1–18 (2018a)

Souppez, J.B.R.: Structural analysis of composite search and rescue vessels under the new BS EN ISO 12215-5. In: RINA Surveillance, Search and Rescue Craft Conference, London, (April 2018b)

Souppez, J.B.: Designing the next generation of small pleasure and commercial powerboats with the latest ISO 12215-5 for hull construction and scantlings. In: 1st SNAME/IBEX Symposium. Society of Naval Architects and Marine Engineers (2019)

Souppez, J.B., Dewavrin, J.M., Gohier, F., Labi, G.B.: Hydrofoil configurations for sailing superyachts: Hydrodynamics, stability and performance. In: Design & Construction of Super & Mega Yachts. Royal Institution of Naval Architects (2019a)

Souppez, J.B.R., Arredondo-Galeana, A., Viola, I.M.: Recent advances in numerical and experimental downwind sail aerodynamics. J. Sail. Technol. **4**(01), 45–65 (2019b)

Souppez, J.B.R., Begovic, E., Sensharma, P., Wang, F., Rosén, A.: Comparative assessment of rule-based design on the pressures and resulting scantlings of high speed powercrafts. In: HSMV 2020, pp. 263–275. IOS Press (2020)

Souppez, J.B.: Experimental testing of scarf joints and laminated timber for wooden boatbuilding applications. Int. J. Mar. Eng. **163**(A3) (2021). https://doi.org/10.5750/ijme.v163iA3.16

Souppez, J.B., Laci, J.: Ultimate strength of quasi-isotropic composites: ISO 12215-5: 2019 validation. Int. J. Mar. Eng. **164**(A2), 237–246 (2022)

Souppez, J.B.R.: Structural design of wooden boats. In: Historic Ships Conference 2023: Historic Vessels: Sustainable Futures. Royal Institution of Naval Architects (2023)

Souppez, J.-B.R.G., Viola, I.M.: Water tunnel testing of downwind yacht sails. Exp. Fluids. **65**, 15 (2024)

Su, S.J., Han, J., Xiong, Y.: Optimization of unmanned ship's parametric subdivision based on improved multi-objective PSO. Ocean Eng. **194**, 2019 (2019)

Su, S.J., Liu, C.B., Zhang, X., Wang, G.H.: Multi object optimization design of cargo hold structure of NOBS tanker based on RBF agent model. J. Wuhan Univ. Technol. **43**(11) (2021)

Su, S.J., Wang, G.H., Zhang, X.: Topology optimization of midship section of cargo hold of V-type non-ballast water tanker. Ship Eng. **44**(4), 58–63 (2022)

Sun, Z., et al: Characteristics of slamming pressure and force for trimaran hull. J. Mar. Sci. Eng. **9**(6), 564 (2021)

Sun, Z., Deng, Y.Z., Zou, L., Jiang, Y.C.: Investigation of trimaran slamming under different conditions. Appl. Ocean Res. **104**, 102316., ISSN 0141-1187 (2020). https://doi.org/10.1016/j.apor.2020.102316

Suominen, M., Toroody, A.B., Valdez, B.O.: Empirical Mode Decomposition for Noise Detection and Filtration of Ice-Induced Load Measurements. OMAE2023-102504 (2023)

Suominen, M., Lu, L., Kujala, P., Bekker, A.: Antarctic Sea Ice Properties on Zero Meridian Side During Austral Summers 2012–14 and 2018–19. POAC21-074 (2021)

Tabri, K., Tõns, T., Suominen, M., Kõrgesaar, M.: Ice-Induced Loads on Offshore Wind Turbines in the Baltic Sea. OMAE2022-79035 (2022)

Terui, C., Hino, T., Takagi, Y.: Numerical analysis of wake wash reduction for catamaran with hydrofoils. Appl. Ocean Res. **135**, 103556 (2023)

Thijssen, J., Fuglem, M., King, T., Paulin, M.: Evaluation of Floating Wind Turbine Platform Designs for Sea Ice Loads, p. 22111. IAHR Ice 2022 (2022)

Tillig, F., Ringsberg, J.W.: Design, operation and analysis of wind-assisted cargo ships. Ocean Eng. **211**, 107603 (2020)

TNO: Offshore Wind Access Report 2020. TNO' Internal R&D Instrument 'Kennis en Innovatie Project' – Programme 2020 (KIP), Pette, The Netherlands (2020) https://publications.tno.nl/publication/34637592/uSPDJu/TNO-2020-R11992.pdf

Truelock, D., et al: Committee V.5 special vessels. In: 21st International Ship and Offshore Structures Congress (ISSC 2022), September 11–15 2022, Vancouver, Canada (2022)

Truelock, D., et al: Committee V.5 special craft. In: 20th International Ship and Offshore Congress, Leige-Delft, September 2018 (2018). https://doi.org/10.3233/978-1-61499-864-8-279

Tusar, M.I.H., Sarker, B.R.: Developing the optimal vessel fleet size and mix model to minimize the transportation cost of offshore wind. Ocean Eng. **274** (2023). https://doi.org/10.1016/j.oceaneng.2023.114041

Uğurlu, F., Yıldız, S., Boran, M., Uğurlu, O., Wang, J.: Analysis of fishing vessel accidents with Bayesian network and Chi-square methods. **198**(2020), 106956 (2020). https://doi.org/10.1016/j.oceaneng.2020.106956

van Walree, F., Thomas, W.L.: Validation of simulation tools for a RHIB operating in heavy seas, in contemporary ideas on ship stability. In: Fluid Mechanics and Its Applications, vol. 134, pp. 690–707. Springer (2023)

van Walree, F., Sgariato, D.: Impulsive loads on and water ingress in a landing craft: model tests and simulations, contemporary ideas on ship stability. In: Fluid Mechanics and Its Applications, vol. 134, pp. 708–725. Springer (2023)

van der Kolk, N.J., Akkerman, I., Keuning, J.A., Huijsmans, R.H.M.: Low-aspect ratio appendages for wind-assisted ships. J. Mar. Sci. Technol. **26**, 1126–1143 (2021)

Valtonen, V., Bond, J., Hindley, R.: Improved method for non-linear FE analysis of polar class ship primary structures. Mar. Struct. **74**, 102825 (2020)

Veltheim, O., Suominen, M., Ikonen, T., Kujala, P.: Validation of an Inverse Model to Determine Ice Load Magnitude and Load Patch on a Ship Hull. IAHR Ice 2022 (2022)

Vergassola, G., Pais, T., Boote, D.: Numerical tools and experimental procedures for the prediction of noise propagation on board superyachts. RINA Trans. Pt. B Int. J. Small Craft Technol. **160**(Part B1), B-9, B-16, (Jan–Jun 2018) https://doi.org/10.3940/rina.ijsct.2018.b1.207.

Vergassola, G., Pais, T., Boote, D., Rasori, S., Tonelli, A.: On the structural assessment of tall ship rigs. Ships Offshore Struct. **2023** (2023). https://doi.org/10.1080/17445302.2023.2226506

Vergassola, G., Boote, D., Pais, T.: Hull–furniture interaction in the primary response to global loads of a carbon fibre sailing yacht. Ships Offshore Struct. **14**(3), 281–294 (2019a). https://doi.org/10.1080/17445302.2018.1498210

Vergassola, G., Pais, T., Boote, D.: Low-frequency analysis of super yacht free vibrations. Ocean Eng. **176** (2019b). https://doi.org/10.1016/j.oceaneng.2019.02.037

Vergassola, G.: The prediction of noise propagation onboard pleasure crafts in the early design stage. J. Ocean Eng. Mar. Energy. **6**(1), 15–301 (2020) https://link.springer.com/article/10.1007/s40722-019-00149-4

Vergassola, G., Boote, D.: A simplified approach to the dynamic effective thickness of laminated glass for ships and passenger yachts. Int. J. Interact. Des. Manuf. (IJIDeM). **14**, 123–135 (2020). https://doi.org/10.1007/s12008-019-00614-2

Wang, Q., Li, Z., Tan, B.: The Effects of Porosity on the Flexural and Uniaxial Compressive Strength of Sea Ice in the Summer Arctic. POAC, 2023, paper 67 (2023a)

Wang, J., Chen, X., Ji, S.: Analysis on the Influence of Measuring Point Failure in Ice Load Identification of Ship Structures. POAC, 2023, Paper106 (2023b)

Warren, M., Ali-Lavroff, J., McVicar, J., Magoga, T., Shabani, B., Holloway, D.: Fatigue estimation on a high-speed catamaran during normal operations. Int. J. Marit. Eng., A55–A67 (2022). https://doi.org/10.5750/ijme.v164i1.736

Warren, M., Ali-Lavroff, J., Holloway, D., Magoga, T.: Comparing linear damage hypothesis to linear elastic fracture mechanics in the estimation of fatigue life in a high speed light craft. Ocean Eng. **290**, 1116338 (2023). https://doi.org/10.1016/j.oceaneng.2023.116338

Wen, X., Del Buono, A., Liu, P., Qu, Q., Iafrati, A.: Acceleration effects in slamming and transition stages for the water entry of curved wedges with a varying speed. Appl. Ocean Res. **128** (2022). https://doi.org/10.1016/j.apor.2022.103294

Windyandari, A., Sulardjaka, K.O., Tauviqirrahman, M.: Bow structure damage analysis for hybrid coir-glass fiber composite fishing boat hull subjected to front collision load. Curved Layer. Struct. **9**, 236–257 (2022)

Wium, D., Lataire, E., Belis, J.: Considerations for the integration of glass in superyacht structures. In: Belis, J., Bos, F., Louter, C. (eds.) Challenging Glass Conference Proceedings, vol. 8, (2022). https://doi.org/10.47982/cgc.8.442

Wu, J., Sun, Z., Jiang, Y., Zhang, G., Sun, T.: Experimental and numerical study of slamming problem for a trimaran hull. Ships Offshore Struct. **16**(1), 46–53 (2021)

Xie, C., Zhou, L., Ding, S., Liu, R., Zheng, S.: Experimental and numerical investigation on self-propulsion performance of polar merchant ship in brash ice channel. Ocean Eng. **269** 113424 (2023)

Xin, J., Shi, F., Fan, S., Jin, Q.: A sharp interface multiphase flow model for two-dimensional water impact of a symmetric and asymmetric wedge. Appl. Ocean Res. **119**(2022), 102988 (2022). https://doi.org/10.1016/j.apor.2021.102988

Xu, S., Tang, Y., Chen, K., Zhang, Z., Ma, T., Tang, W.: Numerical investigation on pressure responsiveness properties of the skirt-cushion system of an air cushion vehicle international journal of naval architecture and ocean. Engineering. **12**(2020), 928–942 (2020). https://doi.org/10.1016/j.ijnaoe.2020.09.006

Yang, J., Sun, H., Li, X., Wu, D.: Experimental study on wave motion of partial air cushion supported catamaran. Int. J. Nav. Archit. Ocean Eng. **15**, 100525 (2023). https://doi.org/10.1016/j.ijnaoe.2023.100525

Yoon, S., Herrnring, H., Müller, F., von Bock und Polach F.: Numerical Analysis of Ice Blocks Impact on Stiffened Plates According to a Mohr-Coulomb Material and Node Splitting Technique. OMAE2023-102584 (2023)

Yu, Z., Lu, W., van Den Berg, M., Jørgen, A., Sveinung, L.: Glacial ice impacts: part ii: damage assessment and ice-structure interactions in accidental limit states (als). Mar. Struct. **75**, 102889 (2021)

Zambon, A., Moro, L., Biot, M.: Vibration analysis of super-yachts: validation of the Holden method and estimation of the structural damping. Mar. Struct. **75**, 102802., ISSN 0951-8339 (2021). https://doi.org/10.1016/j.marstruc.2020.102802

Zhang, Z., Shu, C., Khalid, M.S.U., Yuan, Z., Liu, W.: Investigations on the hydroelastic slamming of deformable wedges by using the smoothed particle element method. J. Fluids Struct. **114** (2022a). https://doi.org/10.1016/j.jfluidstructs.2022.103732

Zhang, G., Zhao, Y., Yang, B., Sun, Z.: Experimental and Numerical Investigation of the Hydrodynamic Effect during the Ice Floe-structure, p. 1325. ISOPE 2022 (2022b)

Zhao, W., et al: On the Structural Analysis of Icebreakers Due to Ramming of First-Year Ice Ridges. OMAE2022-79661 (2022a)

Zhao, W.H., Tian, Y., Yu, C., Gang, X., Lu, P.: Numerical Investigation on Center Loading on a Circular Plate of Model Ice. POAC, 2022, Paper36 (2022b)

Zhou, M., Jian, J., Wang, Q., Yang, M.: Ice-Pod Interaction Analysis for Ice-Classed Azimuth Propullsor. ISOPE2023, 1952 (2023)

Zhu, X., Liang, C., Qian, H.: Design and implementation of a novel adaptive multihull sailboat with liftable side hulls. In: 2023 IEEE International Conference on Robotics and Biomimetics (ROBIO), pp. 1–6. IEEE (2023)

Zhu, L., Cai, W., Chen, M.S., et al.: Experimental and numerical analyses of elastic-plastic responses of ship plates under ice floe impacts. Ocean Eng. **218**, 108174 (2020)

Open Access This chapter is licensed under the terms of the Creative Commons Attribution-NonCommercial-NoDerivatives 4.0 International License (http://creativecommons.org/licenses/by-nc-nd/4.0/), which permits any noncommercial use, sharing, distribution and reproduction in any medium or format, as long as you give appropriate credit to the original author(s) and the source, provide a link to the Creative Commons license and indicate if you modified the licensed material. You do not have permission under this license to share adapted material derived from this chapter or parts of it.

The images or other third party material in this chapter are included in the chapter's Creative Commons license, unless indicated otherwise in a credit line to the material. If material is not included in the chapter's Creative Commons license and your intended use is not permitted by statutory regulation or exceeds the permitted use, you will need to obtain permission directly from the copyright holder.

Committee V.6: Ocean Space Utilization

Chao Tian[1(✉)], Harry Bingham[2], Ciro Busiello[3], Ingo Drummen[4], Nuno Fonseca[5], Zhiqiang Hu[1], Debabrata Karmakar[6], Ekaterina Kim[7], Sarat Mohapatra[8], Motohiko Murai[9], and Robert Sielski[10]

[1] China Ship Scientific Research Center (CSSRC), Wuxi, China
ctian@cssrc.com.cn
[2] Technical University of Denmark (DTU), Lyngby, Denmark
[3] Fincantieri, Trieste, Italy
[4] Maritime Research Institute Netherlands (MARIN), Wageningen, Netherlands
[5] SINTEF Ocean, Trondheim, Norway
[6] National Institute of Technology Karnataka (NITK), Surathkal, India
[7] Norwegian University of Science and Technology (NTNU), Trondheim, Norway
[8] University of Lisbon (CENTEC), Lisbon, Portugal
[9] Yokohama National University, Yokohama, Japan
[10] Sielski and Associates, Melbourne Beach, Florida, USA

Committee Mandate. Concern for the application of offshore structures for ocean space utilization purposes, such as VLFS, seabed mining and offshore aquaculture structures. Focus should be given to fluid-structure interaction induced by the large size, seabed bathymetry and structure flexibility. Due consideration should be given to the comparison of simple and more refined theories to determine the dynamic response of structures with connectors, mooring systems, etc. The engineering applicability should be discussed based on available tank testing and full-scale measurements in the actual engineering structures. General requirements, interpretations and standards used in the design for safety, reliability and serviceability of the offshore structures for different ocean space utilization purposes shall be discussed.

Keywords: Ocean Space Utilization Structures (OSUS) · VLFS · Hydroelasticity · Mooring Analysis · Connectors · Multi-body Hydrodynamics · Offshore Aquaculture · Offshore Floating Photovoltaics (OFPV) · Floating offshore wind turbine · Deep Sea Mining (DSM) · Risk Assessment · Rules and Regulations

1 Introduction

The oceans, covering 71% of the earth's surface, provide human civilization with tremendous and indispensable resources related to biology, energy, minerals, food, medicine, transportation, living and pleasure, etc. (see Fig. 1). About 40% of the world's population lives within 100 km of the coast, 70% of economic activities occur in coastal zones, and more than 80% of international trade is carried out by sea. Human activities of exploitation and utilization of marine resources are expanding from coastal zones and continental shelf regions, islands and reefs to deep sea and polar regions. Figure 1 shows an example of modern ocean space utilization in Hangzhou Bay, China.

In recent years the demand for developable property near coastal cities has surged, driven by the need for residential, commercial, industrial, and research spaces (Wang and Tay 2011). Ocean Space Utilization Structures (OSUS) are utilized for various purposes and are typically constructed in coastal or offshore areas. They have several advantages, such as slight environmental impact, flexibility of construction site, strong wave resistance, convenient construction, simpler installation, and easy expansion (Wang and Wang 2015; Xia et al. 2016). Because of the lack of available nearshore production sites, the industry has moved to more exposed sites with large amplitude waves over the past decade. Significant design challenges have been encountered. In order to give a systematic, comprehensive broad view of ocean space utilization, the publications reviewed here cover a time frame greater than the previous three years.

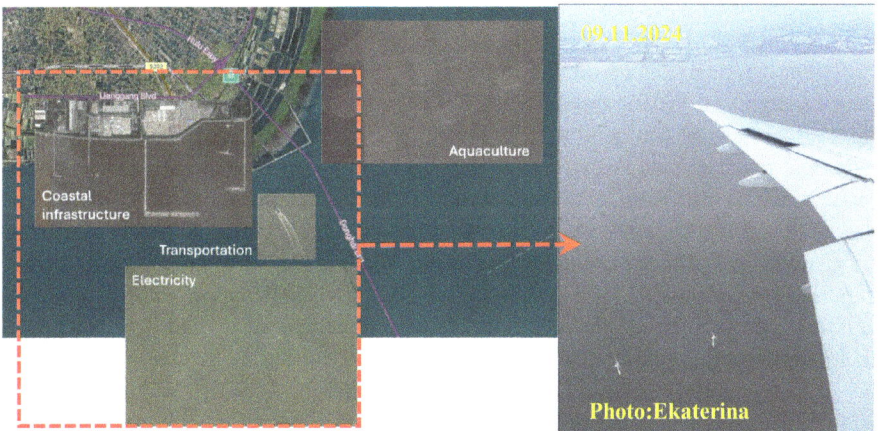

Fig. 1. An example of modern ocean space utilization (Hangzhou Bay, China).

Generally speaking, OSUS have applications in several domains, including offshore marine energy utilization, commercial navigation, fisheries and aquaculture, offshore oil and gas, meteorological monitoring, and environmental protection.

Traditionally, OSUS are utilized in specialized applications to support transportation and infrastructure development in marine environments. This includes offshore ports, pontoons, offshore airports, and other facilities that enhance maritime trade, connectivity, and logistics. A notable example is the Mega-Float project in Japan, which featured a VLFS measuring 4.75 km in length and enabled small aircraft to land and take off on its surface (Kyozuka et al. 2001; Suzuki 2005). In addition, floating bridges are a transportation channel widely used to cross rivers, lakes, fjords, and seas. When developing ocean space, floating bridges provide an effective transportation channel that can connect straits, fjords, or offshore islands and reefs. Compared to traditional bridges, floating bridges have many advantages, such as easy construction, convenient migration, and strong environmental adaptability, and they can be installed without the need for large mechanical equipment, so they are widely used in coastal areas and offshore islands. This has received special attention from government departments and research institutions, especially in countries or regions with numerous mountains, rivers, and straits.

The global demand for seafood has increased sharply, so fishery production is becoming increasingly prevalent. In recent years, with the shrinking of nearshore aquaculture space, deep-sea aquaculture has received attention from many enterprises and research institutions, and different forms of deep-sea aquaculture equipment have been proposed. Norway was the first to propose and design Ocean Farm. China has also designed and built multiple types of deep-sea aquaculture equipment. The world's first semi-submersible offshore fish farm (SOFF)-Ocean Farm 1 - arrived at Frohavet in 2017, and the Deep Blue 1 was installed in the Yellow Sea in 2018. This has offered significant design challenges. OSUS are also designed to support aquaculture operations by providing platforms for fish farming (Xie et al. 2023), shellfish cultivation, and seaweed cultivation within a controlled environment. The fish cages typically consist of flexible nets and circular floating collars made of high-density polyethylene, making them susceptible to deformation under wave action (Shuaishuai Wang et al. 2023; Yang et al. 2023).

In recent years, OSUS have been used to harness renewable energy sources such as offshore wind (Li 2022), tidal (Feng et al. 2023), wave energy (Nguyen et al. 2019, 2020; Zhang et al. 2019), and solar energy (Kumar et al. 2021; Zhang et al. 2022a). These structures are equipped with turbines, generators, and other energy conversion equipment to harness the ocean's natural energy and produce electricity. Ocean photovoltaic farms can fully utilize the vast space of the ocean and choose to deploy FPV in areas with mild sea conditions, providing more green energy for humanity's growing energy demand and more technological means and equipment to reduce carbon emissions. The expansion of FPV systems into marine environments presents a significant shift from reservoir-based installations, offering abundant space and new challenges due to complex environmental conditions. Up to now the technological innovations of FPV systems continue, particularly for offshore applications being the frontier, due to insufficient technological maturity and exposure to harsher conditions (Shi et al. 2023b).

Additionally, seabed nodule deposits contain significant amounts of metallic elements such as manganese, copper, cobalt, and nickel, which are regarded as an important alternative resource to land-based minerals (Sharma 2017). Deep-sea minerals are considered vital for high-tech industries that support a low-carbon economy. These minerals include gas hydrates, polymetallic nodules, cobalt-rich crusts, polymetallic sulfides, and rare earth elements, which are essential for various applications such as electric vehicles, renewable energy, and aerospace components. Consequently, deep-sea mining projects for the exploitation of seabed minerals have attracted much interest over the past few decades owing to the abundance of seabed polymetallic nodules.

Therefore Ocean Space Utilization Structures (OSUS) are crucial to the sustainable development, playing an increasingly essential role for ocean space utilization, resources exploitation and energy recovery (Li et al. 2022a). The main OSUS examples include seaports, VLFS, artificial islands, cross-sea bridges, sub-sea tunnels, coastal protection structures, seabed mining, mariculture pastures, etc. A thorough and complete review of technical demands and challenges of all the above items surpass the background and competitiveness of the committee members. Therefore only typical ocean space utilization structures, such as VLFS, seabed mining and offshore aquaculture structures will be mostly touched in the present and the main technical conditions and challenges for the current ocean space utilization structures will be given in this report.

The near-shore ocean space is constantly decreasing. Owing to the limited space in the nearshore, the offshore aquaculture platforms, marine floating photovoltaic (FPV) farms, offshore wind farms, etc. are expected to experience a transition from nearshore to offshore applications, moving from shallow to intermediate and deep water (Jiang et al. 2023). When OSUS is designed to operate in complex and harsh environments, it must harness the resources and potential of the ocean environment. Failures may cause serious casualties, economic losses and environmental pollution. Taking this into account, the wind, wave, and current environmental conditions for these very large floating structures are first addressed in Sect. 2. In Sect. 3 we address fluid-structure interaction in terms of global loads, motions and structural responses. Since floating solar power plants and mobile offshore bases are sequentially connected multi-modular floating structures, research trends in the design and treatment of connectors is covered in Sect. 4. Section 5 discusses the positioning and mooring of OSUs which are critical to their operational efficiency and safety. In Sect. 6, based on a review of tank testing and full-scale measurements of OSUS, challenges on engineering applicability are provided. Section 7 gives a review on general requirements, interpretations and standards used in the design for safety, reliability and serviceability of offshore structures for different ocean space utilization purposes. After risk assessments and reliability analysis is discussed in Sect. 8, conclusion and recommendations are given in the last section.

2 Environmental Conditions

2.1 Introduction

In this Section, we highlight some of the most relevant aspects of environmental conditions on the topic of Ocean Space Utilization. In particular, we focus on those aspects of wind, wave, and current conditions that present special challenges for the very large, and generally nearshore, structures that are of concern here. Table 1 from Committee I.1 Environment Ch. 1 summarizes the required environmental input for a broad range of marine structures. As noted there, the special environmental features of relevance to very large structures are mainly associated with variable wind, wave and current conditions along the structure, bathymetry and other nearshore effects including reflection from coastal boundaries and generation of infra-gravity waves. Since we do not currently anticipate the deployment of very large structures in ice-impacted regions, this topic is not considered here.

Table 1 Main wave models and software

Scale	Type of model	Model	Solver or Software
Large-Medium Scale	Phase-averaged models (Wind-wave models)	The third generation wave model (with ST1, ST2, ST3, ST4, ST6 source term)	SWAN WAVEWATCH III MIKE21 SW ……

(*continued*)

Table 1 (continued)

Scale	Type of model	Model	Solver or Software
Medium-Small Scale	Phase-resolved models	Nonlinear shallow water model Mild-slope equation, Boussinesq-type model High-order spectral method Green-Naghdi model	SWASH REF/DIF FUNWAVE BOSZ NHWAVE
Small Scale or localized region	CFD or potential flow models	N-S equation Potential flow theory	OpenFOAM STARCCM+ AQWA WAMIT

2.2 Wave Models

Owing to the random nature of sea surface waves, this is one of the most complex phenomena to predict (Thomas and Dwarakish 2015). Validated wave evolution models are needed for different environmental characteristics. For application to the needs of Ocean Space Utilization, wave modeling at large and medium scales mainly consists of phase-averaged wind-wave models for the open ocean, and phase-resolved models that can describe the evolution of specific wave surfaces. The main wave models and software currently employed are shown in Table 1.

2.2.1 Phase-Averaged Models for the Open Sea

Assessment of the wave environment in the open ocean is often driven by the surface wind field and requires the ability to model wave fields on the scale of thousands of square kilometers. The wind-wave models are based on a fully spectral representation of the energy or action balance equation with all physical process modelled explicitly, and sometimes extended to coastal regions by adding the finite-depth effects of shoaling, refraction and bottom friction. (Tolman 1991; Booij et al. 1996, 1999) To establish the energy or action balance equation, Miles (1957) and Phillips (1957) proposed the wave generation theories, and Hasselman et al. (Hasselmann 1962) quantified the interaction and conservative exchange of spectral energy. With the introduction of the reduced discrete interaction approximation parameterization of the nonlinear interaction, the third generation of wave models were eventually formulated without a priori limitations for the spectral evolution (Hasselmann 1962). In addition to the third generation wave model WAM, originally built by Komen et al. (Komen et al. 1996), wave models such as WAVEWATCH III (Tolman 1991), SWAN (Booij et al. 1999), MIKE21 SW and others have been built based on the same theory.

The third generation wave model offers a diverse range of source terms for various environmental conditions, mainly considering different wind input and wave energy dissipation. For convenience of selection and utilization, these source terms have been combined to form several pattern packages, including WAM Cycle 3 (ST1) package presented (Komen et al. 1984) and (Snyder et al. 1981), the ST2 package (Tolman and Chalikov 1996), the WAM Cycle 4 (ST3) package presented by Janssen et al. (1991) and Bidlot et al. (2007), the ST4 package (Ardhuin et al. 2010), and the ST6 source term (Zieger et al. 2015). There are still extensive studies for optimization of the empirical parameters in the wave models (Mojtahedi et al. 2024).

Committee I.1 Environment, Ch. .3.2 summarizes the progress made in phase-averaged spectral wave modelling over the past four years. A number of studies have been carried out to characterize the performance of the available source term parametrizations ST1 – ST6 applied to the most commonly used third generation model WAM, WAVE-WATCH III and SWAN. These models generally perform well when the spatial and temporal resolution is adequate to resolve the local regional variations. There is however some disagreement on which source term parametrization performs best with some recommending ST3 and ST4, and others recommending ST6. Some attempts at using Machine Learning (ML) to boost the accuracy of these models show promise. In terms of application to predicting the loading on very large structures, these models can be used to determine whether or not the wave climate is homogeneous along the structure, but it's less clear how to then determine the appropriate forcing in the context of standard linear radiation/diffraction theory.

Wind-wave models have been developed and applied extensively for wave forecasting in various ocean and nearshore regions. For example, the models have been applied to computing the Hurricane Ivan process around Gulf of Mexico (Kalourazi et al. 2021), the wave evolutions around the northern Tyrrhenian Sea and off the Mediterranean Spanish coast (Mentaschi et al. 2015), and in the entire West African region (Foli et al. 2022). The efficiency and robustness of these third generation wave models has been fully proved, and the wave prediction methods in deep water have achieved significant progress and provided generally satisfactory wave simulations. As a result, third-generation wave models have continued to be widely employed in recent years for wave assessment (Churchill et al. 2024; Xu et al. 2024a, b).

Cavaleri et al. (2020) summarize the development of wind-wave models and presents a developmental outlook. They consider that although the present wave modeling activity shows good results for significant wave heights, the only fundamental conceptual improvement for the wind-wave models in the last 30 years has been the recognition by Janssen and Peter (1991) of the two-way interaction with the atmosphere, and the consequent need to use a coupled model. In addition, the approximation for higher peak wave heights is still not as good as the general performance. For the spectra, the maximum heights and the corresponding wave shape in a storm are poorly estimated. They suggest that the traditional opposing processes of input by wind and dissipation by white-capping should be considered as part of a single process.

As Cavaleri et al. (2020) suggested, Leung et al. (2024) have coupled the Weather Research and Forecasting Model, Regional Ocean Modelling System and the third generation wave model SWAN to build the Ocean-Atmosphere-Wave-Sediment Transport

Modelling System, which can be used to evaluate the effect of atmosphere-ocean coupling on sea level. Ditching the spectral models, Hell et al. (2024) constructed the Particle-in-Cell for Efficient Swell wave model to couple atmosphere, ocean, and sea ice. The model can solve for the growth and propagation of a parametric wave spectrum with a reduction in the size of the state vector size by a factor of 50–200 compared to spectral models.

Furthermore, the third generation wave models are being coupled and developed with other wave models to adapt to the requirements of wave simulation under different conditions. For example, the National Oceanic and Atmospheric Administration (NOAA) developed a coupled system between WAVEWATCH III and the surface gravity wave model for the global ocean surface wave simulation (Tolman et al. 2002). Fan et al. (2012) also simulated a 29-yr (1981–2009) period of global ocean surface gravity waves by using this coupled atmosphere–wave Model.

As a wind-wave model, the third generation wave model has also been inverted to evaluate the wind stress. Zhao et al. (2024) utilize WAVEWATCH III to discuss the wave blocking impacts for wind stress modification on offshore wind turbines, and conclude that the wind turbine foundation can change the downstream wind stress by blocking the ocean surface waves.

With advances in computing resources, measurement methods, and emerging technologies, the wave modeling has also emerged as a new topic to improve efficiency and accuracy. Hu et al. (2024) employed the remotely sensed products to improve the drag coefficient parameterization of WAVEWATCH III. Zhang et al. (2024) optimized the third generation wave models by using machine learning. Escobar et al. (2024) ported the wave model to GPU architectures for computational efficiency. Siadatmousavi et al. (2024) improved the wave model results by using data assimilation.

2.2.2 Phase-Resolved Models for Near-Shore Regions

In contrast to phase-averaged models, phase-resolved models resolve individual waveforms. This allows them to more accurately capture wave diffraction, phase-sensitive reflections, wave refraction, nonlinear wave-wave interaction, wave breaking, and wave-induced currents (Kirby 2016). However, phase-resolved model require a grid resolution fine enough to capture the shortest wave length and highest frequency waves of interest, which means that it calls for more computational resources and therefore is able to compute a much smaller range than the phase-averaged models.

Phase-resolved wave models are built based on equations under different assumptions, the more classical of which include linear wave model, Stokes wave model, nonlinear shallow water model, mild-slope equation, Boussinesq-type models, high-order spectral methods, and Green-Naghdi models. Considering the actual demand of simulation, the most commonly used models today are the mild-slope equations, Boussinesq-type models, high-order spectral methods, non-hydrostatic models, and the Green-Naghdi model. A number of open-source solvers have also been released including: the SWASH models (Zijlema et al. 2011), REF/DIF (Kirby and Dalrymple 1983), FUNWAVE (Kirby 1998), BOSZ (Roeber and Cheung 2012), NHWAVE (Shi et al. 2012), which are widespread examples of this type of model, very often mentioned in the wave modelling literature.

Among the available classes of phase-resolved wave model, Boussinesq-type wave models are less time-consuming than non-hydrostatic wave models using a pressure Poisson solver. By employing parallel computing and adaptive mesh refinement techniques, the model can be used to simulate surface wind waves over domains with dimensions of tens of kilometers (Chakrabarti et al. 2017). Therefore, this type of model has also been practically applied to wave simulations for near-shore regions and near-island reefs.

Boussinesq equations are developed from the KdV equation, and the key concept is to make a Boussinesq approximation to the flow field. The most commonly used Boussinesq approximation was derived by Peregrine (1967). Boussinesq equations can improve the non-linearity of the simulated waves by retaining higher order nonlinear terms, and can more accurately represent the dispersion aspects of the waves by modulating reference layers (Madsen et al. 1991; Nwogu 1993; Agnon et al. 1999; Kennedy et al. 2001). With improvements in wave breaking theory, processing techniques of dynamic boundary and wave-making conditions, Boussinesq-type equations have now evolved into a more computationally complete model for coastal waves (Fang et al. 2015; Shirai et al. 2023; Ataie-Assehtiani and Najafi-Jilani 2024). Models for wave evolution on an impermeable seabed have been fully developed (Gao et al. 2023).

Moreover, as in the case of phase-averaged models, there is a trend towards combining phase-resolved models with new techniques such as GPU parallelism and machine learning in order to improve the efficiency of the solving process. Yuan et al. (2020) developed a multiple GPU model for the FUNWAVE solver and established the GPU-accelerated form of a Boussinesq-type equation. Duan et al. (2020) have proposed an artificial neural network-based wave prediction (ANN-WP) model, and achieved phase-resolved wave prediction for long-crested waves based on machine learning.

As detailed in Committee I.1 Environment, Chapter 3.2, the most relevant advances over the past four years in phase-resolved numerical modelling of the wave climate in coastal regions are in the area of nonlinear potential flow theory. Significant progress has been made on physics-based wave breaking models which can be applied to solvers such as REEF3D: NLPF Pákozdi et al. (2022) or HOS-NWT Ducrozet et al. (2023) to cover large domains and capture the influence of breaking on the wave statistics. Although these models are much more computationally intensive than spectral wave models, they can provide more accurate estimates of the nearshore wave climate around a very large floating structure.

2.2.3 Coupled Wave Models for Near-Island or Reef Regions

Near-island or reef lagoon waves have their own unique complexities. Wave modeling becomes a challenge due to the wave transition in sea regions near islands or in reef lagoons with complicated bathymetry, and the mixture of swell from open sea with locally generated wind waves (Cavaleri and Bertotti 2004; Jouon et al. 2009; Zong et al. 2019). In assessing wave propagation and evolution in the region near islands or a reef lagoon, it is necessary to take into account both the wave conditions in the surrounding open sea at the 100-km scale and the refraction and diffraction of waves near the islands. Due to this situation, neither phase-averaged models nor phase-resolved models alone can be used directly to obtain sufficiently satisfactory wave simulations.

For local regions, phase resolved models based on Boussnesq-type equations or the mild-slope equation can also be applied. For example, Zhang et al. (2022) use a time-domain Boussinesq model to discuss the wave interference in San Francisco Bay and the adjacent Ocean Beach. Yao et al. (2023) simulate the wave evolution process in a reef-lagoon-channel system based on a Boussinesq-type model. Wang et al. (2023) also modelled the beach profile evolution based on a Boussinesq wave solver. The solver FUNWAVE, based on Boussinesq-type equations is well developed, and has in recent years been applied to a large number of wave propagation and evolution simulations (Gardner 2020; Wan et al. 2021; Nguyen et al. 2023), as well as wave evolution patterns (Benvenuto et al. 2022; Kwak 2024).

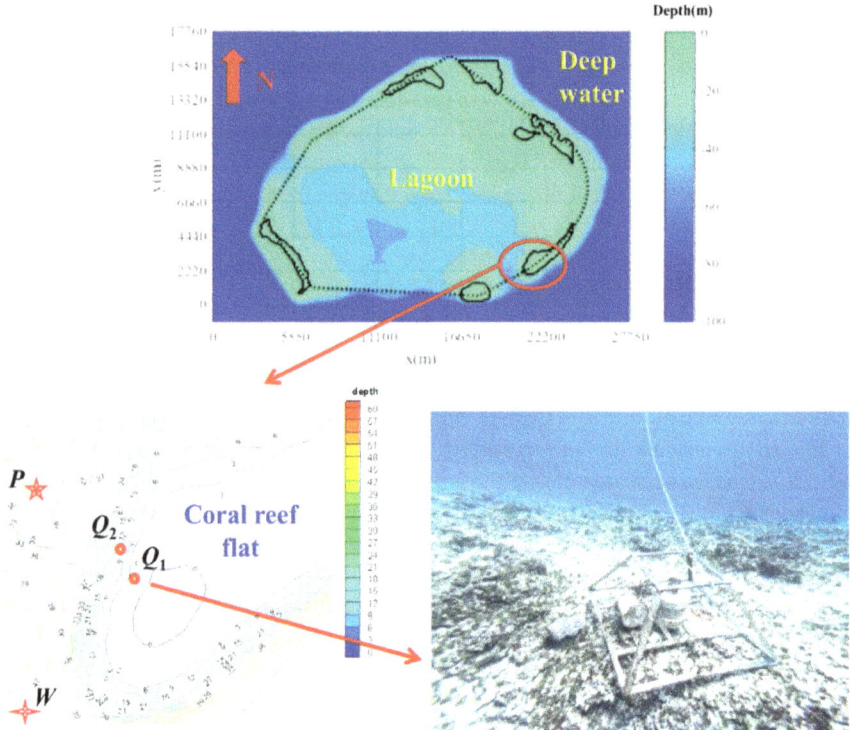

Fig. 2. Bathymetry, layout of the instrument locations and Coral Sea bed (Chen et al. 2023a)

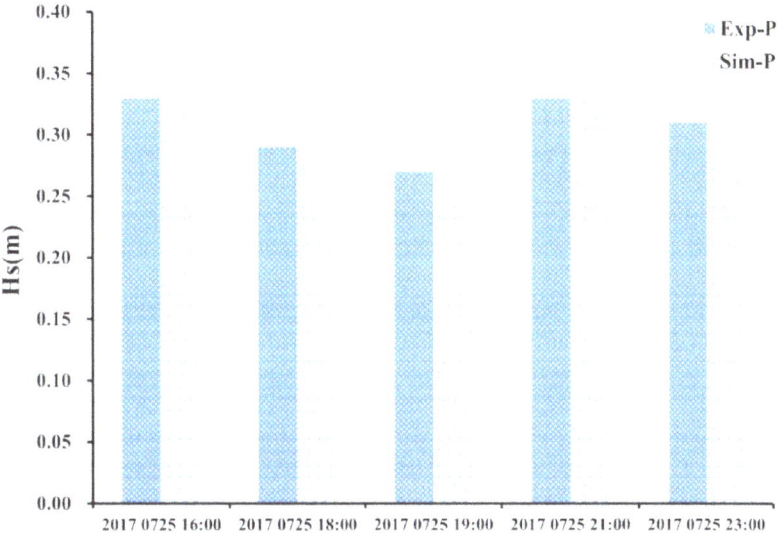

Fig. 3. Comparisons of measured and simulated significant wave heights in the coral reef region (Chen et al. 2023a)

In addition, based on pure phase-averaged models, multi-scale zone and multi-scale grid techniques have been targeted and developed to provide a reliable tool for wave forecasting in complicated geographical environments (Sun et al. 2021b). Meanwhile, the wind input-dissipation source terms were tested to find the most suitable parameters for the near-island regions (Zou et al. 2023).

Furthermore, combining phase-averaged models employing a multi-scale grid with phase-resolved models to establish a multi-size analysis system is an effective approach for solving near-island wave evolution problems when considering larger-scale environmental impacts and evolution. Phase-averaged models employing a multi-scale grid can be used to simulate wave energy propagation from large to relatively localized regions. By taking the results obtained from a large-scale phase-averaged model as the boundary condition, phase-resolved models can be used for simulations of wave refraction and diffraction in shallow water and near-island regions. Based on this approach, a wave modeling system applicable to near-island wave evolution has been developed (Sun et al. 2019, 2021a; Cai et al. 2022). The resulting simulations have been tested by *in-situ* measurement data and are satisfactory (Cai et al. 2019, 2021b; Zou et al. 2023; Chen et al. 2023), see Figs. 2 and 3 for example.

2.2.4 Challenges and Recommendations

- With the development of theoretical and computational methods, existing wave models are able to deal with wave evolution in most environmental conditions from large to small scales. It is considered that the ability to use a single wave model, or to couple multiple wave models, meets most of the needs of wave environment assessment. However, there are still some mechanistic aspects and challenges to wave modeling.

- Simulation methods for detailed waves and their spatial inhomogeneities are not yet well developed. Current wave models impose restrictive assumptions, and they are mainly designed to capture the characteristics of the overall wave field, while the simulation of detailed waves is often limited to traditional CFD methods. However, the large spatial dimensions and the long evolution time of the waves that characterize them require more computational resources if they are simulated by CFD methods. Non-hydrostatic models or higher-order spectral models combined with adaptive meshing are considered to be potentially effective methods for dealing with refined waveforms in the future.
- Because of our current incomplete understanding of the true physics of water-air coupling, in particular during wave breaking, the simulation of some extreme wave events can be difficult for existing large- and mesoscale wave models.
- For shallow water regions, especially coral reefs, it was found that the limiting wave heights could potentially exceed the wave model limitations on the wave height to depth ratio. Such shallow water large wave height events may have a greater impact on the construction of offshore structures and should be given more attention in the future.

2.3 Current

Committee I.1, Chapter 3.3 summarizes recent progress on ocean circulation and coastal current modelling, hindcasting and forecasting. Of most relevance to our topic are the developments in coastal current modelling using either the Finite Volume Coastal Ocean Model (FVCOM) or the Regional Ocean Modelling System (ROMS) which is based on the finite difference method. As for the phase-averaged spectral wave models discussed above, these models show generally good agreement with measurements as long as the forcing is accurately imposed. Machine Learning is also being applied here to improve forecasting.

2.4 Wind

Committee I.1, Chapter 3.4 presents recent progress on surface wind modelling, hindcasting and forecasting. Spurred by the rapidly expanding wind energy sector, wind models have seen correspondingly rapid development over the past four years. The main tool for research on this topic is the open-source Weather Research Forecasting (WRF) model which currently has more than 57,000 users from over 160 countries. Reanalysis of calculations using WRF (or other weather prediction models) by comparison with measurements is widely used to tune parameters and improve the accuracy, as is done in the ERA5 data for example. Significant progress has been made in understanding and quantifying the complex interaction between wind, waves and currents which gives guidance on the behaviour of coupling parameters in different conditions.

2.5 Conclusions and Recommendations

The state of the art in predicting current and future trends in wind and wave climate, on both a global and a regional scale, is to use spectral wave models forced by the Coupled

Model Intercomparison Project (CMIP) General Circulation Models (GCM) of 10 m surface wind speeds. Reanalysis and hindcasting of these simulations based on local point measurements is considered to be quite accurate for characterizing the present and historical conditions, though uncertainties are higher closer to shore. Predicting future trends naturally involves much larger uncertainties which are difficult to quantify but significant progress is being made in this direction through publicly available climate simulation ensembles.

The impact of climate change on the mean and extreme wind and wave environment over the next century is an important consideration for the design of all marine and offshore structures. The newest CMIP6 projections based on a high carbon emissions scenario, predict much larger changes in the mean wind resources by the end of the 21st century compared to the predictions from the earlier CMIP5 model. Mean wind speeds and significant wave heights are predicted to increase by up to 10% in the Southern Ocean and the Eastern tropical South Pacific, and by more than 100% in the Arctic. A decline of up to 10% is predicted in the North Atlantic and North Pacific. Uncertainties in these predictions are however generally high.

3 Fluid Structure Interaction

3.1 Introduction

Unlike regular ships, OSUS typically have a vast volume and surface area, with a relatively thin vertical depth and a large horizontal plane, making them suitable for transporting heavy machinery and large quantities of supplies. Due to their enormous size, these structures have low stiffness, and their elastic deformation often needs to be considered. As a result, OSUS are commonly constructed in a multi-modular manner, with connectors linking the individual modules. Each floater is generally considered a rigid body, while the interconnections provide the necessary flexibility.

Consequently, the dynamic behavior of OSUS is often approached as a hydroelastics and multi-body dynamics problem. One notable challenge is the nonlinear multi-body dynamic characteristics, which involve not only the floating body itself, but also additional structures installed onboard, such as floating offshore wind turbines and wave energy converters. These facilities interact with the primary structure, creating a more complex multi-body dynamics scenario. Furthermore, water resonance in the narrow gaps between modules can significantly impact the hydrodynamic response and must be considered during analysis. This challenge has aroused many research interests due to its high nonlinear characteristics. Waals et al. (2018) provided an example of a multi-modular structure for a floating mega island composed of triangle floaters. Ding et al. (2020a, b, 2021) and Yang et al. (2019a) investigate the hydroelastic responses of a 3-module and 8-module VLFS numerically and experimentally. Several scholars have conducted hydrodynamic analyses on fully flexible floating bodies that are considered monolithic floating structures (Suzuki 2005; Song et al. 2022). Moreover, due to the large size of OSUS, the seabed where they are located is often uneven. The wind, wave, and current environments can vary significantly along the length of the OSUS due to changes in seabed topography and the influence of nearby islands and reefs. Therefore, the effects of spatially varying wave fields on the dynamic response of OSUS must be carefully considered.

3.2 Global Loads and Hydrodynamic Characteristics

This section summarizes the recent development in global loads and hydrodynamic characteristics of OSUS in three aspects, including hydroelasticity modelling, inhomogeneous incident waves, and seabed topology influences and shallow water effect.

3.2.1 Development of Hydroelasticity Modelling

Considering the scale characteristics of OSUS, the elastic deformation of these structures is significant, and the hydrodynamic coefficients vary greatly under wave action. The traditional potential flow theory, which assumes a rigid body, struggles to accurately predict the hydrodynamic loads on OSUS. Consequently, researchers have increasingly adopted hydroelasticity theories that account for the elastic deformation of floating structures to predict the hydrodynamic loads of large floating bodies. The development of hydroelasticity modelling has advanced significantly over the years, driven by improvements in computational methods and a deeper understanding of fluid-structure interaction phenomena. Hydroelasticity theory has evolved from two-dimensional to three-dimensional approaches, and the complexity of the problems addressed has expanded from linear to nonlinear categories. Below are some key milestones and methodologies in the evolution of hydroelasticity modelling:

1). Fluid Dynamics Modelling

The fundamental aspect of hydroelasticity modeling for OSUS is the accurate representation of fluid flow and the prediction of wave loads around the elastic structure. Betts et al. (1977) and Bishop (1979) were the first to propose a two-dimensional linear hydroelasticity theory based on strip theory, which simplifies the flow field into a two-dimensional plane and models the structure as a non-uniform Euler beam or Timoshenko beam. The theory uses a linear superposition of basic functions to describe the overall motion and deformation of ships in waves. Liu and Sakai (2002) and Qiu (2007) applied the boundary element method in the fluid domain and the finite element method in the structural domain to analyze the two-dimensional behavior of flexible floating structures under regular, irregular, and solitary waves, as well as under moving loads. Their results indicated that linear theory might not be applicable in extreme ocean conditions, where wave nonlinearity could cause significant differences in response. Pham et al. (2008) further evaluated the hydrodynamic diffraction and radiation forces, studying the hydroelastic responses of a pontoon-type VLFS with a horizontal submerged plate using the eigenfunction expansion matching method.

Since the two-dimensional hydroelasticity theory analyzes each strip individually and performs an integral solution along the ship's length, it overlooks the mutual interference between fluid motions in the longitudinal direction. Consequently, this theory is only applicable to slender structures and cannot effectively address large floating bodies with comparable length and width. To meet the computational requirements for the hydroelasticity of a wider range of structures, it is essential to extend two-dimensional hydroelasticity to three-dimensional hydroelasticity. Wu (1984) and Price and Wu (1985) proposed a generalized fluid-structure coupling interface condition, developing a new three-dimensional hydroelasticity theory by integrating three-dimensional seakeeping theory with three-dimensional structural dynamics theory. Aksu et al. (1991) employed both two-dimensional and three-dimensional methods to calculate the response of slender and non-slender elastomers under irregular waves. Their study found that while both methods accurately predict the hydroelasticity of slender bodies in heading waves, the three-dimensional method outperforms the two-dimensional method for non-slender bodies. Fluid dynamics modeling solutions have evolved and matured since the early 20th century, addressing a range of problems through the Morison equation and potential flow theory (Faltinsen et al. 1995). However, these methods are limited by their assumptions regarding linear wave conditions, small wave heights, and inviscid fluid. More recently, the combination of higher-order Boundary Element Method (BEM) with potential flow theory for simulating nonlinear fluid dynamics has gained popularity (Cheng et al. 2016; Shirkol et al. 2016; Shirkol and Nasar 2018).

Nevertheless, the application of the Boundary Element Method (BEM) still encounters challenges due to its discretization along the body's boundary rather than within the fluid volume. This limitation prevents the resolution of strong nonlinearities, such as wave slamming and green water effects. To address these issues, techniques from CFD are employed. By utilizing appropriate boundary conditions and turbulence models, CFD numerically solves the Navier-Stokes equations to simulate fluid flow around structures. Seng (2012) developed a coupling method to calculate the hydroelastic response using a beam model, demonstrating that the CFD method accurately predicts the hydroelastic response of structures. Jason et al. (2018) predicted the bending moment of a segmented wave-piercing catamaran resulting from wave slamming through both one-way and two-way fluid-structure coupling. Despite its high computational demands, the CFD approach is essential for many studies.

2). Structural Dynamics Modelling
Simultaneously, it is essential to model the dynamic behavior of the elastic structure. This involves formulating the equations of motion for the structure, taking into account its geometry, material properties, and boundary conditions. The two primary methods for analyzing structural elasticity are modal superposition and the direct integration method (Fu et al. 2007; Zhang et al. 2018; Chen et al. 2023b). Modal superposition is a valuable and computationally efficient tool for evaluating the dynamic response of linear structures, as it allows calculations to focus on the dominant modes. By superposing a selected number of eigenmodes, a structure's dynamic response can be effectively simulated, particularly in investigations conducted in the frequency domain. For instance, Ling et al. (2022) studied wave loads on various sections of a tourism platform using the elastic modal superposition method. Similarly, Wang et al. (2024) examined the characteristics of overall elastic deformation and wave loads on a multiply-connected domain tourism platform. However, it is important to note that modal superposition assumes linearity in the system response. In cases where the structure exhibits significant nonlinear behavior, such as large displacements or material nonlinearity, modal superposition may yield inaccurate results.

The direct integration method, also known as time history analysis, involves directly integrating the equations of motion of the structure over time to determine its dynamic response. This approach is particularly well-suited for analyzing structures subjected to complex loading patterns or time-varying excitations (Kim et al. 2013; Pal et al. 2018; Lakshmynarayanana and Temarel 2019).

For modeling structures, several models can be employed based on the degree of simplification required: beam model, plate model, beam-plate model, and 3D structural model. The beam model is the simplest structural representation, commonly used for slender bodies. This includes the Euler-Bernoulli beam model (Riyansyah et al. 2010; Sengupta et al. 2017; Lu et al. 2019; Zhang et al. 2021a; Wei et al. 2024) and Timoshenko beam model (Huang et al. 2023). To address the limitations of the beam model for pontoon-type VLFS, plate and beam-plate models have been developed. These models assume the structures to be uniform rectangular plates or combinations of beams (referred to as grillages) and plates (Sun et al. 2002; Gao et al. 2011; Karmakar and Soares 2018; Karperaki and Belibassakis 2021), as their horizontal dimensions are significantly larger than their thickness. The most comprehensive model, capable of accurately representing the distribution of mass and stiffness for global bending, shear, and torsional stresses, is the 3D structural model (Song et al. 2022). This model requires the generation of complex structural shapes and the provision of true stiffness; however, it is the most complicated to design and produce, as it must meet the similarity criteria for both hydrodynamic and structural models simultaneously. Therefore, the beam model or the beam-plate model is typically utilized during the initial design stage to estimate global loads, while the 3D structural model is further developed for assessing local stress and fatigue limits to ensure system safety.

3). Fluid-Structure Interaction (FSI) Modelling

The next step is to couple the fluid dynamics model with the structural dynamics model to account for the interaction between the fluid and the structure. Hydroelastic responses of floating structures can be predicted in both the frequency domain and the time domain. In the frequency domain, linear fluid dynamics models are typically linked and solved alongside structural dynamics models (Wei et al. 2017; Zhang et al. 2023a; Shi et al. 2023c). However, this type of coupling is limited to analyzing the steady state of hydroelastic calculations and cannot investigate transient and nonlinear phenomena, such as bow slamming. In recent years, advancements in computational resources have enabled researchers to conduct time-domain hydroelastic analyses for a variety of structures using time-domain methods (Kang and Kim 2017; Yang et al. 2019a; Li et al. 2022b; Huang et al. 2023). The structural dynamic models are mostly coupled with nonlinear hydrodynamic models.

The process of fluid-structure interaction can be categorized into two approaches: the monolithic approach and the partitioned approach. The monolithic approach involves solving the fluid and structural components together within a single, large matrix, offering the advantage of full implicit stability (Bungartz et al. 2016). However, this coupling method is challenging to establish due to the differing discretization schemes and grid types of each solver. In contrast, the partitioned approach utilizes two separate solvers that communicate at an interface without altering their respective codes. An example of the partitioned approach in the hydroelastic study of a flexible segmented barge in waves was presented by Senjanović et al. (2008). A 3D finite element method (FEM) structural model was combined with a 3D potential flow code based on radiation-diffraction theory in their fluid-structure interaction (FSI) technique. Tang et al. (2023) proposed a fully nonlinear hydroelastic numerical method for the fluid-structure interactions of elastic ships, utilizing BEM-beam coupling. The partitioned approach can be further divided into one-way coupling and two-way coupling. The primary distinction between the two is that in the two-way coupling process, structural deformation is fed back into the flow field, and the added mass effect is taken into account (as illustrated in Fig. 4). Two-way coupling approaches, such as the Arbitrary Lagrangian-Eulerian (ALE) method (Duarte et al. 2004), are often employed. This method allows the fluid and structure to exchange information at each time step, accounting for the feedback effects between them. Lakshmynarayanana et al. (2019) and Takami et al. (2018) investigated the influences of one-way and two-way algorithm on ship hydroelasticity based on a commercial co-simulation interface. Benra et al. (2011) investigated the vortex shedding process and developed a model of the pump using one-way and two-way coupling methods. While the one-way coupling approach is computationally efficient, it may not be suitable for investigating the springing and whipping of structures, as it ignores the effects of added mass, potentially leading to an underestimation of wave loads. Consequently, due to the strong interaction between structures and waves, the two-way coupling approach is the most appropriate technique for hydroelasticity research within the partitioned coupling framework.

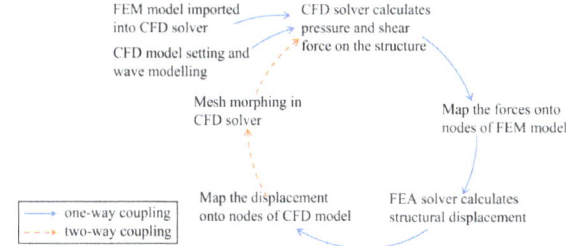

Fig. 4. Difference of one-way and two-way FSI coupling (Huang et al. 2023)

4). Advanced Modelling Techniques

As illustrated above, the development of fluid-structure coupling methods necessitates the simultaneous calculation of both the fluid and structural components, which places significant demands on computational resources. This is particularly challenging for large-scale OSUS systems, where hydroelasticity simulations may encounter computational difficulties. Researchers are continually advancing modeling techniques to enhance the efficiency and accuracy of hydroelasticity simulations. These advancements include meshless methods, reduced-order modeling, and improved fluid-structure coupling algorithms. Additionally, efforts are being made to integrate more complex physical phenomena—such as nonlinearities, fluid-structure contact, and material failure—into the coupling models.

Meshless methods, such as the Smoothed Particle Hydrodynamics (SPH) method, offer alternative approaches to traditional mesh-based techniques. These methods do not require a structured mesh, making them particularly suitable for problems involving large deformations or complex geometries. Zhang et al. (2019) coupled the Moving Particle Semi-Implicit (MPS) method with the finite element method (FEM) for numerical investigations of the interaction between regular waves and horizontally suspended structures. Additionally, MITSUME et al. (2014) proposed a partitioned coupling strategy for fluid-structure interaction problems involving free surfaces and moving boundaries, employing both the moving particle simulation method and the finite element method. Overall, meshless methods can enhance computational efficiency and simplify the handling of fluid-structure interfaces.

Researchers are continually developing improved coupling algorithms to enhance the accuracy and efficiency of fluid-structure interaction simulations. These algorithms aim to ensure the proper transfer of forces and displacements between the fluid and structural domains, effectively handle nonlinearities, and capture complex physical phenomena. In implicit coupling systems that utilize a two-way fluid-structure interaction (FSI) approach, this mathematically results in fixed-point equations at the coupling interface. Acceleration methods are required for straightforward fixed-point iteration to stabilize and expedite the FSI process. Davis et al. 2022 (2022) proposed two enhancements to quasi-Newton methods to accelerate coupling iterations for partitioned fluid-structure interaction. Kang et al. (2017) proposed an enhanced turbulence model through the modification of the turbulent kinetic energy dissipation term, thereby achieving more accurate predictions for strongly separated flow fields.

The viscous potential flow coupling method is an effective approach to enhance hydrodynamic calculation efficiency in CFD. The High-Order-Spectral Method (HOS) is employed as an external potential flow theory to generate waves quickly and accurately, allowing the CFD solver to freely select the range and occurrence time of the wave field (Aliyar et al. 2022; Lu et al. 2022; Yali Zhang et al. 2023). By improving the computational efficiency of the fluid dynamics component, this method also optimizes the speed of hydroelasticity calculations.

In recent years, the combination of artificial intelligence and CFD has emerged as a novel approach to improving calculation efficiency in fluid-structure interaction (FSI) simulations. Surrogate models can be developed to approximate CFD simulation results, trained using datasets from pre-computed CFD simulations. This allows for rapid estimation of fluid flow around structures under various wave conditions. Surrogate models, such as artificial neural networks or Gaussian process models, can significantly reduce computational costs by eliminating the need to run full-scale CFD simulations for every design iteration or scenario (Bublík et al. 2023; Mazhar et al. 2023; Zhang et al. 2023a; Zhou et al. 2023; Ni et al. 2024). In summary, the development of hydroelasticity modeling techniques is an active area of research, aiming to provide engineers and researchers with more accurate and efficient tools for predicting the behavior of OSUS. This advancement aims to enhance design and analysis capabilities across a wide range of applications.

3.2.2 Inhomogeneous Incident Waves

For very large structures and systems built offshore or near islands and reefs, the seabed in these locations is often uneven. Due to variations in seabed topography and the influence of nearby islands and reefs, the wave environment around these structures becomes unstable. Consequently, changes in wave conditions along the length of the floating structures are likely to occur. It is essential to consider the effects of these spatially varying wave fields on the dynamic response of very large floating bodies.

Fu et al. (2007) and Wei et al. (2017, 2018) investigated the hydroelastic responses of freely floating structures in inhomogeneous regular and irregular waves. Zhang et al. (2023a) innovatively developed a frequency-domain hydroelastic stress analysis of floating structures under spatially inhomogeneous wave field. Dai et al. (2020) comprehensively examined various effects of wave inhomogeneity on the dynamic responses of bridges. Li et al. (2023) investigated the hydroelastic responses of floating bridge in inhomogeneous waves and predicted the time series of first- and second-order wave forces using linear transfer functions and Newman's approximation, respectively. Cui et al. (2023) studied the long-term extreme responses of a very long floating bridge under various inhomogeneous wave fields. These studies demonstrated that a correlated inhomogeneous wave field can induce significantly large extreme axial forces, and that the bending moments, shear forces, and torsional moments of the structure are highly influenced by wave inhomogeneity. A simplified method for estimating the hydroelasticity of OSUS in inhomogeneous incident waves was also established and applied to an 8-module very large floating structure exceeding 2,400 m in length (Ding et al. 2019).

3.2.3 Seabed Topology Influences and Shallow Water Effect

The complex geographical environment, particularly the depth-varying seabed, can significantly influence ocean waves and the hydroelasticity responses of OSUS, whether positioned near an island or close to the coast. Consequently, it is crucial to account for shallow water effects when analyzing the elastic response of OSUS.

Song et al. (2005) investigated the hydroelastic response of VLFS under varying water depths using both experimental and numerical methods. To simulate an uneven seabed, several shoals were placed at the bottom of the wave basin. Their findings indicated that the non-uniform seabed environment has a notable effect on the hydroelastic response of VLFS. Kyoung et al. (2005) also studied the influence of seabed topography on the hydroelastic behavior of VLFS using experimental and numerical approaches, where the fluid domain was modeled using a finite-element method based on the variational principle. Karperaki et al. (2016) explored the effects of variable bathymetry, stiffness, and damping coefficients on the hydroelastic responses of a VLFS that is elastically connected to the seabed. Yang et al. (2019a) and Yang et al. (2019b) compared numerical results of the hydroelastic response of a floating structure subjected to waves in shallow water over a specified seabed profile with model test results. Their studies concluded that significant differences in wave statistics are observed as waves propagate and approach an island. The complex seabed profile appears to have a considerable impact on the hydroelastic responses of VLFS.

In addition, seabed topography and shallow water effects present challenges to the design of mooring systems for OSUS (Karmakar and Soares 2012). The presence of seabed features, such as sandbars and bathymetric variations, can influence the optimal placement and alignment of mooring anchors. Seabed irregularities may necessitate adjustments in anchor positioning to ensure proper load distribution and to avoid interference with other structures or subsea infrastructure. Furthermore, shallow water conditions can impact the required length and scope of mooring lines, as the reduced water depth alters the catenary profile and tension distribution along the mooring lines.

Karmakar and Soares (2012) employed the ordinary eigenfunction expansion method to study wave scattering by a floating plate connected to mooring lines in shallow and finite water depths. Wei et al. (2024) developed a time-domain hydroelastic-mooring coupling model by integrating a quasi-static mooring module into a fully coupled Computational Fluid Dynamics (CFD) and discrete-module-beam (DMB) approach, demonstrating that the structural elasticity of VLFS has a significant impact on mooring tensions. Similarly, Nguyen et al. (2018) found that the hydroelastic responses of pontoon-type VLFS can be significantly reduced using vertical elastic mooring lines and proposed an optimization procedure to determine the optimal mooring line stiffness. Liang et al. (2019) introduced a shallow water mooring system design methodology that combines the NSGA-II algorithm with a vessel-mooring coupled model to optimize mooring system performance while accounting for practical design requirements.

3.3 Motion Characteristics

3.3.1 Multibody Hydrodynamics

OSUS typically consist of a large deck supported by pontoons or other buoyant structures, owing to their significantly larger size compared to conventional floating structures. To facilitate construction and maintenance, OSUS is often composed of multiple modules. During the calculation process, these structures can be treated as interacting rigid bodies. Multibody hydrodynamics is a field dedicated to analyzing and simulating the complex interactions between multiple bodies in fluid flow.

In multibody hydrodynamics, a deformable structure is treated as several separate floating modules that move in tandem while being connected by a beam or connector with rigidity equal to that of the structure itself. Compared to the finite element method, multibody hydrodynamics employs a beam-connected-discrete-modules (BCDM) hydroelastic method to address the hydroelasticity problem with a lower computational burden. Additionally, the BCDM method's unique discrete approach to hydrodynamic calculation can naturally integrate with the treatment of inhomogeneous wave field by discretizing it into multiple homogeneous wave fields (Fu et al. 2017), providing a distinct advantage in handling inhomogeneous wave conditions. Zhang and Lu (2018) demonstrated that this approach can be extended to more complex floating structures by converting the structural stiffness, using the finite element method (FEM), into the stiffness of the corresponding end node in the BCDM framework. An example of discretized VLFS into M rigid bodies and M-1 beam-theory-based elements is depicted in Fig. 5.

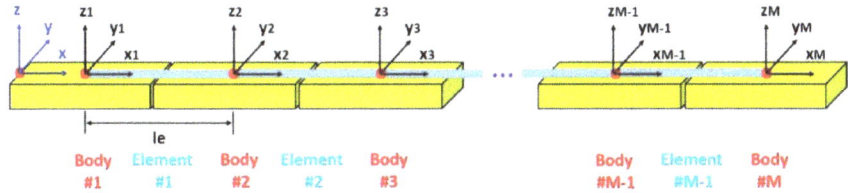

Fig. 5. Multibody-based hydroelastic analysis (Jin et al. 2020)

To account for nonlinear and transient load effects, Wei et al. (2018) integrated the convolution principle with the linear frequency-domain BCDM method, extending it into the time domain. Li et al. (2022) further extended the BCDM method to second-order by incorporating linear hydroelastic responses into multi-body hydrodynamics. The second-order BCDM method is the first to enable hydroelastic analysis that couples complete second-order wave loads with inhomogeneous wave fields. The above studies on the BCDM method primarily focus on analyzing the overall hydroelastic responses of OSUS. In order to achieve a more comprehensive local stress analysis of OSUS, Zhang et al. (2023) investigated a two-step stress analysis method based on the quasi-static stress analysis approach. The modified approach incorporates the effects of both overall elastic deformation and local response into the quasi-static analysis, making the stress analysis of OSUS that considers the influence of local response the first time possible.

Miao et al. (2024c) shows a new type of pontoon bridge, which is connected by multiple pontoon-truss composite modules, called pontoon-truss composite floating bridge (PTCFB), and the three-dimensional hydroelastics theory is used to calculate the hydroelastic performances of the PTCFB, and the motion responses of the modules were verified by the experimental results. The results show that the water depth has a greater impact on the motion response of the floating bridge while $\omega < 1.0$ rad/s, the wavelength has an obvious influence on the section loads of the PTCFB. In addition, Miao et al. (2024b) improves a numerical method to study the dynamic performances of floating structures in the time domain based on the three-dimensional hydroelastic theory, which can calculate the dynamic properties of floating structures with or without considering the effect of the various nonlinear factors and elastic deformation. Based on the numerical method, the computational program THAFTS-NATD was developed. This program accounts for the effects of nonlinear generalized restoring forces and nonlinear generalized incident wave forces, which arise from the large motions of a floating body and instantaneous changes in its wetted surface. Using THAFTS-NATD, the motion responses of a new offshore floating bridge under regular and random waves were calculated and compared with the experimental results.

For floating multi-body systems with a limited number of floaters, many scholars have investigated hydrodynamic interactions, gap resonance, and mechanical couplings. Feng and Bai (2017) developed a numerical model of two floating barges, both with and without connections, to study the coupling effects between multiple floating bodies and their interconnections. Zhao et al. (2018) presented a time-domain analysis to predict the integrated dynamic responses of a side-by-side offloading floating liquefied natural gas (FLNG) system. Pessoa et al. (2016) examined the side-by-side moored system involving a FLNG and a shuttle tanker, considering low-frequency relative motions, and found that second-order wave forces are crucial for the stability of the side-by-side floating system. Lu et al. (2010) investigated fluid resonance in the narrow gaps between three identical fixed rectangular structures subjected to incident waves normal to those gaps, revealing that the maximum resonant wave height in the gaps can be up to four times the incident wave height.

The multibody approach has also been used by many scholars to study hydroelastic behavior. Lee et al. (2018) analyzed the FEM-based mooring line model applicable to a multibody system based on flexible multibody dynamics. Wei et al. (2017) investigated a VLFS's hydroelastic behavior in inhomogeneous sea conditions. Jin et al. (2020) implemented the DMB method to solve a floater-connector-mooring coupled system. Zhang et al. (2021) explored the hydroelasticity of a VLFS with a certain number of wind turbines. Wei et al. (2024) investigated the coupled effects between structural hydroelasticity and loose-type mooring systems on a deformable VLFS in waves based on BCDM approach. Potential flow coupled with multi-body dynamic method (MBD) has been established by Lu and Zhang and applied to study the transient response of a VLFS (Zhang and Lu 2018; Lu et al. 2019). Al-Solihat and Nahon (2018) developed a coupled rigid-flexible multibody dynamic of a FOWT model. Leng et al. (2023) proposed a new numerical method to analyze turbine aeroelastic performance and fluid–structure interaction based on flexible multibody dynamics and large eddy simulation with anisotropic actuator line method. Jin et al. (2023) extended a discrete-module-beam (DMB) method based on 3D potential flow theory to include second-order wave forces and evaluated the hydro-elastic behaviors of a moored submerged floating tunnel.

3.3.2 Fluid Structure Interaction

As previously mentioned, fluid-structure interaction (FSI) plays a crucial role in the hydroelasticity analysis of OSUS. FSI refers to the study of interactions between fluid flow and the elastic response of structures, taking into account the dynamic behavior of both the fluid and the structure. This section presents some motion characteristics calculated using the FSI approach. During the early design stage, simplified structural models such as beams and plates are often employed to perform FSI analysis on floating bodies. Seto et al. (2005) developed a three-dimensional hydroelastic response analysis method and applied it to realistic Mega-Float models consisting of a deck, bottom, and bulkheads. The analysis was conducted for models situated both in open sea conditions and in environments protected by breakwaters and shore, while also considering varying bottom topography. Watanabe et al. (2006) analyzed the hydroelastic responses of a circular Mindlin plate, providing benchmark solutions for validating numerical methods used in the hydroelastic analysis of VLFS. Xu and Wellens (2022) studied the nonlinear fluid–structure interaction (FSI) of free surface waves with large-scale polymer offshore floating photovoltaics (LPOFPV). The floating structure was modeled as a nonlinear Euler-Bernoulli–von Kármán (EBVK) beam coupled with the water beneath. Li et al. (2023) conducted a fluid-structure-material coupling analysis for a floating laminated structure, calculating various hydrodynamic quantities such as reflection and transmission coefficients, deflection, shear force, and bending moment. The results indicated that the edges of the laminated structure significantly influenced wave reflection and transmission, similar to the behavior observed in a single-layer homogeneous plate. The highest reflection coefficient peak was typically observed when the structural edges were free.

For structures with complex geometrical shapes and local deformations, three-dimensional finite-element (FE) models are favoured over basic beam and plate theories in order to account for the nonlinear effects. Kim et al. (2013) applied continuum mechanics-based FEM to model floating structures with arbitrary geometries while the fluid was solved using BEM. Related works on the hydroelasticity can also be found in Chen et al. (2023) and Hamamoto and Fujita (1995), which combined 3D FE models with linear hydrodynamic models. Tuitman and Malenica (2009) presented a methodology to solver the seakeeping, slamming and whipping problems using a beam model or full three-dimensional finite element model of the structure.

Fully nonlinear hydrodynamic models are essential for capturing the hydroelastic effects of VLFS subjected to large, steep waves or other significant nonlinear phenomena. Huang et al. (2023) investigated the hydroelastic responses of a single-module VLFS in extreme waves using a CFD-FEA coupling method. Their findings indicated that freak waves can significantly increase the vertical bending moment of the VLFS. Lakshmynarayanana and Temarel (2019) applied a two-way coupling between RANS/CFD and finite element software to predict the symmetric distortions and bending moments of a highly elastic rectangular stationary barge in regular head waves. The results showed that nonlinearities contributed to the discrepancy between the present coupling approach and traditional 2D linear methods.

3.3.3 Gap Resonance

In the study of multi-body hydrodynamics of floating structures, special attention must be paid to fluid resonance within narrow gaps. When the incident wave frequency is close to the natural frequencies of fluid resonance within these gaps, gap resonant motions may occur, leading to significant wave elevation (Ekerhovd et al. 2021; Zou et al. 2024). Based on ANSYS-AQWA, Miao et al. (2021b) analyzed the complex multi-body hydrodynamic interactions by introducing artificial damping on the gap surfaces in both frequency-domain and time-domain simulations. Li and Wang (2021) investigated water resonance between two fixed rectangular barges in a 2D numerical wave tank using fully nonlinear potential flow theory. Their study found that for wider gaps, higher-order sloshing modes dominate, while in narrower gaps, piston mode motion is more prevalent. Cheng et al. (2022) optimized the design and layout of an integrated system of modular WEC-type floating breakwaters and a pontoon-type VLFS. Their findings showed that multi-modal wave resonance in the WEC-VLFS gap resulted in several peak efficiencies. Li (2019) also analyzed second-order wave resonance in a gap formed by twin barges using fully nonlinear potential flow theory, discovering that a large gap width-to-draught ratio significantly induces second-order resonance.

Due to the limitations of potential flow theory in fluid modeling, some researchers have developed numerical wave flumes based on the Navier–Stokes equations to account for viscous dissipation in gap resonance simulations. He et al. (2021) investigated fluid resonance in the narrow gap between two rigidly connected heave boxes subjected to waves. They examined the effects of body draft and the distance between the connected boxes. The results showed that the hydrodynamic characteristics of the rigidly connected boxes differ significantly under free heave motion compared to a motion-constrained condition. Yin et al. (2023) explored gap wave resonance in a box-wall system with both impermeable and permeable beds. Their numerical studies revealed that the resonant amplitude in the gap is reduced when a foundation bed is present, and this reduction becomes even more pronounced when the permeability of the bed is considered. Gao et al. (2021) simulated gap resonance between two boxes excited by incident regular waves using OpenFOAM. Their findings indicated that the heave motion of the upstream box increases the fluid resonant frequency while decreasing the resonant wave height.

3.3.4 Offshore Aquaculture Structures

Increasingly attention has been attracted to the fishery and aquaculture equipment. Numerical investigations on both steady and unsteady performance of aquaculture units have been carried out during the past decades.

The dynamic properties of a flexible net sheet exposed to waves and currents were investigated by using a numerical model, where the net was modelled by dividing it into super elements (Lader and Fredheim 2006). The effects of Reynolds number, net solidity and mesh pattern, and flow direction on the drag force on submerged nets of fish cages were studied (Zhan et al. 2006). Kristiansen and Faltinsen proposed a screen type of force model for the viscous hydrodynamic load on nets (Kristiansen and Faltinsen 2012, 2015).

The net cage was modeled using truss elements that represented several parallel twines, and the strength analysis was performed using commercial explicit finite element software ABAQUS to calculate the distribution of loads in the net cage due to current, weights, and gravity (Moe et al. 2010). A three-dimensional numerical model was established to simulate the flow field inside and around the gravity cages in a current, and the realizable k–ε turbulence model was chosen to describe the flow and the governing equations were solved by using the finite volume method (Zhao et al. 2013b). Hydrodynamic drag and wake properties of square aquaculture cage arrays were studied to improve understanding of nutrient dynamics from fish cages to guide the design of integrated multitrophic aquaculture by Turner et al. (2016). An innovative hybrid method was proposed to determine the hydroelastic responses of full-scale floater-and-net systems, and the net for the fish cage was vertically and peripherally divided into similar interconnected sections with different hydrodynamic parameters, which were assumed to be uniformly distributed over each section (Ma et al. 2016).

The semi-submersible offshore fish farm (SOFF) is a new type of offshore aquaculture structure, consisting of pontoons, frame system, and net system. Miao et al. (2021b) proposed a hybrid numerical method of direct time domain and indirect time domain for the interaction between the SOFF and waves. The effects of the wave parameters on the SOFF were studied and several conclusions were obtained. The structural strength of the world's first offshore fish farm, *Ocean Farm 1*, operating in Norway subjected to supply vessel collisions was evaluated by using nonlinear finite element simulations (Yu et al. 2019). Liu et al. (2021) examined the hydrodynamics of a semi-submersible fish farm under current using the porous media method combined with a rigid wall based on the shear stress transport k-omega turbulence model. Martin et al. (2022) presented a CFD approach for the simulation of open ocean aquaculture structures in waves, and the two-way coupled solution for the fluid and floating body dynamics is described by the conservation equations of mass and momentum. Wang et al. (2022a) investigated the hydrodynamics of and nonlinear interaction between the large offshore fish farm "ShenLan 1" and regular waves using the open-source computational fluid dynamics (CFD) toolbox REEF3D. Li et al. (2023a) proposed a novel integrated offshore design, JOWT-SC, which combines a jacket-supported offshore wind turbine structure (JOWT) with an aquaculture steel cage, and the numerical modeling and time-domain simulations were conducted in software SACS, all structural members are slender bodies modeled as Morison tubular member elements. Zhang et al. (2024c) studied the effects of current direction, current velocity, and ten different forms of net breakage on net tension and deformation by using OrcaFlex, and the net is regarded as a Morison unit of a slender rod and simplified by the net concentration method.

When discussing the vessel-shaped fish farm, the hydrodynamic and motion response analysis was conducted using a coupled time-domain method (Lin et al. 2018). A linear mathematical model of a freely floating 2D closed flexible fish cages in waves was developed by Strand and Faltinsen (2019), and it was found that the wave-induced rigid body motion responses of a flexible CFFC in sway, heave, and roll were significantly different from the responses of a rigid CFFC, and non-linear free surface effects must be accounted for inside the tank in realistic sea conditions. Y. Wang Fu, et al. (Wang et al. 2023d) proposed a novel method to make it possible for the simulation of the hydro-elasticity of the very large floating body of a vessel-shaped fish cage and the large deformation of the fish net considering the diffraction and radiation waves fields induced by the floating body. It is the first time to find that the diffraction and radiation waves make a significant difference in the twine tension and connector load. In the method, the hydro-elasticity analysis is realized through a combination of the BCDM method with state space model to simulate the radiation force.

Chen et al. (2025) constructed a fully-coupled aero-hydro-net-mooring time-domain model by coupling AQWA and FAST for preliminary analysis of a turret-moored deep-sea aquaculture vessel powered by wind energy. A novel concept that combines multiple megawatt vertical-axis wind turbines and a solar array with a floating steel fish-farming cage was proposed, and a fully coupled time-domain simulations for the combined wind-solar-aquaculture system were carried out to estimate the extreme responses of this system in survival conditions by Zheng et al. (2020).

3.3.5 Deep Sea Mining

Despite its potential benefits, deep sea mining poses significant environmental risks and operational challenges that necessitate careful management and regulation. Guo et al. (2023) summarizes the challenges with deep sea mining, which relate to extreme environmental conditions (high-pressure and low temperature), complex hydrodynamic environments, complex geology, inaccessibility to human activities, environmental impact, monitoring and data collection.

In deep sea mining (DSM) operation the mother ship and the mining machine are linked through an umbilical system. the coupling dynamic response between the mining vessel and the lifting pipe is a significant challenge, which directly affects the structural design of the lifting system and the safety of field operation. L. Sun et al. (2017)

developed a simulation method that integrates the dynamics of a vessel, riser, and body system specifically for DSM operations. This coupling effect is essential for accurately simulating the operational conditions of DSM, as it allows for a comprehensive analysis of the interactions between these components. The paper emphasizes the incorporation of dynamic positioning systems in the simulation. Song et al. (2021) investigated the dynamic behavior of the lifting pipe and mining vessel during the process of deep-sea mining, considering ocean current, surface wave, pipe dynamics and vessel-pipe contact mechanics. The coupling simulation results show that the coupling effect has a significant effect on the time domain dynamic response of the lifting pipe but has little effect on the average effective tension and longitudinal amplitude along the pipe length.

One of the key challenges in deep-sea mining is developing technologies for transporting minerals from the seafloor to the surface. Vertical transportation systems, a dominant approach for conveying ores, have recently garnered extensive attention due to their high efficiency and stability. Significant research has focused on the lifting of solid-liquid two-phase minerals and the vortex-induced vibration of flexible cantilever risers associated with deep-sea mining ships. Recent studies have explored the internal and external flow dynamics and their coupling effects within the transport system.

For the internal flow of the transport system, solid-liquid multi-phase flow during mineral transportation contributes to the structural dynamics of the mining riser. Liu et al. (2023) conduct both numerical simulations and experiments on the solid-liquid two-phase flow in the DSM risers and summarize the dominating factors such as feeding concentration, flow velocity and pipe inclination for DSM system design. Wang et al. (2024c) investigated the impact of various particle shapes on slurry transport efficiency and flow characteristics at higher concentrations in the hydrodynamic transport of DSM system. A numerical method combining computational fluid dynamics (CFD) and the discrete element method (DEM) within the Euler-Lagrange framework is employed to simulate the flow and hydrodynamic properties of graded heterogeneous coarse particles in a deep-sea hydraulic lift pipe. For the external flow of the transport system, Vortex-induced vibration (VIV) analysis plays a significant role in the evaluation of hydrodynamic performance for the marine riser. Jin et al. (2023b) employ a quasi-3D coupling algorithm based on the discrete vortex method and finite element method to calculate the unsteady hydrodynamic forces of the riser pipeline under a wide range of bottom weights and different current speeds. For the coupling action of the internal and external flow, it is difficult to be simulated directly due to large differences in mechanical properties. Cao et al. (2023) propose a new three-dimensional numerical model of the vertical transport system in deep-sea mining using the vector form intrinsic finite element (VFIFE) method.

Composite materials have received great attention in the design of flexible risers for deep-sea mining due to their significant advantages of high strength and lightweight. Considering the nonlinearity of materials and detailed geometric features, Sun et al. (2024) established a theoretical method for predicting the structural stiffness of a new type unbonded flexible risers under axial tension and internal pressure, By iteratively solving the equilibrium equation, the strain and displacement fields of the flexible riser are obtained.

Additionally, the potential impact of DSM on deep-sea biodiversity and ecosystem functionality must be taken into account. The sediment plumes generated by mining activities in the ocean should be investigated for this problem (Haalboom et al. 2023). Peacock and Ouillon (2023) summarize and outline the research on DSM plumes. Bai et al. (2024) employ the two-fluid Eulerian model to investigate how various parameters affect the initial stages of plume discharge, and identify the key influencing factors, such as discharge velocity, discharge concentration, and pipe diameter in governing the evolution of the plume.

3.4 Structural Responses

The elastic response of OSUS in waves is a fundamental type of structural behavior, arising from the elastic deformation of the floating body itself or the relative motion between modules induced by the connectors. Typically, the magnitude of this elastic deformation can be of the same order as the vertical global motion of the structure. Jiang et al. (2024) presented a coupled high-fidelity numerical approach for accurately modeling the fully nonlinear wave–structure interaction of a floating, highly flexible structure in waves. This study shed light on how structural deformations in waves are influenced by rigid body motions. Song et al. (2023) investigated the dynamic behavioural characteristics of multi-connected offshore floating photovoltaics and found that the corner module in the multi-connected system showed critical responses compared to others, warranting more attention. Chen et al. (2023) studied the hydroelastic response of both narrow and square VLFS using the discrete-module-finite-element hydroelasticity method and calculated the displacement and structural force responses at the interfaces. Zhang et al. (2023) simulated a 15 MW wind turbine using a CFD-FEA Fluid-Structure Interaction (FSI) method at full scale, revealing that stresses and tip deformations of blades are significantly amplified by periodic motions. Chen et al. (2022) analyzed the hydroelastic response of double-segment floating sandwich structures under wave action and optimized its hydroelastic response. Zheng et al. (2023) assessed the structural strength of the thin-walled blade root joint of a floating offshore wind turbine (FOWT), considering the interaction effects between each component of the blade root joint on structural responses. Xiang (2024) studied the hydrodynamic interaction of a pontoon-supported floating bridge.

To ensure structural safety, several scholars have conducted studies aimed at mitigating the structural response of floating bodies. Cheng et al. (2016) investigated the fluid-structure interaction of a pontoon-type VLFS edged with dual horizontal/inclined perforated plates in oblique irregular waves. Singla et al. (2018) analyzed the effectiveness of partial vertical permeable barriers in three different configurations, positioned at a finite distance from a very large floating structure, to mitigate the wave-induced response. Their findings indicated that by carefully selecting the porous-effect parameter and the distance between the barrier and the floating plate, significant reductions in free surface elevation and deflection of the floating plate could be achieved. Hong and Hwang (2023) presented an analytical study examining the hydroelastic responses of pneumatically supported floating structures, finding that pneumatic supports could significantly reduce hydroelastic responses in most cases. Yang et al. (2023) applied boundary control in fluid-structure interaction problems to reduce the hydroelastic response of floating beam systems subjected to irregular waves. Additionally, Gao et al. (2013) employed flexible connectors and gill cells to reduce the hydroelastic response of pontoon-type VLFS.

Stress response is another critical aspect that requires sufficient attention during the detailed design phase. For large structures, estimating dynamic wave-induced stresses in a single step can be impractical due to the considerable computational burden involved. Consequently, a significant section of the structural module is often isolated for analysis, where FEA or multibody analysis is performed based on global responses and stress resultants (such as bending moments, torsional moments, and shear forces) to determine the stress distribution. Wiegard et al. (2021) conducted a detailed stress analysis during an extreme wave event utilizing a partitioned fluid-structure interaction approach. Wang et al. (2023) outlined a robust design process for floating wind turbine hulls, calculating the time-domain structural stress of the floater; the conclusions and suggestions from this study serve as a valuable reference for the analysis and design of semi-submersible floating wind turbine platforms. Additionally, Song et al. (2022) assessed the stress response of a trapezoidal floating body in numerical waves using the 3D hydro-elastic computational platform HOMER.

3.5 Conclusions and Recommendations

It is important to note that with the rapid advancement of engineering applications for OSUS, research on the hydroelastic response of these structures in complex sea environments will encounter increasing challenges. Addressing these difficulties will require significant effort and innovation. Below are some recommendations to help overcome these challenges.

- Accurately predicting hydrodynamic loads is a fundamental aspect of designing OSUS. Potential flow theory and viscous flow theory are the two primary approaches for solving fluid flow and determining structural loads. However, potential flow theory cannot account for fluid viscosity or strong nonlinear phenomena, such as slamming and green water. Consequently, Navier-Stokes equation-based theory in the time domain is gradually becoming the primary method for hydrodynamic prediction. Given its high computational resource demands, there is a pressing need for the development of effective nonlinear hydroelastic approaches that consider wave nonlinearity, seabed topology, complex mooring systems, and module connectors. Future research could explore the combination of CFD with artificial neural networks, efficient coupling algorithms, and multibody dynamic structural modeling.
- The majority of hydrodynamics research concentrates on wave-induced loads and elastic displacement of VLFS under surface waves. This analysis is typically essential during the initial design stage of a structure. However, it is also necessary to conduct a local analysis at stress concentration points, particularly when considering fatigue and fracture during the configuration of welding lines. Additionally, when constructing an underwater space station, it is crucial to explore the interaction of OSUS with internal waves, especially large internal solitary waves.

4 Connectors Between Floating Bodies

4.1 Introduction

Utilization of offshore space requires the construction of onshore structures (which may include landfills) or the installation of floating structures. The floating structures can be classified into point structures and linear or areal structures. The former include floating drilling rigs, FPSOs, and floating offshore wind turbines, while the latter include such as a floating bridge, VLFS, MOB, and floating solar power facility. In terms of the response of floating bodies in waves, the former can be regarded as rigid bodies in main, while the latter behaves more as other-degree-of-freedom floating bodies than as rigid bodies as their one- or two-dimensional extent increases. An example of a two-dimensional floating body is a VLFS, such as a mega-float, which is a one-piece floating body, but is a thin structure compared to its areal size (often several kilometers in length) and has a relatively low bending stiffness, resulting in a flat plate-like elastic response. A mobile offshore base (MOB) can also be considered a multi-degree-of-freedom floating structure composed of a number of rigid floating modules.

In floating solar power plants, which have been actively studied in recent years, it is envisioned that unit floats will be sequentially connected and deployed, and it is imagined that elastic response will occur in other degrees of freedom than in the past.

In this type of space utilization with large floats, floats of a certain size are deployed while being connected on site. Connectors are used to connect the floating units (permanently or temporarily), but the boundary conditions of the connection change depending on whether the connection is rigid or has some degree of freedom. It is easy to imagine that this change in boundary conditions would also change the response of the entire floating unit. This chapter presents research trends and treatment of these connectors.

4.2 Connector Design

4.2.1 Rigid Connectors

Rigid connectors are one of the oldest and simplest types of connectors used in very large floating structures (VLFS). These connectors provide a fixed, inflexible connection between modules and ensure that the modules move as a single unit. The main advantages of rigid connectors are simplicity and structural solidity. However, they also have significant drawbacks, especially in terms of flexibility and adaptability to wave-induced motion.

The use of rigid connectors featured prominently in early VLFS designs, such as the Mega-Float project in Japan in the 1990s. Researchers such as Suzuki (2005) emphasized the effectiveness of rigid connectors in maintaining the structural integrity of pontoon-type VLFS. The rigid connectors used in this project provided the stability and strength needed to support these large structures.

While effective in terms of maintaining structural integrity, the limitations of the rigid connectors became apparent as the need for more flexible and adaptable structures increased. The rigidity of these connectors meant that they could not accommodate dynamic movements caused by waves or other environmental factors. This resulted in increased stresses on the structure and the potential for damage.

In recent years, there has been renewed interest in optimizing rigid connectors to improve their performance. Researchers are exploring new materials and design techniques to improve the durability and adaptability of rigid connectors. For example, advanced composite materials are being investigated because of their potential to provide greater strength and flexibility compared to traditional materials. These innovations aim to address the limitations of rigid connectors and make them suitable for a wider range of applications.

One of the key challenges in optimizing rigid connectors is finding the right balance between strength and flexibility.

4.2.2 Hinged Connectors

Hinged connectors were introduced as a solution to the limitations of rigid connectors, offering greater flexibility and adaptability to wave-induced motions. These connectors allow rotational movement between modules, reducing the stresses caused by wave action and improving the overall dynamic response of the structure. The concept of hinged connectors gained prominence in the early 2000s, with several studies exploring their potential benefits and applications in VLFS.

The early exploration of hinged connectors was driven by the need to address the dynamic challenges posed by the marine environment. Researchers like Xia et al. (2000a) and Khabakhpasheva and Korobkin (2002) were among the first to investigate the use of hinged connectors in VLFS. Their studies demonstrated that hinged connectors could significantly reduce the hydroelastic response of the structure, making them suitable for applications in more dynamic marine environments. These early studies laid the groundwork for further research and development in this area. In the recently, Wang et al. (2023) showed experimental results of the connector loads on hinged connected multi-module floating bodies.

One of the key advantages of hinged connectors is their ability to accommodate rotational movements between modules. This flexibility helps to distribute the stresses caused by wave action more evenly across the structure, reducing the likelihood of structural damage. Zhao et al. (2015) conducted a comprehensive study on the influence of hinged conditions on the hydroelastic response of compound floating structures. Using the Mindlin plate theory and Wiener–Hopf technique, the study analyzed the effects of different torsional stiffness at the joints on the displacement and bending moments of the floating plates. The results provided valuable insights into the optimal design of hinged connections to minimize dynamic stress and enhance the stability of VLFS.

4.2.3 Flexible Connectors

Flexible connectors represent a significant advancement in the design of Very Large Floating Structures (VLFS), offering the highest level of adaptability and flexibility to wave-induced motions. These connectors are designed to absorb and dissipate wave energy, reducing the stresses on the structure and improving its overall dynamic response. The development and implementation of flexible connectors have been driven by the need to enhance the performance and durability of VLFS in increasingly challenging marine environments.

Wang et al. (1991) treated the rigid floating bodies connected by flexible connectors. In the middle of 2000s, Fu et al. (2007) proved that the stiffness of connectors is a critical parameter in determining the hydroelastic response of the structure. Wang et al. (2009) proposed the use of hinge connectors instead of rigid ones since the non-rigid connectors are more effective in reducing the hydroelastic response as comparing to the rigid connectors. Different from the traditional flexible connectors, Shi et al. (2022) proposed a new type of face-contact (FC) connector. The core idea is to disperse the huge connection load to the whole end face of modules, so as to significantly reduce the local stress level at connection areas. These studies demonstrated that flexible connectors could significantly enhance the performance of VLFS, making them suitable for applications in more dynamic and harsh marine environments. The ability of flexible connectors to accommodate large deformations without compromising structural integrity was a key factor in their adoption.

One of the primary advantages of flexible connectors is their ability to absorb and dissipate wave energy. This capability helps to reduce the dynamic loads on the structure, minimizing the risk of structural damage and extending the lifespan of the VLFS. Zhang et al. (2016) proposed a type of hybrid flexible connectors embedded with wave energy harvester.

The use of flexible connectors has been particularly beneficial in applications where the structure is exposed to high wave energy, such as deepwater resource exploitation and offshore wind farms. Studies by Gao et al. (2011) and Michailides et al. (2013) have shown that flexible connectors can significantly reduce the wave-induced stresses and enhance the stability and performance of VLFS.

Despite their advantages, flexible connectors also present some challenges, particularly in terms of their design and maintenance. (Zhao et al. 2015) proposed an generalized optimization method for stiffness configuration of flexible connectors for multi-modular floating systems, which is capable to handle quite large number of influential factors in a decision making process. Based on the optimization method, some scholars conducted stiffness design for the different connectors (Shi et al. 2018; Lu et al. 2022). The materials used in flexible connectors must be able to withstand the harsh marine environment, including exposure to saltwater, UV radiation, and mechanical wear. For instance, advanced composite materials and elastomers are being investigated for their potential to provide greater flexibility and resistance to environmental degradation.

In recent years, there has been a growing interest in optimizing the design of flexible connectors to further enhance their performance. Advanced simulation tools and modeling techniques are being used to better understand the behavior of flexible connectors under different conditions. Xu et al. (2014) introduced a nonlinear network modeling approach for multi-module floating structures with arbitrary flexible connections. Based on the network dynamic theory, Zhang et al. (2015) proposed an amplitude death stability design method for the flexible connector of the VLFSs. Subsequently, the flexible connection configurations were studied by Zhang et al. (2017) using the amplitude death stability design method.

One of the key areas of focus in the optimization of flexible connectors is the development of new materials that can provide greater strength and durability. Composite materials, for example, are being investigated for their potential to offer superior performance compared to traditional materials. These materials can provide the necessary flexibility while also enhancing the overall strength and durability of the connectors (Shi et al. 2022). Additionally, researchers are exploring the use of smart materials that can adapt to changing environmental conditions, further enhancing the performance of flexible connectors.

Another important aspect of flexible connector design is the integration of damping mechanisms to further reduce the stresses caused by wave action. By incorporating damping elements into the design of flexible connectors, it is possible to absorb and dissipate wave energy more effectively, reducing the overall dynamic response of the structure. This approach has been shown to significantly enhance the performance of flexible connectors in dynamic marine environments.

The use of flexible connectors is also being explored in combination with other types of connectors to create hybrid solutions.

4.2.4 Semi-Rigid Connectors

Semi-rigid connectors offer a compromise between the rigidity of fixed connectors and the flexibility of hinged or flexible connectors. These connectors provide a certain degree of flexibility while maintaining a strong connection between modules, making them suitable for a wide range of applications.

Wang et al. (2009) and Riyansyah et al. (2010) demonstrated that semi-rigid connectors could effectively reduce the hydroelastic response of VLFS while maintaining structural integrity.

One of the key advantages of semi-rigid connectors is their ability to adjust the stiffness of the connection based on the environmental conditions. This adaptability helps to distribute the stresses caused by wave action more evenly across the structure, reducing the likelihood of structural damage. Wei et al. (2017) proposed a discrete-modules-based hydroelasticity method for predicting the response of floating structures in inhomogeneous sea conditions.

Studies by Michailides et al. (2013) and subsequent research have shown that semi-rigid connectors can significantly enhance the performance of VLFS, making them suitable for a wide range of marine applications.

In recent years, there has been a growing interest in optimizing the design of semi-rigid connectors to further enhance their performance. One of the key areas of focus in the optimization of semi-rigid connectors is the development of new materials that can provide greater strength and durability. Composite materials, for example, are being investigated for their potential to offer superior performance compared to traditional materials.

Another important aspect of semi-rigid connector design is the integration of damping mechanisms to further reduce the stresses caused by wave action. By incorporating damping elements into the design of semi-rigid connectors, it is possible to absorb and dissipate wave energy more effectively, reducing the overall dynamic response of the structure.

4.2.5 Adaptive Connectors

Adaptive connectors represent offering the ability to dynamically adjust their stiffness in response to changing wave conditions. Xia et al. (2016) proposed a special active-control connector system with air-springs and showed that tuning the connector stiffness in response to variable wave conditions, the method aimed to reduce oscillations and improve the stability of the floating structure. Numerical simulations verified the feasibility and effectiveness of this approach.

One of the key advantages of adaptive connectors is their ability to dynamically adjust their stiffness based on real-time environmental conditions. This adaptability helps to distribute the stresses caused by wave action more evenly across the structure, reducing the likelihood of structural damage.

Recent research has continued to explore innovative solutions to enhance the performance and applicability of adaptive connectors. Kou et al. (2019) demonstrated the potential of adaptive connectors to significantly reduce the dynamic response of VLFS, improving their stability and operational safety in random sea conditions. These studies provided valuable insights into the design and optimization of adaptive connectors.

One of the key areas of focus in the optimization of adaptive connectors is the development of new materials that can provide greater strength and durability.

Another important aspect of adaptive connector design is the integration of advanced control systems to dynamically adjust the stiffness of the connectors in real-time. By incorporating sensors and actuators into the design of adaptive connectors, it is possible to monitor environmental conditions and adjust the stiffness of the connectors accordingly.

Xia et al. (2022) proposed a stiffness-adjustable connector. The proposed adaptive connector consists of a cylindrical spring with an embedded actuator, making its stiffness adjustable. In a case study, a layout for such connectors is suggested to reduce the surge, heave, pitch, and yaw motions of the VLFS in random seas. In order to control the vibration responses of the VLFS, a state feedback control scheme is developed using Sequential Quadratic Programming, which is able to adapt to varying wave conditions. Numerical studies indicate that the control method based on the stiffness-adjustable connectors was able to greatly reduce the responses of the modules when compared to flexible connectors and was also able to reduce the connector loads when compared to hinged connectors.

4.3 Connector Loads

OSUSs are typically constructed from multiple floating bodies connected by flexible or rigid connectors. However, these connectors are often the most critical and vulnerable components of the entire structure. Under harsh sea conditions, connectors endure significant loads, and if these loads exceed their design limits, the connectors can fail, jeopardizing the integrity of the entire structure. Therefore, accurately predicting the loads on the connectors is essential to ensure the safety and reliability of the overall system.

In hydroelastic analysis, both floating bodies and connectors can be modeled as either rigid or flexible. There are four primary categories of models: (1) rigid module and rigid connector (RMRC); (2) rigid module and flexible connector (RMFC); (3) flexible module and rigid connector (FMRC); and (4) flexible module and flexible connector (FMFC) (Fu et al. 2007). Among these, the RMFC model is the most widely used for predicting the hydroelastic responses of VLFS, where the deformation of the VLFS is represented by the rigid body displacements of the floating modules, along with flexible connectors. A comparative study by Riggs et al. (2000) on the linear, wave-induced responses of a five-module mobile offshore base (MOB) based on RMFC and finite element analysis (FEA) models found that when the natural frequencies and corresponding normal modes align with those of the FEA model, the simplified RMFC model can predict responses with high accuracy. The findings demonstrate that if the natural frequencies and related normal modes match the frequencies and modes of the FEA model, the simplified RMFC model can predict the response very well. Wang (2022) explored the coupling effects between cage motion, disturbed waves, and loads on nets, analyzing the load effects on nets, steel structures, and connectors induced by diffraction and radiation waves. Ding et al. (2020) examined the connector loads of a three-module VLFS deployed in shallow waters using the 3D hydroelasticity theory and the RMFC model, discovering that the x-connector force was greater than the other forces. Yang et al. (2019) calculated the motions of each module and connector loads using the RMFC model while considering the depth-varying seabed and inhomogeneous waves. Their results indicated that the two-node horizontal bending mode significantly influences the connector loads in the x and y directions. Additionally, different torsional modes and coupled modes resulting from relative pitch and surge across modules also affect the connector load in the vertical direction.

Several researchers have investigated the hydroelastic responses of VLFS using models where both the connectors and the floating modules are considered flexible. Zhu et al. (2022) employed the FMRC modeling method to connect eight identical single modules along the longitudinal direction through fixed hinges. Their research demonstrated that the floating body of maritime airports constructed using the FMRC method can effectively reflect the dynamic response characteristics of similar structures. Wu et al. (1993) utilized the FMFC model to study a five-module system. Their calculations revealed that the longitudinal connector force was greater than the transverse and vertical forces. Additionally, Ertekin et al. (1993) analyzed a floating system composed of 16 modules using both RMFC and FMFC models. Their findings indicated that both approaches maintain high computational efficiency when applied to large floating structures.

4.4 Conclusions and Recommendations

FPV systems consist of three main components: the photovoltaic module, the floating structure, and the mooring. At this time, offshore floating photovoltaics are in the preliminary stages of development and exploration, with few successful installation projects. Numerous studies have focused on its potential applications. For example, studies by Kumar et al. (2021) and Claus and López (2022) highlight the importance of innovative design and engineering for successful deployment of offshore FPV systems.

Offshore photovoltaic modules are considered to have several advantages, including improved power generation efficiency due to the cooling effect of water, and extensive installation on the water's surface to reduce evaporation of water and inhibit algae growth. There have also been studies on how the tilt angle of photovoltaic panels affects the efficiency of power generation systems, finding no significant difference in power generation efficiency when the tilt angle of the photovoltaic modules is between 30° and 12°.

On the other hand, the marine environment envisioned for FPV systems is harsh, with high salinity, strong currents, and severe waves. For modular floating platforms, semi-submersible designs, which can be cost-effectively manufactured and installed, generally demonstrate good hydrodynamic performance, but larger platforms can increase the size and cost of the mooring system. In addition, materials used in PV systems should be selected for their resistance to corrosion, fatigue, and biofouling; the work of Jiang et al. (2021). Shi et al. (2023) also point out the importance of improving connector life and reliability of FPV systems through the use of advanced materials such as composite materials and high-strength alloys. In order to ensure the durability and efficiency of FPV systems, research is underway to optimize the design of these systems, including the placement of PV modules, the materials used in the floating structure, and the configuration of the mooring system. Song et al. (2022, 2023) indicated that the dynamic responses of hinged connectors were significantly higher than those of fixed connectors under identical sea conditions. Luo et al. (2024) indicated that the airgap performance deteriorates rapidly when single module draught surpasses half of pontoon diameter, and improving soft rope stiffness is more effective and produces less disturbance to other behaviors than enhancing mooring stiffness via experimental results.

In addition, for numerical analysis of the response of FPV systems in waves, which are very flexible and have an areal extent, and the underwater shape of the floating part is a slightly complex module, connected by connectors, there are an increasing number of examples of analysis using commercial software such as ANSYS-AQWA, as well as open source codes and other extensions and combinations. Such an approach has the potential to significantly reduce the computational resources required for large-scale simulations and could be a valuable tool for optimizing the design and performance of FPV systems.

5 Positioning and Mooring

5.1 Introduction

The effective positioning and mooring of ocean structures are critical to their operational efficiency and safety. A well-designed positioning and mooring system must accommodate the environmental loads while considering economic viability and environmental sustainability. Such systems are diverse in design and function, however they can be classified into three main groups: caisson or pile type dolphins, mooring lines and tendons, and dynamic positioning, see the ISSC 2022 Committee V.6 report. While traditional catenary mooring configurations remain common, the emergence of new materials and technologies has expanded possibilities for new technologies. This is especially relevant for mooring of structures for ocean space utilization which often present new challenges in terms of design requirements.

This chapter reviews recent developments in design analysis methods and on technological solutions. The focus is on floating infrastructures for ocean space utilization, including very large floating structures (Sect. 5.2) and aquaculture facilities (Sect. 5.3). Conventional oil & gas platforms and renewable energy floating systems are mostly not included as they are covered by other ISSC Committees. The Chapter ends with a summary and recommendations for further research and development work.

5.2 Floating Infrastructures and Cities

This Section addresses very large floating structures (VLFS), reference is made to a couple of relevant publications on ocean energy systems and dynamic positioning. The focus is on the station keeping of these structures and on summarizing the research work performed during the past period (Fig. 6).

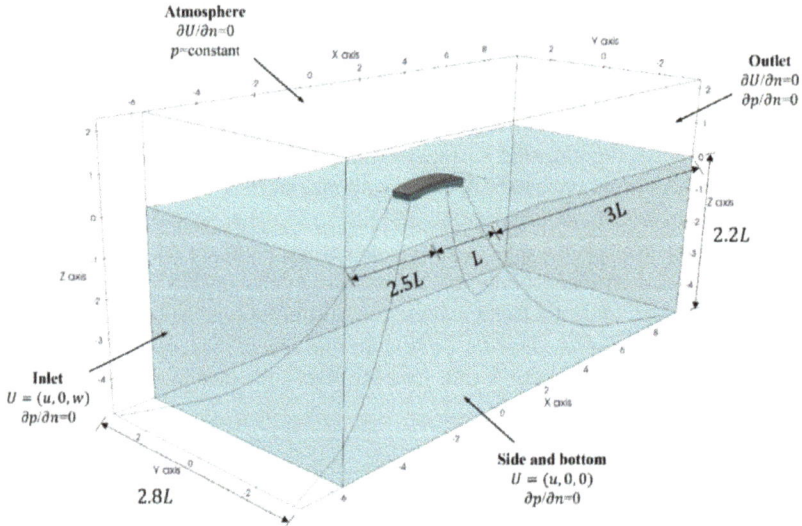

Fig. 6. Computational domain and boundary conditions (Wei et al. 2024)

Wei et al. (2024) performed a coupled analysis between VLFS and catenary mooring considering structural hydroelasticity in waves. The methodologies employed are designed to accurately simulate the interactions between the fluid dynamics and the structural responses of the VLFS. Validation against experimental data confirms the accuracy of the numerical results, highlighting the influence of mooring systems on structural dynamics. The study examines the interaction between mooring systems and the hydroelastic behavior of the VLFS, focusing on the effects of different mooring configurations. The analysis highlights the differences in dynamic responses between taut and catenary mooring systems, emphasizing their impact on structural stability. The findings suggest that the choice of mooring type can significantly influence the hydroelastic behavior of the VLFS, particularly in wave conditions. The study successfully integrates

CFD and multibody dynamics to evaluate the hydroelastic behavior of flexible VLFSs, demonstrating good agreement with experimental data (Fig. 7).

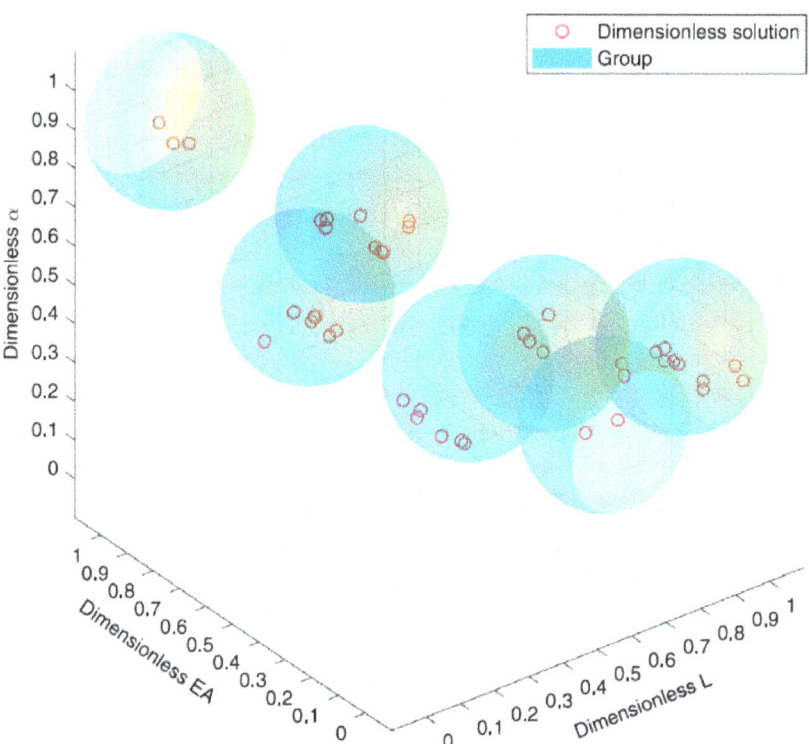

Fig. 7. Illustration of the optimal solutions grouping process (Liang et al. 2020).

The complexity and reliability issues associated with scale model experiments for very large floating structures (VLFS) due to excessive mooring lines are evaluated by Liang et al. (2020). A mooring system simplification methodology is proposed, which was previously validated through numerical studies, and is further evaluated through model tests involving a VLFS with 20 mooring chains. Non-moored decay tests are used to calibrate the VLFS model, ensuring that its motion responses align with simulation results. The decay test assesses the natural period and damping coefficient of the VLFS model with both mooring systems, confirming their equivalence. The equivalence of the mooring systems is further validated through irregular wave tests, with motion response data analyzed for both operational and survival conditions. The study successfully demonstrates the effectiveness of the mooring system simplification methodology, with a 10-chain system validated against a 20-chain original system through comprehensive testing. The findings highlight the potential for simplified mooring systems to enhance experimental reliability and reduce setup challenges in offshore engineering research (Fig. 8).

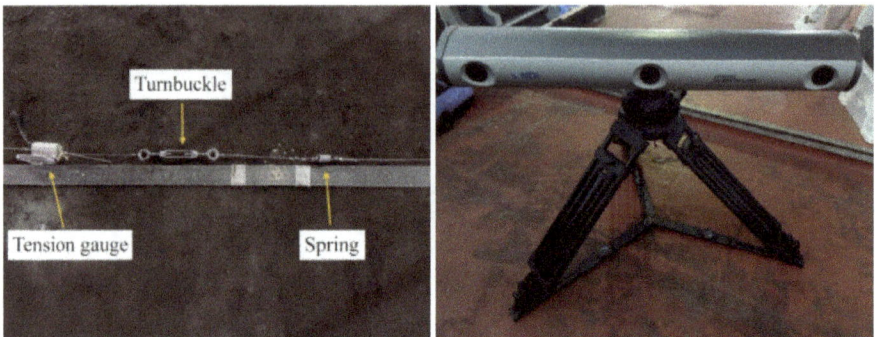

Fig. 8. Diagram of mooring line and schematic diagram of Optotrak Certus (Feng et al. 2022).

The impact of different mooring types on the hydroelastic response and other parameters of a single module pontoon type Very Large Floating Structure (VLFS) through physical model tests are investigated by Feng et al. (2022). Four mooring types are analyzed, focusing on mooring conditions' effects on mooring force, incident coefficient, reflection coefficient, and energy dissipation coefficient. The study concluded that different mooring types significantly affect the hydroelastic response, mooring force, and wave interaction parameters of VLFS. Initial tension emerged as a critical factor influencing the hydroelastic response, with bollard mooring generally resulting in larger responses than seabed anchoring. The reflection, transmission, and energy dissipation coefficients exhibited consistent trends with wavelength changes, highlighting the complex dynamics of mooring systems. The results emphasized the need for careful consideration of these coefficients in the design of mooring systems for VLFS (Fig. 9).

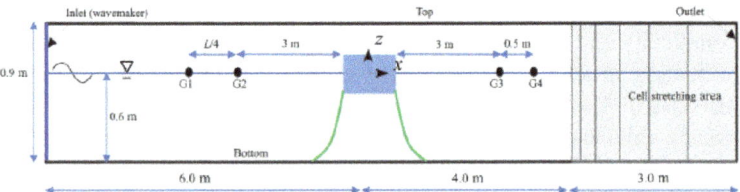

Fig. 9. Numerical model setup for the physical experiment of a box-type floating breakwater (Chen et al. 2024).

A framework for assessing the feasibility of a multi-purpose platform that integrates wind turbines, wave energy converters (WECs), and aquaculture facilities, highlighting the cost challenges associated with mooring systems in emerging technologies proposed by Chen et al. (2024a) identifies the optimization of compact WEC arrays and shared moorings as a potential strategy to reduce energy costs, while recent experiments reveal challenges in shared mooring system designs. The study indicates that activating turbulence modeling does not significantly affect numerical results, as shown in previous mesh-based Computational Fluid Dynamics (CFD) studies. The volume of fluid (VOF) method is employed to approximate the air-water interface, with density and viscosity calculated as α-weighted averages in the interface cells. The paper reviews three mooring analysis codes i.e., MAP++, MoorDyn, and Moody, detailing their features and applications in floating structure simulations. It is seen that the development of an open-access mooring restraint library enhances the capabilities of OpenFOAM's rigid body motion solvers, allowing for more complex simulations of floating structures. The study encourages the use of the open-access library for reproducible research and further exploration of floating structure dynamics in diverse applications (Fig. 10).

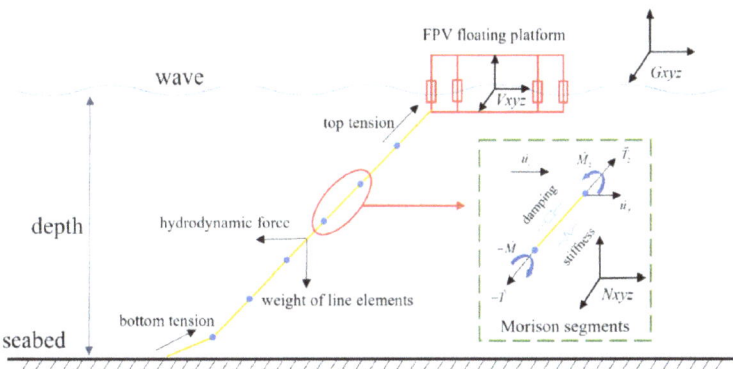

Fig. 10. The schematic diagram of the mooring system model based on the finite element analysis method (Tang et al. 2024).

A nonlinear coupled dynamic response analysis using the Finite Element Method (FEM) in OrcaFlex to evaluate the performance of Floating Photovoltaics (FPV) systems is deployed by Tang et al. (2024). A comprehensive research framework is established to assess the dynamic interactions between hydrodynamic forces and structural responses of the floating platforms. The study evaluates the effectiveness of additional viscous damping corrections on the dynamic responses of the platforms, demonstrating improved accuracy in motion predictions. The findings suggest that extreme tension forces increase with longer return periods and greater water depths, highlighting the need for robust design considerations in mooring systems. Key findings include the effectiveness of damping corrections, the comparative stability of platform shapes, and the implications of tensioner stiffness on operational safety. Future research directions are suggested, focusing on the integration of additional parameters and the exploration of innovative mooring solutions to enhance the resilience of FPV systems (Fig. 11).

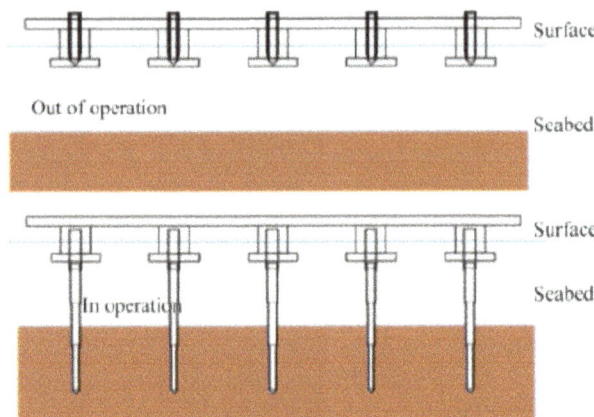

Fig. 11. Two operational conditions of the novel positioning pile (Xu et al. 2020).

A conceptual design for novel telescopic piles intended to position a multi-modular very large floating structure (VLFS) that functions as a movable floating airport is proposed by Xu et al. (2020). These telescopic piles can automatically engage with the seabed to withstand environmental loads and can be retracted for relocation. The implementation of a Genetic Algorithm-Finite Element Analysis (GA-FEA) method led to a 31% reduction in pile weight while maintaining structural efficiency. The study successfully demonstrates the feasibility of using novel telescopic positioning piles for VLFS applications, particularly in floating airport scenarios. The structural analysis confirms that the piles can withstand external loads while providing high positioning accuracy. The optimized design maintains structural integrity and safety, showcasing the effectiveness of the GA-FEA method compared to traditional optimization techniques (Fig. 12).

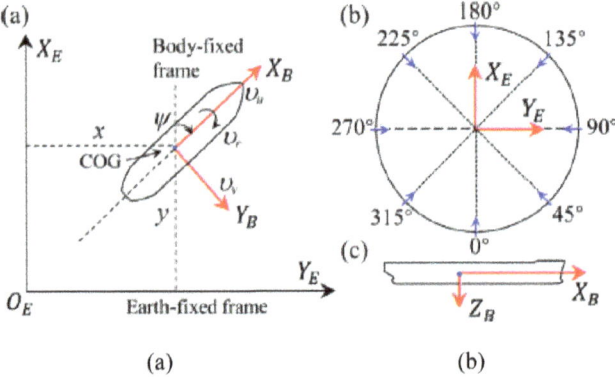

Fig. 12. Definitions of coordinate systems and sea load directions. (a) Top view of earth-fixed frame and body-fixed frame. (b) Top view of the sea load directions (Shi et al. 2023).

Shi et al. (2023a) discussed the development of Dynamic Positioning (DP) technology for vessels, which allows for automatic control against marine environmental loads to maintain desired positioning and heading. The authors emphasize the need for control strategies that do not overly rely on prior knowledge of hydrodynamic parameters, especially in the context of varying ocean disturbances. The findings highlight the significant impact of environmental disturbances on control performance, with the proposed strategy exhibiting less degradation compared to the baselines. The study concludes that the proposed DP control strategy successfully addresses uncertainty and disturbance estimation (UDE) problems and enhances control performance, ensuring safety-related constraints are met. It emphasizes the robustness of the strategy in varying sea states and its practical relevance for offshore engineering applications (Fig. 13).

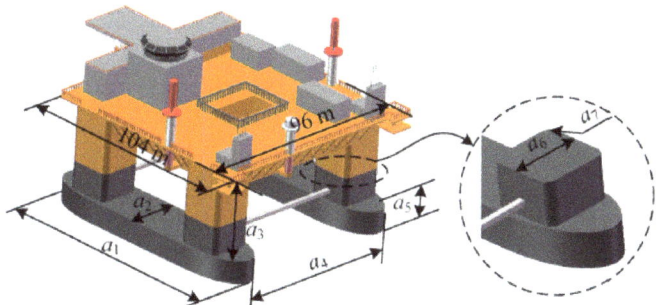

Fig. 13. Structure parameters of the SEMI platform (Yuan et al. 2023).

Yuan et al. (2023) studied the Multidisciplinary design optimization of dynamic positioning system for semi-submersible platform. A simultaneous dynamic optimization approach is proposed to address the challenges of achieving a global optimal solution amidst the discoordination between structural and dynamic performance parameters. The radial basis function (RBF) model is extended to include constraint conditions, and the alpine skiing optimization (ASO) algorithm is utilized to solve the multidisciplinary design optimization problem. The results indicate that the co-design strategy is a viable method for optimizing dynamic positioning systems in complex marine environments. This study presents a pioneering simultaneous optimization approach for the DPS, addressing the limitations of conventional strategies that often overlook inter-disciplinary relationships. The co-design strategy effectively minimizes operational costs while improving dynamic performance metrics, such as rolling and yaw responses. Limitations include assumptions regarding thruster capabilities and the need for further enhancements to the ASO algorithm's efficiency (Fig. 14).

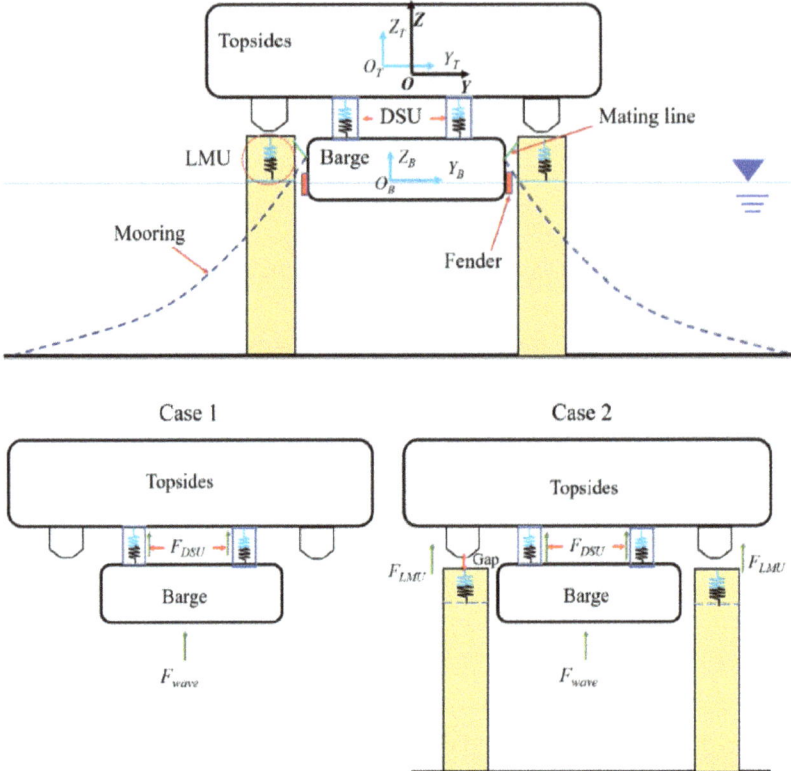

Fig. 14. Configuration of a typical float-over system and schemes of the Heave-only CPTDM with two different configurations (Chen et al. 2024).

Chen et al. (2024b) introduced a Constant Parameter Time-Domain Model (CPTDM) to analyze the nonlinear dynamics of float-over installations, focusing on rigid body motions, impact-absorbing systems, and mooring dynamics. The model employs the Cummins equation to address wave-induced motions of floating bodies and incorporates state-space models for improved computational efficiency. Validation against the AQWA hydrodynamic analysis package shows that the CPTDM effectively captures the nonlinear physics relevant to float-over operations, with potential applications in offshore wind maintenance. The adaptive time step method is shown to improve the capture of nonlinear behaviors compared to fixed time step methods. Despite some discrepancies, the Fully Coupled CPTDM demonstrates the ability to capture the main characteristics of the dynamic responses associated with float-over systems. The selection of state-space model orders is discussed, based on Hankel singular values, to optimize the model's performance in simulating the float-over system dynamics.

5.3 Aquaculture Facilities

Due to decreasing fish stocks and population growth, there is a need to study and develop aquaculture offshore facilities that employ multiple net cages globally to fulfil the world's demand. Therefore, apart from the dynamic analysis of floating flexible fish cages, studying different mooring systems associated with aquaculture facilities (net cage supported platform, floating offshore base to support aquaculture, and fish farm consisting of an array of flexible net cages) is also essential. The following paragraphs explore recent research addressing various facets of mooring system analysis in offshore aquaculture.

Sulaiman et al. (2013a) presented a mooring system design for a large offshore aquaculture ocean plantation floating structure under wind and current loadings on mooring components were conducted. Further, Sulaiman et al. (2013b) performed a qualitative risk analysis of a floating aquaculture plant on the mooring system and concluded that "an integrated approach to risk analysis will assist the aquaculture sector in reducing risks to successful operations.

Li et al. (2019) optimized the mooring system for a vessel-shaped aquaculture system based on the FEM model and they suggested that the fatigue life of the mooring system can be treated as a target in the further calculation and uncertainties should be considered. Liu et al. (2022a) studied a novel open ocean aquaculture ship coupled with a wind turbine connected with turret mooring system. They observed that aquaculture ships can avoid enormous wind, wave, and current loads with the help of the weathervaning effect by adopting a turret mooring system. Liu et al. (2020) studied numerically the hydrodynamic behaviour of a semisubmersible aquaculture facility and mooring line tension is beneficial to the stability of the semisubmersible aquaculture facility and helps avoid large motion of the aquaculture facility in severe waves through increasing the draught.

The multi-module aquaculture platform through a series of physical model experiments under regular waves was developed by Bi et al. (2021). The tensioned mooring chains restrain the modules on the weather side on the front wave side, and the 1st-order wave force of a high frequency practically does not cause longitudinal and vertical motion of the modules of the aquaculture platform. The floating aquaculture platform featuring a hinged multi-body design is studied (Ma et al. 2022), hey observed that the amplitudes of the mooring force exhibit a progressive reduction and demonstrated that the flexible hinge connection is more beneficial to the aquaculture platform design than the rigid connection.

Hou et al. (2020) studied the annual fatigue damage of the mooring system in an offshore fish farm system by the generalized probability density evolution method and a virtual stochastic process. Further, Hou et al. (2022) also calculated the mooring force distribution of a mooring system with one rope failure. They found that one rope failure increases the maximum mooring force significantly.

Tang et al. (2021) built physical and numerical models to investigate the situation of the mooring system when one rope breaks for the gravity cage system, two mooring ropes break continuously, and studied in irregular waves for the failure of mooring lines {Formatting Citation}. They observed that the maximum mooring force was increased by up to two times when one rope broke and the second failure of the ropes can result from a great increase in the mooring force. Additionally, it was found that the difference between the physical model and the numerical model becomes non-negligible when the current velocity reaches 0.2 m/s.

Xu et al. (2012) developed different arrangements of cage arrays and calculated the mooring forces and they applied different designs of mooring systems and found that the amplitudes of mooring forces for a twin mooring system are higher than the ones for an orthogonal system (Xu et al. 2013) performed time-domain simulations to analyze the structural responses of both a single-cage and a 1 × 4 multi-cage fish farm in response to grid mooring line failures by employing the program FhSim (Cheng et al. 2021) and concluded that the failure of a single mooring line is unlikely to trigger a cascading collapse of the fish farm under current velocity.

A large fish farm consisting of 16 net cages arranged in a 2 by 8 layout based on the lumped mass model was modeled to examine nonlinear dynamics of the aquaculture cage array induced by severe conditions (Shen et al. 2023) and surge motion is highly sensitive to changes in bending stiffness. Wang et al. (2024a) numerically studied the dynamic of grid mooring system for an array of gravity cages under wave-current loads based on lumped mass method and Morison equation. They examined that enlarging the frame line length can equalize the tension distribution within the mooring system and minimize the peak system. They recommended that to reduce the mooring line breakage not to use 90° of environmental load to the cage system. A fish farm of six cages with a complete mooring system was built by Liu et al. (2022b) and they found that the mooring forces of the downstream frame ropes are very small and the ropes do not play a structural role.

An analytical model of wave-current interaction with an array of moored floating flexible net cages of cylindrical shapes was developed by Mohapatra and Soares (2024) under the linear water wave theory. The results indicate that as mooring stiffness and current increase, both wave loads and cage displacements also rise, indicating that elastic collars with different mooring system are recommended to reduce the displacement of cages in the future. Recently, Mohapatra and Guedes Soares (2024) presented a review of the hydroelastic theoretical models of floating flexible net cages and elastic floaters for offshore aquaculture applications and it was suggested to consider the array of different cage geometries (hexagonal) with mooring systems and elastic floaters to prevent damage of cages in harsh wave conditions (Fig. 15).

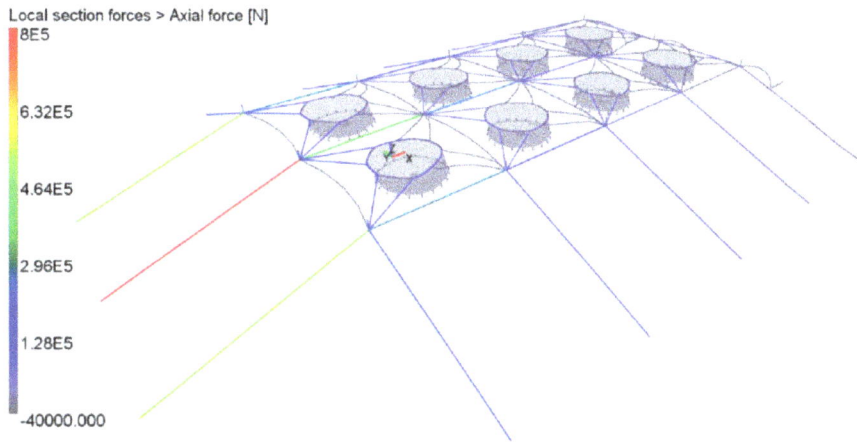

Fig. 15. A snapshot of the simulation of the fish farm subjected to waves and current (Nasyrlayev et al. 2023)

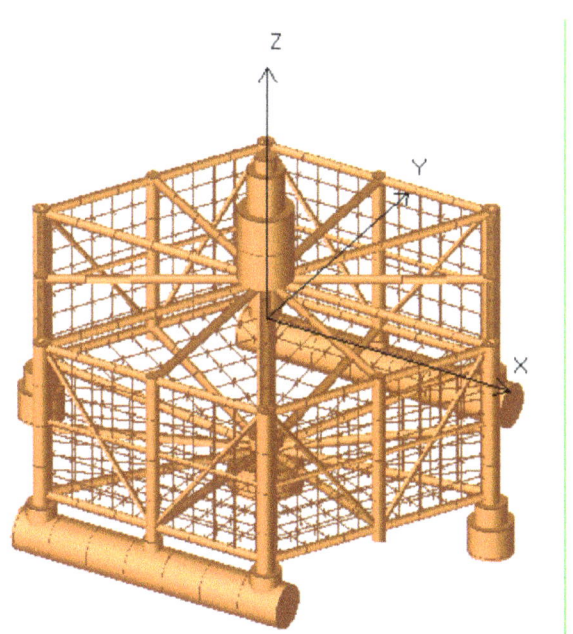

Fig. 16. Hydrodynamic model of the aquaculture platform (Yu et al. 2023)

Nasyrlayev et al. (2023) investigated the environmental loads on slender gravity salmon fish cages and the related loads on mooring lines, anchors and other structural elements. Different configurations of a multi-cage fish farm were considered under several high-energy offshore conditions. A structural finite element method (FEM) was used to simulate the responses to the wind and current conditions. The extensive sensitivity analysis provides insights into the range of possible mooring line loads for different configurations. Both the current magnitude and angle have a significant impact on the mooring line tensions. Farms with a symmetrical configuration minimize the mooring line loads (Fig. 16).

Yu et al. (2023) present numerical and experimental studies on the mooring system loads of an aquaculture structure under several waves and current conditions for the operational draft and for the bottom-sitting survival condition. The floater consists of an assembly of slender elements into a hexagonal prism with a diameter of 110 m and height of 78.5 m The numerical method is based on a combination of potential flow loads and Morison type of loads and the solution obtained in the time domain. The paper describes a procedure to calculate the Morison loads on the net, including shadowing effects. A spread mooring system with a hybrid mooring line composition, containing steel chain and fibre rope in each line, was designed and its performance was assessed. The authors conclude that the numerical model provides results in good agreement with the experiments.

5.4 Conclusions and Recommendations

Floating structures for ocean space utilization are kept in station mostly by mooring systems. Most common are catenary (soft) mooring systems, although inclined taut systems and vertical tendons are also applied, depending on the design requirements. One alternative for shallow water installations is the use of piles, or caissons. Dynamic positioning is rarely applied for the type of installations considered in this report.

For VLFS it may be important to solve the coupled elastic-hydrodynamic-mooring problem, as it has been shown that the mooring system influences the structural dynamics (and vice-versa). A careful design of the mooring system has the potential to minimize the structural responses. More generally, modelling of complex coupling effects within the evaluation of mooring system loads is a need and a challenge common to many of the installations for ocean space utilization. In this respect, the report presents examples for VLFS, multipurpose platforms, arrays of similar floaters (e.g. wave energy converters), floating islands for harnessing solar energy and aquaculture platforms.

Research needs are identified in several topics. A non-exhaustive list includes:

- Shared mooring systems, which offer a promising solution to save materials and reduce the cost of mooring systems, particularly for arrays of floating structures.
- Optimizing the distribution of mooring loads on Very Large Floating Structures (VLFS) and photovoltaic solar islands, while ensuring the system remains economical and simple for installation and maintenance.
- Mooring in shallow water continues to pose challenges in wave-energy-intensive locations due to the tendency of catenary systems to exhibit slack-tensioned behavior, leading to high dynamic loads.
- Accurate prediction of wave drift loads for complex multibody systems. Conditions with current add complexity and the related predictions are beyond the existing design analysis tools capabilities. Additionally, viscous wave drift effects are often significant and are typically represented by Morison drag elements, which are incorporated into

the equations of motion alongside potential flow terms. However, the drag coefficients are often oversimplified, disregarding their dependency on the Keulegan-Carpenter number and the influence of free surface effects.

6 Tank Testing and Full Scale Measurements

6.1 Introduction

The utilization of ocean space is of great significance to humanity. Last several decades have witnessed the rapid growth of ocean resources and space demand. The structural characteristics of OSU structures, such as floating bridges, floating fish farms, floating photovoltaic (FPV) equipment and floating offshore wind turbines (FOWT), are quite different from traditional ocean platforms. Since there exists limitations in the theory assumptions and numerical modeling, Model experimental and full scale measurement play an important role in the research, design and development of OSU structures. In addition, model tests and full scale measurements also provide useful data for verification and validation of numerical methods from different codes and software.

6.2 Model Scale Measurements

6.2.1 Floating Bridge

Numerous researchers have also conducted studies on floating bridges, it is found that the most important factors affecting the dynamic responses of pontoon bridges include wave and current parameters (Kvåle et al. 2016; Sha et al. 2018; Viuff et al. 2019, 2023a; Xu et al. 2020b; Dai et al. 2022; Xiang et al. 2022a), vehicle loads (Wang and Jin 2016; Jun and Wei 2021; Miao et al., 2021a), water depth (Zhang et al. 2008), mooring system (Xiang et al. 2022), rotation stiffness of connectors (Huang et al. 2021, 2022) and so on (Manisha and Sahoo 2019; Wei et al. 2019; Wang et al. 2023).

Xiang (2023, 2024) researched a long floating bridge crossing the deep and wide fjords located west of Norway and analyzed the hydrodynamic and hydroelastic characteristics and hydrodynamic interaction between the pontoons of the floating bridge using numerical methods and physical model tests. The influences of various environmental loads and homogeneous wave conditions on the hydrodynamic responses and structural responses of an end-anchored floating bridge were calculated using the codes SIMO/RIFLEX (Cheng et al. 2018a, b, 2019, 2020). Fenerci et al. (2022) investigated the influence of the hydrodynamic interaction between the adjacent pontoons on the global responses of a long floating bridge, and such interactions cannot be ignored while calculating the dynamic responses. The extreme responses of an end-anchored floating bridge were investigated through a numerical parameter study based on an extensive amount of time domain simulations (Viuff et al. 2019), the structural responses were calculated under several wave conditions, and the wave- and current-induced responses of a floating bridge were investigated based on small-scale model tests and compared with the numerical simulation results by Viuff et al. (2020, 2023b). A new machine learning-based Monte Carlo framework was developed to predict the cumulative distribution function of the long-term extreme response of a floating suspension bridge under wind and wave actions (Xu et al. 2020b).

Since the floating bridge usually comprises multiple modules connected by articulations, usually reaching hundreds of meters or even several kilometers, it is generally necessary to consider the connector loads of the floating bridge reduced by waves, currents, and other loads. The effects of the nonlinear properties of connectors and moving loads effects on the motion responses and connector forces of a nonlinearly connected ribbon floating bridge were proposed by Shixiao et al. (2005), Wang et al. (2009) and Fu and Cui (2012) based on the numerical simulations and model tests, and it is found that nonlinearity and initial gap of the connectors are important for the motion responses of the multi-module floating bridge. The hydrodynamic responses of a newly designed pontoon-type floating bridge under regular waves were also investigated by experiments and hydroelastic analysis (Sun et al. 2018). Huang et al. (2021, 2022) proposed a calculation method that can be widely used in the response analysis of floating bridges with elastic-hinged irregular spans, and studied the influence of elastic hinge stiffness and other parameters on the motion response, bending moment and shear force of floating bridges.

Physical model tests in the wave basin of the floating bridges play an important role in the design stage, and the experimental results are often used to verify the rationality of the floating bridge configuration design (Sun et al. 2018; Huang et al. 2023), and the hydro-elasticity tests can also verify the safety of the floating bridge structure design (Fu and Cui 2012; Rodrigues et al. 2020, 2022; Viuff et al. 2023a, b; Xiang 2023; Xiang et al. 2023). At the same time, the test results are often used to verify the accuracy of numerical calculation methods (Sun et al. 2018; Huang et al. 2023). Otherwise, the physical model test is also often used to research the new floating structure to determine the impact of specific factors on the floating bridge in the research and initial design stage (Chen et al. 2017), and it also provides a reference for the research, development, design, and improvement of floating structures. Chen et al. (2017) carried out both numerical and physical model tests to investigate the hydrodynamic performances of a moored irregularly shaped pontoon in finite-depth water, the physical model tests were conducted with model-scale 1:16. Sun et al. Sun et al. (2018) carried out experiments of a newly designed type of floating bridge in the deep-water tank of the Shanghai Institute of Shipping and Transportation, and the floating bridge model was made of aluminum alloy, high-strength steel, and rubber cloth with scale ratio 1:6. The model tests of a three-span continuous beam bridge were carried out in the circulating water channel of State Key Laboratory of Ocean Engineering in Shanghai Jiao Tong University with a scaling ratio of 1:70 by Xiang et al. (2022b, 2023). Rodrigues et al. (Rodrigues et al. 2020; Rodrigues et al. 2022) performed the model tests on a straight-side anchored bridge by the truncated method, and the 1:33.3 scaled model was built in aluminum, PVC foam, and steel. Xiang (2023, 2024) provided an overview of the background, design, execution, and correlation work of extensive hydro-elastic tests of a long floating bridge in the Ocean basin of SINTEF Ocean, including three groups: the single pontoon tests under the scale of 1:25, the three pontoons tests under the scale of 1:25, and the high bridge tests under the scale of 1:31 (as illustrated in Fig. 17). It can be seen from the above literature that when designing a new type of floating bridge, physical model tests are essential, but such reports are not as numerous as numerical calculations.

Fig. 17. Floating bridge model tests - high bridge tests

A new type of pontoon-truss composite floating bridge was designed and investigated by Miao et al. (2024a), and the module of the floating bridge is composed of pontoons and trusses (as shown in Fig. 18), there are connectors between the pontoons and trusses, the most prominent feature distinguishing it from traditional pontoon bridges. The physical model tests of the typical module and a 4-module floating bridge were carried out in the wave basin to investigate the variation of the motion response of the modules and the connector loads between the modules with wave characteristics. In the experiments, the motion of the first and second modules of the floating bridge was measured, and the three-component balance was used for measurement of the load on the connector between the modules. The time histories of the motion responses and connectors are obtained, then the motion RAOs and transfer functions of the connector load are obtained, and the several important conclusions were obtained.

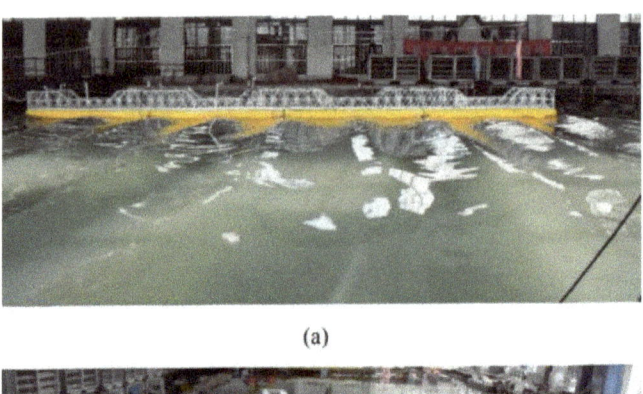

Fig. 18. Test model of the floating bridge. (a) The wave direction is 0°; (b) The wave direction is 15°. (This figure is available in colour online.)

6.2.2 Offshore Fish Farm

The traditional fish cage has been investigated by numerical simulations and physical models by many researchers. With the increasing improvement of new type fishery and aquaculture equipment, such as SOFF and vessel-shaped offshore fish farm, the physical experimental method for above new type aquaculture equipment also need improvements and estimates of the performance of them in waves and current.

The mooring loads of a realistic aquaculture fish farm system in both regular and irregular waves were investigated by numerical simulations and model tests (Shen et al. 2018). A wake model for the net structure was described in detail, and a method for calculating the current forces experienced by the net structure and the resulting deformation was derived and compared with model tests (Aarsnes et al. 1990; LøLand 1993). The hydrodynamic loads acting on a three-dimensional highly flexible simplified net cage structure in a uniform flow and the associated deformation of the net cage were investigated experimentally (Moe-Føre et al. 2016). A method based on linear hydrodynamic theory and physical model tests have been developed to estimate wave loads, mooring displacement, and mooring forces of a sea cage used in aquaculture (Ito et al. 2014). The dynamic properties of a dual pontoon floating structure with and without a fish net were investigated using physical and numerical models (Tang et al. 2011). A numerical analysis of a cage based on a Morison-force model under combinations of current and waves was carried out to estimate the loads on the cages and the numerical results were compared with experimental data (Cifuentes and Kim 2017). The hydrodynamic characteristics of a floater-net system in oscillatory and steady flows were investigated through forced oscillation experiments in a towing tank, and the effects of Keulegan-Carpenter number, Reynolds number, and reduced velocity were studied (Fu et al. 2014). The dynamic behavior of a box-shaped net cage was investigated in pure waves and steady current by numerical model and physical experiments (Zhao et al. 2013).

A series of physical model experiments was performed by Zhao et al. (2019) to investigate the hydrodynamic responses of a semi-submersible offshore fish farm in waves. The mooring line tension and motion response of the fish farm were measured at three draughts. The study indicated that the tension on the windward mooring line is greater than that on the leeward mooring line. As the wave height increases, the mooring line tension and motion responses including the heave, surge, and pitch exhibit an upward trend. The windward mooring line tension decreased slightly with increasing draught. The existence of net resulted in approximately 42% reduction in mooring line tension and approximately 51% reduction in surge motion. However, the heave and pitch of the fish farm increased slightly with the existence of net. It was found that the wave parameters, draught, and net have noticeable effect on the hydrodynamic response.

The model tests of a SOFF (Fig. 19) were carried out by Miao et al. (2021b) in the wave basin, and the heave and pitch motions with wave periods were obtained. The numerical results were verified by the experimental results. The effects of some key parameters on the motion responses were investigated by the numerical model and model tests. The results show that the draught of SOFF has an obvious influence on the heave and pitch motion. Heave and pitch response amplitudes are positively correlated with wave amplitude.

Fig. 19. Photograph of SOFF models in waves

A physical model of a semi-submersible offshore fish farm with a scale of 1:30 has been tested by Huang et al. (2020) in a wave-current tank. The mooring force of the fish farm in waves was measured and compared for different single-point mooring arrangements, and henceforth, a safe and reliable mooring form is chosen as the mooring system for the fish farm. On this basis, the tested results for the mooring force, heave, pitch and roll of the fish farm in waves and current are presented. The performance of blocking water flow for the fish farm in pure current was also evaluated. It has been found that increasing the anchor chain length and adding a sinker suspending from the chain could reduce the mooring force significantly despite the fluctuations in the mooring force. During the experiments, the tested value for the heave and pitch as well as the roll has been small, which indicates that the fish farm has a good stability. The mooring force has become larger, while the heave became smaller as the drought increased.

6.2.3 Offshore Floating Photovoltaic

As the global demand for renewable energy escalates, FPV systems are gradually expanding from coastal waters to offshore areas. The expansion of Floating Photovoltaic (FPV) systems into marine environments presents a significant shift from reservoir-based installations, offering abundant space and new challenges due to complex environmental conditions. Therefore a enhanced environmental adaptability and safety requirements for FPV system are needed during the design stage.

Zhang et al. (2024a) improved upon the successful modular floating PV development at Tengeh Reservoir, adapting it to withstand harsher marine conditions and establishing one of the world's largest nearshore floating photovoltaic farms near Woodlands, Singapore. The FPV system, composed of standardized floating modules made of high-density polyethylene (HDPE), is designed to follow wave motion, showcasing its adaptability to marine environments. Through full-scale experimental testing and numerical simulations on a representative subsystem of the floating PV farm, the hydrodynamic performance of the system is verified, focusing on its ability to withstand the combined effects of wind, waves, and currents.

Jiang et al. (2023) proposed an FPV concept capable of withstanding extreme sea conditions with wave heights over 10 m, utilizing standardized lightweight semi-submersible floats and forming an FPV array through soft connections. Model tests on a 2 × 3 FPV array under regular and irregular wave conditions demonstrated excellent hydrodynamic performance and no contact between adjacent modules under extreme wave conditions. In addition to using semi-submersible floating platforms, researchers at the University of Oviedo developed the HelioSea system, an offshore solar concept combining a dual-axis tracking system with a Tension Leg Platform (TLP), offering advantages such as mobility, minimal vertical movement, low cost increase with water depth, and deepwater capabilities.

Claus et al. (2024) outlined the first experimental proof of concept for the HelioSea system, with tests conducted on a 1:30 scale model in the wave basin of the University of Porto to assess responses to regular and irregular waves. A total of 27 regular wave tests were conducted to establish the Response Amplitude Operators (RAOs) of the structure, confirming the stability of the proposed concept in more realistic scenarios.

Luo et al. (2024) categorized FPV into zero-airgap type and positive-airgap type. The zero airgap type, while offering better water cooling effects, is more exposed to wave loads, posing significant threats to structural safety. Consequently, the positive-airgap type is considered more practical. Luo et al. proposed a pontoon-truss floating platform, connecting four such platforms with flexible connection ropes to form a multi-module FPV system. Model experiments assessed its feasibility under the combined action of wind, wave, and current, discussing the impact of several key design parameters on the system's hydrodynamic performance, with a focus on airgap response. Beyond investigating the safety and hydrodynamic performance of FPV systems, researchers have also extended their interest to the economic feasibility of these systems.

Huang et al. (2024) innovatively installed a solar simulator on top of a wave tank to study the impact of wave-induced motion on the power output of FPV systems. The results indicated significantly lower power output in waves compared to calm water, establishing a clear link between wave-induced power loss and the panel's rotational motion. An empirical equation was derived to predict power loss through rotational amplitude, highlighting the importance of implementing wave attenuation technologies such as breakwaters to minimize the impact of waves on floating solar systems. The development of FPV systems is transitioning from nearshore to offshore regions, with experimental studies focusing on environmental adaptability, structural safety, hydrodynamic performance, and economic viability. As the global demand for renewable energy grows, FPV systems offer a sustainable solution that must address the challenges posed by marine environments to realize their full potential.

6.3 Full-Scale Measurements

6.3.1 VLFS

To further verify the numerical methods and investigate the performance of such a very large floating structure (VLFS), a "Scientific Research and Demonstration Platform (SRDP)" was built and deployed in 2019 at the site about 1,000 m off an island with water depth around 40 m in the South China Sea (as illustrated in Fig. 20). It is a simplified small model of a two-module semi-submersible-type VLFS.

Fig. 20. The SRDP system deployed near island and reefs (Ding et al. 2021a)

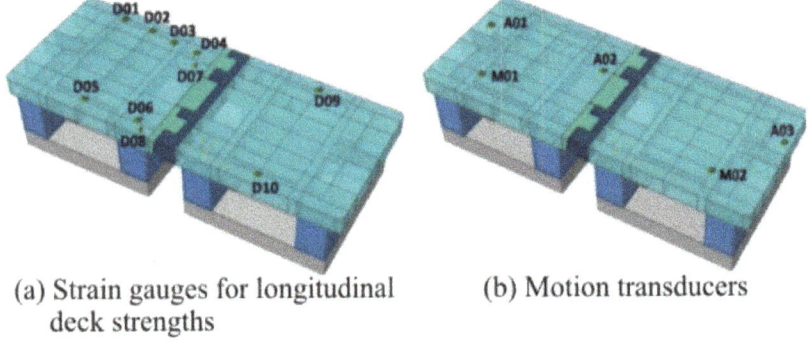

(a) Strain gauges for longitudinal deck strengths

(b) Motion transducers

Fig. 21. Layout of the measuring points (Ding et al. 2021a)

The numerical simulation of its responses on severe waves with a focus on motions and connector forces is conducted by DCAM, and compared with the on-site measurements, as illustrated in Fig. 21 (Ding et al. 2021). Experimental and numerical studies were carried out by Miao et al. (2022) to investigate the coupled behavior of the two connected semi-submersibles with the emphasis on the motion responses of the two modules and the forces in the connectors. A series of wave basin model tests for a single module and two connected modules have been conducted, and the measured data were compared with the corresponding numerical simulations which showed a good agreement.

A mooring system has been designed for the position-keeping of the SRDP which is connected by hinge-type connectors. An analysis method has been developed by Ni et al. (2021) based on the three-dimensional frequency domain hydroelasticity theory in conjunction with the time domain quasi-static analysis method of mooring actions, which takes into account of the coupling effect of the platform motion and mooring tension.

A new approach referred as "the network modeling method" was developed by R. Ding et al. (2021) to analyze the behaviors of marine structures, and the method is briefly described and applied to predict the loads acting on the connectors between the SRDP. Based on this method, the response amplitude operators (RAOs) of the connector loads of the SRDP in regular waves, and the time variations of the connector loads of the SRDP in an on-site measured random sea state are predicted and presented. The significant stresses at 20 spots of the local connection structure induced by the connector loads in the sea state are further calculated.

6.3.2 Floating Bridge

The dynamic behaviour of long-span bridges is governed by stochastic loads from typically ambient excitation sources. In real life, these loads cannot be measured directly at full scale. However, inverse methods can be utilised to identify these unknown forces using response measurements together with a numerical model of the relevant structure. The full-scale floating bridge tests are even rarer, a long-span pontoon bridge that has been monitored since 2013 by Petersen et al. (2019), and they presented a case study of full-scale identification of the wave forces on the floating bridge, Bgsoysund bridge. First, a numerical model of the structure is formed, resulting in a reduced-order state-space model that takes into account the frequency-dependent hydrodynamic mass and damping from the fluid, based on fitting of rational transfer functions. Using acceleration data of the structure measured during several events of moderate and strong seas, the wave forces are identified using stochastic-deterministic methods for combined state and input estimation.

6.3.3 FPV

Low displacement floating structures are usually the standard solution for offshore floating photovoltaic equipment. These structures are light, cost effective and easy to install, render-ing the solution suitable for floating solar park in closed waters where the wave conditions are much more benign than offshore. Moving low displacement floaters offshore brings along a number of challenges mainly related to the metocean condition and the water depth of the installation. Currently many pilot projects are being developed with aim at collecting full scale data on the behaviour of FPV in terms of mooring response and hydrodynamic performance.

6.3.4 FOWT

Floating offshore wind turbine is a sector which is rapidly evolving aiming to become a key market player in the coming decade. Currently, there are a few floating wind farms in operation (e.g. Hywind, Provence Grand Large, Windfloat Atlantic, Kinkardine) and considerably more demonstrators (e.g. TetraSpar, Floatgen, DemoSATH). New projects are planned worldwide.

Due to its floating nature, FOWT presents new and additional dynamics to bottom fixed off-shore wind turbines due to hydrodynamic and aerodynamic motions which have an impact on wind turbine accelerations, therefore, monitoring the dynamics (i.e.

position, velocities and accelerations) are crucial for floating offshore wind platforms. Nowadays, there is a significant availability of high-resolution GNSS systems and MRU sensors for low and wave frequency monitoring respectively.

To ensure station keeping, floating wind turbines are equipped with mooring systems are exposed to a wide variety of environmental conditions. Mooring systems are not always as reliable as we want them to be and therefore mooring failures do happen. This is an attention point for floating wind farms as hundreds of mooring lines will be installed and ensuring their integrity in a cost effective way is a challenge. One of the integrity measures taken by the oil&gas industry is to install loadcells (direct mooring line tension monitoring) or incli-nometers (in-direct mooring line tension monitoring) in the mooring lines. This approach is feasible for a single highly profitable floater of the oil&gas industry but not for every moor-ing line and wind turbine in a wind farm. MARIN recommends to only instrument the moor-ing lines of the demonstrators and a few wind turbines of a wind farm and use a mooring integrity framework to calculate the mooring line tensions of the remaining floating wind turbines. This mooring integrity framework comprises a numerical model of the mooring lines and uses the measured motions at the fairlead to determine the mooring line tensions. MARIN has instrumented the TetraSpar floating wind turbine extensively to validate the mooring integrity framework. Together with other floating wind turbines, this work is being done with the industry as part of the MoniMoor Joint Industry Project.

6.4 Conclusions and Recommendations

Compared to numerical methods, model testing offers significant advantages in terms of intuition and reliability, yet it also encounters numerous challenges. Model tests systems are typically carried out at scales ranging from 1:10 to 1:100. Results are Froude scaled. Compared to numerical methods model tests will typically have a better representation of physics. For large systems, simplification of the model may be necessary, particularly when dealing with very small scales, where accurately scaling individual floating bodies becomes challenging. Additionally, the actual ocean conditions, which include high temperatures, wind waves, salt spray, ocean currents, corrosion, biological attachment, and extreme natural disasters, are extremely complex. Therefore, it is difficult to achieve comprehensive and accurate simulations in a test tank. Full scale measurements include all relevant physics as well as the actual structure. An important drawback here is that the conditions under which the measurements are done cannot be chosen and are more difficult to determine than is the case for numerical methods or model tests. The triangle between numerical methods, model testing and full scale measurements offers the full learning potential. Given the importance of ocean space utilization and the relative inexperience in the topic it is highly recommended to start up (cooperation) projects that include all three aspects of this triangle.

Scholars have conducted experimental research and numerical calculations on different types of floating bridges, analyzing their motion response, connector load, and mooring cable tension. However, in the design and testing of floating bridges, terrain and geological conditions are rarely considered, and the impact of terrain changes on floating bridges is not taken into account. Especially in nearshore areas where terrain changes greatly and water depth gradually becomes shallower, wave parameters will undergo significant changes, and the impact on the motion response of floating bridges and mooring cable tension cannot be ignored. There are few reports on the model test of floating bridges under the combined action of vehicle load and wind, wave and current loads, mainly due to the difficulty of controlling vehicle load in the test.

As a new type of floating structure, the characteristics of aquaculture platforms are different from traditional marine platforms. In order to solve the problem of predicting the overall dynamic response of net and floating bodies, model experiments were conducted on different types of aquaculture platforms to explore the motion response and mooring cable tension of aquaculture platforms under different wave environmental conditions. However, there has always been a challenge in the scaling test of aquaculture platforms, as the scaling ratio of the net and floater is difficult to maintain consistency, resulting in deviations when converting to the actual structure. In addition, the connection between the net and frame structure of the aquaculture platform is difficult to simulate finely in model experiments. These issues need to be taken seriously in the next step of research, and new model testing techniques or full-scale experiments should be developed to solve the above problems.

Structural strength and fatigue life are also crucial factors for the system. In arrays, FPVs are connected via connectors, and wave-induced motion can cause fatigue on these connect-ors. How to appropriately simulate these smaller-scale connectors in scale tests is a worth-while consideration. Furthermore, power output is a vital component of the energy system. Most FPV experimental studies solely focus on the installed power capacity calculated based on the number of photovoltaic panels. Taking into account the actual dynamic power output under the influence of shelter and motion effects is also a challenging work.

In the future, the technical feasibility of tank test design needs to be further improved by considering a larger array model, selecting recyclable composites with higher strength and durability, integrating system components including mooring and anchoring, and evaluating the energy performance of the array under operating conditions. For a large FPV system, it may be necessary to simplify the model. In this case, it is recommended to reduce the num-ber of floating bodies in the model, rather than completely scale according to the full-size structure.

Given that OSUS typically operate at sea for extended periods and are generally large in size, long-term structural health monitoring is essential. This monitoring helps personnel take timely measures to prevent accidents. An effective monitoring system can enhance overall management standards and extend the operational lifespan of structures by tracking their real-time response to various loads. The subsequent research work will give priority to collecting actual data through the implementation of on-site monitoring system to verify the effectiveness of OSUS designs.

7 General Requirements, Interpretations and Standards

7.1 Introduction

General requirements, interpretations and standards used in the design for safety, reliability and serviceability of the offshore structures for different ocean space utilization purposes have been issued by many classification societies. In some cases, they are specific to particular types of structures, such as floating fish farms, and in other cases are general, and refer to their rules for ships or for drilling units. Some countries have developed national regulations and standards for these unusual types of structures, but there are no international regulations. Because of the diverse nature of structures for ocean space utilization, there is no single industry that has developed standards, although there are efforts underway within sectors of the field.

This chapter cites some references that were published prior to the 2020 ISSC because they were not previously cited in a report of the congress as applicable to ocean space utilization.

7.2 General Requirements

The diversity of this field precludes any single standard that would be applied to all structures used for ocean space utilization. In almost all cases, prescriptive rules do not apply, and the design of these structures requires an analysis using first principles, which is then reviewed by the appropriate regulatory body.

7.3 Classification Society Rules and Guidelines

7.3.1 American Bureau of Shipping

ABS has published the Rules for Building and Classing Mobile Offshore Units (MOU Rules) specifically, Part 3, Hull Construction and Equipment (ABS 2024a). These rules contain specific formulae for determination of scantlings for conventional structure, such as watertight bulkheads and flats. Otherwise, more general guidance is provided for determination of overall loads, identification of critical structural areas, and fatigue analysis. Wave loads are to be based on acceptable calculations, model tests or full-scale measurements, and two indices are provided covering shallow-water wave theory and deep-water wave theory.

ABS has published their Guide for Building and Classing Aquaculture Installations (ABS 2022), which provides class requirements for the design, construction, installation and survey of non-self-propelled, sited aquaculture installations.

ABS recently revised their Guide for Building and Classing Floating Offshore Wind Turbines (ABS 2013). This guide provides criteria for the design, construction, installation and survey of permanently sited floating offshore wind turbines, addressing the floating substructure, the station keeping system, and onboard machinery. ABS has also published their Requirements for Nuclear Power Systems for Marine and Offshore Applications (ABS 2024c), which has been developed to provide requirements for design, construction, and survey for class review and approval of ships, barges, offshore units, and installations having onboard nuclear power system installations. The ABS Rules for

Building and Classing Offshore Installations (ABS 2023a) provides structural requirements for offshore electric generating plants, but not the requirements for the electrical generation facilities. These are covered by Part 4 of the ABS Rules for Building and Classing Marine Vessels (ABS 2024d) and the Rules for Building and Classing Mobile Offshore Units. The ABS Requirements for Use of Supercapacitors in the Marine and Offshore Industries (ABS 2022) also apply to power generation. The ABS Requirements for Offshore Substations and Electrical Service Platforms (ABS 2023b) provides requirements on the design, construction, and survey of offshore installations with equipment installed onboard primarily for the transmission of power to an onshore substation or power grid serving other assets or locations, but not for transmission of power to shore from wind farms or other power-generation offshore installations.

7.3.2 Bureau Veritas

Bureau Veritas has published their Rules for the Classification of Offshore Units (Veritas 2023). In these rules an offshore unit is defined as a floating unit for use in connection with offshore recovery of subsea resources including but not limited to hydrocarbons and other units may be also considered as offshore units where deemed appropriate by the Society. The BV NR572, Classification and Certification of Floating Offshore Wind Turbines (Veritas 2024) covers floating substructures supporting vertical axis wind turbines and multiple wind turbines. Additionally, BV rules include Classification of Mooring Systems for Permanent and Mobile Offshore Units (Veritas 2024). BV also has NR580 Rules for the Classification of Floating Establishments (Veritas 2012), which applies to stationary berthed non-propelled floating units equipped for missions such as activities intended for public and accommodation facilities (e.g. hotel, restaurant, hospital, museum, or sport center) moored or anchored in smooth stretches of water.

7.3.3 China Classification Society

The China Classification Society has published Rules for Classification of Mobile Offshore Units, (CCS 2023a), which primarily addresses installations intended for oil and gas exploration, production, and storage. However, there are sections covering leisure units, geological survey units, riprap levelling units, scientific research units, rocket launch/recycling units, power generating units, and offshore wind turbine service units, The bulk of these rules contain specific requirements for structure, structural connections, loads, mooring, and equipment requirements. The special sections provide a few special requirements for these structures, but refer back to other parts of the rules for most requirements. The CSS Rules for Classification of Offshore Floating Installation (CCS 2023b) primarily addresses installations intended for oil and gas exploration and production. However, it contains sections Part X: Special Requirements for Floating Fisheries, Part XI: Special Requirements of Floating Leisure Installations, including underwater sightseeing cabins, Part XII: Special Requirements of Floating Power Generation Installation, including wind power, wave energy, photovoltaic power, tidal current, and tidal energy, Part XIII: Special Requirements of Floating Substation Installation, Part XIV: Special Requirements of Floating Parking Installation, and Part XV: Special Requirements of Floating Observation and Communication. These special parts contain a few special rules pertaining to these structures, but mostly refer back to other places in the rules for requirements.

7.3.4 Class NK

Class NK has published under their Rules for the Survey and Construction of Steel ships Part P, Mobile Offshore Drilling Units and Special Purpose Barges and Part PS, Floating Offshore Facilities for Crude Oil Petroleum Gas Production, Storage, and Offloading. The definition of a special purpose barge is, "steel-made ships and floating structures, and those are generally positioned for a long period of time or semi-permanently at a specific sea area or fixed at a specific sea area." It would appear that these rules were developed with oil and gas facilities but could be adopted for other offshore structures for ocean space utilization. General guidance is given for loads, including the design wave, and other guidance for design of structure.

7.3.5 Croatian Register of Shipping

The Rules for the Classification of Ships of the Croatian Register of Shipping (CRS 2023) define a ship as "a ship, vessel, unit or offshore structure of any kind, whether or not connected to the shore or sea/riverbed." Therefore, offshore structures intended for ocean space utilization fall under these rules. However, the rules are written for the design of conventional ship structure, so interpretation would be needed in applying them to unusual offshore structures for ocean space utilization. Perhaps the section for barges and pontoons could apply, or else there are provisions in the rules for design by direct calculation using the finite element method.

7.3.6 DNV

DNV has several rules that could pertain to offshore structures intended for ocean space utilization. These include self-elevating units, wind turbine installation units and lift-boats; floating fish farming units and installations; and floating wind installations. Of special note are the rules for Floating Infrastructure Units and Installations DNV-RU-OU-0503 (DNV 2024). This document covers classification of floating infrastructure installations and buoy installations of the following design types: ship-shaped installations, barge/pontoon installations, column-stabilised installations, cylindrical installations, self-elevating installations, tension leg installations, and hull shapes other than the above. The infrastructure for which the rules are intended includes housing, office spaces, hotels, parking, parks, and industrial plants of various sorts, for example solar energy systems, wave energy systems, and desalination of sea water plants. Specific rules or guidance is not given for the design and fabrication of structure. Rather cross-reference is given to other DNV rules for specific types of offshore structure. These rules have recently added a section on floating spaceport. DNV-ST-0119 Floating Wind Turbine Structures (DNV 2021) specifies general principles and requirements for the structural design of floating wind turbine structures. Other DNV rules include DNV-RU-OU-0503, Floating Fish Farming Units and Installations (DNV 2022) and Recommended Practice DNV-RP-0584, Design, Development and Operation of Floating Solar Photovoltaic Systems

(DNV 2021). These are structures where solar panels are placed on a body of water. The DNV recommended practice focuses on systems located in sheltered, inland water bodies, while still being applicable for near-shore locations. DNV Rules for Classification of Offshore Units, DNV-RU-OU-0104, Self-elevating Units, including Wind Turbine Installation Units and Liftboats (As 2017) are intended for fixed offshore structures, including wind turbines and accommodation structures. DNV rules for classification of offshore units, DNV-RU-OU-0512, Floating Wind Installations (DNV 2023) covers most types of floating wind turbine installations, including collum-stabilized units, but are not to be manned structures.

7.3.7 Indian Register

The Indian Register of Shipping has published Rules and Regulations for the Construction and Classification of Floating Offshore Units (IR 2024). These rules are applicable to ship-shaped offshore units that are engaged for operations such as the production, storage and offloading of oil but excluding the ship types defined the Rules and Regulations for Construction and Classification of Steel Ships and Rules for Bulk Carriers and Oil Tankers. As such, they are not applicable to the offshore structures that will be used for ocean space utilization.

7.3.8 Korean Register

The Korean Register of Shipping has published their Guidance for Floating Structures (KR 2024) for floating hotel, floating restaurants, and floating performing place, etc. No specific rules and guidance are provided for structure. Reference is made to the relevant chapters of the KR rules. This guide specifically covers mooring and anchoring of these floating structures.

7.3.9 Lloyds Register

Lloyds Register has published their Rules and Regulations for the Classification of Offshore Units (LR 2023). However, these rules apply only to offshore units used in oil and gas, and there are no rules applicable to the offshore structures that will be used for ocean space utilization. They have also published their Recommended Practice for Floating Offshore Wind Turbine Support Structures (LR 2024), which covers floating offshore wind turbine support structures and dynamic cables but not the remainder of the turbine structure. These guidelines apply only if the unit response and design of the support structure are likely to be dominated by the presence of a wind turbine, rather than other equipment such as that associated with hydrogen production, the unit is normally unattended, and the unit does not have an increased importance or criticality (for example, the unit is a substation).

7.3.10 Polish Register of Shipping

The Polish Register of Shipping has published their Rules for the Classification and Construction of Mobile Offshore Drilling Units (PRS 2023) but these only apply to vessels intended for oil and gas or undersea mining.

7.3.11 RINA

RINA has Rules for the Classification of Steel Fixed Offshore Platforms (RINA 2015) and Rules for the Classification of Floating Offshore Units Intended at Fixed Locations and Mobile Offshore Drilling Units. (RINA 2024). However, according to those rules a "Floating unit means a unit permanently moored on the site by means of a fixed or disconnectable mooring system and intended for the storage, production and offloading of hydrocarbons." Therefore, they do not apply to units for ocean space utilization.

7.3.12 International Association of Classification Societies

There are no specific IACS documents intended for vessels used in ocean space utilization.

7.4 National and International Standards

The Norwegian standard NS 9415 was developed to specify technical requirements for dimensioning, design, installation, and operation of floating fish farming installations. Although it primarily pertains to nearshore fish farming installations, it provides an approach for offshore sites (Standard 2003). The Scottish government developed a technical standard to minimize fish escapes (Group 2015), which covers design, construction, materials, manufacture, installation, maintenance and size of equipment. ISO 16488 (2015) suggests a general method that can be used for the systematic analysis, design, and evaluation of net cage marine finfish farms. The methodology presented in this international standard allows assessing the suitability of floating structures, nets, and mooring equipment for a given finfish farm and its environment.

In January 2023, the U.S. Department of Interior finalized the transfer of regulations governing offshore renewable energy activities from the Bureau of Ocean Management to the Bureau of Safety and Environmental Enforcement (BSEE). This transfer includes evaluating and overseeing facility design, fabrication, installation, safety management systems and oil spill response plans. API RP 2A-WSD, Twenty-first Edition, has again been incorporated by reference into 30 Code of Federal Regulations Part 285 – Renewable Energy and Alternate Uses of Existing Facilities on the OCS (Watkins et al. 2024).

Regulation of floating cities faces a lack of a legal definition of a floating city. Is it legally part of the adjacent city and how should it be considered when deployed offshore in marine areas beyond national jurisdiction? Should it be treated as real property (land) or personal property (chattel)? (Wang 2023) addresses these and other issues that have implications for the requirements and standards to be applied to these structures.

The International Organization for Standards (ISO) has published ISO 29400 Ships and Marine Technology — Offshore Wind Energy - Port and Marine Operations (ISO 2020), which provides comprehensive requirements and guidance for the planning and engineering of port and marine operations of offshore wind farms. An ISO committee is currently producing a draft of ISO/WD 25249-1, Corrosion protection of offshore wind structures.

7.5 Industry Standards

Because of the many different activities and diverse structures employed for ocean space utilization there is no central industry to organize all applications and therefore no unified industry standards. However, for certain aspects of the broad field, there are industries developing and there are also allied industries that have developed or are currently developing industry standards. One example is the offshore oil and gas industries, whose standards for offshore structures can be applied to other large floating structures. The International Association of Oil and Gas Producers has developed Joint Industry Project JIP-35, which aims to achieve industry-level standardization. Its focus is on offshore structures design specifications (IOGP 2024). Eleven IOGP member companies participate in developing these specifications, with the objective to leverage and improve industry level standardisation for projects globally in the oil and gas sector. The work has developed a minimised set of supplementary requirements for the design and operation of offshore structures based on a critical review of the participating members' company specifications, building on recognised industry and international standards. The task covers 11 subdisciplines, or which three, weight management, station keeping, and metocean could pertain to large floating offshore structures.

The American Petroleum Institute has formed Climate Action Framework (CAF) initiative to accelerate technology to reduce emissions while meeting growing energy needs. Within that initiative with the Offshore Operators Committee, they have developed Recommended Practice 75 W, Safety and Environmental Management System for Offshore Wind Operations and Assets (API 2024). This recommended practice provides companies engaged in offshore operations in the offshore wind energy sector with a framework for the establishment, implementation, and maintenance of a Safety and Environmental Management System (SEMS) to manage and reduce risks associated with safety and the environment to prevent incidents and unplanned events. This recommended practice applies, in part or whole, to companies engaged in offshore operations, through the project life cycle.

Hogan-Lovells. (2022) reviewed the regulatory framework for offshore wind turbines in fourteen countries. The document addresses issues such as business aspects, leasing of offshore sites, tariffs, and offtake of power to shore. There is also limited discussion of the regulation of structural issues.

7.6 Interpretation of Requirements, Standards, Rules, and Regulations

Many of the requirements, standards, rules and regulations that may pertain to an offshore structure intended for ocean space utilization purposes were not written with that specific use in mind. Therefore, interpretation of existing documents is needed on the part of designers, builders, and regulators is needed.

A review of standards and guidelines for design and analysis of offshore fish farms was made by (Chu et al. 2023). They cite rules and guides published by ABS and DNV, and cite rules published by BV for nearshore waters where wave height does not exceed two meters. They state that BV provides classification services for offshore fish farms based on their existing rules and standards. Chu et al. state that LR is developing rules for offshore fish farming that at the time of their writing were not published. They cite a book chapter by Barker (1990) which presented LR's principles of approval for offshore fish pens that takes a similar approach to other maritime classification groups.

Although floating bridges either or permanent have been used for centuries, no specific rules or regulations have been developed for them. Rather, modifications of existing standards have been used by designers. Lwin (2000) reports that the replacement bridge over Lake Washington in Washington state in the U.S. was designed in accordance with (Officials 1973) requirements (latest update in 2002 cited in references below), but because AASHTO had nothing specific for floating bridges, the requirements were modified by the bridge designers in several ways, including wind and wave forces and damage from allision or flooding. The International Federation for Structural Concrete (2020) has published Guidelines for Submerged Floating Tube Bridges (FIB 2020), which gives wide information on the design, construction and management of these structures.

ABS reviewed the process by which they classified a floating wind farm offshore of Portugal (de Almeida et al. 2022). For classification of these newer-type structures, a combination of several sources was used, including in-house research, existing rules and guides, international standards, industry feedback, offshore industry best practice, and project experience. For this project, the ABS Guide for Building and Classing Floating Offshore Wind Turbines (ABS 2024) was the main document, supported by several other ABS guides and guidance notes. In this case, several other rules and guides were used because the floating base and the topsides units were fabricated separately, and the bases had to be treated as barges while being towed to the assembly site. ABS also worked with the Portuguese flag administration, "Direção Geral de Recursos Naturais, Segurança e Serviços Marítimos" (DGRM) to comply with their requirements throughout the project.

Rawson and Huang et al. (2022) reviewed the process by which the U.S. Coast Guard evaluates and approves vessels intended for the support of offshore wind turbines. Floating wind turbines are not specifically addressed because none have been or are intended for installation in U.S. waters, but the process used for approval described would be applied for those structures. In addition, the paper describes interaction with ABS, the U.S. Bureau of Ocean Energy Management (BOEM) the U.S. Department of Energy, U.S. Department of Commerce, and industry stakeholders.

7.7 Conclusions and Recommendations

The field of ocean space utilization is rapidly evolving as new structures are being designed and built. In many cases there are not existing rules and regulations to govern their design and construction, certainly not to the extent of conventional ships such as containerships and tankers. In many cases rules developed for other applications such as oil and gas exploration must be modified to suit the newer units. Often, a first-principles design approach is required, with some guidance as loads, mooring requirements, and similar areas. In other cases, such as floating hotels, recognition of these structures is made in the rules, but reference is made to other rules that need to be modified to suit these newer types of structures. Trends of new rules being issued every year indicate that as the industry develops, new rules will emerge from classification societies and other organizations. Encouragement should be made to all such organizations to develop the rules and regulations for structures used for ocean space utilization so that designers will have solid guidance to their development.

8 Risks and Reliability

8.1 Introduction

In recent years, the development of coastal land has expanded. Therefore, some conceptual designs involving floating bodies have occurred, such as floating airports and ports, floating ecopolises, floating aquaculture farms and the floating tourism platforms) (Xia et al. 2000b). The new technical challenges have emerged as the research and application of OSU floating bodies in depth, including (1) the description of complicated environment; (2) the analyses of hydroelastic responses of multi-module floating structures in inhomogeneous waves; (3) the design and safety assessment of the weakest parts of the OSU floating bodies, the connectors between modules; (4) the design and safety insurance of economic and effective mooring system; (5) the long-term anti-corrosion measures (Wu et al. 2021). Those challenges all focus on the safety and reliability of OSU floating bodies. Therefore it is necessary to review risk assessments, reliability analyses, and structure safety design criteria in the design and construction of floating bodies for ocean space utilization (OSU).

8.2 Risk Analysis for OSU Floating Bodies

Risk analysis method was originally introduced from nuclear industry to marine structures since 1990s. The Piper Alpha disaster in 1988 was a wake-up call for offshore industry practitioners. Since then, risk analysis was started to be used in offshore platforms to avoid such accidents. The offshore Safety Case published by the UK Health and Safety Executive was notably a type of risk analysis for preparing offshore installation (Wang and Gu 2021). In re-cent years, International Maritime Organization (IMO) have used the risk analysis method as a tool for rule-making process and established the formal safety assessment (FSA) methodology (IMO 2015).

Basically FSA involves five steps: identification of hazards, risk analysis, risk control options, cost-benefit assessment and recommendations for decision making (IMO 2015). Although the steps for risk analysis are common procedures for ships and marine structures, the framework of risk analysis for OSU floating bodies also need specific considerations. Partly because the OSU floating bodies with multi-purposes have a large number of crews and people on board, which is different from cargo ships and offshore structures. Another reason is that there exists novel designs and the traditional design approaches may not be sufficient for OSU floating bodies.

Risk can be expressed as a combination of the frequency and severity of a consequence. Generally the risk assessment ranged from operational risk to floating body performance risk (stability failure, structure failure etc.) based on accident scenarios. The risk assessment related to floating body performance contributes to the risk-based design. Wang and Gu (2021) established an risk-based approach to achieve the ultimate strength design criteria of VLFS airports. In this paper, event tree risk analysis methods and reliability assessment methods were used as tools to calculate ultimate strength design criteria. A reasonable target reliability range for VLFS was determined by risk assessment methods and principles considering the structure failure probability and consequences. The longitudinal ultimate strength for the typical sections of VLFS was calculated by applying the simplified progressive failure method. When the structure safety level and target reliability are determined, the ultimate strength design criteria can be calculated by limit state equation of ultimate strength failure mode.

In order to guide design for offshore wind industry and offshore aquaculture industry, Gudmestad (2023)discussed the safety levels for marine fish farms and offshore wind turbines platform bed on life safety and consequence analysis according to ISO 19902 (2007) and ISO 19906 (2010). The wind turbines and transformer stations safety level is suggested to be designed to exposure level 10^{-3} and the offshore substation is recommended to be 10^{-4}. Considering the storms survival and environmental damage, the fish cage safety level set to be 10^{-4}. This paper also suggested to carry out a qualitative risk analysis to determine the safety level for an offshore platform.

Zhang et al. (2021) conducted storm damage risk assessment for offshore cage culture based on damage mechanisms of sea cages subjected to storm waves. Risk levels were classified into three grades based on the load/strength ratio, which corresponds to severe dam-age, moderate damage and slight or no damage.

Yeter et al. (2020) provide a framework to conduct a life-cycle cost assessment accounting for the optimal inspection and maintenance policy of monopile offshore wind turbines based on risk and reliability assessment. The fatigue reliability was chosen as the indication for life-cycle inspection. The event tree method is employed to assess the expected cost of failure to be included in the capital investment as the structural risk premium, and the total expected cost to be included in the operational cost. Tremps et al. (2024) applies the Failure Mode, Ef-fects, and Criticality (FMECA) methodology to identify the most critical failure modes for monopile offshore wind structures. Despite different failure modes with different causes and mechanisms, all could be classified into four categories: corrosion, fatigue, connection failure and deformation, buckling and displacement.

In order to solve the problem of correlated failures for floating offshore wind turbines, Sun et al. (2023)proposed a correlated-Failure Mode and Effect Analysis (FMEA) method to identify critical failures of floating offshore wind turbines with correlated failures by including a model of correlated failures derived from the analysis of real data. The Risk Priority Numbers of failure items, including systems, components, failure modes, and failure causes of floating devices were calculated using the proposed model. Results shows that the wind turbine is more critical than other systems, with a contribution of over 43% RPN to the total; "Broken mooring lines" and the floating foundation are the top failure mode and components; 38 cru-cial failure causes such as wear, fatigue, lubrication fault are identified.

Claus and López (2022) discussed the key issues in the design of marine floating photovoltaics (FPV) systems under marine environment. Some risk factors are identified for floating photovoltaics, such as extreme environmental loads (mainly wind, waves and currents), saltwater corrosion and biofouling. The measures to avoid those risk factors are also suggested, in-cluing using plastic materials, Antifouling coatings and sustainable materials and Fibre-reinforced polymers (FPR) to construct marine FPV.

The use of floating photovoltaic systems in freshwater and marine environments is forecast to increase dramatically worldwide. However, there are few studies focusing on the environment impact of floating solar photovoltaic systems. Benjamins et al. (2024) studied the potential environmental impacts of floating solar photovoltaic systems, which may be used in severity assessment of consequence during risk analysis. Five potential environmental impacts are as follows:

1) abrupt changes to light levels in the water column below the FPV structures
2) impacts on hydrodynamics and water-atmosphere interchange
3) energy emissions
4) impacts on benthic communities, including artificial reef effects and accommodation of invasive non-native species
5) impacts on mobile species (fish, marine mammals etc.).

8.3 Reliability Assessment for OSU Floating Bodies

8.3.1 Structure Failure Modes

Structure failure modes is different due to the various types of OSU floating bodies. Some theoretical and experimental studies on ultimate strength for semisubmersible-type VLFSs were performed to research the structure failure modes. The semisubmersible-type VLFSs is subjected to multi-axial loads in complex marine environment, which makes brace strut and connector foundation structure with weak stiffness easy to damage, thus affecting the safety and reliability of the very large floating structure. Model test of the brace strut of semisubmersible-type VLFS ultimate strength under bending and shear loads was carried out (Zhao et al. 2021, 2022). Based on the mixed similarity theory of model design, an ultimate strength test model of the brace strut structure under bending and shearing load was designed. By comparison and analysis of results from model test and numerical simulation method, the correctness of the numerical simulation method was verified. As shown in Figs. 22, 23, and 24. The research results can provide support for the design and safety reliability assessment of the VLFS under bending and shear loads. Another type of VLFSs is large floating tourism platforms, China Ship Scientific Research Center (CSSRC) have designed the ocean heart large floating tourism platforms. This type of VLFS also encounter multi-axial loads due to large horizontal and longitudinal scale, which results in ultimate strength under bending and torsion loads (X. Wang et al. 2022).

Fig. 22. Semi-submersible-type VLFS Brace strut model test

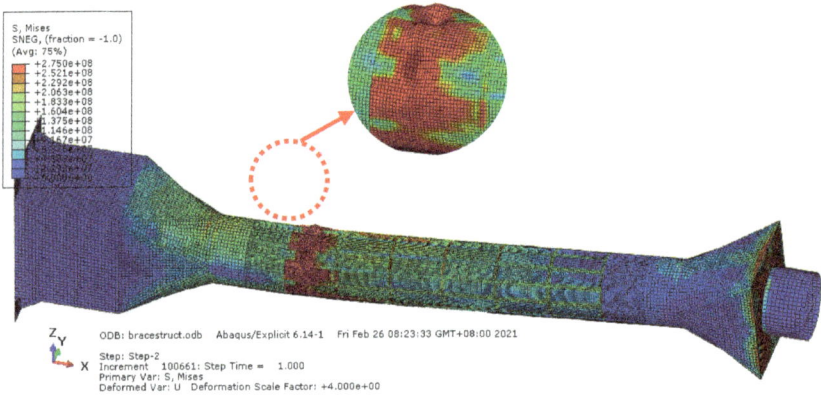

Fig. 23. Failure mode of the brace strut ultimate strength under bending and shear loads

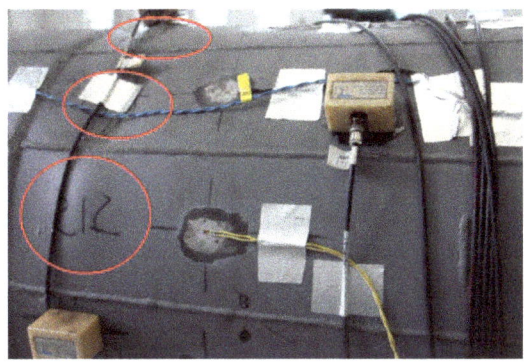

Fig. 24. Deformation results for VLFS Brace strut model test

As for other OSU floating bodies, Aryai et al. (2021)presented the common failure modes used for the structural reliability assessment of aquaculture equipment and offshore wind turbines. Aquaculture equipment, the offshore renewable energy platforms, Semi-submersibel VLFS airport and large floating tourism platforms have their own failure modes, as show in Table. Those failure modes for sub-system may have co-relations with each other. Therefore there is a tough task for calculating the joint-failure probability among the co-related failure modes (Table 2).

Table 2 Common failure modes used for typical OSU floating bodies: (Aryai et al. 2021; Li and Soares 2022; Wang et al. 2022b)

Floating bodies	Sub-system	limit-states
Aquaculture	Grid mooring system for net cage system	Fatigue
	Floating collar	Tensile strength Buckling Fatigue
	Mooring chains	Ultimate strength
Offshore Wind Turbine	Foundation	Fatigue Buckling Ultimate Strength
	Anchors	Crack growth Ultimate Strength
	Blade	Excessive displacement of blade tip Overload of blade root Fatigue
	Gears	Fatigue Corrosion

(*continued*)

Table 2 (*continued*)

Floating bodies	Sub-system	limit-states
	Mooring line	Fatigue Ultimate strength
Semi-submersibel VLFS airport, large floating tourism platforms	Floating body	Multi-axial ultimate strength under complex loads
	Connecter	Fatigue Ultimate strength

8.3.2 Uncertainty Analysis

Based on the uncertainties sources, uncertainties could be classified into two categories: (a) aleatory uncertainty and (b) epistemic uncertainty (Parunov et al. 2022). Some OSU floating bodies were deployed in the region near islands and reefs, which may induce wave-induced loads variability with the complicated environments surrounding. The wave-induced loads variability assessment directly influence the reliability-based method in Structural Reliability Analysis (SRA). Wu et al. (2021)described the new technical challenges facing by floating structures deployed near islands and reefs, including the description of inhomogeneous waves and the analyses of hydroelastic responses. The wave conditions near islands and reefs are influenced by the complicated geographic surroundings, which are different from those in open seas. Ding et al. (2021) conducted an 8-module VLFS basin model test in the waters near islands and under complex seabed, which shows that the inhomogeneous encounter waves near islands and reefs have significant impact on the responses of multi-module VLFS. Chen et al. (2020) observed wave bimodal spectrum in the lagoon near islands and reefs, which further verified the statement by by illustrating the features of a mixture of unimodal and bimodal spectrum near islands and reef. The wave conditions brings uncertainties of wave loads prediction in terms of wave spectrum and wave scatter diagram, which have an impact on short-term and long-term wave loads in spectral analysis.

Except for the uncertainties from wave conditions, the floating body itself also causes uncertainties due to the assumptions of rigid body or large flexibility. A rigid body assumption is suitable for most cases in small or middle scale of ships and floating structures. However, the large floating bodies show characteristics of flat expand type with large length and width, large flexibility and discontinuous multi-free-surface water region within the floating body (Ling et al. 2022; Wang et al. 2024b). Those characteristics trigger uncertainty of wave loads that have not been studied in detail.

Ramezani et al. (2023) review the uncertainties for the structural design of floating offshore wind turbines. Aerodynamic and hydrodynamic uncertainties must be taken into account for the structural design of FOWTs. Uncertainties in structures, materials, manufacturing, and construction needs to be clarified for assessing probability of failure. Growing uncertainties over time, such as fatigue and corrosion damages, have not taken into account in the prevailing practice. The long-term environmental loads and fatigue reliability are commonly assessed without accounting for the changing nature of uncertainties over time.

8.3.3 Reliability Analysis Methods

Structural reliability may be expressed as the probability that a structure works with a predefined extent of safety throughout its service-life (Aryai et al. 2021). After obtaining the limit-state functions and stochastic parameters, the reliability of the structure for a predefined failure mode can be obtained by using reliability analysis methods such as First-Order Reliability Method (FORM), Second-Order Reliability Method (SORM), Outcrossing methods, and Monte Carlo Simulation (MCS).

With the long-term serviceability of large floating structures, general corrosion may occur in stiffened plates and cause degradation of longitudinal or local strength. Therefore the ultimate strength is decreasing in the long-term serviceability of large floating structures. Wang et al. (2022) conducted a studies on time-variant reliability under corrosion-caused strength degradation for large floating tourism platforms. Ultimate strength under discrete stochastic extreme load events for the large floating structures is chosen as time-dependent limit state. Results show that severe corrosion deterioration degrades ultimate strength and reduces time-variant reliability over time.

Aryai et al. (2021) present a review on reliability of multi-purpose offshore platforms (MPOPs), mainly focusing on Australian aquaculture, offshore engineering and renewable energy industries. This paper points out that the offshore floating bodies facing harsh offshore environment is one of the main challenges in developing the MPOPs. Employing structural reliability analysis methods for assessing the structural safety of the novel aquaculture-MPOPs comes with different limitations. The factors affecting the reliability of the MPOPs are as follows:

- Data scarcity
- Load prediction with different methods
- Uncertainties within the models, environmental and design parameters
- Failure modes identification specific to the novel design
- Design standards and current safety level

Zhao et al. (2024) studied fatigue reliability of floating offshore wind turbines under the random wave environmental conditions based on surrogate model. Short-term fatigue damages are estimated by the surrogate model. And Monte Carlo simulation is employed to assess the fatigue reliability. Yu and Xu (2024) studied fatigue reliability assessment of the tower floating offshore wind turbines under correlated wind-wave-current loads. The fatigue risk of the floating offshore wind turbines increases when considering the parametric correlation.

8.4 Conclusions and Recommendations

The OSU floating bodies are novel marine structures with specific environmental and design parameters. The specific failure modes for those OSU floating bodies can be identified by FMEA or FMECA under risk analysis framework. The attempt to co-relate different failure modes has been adopted by some researchers. However, multi-failure and system reliability assessment for the floating body as a whole is still a problem to be explored. In addition, the wave load and strength uncertainties used in reliability analysis need to be quantified considering environment conditions and other influential factors. Several recommendations for research are as follows:

- Establishing a risk framework for OSU floating bodies, with merging reliability methods to assist design and engineering application of OSU floating bodies.
- Collecting failure data from experiments and OSU floating structure in service.
- Ascertaining the safety level and target reliability for different OSU floating bodies and setting the target safety level at a sub-system that is in line with the total risk acceptance criteria of the OSU floating bodies.
- Identifying design-specific failure modes and quantifying the load and structure uncertainty.
- Proposing a system-based reliability analysis approach to assess OSU floating bodies reliability as a whole.

9 Conclusions and Recommendations

9.1 Introduction

Ocean Space Utilization Structures (OSUS) serve as foundational infrastructure supporting the development of marine resources. OSUS are typically composed of multiple floaters, with their relative motions are constrained by connectors. Currently, there are several prevailing types of OSUS. For example, infrastructures such as very large floating structures, floating bridges, floating breakwaters have been widely used in coastal areas and near islands due to their convenience, and many scholars have conducted in-depth research on the performance by tank testing and full scale measurements. Meanwhile, offshore fish farm and offshore floating photovoltaic equipment have become the current research focus due to the growing demand for resources.

Compared to the FSI analysis of conventional floating structures, predicting and analyzing the dynamic responses of OSUS remains a challenging task for both industry and academia. The elasticity effects plays a more significant role in the FSI analysis of OSUS. Additionally, external environmental factors, such as the inhomogeneity of incident waves, seabed topology, and shallow water effects, are critical considerations during the initial design stages. Consequently, particular attention must be given to multi-body motion behavior, gap resonance, and the safety of connectors.

9.2 Conclusions

The environmental conditions experienced by structures for ocean space utilization depends on the location of the structure. For structures moored in deep water far offshore, the environment is no different from that of conventional structures. However, large structures moored closer inshore have a greater concern for variable wind, wave, and current conditions along the structure, bathymetry and other nearshore effects including reflection from coastal boundaries and generation of infra-gravity waves. Nonlinear potential flow theory, although more computationally intensive than spectral wave models provide estimates that are more accurate of the nearshore wave climate around a very large floating structure. Coastal current modelling using the finite difference method with either the finite volume coastal ocean model or the regional ocean modelling system can show good agreement with measurements.

The flexibility of many large structures used for ocean space utilization precludes the use of potential flow theory for loads calculation because it assumes a rigid body. The combination of the higher-order boundary element method with potential flow theory for simulating nonlinear fluid dynamics does not resolve strong nonlinearities, such as wave slamming and green water effects. The best results are seen using computationally intensive computational fluid dynamics analysis.

With the rapid advancement of engineering applications for ocean space utilization structures, research on the hydroelastic response of these structures in complex sea environments will require significant effort and innovation. Accurately predicting hydrodynamic loads is a fundamental aspect of designing nonlinear ocean space utilization structures. Potential flow theory and viscous flow theory are the two primary approaches for solving fluid flow and determining structural loads. However, potential flow theory cannot account for fluid viscosity or strong nonlinear phenomena, such as slamming and green water. Consequently, Navier-Stokes equation-based theory in the time domain is gradually becoming the primary method for hydrodynamic prediction.

Most of the hydrodynamics research concentrates on wave-induced loads and elastic displacement of very large floating structures under surface waves. This analysis is typically essential during the initial design stage of a structure. However, it is also necessary to conduct a local analysis at stress concentration points, particularly when considering fatigue and fracture during the configuration of welding lines. Additionally, when constructing an underwater space station, it is crucial to explore the interaction of ocean space utilization structures with internal waves, especially large internal solitary waves.

Given that ocean space utilization structures typically operate at sea for extended periods and are generally large in size, long-term structural health monitoring is essential. This monitoring helps personnel take timely measures to prevent accidents. An effective monitoring system can enhance overall management standards and extend the operational lifespan of structures by tracking their real-time response to various loads.

Connectors joining multi-body large floating structures are of five types: rigid connectors, hinged connectors, flexible connectors, semi-rigid connectors, and adaptive connectors. Various factors in the overall geometry of the structure will make one type of connector better than others, but research continues in all types of connectors to improve their performance. Much of this research addresses the needs of large floating photovoltaic structures, which consist of many separate floating platforms that extend in both directions to form a large surface.

It may be important to solve the coupled elastic-hydrodynamic-mooring problem, as it has been shown that the mooring system influences the structural dynamics (and vice-versa). A careful design of the mooring system has the potential to minimize the structural responses.

The model test and full-scale test of marine structures are important aspects of their design and development, especially for understanding the overall performance of new marine platforms, including motion response, mooring line tension, and structural loads. The model tests and full-scale measurements of floating bridges, aquaculture platforms, floating photovoltaic platforms, and other marine structures not only revealed the dynamic response laws of these floating structures, but also verified the proposed numerical analysis methods, providing reference for the research and design of similar structures.

Because the field of ocean space utilization is relatively new and is evolving, few rules, regulations and standards exist. Regulatory bodies and other organizations are working to develop such documents, so the designer of a structure must be particularly alert to any new documents or changes in existing documents. For the present, an amalgam of guidance developed for different structures as well as first-principles calculations must be used until specific guidance for a particular type of structure is developed.

Risk and reliability analysis is essential for design and manufacture of OSU floating bodies. The novel marine structures with specific environmental and design parameters bring new failure modes. Multi-failure and system reliability assessment for the floating body as a whole is still a problem to be explored. The wave load and strength uncertainties used in reliability analysis need to be quantified.

9.3 Recommendations

- The rapid changes in the environment and their effect on the loading of very large floating structures moored closer inshore need to be understood better.
- Methods are needed to reduce the computational demands of computational fluid dynamics analysis needed for computing hydrodynamic loads on very large floating structures.
- High computational resource demands for Navier-Stokes equation-based theory in the time domain presents a pressing need for the development of effective nonlinear hydroelastic approaches that consider wave nonlinearity, seabed topology, complex mooring systems, and module connectors. Future research could explore the combination of computational fluid dynamics with artificial neural networks, efficient coupling algorithms, and multibody dynamic structural modelling.
- The technical feasibility of tank test design needs to be further improved by considering a larger array model, selecting recyclable composites with higher strength and durability, integrating system components including mooring and anchoring, and evaluating the energy performance of the array under operating conditions. For a large Floating Photovoltaic system, it may be necessary to simplify the model. In this case, it is recommended to reduce the number of floating bodies in the model, rather than completely scale according to the full-size structure.
- Mooring in shallow water continues to pose challenges in wave-energy-intensive locations due to the tendency of catenary systems to exhibit slack-tensioned behavior, leading to high dynamic loads. Optimizing the distribution of mooring loads on OSUS and photovoltaic solar islands, while ensuring the system remains economical and simple for installation and maintenance. Shared mooring systems, which offer a promising solution to save materials and reduce the cost of mooring systems, particularly for arrays of floating structures.
- The characteristics of new type floating structures are different, and model tests and full-scale measurements need to be carried out. For floating bridge tests, it is necessary

to consider the effects of uneven waves and terrain changes in fjords or nearshore areas. Aquaculture platforms need to consider the coupling response between the net and the floating body, as well as the precise measurement of the connection stress between the net and the frame. When testing floating photovoltaic platforms, it is necessary to consider the coupling response between floating photovoltaic and mooring systems, as well as the mutual interference hydrodynamic of multiple floating photovoltaic platforms.

- Regulatory bodies, classification societies and industry groups should continue to develop the rules, regulations and guidance documents for all types of structures being built or contemplated for ocean space utilization.
- Establishing a risk framework for OSU floating bodies, with merging reliability methods to assist design and engineering application of OSU floating bodies may be a new approach to achieve the goal of risk-based design.

References

Aarsnes, J.V., Rudi, H., Løland, G.: Current forces on cage, net deflection. In: Engineering for Offshore Fish Farming, pp. 137–152. Thomas Telford Publishing (1990)

ABS, A.: Guide for Building and Classing Floating Offshore Wind Turbine Installations. The American Bureau of Shipping, Houston (2013) [Preprint]

ABS: Guide for Building and Classing, Aquaculture Installations. American Bureau of Shipping (2022a)

ABS: Requirements for Use of Supercapacitors in the Marine and Offshore Industries. American Bureau of Shipping (2022b)

ABS: Rules for Building and Classing Mobile Offshore Units (MOU Rules): Part 3, Hull Construction and Equipment. American Bureau of Shipping (2024a)

ABS: Requirements for Nuclear Power Systems for Marine and Offshore Applications. American Bureau of Shipping (2024b)

ABS: Rules for Building and Classing Offshore Installations. American Bureau of Shipping (2023a)

ABS: Requirements for Offshore Substations and Electrical Service Platforms. American Bureau of Shipping (2023b)

Agnon, Y., Madsen, P.A., Schäffer, H.A.: A new approach to high-order Boussinesq models. J. Fluid Mech. **399**, 319–333 (1999)

Aksu, S., Price, W.G., Temarel, P.: A comparison of two-dimensional and three-dimensional hydroelasticity theories including the effect of slamming. Proc. Inst. Mech. Eng. Part C J. Mech. Eng. Sci. **205**(1), 3–15 (1991)

Al-Solihat, M.K., Nahon, M.: Flexible multibody dynamic modeling of a floating wind turbine. Int. J. Mech. Sci. **142**, 518–529 (2018)

Aliyar, S., et al.: Numerical coupling strategy using HOS-OpenFOAM-MoorDyn for OC3 Hywind SPAR type platform. Ocean Eng. **263**, 112206 (2022)

de Almeida, P.L., Samuelsson, L.H., Santos, F.B.: Classification of an offshore wind farm. In: Offshore Technology Conference, p. D021S016R005. OTC (2022)

API: RP 75W: Safety and Environmental Management System for Offshore Wind Operations and Assets. American Petroleum Institute (2024)

Ardhuin, F., et al.: Semiempirical dissipation source functions for ocean waves. Part I: definition, calibration, and validation. J. Phys. Oceanogr. **40**(9), 1917–1941 (2010)

Aryai, V., et al.: Reliability of multi-purpose offshore-facilities: present status and future direction in Australia. Process Safety Environ. Protect Transact. Inst. Chem. Eng. Part B. **148**, 437–461 (2021)

Ataie-Assehtiani, B., Najafi-Jilani, A.: A higher-order two-dimensional Boussinesq wave model. J. Coast. Res. **50**(sp1), 1183–1187 (2024)

Bai, T., et al.: Numerical simulation and analysis of initial plume discharge of deep-sea mining. Ocean Eng. **310**, 118794 (2024)

Barker, C.J.: Classification society rules for fish farms. In: Engineering for Offshore Fish Farming, pp. 31–38. Thomas Telford Publishing (1990)

Benjamins, S., et al.: Potential environmental impacts of floating solar photovoltaic systems. Renew. Sust. Energ. Rev. **199**, 114463 (2024)

Benra, F.-K., et al.: A comparison of one-way and two-way coupling methods for numerical analysis of fluid-structure interactions. J. Appl. Math. **2011**(1), 853560 (2011)

Benvenuto, J., et al.: USCG Buffalo sector Boat Basin wave study: optimization of wave attenuators using FUN-WAVE numerical modeling. Ports. **2022**, 681–688 (2022)

Betts, C.V., Bishop, R.E.D., Price, W.G.: The symmetric generalised fluid forces applied to a ship in a seaway. RINA Suppl. Papers. **119** (1977)

Bi, C.W., et al.: Physical model experimental study on the motion responses of a multi-module aquaculture platform. Ocean Eng. **239**, 109862 (2021) [Preprint]

Bidlot, J.-R. et al.: A revised formulation of ocean wave dissipation and its model impact (2007)

Bishop, R.E.D., Price, W.G.: Hydroelasticity of Ships. Cambridge University Press (1979)

Booij, N., Holthuijsen, L.H., Ris, R.C.: The "swan" wave model for shallow water. Coast. Eng. **1** (1996)

Booij, N., Ris, R.C., Holthuijsen, L.H.: A third-generation wave model for coastal regions: 1. Model description and validation. J. Geophys. Res. Oceans. **104**(C4), 7649–7666 (1999)

Bublík, O., et al.: Neural-network-based fluid–structure interaction applied to vortex-induced vibration. J. Comput. Appl. Math. (2023). https://doi.org/10.1016/j.cam.2023.115170

Bungartz, H.-J., et al.: preCICE–a fully parallel library for multi-physics surface coupling. Comput. Fluids. **141**, 250–258 (2016)

Cai, Z., et al.: Waves enter a reef lagoon with double barriers in South China Sea: in-situ measurement and simulation. In: ISOPE International Ocean and Polar Engineering Conference, p. ISOPE-I. ISOPE (2019)

Cai, Z., et al.: Simulation of inhomogeneous inlet waves by the Boussinesq equations. In: ISOPE International Ocean and Polar Engineering Conference, p. ISOPE-I. ISOPE (2022)

Cao, Y., et al.: Modeling and dynamic analysis of integral vertical transport system for deep-sea mining in three-dimensional space. Ocean Eng. **271**, 113749 (2023)

Cavaleri, L., Barbariol, F., Benetazzo, A.: Wind–wave modeling: where we are, where to go. J. Mar. Sci. Eng. **8**(4), 260 (2020)

Cavaleri, L., Bertotti, L.: Accuracy of the modelled wind and wave fields in enclosed seas. Tellus A Dyn. Meteorol. Oceanogr. **56**(2), 167–175 (2004)

Chakrabarti, A., et al.: Boussinesq modeling of wave-induced hydrodynamics in coastal wetlands. J. Geophys. Res. Oceans, 3861–3883 (2017). https://doi.org/10.1002/2016JC012093

Chen, H., Medina, T.A., Cercos-Pita, J.L.: CFD simulation of multiple moored floating structures using OpenFOAM: an open-access mooring restraints library. Ocean Eng. **303**, 117697 (2024a)

Chen, M., et al.: A fully coupled time domain model capturing nonlinear dynamics of float-over deck installation. Ocean Eng. **293**, 116721 (2024b)

Chen, M., et al.: A novel SPM wind-wave-aquaculture system: concept design and fully coupled dynamic analysis. Ocean Eng. **315**, 119798 (2025)

Chen, W., et al.: Study on applicability of Bi-modal wave spectrum in waters near islands and reefs. Huadong Chuanbo Gongye Xueyuan Xuebao/J. East China Shipbuild. Inst. **61**(3), 120–130 (2020)

Chen, W., et al.: Study on wave attenuation on a coral reef flat in SCS by in-situ measurement and simulation. Ocean Model. **182**, 102171 (2023a)

Chen, X., et al.: Numerical and experimental analysis of a moored pontoon under regular wave in water of finite depth. Ships Offshore Struct. **12**(3), 412–423 (2017)

Chen, Y., et al.: Hydroelastic analysis of double-segment floating sandwich structures under wave action. Ocean Eng. **260**, 111993 (2022)

Chen, Y., et al.: A discrete-module-finite-element hydroelasticity method in analyzing dynamic response of floating flexible structures. J. Fluids Struct. **117**, 103825 (2023b)

Cheng, H., et al.: Effects of mooring line breakage on dynamic responses of grid moored fish farms under pure current conditions. Ocean Eng. **237**, 109638 (2021) [Preprint]

Cheng, Y., et al.: Dual inclined perforated anti-motion plates for mitigating hydroelastic response of a VLFS under wave action. Ocean Eng. **121**, 572–591 (2016)

Cheng, Y., et al.: Wave energy extraction and hydroelastic response reduction of modular floating breakwaters as array wave energy converters integrated into a very large floating structure. Appl. Energy. **306**, 117953 (2022)

Cheng, Z., Gao, Z., Moan, T.: Hydrodynamic load modeling and analysis of a floating bridge in homogeneous wave conditions. Mar. Struct. **59**, 122–141 (2018a)

Cheng, Z., Gao, Z., Moan, T.: Wave load effect analysis of a floating bridge in a fjord considering inhomogeneous wave conditions. Eng. Struct. **163**, 197–214 (2018b)

Cheng, Z., Gao, Z., Moan, T.: Numerical modeling and dynamic analysis of a floating bridge subjected to wind, wave, and current loads. J. Offshore Mech. Arct. Eng. **141**(1), 11601 (2019)

Cheng, Z., Gao, Z., Moan, T.: Extreme responses and associated uncertainties for a long end-anchored floating bridge. Eng. Struct. **219**, 110858 (2020)

Chu, Y.-I., et al.: Offshore fish farms: a review of standards and guidelines for design and analysis. J. Mar. Sci. Eng. **11**(4), 762 (2023)

Churchill, B., et al.: Modelling the sea surface of Cardigan Bay. Emerg Minds J Stud Res. **2**, 112–129 (2024)

Cifuentes, C., Kim, M.H.: Hydrodynamic response of a cage system under waves and currents using a Morison-force model. Ocean Eng. **141**, 283–294 (2017)

Claus, R., et al.: Experimental proof-of-concept of HelioSea: a novel marine floating photovoltaic device. Ocean Eng. **299**, 117184 (2024)

Claus, R., López, M.: Key issues in the design of floating photovoltaic structures for the marine environment. Renew. Sust. Energ. Rev. **164**, 112502 (2022)

CRS: Rules for the Classification of Ships. Croatian Register of Shipping (2023)

CCS: China Classification Society Rules for Classification of Mobile Offshore Units. China Classification Society (2023a)

CCS. (2023b). Rules for Classification of Offshore Floating Installation. China Classification Society.

Cui, M., Cheng, Z., Moan, T.: Effects of inhomogeneous wave modeling on extreme responses of a very long floating bridge. Appl. Ocean Res. (2023) [Preprint]

Dai, J., et al.: Inhomogeneous wave load effects on a long, straight and side-anchored floating pontoon bridge. Mar. Struct. **72**, 102763 (2020)

Dai, J., et al.: Effect of wave-current interaction on a long fjord-crossing floating pontoon bridge. Eng. Struct. **266**, 114549 (2022)

Davis, K., Schulte, M., Uekermann, B.: Enhancing quasi-Newton acceleration for fluid-structure interaction. Math. Comput. Appl. **27**(3), 40 (2022)

Ding, J., et al.: A simplified method to estimate the hydroelastic responses of VLFS in the inhomogeneous waves. Ocean Eng. **172**, 434–445 (2019)

Ding, J., Wu, Y., et al.: Investigation of connector loads of a 3-module VLFS using experimental and numerical methods. Ocean Eng. **195**, 106684 (2020a)

Ding, J., Xie, Z., et al.: Numerical and experimental investigation on hydroelastic responses of an 8-module VLFS near a typical island. Ocean Eng. **214**, 107841 (2020b)
Ding, J., Wu, Y.-s., et al.: A direct coupling analysis method and its application to the scientific research and demonstration platform. J. Hydrodyn. **33**, 13–23 (2021a)
Ding, J., et al.: Experimental study on responses of an 8-module VLFS considering different encounter wave conditions. Mar. Struct. **78**, 102959 (2021b)
Ding, J., Wu, Y., et al.: Overview: research on hydroelastic responses of VLFS in complex environments. Mar. Struct. **78**(4), 102978 (2021c)
Ding, R., et al.: An application of network modeling method to scientific research and demonstration platform—connector load analysis. J. Hydrodyn. **33**, 33–42 (2021d)
DNV: Rules for classification offshore units Self-elevating units, including wind turbine installation units and liftboats (2017)
DNV: Recommended Practice DNV-RP-0584, Design, Development and Operation of Floating Solar Photovoltaic Systems. DNV (2021)
DNV: Floating Fish Farming Units and Installations, DNV-RU-OU-0503 Edition July 2022. DNV (2022)
DNV: Rules for Classification of Offshore Units, Floating Wind Installations DNV-RU-OU-0512 Edition July 2023. DNV (2023)
DNV: DNV Rules for Classification, Offshore Units, DNV-RU-OU-0571, Floating Infrastructure Units and Installations. DNV (2024)
Duan, W., et al.: Phase-resolved wave prediction model for long-crest waves based on machine learning. Comput. Methods Appl. Mech. Eng. **372**, 113350 (2020)
Duarte, F., Gormaz, R., Natesan, S.: Arbitrary Lagrangian–Eulerian method for Navier–Stokes equations with moving boundaries. Comput. Methods Appl. Mech. Eng. **193**(45/47), 4819–4836 (2004)
Ducrozet, G., Wang, Y., Derakhti, M.: Enhanced wave breaking modelling in a high-order spectral model. In: B'Waves 2023 (2023)
Ekerhovd, I., et al.: Numerical study on gap resonance coupled to vessel motions relevant to side-by-side offloading. Ocean Eng.. [Preprint]. **241**, 110045 (2021)
Ertekin, R.C., et al.: Efficient methods for hydroelastic analysis of very large floating structures. J. Ship Res. **37**(01), 58–76 (1993)
Escobar, J. et al.: Porting the Meso-NH atmospheric model on different GPU architectures for the next generation of supercomputers (version MESONH-v55-OpenACC) (2024)
Faltinsen, O.M., Newman, J.N., Vinje, T.: Nonlinear wave loads on a slender vertical cylinder. J. Fluid Mech. **289**, 179–198 (1995)
Fan, Y., et al.: Global ocean surface wave simulation using a coupled atmosphere–wave model. J. Clim. **25**(18), 6233–6252 (2012)
Fang, K., Liu, Z., Zou, Z.: Efficient computation of coastal waves using a depth-integrated, non-hydrostatic model. Coast. Eng. **97**, 21–36 (2015)
Fenerci, A., et al.: Hydrodynamic interaction of floating bridge pontoons and its effect on the bridge dynamic responses. Mar. Struct. **83**, 103174 (2022)
Feng, B., et al.: Design, modeling and experiments of swing L-shape piezoelectric beam applied to tidal and wave energy harvesting. Ocean Eng. **289**, 116193 (2023)
Feng, M., et al.: Experimental study of mooring type effect on the hydrodynamic characteristics of VLFS. China Ocean Eng. **36**(1), 155–166 (2022)
Feng, X.Y., Bai, W.: Hydrodynamic analysis of marine multibody systems by a nonlinear coupled model. J. Fluids Struct. **70**, 72–101 (2017)
FIB: Guidelines for Submerged Floating Tube Bridges. Guide to Good Practice (ISBN 978-2-88394-144-1). International Federation for Structural Concrete (2020)

Foli, B.A.K., et al.: A WAVEWATCH III® model approach to investigating ocean wave source terms for West Africa: input-dissipation source terms. Remote Sensing Earth Syst. Sci. **5**(1), 95–117 (2022)

Fu, S., et al.: Hydroelastic analysis of flexible floating interconnected structures. Ocean Eng. **34**(11–12), 1516–1531 (2007)

Fu, S., et al.: Experimental investigation on hydrodynamics of a fish cage floater-net system in oscillatory and steady flows by forced oscillation tests. J. Ship Res. **58**(01), 20–29 (2014)

Fu, S. et al.: A time-domain method for hydroelastic analysis of floating bridges in inhomogeneous waves (2017)

Fu, S., Cui, W.: Dynamic responses of a ribbon floating bridge under moving loads. Mar. Struct. **29**(1), 246–256 (2012)

Gao, J., et al.: Effects of free heave motion on wave resonance inside a narrow gap between two boxes under wave actions. Ocean Eng. **224**, 108753 (2021)

Gao, J., et al.: Mechanism analysis on the mitigation of harbor resonance by periodic undulating topography. Ocean Eng. **281**, 114923 (2023)

Gao, R.P., et al.: Hydroelastic response of very large floating structure with a flexible line connection. Ocean Eng. **38**(17–18), 1957–1966 (2011)

Gao, R.P., Wang, C.M., Koh, C.G.: Reducing hydroelastic response of pontoon-type very large floating structures using flexible connector and gill cells. Eng. Struct. **52**, 372–383 (2013)

Gardner, M.: Assessing the Impact of an Artificial Reef to Mitigate Coastal Erosion Using the Phase Resolving Wave Model FUNWAVE. University of Rhode Island (2020)

Group, S.M.G. for S.A.S.T.S.S.: A Technical Standard for Scottish Finfish Aquaculture. Scottish Government (2015)

Gudmestad, O.T.: Selection of safety level for marine structures. Procedia Struct. Integrity. **48**, 113–118 (2023)

Guo, X., et al.: Deep seabed mining: Frontiers in engineering geology and environment. Int. J. Coal Sci. Technol. **10**(1) (2023)

Haalboom, S., et al.: Monitoring of a sediment plume produced by a deep-sea mining test in shallow water, Málaga Bight, Alboran Sea (southwestern Mediterranean Sea). Mar. Geol. **456**, 106971 (2023)

Hamamoto, T., Fujita, K.: Three dimensional BEM-FEM coupled dynamic analysis of module-linked large floating structures. In: ISOPE International Ocean and Polar Engineering Conference, p. ISOPE-I. ISOPE (1995)

Hasselmann, K.: On the non-linear energy transfer in a gravity-wave spectrum part 1. General theory. J. Fluid Mech. **12**(4), 481–500 (1962)

He, G., et al.: Two-dimensional numerical study on fluid resonance in the narrow gap between two rigid-connected heave boxes in waves. Appl. Ocean Res. **110**, 102628 (2021)

Hell, M., Chapron, B., Fox-Kemper, B.: A particle-in-cell wave model for efficient sea-state and swell estimates in earth system models-PiCLES. Authorea Preprints. (2024) [Preprint]

Hong, S., Hwang, W.: Mitigation of hydroelastic response in pneumatically supported floating structures. Appl. Ocean Res. **138**, 103640 (2023)

Hogan-Lovells.: Review of Offshore Wind Worldwide: Regulatory Framework in Selected Countries. Hogan Lovells International LLP (2022)

Hou, H.M., et al.: Reliability assessment of mooring system for fish cage considering one damaged mooring line. Ocean Eng. (2022) [Preprint]

Hou, H.M., Dong, G.H., Xu, T.J.: Analysis of probabilistic fatigue damage of mooring system for offshore fish cage considering long-term stochastic wave conditions. Ships Offshore Struct., **17**(2), 398, 1–12 (2020)

Hu, Y., et al.: Improvement of drag coefficient parameterization of WAVEWATCH-III using remotely sensed products during tropical cyclones. Ocean Dyn. **74**, 843–858 (2024)

Huang, H., et al.: A theoretical calculation method of irregular bridge span pontoon connected by elastic hinges. Mar. Struct. **76**, 102929 (2021)

Huang, H., et al.: Structural analysis method of a pontoon-separated floating bridge connected by elastic hinges. Ships Offshore Struct. **17**(9), 2045–2057 (2022)

Huang, H., et al.: Experimental and numerical study on dynamic responses of floating bridge under the shielding effect of a floating platform. Mar. Struct. **89**, 103379 (2023a)

Huang, L., et al.: Floating solar power loss due to motions induced by ocean waves: an experimental study. Ocean Eng. **312**, 118988 (2024)

Huang, S., Hu, Z., Chen, C.: Application of CFD–FEA coupling to predict hydroelastic responses of a single module VLFS in extreme wave conditions. Ocean Eng. (2023b). https://doi.org/10.1016/j.oceaneng.2023.114754

Huang, X.H., et al.: Hydrodynamic performance of a semi-submersible offshore fish farm with a single point mooring system in pure waves and current. Aquac. Eng. (2020). https://doi.org/10.1016/j.aquaeng.2020.102075

IMO: Revised Guidelines for Formal Safety Assessment (FSA) for Use in the IMO Rule-Making Process (2015)

IOGP: Joint industry project JIP-35, standardization of offshore structures specifications (2024). Retrieved from https://www.iogp.org/workstreams/engineering/standards/jip35

IR: Rules and Regulations for the Construction and Classification of Floating Offshore Units. Indian Register of Shipping (2024)

ISO: ISO 16488, Marine finfish farms: open net cage-design and operation (2015). Retrieved from https://www.iso.org/standard/56852

ISO: ISO 29400, Ships and marine technology — offshore wind energy – port and marine operations (2020). Retrieved from https://www.iso.org/standard/71100.html

Ito, S., Kinoshia, T., Bao, W.: Hydrodynamic behaviors of an elastic net structure. Ocean Eng. **92**, 188–197 (2014)

Janssen, P.A.E.M.: Quasi-linear theory of wind-wave generation applied to wave forecasting. J. Phys. Oceanogr. **21**(11), 1631–1642 (1991)

Jason, M.V., et al.: Fluid–structure interaction simulation of slam-induced bending in large high-speed wave-piercing catamarans. J. Fluids Struct. **82**, 35–58 (2018)

Jiang, C., El, M.O., Zhang, G.: Nonlinear modeling of wave–structure interaction for a flexible floating structure. Ocean Eng. **300**, 1.1–1.14 (2024)

Jiang, D., et al.: Research and development in connector systems for very large floating structures. Ocean Eng. **232**, 109150 (2021)

Jiang, Z., et al.: Design and model test of a soft-connected lattice-structured floating solar photovoltaic concept for harsh offshore conditions. Mar. Struct. **90**, 103426 (2023)

Jin, C., et al.: Discrete-module-beam-based hydro-elasticity simulations for moored submerged floating tunnel under regular and random wave excitations. Eng. Struct. **275**, 115198 (2023a)

Jin, C., Bakti, F.P., Kim, M.: Multi-floater-mooring coupled time-domain hydro-elastic analysis in regular and irregular waves. Appl. Ocean Res. **101**, 102276 (2020)

Jin, G., et al.: Numerical analysis of vortex-induced vibration on a flexible cantilever riser for deep-sea mining system. Mar. Struct. **87**, 103334 (2023b)

Jouon, A., et al.: Wind wave measurements and modelling in a fetch-limited semi-enclosed lagoon. Coast. Eng. **56**(5–6), 599–608 (2009)

Jun, Z., Wei, L.: Dynamic model of a hinged-girder floating bridge subjected by moving loads. Appl. Ocean Res. **114**, 102804 (2021)

Kalourazi, M.Y., et al.: WAVEWATCH-III source terms evaluation for optimizing hurricane wave modeling: A case study of Hurricane Ivan. Oceanologia. **63**(2), 194–213 (2021)

Kang, H.Y., Kim, M.H.: Time-domain hydroelastic analysis with efficient load estimation for random waves. Int. J. Nav. Archit. Ocean Eng. **9**(3), 266–281 (2017)

Kang, Z., Ni, W., Sun, L.: A numerical investigation on capturing the maximum transverse amplitude in vortex induced vibration for low mass ratio. Mar. Struct. **52**, 94–107 (2017)

Karmakar, D., Soares, C.G.: Scattering of gravity waves by a moored finite floating elastic plate. Appl. Ocean Res. **34**, 135–149 (2012)

Karperaki, A.E., Belibassakis, K.A.: Hydroelastic analysis of very large floating structures in variable bathymetry regions by multi-modal expansions and FEM. J. Fluids Struct. **102**, 103236 (2021)

Karperaki, A.E., Belibassakis, K.A., Papathanasiou, T.K.: Time-domain, shallow-water hydroelastic analysis of VLFS elastically connected to the seabed. Mar. Struct. **48**, 33–51 (2016)

Kennedy, A.B., et al.: Boussinesq-type equations with improved nonlinear performance. Wave Motion. **33**(3), 225–243 (2001)

Khabakhpasheva, T.I., Korobkin, A.A.: Hydroelastic behaviour of compound floating plate in waves. J. Eng. Math. **44**, 21–40 (2002)

Kim, K., Lee, P., Park, K.C.: A direct coupling method for 3D hydroelastic analysis of floating structures. Int. J. Numer. Methods Eng. **96**(13), 842–866 (2013)

KR: Guidance for Floating Structures. Korean Register of Shipping (2024)

Kirby, C.: The restrictions on predictability implied by rational asset pricing models. Rev. Financ. Stud. **11**(2), 343–382 (1998)

Kirby, J.T.: Boussinesq models and their application to coastal processes across a wide range of scales. J. Waterw. Port Coastal Ocean Eng. **142**(6), 3116005 (2016)

Kirby, J.T., Dalrymple, R.A.: Propagation of obliquely incident water waves over a trench. J. Fluid Mech. **133**, 47–63 (1983)

KM, P., Karmakar, D., Soares, C.G.: Hydroelastic analysis of articulated floating elastic plate based on Timoshenko–Mindlin plate theory. Ships Offshore Struct. **13**(sup1), 287–301 (2018)

Komen, G.J., et al.: Dynamics and Modelling of Ocean Waves. Cambridge University Press (1996)

Komen, G.J., Hasselmann, S., Hasselmann, K.: On the existence of a fully developed wind-sea spectrum. J. Phys. Oceanogr. **14**, 1271–1285 (1984)

Kou, Y., et al.: Performance characteristics of a conceptual ring-shaped spar-type VLFS with double-layered perforated-wall breakwater. Appl. Ocean Res. **86**, 28–39 (2019)

Kristiansen, T., Faltinsen, O.M.: Modelling of current loads on aquaculture net cages. J. Fluids Struct. **34**, 218–235 (2012)

Kristiansen, T., Faltinsen, O.M.: Experimental and numerical study of an aquaculture net cage with floater in waves and current. J. Fluids Struct. **54**, 1–26 (2015)

Kumar, M., Niyaz, H.M., Gupta, R.: Challenges and opportunities towards the development of floating photovoltaic systems. Sol. Energy Mater. Sol. Cells. **233**, 111408 (2021)

Kvåle, K.A., Sigbjörnsson, R., Øiseth, O.: Modelling the stochastic dynamic behaviour of a pontoon bridge: A case study. Comput. Struct. **165**, 123–135 (2016)

Kwak, M.S.: Estimate of wave overtopping rate on Armoured slope structures using FUNWAVE-TVD model. J. Korean Soc. Coastal Ocean Eng. **36**(1), 11–19 (2024)

Kyoung, J.H., et al.: Hydroelastic response of a very large floating structure over a variable bottom topography. Ocean Eng. **32**(17–18), 2040–2052 (2005)

Kyozuka, Y., Kato, S., Nakagawa, H.: A numerical study on environmental impact assessment of mega-float of Japan. Mar. Struct. **14**(1–2), 147–161 (2001)

LøLand, G.: Current forces on, and water flow through and around, floating fish farms. Aquac. Int. **1**(1), 72–89 (1993)

Lader, P.F., Fredheim, A.: Dynamic properties of a flexible net sheet in waves and current—A numerical approach. Aquac. Eng. **35**(3), 228–238 (2006)

Lakshmynarayanana, P.A., Temarel, P.: Application of CFD and FEA coupling to predict dynamic behaviour of a flexible barge in regular head waves. Mar. Struct. **65**, 308–325 (2019)

Lee, H.-W., et al.: Coupled analysis method of a mooring system and a floating crane based on flexible multibody dynamics considering contact with the seabed. Ocean Eng. **163**, 555–569 (2018)

Leng, J., et al.: A fluid–structure interaction model for large wind turbines based on flexible multibody dynamics and actuator line method. J. Fluids Struct. (2023). https://doi.org/10.1016/j.jfluidstructs.2023.103857

Leung, N.-C., et al.: WRF-ROMS-SWAN coupled model simulation study: effect of atmosphere–ocean coupling on sea level predictions under tropical cyclone and northeast monsoon conditions in Hong Kong. Atmosphere. **15**(10), 1242 (2024)

Li, H., et al.: Demands and challenges for construction of marine infrastructures in China. Front. Struct. Civ. Eng. **16**(5), 551–563 (2022a)

Li, H., Soares, C.G.: Assessment of failure rates and reliability of floating offshore wind turbines. Reliab. Eng. Syst. Saf. **228**, 108777 (2022)

Li, L., et al.: Design optimization of mooring system: an application to a vessel-shaped offshore fish farm. Eng. Struct. **197**, 109363 (2019)

Li, L.: Full-coupled analysis of offshore floating wind turbine supported by very large floating structure with consideration of hydroelasticity. Renew. Energy. **189**, 790–799 (2022)

Li, N., et al.: Dynamic analysis of an integrated offshore structure comprising a jacket-supported offshore wind turbine and aquaculture steel cage. Ocean Eng. **274**, 114059 (2023a)

Li, S., et al.: Second-order hydroelastic analysis of a flexible floating structure under spatially inhomogeneous waves. Mar. Struct. **86**, 103306 (2022b)

Li, S., et al.: Hydroelastic analysis of a floating bridge under spatially inhomogeneous waves, with emphasis on the effect of drift force modeling. Appl. Ocean Res. **139**, 103666 (2023b)

Li, Y.: Fully nonlinear analysis of second-order gap resonance between two floating barges. Eng. Anal. Bound. Elem. **106**, 1–19 (2019)

Li, Y., Wang, X.: Numerical study of effects of gap and incident wave steepness on water resonance between two rectangular barges. Eur. J. Mechan.-B/Fluids. **86**, 157–168 (2021)

Liang, M., et al.: A shallow water mooring system design methodology combining NSGA-II with the vessel-mooring coupled model. Ocean Eng. **190**, 106417 (2019)

Liang, M., et al.: Experimental evaluation of a mooring system simplification methodology for reducing mooring lines in a VLFS model testing at a moderate water depth. Ocean Eng. **219**, 107912 (2020)

Lin, et al.: Numerical analysis of a vessel-shaped offshore fish farm. J. Offshore Mech. Arct. Eng. (2018) [Preprint]

Ling, H., et al.: Complex wave loads and hydroelastic responses of a very large tourism floating platform "Ocean Diamond". Ocean Eng. **265**, 112608 (2022)

Liu, H.-F., et al.: Hydrodynamic assessment of a semi-submersible aquaculture platform in uniform fluid environment. Ocean Eng. **237**, 109656 (2021)

Liu, H.-F., Bi, C.-W., Zhao, Y.-P.: Experimental and numerical study of the hydrodynamic characteristics of a semisubmersible aquaculture facility in waves. Ocean Eng. **214**, 107714 (2020)

Liu, H., et al.: Feasibility study of a novel open ocean aquaculture ship integrating with a wind turbine and an internal turret mooring system. J. Mar. Sci. Eng. **10**(11), 1729 (2022a)

Liu, L., et al.: Numerical investigation on dynamic performance of vertical hydraulic transport in deepsea mining. Appl. Ocean Res. **130**, 103443 (2023)

Liu, X., Sakai, S.: Time domain analysis on the dynamic response of a flexible floating structure to waves. J. Eng. Mech. **128**(1), 48–56 (2002)

Liu, Z., Wang, S., Guedes Soares, C.: Numerical study on the mooring force in an offshore fish cage array. J. Mar. Sci. Eng. **10**(3), 331 (2022b)

LR: Rules and Regulations for the Classification of Offshore Units. Lloyd's Register of Shipping (2023)

LR.: Recommended Practice for Floating Offshore Wind Turbine Support Structures. Lloyd's Register of Shipping (2024)

Lu, D., et al.: A method to estimate the hydroelastic behaviour of VLFS based on multi-rigid-body dynamics and beam bending. Ships Offshore Struct. **14**(4), 354–362 (2019)

Lu, L., et al.: Numerical investigation of fluid resonance in two narrow gaps of three identical rectangular structures. Appl. Ocean Res. **32**(2), 177–190 (2010)

Lu, X., Dao, M.H., Le, Q.T.: A GPU-accelerated domain decomposition method for numerical analysis of nonlinear waves-current-structure interactions. Ocean Eng. **259**, 111901 (2022a)

Lu, Y., et al.: A new type connection with optimum stiffness configuration for very large floating platforms. Proc. Inst. Mech. Eng. Part M: J. Eng. Marit. Environ. **236**(3), 764–778 (2022b)

Luo, W., et al.: Conceptual design and model test of a pontoon-truss type offshore floating photovoltaic system with soft connection. Ocean Eng. **309**, 118518 (2024)

Lwin, M.M.: Floating bridge. In: Chen, W.F., Duan, L. (eds.) Bridge Engineering Handbook, p. Chapter 22. CRC Press (2000)

Ma, C., et al.: Dynamic behaviors of a hinged multi-body floating aquaculture platform under regular waves. Ocean Eng. **243**, 110278 (2022)

Ma, L., et al.: A hybrid empirical-numerical method for hydroelastic analysis of a floater-and-net system. Transactions, 124 (2016) [Preprint]

Madsen, P.A., Murray, R., Sørensen, O.R.: A new form of the Boussinesq equations with improved linear dispersion characteristics. Coast. Eng. **15**(4), 371–388 (1991)

Manisha, K., R.B., Sahoo, T.: Effect of bottom undulation for mitigating wave-induced forces on a floating bridge. Wave Motion. **89**, 166–184 (2019)

Martin, T., et al.: Modelling open ocean aquaculture structures using CFD and a simulation-based screen force model. J. Mar. Sci. Eng. **10**(3), 332 (2022)

Mazhar, F., Javed, A., Altinkaynak, A.: A novel artificial neural network-based interface coupling approach for partitioned fluid–structure interaction problems. Eng. Anal. Bound. Elem. **151**, 287–308 (2023)

Mentaschi, L., et al.: Performance evaluation of Wavewatch III in the Mediterranean Sea. Ocean Model. **90**, 82–94 (2015)

Miao, Y., et al.: Numerical modeling and dynamic analysis of a floating bridge subjected to wave, current and moving loads. Ocean Eng. **225**, 108810 (2021a)

Miao, Y., et al.: Investigation on hydrodynamic performance of a two-module semi-submersible offshore platform. Ships Offshore Struct. **17**(3), 607–618 (2022)

Miao, Y., Chen, X., Tang, M., et al.: Experimental investigation on the dynamic properties of a floating bridge under regular waves: model design, implementation, and analysis. Ships Offshore Struct., 1–16 (2024a)

Miao, Y., Chen, X., Xu, L., et al.: Numerical studies of hydroelastic performances of a pontoon-truss composite floating bridge under regular waves. J. Mar. Sci. Technol. **30**, 1–17 (2024b)

Miao, Y., Chen, X., Cheng, X.: Research on the dynamic responses of a multi-module offshore floating bridge based on the three-dimensional hydroelastic theory in the time domain. Ocean Eng. **313**, 119314 (2024c)

Miao, Y.J., et al.: Experimental and numerical study of a semi-submersible offshore fish farm under waves. Ocean Eng. **225**(4), 108794 (2021b)

Michailides, C., Loukogeorgaki, E., Angelides, D.C.: Response analysis and optimum configuration of a modular floating structure with flexible connectors. Appl. Ocean Res. **43**, 112–130 (2013)

Miles, J.W.: On the generation of surface waves by shear flows. J. Fluid Mech. **3**(2), 185–204 (1957)

Mitsume, N., et al.: MPS–FEM partitioned coupling approach for fluid–structure interaction with free surface flow. Int. J. Comput. Methods. **11**(04), 1350101 (2014)

Moe-Føre, H., et al.: Structural response of high solidity net cage models in uniform flow. J. Fluids Struct. **65**, 180–195 (2016)

Moe, H., Fredheim, A., Hopperstad, O.S.: Structural analysis of aquaculture net cages in current. J. Fluids Struct. **26**(3), 503–516 (2010)

Mohapatra, S.C., Guedes Soares, C.: A review of the hydroelastic theoretical models of floating porous nets and floaters for offshore aquaculture. J. Mar. Sci. Eng. **12**(10), 1699 (2024)

Mohapatra, S.C., Soares, C.G.: A semi-analytical model of an array of moored floating flexible offshore net cages under current loads. Ocean Eng. **291**, 116309 (2024)

Mojtahedi, A., et al.: The role of different wind input, whitecap dissipation and quadruplet wave-wave interaction terms in wave evolution in Lake Michigan. Ocean Eng. **307**, 118149 (2024)

Nasyrlayev, N., et al.: Modelling the response of offshore aquaculture fish pens to environmental loads in high-energy regions. Appl. Ocean Res. (2023) [Preprint]

Nguyen, H.P., et al.: Reducing hydroelastic responses of pontoon-type VLFS using vertical elastic mooring lines. Mar. Struct. **59**, 251–270 (2018)

Nguyen, H.P., Wang, C.M., Luong, V.H.: Two-mode WEC-type attachment for wave energy extraction and reduction of hydroelastic response of pontoon-type VLFS. Ocean Eng. **197**, 106875 (2020)

Nguyen, H.P., Wang, C.M., Pedroso, D.M.: Optimization of modular raft WEC-type attachment to VLFS and module connections for maximum reduction in hydroelastic response and wave energy production. Ocean Eng. **172**, 407–421 (2019)

Nguyen, Q.T., Mao, M., Xia, M.: Numerical modeling of nearshore wave transformation and breaking processes in the Yellow River Delta with FUNWAVE-TVD wave model. J. Mar. Sci. Eng. **11**(7), 1380 (2023)

Ni, W., et al.: Power prediction of oscillating water column power generation device based on physical information embedded neural network. Energy. **306** (2024)

Ni, X., et al.: Performance analysis of the mooring system of a two-module scientific research and demonstration platform. J. Hydrodyn. **33**(5), 901–914 (2021)

Nwogu, O.: Alternative form of Boussinesq equations for nearshore wave propagation. J. Waterw. Port Coast. Ocean Eng. **119**(6), 618–638 (1993)

Officials, A.A. of S.H.: Standard Specifications for Highway Bridges. American Association of State Highway Officials (1973)

NS.: Marine Fish Farms: Requirements for Design, Dimensioning, Production, Installation and Operation (Norwegian Standard 9415). Norwegian Standardisation Association (2009)

Pákozdi, C., et al.: Joint-Industry Effort to Develop and Verify CFD Modeling Practice for Predicting Wave Impact, vol. 1. Offshore Technology (2022) [Preprint]

Pal, S.K., Datta, R., Sunny, M.R.: Fully coupled time domain solution for hydroelastic analysis of a floating body. Ocean Eng. **153**, 173–184 (2018)

Parunov, J., et al.: Uncertainties in modelling the low-frequency wave-induced global loads in ships. Mar. Struct. **86**, 103307 (2022)

Peacock, T., Ouillon, R.: The fluid mechanics of deep-sea mining. Annu. Rev. Fluid Mech. **55**(1), 403–430 (2023)

Peregrine, D.H.: Long waves on a beach. J. Fluid Mech. **27**(4), 815–827 (1967)

Pessoa, J., Fonseca, N., Soares, C.G.: Side-by-side FLNG and shuttle tanker linear and second order low frequency wave induced dynamics. Ocean Eng. **111**, 234–253 (2016)

Petersen, Ø.W., Øiseth, O., Lourens, E.: Full-scale identification of the wave forces exerted on a floating bridge using inverse methods and directional wave spectrum estimation. Mech. Syst. Signal Process. **120**, 708–726 (2019)

Pham, D.C., Wang, C.M., Utsunomiya, T.: Hydroelastic analysis of pontoon-type circular VLFS with an attached submerged plate. Appl. Ocean Res. **30**(4), 287–296 (2008)

Phillips, O.M.: On the generation of waves by turbulent wind. J. Fluid Mech. **2**(5), 417–445 (1957)

Price, W.G., Wu, Y.: Hydroelasticity of marine structures. In: Theoretical and Applied Mechanics, pp. 311–337. Elsevier (1985)

PRS: Rules for the Classification and Construction of Mobile Offshore Drilling Units. Polish Register of Shipping (2023)

Qiu, L.: Numerical simulation of transient hydroelastic response of a floating beam induced by landing loads. Appl. Ocean Res. **29**(3), 91–98 (2007)

Ramezani, M., et al.: Uncertainty models for the structural design of floating offshore wind turbines: A review. Renew. Sust. Energ. Rev. **185**, 113610 (2023)

Rawson, C., Huang, W.: Navigating the path to energy transition: understanding the U.S. regulations for offshore wind vessels [Paper Presentation]. Offshore Technology Conference, Houston, TX, USA (2023, May 2–5)

Rawson, C., Huang, W.: Navigating the path to energy transition: understanding the US regulations for offshore wind vessels. In: Offshore Technology Conference, p. D031S031R004. OTC (2022)

Riggs, H.R., Ertekin, R.C., Mills, T.R.J.: A comparative study of RMFC and FEA models for the wave-induced response of a MOB. Mar. Struct. **13**(4–5), 217–232 (2000)

Riyansyah, M., Wang, C.M., Choo, Y.S.: Connection design for two-floating beam system for minimum hydroelastic response. Mar. Struct. **23**(1), 67–87 (2010)

RINA: Rules for the Classification of Steel Fixed Offshore Platforms. RINA S.p.A. (2015)

RINA: Rules for the Classification of Floating Offshore Units Intended at Fixed Locations and Mobile Offshore Drilling Units. RINA S.p.A (2024)

Rodrigues, J.M., et al.: Design and verification of large floating coastal structures: floating bridges for fjord crossings. In: ISOPE International Ocean and Polar Engineering Conference, p. ISOPE-I. ISOPE (2020)

Rodrigues, J.M., Viuff, T., Økland, O.D.: Model tests of a hydroelastic truncated floating bridge. Appl. Ocean Res. **125**, 103247 (2022)

Roeber, V., Cheung, K.F.: Boussinesq-type model for energetic breaking waves in fringing reef environments. Coast. Eng. **70**, 1–20 (2012)

Scottish Government: A Technical Standard for Scottish Finfish Aquaculture (ISBN 9781785443725). The Scottish Government (2015)

Seng, S.: Slamming and Whipping Analysis of Ships. Technical University of Denmark, Copenhagen (2012)

Sengupta, D., Pal, S.K., Datta, R.: Hydroelasticity of a 3D floating body using a semi analytic approach in time domain. J. Fluids Struct. **71**, 96–115 (2017)

Senjanovic, I., Malenica, S., Tomasevic, S.: Investigation of ship hydroelasticity. Ocean Eng. **35**(5–6), 523–535 (2008)

Seto, H., et al.: Integrated hydrodynamic–structural analysis of very large floating structures (VLFS). Mar. Struct. **18**(2), 181–200 (2005)

Sha, Y., et al.: Numerical investigations of the dynamic response of a floating bridge under environmental loadings. Ships Offshore Struct. **13**, 113–126 (2018)

Sharma, R.: Deep-sea mining: current status and future considerations (2017)

Shen, H., et al.: Nonlinear dynamics of an aquaculture cage array induced by wave-structure interactions. Ocean Eng. **269**, 113711 (2023)

Shen, Y., et al.: Numerical and experimental investigations on mooring loads of a marine fish farm in waves and current. J. Fluids Struct. **79**, 115–136 (2018)

Shi, F., et al.: Non-hydrostatic wave model NHWAVE user's guide for modeling submarine landslide tsunami (version 1.1). Research Report NO. CACR-12-04 [Preprint] (2012)

Shi, Q., et al.: A face-contact connector for modularized floating structures. Mar. Struct. **82**, 103149 (2022)

Shi, Q., et al.: Finite-time adaptive anti-disturbance constrained control design for dynamic positioning of marine vessels with simulation and model-scale tests. Ocean Eng. **277**, 114117 (2023a)

Shi, Q.J., et al.: Optimized stiffness combination of a flexible-base hinged connector for very large floating structures. Mar. Struct. **60**, 151–164 (2018)

Shi, W., et al.: Review on the development of marine floating photovoltaic systems. Ocean Eng. **286**, 115560 (2023b)

Shi, Y., et al.: Hydroelastic analysis of offshore floating photovoltaic based on frequency-domain model. Ocean Eng. **289**, 116213 (2023c)

Shirai, H., Onda, S., Hosoda, T.: Boussinesq models with moving boundaries and their applicability to waves generated by lateral oscillation and bottom deformation. J. Hydraul. Eng. **149**(8), 4023023 (2023)

Shirkol, A.I., Nasar, T.: Coupled boundary element method and finite element method for hydroelastic analysis of floating plate. J. Ocean Eng. Sci. **3**(1), 19–37 (2018)

Shirkol, A.I., Nasar, T., Karmakar, D.: Wave interaction with very large floating structure (VLFS) using BEM approach–revisited. Perspect. Sci. **8**, 533–535 (2016)

Shixiao, F., et al.: Hydroelastic analysis of a nonlinearly connected floating bridge subjected to moving loads. Mar. Struct. **18**(1), 85–107 (2005)

Siadatmousavi, S.M., Kalourazi, M.Y., Kholgh, A.K.: Improving the WAVEWATCH-III wave model results using data assimilation in the Persian Gulf. Ocean Eng. **300**, 117460 (2024)

Singla, S., Martha, S.C., Sahoo, T.: Mitigation of structural responses of a very large floating structure in the presence of vertical porous barrier. Ocean Eng. **165**, 505–527 (2018)

Snyder, R.L., et al.: Array measurements of atmospheric pressure fluctuations above surface gravity waves. J. Fluid Mech. **102**, 1–59 (1981)

Song, H., et al.: Hydroelastic response of VLFS on uneven sea bottom. In: International Conference on Offshore Mechanics and Arctic Engineering, pp. 433–443 (2005)

Song, J., et al.: Dynamic response of multiconnected floating solar panel systems with vertical cylinders. J. Mar. Sci. Eng. **10**(2), 189 (2022a)

Song, J., et al.: Wave-induced structural response analysis of the supporting frames for multi-connected offshore floating photovoltaic units installed in the inner harbor. Ocean Eng. **271**, 113812 (2023)

Song, Q.H., et al.: Analysis of longitudinal coupling dynamic characteristics of deep sea mining vessel and stepped lifting pipe. Sci. Rep. **11**(1), 19190 (2021)

Song, X., Liu, W., Zhang, G.: Research on structural response characteristics of trapezoidal floating body in waves. J. Mar. Sci. Eng. **10**(11), 1756 (2022b)

Standard, N.: Marine Fish Farms. Requirements for Design, Dimensioning, Production, Installation and Operation. NS (2003)

Strand, I.M., Faltinsen, O.M.: Linear wave response of a 2D closed flexible fish cage. J. Fluids Struct. **87**, 58–83 (2019)

Sulaiman, O.O., Magee, A., et al.: Mooring analysis for very large offshore aquaculture ocean plantation floating structure. Ocean Coastal Manag. **80**, 80–88 (2013a)

Sulaiman, O.O., Sakinah, N., et al.: Qualitative risk analysis study of offshore aquaculture ocean plantation system. Glob. J. Sci. Front. Res. Environ. Earth Sci. **13** (2013b)

Sun, H., et al.: On the interaction of surface waves with an elastic plate of finite length in head seas. China Ocean Eng. (2002) [Preprint]

Sun, H., et al.: Theoretical and numerical methods for predicting the structural stiffness of unbonded flexible riser for deep-sea mining under axial tension and internal pressure. Ocean Eng. **310**, 118672 (2024)

Sun, J., et al.: An experimental investigation on the nonlinear hydroelastic response of a pontoon-type floating bridge under regular wave action. Ships Offshore Struct. **13**(3), 233–243 (2018)

Sun, L., et al.: Coupled dynamic analysis of deep-sea mining support vessel with dynamic positioning. Mar. Georesour. Geotechnol., 1064119X.2017.1391901 (2017)

Sun, Y., et al.: Failure analysis of floating offshore wind turbines with correlated failures. Reliab. Eng. Syst. Safety. **238**, 109485 (2023)

Sun, Z., et al.: A comparison of WAVEWATCH III grid models for a typical reef lagoon. In: ISOPE International Ocean and Polar Engineering Conference, p. ISOPE-I. ISOPE (2019)

Sun, Z., Xu, D., et al.: Observation and simulation of wind waves near a typical reef lagoon in South China Sea. J. Hydrodyn. **33**, 24–32 (2021a)

Sun, Z., Zhang, H., et al.: Wave energy assessment of the Xisha Group Islands zone for the period 2010–2019. Energy. **220**, 119721 (2021b)

Suzuki, H.: Overview of Megafloat: concept, design criteria, analysis, and design. Mar. Struct. **18**(2), 111–132 (2005)

Takami, T., et al.: A numerical simulation method for predicting global and local hydroelastic response of a ship based on CFD and FEA coupling. Mar. Struct. **59**, 368–386 (2018)

Tang, H.-J., Huang, C.-C., Chen, W.-M.: Dynamics of dual pontoon floating structure for cage aquaculture in a two-dimensional numerical wave tank. J. Fluids Struct. **27**(7), 918–936 (2011)

Tang, H.-J., Yang, R.-Y., Yao, H.-C.: Experimental and numerical investigations of a mooring line failure of an aquaculture net cage subjected to currents. Ocean Eng. **238**, 109707 (2021)

Tang, H.-J., Yang, R.-Y., Yao, H.-C.: Experimental and numerical study on the hydrodynamic behaviors of mooring line failure on a net cage in irregular waves and currents. Front. Mar. Sci. **10**, 1122855 (2023a)

Tang, Y., et al.: A fully nonlinear BEM-beam coupled solver for fluid-structure interactions of flexible ships in waves. J. Fluids Struct. **121**, 103922 (2023b)

Thomas, J.T., Dwarakish, G.S.: Numerical wave modelling–a review. Aquat. Procedia. **4**, 443–448 (2015)

Tolman, H.L.: A third-generation model for wind waves on slowly varying, unsteady, and inhomogeneous depths and currents. J. Phys. Oceanogr. **21**(6), 782–797 (1991)

Tolman, H.L., et al.: Development and implementation of wind-generated ocean surface wave Modelsat NCEP. Weather Forecast. **17**(2), 311–333 (2002)

Tolman, H.L., Chalikov, D.: Source terms in a third-generation wind wave model. J. Phys. Oceanogr. **26**(11), 2497–2518 (1996)

Tremps, L., Yeter, B., Kolios, A.: Review and analysis of the failure risk mitigation via monitoring for monopile offshore wind structures. Energy Rep. **11**, 5407–5420 (2024)

Tuitman, J.T. and Malenica, ? (2009) 'Fully coupled seakeeping, slamming, and whipping calculations', Proc. Inst. Mech. Eng. Part M J. Eng. Marit. Environ., 223(3), pp. 439–456.

Turner, A.A., Jeans, T.L., Reid, G.K.: Experimental investigation of fish farm hydrodynamics on 1: 15 scale model square aquaculture cages. J. Offshore Mech. Arct. Eng. **138**(6), 61201 (2016)

Veritas: NR580 Rules for the Classification of Floating Establishments. Bureau Veritas (2012)

Veritas: Classification of Mooring Systems for Permanent and Mobile Offshore Units. Bureau Veritas (2021)

Veritas: Rules for the Classification of Offshore Units. Bureau Veritas (2023)

Veritas: NR572, Classification and Certification of Floating Offshore Wind Turbines. Bureau Veritas (2024)

Viuff, T., et al.: Effects of wave directionality on extreme response for a long end-anchored floating bridge. Appl. Ocean Res. **90**, 101843 (2019)

Viuff, T., et al.: Model uncertainty assessment for wave-and current-induced global response of a curved floating pontoon bridge. Appl. Ocean Res. **105**, 102368 (2020)

Viuff, T., et al.: Experimental study of floating bridge global response when subjected to waves and current. Appl. Ocean Res. **138**, 103588 (2023a)

Viuff, T., et al.: Model test of a hydroelastic truncated floating bridge with a stay-cable tower. Appl. Ocean Res. **135**, 103539 (2023b)

Waals, O.J., Bunnik, T.H.J., Otto Offshore and Arctic Engineering, W.J.B.T.-A. 2018 37th I.C. on O: Model Tests and Numerical Analysis for a Floating Mega Island (2018)

Wan, W., et al.: Simulation of storm wave transformation in macro-tidal environment based on FUNWAVE-GPU. In: ISOPE International Ocean and Polar Engineering Conference, p. ISOPE-I. ISOPE (2021)

Wang, B.T.: Development of the legal definition of the floating city: judicial interpretation of structural characteristics of floating homes and developments. In: World Conference on Floating Solutions, pp. 139–157. Springer (2023)

Wang, C., Fu, S., Cui, W.: Hydroelasticity based fatigue assessment of the connector for a ribbon bridge subjected to a moving load. Mar. Struct. **22**(2), 246–260 (2009a)

Wang, C.M., Riyansyah, M., Choo, Y.S.: Reducing hydroelastic response of interconnected floating beams using semi-rigid connections. In: International Conference on Offshore Mechanics and Arctic Engineering, pp. 1419–1425 (2009b)

Wang, C.M., Tay, Z.Y.: Very large floating structures: applications, research and development. Procedia Eng. **14**, 62–72 (2011)

Wang, C.M., Wang, B.T.: Large floating structures. Ocean Engineering & Oceanography. **3** (2015)

Wang, D., Ertekin, R.C., Riggs, H.R.: Three-dimensional hydroelastic response of a very large floating structure. Int. J. Offshore Polar Eng. **1**(04), 307 (1991)

Wang, G., et al.: A numerical study of the hydrodynamics of an offshore fish farm using REEF3D. J. Offshore Mech. Arct. Eng. **144**(2), 21301 (2022a)

Wang, H., et al.: Dynamic performance analysis of grid mooring system for gravity cages. Ships Offshore Struct. **20**, 1143–1160 (2024a)

Wang, H., Jin, X.: Dynamic analysis of maritime gasbag-type floating bridge subjected to moving loads. Int. J. Nav. Archit. Ocean Eng. **8**(2), 137–152 (2016)

Wang, P., Fang, K., et al.: A beach profile evolution model driven by the hybrid shock-capturing Boussinesq wave solver. Water. **15**(21), 3799 (2023a)

Wang, Q., et al.: Numerical and experimental studies on hydroelastic responses of a multiply-connected domain tourism platform in head waves. Ocean Eng. **294**, 116795 (2024b)

Wang, S., et al.: Design, local structural stress, and global dynamic response analysis of a steel semi-submersible hull for a 10-MW floating wind turbine. Eng. Struct. **291**, 116474 (2023b)

Wang, S., et al.: Research on the connector loads of a multi-module floating body with hinged connector based on FMFC model. Appl. Sci. **13**(10), 6212 (2023c)

Wang, X., et al.: Time-variant reliability under corrosion-caused strength degradation for floating structures in the South China Sea. Int. J. Offshore Polar Eng. **32**(04), 457–462 (2022b)

Wang, X., Gu, X.: Risk-based ultimate strength design criteria for very large floating structure. Ocean Eng. **223**, 108627 (2021)

Wang, Y., et al.: Loads on a vessel-shaped fish cage steel structures, nets and connectors considering the effects of diffraction and radiation waves. Mar. Struct. **86**, 103301 (2022c)

Wang, Y., Fu, S., et al.: Load effects on vessel-shaped fish cage steel structures, nets and connectors considering the effects of diffraction and radiation waves under irregular waves. Mar. Struct. **91**, 103468 (2023d)

Wang, Y., et al.: Simulation of special-shaped graded particulate hydraulic transport in deep-sea mining scenarios. Powder Technol. **448**, 120344 (2024c)

Wang, Z., Sha, Y., Jakobsen, J.B.: Floating bridge response under combined ship collision, wind and wave loads. Ships Offshore Struct., 1–18 (2023e)

Watanabe, E., Utsunomiya, T., Wang, C.M.: Benchmark hydroelastic responses of a circular VLFS under wave action. Eng. Struct. **28**(3), 423–430 (2006)

Watkins, P.D., Roueche, J.N., Miller, D.L.: American petroleum institute standards and other initiatives in support of the energy transition. In: Offshore Technology Conference, p. D031S038R002. OTC (2024)

Wei, W., et al.: A discrete-modules-based frequency domain hydroelasticity method for floating structures in inhomogeneous sea conditions. J. Fluids Struct. **74**, 321–339 (2017)

Wei, W., et al.: A time-domain method for hydroelasticity of very large floating structures in inhomogeneous sea conditions. Mar. Struct. **57**, 180–192 (2018)

Wei, W., et al.: A time-domain method for hydroelasticity of a curved floating bridge in inhomogeneous waves. J. Offshore Mech. Arct. Eng. **141**(1), 14501 (2019)

Wei, Y., et al.: Coupled analysis between catenary mooring and VLFS with structural hydroelasticity in waves. Mar. Struct. **93**, 103516 (2024)

Wiegard, B., et al.: Fluid-structure interaction and stress analysis of a floating wind turbine. Mar. Struct. **78**, 102970 (2021)

Wu, Y.: Hydroelasticity of Floating Bodies. University of Brunel (1984)

Wu, Y., et al.: Composite singularity distribution method with application to hydroelasticity. Mar. Struct. **6**(2–3), 143–163 (1993)

Wu, Y., et al.: The progress in the verification of key technologies for floating structures near islands and reefs. J. Hydrodyn. **33**, 1–12 (2021)

Xia, D., Kim, J., Ertekin, R.C.: Review on conceptual designs and key technologies of very large floating structures. J. Ship Mech. **24**, 135–148 (2000a)

Xia, D., Kim, J.W., Ertekin, R.C.: On the hydroelastic behavior of two-dimensional articulated plates. Mar. Struct. **13**(4–5), 261–278 (2000b)

Xia, S., et al.: Vibration control of multi-modular VLFS in random sea based on stiffness-adjustable connectors. Appl. Sci. (2022) [Preprint]

Xia, S.Y., et al.: On retaining a multi-module floating structure in an amplitude death state. Ocean Eng. **121**, 134–142 (2016)

Xiang, S., Cheng, B., Tang, M., Abdelbaset, H.: Effects of spatial inhomogeneity of wave excitations on structural behaviors of multi-span floating bridges. Ocean Eng. **243**, 110340 (2022a)

Xiang, S., Cheng, B., Tang, M., Zhang, S.: Hydrodynamic characteristics of deep-water bridge floating foundations with different mooring systems. Ocean Eng. **257**, 111635 (2022b)

Xiang, S., et al.: Structural dynamic performance of floating continuous beam bridge under wave and current loadings: an experimental study. Appl. Ocean Res. **137**, 103604 (2023)

Xiang, X.: Extensive hydro-elastic floating bridge tests: planning, design, implementation, and numerical comparison. Appl. Ocean Res. **140**, 103741 (2023)

Xiang, X.: Floating bridge global responses with hydrodynamic interaction. Ocean Eng. **291**, 116420 (2024)

Xie, W., et al.: Dynamic responses of a large fishing net under waves with boundaries deforming with time. Ocean Eng. **276**, 114245 (2023)

Xu, D.L., et al.: On study of nonlinear network dynamics of flexibly connected multi-module very large floating structures. In: Vulnerability, Uncertainty, and Risk: Quantification, Mitigation, and Management, pp. 1805–1814 (2014)

Xu, P., Wellens, P.R.: Theoretical analysis of nonlinear fluid-structure interaction between large-scale polymer offshore floating photovoltaics and waves. Ocean Eng. **249**, 110829 (2022)

Xu, S., et al.: A novel conceptual telescopic positioning pile for VLFS deployed in shallow water: structure design. China Ocean Eng. **34**(4), 526–536 (2020a)

Xu, T.J., et al.: Numerical investigation of the hydrodynamic behaviors of multiple net cages in waves. Aquac. Eng. **48**(none), 6–18 (2012)

Xu, T.J., et al.: Analysis of hydrodynamic behavior of a submersible net cage and mooring system in waves and current. Appl. Ocean Res. **42**(Complete), 155–167 (2013)

Xu, X., Sasmal, K., Wen, Y., et al.: An integrated approach for the decision of wave energy converter deployment based on forty-five-years high-resolution wave climate modeling. Energy. **305**, 132238 (2024a)

Xu, X., Sasmal, K., Xu, H., et al.: Ocean wave modelling for indo-Pacific region under extreme weather conditions using the third-generation wave models. In: International Conference on Offshore Mechanics and Arctic Engineering, p. V002T02A005. American Society of Mechanical Engineers (2024b)

Xu, Y., et al.: Efficient prediction of wind and wave induced long-term extreme load effects of floating suspension bridges using artificial neural networks and support vector machines. Ocean Eng. **217**, 107888 (2020b)

Yang, J.S., Yang, J.L., Wang, B.: Boundary control for floating beam system in irregular waves under one end pinned to reduce hydroelastic response. Ocean Eng. **270**, 113586 (2023)

Yang, P., Li, Z., et al.: Boussinesq-Hydroelasticity coupled model to investigate hydroelastic responses and connector loads of an eight-module VLFS near islands in time domain. Ocean Eng. **190**, 106418 (2019a)

Yang, P., Liu, X., et al.: Hydroelastic responses of a 3-module VLFS in the waves influenced by complicated geographic environment. Ocean Eng. **184**, 121–133 (2019b)

Yao, Y., et al.: Modeling wave processes in a reef-lagoon-channel system based on a Boussinesq model. Ocean Eng. **268**, 113404 (2023)

Yeter, B., Garbatov, Y., Soares, C.G.: Risk-based maintenance planning of offshore wind turbine farms. Reliab. Eng. Syst. Saf. **202**, 107062 (2020)

Yin, Y., Jiang, S.-C., Liu, H.: Fluid resonance in the narrow gap for a ship close to the vertical caisson on impermeable and permeable beds. Ocean Eng. **286**, 115422 (2023)

Yu, J., et al.: Mooring design of offshore aquaculture platform and its dynamic performance. Ocean Eng. **275**, 114146 (2023)

Yu, Q., Xu, J.: Fatigue reliability assessment of floating offshore wind turbines under correlated wind-wave-current loads. Ocean Eng. **313**, 119442 (2024)

Yu, Z., et al.: Numerical analysis of local and global responses of an offshore fish farm subjected to ship impacts. Ocean Eng. **194**, 106653 (2019)

Yuan, Y., et al.: FUNWAVE-GPU: multiple-GPU acceleration of a Boussinesq-type wave model. J. Adv. Model. Earth Syst. **12**(5), e2019MS001957 (2020)

Yuan, Y., et al.: Multidisciplinary design optimization of dynamic positioning system for semi-submersible platform. Ocean Eng. **285**, 115426 (2023)

Zhan, J.M., et al.: Analytical and experimental investigation of drag on nets of fish cages. Aquac. Eng. **35**(1), 91–101 (2006)

Zhang, C., et al.: Development of compliant modular floating photovoltaic farm for coastal conditions. Renew. Sust. Energ. Rev. **190**, 114084 (2024a)

Zhang, D., et al.: Hydrodynamic modelling of large arrays of modularized floating structures with independent oscillations. Appl. Ocean Res. **129**, 103371 (2022a)

Zhang, G., Chen, X., Wan, D.: MPS-FEM coupled method for study of wave-structure interaction. J. Mar. Sci. Appl. **18**, 387–399 (2019)

Zhang, H.C., et al.: Network dynamic stability of floating airport based on amplitude death. Ocean Eng. **104**, 129–139 (2015)

Zhang, H.C., et al.: Wave energy absorption of a wave farm with an array of buoys and flexible runway. Energy. **109**, 211–223 (2016)

Zhang, H.C., et al.: Connection effect on amplitude death stability of multi-module floating airport. Ocean Eng. **129**, 46–56 (2017)

Zhang, J., et al.: Analytical models of floating bridges subjected by moving loads for different water depths. J. Hydrodyn. Ser. B. **20**(5), 537–546 (2008)

Zhang, L., et al.: Optimized WAVEWATCH III for significant wave height computation using machine learning. Ocean Eng. **312**, 119004 (2024b)

Zhang, S., et al.: Frequency-domain hydroelastic stress analysis of floating structures under spatially inhomogeneous wave field. Ocean Eng. **288**, 115937 (2023a)

Zhang, X., et al.: A time domain discrete-module-beam-bending-based hydroelasticity method for the transient response of very large floating structures under unsteady external loads. Ocean Eng. **164**, 332–349 (2018)

Zhang, X., et al.: Feasibility of very large floating structure as offshore wind foundation: effects of hinge numbers on wave loads and induced responses. J. Waterw. Port Coast. Ocean Eng. **147**(3), 4021002 (2021a)

Zhang, X., Lu, D.: An extension of a discrete-module-beam-bending-based hydroelasticity method for a flexible structure with complex geometric features. Ocean Eng. **163**, 22–28 (2018)

Zhang, Y., et al.: Storm damage risk assessment for offshore cage culture. Aquac. Eng. **95**, 102198 (2021b)

Zhang, Y., et al.: Phase-resolved modeling of wave interference and its effects on nearshore circulation in a large ebb Shoal-Beach system. J. Geophys. Res. Oceans. **127**(10), e2022JC018623 (2022b)

Zhang, Y., et al.: Aerodynamic and structural analysis for blades of a 15MW floating offshore wind turbine. Ocean Eng. **287**, 115785 (2023b)

Zhang, Y., et al.: Hydrodynamic analysis and validation of the floating DeepCwind semi-submersible under 3-h irregular wave with the HOS and CFD coupling method. Ocean Eng. **287**, 115701 (2023c)

Zhang, Y., et al.: A study of tension and deformation of flexible net under the action of current. Ships Offshore Struct. **19**(1), 27–44 (2024c)

Zhao, B., et al.: Wind stress modification by offshore wind turbines: A numerical study of wave blocking impacts. Ocean Eng. **313**, 119651 (2024a)

Zhao, C., et al.: Influence of hinged conditions on the hydroelastic response of compound floating structures. Ocean Eng. **101**, 12–24 (2015)

Zhao, D., et al.: Coupled analysis of integrated dynamic responses of side-by-side offloading FLNG system. Ocean Eng. **168**, 60–82 (2018)

Zhao, G., Dong, S., Zhao, Y.: Fatigue reliability analysis of floating offshore wind turbines under the random environmental conditions based on surrogate model. Ocean Eng. **314**, 119686 (2024b)

Zhao, N., et al.: Ultimate strength of brace strut structure of VLFS under bending and shear loads. Chuan Bo Li Xue/J. Ship Mechan., 1344–1353 (2022). https://doi.org/10.3969/j.issn.1007-7294.2022.09.009

Zhao, N., Gu, X.K., Li, Z.J.: Research development of ultimate strength of very large floating structures. Chuan Bo Li Xue/J. Ship Mech., 1412–1426 (2021). https://doi.org/10.3969/j.issn.1007-7294.2021.10.013

Zhao, Y.-P., Gui, F., et al.: Numerical analysis of dynamic behavior of a box-shaped net cage in pure waves and current. Appl. Ocean Res. **39**, 158–167 (2013a)

Zhao, Y.-P., Bi, C.-W., et al.: Numerical simulation of the flow field inside and around gravity cages. Aquac. Eng. **52**, 1–13 (2013b)

Zhao, Y., et al.: Experimental investigations on hydrodynamic responses of a semi-submersible offshore fish farm in waves. J. Mar. Sci. Eng. **7**(7), 238 (2019)

Zheng, T., Chen, N.-Z., Yuan, L.: Structural strength assessment for thin-walled blade root joint of floating offshore wind turbine (FOWT). Thin-Walled Struct. **191**, 111057 (2023)

Zheng, X., et al.: An offshore floating wind–solar–aquaculture system: concept design and extreme response in survival conditions. Energies. (2020). https://doi.org/10.3390/en13030604

Zhou, Z., et al.: Real-time hybrid simulation incorporating machine learning for deep-water bridges subjected to seismic ground motion with fluid-structure dynamic interaction. Soil Dyn. Earthq. Eng. **175**, 108263 (2023)

Zhu, T., et al.: WRF-CFD/CSD analytical method of hydroelastic responses of ultra-large floating body on maritime airport under typhoon-wave-current coupling effect. Ocean Eng. **261**, 112022 (2022)

Zieger, S., et al.: Observation-based source terms in the third-generation wave model WAVEWATCH. Ocean Model., 2–25 (2015). https://doi.org/10.1016/j.ocemod.2015.07.014

Zijlema, M., Stelling, G., Smit, P.: SWASH: an operational public domain code for simulating wave fields and rapidly varied flows in coastal waters. Coast. Eng. **58**(10), 992–1012 (2011)

Zong, Z., et al.: Proposal of a rational function wave Spectrum (RFWS) form for atoll waves. Ocean Eng. **189**, 106402 (2019)

Zou, L., et al.: Evolution wave condition using WAVEWATCH III for island sheltered area in the South China Sea. J. Mar. Sci. Eng. **11**(6), 1158 (2023)

Zou, M., et al.: Experimental and numerical investigation of gap resonances between side-by-side fixed barges under beam sea excitation. Ocean Eng. **297**, 117150 (2024)

Open Access This chapter is licensed under the terms of the Creative Commons Attribution-NonCommercial-NoDerivatives 4.0 International License (http://creativecommons.org/licenses/by-nc-nd/4.0/), which permits any noncommercial use, sharing, distribution and reproduction in any medium or format, as long as you give appropriate credit to the original author(s) and the source, provide a link to the Creative Commons license and indicate if you modified the licensed material. You do not have permission under this license to share adapted material derived from this chapter or parts of it.

The images or other third party material in this chapter are included in the chapter's Creative Commons license, unless indicated otherwise in a credit line to the material. If material is not included in the chapter's Creative Commons license and your intended use is not permitted by statutory regulation or exceeds the permitted use, you will need to obtain permission directly from the copyright holder.

Committee V.7: Structural Assessment During Operations

J. M. Underwood[1(✉)], A. Barbato[2], A. Bekker[3], M. Braun[4], M. A. Eder[5], P. Hess[6], C. Jochum[7], D. Morato[8], V. Nilva[9], N. Osawa[10], P. Pahlavan[11], D. Sarsoza Burgos[12], Â. P. Teixeira[13], I. Thompson[1], and Y. Wang[14]

[1] Southampton, United Kingdom
james.underwood@uk.bmt.org
[2] Fincantieri, Trieste, Italy
[3] Stellenbosch University, Stellenbosch, South Africa
[4] German Aerospace Center (DLR), Geesthacht, Germany
[5] Technical University of Denmark (DTU), Roskilde, Denmark
[6] Naval Surface Warfare Center Carderock Division, West Bethesda, USA
[7] Brest, France
[8] Technical University of Munich (TUM), Munich, Germany
[9] Odessa, Ukraine
[10] Kanazawa Institute of Technology, Kanazawa, Japan
[11] Delft University of Technology, Delft, Netherlands
[12] São Paulo, Brazil
[13] Lisbon, Portugal
[14] Changsha, Hunan, China

Committee Mandate. Concern for the structural assessment of ship and offshore structures during operations, including unmanned operations. The focus shall be on methodologies translating monitoring and inspection data into operational and life-cycle management advice, with associated criteria for decision making. This shall include diagnosis and prognosis of structural health, prevention of structural degradation and failures, and structural renewal and reuse.

The research and development in passive, latent and active systems including their sensors and actuators shall be addressed. Special attention is to be given to structural digital twin technology and methods including reduced order analysis, inverse modelling, and AI technology application, combined with the use of monitoring systems and inspection data, to provide real-time advice for safe operation during the structural life-cycle.

Keywords: Artificial Intelligence · Autonomous Systems · Acoustic Monitoring · Corrosion · Digital Shadow · Digital Twin · Fatigue · Machine Learning · Motion Monitoring · Offshore Structure · Predictive Maintenance · Ship Structure · Strain Measurement · Structural Assessment · Structural Health Monitoring · Wind Turbine Structure

1 Introduction

1.1 Introduction and Background to Mandate

This report presents the first review of the new ISSC Committee V.7 on Structural Assessment During Operations. The Committee's mandate has many similarities to that of the previous ISSC Committee V.7 on Structural Longevity (ISSC 2022), in looking to monitor and assess the performance of ship and offshore structures, and the aim of assessing the structural health of the platform. It is the view of this committee, that the primary differentiator lies in the timescales associated with the processing and use of the data available from structural monitoring.

Whilst assessment for structural longevity may take many months to process, often away from the asset, this committee focuses on the ability to use and respond to collated data within the timescales of the operation. For reasons such as speed, as much as limitations of moving large amounts of data from offshore to shore-based computational facilities, processing of data may be more likely conducted onboard the asset; therefore, being able to inform operators (or operating systems where a level of autonomy is present), to influence the operation of the platform during the timescales of its operation. However, the requirement to process date onboard the asset has not been considered a requirement for literature included within the review.

The definition of what may be considered an operation and associated timescales have not been narrowed in the review and reporting, and so this has been contextualised throughout the report in reference to the topic being discussed. Monitoring structural health for corrosion or fatigue would anticipate a need for longer monitoring timescales than, for example, stress monitoring of a crane during a lifting operation. For these examples, the operational timescale may be defined as between docking periods or limited to the duration of a voyage for the former, whilst the latter may be limited to the number of hours that a lifting operation is being conducted.

In all monitoring processes there is a level of similarity, beginning with the collation of data from the monitoring system, processing of that data, and responding to the data to ensure operation continues within the operational limits of the structure. However, how this is achieved may vary significantly. At a simplistic level, the readout of loadcells on a crane hook to the operator may be considered structural assessment during the operation, on the basis that the operator has an understanding of the planned lifting evolution and the operational limits of the crane, and therefore will react to changing information that may indicate the safety of the lift is in doubt. At a more complex level an autonomous vessel on transit may need to be trained through a digital representation of the vessel, from which it can collate and process structural data to influence its operation; for example, in terms of speed or heading against the prevailing conditions to keep slamming loads below required limits. Both ends of the spectrum are considered within the scope of the Committee's mandate.

1.2 Digital Representations

Digital transformation has enabled a variety of digital-physical entanglements for ship structures that springboard operational hindsight, insight and foresight. Reilly & Jorgensen (2016) refer to "smart" ships as highly instrumented, automated and interconnected vessels where onboard sensors monitor the state of the vessel towards maturing developments for autonomous ships. Where processing of data onboard the vessel or platform is conducted, often a digital representation of the structure is employed to assess structural capacity and demand for decision making, with or without human engagement. Such representation may be in the form of a calculation algorithm, through to full Finite Element Analysis (FEA) representation. The term "digital-twin" has been widely used in recent years, with common understanding being that of a digital counterpart of a physical object (Kritzinger et al. 2018). NASA originally defined the term digital-twin as an integrated multi-physics multi-scale probabilistic simulation of a physical system, that uses the best available physical models, sensor updates, fleet history, etc. to mirror the life of its twin (Shafto et al. 2010). In this regard the integration may be considered full in nature, with the flow of data both ways between the digital model and physical object (Kritzinger et al. 2018). Within the literature the term digital-twin often used to define digital representation with one-way or two-way influence on the physical asset. More recently the term "digital-shadow" has also entered use, being used to define a digital representation of the physical object integration with one-way data flow between the physical object, and any return influence requiring "manual" intervention (Kritzinger et al. 2018). Therefore, the following definitions may be considered applicable:

- Digital-Model: A design stage digital representation of the structure during phases when there is no physical product available, or of an existing structure with no exchange of data between them. Developed for the purposes of assessing the capability or capacity of the structure in reference to predicted demand loading through-life from operational loads, extreme and accidental loads as deemed applicable. Some literature has adopted terminology by Kritzinger et al. (2018) who classifies DTs according to the level of data integration. i.) A *digital model* relies on manual data transfer between physical and digital entities.
- Digital-Shadow: Changes in the physical object automatically affect changes in the digital object, but not vice versa (Kritzinger et al., 2018).
- Digital-Twin: A widely adopted definition (Niederer et al., 2021) states that a "DT is a set of virtual information constructs that mimics the structure, context and behaviours of an individual or unique physical asset, that is dynamically updated with data from its physical twin throughout its life-cycle, and that ultimately informs decisions that realize value". In the maritime context this is echoed by the latest definition by Van Der Horn & Valluri (2024) where a DT is defined as "a virtual representation of a physical asset, along with its environment and processes, comprised of integrated models that are updated through data exchange to provide decision-making support over its lifecycle."

The digital-model is not considered to be within the scope of the mandate of this Committee, being sufficiently covered by other ISSC Committees. Digital-shadows and digital-twins are considered within the scope of the mandate of this Committee, the distinction between digital twins and digital shadows is not consistently adhered to in literature.

Other emerging DT concepts for operational assets include that of a digital twin instance (DTI), referring to a single physical to virtual instance, whereas digital twin aggregates (DTA) combine information from DTI and DTAs. Such aggregations being referred to as a Digital Twin Ecosystem by Rigby and Christmas (2023). The individual reflection of multiple assets may lead to significant cost and complexity. Large systems often require a composite representation to manage information and knowledge among multiple DTs. A *federated digital twin* (FDT) is a potential solution to establish communication between multiple DTs, to achieve efficient and optimized cooperation in a virtual space. A FDT comprises the integration of autonomous DTs interconnected in a virtual environment, governed by a set of rules that determine interoperability, coordination and secure communications.

1.3 Report Content

The report is split into four primary chapters. Section 2 introduces the major themes and underlying approaches referred to within the committee's mandate, for example discussing Artificial Intelligence, Digital Twins and Machine Learning technologies for structural assessment and associated operational decision making. Section 3 introduces inspection and monitoring methods and techniques, including available sensor technologies that may be used within a structural assessment approach. Sections 4 and 5 present the application of approaches to measurement and use of structural assessment data in the context of offshore structures and ships respectively. Section 6 concludes the report with recommendations for the future direction of ISSC Committee V.7.

2 Condition, Performance, and Safety Assessment Techniques During Operations

2.1 Introduction

The use of material and environmental state awareness information is increasing to support operation and maintenance of ships and offshore structures throughout their service life. The information is used by both human and machine-based, operators and decision-makers, onboard and onshore, to meet performance and safety requirements without significant degradation of the material condition. Overuse or overloading of the structure could require immediate operational change, or lead to a reduction in service life and/or increased maintenance and repair costs. Traditionally, structural health monitoring (SHM) has supported overall life-cycle management of a structure to better understand maintenance, repair and recapitalization requirements (remaining useful life). The previous ISSC V.7 Structural Longevity committees (2012–2015, 2015–2018, 2018–2022) focused on structural life-cycle management without direct connection streaming and analysis of immediate or near-term vessel operational information.

Tools including Artificial Intelligence, Digital Twins, and Machine Learning, are increasingly used to better utilise and apply the acquired hull, environment, and other relevant source data. Data from a wide range of sources are being collected continuously and intermittently to inform human and machine decision-making, including direct structural measurements (e.g. strain, vibration), indirect structural measurements (e.g. acceleration), operating conditions (e.g. speed), and environmental data. Data processing requirements depend on the timescale of a resulting decision. For on-board operational guidance, data must be processed near or "faster than real-time" (through predictive algorithms) to inform operations, whereas for maintenance, the data processing can occur over a much longer timescale.

This chapter introduces the major themes and underlying approaches of this committee, which will be expanded upon in later chapters. Section 2.2 describes opportunities afforded by digitalization, Digital Twins (DTs), Artificial Intelligence (AI) and Machine Learning (ML). Definitions and descriptions of failure and degradation for operationally relevant limit states are introduced in Sect. 2.3. The relationship between operational and life-cycle perspectives are described in Sect. 2.4. Section 2.5 explores the operational system health monitoring aspects of autonomous/unmanned-to-manned vehicles and platforms, including Class Society perspectives.

2.2 Digitalization and Digital Engineering

2.2.1 Digital Twins and Machine Learning

DT comprises three main elements: (i) The physical reality (including the physical system, physical environment and physical processes), (ii) the virtual representation (an idealized, digital form of the physical reality), and (iii) the link which enables communication between the real asset and virtual representation which is termed entanglement (Minerva et al., 2020). Entanglement is characterized by i.) Connectivity - the direct or indirect means through which updates in state and/or behaviours can be realized between real and digital assets. ii.) Promptness - quantifies how timely communication between the real asset and its digital representation takes place. The latency of the digital reflection should be negligible compared to the needs of the intended use of the DT. iii.) Association - refers to the direction of communication between the real asset and its digital counterpart. Communication can be uni-directional (typically from the physical asset to digital counter-part) or bi-directional where virtual to physical information flow may additionally be enabled. Entanglement can be defined as weak (where information is inferred by in-direct observations), simple (uni-directional link, not necessarily real-time with links that may be interrupted) or strong where a constant bi-directional link is established, and the digital representation may be the controlling instance and may be ultimately automated.

In their review Liu et al. (2021a) state the commonalities between DT definitions as being: i.) Individualized: A DT is one-to-one with the individual physical twin. ii.) High-fidelity: The DT can accurately simulate the desired response of the physical in the virtual space. iii.) Real-time: This means that the DT responds to physical twin with appropriate latency to deliver value. iv.) Controllable: Changes on DT or physical twin control the other twin.

Other emerging DT concepts for operational assets include that of a DT instance, referring to a single physical to virtual instance, whereas DT aggregates combine information from DTI and DTAs. The individual reflection of multiple assets may lead to significant cost and complexity. Large systems often require a composite representation to manage information and knowledge among multiple DTs. A federated digital twin (FDT) is a potential solution to establish communication between multiple DTs, to achieve efficient and optimized cooperation in a virtual space. An FDT comprises the integration of autonomous DTs interconnected in a virtual environment, governed by a set of rules that determine interoperability, coordination and secure communications.

In a critical review of DT applications in Structural Integrity Management (SIM), Li and Brennan (2024a) define a DT for SIM as "a virtual representation of a physical structure that mirrors the same structural conditions in real-time". An integrated DT-based time-domain fatigue framework was proposed for SIM where the link between the physical domain (i.e. physical structure in the real world) and the digital/virtual domain (computer-based model) is established through the flow of monitoring data. The key enabling technologies are i.) model updating, ii.) real-time simulation and iii.) data-driven forecasting.

In related works, Li and Brennan (2024b) focus on feedback from the digital domain whereby DT-enabled virtually monitored data is incorporated into ship inspection planning based on four elements: i.) virtual monitoring - entailing modal- decomposition and expansion, ii.) data-driven forecasting through Bayesian updating (based on Markov-Chain Monte Carlo simulations) to update long-term stress range distributions, iii.) fatigue reliability using a probabilistic fracture mechanics formulation and iv.) inspection planning based on the calculated time-variant reliability index and the target reliability.

In the context of an industry-academia joint R&D project on DT for ship structures (DTSS), Fujikubo et al. (2024) defined a ship DT as "a virtual representation of the physical object that is constructed and maintained in cyberspace, incorporating actual monitoring/inspection and environmental data". The DTSS comprises a Finite Element structural model and includes data analysis and physical simulation capabilities to provide real-time estimates of ship performance. The performance of the object in cyberspace is then fed back to physical space for decision support at any stage of the ship's lifetime, from design, construction, and operation to recycling and disposal.

Schirmann et al. (2019) used DT-based predictions to generate cumulative fatigue damage for four ship route scenarios. The DT enabled insights such as an understanding that large head winds were correlated with excessive fatigue damage along one route. The authors state that DT-based insights may be considered alongside operational needs to assist decisions across a ship fleet.

Considering a DT as a digital model of its physical twin that includes a fully coupled information feedback loop between the real world and the virtual world (Papanikolaou et al. (2024), Giering and Dyck (2021), Minerva et al. (2020) and Kritzinger et al. (2018)), through a collection or cluster of coupled numerical models of its physical twin, the DT has the capability to make accurate and timely predictions of the effects a specific damage state has on the structural response or performance as e.g. presented in Haghshenas et al. (2023), Fahim et al. (2022), Stadtmann et al. (2023), Branlard et al. (2024) and Mahmoud et al. (2024).

A multimodal SHM system combines both, in-situ (locally embedded) and ex-situ (remote) sensor data. A general review of SHM methods in composite structures is provided by Wilson et al. (2017) and the importance of Machine Learning (ML) in

SHM is discussed by Shibu et al. (2023). Recent advancements in SHM are presented by Alokita et al. (2018). A relevant example for a multimodal application in SHM is presented by Samareh-Mousavi et al. (2024). In-situ sensors can be locally mounted onto the surface of the structure or be embedded in the material. Ex-situ NDT methods can be classified as either tactile or remote. Tactile SHM refers to ultrasound, eddy-current or shearography NDT methods that are applied and conducted directly on the structure by e.g. robot crawlers as proposed by Marques and Sattar (2023), or underwater drones. Remote SHM refers to visual-spectrum and infrared-spectrum based inspection. Examples for IR based damage reconstruction methods for SHM in composite materials can be found in von Houwald et al. (2024), Demleitner et al. (2024), Sarhadi et al. (2022) and Albuquerque et al. (2024). Most typically such methods are deployed by airborne drones or other platforms as presented by Chen et al. (2023a), Attallah et al. (2023) and Petrosjan (2021).

A SHM system must be capable of detecting the exact location, size, depth and shape of a damage and needs to be able to track damage evolution with sufficient temporal and spatial resolution (≤ 10 mm). Inferring from the performance of current state-of-the-art SHM technology discussed by Tonelli et al. (2021) it appears that none of the existing sensor systems and measurement technologies alone are currently living up to this requirement. One step towards overcoming this limitation is to combine in-situ and ex-situ data and to integrate them into a multimodal SHM system as proposed by Khana et al. (2014). However, the integration of vastly different and not ad-hoc compatible sensor data to render an accurate and meaningful damage characterization in a multidimensional parameter space is complex. The two main bottlenecks relate to: 1) managing, storing, and accessing large data sets; and 2) filtering and interpreting data.

Real time analytics approaches are therefore becoming increasingly important to filter big data before processing to facilitate prompt detection. An introduction to the concept of real time analytics can be found in Steele et al. (2016) and Chen et al. (2023b). ML models have proven their suitability to handle this type of problems and have been successfully applied to different NDT methods albeit mostly on the laboratory scale. A review of ML applications in defect detection is provided by Liu et al. (2024). The common limitation when applying ML models is the training process which requires many training cases, which for a multimodal SHM system can be challenging to establish – both in terms of number and fidelity. Moreover, the training process of a multi-parameter ML model can be computationally demanding. More research and development are required for identifying and training suitable ML models which can handle several different multimodal sensor data feeds.

2.2.2 Applications and Limitations of the DT

In the framework of this section, a DT specifically refers to a finite element model or a collection thereof in which the structure and the damage are discretized with finite elements such as the concept introduced in the ReliaBlade project (2024). The DT numerically solves the governing coupled non-linear partial differential equations (this includes coupled multi-physics problems e.g. thermo-mechanics, electro-mechanics etc.) by shape function approximations and provides the internal field variable histories such as stress, strain and displacement as a function of the loads, boundary conditions and the damage state. When interconnected with SHM, the DT provides condition and performance metrics - such as the remaining lifetime and power production - with respect to the different operation parameters (e.g. rotor speed, turbulence level, wind speed, wave conditions etc.) based on the current damage state. In this way the DT generates a set of possible scenarios that can be used as training data for the AI. For example, the DT can predict the effect on the lifetime if operation continues as is, or if the current most critical damage is repaired or if the structural loads are ameliorated by rotor speed reduction. This capability is a key element for establishing an Autonomous System (AS).

The first hurdle in the realization of the DT is the development of a suitable preprocessor designed to convert the multimodal SHM data into an accurate geometrical and temporal discretization of the as-measured damage state in the finite element mesh. A treatise of the challenges of damage identification in SHM is presented by Burgos et al. (2020). To provide a concrete example: If delamination damage is detected by the SHM in a rotor blade, the sensor data must first be converted into information regarding the exact location and shape of the delamination in a global structural coordinate system. Additionally, its location in the thickness direction must be determined in a local layup material coordinate system. Subsequently, this information must be further converted into finite element mesh information, determining which nodes of the intact structure need to be locally disconnected to discretize the delamination crack geometry in the DT. In other words, SHM sensor data cannot be directly fed into a DT but requires data preprocessing in several stages.

The second limitation preventing realization of a DT pertains to the mutually competing requirements between accuracy and computational efficiency, as an increase of the former entails an exponential decrease of the latter. Computational speed is required to increase the fidelity of the AI, surrogate model, which is proportional to the number of training sets generated per unit of time. On the other hand, higher accuracy of the predictions requires an increase of the finite element discretization level which leads to soaring solution times. For a DT to make meaningful predictions, requires the implementation of material damage models that add a considerable computational demand into the finite element framework. Some relevant examples of such numerical damage modelling approaches for composite materials can be found in Miao et al. (2022), Eder et al. (2019) and Erives Anchondo (2022).

While research in the past decades has mainly focused on improving the accuracy of such numerical damage models, the aspect of computational efficiency especially in terms of large-scale structural applications has been mostly neglected as pointed out by Chen and Eder (2020). More research and development are required for finding finite element analysis approaches which can reduce the solution time whilst maintaining the required accuracy. In the following, a few promising improvements are provided. One approach is the utilization of e.g. sub-models or other techniques as proposed by Haider (2024). The application of surrogate modelling is also a developing field with promising results as presented by Silionis et al. (2024). Another approach, for example, to increase the speed of discrete fatigue crack growth analyses is to decouple the computationally expensive fracture analysis from the crack propagation simulation. An application of this approach to adhesive bondlines in wind turbine rotor blades – coined FASTIGUE - was proposed by Eder and Chen (2020). It is expected that the rapidly evolving ML technology will entail a significant reduction of computation time in DTs of large structures through co-simulation capabilities. ML models can be implemented in fatigue growth simulations to replace the computationally expensive parts which involve repeating identical computation steps. In this way, the solution is predicted by a ML model, where rerunning the same type of simulation repeatedly can be omitted.

Van Der Horn and Valluri (2024) describe a verification and validation (V&V) framework for DTs to establish trust for the users. Verification is described as "Did I build the thing right?" and Validation is described as "Did I build the right thing?". The authors first define terms of reference as they apply to DTs then proceed to provide a framework to define, determine, generate, and document a Digital Twin V&V to establish credibility and support use in specific applications. While not prescriptive, the paper covers important aspects to ensure trustworthy use of a digital representation of a physical asset for decision-making. A future American Bureau of Shipping Guide on the topic is forthcoming and will include modelling, simulation and DT V&V.

2.2.3 Applications and Limitations of AI

AI takes the role of making automated decisions on different levels of operation by minimizing a given loss function. For this purpose, the AI not only uses information on the current structural health state from the SHM but receives prognostic data of the structural health index predicted by the DT for different virtual scenarios of operation. In addition, it considers information from SCADA supplemented by the predicted wind speed, turbulence levels and wave conditions from meteorological forecasts and considers predictions of the price of electricity, among other factors.

It should be emphasized that the role of the AI should be strictly distinguished from the role and purpose of a DT (Mahmoud et al. 2024). If well trained, the AI can obtain a solution of the underlying discrete optimization problem for e.g. maximizing the total power output over the total lifetime of each individual wind turbine in a windfarm. Subsequently, on Level-1 the AI will continuously change the control parameters of the physical twin in quasi-real time during operation – which closes the feedback loop of the AS. On Level-2 the AI aids the decision regarding the scheduling of the upcoming predictive maintenance interval.

The first notable advantage of the AI is its capability to control the operation of several assets simultaneously by considering the complex coupled dependencies of a multidimensional parameter space in quasi real time. These rather small changes made during short time increments in the measurement-prediction-control procedure have a cumulative effect which eventually facilitates a net increase of the asset's longevity. The second advantage of the AI lies in a significant transition from reactive maintenance towards predictive maintenance. However, despite the power of AI and ML models to solve ill-posed and complex structural analysis problems efficiently, the often overlooked downside is the training effort. Wang et al. (2022a) provide a comprehensive review on this subject. The accuracy and reliability of AI and ML models alike are strongly dependent on the quality of training and their inability of these techniques to extrapolate solutions beyond the trained parameter space. A treatise of current limitations of deep learning are provided by Thompson et al. (2022). Apart from increasing computational procedures to increase the learning speed, another issue is the supply of sufficient training data.

Training of an AI requires a large number of training sets comprising ground truths (GTs) and the associated input parameters. In terms of the AI envisioned in this chapter, the GT would be the structural health index, and its associated state parameters comprising SHM sensor data, environmental data and operational parameters. In the context of offshore wind turbines, SHM sensor data is complimented by wave loads, aeroelastic loading conditions as a function of rotor speed to name only a few. A current limitation is the high demand for accuracy of the training data between the GT and the state parameters of the AI input, where any uncertainty in the training data will inevitably increase the uncertainty of the AI predictions. Meaning, that the generation of training data with sufficient accuracy and repeatability i.e. precision can be challenging. This situation is exacerbated by the need for a vast number ($>1e^6$) of high-quality training cases whose number increases with the statistically varied linear combination of the number of AI input parameters. This should highlight the research demand on new ways to continuously provide training data for the AI with sufficient quality and quantity.

The generation of training sets for structural health problems requires both, experimental data of known initial defects grown under controlled laboratory conditions in testing machines supplemented with high-fidelity numerical simulations of the same. The reason being that experimental data is arguably most accurate; however, synthetic training data is required to cope with the high demand for training cases. Regarding experiment-based training sets, new experimental approaches are required in which the damage is tracked with multiple integrated SHM sensors instead of employing a single NDT method as is currently most typical. While experimentally generated training sets arguably have high fidelity in terms of being realistic, they cannot be provided in the required abundance due to cost and time limitations. This emphasizes the role of DTs to generate training sets through virtual test simulations to supplement training cases for the AI.

An important aspect of virtual- or synthetic training set generation is the development of more advanced co-simulation techniques, and an extension of current data augmentation techniques as treated by Maharana et al. (2022). The term "data augmentation" refers to the process of perturbing the GT and associated parameters (e.g. by flipping, rotating or translation) to generate new – physically meaningful – training sets while avoiding to re-run a computationally expensive simulation.

2.2.4 Concluding Remarks

The advantages of advancing to partial or full autonomy in operating systems are undoubtedly striking and tempting. The endeavours to attain this goal started recently due to major advances in ML, AI, big data handling accompanied by an increase of computational power. ML will increasingly play a central role in structural applications to overcome some of the hurdles mentioned in this section by their ability to solve ill-posed inverse problems which many well established numerical approaches such as the finite element method have difficulties solving. However, more highly interdisciplinary and collaborative research will be required on system integration and on interfacing different realms which have traditionally developed rather independently – such as mechanical testing, NDT, ML and numerical analysis of large structures.

2.3 Platform Operator Definitions of Failure and Degradation

During operations, crews and their broader organizations manage degradation and failures related to loss of material thickness, fatigue crack initiation and crack growth, panel buckling and ultimate bending collapse of the hull girder, which may involve multiple sets of these aspects. Collette et al. (2022) interviewed ship crews to understand how decisions are made in support of future uncrewed vessel research. Although operators constantly assess the current state and the vessel's capabilities, structural aspects tend to have little influence on immediate operational decisions. Interviewed operators considered structural concerns as issues best addressed while in port, except for coating to control corrosion. However, extreme structural risks remain present and the effects of gradual degradation demand attention through the duration of lifetime operation.

2.3.1 Corrosion

Loss of material thickness due to corrosion reduces structural strength and the resistance to other degradation and failure modes. Vukelic et al. (2021) quantify the corrosion damage of butt-welded shipbuilding steel, including mass change, surface morphology, and pit development and growth. Kim et al. (2020a) propose an approach to fit corrosion measurements to relationships that account for nonlinear time-dependent corrosion. A follow-on study uses a series of statistical tests to represent corrosion development, including a validating case study of a ballast tank structure (Kim et al. 2020b). The results show how corrosion depth varies with time under the influences of different assumptions. Esfidan & Ranji (2024) take a similar approach, but they propose corrosion models that vary depending on the type of ship structure.

Data-driven approaches have been popular to detect corrosion and model thickness reduction. Imran et al. (2023) review such approaches to detect or predict corrosion damage along with associated predictive maintenance models. Khayatazad et al. (2020) and Yao et al. (2019) show examples of image processing to detect and locate corrosion damage. The former paper relied on a colour and roughness detection algorithm, while the latter trained a neural network. Using Bayesian inference and previous corrosion depth measurements on deck plates of ballast tanks in an oil tanker, Kim et al. (2022) developed a probabilistic model of corrosion wastage. Diao et al. (2021) use low-alloy steel chemical composition and environmental parameters taken during corrosion experiments as input features to a machine learning approach to estimate corrosion rate. The results generalized to accurately predict corrosion rates in low-alloy steels containing nickel, despite using training samples that lacked nickel.

In many cases, corrosion risks are assessed considering individual components. Woloszyk and Garbatov (2024) present a probabilistic approach to represent component corrosion based on a limited set of measurements. Taking a broader, system-based view, Gong and Frangopol (2020) assess corrosion of multiple bay sections in an ultimate strength reliability analysis of a tanker. They show that the typical assumption of corrosion at one section results in an underestimate of failure probability.

2.3.2 Fatigue

Crack initiation due to cyclic stresses tends not to concern ships' crews, but the long-term effects on operations require attention from the larger operating organization, especially when considering life extension. The growth of those cracks to the point that they reduce safety and require unplanned repairs demands additional short-term attention.

Typical fatigue crack initiation analysis applies spectral analysis, representing operations and the wave environment with a probabilistic operational profile. Damage accumulation calculation often relies on the linear Palmgren-Miner rule and fatigue resistance as represented using appropriate stress-life S-N curves. Assuming fatigue crack initiation when the accumulated damage reaches unity provides a way to quantify the fatigue state of components. Yosri et al. (2022) present a case study of side details for a tanker as a typical application of these methods.

Cracks growing large could interrupt operations, with high economic costs for commercial vessels and a risk to meeting mission objectives for military vessels. In some cases, their presence can reduce a ship's ultimate strength, as Firdaus & Adiputra (2024) review. To reduce associated risks, Wang et al. (2019) examined 1 year of a container vessel's operations; they found that voyage optimization could limit crack growth by more than 50%. Makris et al. (2023) explain an approach to model crack growth considering a large collection of possible crack paths due to random sea state sequences. The authors use the results afterwards in a reliability analysis for a container ship. Their focus on component failure is common; studies that examine the larger structural system may include crack growth, but tend to evaluate ultimate strength failures (Kwon and Frangopol, 2012).

Fang et al. (2022) outline a way to predict real-time crack growth of an offshore platform within a digital twin to mitigate the risk of structural failure. They include complexities related to multi-mode loading, while limiting stress intensity factor calculation time with a surrogate finite element model. They trained a dynamic Bayesian network rack to further limit the time for growth calculations based on installed sensors.

2.3.3 Ultimate Hull Girder Strength

The ultimate compressive strength of stiffened plates, which may fail from yielding, panel buckling, or ultimate collapse of the hull girder, is complicated by multiple material, loading, and geometry factors. The International Association of Classification Societies (IACS) common structural rules (Common Structural Rules for Bulk Carriers and Oil Tankers, IACS, 2024) provide analysis guidelines. However, as Ringsberg et al. (2021) explain in a benchmark study, various developments in shipbuilding and complications in analysis challenge analysts to predict ultimate strength accurately. They outline historical developments and current approaches, mentioning that nonlinear finite element analysis is the preferred approach to calculate structural capacity despite its time-consuming nature. Further, the study addresses topics related to mesh density, residual stresses, geometric imperfections, material models, and finite element solver details.

Tekgoz et al. (2020) review developments in ultimate strength analysis. They explain the literature related to modelling generalized and pitting corrosion, addressing material properties degraded by corrosion, and the general effects of corrosion, dents, cracks, and cyclic loads on structures and their ultimate strength. In a review of corrosion effects on ultimate strength, Woloszyk & Garbatov (2022) lay out commonly used models to represent mean thickness loss and consideration of reduced material properties from corrosion. They review the effects of pitting corrosion, uniform and nonuniform general corrosion, and reliability analysis for corroded structures. Reliability analyses, such as those mentioned above by Gong and Frangopol (2020) and Kwon and Frangopol (2012), are the typical way to evaluate the risk of ultimate strength failure.

While most studies focus on steel structures, Liu et al. (2020) review developments in analysing ultimate strength for stiffened aluminium structures. Analysis of aluminium stiffened structures differ from that of steel because of yield strength reduction in the heat affected zone due to welding. As such, manufacturing method plays a significant role in the structure's ultimate load capacity.

2.4 Through Life Planning Methodologies and Tools

The concept of structural longevity focuses mostly on the hull integrity life-cycle question, as explored in the three previous ISSC V.7 committees (ISSC 2015, ISSC 2018, and ISSC 2022). Inspection techniques (global and local) are presented with their limitations. There is a need for connecting this information and data to answer the question of structural integrity, following the strategy of risk-based inspection to minimize the inspection costs. As for inspection, the intermittent acquisition of relevant data for further decision making, material science is a key point. This relates to the concept of degradation and its modelling (corrosion, pitting, coating degradation, fatigue and crack, thickness reduction). The authors explore different techniques and highlight limitations due to lack of cleanliness, inspection accuracy and reproducibility, skills of the operator, accessibility and recommend design improvement to ease the use of inspection techniques.

Digital twins support the connection between data and decision-making, but their use is not yet applied widely in the ship industry due to insufficient technology readiness. This aspect was highlighted by Erikstad (2019) by proposing a digital maturity service index. The author concluded that there is still a long way before this field (and especially digital twin) is mature and recommend, starting with development of digital services, i.e. a service driven perspective starting from the end-user needs for decision support in operation. This is similar to a typical engineering design process. Such an approach avoids the tough question of how to set up a relevant model, fed by the data (and which data) to twin a ship and perform integrity analysis and prediction while in operation.

Gathering data is easier than managing and making use of the data. Rovelli et al. (2022) present a general methodology for a modular tool tested on a steelmaking plant. The needs (or issues stated) are: capturing relevant temporal dynamics of an industrial process, management of multiple datasets, and provision of a correct Life-Cycle Analysis (LCA) interpretation to a non-expert audience.

The literature contains numerous papers related to systems and safety assessment (Wang, 2017). To some extent this can be a path to life planning methodology, as life planning also requires structural safety. The digital twin approach is also used in railway transportation, probably because its environment is milder than for maritime transportation. Bekker and Erikstad (2023) presented a conference paper entitled "Digital twin service patterns for maintenance management of operational rail assets". This is an example of how the digital service concept, aiming toward end-user needs, is more mature, and is based on engineering sensors to communicate the state and behaviour of an asset.

Consequently, virtual monitoring approaches offer to mitigate challenges associated with large volumes of instrumentation, typically, at the cost of accuracy. Several studies examined how ship tracks could be combined with measured or hindcast wave data and spectral analysis to calculate stress responses at any structural location (Thompson, 2018; Aarsnes et al., 2019). Van Der Horn et al. (2022) build on that approach to predict future damage with Monte Carlo simulations based on a mix of design and encountered conditions. Finally, as stated by Mogeke and Magoga (2023) virtual monitoring of a hull becomes more and more promising for life planning based on ship structural performance.

2.5 Autonomous/Unmanned-to-Manned Vehicles and Platforms

The progression towards increased autonomy prevails across the transport sectors, and the marine sector is no different. Interest spans the spectrum of marine vessels which have been categorized by Im et al. (2018) as shown in Fig. 1. Within these definitions there is inference to whether the vessel is crewed or uncrewed, and whether the vessel is controlled, remotely controlled or autonomous in its operation. Similarly, the ABS Smart Functions (ABS 2022) define decision control levels progressing from Manual, through Smart (system and human augmentation), Semi-Autonomy (increased system control with human augmentation), to Full Autonomy. Demonstration of an uncrewed vessel can be seen from small craft through to much larger container vessels. The Ocius' Bluebottle Uncrewed Surface Vessel (USV), has a use case to collect data supporting offshore wind turbine operators with environmental data and a provision for mine counter measure path planning. These vessels operate at up to 5kts using conventional sail or solar-powered propeller propulsion. At a larger scale the 80 m 120TEU container vessel Yara Birkeland (Yara 2024) has been undergoing autonomous (though crewed) trials to train the control and navigation sensors to enable future autonomous operations.

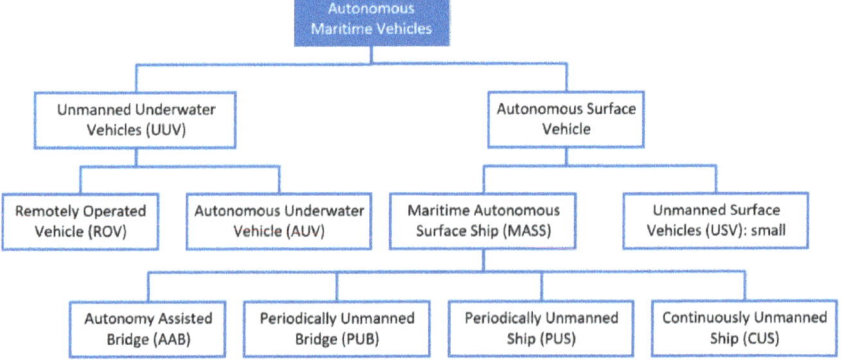

Fig. 1. Autonomous vessel definitions (Im 2018 - edited).

In published literature, autonomous vessel research primarily focuses on the navigation and manoeuvring control of the vessel. However, in crewed vessels, the operational limitations are often related to the human and not the structural capacity of the vessel. For example, a vessel with a short roll period may cause motion sickness as much as one with excessive roll; or, under slamming loads there are well published limits to the extent of slamming induced vibration the human body can safely experience over a given time period that may limit the vessel's operational profile. The move to uncrewed vessels may therefore present opportunities to operate closer to the structural capacity of the vessel if this can be assessed within the timeframe of the operation. In recent rule changes, reference to structural assessment during operation has been included.

Within the MCA Workboat Code (MCA 2024), additional requirements have been included in Annex 2 to Edition 3 of the code to include "vessels with no persons on board operated from a remote operation centre", or a "Remotely Operated Unmanned Vessel", and is applicable to vessels less than 24 m in length. The Annex does not remove any of the structural requirements compared to those required for a crewed vessel; however, the Annex introduces an additional requirement for the recording, both onboard and at the remote operation centre, of hull stresses, with the possibility that these are also to be displayed to the operator. It is not further elaborated by what means or how extensively hull stress should be assessed and recorded.

DNV's guidance for autonomous and remotely operated ships (DNV 2021) Appendix D also notes the requirement for the vessel to monitor movement and hull stress, but again without further elaboration as to the end goal of this assessment, or its potential active use in the operation of the vessel. There is also no further information provided as to the extent or methodology that would be acceptable to achieve such assessments.

Lloyd's Register's code for unmanned marine systems (LR 2017) does not include additional requirements for monitoring of hull stresses, but as per the MCA Workboat Code and DNV code, does not exempt any structural requirements otherwise applicable to a crewed vessel.

Whilst it may be considered beneficial to increase the level of information available to those monitoring or controlling remotely operated or autonomous vessels, a balance needs to be struck. An influencing factor in the move towards increased autonomy is to reduce the risk of human error in vessel operations. However, as discussed by Whalstrom et al. (2015), a balance is needed to avoid information overload, but also boredom of the operator that may lead to the need for intervention being missed.

For both remotely operated and autonomous vessels (where remote monitoring would be anticipated as a minimum), the ability to process sensor data onboard or elsewhere is an important consideration. Sensors continuously collecting data can generate large amounts of data very quickly, which may be unfeasible to transmit to shore-based processing facilities and relay commands back to the vessel. Therefore, a level of on-board data processing is required to filter or edge process the collected data, relaying a reduced data sample or only relaying data after detecting an anomaly from the accepted baseline profile.

Integrated application of DTs and Artificial Intelligence (AI) into an Autonomous System (AS) presents a vision going beyond current state-of-the-art capabilities. A DT for offshore wind energy applications might feature a structural finite element model as proposed by Branner et al. (2022) which discretizes the entire structure usually with higher fidelity sub-models in regions of specific interest.

In a pre-processing step Machine Learning (ML) is used to characterize the damage state in the structure by using a multimodal Structural Health Monitoring system (SHM). Multimodal refers to the combination of manifold SHM data e.g. from embedded sensors and remote inspection data from airborne based thermography and NDT from robot crawlers. Subsequently, the damage characterised by of type, location, shape and size is discretized in the DT model where their possible effects on the structural performance can be predicted for different loading scenarios. These predictions are then transferred to an AI model which optimises control commands fed back to the physical twin. Moreover, the AI provides reliable recommendations that aid the decision-making process of the human operators.

Such an integrated AS, as envisioned by Chen et al. (2021a) facilitates the transition from time interval-based maintenance towards predictive maintenance with significant savings in Operations and Maintenance (O&M) costs. While ML is already used in predictive maintenance, as discussed by Liu et al. (2023a), humans still decide on the maintenance intervals and repair strategies. Moreover, these decisions are based on information with a significant time lag and do not pertain to immediate adaptations of the operational control parameters.

According to Chen et al. (2021) and Branner et al. (2021), the AS would be able to make decisions and actively aid the decision-making process by considering a multidimensional parameter space in near real time, which would be far beyond current human based O&M capabilities. Figure 2 depicts a schematic representation of the AS, showing the three main development technologies highlighted red. Applied to offshore wind, the AS optimizes the operation of individual assets within an entire windfarm by controlling individual wind turbines. A direct feedback loop exists between the physical twin and the AI. A secondary loop exists where a DT provides the sensitivities or the weights for the parameters in the convolutional neuronal network. The human takes a supervisory role by evaluating the decisions made by the AI. The AI aides the decision making for predictive maintenance, leaving an option for the AI autonomy to be overruled.

Through the continuous feedback from the SHM – DT loop, it is always up to date regarding the health status and risk level of individual turbines and individual components within the turbine. It can distinguish between turbines being exposed to different turbulence levels in the wake field and can consider the effect of wind gusts on specific wind turbines. For instance, the AS could decide to pitch the rotor of individual wind turbines at a certain wind speed which was predicted to be critical for their present structural health state. The AS could also try to maximize the lifetime of the asset while accepting a decreased power output during this period or try to maximize the power output whilst keeping the structural health above a certain acceptable threshold within the upcoming maintenance interval. These turbine control decisions would be based on accurate predictions of their effects on the structure based on the current health state at a much higher frequency than is currently feasible.

Realising such an AS requires major advances of three different key enabling technologies: (1) the integration of advanced multisensory SHM systems and management of sensory data, (2) the development of high-fidelity DTs and (3) the development of an AI and adequate training thereof.

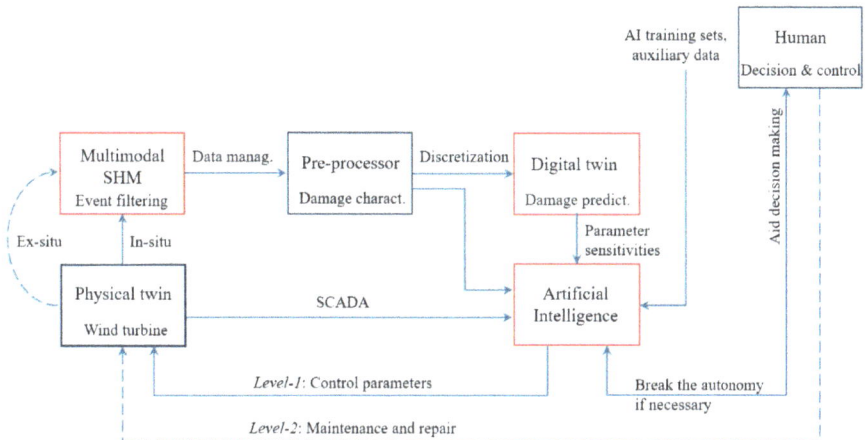

Fig. 2. Schematic depiction of the AS, as outlined above showing the three main key development technologies indicated in red.

2.6 Concluding Remarks

Material and environmental state awareness information is being used to inform human and machine operation and maintenance of ships and offshore structures throughout their service life. Continued development and experimentation is underway to more effectively employ AI, digital twins and Machine learning to digest the acquired hull, environment and other relevant data sources. The maturity of operational structural assessment methods continues to evolve on crewed and uncrewed platforms, but currently is very limited on uncrewed vessels where other onboard systems take the majority of the monitoring focus.

3 Inspection and Monitoring During Operations

3.1 Introduction

The essence of a system for structural assessment is the means to collect the required data from which assessments and decisions can be made. This chapter presents discussion of a variety of sensor types for collation of a broad range of data types. Some sensors and associated algorithms for collating and processing the collected data are presented in relation to their real-world application, where others presently remain proven within laboratory conditions, requiring further development to reach their real-world potential. It is often a combination of sensors that enable the health of a structure to be ascertained.

3.2 Inspection Techniques During Operation

This section describes inspection methods for ship and offshore structures that can be applied during operations, and includes a review of conventional and AI-assisted inspection methods. Recent developments of remotely-operated and autonomous vehicles (onboard, underwater, and aerial) applicable to ship and offshore structures during operation are reviewed.

3.2.1 Conventional and AI-Assisted Methods of Inspection

The use of Artificial Intelligence (AI) presented opportunities to remove the human requirement in the processing of inspection data, or to process and label types of data or quantities of data that would otherwise be overwhelming or impossible to process. As shown by the following studies, data sources can be varied and the AI used to process data may also differ.

Nicard et al. (2023) investigated ultrasonic Coda Wave Interferometry (CWI) to determine whether it can be used to continuously and quantitatively measure thickness loss due to corrosion on an S355 grade steel. Their study involves laboratory experiments where a steel sample was immersed in a 3% NaCl solution to induce uniform corrosion. The change in response to an emitted ultrasonic signal was evaluated over time and correlated to the loss in thickness. The authors conclude that CWI can be used to evaluate thickness loss under controlled laboratory conditions, but the method is very sensitive to any minor physical change in the material including as temperature or deformation.

Brijder et al. (2023) proposed an algorithm for corrosion detection and three algorithms for corrosion prognosis by using Bayesian filtering approaches. The authors quantitatively compare the algorithm accuracy against synthetic datasets having characteristics typical for wall thickness measurements using ultrasound sensors. The results show that a corrosion prognosis algorithm based on the Pourbaix corrosion model using unscented Kalman filtering outperforms the algorithms based on a linear corrosion model and the bimodal corrosion model introduced by other studies.

Han et al. (2024) proposed a novel inspection method for composite hull plates that combined ultrasonic testing with a correction for the wave propagation velocity dependent on the composite plate characteristics. The authors tested glass fibre reinforced polymer panels with different ply layer counts, glass fibre weight fractions and fabric types using an ultrasound probe to determine the propagation speed. Statistical analysis of the resulting data yielded a regression model of the wave propagation speed which could be used to correct thickness measurements. The authors conclude that the method can reduce measurement errors from 20%–30% to 1%–2%.

Anyfantis (2021) presented two condition monitoring techniques for structural damage in ships based on a numerical investigation in a simplified hull girder. The first method was an optimization scheme which used FE model iteration to locate damage from a number of strain readings by minimizing the difference between the measured strains and the computed strains. The second method used an artificial neural network trained using FE analysis data to locate structural damage given the strain in 40 locations on the ship. The author concluded that the optimization scheme was slightly more accurate than the ANN method; however, the computational cost is much higher as it requires multiple iterations of a large numerical simulation.

3.2.2 Remotely-Operated Vehicles for Inspection

Multiple studies demonstrate the increasing and broad use of Remotely-Operated Vehicles (ROVs) for inspection. Antony Jacob et al. (2021) introduce a pulse eddy current (PEC) technique to estimate the thickness in steel structures covered by marine growth. Their methodology is based on PEC transducers mounted on an ROV. Case studies are presented using field measurements. Applications, advantages, and limitation of these technique are discussed in a comparison to a diver-based inspection technique.

Waszak et al. (2023) present a large-scale dataset of annotated underwater inspection images for the training of autonomous ship inspection algorithms. The images were collected from ROV lifecycle inspections of ships and annotated by experts. Currently available datasets aimed to classify fish and marine growth whilst datasets aimed specifically at inspection remain undisclosed. It is suggested that because corrosion, paint peel and marine growth are difficult to annotate as they appear together and overlap, fusing their classes and running segmentation in post-processing would be more effective.

Kim et al. (2022) propose a hull surface condition inspection method that uses pictures taken by an ROV. The images are classified using a combination of six convolution neural networks. Each network was trained on a dataset of underwater images. The trained classifiers were then subjected to a new set of inspection images for hyperparameter tuning after which the classifiers were combined in a voting ensemble to more reliably determine the hull surface condition. The authors conclude that the method can reach classification accuracy upwards of 98%. They recommend to extend the training set to a wider variety of hulls and visibility conditions.

Jiao et al. (2024) present an underwater inspection system for concrete structures that uses a remote operated vehicle. The inspection vehicle is equipped with a camera for visual inspection of the structure. A classifier is used to recognise different types of damage such as cracks, corrosion, and exposed reinforcement. The system has been tested to a limited degree and more validation data is needed to properly assess the performance.

3.2.3 Autonomous Robotic Inspection

Autonomous inspection methods continue to be a future aspiration, with examples primarily within the research and development stages.

Lin and Dong (2023) addresses the imperative of enhancing ship navigation safety through efficient hull inspections. They explore the integration of computer algorithms and robots for streamlined data processing. Categorizing approaches into in-dock, underwater, and health monitoring-based methods, the study reviews manual and automated techniques. The paper concludes with a comprehensive discussion on hull inspection publications, suggesting potential directions for future development.

Kalaitzakis et al. (2021) investigate the feasibility of an inspection method that uses a drone to perform digital image correlation (DIC) measurements autonomously. The authors experimentally tested the ability of the drone to maintain a position in front of a target and to accurately carry out DIC measurements. The drone could successfully evaluate stresses and detect damage in a four-point bending test of a concrete beam while keeping on target autonomously.

3.3 Monitoring Techniques During Operation

This section focuses on sensing technologies for monitoring of ship and offshore structures during operation. State-of-the-art systems and implementations for monitoring strain, vibration, motion, acoustic emissions, and guided waves are assessed. Methods proposed for fusion of data from various sensor types are also reviewed.

3.3.1 Emerging Sensor Technologies and Applications

State-of-the-art technologies can be used for a variety of 'fleet needs' such as ship operation reliability, navigation check and control (ship classification in moderate-resolution synthetic aperture radar (SAR) images, automatic identification system (AIS) "duplication", usage of unmanned aerial vehicles (UAV)) and even crew's health care (usage of wearable devices and wireless biomonitoring sensors). Ship positioning on the water surface with sufficient accuracy is one of the fundamental requirements for safe navigation, which under "standard conditions" is achieved using GPS and the ship's radar. However, these solutions may not be enough in coastal areas, in narrow areas with a significant concentration of ships, where the risk of accidents increases. SAR images, monitoring of AIS integrity, usage of multiple UAVs can be used as additional tools that increase ship's operation reliability.

SAR is widely used in ship monitoring due to its all-time and all-weather observing ability with the swath of SAR images proportional to image resolution. High resolution SAR images (about 1 m) usually corresponds to a narrow swath of 5 km, whereas current satellite-borne SAR can provide swaths of 100–450 km at image resolution of 10–30 m. In this context, a reliable and operable method for ship classification in moderate-resolution SAR images is proposed by Zhou et al. (2021). Furthermore, the results of the comparative experiments show a fairly good performance for ship targets classification.

3.3.2 Strain Monitoring

Strain monitoring is the process of measuring and analysing mechanical strains or deformations experienced by a structure under operational or representative loading conditions. This process helps in assessing the structural integrity and performance, by monitoring how a structure deforms in response to applied forces. Measuring strain of ship structures provides insights into the stresses within components, which can be linked to material behaviour and ultimate strength. Sensors placed at critical points on the hull can accurately record strain changes due to loads from waves or manoeuvres, allowing for precise mapping of the supporting structure's dynamic stress state behaviour.

Strain gauges are the primary sensor type used by hull strain monitoring system (HSMS) and these are broadly separated into long base strain gauges (LBSGs) and short base strain gauges (SBSGs). By virtue of their length, LBSGs are relatively insensitive to local stress concentrations as they measure the changes in displacement between the two ends of the sensor, effectively averaging high localised strains. Long baseline strain gauges measure global hull girder strains, and are typically positioned along the ship's length to capture overall bending moments. Short baseline strain gauges measure local shear stresses, especially in areas with complex loading conditions like deck openings and confined locations (Phelps & Morris, 2013; Silva-Campillo et al., 2023). SBSGs are considerably more compact than LBSGs. Four different sensor technologies are notable: bondable or weldable foil gauges, fibre-optic sensors, strain rings, and short length LVDT type sensors.

Recently, optical strain measurements have gained importance due to their immunity to electromagnetic interference, ability to be embedded in structures, and long-term stability. Fiber Bragg Gratings sensors can measure global hull girder strains and local strains or accelerations from bow or bottom slamming or other external loads (Güemes, 2014). These sensors require permanent temperature measurement for compensation and an interrogator to translate optical signals into an electrical equivalent for post-processing (Güemes, 2014).

Liu et al. (2023b) describe the advent of wireless and passive RFID technologies with high efficiency and inexpensive hardware equipment as a new era of next-generation intelligent strain monitoring systems for engineering structures.

Experimental validation of wireless strain monitoring has been conducted by Gregori et al. (2019). Wireless strain gauges having a hybrid form between an RFID tag and a usually thin-film resistive strain gauge have been experimented. Installation and maintenance problems of the wireless sensor networks are overcome allowing high measurement accuracy, comparable to that of wired strain sensors, together with a long measurement transmission distance.

The huge amount of strain data acquired by advanced sensors should be processed to evaluate the health status of the ship structures, allowing optimized scheduled maintenance interventions. For this aspect, Liu and Bao (2023) presented a method for automatic interpretation of strain distributions measured from distributed fibre optic sensors for crack monitoring.

The choice of sensor type and placement depends on monitoring objectives, such as detecting overloading, fatigue damage, and other structural issues. Combining different sensor types can provide comprehensive structural health monitoring, ensuring the vessel's safety and integrity (Phelps & Morris 2013).

Silionis and Anyfantis (2023) delve into predictive maintenance for ship hull structures by detecting corrosion-induced thickness loss through strain measurements from strategically placed in-situ sensors. A detailed Finite Element model, subjected to quasistatic loading conditions and considering operational variability, facilitates data collection. The deterministic nature of the resulting signals enables the construction of a detector, evaluated for performance on a 10% thickness loss scenario. This work claims to mark a shift from preventive to predictive maintenance in the maritime sector, enhancing hull structure integrity.

Shen et al. (2023) propose a distributed optical fibre sensor beneath an anti-corrosion wrapping system to monitor the non-uniform corrosion of steel piles in a corrosive environment. The methodology is based on measuring strain variation induced by corrosion products and analysing non-uniform corrosion influenced by different environmental factors. Two analytical models considering build-ups and run-offs of corrosion products are established to quantify non-uniform corrosion-induced mass loss at any location on the surface of the steel pile. Visualization through a colour cloud map is applied to illustrate the severity of non-uniform corrosion on the pile surface.

Sireta and Storhaug (2022) present a method that uses a 3D finite element model to reconstruct the full structural response of a container ship from the data of strain gauges mounted on board. The method makes use of a modal decomposition of the measured structural responses at all of the strain gauges. Once the response modes were established from FE simulations of specific load cases, the proposed method uses the measured time series from the strain gauges to reconstruct the response from a linear combination of those modes using a conversion matrix. For validation, full scale measurement data from 6 strain gauges on a 19000 TEU container ship were used. Error levels on the method were 1% to 20% on maximum stresses and 3% to 80% for fatigue damage estimations. The authors conclude that the error levels could be further reduced if their proposed method for optimal sensor placement was employed.

3.3.3 Dynamic Response Monitoring

Vibration monitoring involves the measurement and analysis of the oscillations of a structure to assess its health and performance. This technique helps in understanding how a structure responds to dynamic loads and identifying any changes or abnormalities that might indicate structural issues.

Sousa et al. (2022) propose a Fiber Bragg grating-based (FBG) sensing technique to detect the initial stages of corrosion development in carbon steel. The thickness loss of steel plating is monitored using embedded FBG sensors. The proposed method consists in monitoring the change in natural frequency of the plates of the monitored structure due to thickness loss. Laboratory experiments were carried out on 1020 carbon steel under accelerated corrosion by immersion in a 0.1 M NaCl solution. The actual thickness loss was measured using an ultrasonic thickness gauge for validation. A link was found between the loss of mass and the natural frequency of the plate.

Chen and Cui (2023) present an FBG strain sensor designed for structural health monitoring in ship constructions under long-term alternating loads. Using a genetic algorithm, the sensor's structural parameters were optimized, achieving a sensitivity of 3.79 pm/µε and a 5.34-fold increase in strain sensitivity. The optimized sensor was successfully used to identify the modal parameters of a reinforced plate structure, confirming its effectiveness through comparison with numerical simulations.

Roberts et al. (2021) integrated GPS and fibre optic sensors to measure the dynamic displacements of the Smyril ferry's hull in the North Atlantic. FBG sensors, placed in the bow and engine room, gathered data at 1 kHz, while GPS tracked movements. Dominant frequencies, such as 12.25 Hz from the engine, were detected, correlating with both FBG and GPS data. The results provide insights into long-term ship health monitoring and vibration characteristics.

Szeleziński et al. (2021) investigate the use of laser vibrometry for detecting defects in welded joints of thin-walled plates. Laser vibrometry and piezoelectric accelerometers were used to measure vibrations induced by a modal hammer. The results, validated through a mathematical model, suggest laser vibrometry as a complementary method to accelerometers in non-destructive testing and structural health monitoring, especially in environments with electromagnetic interference.

Jang et al. (2020) examined the vibration characteristics of underwater rotating propellers using a tracking laser Doppler vibrometer (LDV). The study simulates environmental conditions such as cavitation and turbidity to improve measurement accuracy. Principal component analysis (PCA) was used to filter noise and determine the propeller's natural frequency, confirming LDV's suitability for monitoring underwater propellers.

Rhakasywi et al. (2020) developed a vibration monitoring system for centrifugal pumps on ships, using six accelerometers to track vibrations. Measurements exceeding the ISO 10816-3 threshold frequency of 20 Hz suggest potential structural damage, highlighting the importance of monitoring to prevent failures in pump systems.

Peeters et al. (2024) used different accelerometers, i.e. modal ICP accelerometers, seismic ICP accelerometers, and MEMS DC accelerometers, for modal testing of large offshore wind turbine blades and characterization of their dynamic properties. The study identifies modal parameters such as eigenfrequencies and damping ratios. The findings are intended to inform industrial guidelines for blade testing and validate structural models.

3.3.4 Motion Monitoring

Motion monitoring involves continuous or periodic measurement and analysis of movement or displacement of a structure over time. This process can be used for assessing the dynamic behaviour of structures and detecting changes that may indicate potential issues.

Nelli et al. (2021) estimated sea state conditions from ship motion data using a response amplitude operator (RAO) during the Antarctic Circumnavigation Expedition. They calculated the RAO with NEMOH and HydroSTAR models and compared the reconstructed sea states with wave energy spectrum measurements from the WaMoS-II marine radar. Their results demonstrated a strong agreement between the reconstructed sea states and direct observations.

Min et al. (2021) discussed the use of Fiber Optic Gyroscopes (FOG) for measuring ship motion. The FOG technology is highlighted for its high accuracy in capturing angular velocity and rotational displacements, which are critical for real-time monitoring and navigation of marine vessels. The review emphasizes the robustness and precision of optical fibre sensors in harsh marine conditions, making them ideal for long-term monitoring.

Chen et al. (2023c) focus on the use of optical fibre sensors for monitoring the health of marine structures, with a specific section dedicated to LIDAR-based systems. These systems are used for high-resolution measurement of wave-induced displacements on offshore platforms. The paper discusses the advantages of LIDAR in providing precise, real-time displacement data, which is crucial for the dynamic stability analysis of these structures. The use of LIDAR technology is presented as a critical tool for ensuring the safety and operational efficiency of offshore installations.

Kargar and Hao (2022) present the development and application of a drifter-based piezoelectric sensor system designed for measuring ocean wave-induced motions. The system integrates Inertial Measurement Units (IMUs) to track both linear and angular motion of the drifter, which simulates the behaviour of offshore structures under wave loads. The research demonstrates the effectiveness of IMUs in capturing detailed motion data, including pitch, roll, and yaw, which are essential for assessing the stability and dynamics of offshore platforms.

3.3.5 Impedance-Based Methods

Impedance-based methods involve the measurement and analysis of the electro-mechanical impedance of a structure. The principle behind these methods is that changes in the structural condition, such as damage or degradation, alter the electro-mechanical impedance of the structure. By monitoring these changes, it is possible to detect and locate faults or assess the overall condition of the structure.

Huynh et al. (2017) review the latest advancements in impedance-based SHM technologies, particularly for marine applications. It discusses improvements in sensor design and data interpretation methods, as well as ongoing challenges like environmental compensation and long-term durability. The paper also outlines future directions for research to enhance the effectiveness of impedance-based SHM in maritime settings.

Le et al. (2022) investigate the use of piezoelectric impedance sensors for monitoring marine structures. These sensors are found to be effective in detecting early-stage damage, particularly in harsh maritime environments. The study highlights the potential adaptability of the technology to complex marine structures, emphasizing its potential for real-time damage assessment and long-term monitoring.

3.3.6 Acoustic Emission Monitoring

Acoustic emission (AE) monitoring is a technique used to detect and analyse transient elastic waves emitted by the rapid release of energy within a material or structure. This energy release typically occurs due to processes such as damage formation and growth.

Yao et al. (2022) investigate characterization of AE sources during fatigue crack growth in Al 2024-T3 specimens. They aim to establish a linear relationship between count rate and crack growth rate based on waveform analysis, particularly in 2024-T3 aluminium alloys. The study synchronizes AE waveforms with load levels to identify dominant frequency ranges. AE waveforms are collected during fatigue crack growth in edge-notched specimens at different load ratios. The results show a higher AE count rate at lower load ratios, indicating an impact of the load ratio on the total AE activities collected. Dominant frequency ranges related to crack growth and crack closure are identified. By selecting waveforms based on proposed frequency ranges, a better linear relationship between count rate and crack growth rate is observed. This research enhances understanding of AE sources and waveforms for future structural health monitoring applications.

Bhange et al. (2022) studies the crack growth in a high-strength low-alloy DMR 249A ship steel under real sea state conditions (value of 4). The experimental setup includes a fatigue loading machine, AE sensors and nodes for data preprocessing. The researchers developed a methodology to identify crack propagation independently of AE parameters, proving useful in distinguishing noise and crack information in ship steel. This is achieved by creating phase portraits of the time domain signal and overlaying crack information onto these portraits.

Shi et al. (2023) investigate AE behaviour and its correlation with fatigue crack growth in Hadfield steel during bending fatigue tests. They analyse the probability density function for AE parameters based on the power law distribution and observe a sharp increase in the moving average and cumulative sum of AE parameters as early warning signs of final failure. Two distinct parts are identified in the stable fatigue crack growth process, with differing slopes of duration rate vs. stress intensity factor.

In Schoone et al. (2023), the potential of AE technology for characterizing flow in hydraulic transport processes, particularly in deep-sea mining, is highlighted. The selection of structural materials for system components is critical for corrosion resistance and optimal AE signal transmission. This experimental study evaluates the AE attenuation behaviour of four materials suitable for marine environments using thin plates. The results identify materials with the highest and lowest attenuation, offering insights into the propagation of different wave modes on the plates.

Sheikh et al. (2021) propose a corrosion monitoring technique that uses acoustic emission (AE) sensors to detect corrosion damage, and uses features of the measured signals to predict the extent of the damage using a neural network. In a laboratory experiment, mild steel samples were exposed to a corrosive solution combined with anodic polarisation. AE sensors were used to measure signals induced by the corrosion process. Three types of neural networks were used to attempt he prediction of corrosion damage, namely, a back propagating neural network, a radial basis function network and a Naive Bayes classifier. The corrosion severity, quantified by the mass loss was predicted most accurately by the RBF network.

Wu et al. (2021) investigate the AE from the corrosion of stainless steel on a more fundamental level based on small scale laboratory experiments. The goal of the experiments was to correlate acoustic emission signals with visual observations taken with a microscope. The authors managed to attribute different AE signals to hydrogen bubble formation and electrolyte solution perturbation. The characterization of these signals can help in better identification of corrosion in AE based monitoring systems.

Scheeren et al. (2024) present a methodology for detecting damage in low-speed bearings in offshore applications using waveform-similarity-based clustering of AE signals under fatigue loading. The approach identifies consistent AE sources by analysing waveform cross-correlation, with validation through data from a low-speed run-to-failure experiment. The results demonstrate the high potential of the method for assessing damage in low-speed roller bearings in offshore applications.

Riccioli and Pahlavan (2023) evaluated the feasibility of detecting and localizing corrosion-fatigue damage in submerged steel structures using non-contact AE. Accelerated corrosion-fatigue experiments were performed by applying both corrosion and cyclic loading. Underwater transducers measured AE signals, and a source localization algorithm successfully identified corrosion-fatigue signals. The results show that non-contact AE can effectively detect damage signals with good signal-to-noise ratio and can aid in damage characterization.

3.3.7 Guided Waves Monitoring

Guided waves are specific types of elastic waves that travel through a medium, such as a solid structure, in a constrained manner due to the free surfaces of thin-walled structures. These waves can travel over long distances with relatively low attenuation and are influenced by the structure geometry and material properties.

Zima et al. (2024) present a method of the assessment of corrosion damage by differential time of flight of guided ultrasound waves. The paper investigates wavefront distortions caused by the variation in thickness of damaged specimens. A combination of numerical simulations and experiments were carried out to evaluate how the wavefront of a signal from one source reaches a circular array of sensors around that source under different levels of surface degradation. The authors conclude that the method presented can serve as a robust base for a corrosion assessment method.

Roch et al. (2023) present a feasibility study on the application of guided waves in the detection of corrosion related thickness loss in stiffened ship plates. The study is composed of numerical tests and experimental tests on a portion of a hull plate with a stiffener. The method used by the authors exploits the dispersion relation of lamb waves and the wave speed dependence on plate thickness. By evaluating the change in time of flight of signals from one emitting transducer to receiving transducers, the average plate thickness can be estimated. A numerical model was made of a stiffened plate, where the corrosion damage was simulated by a general thickness loss over the entire plate. The authors conclude that the use of time of flight of guided waves is a promising technique for the detection of corrosion damage.

Hu et al. (2022) present an experimental and numerical study into a novel method of early fatigue damage detection using guided waves in metallic plates with one side exposed to water. The proposed method exploits the tendency of leaky S0 Lamb waves to generate second harmonics when they interact with a change in microstructural features due to fatigue. The authors conclude that the increased nonlinearity in the material caused by the onset of fatigue damage can be detected by the second harmonic generation.

Saccone and Pahlavan (2024) investigated how ultrasound-guided waves transmit through ship hull structures and how their transmission depends on the angle of incidence. The study combines spectral finite element simulations with experimental measurements to derive an expression for the transmission coefficient at various angles. The findings aim to extend the coverage of the monitoring system.

Adams et al. (2024) propose a method to assess fibre-reinforced composite stiffness using ultrasonic guided waves. By comparing wave speeds with a reference dataset, the method estimates stiffness properties through an inverse approach, achieving accuracy of about 10%. Experimental tests on a glass fibre-reinforced composite plate revealed larger discrepancies compared to mechanical tests, attributed to material property variations and averaging effects.

3.3.8 Data Fusion

Xin et al. (2024) present a SHM method that uses measurements from strain gauges, accelerometers and displacement sensors as input to a deep learning network to detect damage in offshore jacket platforms. The deep learning algorithm was trained to relate environmental conditions to normal dynamic responses of the platform using finite element model data. Scale model tests were then carried out with simulated environmental excitations on a structure equipped with sensors. The algorithm fuses the data from the sensors and evaluates whether the dynamic response matches expected behaviour given the known environmental conditions and the response of individual structural members. The experiments yield a damage recognition rate of 88%.

Ding et al. (2022) measured static strains on a model of a propeller blade using surface-mounted FBGs. Reconstructed deformations of the blade were created using an inverse finite element method. The reconstructions were compared to deformation measurements of a three-coordinate measuring arm. Error between reconstructions and deformation measurements was below 7%.

Finnegan et al. (2022) performed static and fatigue experiments on an 8 m long full-scale glass-fibre composite tidal turbine blade. The load was applied in two phases. Firstly it was amplified to simulate a 20-year lifespan in 200e3 cycles using known SN curves. Secondly, the load was further amplified 1.39 times per standard DNVGL-ST-0164 to simulate another 20 years lifetime in 100e3 cycles. They monitored blade stiffness through LVDT measurements and accelerometers. They noted a slight drop in stiffness within the first phase. An insignificant drop was noticed in the second phase.

Davies et al. (2022) describe the performance of static and fatigue tests for a composite tidal turbine. Firstly, static and cyclic coupon tests were performed for coupons made from synthetic or natural composite materials. From these coupons, S-N curves were made for pristine condition and seawater-aged condition. Secondly, a full-scale 5 m CFRP blade was tested under static and cyclic loading. During these tests, the blade was interrogated using strain gauges, embedded optical fibres and surface-mounted acoustic emission sensors.

Waszak et al. (2022) propose a framework for efficient digital twin creation, emphasizing rapid prototyping and affordability. Using the knowledge graphs paradigm, the proposed four-layer architecture integrates data in a digital twin formalism, ensuring compatibility with industry standards. The accompanying software prototype is evaluated in the context of discrete manufacturing and underwater ship inspections, showcasing practical application.

Maetz et al. (2023) propose a method for the monitoring of the grouted connection between an offshore wind turbine transition piece and the monopile foundation. The method makes use of a combination of ground penetrating radar, acoustic emission and strain gauges to detect and localize the fatigue damage. The study includes model scale experiments on monopile-transition piece connections subjected to accelerated life cycle loading. The measurements from the ground penetrating radar were validated against vibration analysis, acoustic emission measurements, and modal analysis from strain gauges.

Momber et al. (2022) describe availability, acquisition and processing of signals/data from meteorological (for atmospheric corrosion) and oceanographic (for immersed seawater corrosion) sensors in the German offshore territory and their integration into an online monitoring and maintenance model for protective systems for marine structures.

Zhang and Wu (2022) combine acoustic emission (AE) measurements and electrochemical noise measurements to monitor stress corrosion cracking damage in 304 stainless steel. Laboratory experiments were conducted on stainless steel samples under strain and high temperature and pressure. The steel sample was equipped with AE sensors and the electrochemical potential fluctuations were measured. The authors conclude that electrochemical noise is effective at detecting crack initiation while acoustic emission measurements are more effective at monitoring its propagation.

3.3.9 Data Communication

Collecting huge amounts of data to represent a ship's response to its environment creates a challenge to effectively communicate the readings. Despite advances in wireless communication on land, at sea, high-speed data transmission remains difficult. Alqurashi et al. (2022) explain that data communication challenges obstruct the maritime industry's ability to make timely and informed decisions. In their review of maritime communications, they note that limited range, coverage, and capacity are among the biggest challenges. The authors also explore developing technologies to address these challenges.

Networks of autonomous vessels provide a means of meeting communications requirements. Nomikos et al. (2023) review the opportunities and challenges associated with using unmanned aerial vessels in wireless networks, including cooperation with autonomous surface and underwater vessels. They explore open issues related to signal transfer and autonomous vessel trajectory determination to avoid collisions and ensure adequate coverage, machine learning, and applications with advances in edge computing.

To manage the challenge of limited bandwidth, Jurdana et al. (2021) propose a way to compress navigation data and limit the size of transferred data. Using actual data from a training ship, the reduced data was about 3% the size of shipboard data, on average.

3.3.10 Virtual Sensing

Virtual sensing in monitoring ship and offshore structures refers to the use of computational models, algorithms, and indirect measurements to estimate physical quantities that are difficult or expensive to measure directly on the structures. Instead of relying entirely on physical sensors, virtual sensing techniques combine data from limited physical sensors with mathematical models to infer the condition or performance of various parts of the structure. Additionally, the pre-selected monitoring point location is not necessarily co-located with the most critically loaded point on a structure (Liu & Ren, 2022).

Data-driven virtual sensing approaches typically involve the development of a dynamic model of the system based on input and output measurements, which inevitably contain uncertainties. The Kalman filter is an optimal state estimator designed to handle these uncertainties, which are modelled as Gaussian noise with associated covariance matrices. Greś et al. (2024) present a direct method for the estimation of noise, based on subspace identification. Traditionally, subspace identification is employed to determine the linear system matrices (A, B, C, D) which approximate system dynamics. Here, Greś et al. (2024) assume that system parameters are known and instead, these matrices are utilized along with input and output system measurements to obtain an estimate of the noise covariance matrices of the system (Q, R, S). These matrices are obtained by showing that estimates of the noise covariance in innovation form (Qe, Re, Se) can be equivalently used in the Kalman filter. The new scheme is validated through both a Monte Carlo simulation and a laboratory experiment. A comparison with the established auto-covariance least-squares (ALS) scheme shows that the subspace-based approach achieves a smaller root-mean-square-error while maintaining numerical efficiency. In the laboratory experiment, the subspace-based approach successfully estimates system outputs at locations where measurements were not used to derive the model, suggesting that the subspace-based approach may be well-suited to virtual sensing applications.

Lee et al. (2024) explore a real-time method for estimating full-field strain distribution in structures using data from strain gauges. This virtual sensing approach is based on finite element analysis and the mode superposition technique. Both numerical simulations and experiments on a lab-scale offshore jacket structure, subjected to water waves, validate the method. The study provides explanations of the prototype's design and experimental setup, including wave loading conditions. Key topics covered include selecting displacement modes and dividing strain signals.

Moynihan et al. (2024) present a Gaussian process regression (GPR) method to predict the bending moments of an offshore wind turbine (OWT). By using SCADA data such as wind speed, power output, and nacelle acceleration, the model forecasts high- and low-frequency dynamic responses. The approach is tested on a 6 MW OWT to predict bending moments in both fore-aft and side-side directions. Its accuracy is confirmed under various operating conditions, with the model proving highly adaptable for other turbines in the same wind farm without needing extra strain gauges.

Partovi-Mehr et al. (2024) assess the remaining fatigue life of a 6 MW GE Haliade OWT's support structure at the Block Island Wind Farm. Using stress data from both instrumented and non-instrumented locations, two finite element models calculate fatigue damage. Virtual sensing techniques estimate stresses at non-instrumented points, showing that the jacket structure has lower fatigue damage than the tower base. The study suggests that fatigue at the jacket's leg joints is below 40% of its capacity over the turbine's 25-year lifespan, indicating a longer life expectancy for the structure's support components compared to the tower.

Mehrjoo et al. (2022) propose an optimal sensor placement (OSP) framework based on information theory for virtual sensing and condition monitoring. Using a Bayesian OSP approach, the framework minimizes information entropy to improve parameter estimation without needing input data. It also considers sensor installation costs at different locations on an offshore wind turbine jacket support structure. The study explores strain time history and structural parameters, including stiffness and foundation properties. Results suggests that variations in input loads, correlation lengths, and cost weights influence OSP outcomes.

Tarpø et al. (2022) use supervised learning and data-driven models to estimate dynamic strain responses in an offshore wind turbine. Principal component analysis (PCA) is applied to 40 min of training data to develop a model that predicts dynamic strain with high accuracy over a two-month period. The study finds that fatigue damage predictions using the model closely match measured data, with an average error of about 1.8%.

3.3.11 Human-System Interface

The human-system interface is a key element for ship structure health assessment and decision-making. It provides an interaction tool between the crew and information from onboard data, and can also be remotely operated from shore. The primary concern relates to human errors in data analysis and decision-making. Numerous studies have examined potential operational errors in human-machine interfaces (Liu et al., 2022). Although these studies often focus on advanced interfaces, the principles of predicting errors using a success likelihood index method and handling uncertainty with interval type-2 fuzzy sets are relevant to human-system interfaces. For instance, Batalden et al. (2019) presented the benefits of fuzzy logic to enhance human-machine interaction. Recent trends (Singh et al., 2023) explore the use of artificial intelligence based on fuzzy systems for collision avoidance to improve human-system interface design, leading to better and safer decision-making processes.

3.4 Data Acquisition, Transfer and Management

A review of state-of-the-art data networks and architectures applicable for ship and offshore structures during operation is performed. Recently developed standards and guidelines are reported. The emerging developments on tiny-machine-learning methods, i.e. sensor nodes with edge-computed AI algorithms, and their applicability under operational conditions are discussed. Technologies for embedded sensors that become an integral part of structures during manufacturing are also reviewed.

3.4.1 Data Management Guidelines and Standards

Recent efforts introduce new standards to standardize access to data from on-board components and systems. ISO 19847:2024 specifies requirements for shipboard data servers that collect data from other shipboard machinery and systems, and to further share the collected data in a safe and efficient manner. Communication protocols are specified with reference to the data structure of ISO 19848:2024 which stipulates requirements and guidance on the capture and processing of data from monitoring sensors pertaining to the ship operational information, structure, machinery and equipment. Fonseca et al. (2022) used existing data standards and web technologies to model and develop a digital twin of a scale model ship, hereby proving the use on an open framework to facilitate services related to visualization, simulation and remote control.

Presently, some guidance on the standardized management of asset and lifecycle data is documented in relevant standards which are not expressly related to the maritime domain. ISO 15926:2018 applies to the integration of life-cycle data for process plants and contains an ontology for asset planning, including information associated with the engineering, construction and operation which should support information requirements among stakeholders over the duration of plant life cycles. ISO/TR 24464 (2020) deals with automation systems and the integration of industrial data through visualization elements of digital twins. The four parts of ISO 23247-4 (2021) address information exchange in automation systems and integration through a digital twin framework within the manufacturing context, but that does not presently extend to operations. ISO 81346-1 (2022) applies to information structures to manage industrial systems, installations, equipment and products. The structuring principles and reference designation systems enables the handling of information across various data processing systems and different technologies in an unambiguous manner. ISO 10303-239 (2024) provides an application protocol for organizations interested in generating product data representation and exchange for product life cycle support. ISO 10303-2 (2024) contains vocabulary to underpin the 10303 series of International Standards, describing computer-interpretable representation of product information and data and mechanisms to detail products throughout their life cycle. DNV-RP-A204 (2023) offers recommended practice to organizations involved in the procurement, development and operation of digital twins with requirements and recommendations to assure the trustworthiness of digital twins from concept to operation. Considerations include i.) the quality of the digital twin, ii.) the risk of relying on information from the digital twin, and iii.) the integration, deployment, operation and maintenance of digital twins.

With increased digitalization the need emerges to organize and use product data from various sources and lifecycle stages in standardized digital representations. Bronson et al. (2024a) highlight that the maintenance of data models throughout the ship lifecycle is disrupted between upstream (ship design, construction and outfitting, commissioning) and downstream (operations and management, decommissioning) activities owing to changes in the ownership of digital product data, non-formalized data governance standards across the lifecycle and deficiencies in data traceability and record linkage. The use of a digital thread for product lifecycle management with cross-ownership linkages, closed-loop lifecycle management and the expansion and integration of data domains through a taxonomy that includes functional, product, schedule, human, geospatial and contextual information. Current gaps towards the development of digital threads include the lack of open standards, non-standardized data models, non-integrated data pipelines, lack of adoption strategies.

Bronson et al. (2024b) point out that the shipbuilding industry currently does not have a standard solution to realize integrated, interoperable, and multi-domain ship product data models that span from design to operation. Four gaps include the lack of a i) Defined ship product data model that incorporates multi-domain ship data ii.) The lack of industry-wide data exchange standards impeding interoperability, iii.) The lack of an Integrated Product Development Environment (IPDE) impedes collaboration and iv) The lack of data management protocols to support the IPDE, modelling standards, and exchange standards. These gaps hamper more efficient ship lifecycle management and design processes. Practices from Building Information Modelling (BIM) are probed for solutions to overcome challenges related to information modelling, integrated design environment, and 4D engineering and planning. The maintenance and management of product data models should afford due attention to i.) data modelling (the structuring and relationships between information), ii.) data exchange (the transfer of data between applications and programs) and iii.) data management (tools that enable data quality, security, and trustworthiness).

Dihan et al. (2024) state that digital twin data will entail storage of physical entity data, virtual model data, service data from different application objects, situations, and scenarios. The sharing, reuse, and exchange of this data will demand standardization. Storage may be facilitated through category theory (involving the abstract representation of mathematical models and their relations) is a complete foundation for modelling, interoperability, and integration of digital twin data throughout a system lifecycle. Cloud services will play a significant role in future data sharing which will transpire at an associated cost. The integrated use of blockchain and federated learning methodologies is proposed to bolster data security across digital twin systems.

3.4.2 Sensor Nodes for Tiny Machine Learning

Tiny Machine Learning (TinyML) refers to the deployment of machine learning (ML) models on small, resource-constrained devices, such as microcontrollers or other embedded systems. These devices typically have very limited computational power, memory, and energy resources, making the task of running ML models challenging.

Disabato (2023) introduces a methodology to enable intelligent processing in constrained environments through TinyML. TinyML models are designed to fit within the limits of pervasive devices. The methodology approaches this challenge in three ways: first, by using inference-based Deep TinyML solutions where models are trained externally but run on-device; second, by implementing on-device learning for TinyML; and third, by creating Wide Deep TinyML solutions that distribute DL processing across multiple connected pervasive devices.

Han and Siebert (2022) offer a comprehensive review of the state-of-the-art in TinyML, synthesizing existing research and identifying key trends and challenges. The study presents a systematic review of TinyML research by synthesizing findings from 47 papers published since 2019, marking the beginning of formal TinyML research. The review categorizes the relevant literature into five key areas: hardware, frameworks, datasets, use cases, and algorithms/models.

Akyuz et al. (2019) explore ML applications and their future impact on maritime transportation. They present comparative research to assess the current state and potential benefits of ML in areas such as voyage optimization, freight rate management, maintenance forecasting, and digitalization of control systems. The research aims to help maritime professionals select appropriate algorithms for specific challenges in the industry.

3.4.3 Embedded Systems

Shamsuddoha et al. (2022) performed fatigue analysis on a full-scale composite hydrofoil with embedded FBG's, surface-mounted LVDT's and surface-mounted piezoelectric accelerometers. The blade was made with automated fibre placement of carbon fibre prepreg around a glass fibre composite core. At a maximum measured strain of 1550 microstrain and $100e^3$ cycles, no or negligible degradation was measured in terms of stiffness, strain hysteresis and modal frequency.

Hamada et al. (2023) embedded piezoelectric line sensors into flexible CFRP propeller blades in operation. Firstly, they determined a piezoelectric transfer function by comparing strain gauge strain rate results to voltages from the piezoelectric sensor. Secondly, blades were tested that were intact or that had chipped edges. They noted that the strain rate amplitudes for the intact blades and chipped blades were different.

Borgelt et al. (2024) propose a new test concept for cyclic axially loaded grouted connections with fibre optical sensors integrated into the grout annulus. The study aims to provide insight into the mechanisms of damage initiation and propagation through real-time crack detection. The FBG sensory inside the grout annulus was capable of detecting cracks in the grout material as they occurred. These cracks measured by the FBG could be correlated with strain gauge measurements, changes in the displacement curve, and crack patterns.

Mieloszyk et al. (2021) present an experimental study in to embedding FBG strain sensors in the skin of a fast patrol boat. The authors investigate the effect of the embedding process on the characteristics of the FBG. The sensor response spectra are evaluated at different stages of fabrication and after 80 h of sea trials to investigate possible changes. Residual stresses from resin curing and installation of heavy components are amongst the main factors that impact the deformation of the FBG sensors. The authors conclude that the angle between the fibre optic and the lamina orientation is a critical factor in the deformation of the FBG during fabrication, the sea trials however had minimal impact on the permanent deformation of the sensors.

Huijer et al. (2024) investigated the use of embedded piezoelectric sensors for monitoring acoustic emissions in submerged flexible composite marine propeller blades. A full-scale glass-fibre reinforced polymer blade with 24 sensors was tested in artificial seawater. The results demonstrate that these sensors effectively measure acoustic emissions underwater. The study also found that the maximum measurable distance between the source and sensor is about 124 mm for frequencies between 100 and 250 kHz at a noise level of 40 dB.

Riccioli et al. (2023) present a novel approach for measuring transient pressure distributions on maritime structures using an elastic matrix layer embedded with piezoelectric sensors. An epoxy layer with 25 sensors arranged in a grid was fabricated and tested. A finite element-based inverse procedure was used to reconstruct the pressure field from the measurements. The system was validated by measuring pressure generated by a traveling wave in a water tank. Results show that the retrofit layer effectively reconstructs the pressure field, with minimal impact from surface disturbances, though non-ideal mounting conditions can affect measurement accuracy.

3.5 Concluding Remarks

Sensor technologies continue to develop from the more well-known accelerometers and strain gauges, to the lesser-known sensors for acoustic emissions assessment and piezoelectric impedance sensors for monitoring electro-mechanical impedance in structures. Strain gauges and accelerometers have been regularly used in structural testing for many years, and are shown to be implemented for longer term structural health monitoring in ship and offshore structures.

Novel monitoring techniques such as impedance-based methods and acoustic emissions monitoring show their potential benefits in detecting early-stage structural damage and degradation when coupled with structural health monitoring systems to process the data being captured over long-term monitoring.

Structural health monitoring techniques are shown to be progressing to combine sensor technologies to collect different types of data, combining the results to detect damage or inform on degradation of the structure in-service, and even progressing to virtual sensing techniques using a more limited det of data coupled with structural models (digital twins or digital shadows) to predict the probability of damage in-service and probable locations. Therefore, informing on inspection and maintenance schedules for the asset.

It is expected that sensor technology will continue to develop, with improvements to existing technologies in terms of speed of data collection, fidelity, reliability, etc., as well as the development of new sensors, or new applications of sensors within structural health monitoring.

4 Assessment of Offshore Structures

4.1 Introduction

Safety and structural reliability of offshore infrastructures operating in the harsh environment of open water are crucial concerns not only in terms of energy production but also with regards to facilitating the necessary transportation logistics. The inevitable need for regular maintenance and repair arises for offshore structures and their components as they age and deteriorate over time to a degree that it begins to impact their performance and safety, despite the risk of external damages inflicted.

Within this chapter, research, development and application of structural assessment technologies and methodologies for offshore structures during operations is reviewed. The deterioration mechanisms and applicable mechanical limit states of offshore structures are explained, and offshore structure-specified theories and technologies for structural integrity management (SIM) and risk-based integrity management (RIM), which include monitoring/inspection/mitigation/requalification, are reviewed. Simulation (including ROM, surrogate model, and machine learning) and data assimilation methods that have been applied to offshore structures are reviewed, and the application examples of the use of DT for SIM and RIM for offshore structures are reviewed.

4.2 Deterioration Mechanisms - Corrosion Fatigue

Corrosion fatigue (CF) has plagued the offshore industry for decades, entailing increased cost due to maintenance and downtime – and catastrophic failure of the entire structure. CF pertains to situations in which a discrete crack is filled with electrolyte and the high cycle fatigue growth of the mechanical crack driving forces are assisted by electrochemical corrosion processes, taking place inside the crack and particularly at the crack tip c.f. Figure 3a. The viciousness of CF manifests itself in (i) a significant reduction or near absence of a fatigue threshold (Suresh et al. 1981) and (ii) a significant effect on the growth rate as compared to the inert fatigue crack growth conditions c.f. Figure 3b (Ma et al. 2021).

CF tests can generally be distinguished into pre-corrosion and in-situ corrosion testing conditions. Pre-corrosion testing (Kang et al. 2013 & Al-Karawi 2022) procedures foresee the exposure of the specimen to a corrosive environment either in a salt-spray or in humidity chamber or by fully immersing it into a corrosive liquid under controlled conditions. While pre-corrosion tests can be used to study the effect of corrosion induced surface alterations on the stress concentration factor, the effects of hydrogen surface diffusion most typically in the form of S-N curves, such tests disregard the electrochemical contributions occurring inside the crack during CF crack growth.

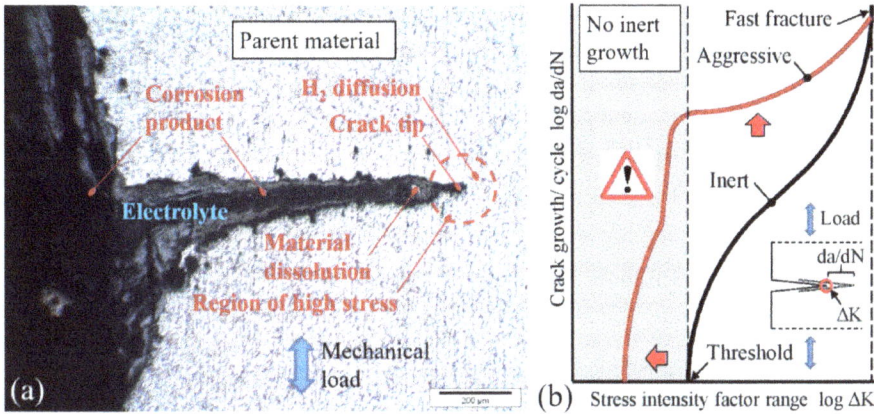

Fig. 3. (a) Micrograph of a CF crack in parent metal with dark corrosion product taken from R-Tech Materials (b) accelerated CF crack growth in aggressive environments (red graph) poses a challenge for safe designs.

Recent studies (Shojai et al., 2023 and 2024) on pre-corrosion testing of structural steel base materials and welded joints for an extensive duration of up to 1 year show that the formation of corrosion pits quickly reduces the fatigue strength of structures that were clean blasted during fabrication; however, almost no difference in fatigue strength was observed for welded specimens that were exposed for 1 month or 1 year. In contrast, the fatigue strength of the base material continuously decreased. In fact, the fatigue strength of the base material and welded specimens was similar after 1 year exposure. A similar fatigue strength was observed in test of base material specimens taken from a ship ballast tank after more than 12 years of service, see Nugroho et al. (2021) and Shojai et al. (2022). In summary, the results indicate that applying a generalized reduction in fatigue strength without accounting for specimen geometry (notch class) as per the guidelines is inappropriate and should be revised to ensure more accurate design of offshore support structures.

Conversely, in-situ CF tests (Suresh et al. 1981; Ma et al. 2021) on the other hand foresee fatigue testing of a fully immersed specimen in a corrosive electrolyte such that both, the mechanical and the electrochemical processes occur simultaneously. The in-situ corrosion fatigue test methodology itself can be split into in-situ-coupon (Zhao et al. 2018 & Ólafsson 2016) and in-situ-fracture testing methodologies (Ma et al. 2021, Huang et al. 2020 & Zhiqiang et al. 1991). The former can be used to investigate crack initiation and the total fatigue life whereas the latter disregards initiation but provides the CF growth parameters for different environmental conditions.

Numerical electro-chemistry models have already been proposed by Turnbull and Thomas (1982), Turnbull and Ferriss (1986) & Bardal (2004), predicting the spatial and temporal states of the electric potential, the pH value and the hydrogen evolution in a crack. However, they were to date not successfully deployed in CF due to the lack of adequate coupling between electrochemistry and the mechanical processes resulting in the development of semi-empirical approaches. CF models can be grossly distinguished between S-N-based and fracture-mechanics based approaches. While SN-based models can be used to consider the effects of surface stress concentration factors (such as pitting), these approaches are insufficient to capture the underlying complex physics of CF and are therefore not further treated herein. Fracture-mechanics based models can be distinguished between continuum-damage models e.g. Phase Field (PF) models (Golahmar et al. 2021, Falkenberg 2019 & Martínez-Pañeda 2021) and discrete fracture-mechanics models. While PF models can handle both crack initiation and propagation, they consider the contribution of the corrosion process by an empirical additional surface energy term which omits modelling of the electrochemical processes.

It can be concluded that a unified model for CF crack growth describing the coupled mechanical and electrochemical processes is currently not available. Most of the available models are semi-empirical and largely ignore the physics governing CF. The complexity of CF is impeded by the considerable parameter space and more research effort is needed to distinguish the influential ones from the less important ones to reduce the complexity of the CF models. The research objective in the future should be to couple the partial differential equations defining fracture mechanics with those defining species diffusion, species transport and the electrochemical reactions by segregating their formulations into a single unified CF theory. This would enable superseding empirical or semi-empirical approaches with physics-based modelling methods. In the authors opinion, as a minimum requirement any relevant CF model should be able to capture the effect of frequency, temperature, salinity, and the crack-tip pH value to be practically applicable to the different offshore sites situated across the globe.

4.3 Mechanical Limit States

4.3.1 Serviceability Criteria

Serviceability limit states (SLS) are criteria focused on conditions affecting a structure's functionality and operational use. SLS considerations include enhancing durability to prevent corrosion and wear, managing deflection, displacement, and accelerations to avoid reducing system performance, ensuring efficient lubrication, and controlling vibrations. These requirements are usually defined qualitatively, aiming to maintain the facility's operational effectiveness and longevity.

Foundations for Offshore Wind Turbines (OWTs) are designed following the limit state philosophy. One of the considered states is the SLS, which verifies that the permanent rotation of the foundation generated from accumulated strains in the soil is below a project specific criterion. Currently, there is limited information about SLS requirements in design standards. Wang and Moan (2024) focuses on evaluating existing methods, criteria, and procedures for SLS design checks for floating wind turbines, with a special emphasis on managing tilt/pitch and nacelle accelerations to optimize power production and minimize its fluctuations. To facilitate the reliability-based design of OWT monopiles with the API p-y design method, Lin et al. (2024) present a two-stage evaluation method of the model bias factor for the p-y method (ε_{py}) of monopiles under the SLS. Page et al. (2021) describes a methodology to estimate the monopile permanent rotation for SLS and discusses its advantages and limitations.

Because offshore wind turbines have stringent SLS requirements and need to be installed in variable and often complex ground conditions, their foundation design is challenging. OWTs are found to have non-negligible probabilities of exceeding the Serviceability at the order of 70% under combined exposure to moderate seismicity and winds close to the rated wind speed (Zhang et al. 2023a). Bhattacharya et al. (2021) provide an overview of the complexities and the common SLS performance requirements for offshore wind turbine. Patra et al. (2022) develop step by step methodology for developing an interaction chart of monopile for the various depth of liquefied soil zone below the mudline level for SLS. And SLS of monopile foundation evaluated for the reliability assessment by considering ground nonuniformity is considered by Sujawat and Kumar (2023). Three dimensional stochastic numerical analyses are carried for analysing the monopile behaviour at the serviceability limit. Using a self-manufactured loading device, Liao et al. (2022) simulate the possible response of the monopile during the serviceability limit state.

4.3.2 Fatigue and Fracture

Fatigue has been considered as one of the causes of catastrophic failures in marine structures. It is also an essential consideration in the design process of offshore structures due to the nature of subjected loadings; wind, waves, and currents. These dynamic effects are even more significant in offshore wind turbine support structures as they are usually subjected to high number of cyclic and wind loads (Yeter and Garbatov 2022). The monopile foundations of offshore wind turbines are also in the scope of concerns for fatigue due to the exposure to long-term dynamic loads (Raed et al. 2020). When it comes to the support structures of OWT, welded-tubular joints are the most threatening parts for the fatigue failure with high-potential of crack initiations due to welding residual stresses and weld profiles. Over the last decades, many studies have been reported regarding the fatigue and fracture mechanic methods for marine structures, and yet the developments are still needed to thoroughly understand the discrepancies between the actual and predicted fatigue strength (Ehlers and Braun 2023). Although fatigue damage may not always be the main reason of severe failure in OWT, it still has a substantial impact on the maintenance cost (Collu et al. 2010).

Using frequency domain criteria, Han et al. (2021) proposed two new formulae which accounts the multiaxial aerodynamic wind forces in fatigue predictions. OWT jackets subjected to multiple random wind loads was considered as a case study and the results obtained from proposed model are compared with those from rain flow counting method which shows the model's reliability.

On the other hand, to obtain more reliable fatigue designs of OWTs, non-linear fully coupling time domain simulations are widely employed. Using this approach, Zhu et al. (2024) proposed a fatigue reliability analysis and design for OWT in the real time domain. The authors predicted probabilistic fatigue life values based on bootstrapping method in combination with Monte-Carlo simulation. Both S-N and damage tolerant approaches are considered for evaluation of fatigue designs. However, with increasing size of OWT, these fully-coupled approach usually needs a high computational efficiency for a large number of load cases. For this reason, Wang et al. (2021a) compared the drivetrain dynamic behaviour of floating wind turbine using fully-coupled and de-coupled approaches, and revealed that de-coupled method could also provide accurate fatigue damage results. Later, Han et al. (2022) also studied the fatigue damage of OWT under combined wave and wind loads using half coupling analysis. By comparing the results with time-domain results based on rain flow cycle counting, the authors proved that half-coupling model could also provide efficient and accurate fatigue damage predictions. Different to these time-domain or frequency-based methods, Farid (2022) proposed an alternative solution by taking advantage of data driven models in machine learning. The author developed a hybrid probabilistic model to predict the real-time fatigue damage under stochastic loadings by the power of fully connected artificial neural network and Gaussian process regression.

Studies investigating the remaining useful life of wind turbines are yet few but the procedures for such assessments of general structures are established in industry standards and guidelines. Therefore, Fajuyigbe and Brennan (2021) performed fitness for purpose assessment of offshore wind turbine monopile with a surface flaw in accordance with British Standard guideline, BS7910. The authors described the issues with load ratio for cracks deeper than 80% of thickness which are having differences in the values obtained from BS7910 and finite element analyses.

Fatigue cracks initiations are usually reported in the high stress concentration regions such as weld toe, notch areas, heat-affected zones and severely corroded areas. Qvale et al. (2021) predicted the fatigue crack initiation and propagation on corroded surface of offshore mooring chain using digital image correlation method. Same approach was also used in combination with 3D surface scans in the work of Shojai et al. (2022) to reveal the impact of corrosion on fatigue strength of welded components. The studies suggested that geometric parameters and residual stresses can influence the fatigue strength of corroded offshore support structures. For the cracks in the heat-affected zone of the material, fatigue crack growth (FCG) of offshore wind monopile weldments were also assessed by Jacob and Mehmanparast (2021). The study compared FCG of weldments in air and seawater, and revealed that FCG in seawater is faster than that in air. It also showed that the corrosion FCG rate was lower when the crack propagation reaches the inner surface of the monopiles. While evaluating FCG, the behaviour of cracks coalescence is also an interesting topics as multiple surface cracks are usually spotted. And, Mishael et al. (2023) recommended for accurate modelling of crack interactions in offshore wind structural connections as the coalescence of nearby cracks can accelerate the FCG rate which may lead to shorter fatigue life.

4.4 Implementation of Methods and Procedures for Safe Operation

Structural Integrity Management (SIM) is a critical aspect of the lifecycle management of offshore structures, ensuring their safety, functionality, and longevity in harsh marine environments. Offshore structures, such as oil platforms and wind turbines, are subjected to extreme environmental conditions, which necessitates robust SIM strategies to prevent catastrophic failures.

Structural health monitoring (SHM) is defined as life-long structural safety evaluation by using sensing techniques and characteristics analysis. Implementation of a SHM system allows localization, recording, and analysis of structural loading and damage to provide risk reduction, maintenance, cost optimization, and lifetime extension. Repair and mitigation methods depend on the type of offshore structures. There are no unified frameworks or guidelines for repairing and mitigation procedures.

The lifetime extension of offshore structures is becoming increasingly significant as many structures are being operated beyond their original design life due to economic and environmental considerations. Recent studies point out the importance of regular condition assessments using advanced non-destructive testing techniques and real-time monitoring systems. These assessments are essential in identifying critical degradation mechanisms, such as corrosion, fatigue, and material embrittlement, which are common in aging offshore platforms.

4.4.1 Structural Integrity Management (SIM)

Recent studies highlight the importance of developing risk-based approaches and advanced inspection techniques in Structural Integrity Management for offshore structures. Risk-Based Inspection (RBI) has emerged as a key methodology, enabling operators to prioritize inspection activities based on the likelihood of failure and the potential consequences (Ozguc, 2020). This approach allows for a more efficient allocation of resources, focusing on areas that pose the highest risk to structural integrity. For example, Tong et al. (2023) demonstrate the digitalization of the Floating Structures Integrity Management Program (FSIMP), which incorporates RBI into its methodology. This integration benefits the FSIMP by enabling structured data evaluation and risk assessment, optimizing inspection and maintenance resources effectively.

Digital twins have also significantly impacted SIM practices in recent years. DTs provide a virtual replica of physical structures, enabling continuous monitoring, predictive maintenance, and enhanced decision-making processes regarding the health of offshore structures (Liu et al., 2021b). According to Li and Brennan (2024), DTs have revolutionized SIM by allowing real-time monitoring and scenario simulation to predict potential issues before they occur. This proactive approach reduces downtime, minimizes maintenance costs, and enhances the safety of offshore operations, offering substantial advantages over traditional SIM practices.

Technological advancements in non-destructive testing (NDT) and sensor technologies have also significantly contributed to improving SIM's effectiveness. Advances in ultrasonic testing, radiography, and acoustic emission techniques have enabled more accurate detection of flaws and defects in offshore structures (Peruń, 2024). When combined with autonomous inspection vehicles (AIVs) and remotely operated vehicles (ROVs), these advancements have revolutionized inspection processes, particularly in deep-water environments where human access is limited. This combination of advanced NDT techniques and robotics offers safer and more effective alternatives for structural inspections.

The application of machine learning and artificial intelligence has also become crucial in enhancing SIM capabilities. Altabey and Noori (2022) discuss the role of AI in improving the accuracy of structural health monitoring systems. Wang et al. (2024) further explored the use of machine learning models to analyse data from sensors on offshore structures, detecting anomalies such as unusual vibrations or stress concentrations that could indicate damage onset. Their study demonstrated that AI-based SIM provides a more accurate and efficient method for assessing structural integrity compared to traditional inspection techniques.

Despite these technological advancements, challenges remain in implementing effective SIM strategies for offshore structures. One major challenge is the uncertainty associated with environmental loads, such as waves, wind, and currents, which can vary significantly over time and space. Another challenge is the aging of offshore structures, as many platforms and wind turbines are operating beyond their original design life, posing significant risks to their structural integrity. Addressing these challenges requires a comprehensive understanding of degradation mechanisms, such as corrosion and fatigue, and the implementation of effective mitigation measures to ensure safety and reliability in offshore operations.

4.4.2 Monitoring, Survey and Inspection Methods

There are various methods for Structural Health Monitoring (SHM) establishment of marine and offshore structures. Pezeshki et al. (2023) categorized current SHM methods into model-based methods, vibration-based methods, and DT methods. Application examples of model-based include Ghasemzadeh et al. (2023), in which the inverse Finite Element Method (iFEM) was applied for the detection of corrosion damage in marine structures. Examples of vibration-based methods include Xiang et al. (2024), which enhanced vibration-based SHM strategies by integrating remote-sensing data and machine learning with Bayesian optimization to minimize environmental and operational condition effects on the modal parameters of OWT towers.

The application examples of DT are presented in the following studies: Fang et al. (2024) proposed a fatigue crack growth prediction method for offshore platforms based on a five-dimensional DT model framework; Zhao and Chen (2022) proposed an acoustic-emission based damage localization method for a DT of OWT blades; Liu et al. (2023c) presented a DT-based framework for SHM system of OWT structures.

Furthermore, vision-based methods are developed to mainly focus on the damages on the local scale of offshore structures, such as fatigue crack and corrosion damage. Jiao et al. (2024) developed a real-time SHM system based on underwater robots, vision-based image processing, and the YOLO algorithm (neural network method). The population-based method, which allows existing information on damage observation to be shared across similar systems, was applied to offshore wind farms (OWFs) by Black et al. (2024).

4.4.3 Repair/Mitigation

Researchers studied the maintenance strategies of OWTs and OWFs. McMorland et al. (2022) reviewed the existing maintenance models for OWTs and discussed the influential factors and inputs for maintenance modelling. Li et al. (2022) proposed a failure mode and effect analysis framework based on a mirrored Bayesian network (BN) to determine the inspection and opportunistic maintenance strategies of OWTs. This model can update the order of failure items based on the operational situations of OWTs to obtain a real-time inspection and opportunistic maintenance strategy. Garcia-Teruel et al. (2022) presented a life cycle assessment using an advanced operation and maintenance model to quantify the environmental impact on floating offshore wind farms. Recently, Xia and Zou (2023) summarized the DT techniques for the operation and maintenance of offshore wind farms and proposed a novel DT-based optimization framework. Promising research directions for maintenance optimization are also identified.

4.4.4 Lifetime Extension Procedures (Requalification)

A recent study by Nguyen and Nguyen (2022) emphasizes the importance of continuous monitoring and dynamic reassessment of structural integrity to extend the service life of fixed steel offshore platforms. By regularly evaluating robustness and redundancy, operators can maintain the safety and functionality of these structures beyond their original design life. Similarly, Singh and Singh (2023) highlight the need for comprehensive reassessments of offshore structures, considering both material and geometrical nonlinearity to understand ultimate strength under incremental loading.

Structural Health Monitoring (SHM) technologies are crucial in extending the life of offshore structures. Yang et al. (2024) discuss advanced SHM technologies, such as wireless sensors and real-time data analytics, which improve fault diagnosis and enable more accurate assessments. These technologies support predictive maintenance strategies, aiding in the requalification process of aging structures.

The integration of AI into life extension strategies is an emerging research area. Ayemere Ukato et al. (2024) review AI applications in optimizing maintenance logistics for offshore platforms. AI models that adapt to changing conditions improve predictive accuracy, helping to minimize downtime and extend equipment lifespan.

In summary, the requalification of offshore structures involves a combination of advanced technologies, material innovations, and AI-driven strategies to ensure the safety, reliability, and longevity of aging offshore assets.

4.4.5 Codes and Guidelines Covering SIM

In general, international standards (e.g., ISO, API, and NORSOK) define requirements and recommendations relative to in-service inspection, condition monitoring, and maintenance. They cover phases such as design, fabrication, transportation, installation and in-service conditions, the latter as a part of the associated integrity management standards. Table 1 shows an overview of relevant ISO standards for offshore structures.

4.5 Risk-Based Integrity Management (RIM): Application of Procedures Based on Structural Reliability Assessment

Offshore Structural Integrity Management (SIM) is the process of demonstrating an offshore structure's fitness for purpose over its entire life and for managing the effects of deterioration, damage, changes in loading, and accidental overloading (API 2014). It involves the systematic assessment, monitoring, and maintenance of offshore structures such as oil rigs, platforms, wind turbines, and subsea installations to ensure their safety, reliability, and longevity.

Following the release of the API-RP-2SIM (API 2014) addressing the issue of Risk-Based Inspection (RBI), standards and guidelines have been developed for implementing risk-based structural integrity management approaches to offshore structures. In particular, the International Organization for Standardization (ISO) has developed a specific

Table 1: ISO standards relevant for SIM and lifetime extension of offshore structures

Standard	Description
ISO 19900:2019	General requirements for offshore structures The standard specifies general requirements and recommendations for the design and assessment of bottom-founded (fixed) and buoyant (floating) offshore structures. It is defined that the durability of a structure shall be achieved by adequate design and assessment, inspection, monitoring, maintenance and repair. Only in-service inspections are mentioned relative to a required strategy (Clause 5.5). The concept monitoring is covered relative to the required SIM scheme (Clause 12.3).
ISO 19901-1:2015	Part 1: Metocean design and operating conditions The standard gives general requirements for the determination and use of meteorological and oceanographic (metocean) conditions for the design, construction, and operation of offshore structures of all types. With respect to monitoring, the standard is intended as an initial reference for operators and asset owners when planning metocean monitoring equipment for offshore installations. Applications such as weather forecasting and climate statistics are included by adopting measurements as a source for, e.g., metocean databases.
ISO 19901-3:2014	Part 3: Topsides structures The standard provides requirements for the design, fabrication, installation, modification and structural integrity management for the topside structure. For in-service inspections and SIM, reference is made to ISO 19902 (and thereby ISO 19901-9), i.e., baseline, periodic and special inspections are mentioned, and requirements are made relative to the level classification. Several considerations are made for corrosion aspects regarding monitoring. However, monitoring is only mentioned two times, i.e., for root-cause analysis of vibration problems and for accidental actions.
ISO 19901-9:2019	Part 9: Structural integrity management The standard primarily covers fixed steel structures. It is defined that a required SIM strategy shall identify the mitigation, monitoring and inspections to be included in the continuous risk reduction requirements for the structure and safety critical components. Different inspection methodologies are provided. However, while basic and periodic (consequence-based) inspection intervals are defined, definitions of RBI and the associated value to be obtained by monitoring are sparse (see Clause 10). Furthermore, monitoring (as a separate field) is only defined in terms of degradation mechanisms and/or integrity enablers (e.g., monitoring of natural frequencies).
ISO 19902:2020	Fixed steel offshore structures The standard covers design of fixed steel offshore structures. For the development and execution of inspection and monitoring strategies, reference is made to ISO 19901-9.
ISO 19904-1:2019	Part 1: Ship-shaped, semi-submersible, spar and shallow-draught cylindrical structures The standard specifies requirements and guidance for structural design and/or assessment of floating offshore platforms. For all entities covered by the standard, the structural integrity management system is the anchorage point with respect to inspection and monitoring (Clause 19), which also includes monitoring of environmental data. Requirements and recommendations regarding inspection and monitoring programmes are provided. These requirements and recommendations include information about techniques, extent and minimum requirements (Clause 19.5). Regarding monitoring, examples of use are provided (Annex A.19.4.5) without further detailing.
ISO 19905-1:2016	Part 1: Jack-ups The standard specifies requirements and guidance for the site - specific assessment of independent jack - up units. In relation to inspection, only project specific in-service inspection programmes are mentioned when the existing in-service inspection programme is outdated (as a result a long-term application). In-service conditions are managed according to a classification by a recognized classification society. Monitoring is specifically recommended with respect to weight control and marine growth.

SIM standard, the ISO/DIS 19901-9 (ISO 2019), that recommends a risk-based approach for developing SIM strategy applied to safety-critical structural items. For those, a performance standard should be established that serves as a basis for appraising their risk level and for defining the SIM strategy. A so-called Major Accident Approach is recommended for selecting critical structural items and typical examples of critical structural items are provided in the appendix of the standard. However, no guideline is provided on how to set up their performance standards.

Along the same lines, Bureau Veritas (2018) and more recently (HSE 2023) provide guidance on the main recommendations and requirements for implementing risk-based structural integrity management (RBSIM) for offshore structures. The documents highlight significant SIM standards from ISO and API to reflect the performance of offshore structures under different requirements.

Overall, offshore RBSIM aims to optimise the management of asset integrity by focusing resources on the most critical risks, thereby enhancing safety, reliability, and operational efficiency. Offshore systems are complex and operate in remote, difficult conditions and corrosive environments; therefore, they are subjected to great hazards and extreme events. Thus, offshore RBSIM has been implemented in such systems with various schemes and processes to tackle such challenges. Some critical processes involve risk-based in-service inspection, risk assessment of structures (significant changes, degradation and damage), structural reliability assessments, maintenance activities and life extension evaluations.

4.5.1 Risk Assessment

Risk Assessment is a key element of offshore RBSIM. It involves identifying potential consequences and the likelihood of occurrence of the main failure modes for various components of offshore systems. Risk assessment methods such as probabilistic risk analysis (PRA) or semi-quantitative approaches are commonly used to quantify and prioritise risks.

Following the recommendation of API-RP2SIM, Ng et al. (2023) proposed a structural integrity management (SIM) approach for a steel gravity-based structure (GBS). This study has identified seven major structural hazards and two non-structural hazards for SIM, and used a qualitative risk assessment approach to establish risk-based underwater inspection (RBUI) and maintenance strategies.

Guédé (2019) has proposed a method for risk assessment and inspection plan development of offshore jacket platforms within the scope of risk-based structural integrity management. Both global and local risk assessments were conducted based on the availability of the data. In particular, semi-quantitative methods are used to assess the local risk ranking of structural components of the platform, and quantitative methods (e.g. structural reliability methods) are adopted for probabilistic fatigue assessment to define inspection plans for selected tubular joints.

Dyer et al. (2022) have provided a platform global risk assessment by employing two machine learning models: a gradient-boosted regression tree and an artificial neural network to forecast the removal age of existing platforms. These models are informed by a unique and extensive dataset containing structural characteristics, incident history, meteorological and oceanographic (metocean) conditions, production information, and geohazard data. Results from this study highlight key factors associated with the platform's remaining useful life (RUL) and can be compared to already established markers of platform integrity.

Qualitative methods can also be improved and used in risk assessments when data is not available, which benefit from accident analysis methodologies and coding taxonomies particularly developed for the offshore context (Bhardwaj et al. 2022). The focus of such methodologies is to highlight major hazards and their representation in a holistic way only as the first step of RBIM.

For low-frequency, high-consequence risks of offshore structures, the conventional bow-tie approach is utilised to manage risks which are mitigated by operational safeguards, such as ice feature overloads, offshore collisions, multi-line mooring failures, and loss of floating stability (Slatnick et al. 2022).

With more data, proper quantitative analyses can be conducted. For example, Bhardwaj et al. (2021) have carried out a comprehensive risk assessment of an FPSO using data from three regulators and assessed the risk using a risk matrix to identify the platform's area at the most risk. They have implemented an evidenced-based Bayesian Network to evaluate the risk of Fire and explosion in FPSOs, but the approach can be applied to structural failure modes.

Due to their capability to handle data uncertainties, incorporate interdependencies and learn from new information, Bayesian Networks have been applied as both expert-driven and data-driven probabilistic approaches in offshore risk and structural integrity assessments (Animah 2024). Furthermore, Hybrid BN models have been developed to solve dynamic and complex risk analysis problems (Alsulieman et al. 2024).

In recent times, structural integrity programs, in particular for floating structures, have been digitalised (Tong et al. 2023). The digitalisation helps ensure the integrity of structures during the service life and provides a holistic overview of assets in a centralised dashboard web-based platform (Tong et al. 2023).

4.5.2 Structural Reliability Assessment

Structural reliability analysis plays a crucial role in offshore structural integrity management by assessing the probability of failure of various components over time. This analysis aids in determining the optimal inspection intervals, maintenance strategies, and overall risk mitigation measures for ageing offshore systems (El-Reedy 2022).

Offshore structural systems are mostly composed of tubular or plate structural elements. The capacity of such structural members can be assessed based on stress-based, strain-based and fracture mechanics approaches. Commonly, accident, ultimate, and serviceability limit states involving load, strength degradation and related uncertainties are formulated to check the design, estimate safety factors or estimate time-varying reliability.

Artificial Intelligence-based surrogate models with high accuracy have also been successfully used to dramatically reduce the computational time required to characterize the hydrodynamic response of floating offshore wind turbines (e.g. Ilardi et al. 2024). However, these continue to be desktop studies away from the asset, but in the future may be able to inform operation of the asset during operation.

To ensure the safety and reliability of deteriorating structural systems, system damage should be detected as early as possible through inspection. Delays in damage detection lead to maintenance or repair delays, which adversely affect the structural integrity. A probabilistic approach to assess the reliability of offshore structural systems with explicit fatigue deterioration and damage detection delays has been developed by Li and Zou (2024). This approach characterized the relationship between the system reliability and the delays in detecting fatigue cracks that depend on the inspection quality, number of inspections, or number of components inspected.

To reduce the computational cost and to increase the feasibility of long-term fatigue evaluations, a novel probabilistic modelling framework is proposed for long-term fatigue reliability analysis of offshore wind turbines (Zhu et al. 2023). The approach adopts a Markov model constrained by environmental conditions to determine the long-term sequence of fatigue loading events and the fatigue reliability is evaluated by subset simulation.

Kurniawan et al. (2023) have investigated the hull structural integrity issues due to corrosion and fatigue-induced cracks. They have developed a time-dependent corrosion model for failure prediction purposes based on the historical data of plate thickness reduction. The hull integrity is assessed in a global failure scenario using a semi-quantitative approach, while local fatigue strength is assessed using reliability analysis to support maintenance planning.

4.5.3 Structural Health Monitoring (SHM)

Effective utilisation of real-time SHM information in the context of structural integrity management (SIM) is crucial for minimizing the expected value of life-cycle costs. Several sensor technologies are implemented on offshore structures; however, data management and optimal inspection & maintenance decisions remain major issues.

The optimal inspection & maintenance decision for the service life of structures changes over time as new information is collected. Therefore, in principle, the optimal strategy for integrity management results from all knowledge and information at any point in time during the life of a structure.

Zhang et al. (2024a) have proposed a framework aimed at optimising the utilisation of real-time SHM information from a life-cycle perspective. The approach builds on the Bayesian posterior decision theory and the concept of Value of Information (VoI) (e.g. Qin 2022) that is used to quantify the benefits (in expected value) related to acquiring additional structural health monitoring information. A comprehensive comparative analysis of various utilisation strategies is conducted, highlighting their respective efficiencies and effectiveness in the context of SIM. Through a case study of welds exposed to fatigue loading, Zhang et al. (2024a) not only showcased the proposed framework but also demonstrated its ability to identify the optimal strategy for real-time SHM information utilisation. This model plays a crucial role in quantifying how the availability of real-time SHM data from a life-cycle perspective impacts the benefits of SIM. Moreover, Sensor Validation Tools (SVTs) have been developed to give insight into the actual condition of the SHM systems (Giordano et al. 2023), extending the classical VoI framework to quantify the additional benefit brought by the information on the state of the SHM system to the decision problems it is meant to support.

4.5.4 Digital Modelling

Xia and Zou (2023) have reviewed the latest research progress on digital twin (DT) technology targeting OWFs operations and maintenance, including failure analysis, operations and maintenance objectives, strategies & optimisation models, DT technology development, as well as DT-based O&M management and optimisation. A DT-based operations and maintenance optimisation framework is proposed to help improve the intelligence level of operations and maintenance. PSA has published a guidance report regarding the use of digital solutions and structural health monitoring for SIM (PSA 2022).

An innovative four-dimensional DT framework, meticulously designed for the fatigue damage assessment of semi-submersible platforms, has been presented (Wu et al. 2024). A key highlight of this study is the introduction of an advanced stress twinning methodology. This novel approach adeptly integrates real-time monitoring data with the results of high-fidelity numerical simulations.

4.5.5 Inspection Planning

RBSIM is likely to lead to a significant improvement in inspection planning. RBI is an integral component of RBSIM, an overview has been provided regarding offshore structures by Qureshi et al. (2023). Cheng et al. (2022) have proposed a Reinforcement

Learning method to optimize the dynamic inspection policy for optimum condition-based maintenance of fatigue-sensitive ship structures. Planning in-service inspection for fatigue cracks is generally carried out for offshore structures.

The optimisation of the inspection time and repair strategy for a generic welded joint in a generic offshore wind turbine structure subject to fatigue is performed based on different types of decision analyses, including the value of information analyses to quantify the expected service life cost encompassing inspection, repair, and fatigue damage for all relevant combinations of inspection time, repair method, and repair time (Farhan et al. 2021). Based on the analysis of the expected service life cost, the optimal inspection time, repair method, and repair time are identified. Possible repair methods for a welded joint in an offshore environment include welding and grinding, for which detailed models are formulated and utilized to update the joint's fatigue performance.

Hlaing et al. (2022) proposed a method composed of dynamic Bayesian networks with Partially Observable Markov Decision Processes (POMDP) solvers to provide optimal inspection and maintenance planning for structural components subject to fatigue deterioration. Heo et al. (2023) have proposed a Markov Decision Process (MDP) framework to deal with operations and maintenance issues. The focus is on using the complex stochastic weather variable on the performance of the MDP model.

4.5.6 Subsea Pipelines

Offshore pipelines are safety-critical structural elements that serve various purposes in offshore facilities. The pipelines are subjected to operational, environmental and accidental loads and their strength is affected by degradation mechanisms.

A new approach has been proposed for risk-based inspection planning of the pipelines (Sözen et al. 2022). This includes the evaluation of the probability of fatigue failure due to internal surface defects and under the effect of variable pressure.

Abdelmaksoud et al. (2024) have introduced a novel framework for characterizing the natural frequency of corroded pipelines conveying fluids. The framework simulates fluid-structure parameters using Gaussian random fields, modified using fuzzy membership functions to account for corrosion effects as identified from field inspection.

Xu et al. (2023) have adopted data-driven models and response surface methods to improve fatigue reliability analysis of deep-water risers. Their approach is more efficient than the use of physics-based models because a large number of complex numerical and iterative solutions are avoided in fatigue reliability analysis.

With the recent wave of digitalization, the DT has been discussed as a powerful technology in a variety of industries, including the oil and gas industry. The emergence of the DT provides an efficient way to realize remote monitoring and control, downtime prediction, and risk reduction of oil and gas subsea pipeline systems (Chen et al. 2022).

4.5.7 Offshore Wind Turbines Supporting Structures

OWFs is a complex system comprising mechanical, electrical and structural components. In particular, the support structure type significantly influences the cost-effectiveness of an offshore wind farm deployment. The offshore wind industry has benefited greatly from the experience obtained with offshore oil & gas platforms. The failure mechanisms and the models developed for both environmental loads and boundary conditions of offshore platforms can also be utilised for the OWT support structures, as reviewed by Yeter and Garbatov (2022).

OWT support structures are exposed to harsh ocean environments with significant uncertainties in soil properties and environmental loads. Wang et al. (2022b) presented the state-of-the-art reliability assessment of OWT support structures, providing a comprehensive review of structural reliability, reliability-based calibration of codes, fatigue reliability and the implementation of reliability assessment.

The main failure modes of OWT support structures include fatigue, buckling, scouring, cracks in welds, excessive deflection, corrosion, vibration and fouling (Martinez-Luengo and Shafiee 2019). Since OWT support structures experience significant cyclic loads, such as wind and wave loads, their design is typically dominated by fatigue reliability (Yeter and Garbatov 2022). In fact, more than 80% of structural failures are caused by fatigue (Mendes et al., 2021).

Failure and maintenance data is available for onshore WT and eventually, for fixed OWT, which can be modified and converted using FMEA and Bayesian theory to make it applicable to floating OWT (Li et al. 2020a, 2021, Sun et al. 2023). Furthermore, Bashir et al. (2022) have proposed a data driven method called Multi-Scale Convolutional Neural Network with Self Attention-based Auto Encoder–Decoder (MSCSA-AED) for damage quantification of floating OWTs. Currently, few studies on the reliability assessment of floating OWT support structures can be found in the literature, as these support structures are still in their early stage of development. Okpokparoro and Sriramula (2021) have proposed a reliability assessment framework by combining FEA and Kriging surrogate models. The framework has been applied to assess the structural reliability of the OC3 Hywind Spar floating wind turbine support structure. The results indicated that the floating wind turbine support structures can be designed at consistent reliability levels using the proposed framework. More studies on the reliability assessment of floating OWT support structures are expected.

As OWFs structures are becoming larger and more distant from the coast, adequate design and health and failure monitoring are needed to ensure their operational integrity, which also benefits from the digital twin (DT) technology, as reviewed by Xia and Zou (2023). In this context, DT technology is regarded as a promising tool to integrate all monitoring data for real-time OWFs failure analysis, understand the mechanisms of failures, and support decision-making on O&M strategies.

The DT-based failure monitoring and remaining life prediction method has many advantages over traditional methods, e.g., integration of data from multiple sources, updating of predictions, and visualization of failures or remaining life, among others (Xia and Zou 2023).

Wang et al. (2021b) have proposed an intelligent DT framework to support real-time monitoring and failure diagnosis of OWFs support structures. Ritto and Rochinha (2021) have integrated machine learning techniques with DT models to achieve rapid identification of structural damages. Tao et al. (2022) developed a five-dimensional DT model for efficient failure monitoring of wind turbines and to guide the reliability assessment of support structures. Other scholars have also used DT to predict the remaining life of OWFs. Yüce et al. (2022) developed a DT framework employing a machine learning algorithm to provide remaining life predictions of OWFs. Moghadam and Nejad (2022) have also developed a dynamic degradation model assisted by DT technology to estimate the remaining life of offshore wind turbine structural components based on the changes in stress.

Although the advantages of DT-based platforms over traditional methods, there are also limitations as currently DT models are applied mainly to the critical elements and components and not comprehensively to the whole system (Xia and Zou 2023).

4.6 Maintenance

Through a diligent execution of probability based appropriate maintenance and repair procedures, it becomes possible to manage and decelerate the rate of deterioration for the purpose of structural life extension. Since maintenance and repair are associated with significant cost and risk for the personnel conducting this operation, research is currently focusing on the development of advanced O&M strategies with the aim to mitigate costs and risk and minimise the need for repair actions. To optimize the operation expenditures (OPEX) and structural integrity of offshore infrastructures, the maintenance and repair plans are strategized in accordance with the guidelines, inspection activities, and decision support systems. The following sections describe different maintenance and repair strategies which are grouped based on decision making approaches, conservative and advanced maintenance approaches.

4.6.1 Decision Making Approaches

For optimal structural integrity management (SIM) crucial for achieving an acceptable OPEX of offshore facilities, an efficient decision-making process for maintenance and repair is arguably one of the most crucial tasks. Generally, these are classified as reactive, open-loop and closed-loop approaches depending on the data type and means of processing the information to reach a decision.

Li et al. (2020b) mentioned that the reactive and open-loop decision making approaches are widely used to manage the offshore windfarms and other infrastructures. The reactive decision-making approach corresponds to the corrective maintenance strategy. In this approach, the decisions are made based on the present state of a component or the infrastructure in a more general sense, avoiding unnecessary repairs and inspections (Nguyen and Chou 2018). On the other hand, the open-loop decision making process is partly related to the planned maintenance strategies that can be distinguished as follows: preventive, opportunistic, age-based, and risk-based. In an open-loop approach, the maintenance strategy is usually optimized after the design and risk assessment stage and implemented throughout the structural lifetime regardless of the change of working condition or the individual status of the asset during operation (Karyotakis & Bucknall 2010). This approach could lead to not being able to update the operational state of infrastructure accurately, and resulting in unreasonable and inefficient maintenance actions.

Alternatively, a closed-loop approach involves a process from collecting information and status to decision-making, taking action, and recording information back into the system (Li et al. 2023). By using this approach, the maintenance plan is iteratively optimized ensuring the reliability and functioning state of the infrastructure. From this perspective, one could state that the closed-loop approach is intertwined with the condition-based maintenance strategies such as predictive and remote-monitoring methods where the state of structures after inspection and repairs are considered to predict structural integrity and next maintenance action. Li et al. (2023) showed that the closed-loop approach can moderate the maintenance model parameter uncertainty which leads to a 3.4% reduction in revenue loss compared to open-loop approaches.

Similarly, Farhan et al. (2021) studied a decision and analysis-based approach to predict the inspection time, the repair schedule and the repair method of a generic welded joint in an offshore wind turbine support structure. The decision-theoretical formulation can provide the basic decisions regarding integrity management and the decisions based on additional information and subsequent actions. At the same time, Value of Information analyses (VoI) can inform the basic decisions while implementing an integrity management strategy, an information acquirement strategy and/or an action strategy. The authors also concluded that VoI analysis using the outcome of system state analysis (SS-A) and predicted information and predicted action decision analysis (PIPA-DA) is more realistic than those using prior decision analysis and predicted action decision analysis

Correspondingly, Markov decision processes (MDP) and partially observable MDP (POMDP) are being applied in the maintenance planning of offshore structures such as do-nothing, inspection, or repair actions. In the work of Morato et al. (2023), a fatigue failure criterion based on different crack growth models were integrated with MDP and dynamic Bayesian Neural Networks (BNNs) for the optimization of maintenance planning. Although the infinite horizon POMDP outperforms the heuristic-based policies, the former sets limits on computational efficiency when it comes to consideration of multiple components and longer timeframes owing to the failure criteria involved. To address this computational problem, Hlaing et al. (2022) proposed optimal scheduling over a 20-years horizon through POMDP and deep reinforcement learning adaptive strategies, for a structural component subject to fatigue loads. These strategies are supported through simulation techniques, offering decision-makers valuable insights, and allowing them to assess and interpret alternate actions without the necessity of retraining neural network models or using value iteration methods. On the other hand, various decision-making tools are also developed and integrated with SIM to be able to consider more influencing parameters and streamline the evaluation process. For instance, Nichols et al. (2015) presented a digitized Technical Limits Weight Control (TLWC) tool based on the global ultimate strength analyses and acceptance criteria for operating region in Malaysian Waters. This TLWC tool can also be integrated with the SIM system of fixed offshore facilities.

4.6.2 Conservative Maintenance Strategies

For the management of offshore facilities, one or more maintenance strategies are usually adopted to ensure the safety through the SIM system. In the maintenance and operation of offshore infrastructure, corrective maintenance, also known as reactive response maintenance, has been widely used for decades (Karyotakis and Bucknall 2010). This strategy makes dynamic decisions to carry out repair or restoration of damaged structures as soon as the failure of a component or equipment occurs. However, with the increased number of offshore infrastructures in deep seas, this approach becomes more difficult especially during harsh weather conditions (Scheu et al. 2012), resulting in increased production downtime and jeopardizing the structural integrity. Therefore, a corrective maintenance strategy is suitable only for less critical components or equipment which will not compromise the safety or operation of the whole infrastructure, and is still used in combination with other maintenance plans.

As a more promising approach, planned strategies (variously known as age-based, risk-based, preventive and opportunistic) and condition-based maintenance strategies have been proposed to ensure the reliability and continuous operation of offshore structures. In planned maintenance, the factors determining the maintenance schedules are the operating parameters, environmental conditions, designs, and deterioration rates. The scheduled inspection, servicing and component replacements are carried out to prevent equipment or structural failures. These tasks are performed regularly at set intervals regardless of the condition of the component to keep the offshore structures in good working order until the next predefined maintenance activity. The age-, risk-, and reliability-based strategies also fall under the preventive maintenance category.

In the age-based approach, components are replaced or retrofitted before reaching their expected lifespan by assuming the likelihood of failure with age (George et al. 2022). However, this approach does not consider the resource availability, deviations on operating parameters, and site constraints. On the other hand, risk-based maintenance is the integration of risk assessments into maintenance planning. It highlights inspection strategy of a component(s) with highest risks considering the probability and consequences of failure based on the gathered data and risk criteria (Chen et al. 2022).

As an advancement in contrast to scheduled planning, the oil and gas offshore industry makes use of Internet of Things (IoT) and in-situ structural health sensing/monitoring technologies for real time risk-based inspection analysis where the inspection schedules can be updated promptly with occurrence changes (Qureshi et al. 2023). To mention a few, com-bination of DT, prognostic model with pressure data from sensors, and high level of AI strategy were used to create an automatic prediction of risk probability and maintenance planning for the subsea oil pipelines by Priyanka et al. (2022). Furthermore, Spahić et al. (2023) trained risk-informed image detection system using a set of synthetic images in a convolutional neural network based on the inspection images remotely acquired by autonomous underwater systems. The system can produce reliable image detection, segment the abnormalities, and ultimately assist in maintenance and inspection tasks of subsea pipelines.

Also in recent years, opportunistic approaches have been practiced in offshore industries to minimize the downtime of the asset due to preventive maintenance activities (McMorland et al. 2023). The triggers for the maintenance opportunity can be distinguished into internal (Erguido et al. 2017) (in case of a need for corrective or preventive maintenance) and ex-ternal (Kennedy et al. 2016) (the weather condition such as low wind speed to generate electricity efficiently in wind farms). The opportunistic strategy takes advantage of those planned or unplanned shutdowns of the system to substitute equipment or components while the resources are available onsite. In recent literatures, opportunistic maintenance is scheduled based on predictive analytics Zhou and Yin

(2019) and the condition monitoring data through DRL (Cheng et al. 2023; Valet et al. 2022). Although this approach is favourable from an economic perspective by reducing the cost of maintenance and production down-time, the remaining useful life of the components will be unutilized.

Although scheduled maintenance strategies could lower the cost of repair and operation downtime to some extent, the planned intervention may not be fully reliable throughout the lifetime of offshore infrastructure. The deterioration rate and mechanism evaluated during design and risk assessments may become invalid as the environmental and operational factors vary through the lifetime of infrastructures. In this regard, condition-based maintenance strategies become promising solutions to address all the variations through the lifespan of the facility and save remaining useful life. The challenge with these strategies is the reliability on smart technologies and sensors, databases with statistical significance, and the accuracy and efficiency of numerical predictive capabilities. According to the recent 4 years period between 2018–2022 of selected literature, preventive maintenance, risk-based maintenance and condition-based maintenance strategies have been predominantly employed to restore the desired functionalities and goals (George et al. 2022).

4.6.3 Advanced Maintenance Strategies

The harsh operating environment makes the maintenance and repair of offshore structures particularly more expensive and difficult. However, digitalization, the development of advanced remote structural health monitoring systems and the technological evolution into Industry 4.0 have become valuable and expedient for the offshore industry. A digital SIM framework can record the inspections, repairs, remote monitoring data, evaluations, and planning outcomes over the entire horizon of an offshore windfarm, and those data are shared for processing by relevant stakeholders (Eichner et al. 2022). DTs have become popular as a smart and safe technology in managing the offshore structures by representing the whole infrastructures in as a digital model with real time data exchange be-tween the target structure and DT (Ambarita et al. 2023).

In a recent work of Hasan et al. (2023), an interactive DT platform, based on the OPC Unified Architecture (UA), Unity3D, and Vuforia, was developed to support in the asset development and optimization. This system, also in combination with Augmented Reality (AR) that enhances the user experience, has been used for both, teaching and research purposes in Norwegian University of Science and Technology (NTNU). Correspondingly, in Haghshenas et al. (2023), a similar system is integrated with the prophet prediction algorithm to investigate the abnormalities and predict the potential failure of wind turbine components in individual wind turbine. The system was attested for failure prediction of mechanical components and bearing and being used for research and teaching.

In the same aspects as DT, data driven models also come into the focus of interest when it comes to more effective and efficient management; those include data processing and pre-dictions through statistical learning, machine learning, and deep learning models (Lv and Fersman 2022). Using BNNs, Hlaing et al. (2023) proposed a virtual load monitoring framework in order to obtain farm-wide load predictions. The findings from application of such a developed framework in an operational wind farm demonstrated that when an epistemic model uncertainty indicator is incorporated, a BNN method has the capability to detect possible inconsistencies with the generated load predictions. This is an informative tool lending itself for optimizing lifecycle decisions through managing the accuracy of the farm-wide virtual load monitoring which provides load information in the place of defunct strain sensors.

4.6.4 Corrosion and Protection

When exposed to the harsh, wet and saline environment, corrosion and corrosion-fatigue (CF) is the leading cause of structural failures in offshore facilities. Countermeasures are taken to reduce the rate of corrosion failures during design, installation and maintenance stages; corrosion allowance, CP, and corrosive coating protection are employed in accordance with regulations (DNV 2021). Momber (2021) also described the active and passive corrosion systems especially for wind power structures in marine environments. However, as the corrosive coating ages, it also deteriorates with time, rendering especially shorter lifetime in the splash zone where CPs are not effective (Brijder et al. (2022)). Moreover, Andresen-Paulsen et al. (2023) concluded that coating age and thickness influence the corrosive damage to welded plates and joints based on the previous literatures. Therefore, as part of the maintenance and repair, corrosive protection systems are regularly inspected and monitored for the necessity to reapply.

With an increasing number of offshore wind farms in the deep sea further from the shore (4C Offshore, (2023)), the visual inspection for corrosion is exacerbated in several different ways: expense, accessibility, resources and man hours. Researchers have been trying to solve that problem by taking advantage of IoT, sensors, and AI. To repair the failed CP systems in oil and gas pipelines before corrosion deterioration becomes critical, a remote VR-enabled system was proposed by Mgbemena et al. (2023) which can monitor and analyse the pipe-to-soil potential readings of the CP system installed. Similarly, Verhelst et al. (2022) developed a software tool compatible with Supervisory Control and Data Acquisition (SCADA) to visualize and translate corrosion-related information for maintenance and repair actions of wind turbines. On the other hand, Thibbotuwa et al. (2022) proposed an ultrasound testbed based virtual monitoring system which can accurately measure the thickness loss up to 1 μm considering coated and bare steel samples, and thus enables predicting the corrosion rate in real-time for practical purposes.

However, when the structure is already severely corroded, the affected areas are clad with corrosion products which can decelerate the corrosion process, or it is replaced partially/completely with a new structure component. Today, such replacements can also be performed with Remotely Operating Vehicles (ROVs). There are commercially available ser-vices for inspection, maintenance, and repair of underwater tasks using ROVs (DeepOcean (2023); TechnipFMC (2023)). For instance, to replace a corroded structure, surface preparation, removal, and replacement of new structure, finally until welding/cladding a new component can be performed using a series of ROVs (OCEANEERING, 2023).

4.7 Co-simulation for Offshore Structures

In our daily lives, the demand for marine operations is steadily increasing, with a projected 52% rise in offshore vessels, including construction vessels for the oil and gas sector and support vessels. This surge has prompted a significant emphasis on the research and development of marine systems, spanning both offshore structures and ships. Liu et al. (2023d) observed that due to similarities in the operating conditions such as high wave loads, unstable and salty environments, offshore structures face the same fundamental issues as ships and therefore, it is becoming increasingly important and necessary for the maritime industry to work on how to increase the safety of offshore structures and ships. Moi et al. (2020) presented that digitization has stepped up as a critical aspect into enabling the maritime industry to be more innovative, efficient and future-proof in the maintenance and operating domain. However, offshore structures and ships are more complex and have an interdisciplinary connection such as mechanical, hydraulics and dynamic fields which leads to a lot of difficulties in creating the multi-domain model. Gomes et al. (2017) came up with one possible solution for this challenge by suggesting using a heterogeneous model-based approach where different conventional models carry out their usual mono-disciplinary analysis, but in addition, the different models can be coupled for simulation (co-simulation), allowing the study of the global behaviour of the system. Hatledal et al. (2021) developed and implemented a high-level fidelity model built with co-simulation framework named 'VICO' in the maritime industry. It has been designed in such a way that physics models and other systems that are not Functional Mock Ups can be integrated using a particular framework into a co-simulation setting. To prove the effectiveness, some case studies were implemented in which vertical displacement of the structures were studied with respect to time which is an important factor for both offshore structure and the ships.

In recent years, significant progress has been made in the domain of co-simulation for offshore structures. Notably, the integration of multi-physics methodologies, precision in hydrodynamic modelling, and the incorporation of machine learning algorithms have collectively augmented our capacity to rigorously evaluate the structural integrity, operational performance, and safety of both offshore structures and ships.

Offshore structures are susceptible to high order environmental loads which induces dynamic pressure and causes the structure to vibrate which is also called vortex-induced vibrations. This vibration leads to significant structural fatigue damage. Zhou et al. (2022) tried to study the frequency of these vibrations using co-simulation approach where they made use of numerical simulation (computational fluid dynamics) and model testing, and with the obtained results, tried to predict the basic parameters needed for fatigue analysis. Similarly, Viswanathan et al. (2023) extended their previous research by implementing a full-scale co-simulation model. In their work, they carried out a multi-physics model simulation of the platform riser tensioner system to understand the heaving motion by using tools like OrcaFlex and SimulationX to model the semi-submersible platform and the hydro-pneumatic tensioner respectively. Chueh et al. (2023) likewise, approached similar problems for offshore wind turbines by implementing dynamic co-simulation models using tools such as SIMPACK, MATLAB/SIMULINK, AeroDyn, HydroDyn, WAMIT, and MAP++ to study the effect of aerodynamics, hydrodynamics and control system dynamics together which resulted in new power control strategy to reduce the oscillations/vibrations of the structures.

Co-simulation approaches have not only been used in structural studies but also in the performance improvements of the offshore power plants. Al-Zareer et al. (2021) showcased their work by integrating photo voltaic systems for offshore power generation using different radiation distribution, optical and photo-electrochemical models and combining them by tools such as FEM and COMSOL. Bagherabadi et al. (2023) demonstrated the application of their model by developing a hybrid power plant model for an offshore vessel. Due to their high-fidelity vessel model, the models were connected in the co-simulation approach to provide a total system simulation with real time capabilities. Also, the integration of real-time co-simulation not only enables dynamic response assessments in actual operational conditions but also supports precise decision-making. Furthermore, recent research endeavours have focused on incorporating machine learning algorithms to enhance the accuracy and efficiency of co-simulation results. Li et al. (2022) proposed a model made by combining tools such as failure mode and effect analysis (FMEA) and Bayesian Network (BN) for the development of predictive maintenance strategies for the offshore structures. They also proposed a risk-based criterion model instead of the cost-based models to achieve real time suggestions for the components that are to be inspected and later verified the FMEA-BN model based on the precisely predicted failure rate data.

The above examples show that co-simulation has proved to be a promising technology in the offshore structure domain by allowing the incorporation of multi-physics and multi-domain models along with the help of real-time data handling clubbed with Machine Learning algorithms. Therefore, it can be said that co-simulation is emerging as a pivotal tool in meeting the evolving demands of offshore operations with a lot of future advancement capabilities.

4.8 Concluding Remarks

The research, development and application of structural assessment technologies and methodologies have been reviewed in the context of offshore structures. In comparison to ship structures, offshore structures often have a much longer service life with increased demands on the need for survey and repair on-site, without the ability to remove the structure to a repair facility.

Within the review, developments in crack and corrosion detection and modelling are discussed, as are advances in in-situ repair techniques. It is noted that for corrosion modelling by numerical methods where the electro-chemistry is simulated have not been successful to date. Therefore, modelling continues to utilise S-N-based or fracture mechanics approaches for corrosion fatigue assessment.

SIM and SHM and the use of DTs for the structural assessment of offshore structures are discussed. Where coupled with machine learning or artificial intelligence for analysing sensor data, the future potential of SIM and SHM is demonstrated. However, challenges related to the uncertainty associated with environmental loads, the aging of the structure need to be overcome to realise the benefits. Data management and optimal inspection & maintenance decisions remain major issues to successful implementation, which will change in time during the life of the asset.

5 Assessment of Ship Structures

5.1 Introduction

The research and development in technologies and methodologies for the structural assessment of ship structure during operations is reviewed.

After outlining the environment, load, and response experienced by the ship structure, functional requirements, benefits, and regulatory requirements of the hull monitoring system (HMS) are explained. A typical HMS configuration is presented, and existing commercial/military/research HMS are reviewed. The ship-specific monitoring technologies, strain measurement/data communication/data processing/human-system interface, are studied.

The difference between HMS and digital twin and DT's benefit is discussed with a special focus on ship structure. The simulation (including ROM, surrogate model, and machine learning) and data assimilation methods that have been applied to ship structure are reviewed, and the application examples of the use of DT for ship structure in decision-making for safe operation and maintenance & repair optimization are studied. Existing/in-development DT systems for ship structure are reviewed.

5.2 Ship Structural Environment, Loading and Response

5.2.1 Environment

Among the listed external weather influences on a ship, notable changes affect ice fields, winds and waves due to global warming in recent decades.

The passage possibility of trans-Arctic routes for two types of ships was assessed (Melia et al. 2016): "open water" ships (OW ships) without specific ice strengthening and ships with low polar class PC6 (summer/autumn operations in medium first-year ice). Early century projections (2015–2029) show that trans-Arctic OW ship transits are possible for at least 30% of September, while PC6 ships have a transit potential of 90% and they can also take advantage of shorter routes impassable to OW ships. Whereas by mid-century (2045–2059), irrespective of RCP OW ships' transit potential is projected to double in the September. And in general, it is predicted that by the end of 21st century average transit times for OW ships may decrease by 9 days (RCP8.5 scenario) or by 4 days (RCP2.6). In 2015–2029, the average OW ships' minimum journey time is 26 days for all European voyages through trans-Arctic route and Suez Canal.

Similar approach of RCP scenario projections (RCP4.5 and RCP8.5) was used for high-frequency extreme ocean wave events (Morim et al., 2021). It was found that global distribution of projected change in high-frequency extreme wave events correlates with changes of extreme surface wind speeds (U10) (Morim et al., 2021) and there is a general increase of Southern Hemisphere U10 (Meucci et al., 2020; Young and Ribal, 2019). 30–40% of the global ocean will be influenced by sustainable changes in extreme waves, but with a strong inter-hemispheric asymmetry. It is predicted that up to 50–100% increases in extreme, persistent wave events over the low and high-latitudes of the Southern Hemisphere while 20–50% decrease across the Northern Hemisphere. Furthermore, wave-driven coastal hazards are expected to increase in the Southern Hemisphere (Morim et al., 2021).

5.2.2 Structural Loadings

Estimating the external loading acting on ships' structures is a complex task that must be done as a part of the assessment of ship structures. Static loads are more straightforward to define based on basic physics concepts. However, slow and rapid loads require solving nonlinear dynamic analysis to calculate the ship's motion and obtain the dynamic pressure on the hull, which is used for defining the level of safety of the ship's structures.

Numerical methods are increasingly used to estimate wave-induced loads on the ship hull. Strip theories, 3D potential theory, and CFD approaches are widely employed to predict the acting wave loads on conventional and nonconventional ship hulls. For example, Wang et al. (2023) have conducted a coupled modelling to study the aerodynamics, hydrodynamics, and structural response of a semi-submersible hull supporting a wind turbine. The hydrodynamic loads on the hull are calculated based on the potential flow theory, including the first-order and second-order wave excitation loads, radiation, restoring, and viscous forces. The main aim of this study was to illustrate an analysis procedure for structural load effect analysis and to shed light on the dynamic characteristics of the support structures from different environmental conditions.

Moreover, experimental tests, either at small or full scale, are involved in estimating the hull loading to calibrate and validate the detailed computational methods. Likewise, class societies' rules exist to guide the implementation of sensor-based direct load measurements on vessels, so-called Hull Condition Monitoring Systems (HCMS) (ABS, 2020; DNV, 2021). Hull monitoring systems are fitted to acquire, display, and record hull responses, and this information is used for decision-making to enhance operational safety. HCMS can provide appropriate information regarding slamming, ship motion, sea state, hull girder stresses, local load, fatigue behaviour, and sloshing. For instance, hull girder monitors show the still water bending moment and wave bending moment and how they vary with time and longitudinal position along the vessel's length.

5.2.3 Structural Responses

As a result of the applied loading, the ship's structure will respond at different levels, with possible consequences on operational functionality and potential failure modes ranging from the global hull-girder level down to the local detail level. The mechanisms through which these failures occur are usually quite different and the consequences vary greatly, from small nuisance level fatigue cracking through to overall collapse of the hull girder under extreme loads.

Application of machine-learning techniques methodologies has been investigated for ship structures fatigue assessment. Lang et al. (2023) investigates the use of a machine learning technique to establish a model for 2800TEU container vessel fatigue assessment. Measurement data from 3 years of cross-Atlantic sailing demonstrated and validated the machine learning model. The fatigue damage amounts predicted using a machine learning model were compared with those obtained from full-scale measurements and direct fatigue calculation. The pros and cons of the methods are compared in terms of their capability, robustness, and prediction accuracy.

5.3 Hull Monitoring System (HMS) Overview

Structural Health Monitoring (SHM) has emerged as a promising branch of engineering, capitalising on the advancements in new sensors, wireless communication networks, and big data analytics coupled with machine learning (ML). The objective of Structural Health Monitoring (SHM) is to assess and predict structural damage and provide decision-making support on safer and more optimal vessel operation, inspection and repairs, and asset integrity management (e.g. ABS 2022, KR 2023, DNV 2024).

It has been conjectured that a Structural Health Monitoring system may allow for optimisation of future designs, operation and/or maintenance, moving from criteria based on experience or conservative estimates to performance-based criteria that take advantage of real-time in-service and behaviour information (Silva-Campillo et al. 2023). Though full realisation may yet to be achieved in practice.

Structural Health Monitoring systems for ship hulls offer significant benefits in terms of safety, cost efficiency, maintenance, environmental protection, and overall operational performance such as: early damage detection by continuous real-time monitoring; load optimization: SHM can help in understanding the load effects on the hull, enabling better distribution and management of cargo loads; optimization of the design, operation and maintenance; supporting the classification process; demonstration of compliance with international safety and environmental regulations, which increasingly require continuous monitoring capabilities; data-driven decisions on the operation of the ship under various conditions.

5.3.1 Functional Requirements

IMO MSC/Circ.646 (IMO, 1994) "Recommendations for the fitting of hull stress monitoring systems" requires that the hardware and software of a hull stress monitoring system be type approved by an Administration, which in practice is usually achieved via a certification of compliance issued by a recognized Class Society. Class societies have also published technical standards and requirements for the use of various hull monitoring systems (ABS 2020a & 2020b).

The functional Requirements for Hull Monitoring Systems typically cover the aspects of the infrastructure for Data Collection and Storage, Data Analysis, Decision-making support functions and Data communication, which are related to specific class notations (KR 2023, NK 2021).

For example, the Functional Requirements for Structural Health Monitoring (SHM) functions defined by (ABS 2022) to obtain the optional SMART (SHM) notation are defined for different SHM tiers, depending on the accuracy and reliability levels of the employed approach for structural health assessment and prediction (Table 2), as follows:

Tier 1: To conduct a vessel-specific load and operation-based structural health estimation using: vessel-specific environmental and sea loads and loading history; vessel-specific operational data such as history of cargo and other payload and loading pattern, vessel speed, heading, draft, and trim; structural strength assessment and damage prediction; accumulated fatigue damage and damage rate estimation using the vessel-specific load and operation history.

Tier 2: Tier 1 approach enhanced with continuous/periodic update on structural conditions (scantlings, repairs and modifications). Finite Element (FE) based or other physics-based or data-driven analytics and simulations incorporated using the vessel-specific loads, operations, and up-to-date structural conditions.
Tier 3: Tier 2 approach calibrated and verified with high-fidelity data for enhanced accuracy and reliability, such as sensor-based measurements.
Tier 4: Integrating the Tier 3 SHM into a risk-based asset management framework, requiring quantification or qualification of uncertainties and risk factors relevant to data collection, data quality, analysis and simulation, construction, and operation.

Table 2: Structural Health Monitoring Tiers SHM (ABS 2022)

SHM Tier	SHM Tier Feature	Coverage	Employed Analytics	Notation
1	Vessel-specific loads and operations	Global hull	Empirical analysis	SMART (SHM) Tier 1
2	Comprehensive structural health assessment	Global hull Local structures	FE-based (or equivalent) analysis	SMART (SHM) Tier 2
3	Comprehensive and calibrated structural health assessment. Sensor package installed for direct monitoring and analysis calibration.	Global hull Local structures Direct sensor-based monitoring on loads and responses for global and identified local structures.	FE-based (or equivalent), analysis Analysis and model calibration	SMART (SHM) Tier 3
4	Comprehensive, calibrated, and risk-based structural health assessment Sensor package installed for direct monitoring and analysis calibration	Global hull Local structures Direct sensor-based monitoring on loads and responses for global and identified local structures	FE-based (or equivalent), analysis Analysis and model calibration Uncertainty and risk assessment	SMART (SHM) Tier 4

5.3.2 Regulatory Requirements

Following the implementation of the Maritime Safety Committee circular, MSC/Circ.646 (IMO, 1994), notable developments have occurred in the design of hull stress monitoring systems, with several recognized Class societies publishing technical standards and requirements for the use of various hull monitoring systems.

It is noteworthy that the term "Hull monitoring system" appears in various forms across different rules and regulations. For instance, the American Bureau of Shipping

(ABS) refers to it as a "Hull Condition Monitoring System," while Bureau Veritas (BV) uses the term "Hull Stress and Motion Monitoring." Similarly, Lloyd's Register (LR) describes it as a "Hull Surveillance System." This diversity in terminology, to a certain extent, reflects the nuanced approaches each society takes towards the specification and implementation of these critical systems.

The leading Classification Societies have integrated a range of considerations regarding structural monitoring systems into their regulatory frameworks (Table 3). DNV (2022) has introduced the HMON notation, indicating the requirement for an approved hull monitoring system to be installed onboard. Component, system and installation requirements are given to ensure that the system complies with the notation. The American Bureau of Shipping (ABS, 2020b) utilizes the HM (Hull Monitoring) notation, which is delineated into three categories: HM1, HM2 and HM3. The condition monitoring systems cover from simple one-motion monitoring systems to sophisticated voyage data recorders including a multitude of hull, systems and machinery parameters. The China Classification Society (CCS, 2023) has introduced the notations HMS for conventional ships and HMS-HSC for high-speed crafts. These notations encompass a wide range of requirements, including system design, data processing methodologies, display standards, and component specifications. Bureau Veritas (BV, 2022) adopted the "MON-HUL" notation, which is specifically designated for ships fitted with a system that can provide real-time data on the longitudinal stresses of the hull girder and vertical accelerations experienced by the vessel. The Nippon Kaiji Kyokai (NK, 2023) has also employed the HMS notation, with the survey and detailed System Requirements and Set-up of Systems elaborated within their documentation. The Lloyd's Register (Lloyd's Register, 2012) has provided the ShipRight SEA (HSS-n) notation for ships with a hull surveillance system that can display in real-time and record the hull stress and motion information. Moreover, ShipRight SEA(ICE) notation is assigned when the ship has been provided with a hull surveillance system that can display in real-time and record the local ice load-induced stresses from a series of strain gauges in the bow region. It is worth noting that Bureau Veritas (BV) and the American Bureau of Shipping (ABS) have developed separate guidelines dedicated to the monitoring of ice loads in their respective regulatory frameworks (BV, 2015 and ABS, 2021).

Table 3: Hull Monitoring class notations

Classification society	Keywords	Notation	Items
ABS	Hull condition monitoring system	HM1 HM2 HM3	1: General 2: System Type Requirements 3: System Requirements 4: Installation 5: Setup and Calibration
BV	Hull Stress and Motion Monitoring	MON-HULL	1. General 2. Sensors design 3. System design 4. Installation and testing

(*continued*)

Table 3: (*continued*)

Classification society	Keywords	Notation	Items
CCS	Hull monitoring system	HMS HMS-HSC	1. General provisions 2. System design 3. Data processing and storage 4. Display and monitoring 5. Component requirements
DNV	Hull monitoring system	HMON	1. General 2. Component requirements 3. System design 4. Installation and testing
LR	Hull surveillance system	HSS-n	1. General 2. System requirements 3. Optional extensions to descriptive note 4. System approval 5. Installation and testing 6. Survey requirements
NK	Hull monitoring system	HMS HMS · R	1. General 2. Survey 3. Hull monitoring systems (System Requirements & Set-up of Systems)

5.3.3 Typical HMS Configuration

The Classification Society's influence has standardized the Health Monitoring Systems (HMS) for marine structures to some extent. Silva-Campillo et al. (2023) conducted a comprehensive review of the essential elements of structural health monitoring for marine structures, analysing current sensor technologies, their theoretical underpinnings, applications, strengths, and limitations. Theories for optimal sensor placement and a historical perspective on sensor use in maritime and offshore applications are also provided.

The foundation of HMS lies in the utilization of sensors that measure strain, acceleration, pressure, temperature, and other critical parameters. In the conceptualization of a HMS, critical considerations include the accuracy and stability of sensor outputs, their durability, and the costs associated with their installation and maintenance (Fujikubo et al. 2024). Fiber Bragg Grating (FBG) sensors (Murawski et al., 2012) offer significant advantages over traditional strain gauges, including resistance to electromagnetic interference, superior corrosion resistance, and the ability to multiplex multiple sensors on a single fibre. These features render FBG sensors particularly well-suited for marine HMS, enhancing stability and durability in tough marine conditions and offering a dependable method for structural health assessment. Despite the advantages, challenges related to technical maturity persist. Chen et al. (2023c) discuss the role of three types of optical fibre sensors in marine Structural Health Monitoring, highlighting their benefits in detecting key structural indicators. There is a call for innovation in sensing technologies, intelligent materials, and machine learning to improve sensitivity and system efficiency. Artificial Intelligence (AI) is highlighted as instrumental in resolving crosstalk issues, ensuring effective signal separation. To address the complexities of deep-sea installations, the paper proposes innovative strategies that leverage Remote Operated Vehicles (ROVs) alongside existing submarine fibre-optic cables. The development of robust fibre encapsulation is identified as an essential step to withstand the marine environment. As technology progresses, the potential for optical fibre sensors in marine SHM is expected to expand significantly (Altabey and Noori, 202).

5.4 Review of HMS Application

5.4.1 Commercial Vessels

Wines et al. (2023) present a ship hull monitoring system for fast ferries to enable operators to ensure operation within the speed wave curve design limits and reduce the risk of exceedance due to variations in operator perception of wave height. A system implementing Fiber Bragg Grating sensors within the midships region is trialled, correlated to FEA, and linked to GPS and accelerometers for speed, position, course and hull motion monitoring.

5.4.2 Military Vessels

Military Systems on board of naval ship are, in principle, in line with the content of chapter 5.3.4 for typical HSMS configuration, according to the descriptions that have been published in recent years.

HSMS have been adopted by different Navies around the world as stated in (Phelps and Morris, 2013), that quotes French Navy, Royal Norwegian Navy and U.S. Navy. Reference (Phelps and Morris, 2013) also reports about key benefits that can be gained through installation of HSMS on naval ship, with respect to those relevant to commercial ships, mainly due the operational profile of naval ships, that is likely to vary over their life and across the fleet.

Detailed description of HSMS by Norwegian Navy is given in reference (Torkildsen et al., 2005), where successful installation of fibre optic sensors has been reported for a Mine Counter Measure Vessel.

Swartz et al. (2010) introduces a hull monitoring systems with wireless sensors into its architectural design, in order to reduce cost installation and complexity, typical for highly compartmentalized layout of naval ships. Furthermore, the fact that they are low-cost can lead to higher sensor densities in a hull monitoring system thereby allowing properties, such as hull mode shapes, to be accurately calculated.

Thompson (2018) proposed a virtual hull monitoring approach for four hull locations. Input data included the ship GPS location, sea loads estimated from linear frequency-domain seakeeping code in combination with linear finite element analysis to estimate the hull stress ranges without requiring an in-situ hull strain or acceleration measurements. These methods delivered insight into the accumulated fatigue damage for 10 naval ships with varying operational profiles. The author stressed that the findings hinge on assumptions and uncertainties in the data used and that real-time data (from a DT) would replace assumptions with facts.

Deployment and subsequent maintenance costs of Instrumented Hull Monitoring (IHM) represent the major limit to its application, although its accuracy and reliability. For this reason, an interesting approach to hull monitoring has found different scientist involved in its development: "Virtual hull monitoring (VHM) is a digital structural health monitoring approach that can be used to calculate the fatigue damage accumulated in a ship's hull. Ship-board data – such as speed, position, and bearing – is enriched with available wave data" (Mogeke et al., 2023).

Virtual hull monitoring has been also investigated in (Hageman and Thompson, 2022; Thompson, 2020) for frigate-size vessel showing that fatigue damage estimates - calculated using wave hindcast data - are in good agreement with values derived from strain measurements.

European Defence Fund project dTHOR on Digital Ship Structural Health Monitoring (dthor.eu), aims to develop a system based on innovative utilisation of large amounts of load and response measurements from robust and advanced sensors with a digital framework for data exchange. dTHOR aims to consolidate end-users military operational requirements based on improved battle damage and structural integrity assessment, reduced hydro-acoustic signatures, and more accurate operation of weapon systems.

5.4.3 Research Systems

Research HMS are typically bespoke systems. The focus of research has mostly been to improve already existing concepts and investigate issues related to ship's structure but also the development of new structural monitoring systems. Typical SHM systems for ships are based on key parameters like hull bending moments, sectional forces, slamming and wave impacts, high and low frequency vibrations and structural deflections and strains (Pezeshki et al., 2023). The specific sensor used depends on the ship type, operational conditions and monitoring requirements (ABS 2020). As DTs and cloud platforms have gained importance in recent years, the focus has gone to visualization of real time sensor data, the remote monitoring and early warning systems. Especially for this reason, acquired data needs to be precise and reliable at all times, which emphasizes the need for a robust measurement system. Marine environments pose unique challenges for data acquisition and transmission due to harsh conditions like corrosion, fouling, and extreme weather events. Ruggedized sensor nodes, fault-tolerant communication protocols, and redundant data storage mechanisms are essential for ensuring reliable data collection and preservation (Svendsen et al., 2022). Also, advanced signal processing and error correction techniques can further enhance data quality and integrity. Sensor data can be integrated with numerical models, such as finite element models (FEM), to create

DTs of ship structures. These virtual representations can be continuously updated with real-time data, enabling accurate simulations and predictions of structural behavior under various operational and environmental conditions. This integration facilitates informed decision-making and optimized asset management. By addressing these key aspects, ship structures can benefit from optimized sensor networks, enabling efficient energy management, robust data acquisition and transmission, advanced damage detection and prediction, seamless integration with DTs, and centralized monitoring and early warning systems (Preethichandra et al., 2023). Collaborative efforts among researchers, industry partners, and stakeholders will drive the continued advancement of these technologies, enhancing the safety, reliability, and sustainability of marine operations.

Bekker & De Koker (2023) proposed a DT to assist navigators with increased situational awareness in ice, by providing insight into the structural margin during operation. The ice-induced propeller moment is rapidly determined through an inverse method and in-direct shaft measurements (sheer strain and high-resolution rpm). The real-time virtual sensing of loads is achieved through inversion based on a modal superposition approach which avoids time-consuming regularization (Nickerson and Bekker, 2022).

Liu and Ren (2023) developed a rapid acquisition method for determining the structural yield strength evaluation stresses at monitoring- and non-monitoring points, using a ship DT model. A three compartment midships section of a 21,000TEU ultra-large container ship was selected. The study considered both, vertical, and horizontal wave bending moments simultaneously, showing high accuracy when assessing points at the monitored locations.

5.4.4 General Application Examples

Zhang et al. (2024b) propose the online identification of wave bending and torsional moment at a critical cross section to monitor hull structural safety against the design threshold. To determine the wave bending moment of a hull structures without large openings, a long baseline strain gauge (LBSG) is installed and optimised using C-optimal and D-optimal design methods, and coupled toan associated FE model. For hull structures with large openings, a joint dual-section monitoring method (JDSMM) is proposed, monitoring two adjacent cross sections of the hull structure for vertical and horizontal bending moments, free torsional moment, and bi-moment. The accuracy of the JDSMM was verified using several time-domain numerical examples.

The determination and prediction of structural dynamic parameters have been investigated on polar vessels. Operational modal analysis (OMA) enables the identification of structural parameters without measurements of the system input forces. However, OMA and its associated assumptions may result in smeared modal estimates which may be improved if known weather parameters could be included in the system identification and tracking analysis. Soal et al. (2019) constructed a statistical regression model to correlate environmental factors (water and air temperature, ship speed, wind velocity and direction) with dynamic structural parameters, determined through OMA on a ship-wide accelerometer array (Fig. 4). A Kalman filter is then employed to combine the predictions from the regression model with OMA Stochastic Subspace Identification (SSI) estimates of the natural frequencies, mode shapes and modal damping of a global ship structure during operation.

Van Zijl et al. (2021) deployed OMA to track five global ship vibration modes is varying environmental conditions (including calm water, storm and ice navigation). The automated OMA analysis comprised the identification, pole cleaning, clustering, and selection of poles to result in a model that tracks changes of modal parameters in the changing vessel environment.

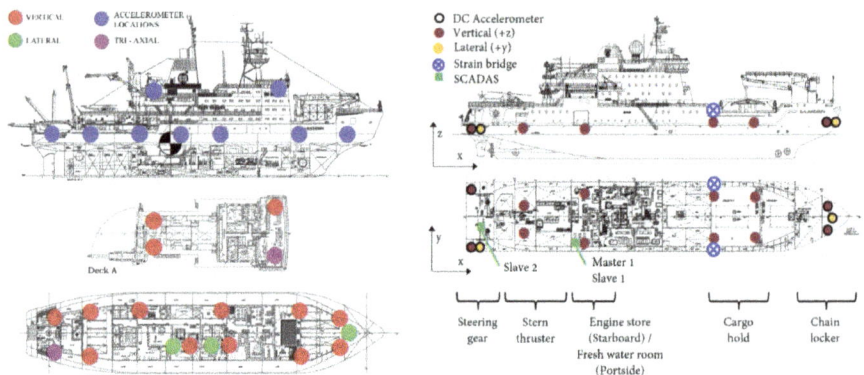

Fig. 4. Accelerometers throughout the structure of the Polarstern. (Soal et al., 2019).

Fig. 5. Hull monitoring system on the SA Agulhas II (Pferdekamper & Bekker, 2024).

The modal responses were quantified by determining the root-mean-square value in the frequency range where the mode shape remained 95% similar to that at a natural frequency. In open water storms bending responses dominated, whereas higher levels of damping were apparent during ice passage. Bossau and Bekker (2022) used six pairs of vertical acceleration sensors (shown in Fig. 5) in the hull on the same vessel to develop an algorithm to detect the location of wave impacts on the hull. The broadband nature of the slamming excitation presents distinct vertical line features on Morlet scalogram images, which may be detected using the Hough transform. Custom thresholds for structurally significant slams were determined from extensive operational data. The wave impacts to whipping responses and hull fatigue through a monitoring system comprising accelerometers and a strain gauge pair in proximity to observed cracks in the mid-ship hull structure. The global bending modes were determined from the accelerometer measurements to determine the relative fatigue contribution of whipping. The low-frequency wave bending and high-frequency whipping vibration were isolated through appropriate filtering, from where rain flow counting and Miner's law quantified fatigue damage from recorded strain-time series. Increased slamming frequency and intensity correspond to increased excitation of the first and second vertical bending modes. In comparison to the head sea test, following waves accrue 46 times more damage, with whipping contributing 98%.

5.5 Digital Twin (DT) for Ship Structures

5.5.1 Benefits

The application of DTs (generally discussed in Sect. 2.2.1) for ship structures present a number of benefits. Gausdal et al. (2018) observe that the intent of the digital transformation is to reduce costs, increase business effectiveness and increase regulation in the maritime industry. Traditionally the prognostic management of asset health has been hampered by uncertainty in the material behaviour, operational conditions and loads that products face in deployment (Uhlenkamp et al. 2019). The DT-enabled coupling between operational data and virtual models, springboards a means through which uncertainties in asset health management (material behaviour, operational context, and loads) can be reduced through detailed and realistic simulation (Katsoulakos et al. 2024). Kutzke et al. (2021) mention that DT technology facilitates reliability and increased robustness by providing increased awareness to system operators and maintainers. A core function of living DT solutions purposes to assist in decision aiding, with the following advantages which enable Condition Based Maintenance (CBM) on a novel scale (Bekker & Erikstad, 2023):

1. DT technology is an aggregator of data and tools with the ability to realize information sharing and convenient communication towards decision support.
2. It is now possible to observe, analyse, and understand real-world interactions and impacts on different objects at a very granular level (Minerva et al., 2020).
3. Optimization of operational setpoints based on the best balance between competing objectives (Katsoulakos et al., 2024).
4. DTs enable automation of manual and repetitive tasks, thereby potentiating increased capacity to conduct higher volumes/more frequent inspections. Higher quality data may lead to better trending models and enhanced predictions.
5. DTs enable remote monitoring and control of operational assets and potentiate digital visibility in hazardous/accessible environments.
6. Given the correct data resolution/fidelity and sufficient calculation efficiency, DTs may capitalize on softwarized expertise. In the same way that specialist software places thousands of man hours' worth of programming at the fingertips of an analyst, advanced diagnostics may be deployed and enhanced throughout the life cycle of a DT as better tools/algorithms evolve.
7. Digital vigilance: Digital systems are impartial, relentless and do not require rest. Algorithmic, rule-based tasks that dispose workers to boredom or unaccommodating hours could be transferred in full/or part to a "digital watchman".

5.5.2 Computational Techniques for Digital Twins

Computation techniques for DTs have not been reviewed by this committee, but is considered to be an area of development that may be reviewed by future committees. Topics here might include reduced order modelling, computational techniques to speed up processing, edge processing, use of surrogate models, etc.

5.5.3 Data Assimilation

Data assimilation is a method used in metrology and other fields to combine empirical data with model predictions. It aims to produce the best estimate of a system's current state by integrating observations into numerical models. Filtering, estimation, smoothing, and prediction are often used in engineering fields, (e.g., Moheimani et al., 1998), as techniques that offer great promise in many DT applications. Another goal of data assimilation is to interpolate sparse observation data using knowledge of the system being observed. Variational (adjoint methods) and sequential methods (Kalman filters) have been successfully applied in data assimilation for DT applications (Brandtstaedter et al., 2018). For a DT of a ship structure (DTSS), it is important not only to perform an accurate prediction of the hull stress response at local measurement points but also to estimate the overall stress response throughout the whole ship. Data assimilation is indispensable in such analyses.

In the Japanese DTSS R&D project (Fujikubo et al., 2024), the three data assimilation methods, shown in Fig. 5.6.4-1, were applied. They include the wave spectrum method, the Kalman filter (KF) method, and the inverse finite element method (iFEM).

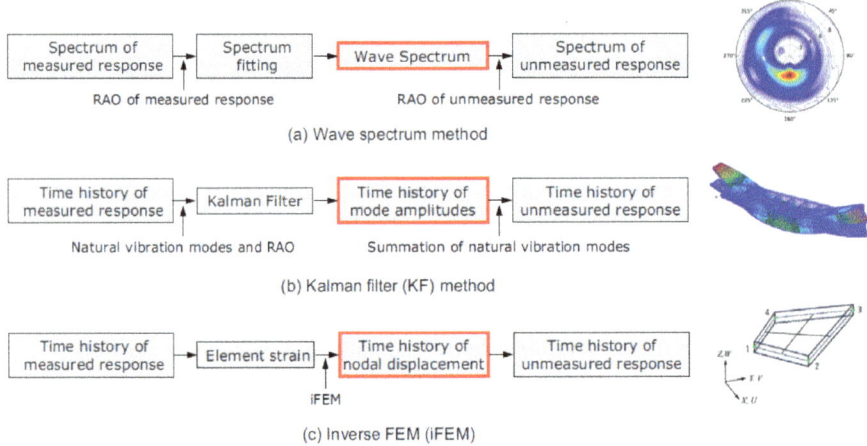

Fig. 6. Three data assimilation methods adopted in the Japanese DTSS R&D project (Fujikubo et al. 2024).

The short-term sea state wave spectrum is estimated using the measured ship response and associated linear transfer functions from the wave spectrum method. Once the wave spectrum is identified, it becomes possible to estimate the spectrum of the unmeasured responses at any location in the ship structure (Fig. 6a).

Researchers studied sea-state estimation from measured ship responses. Nielsen (2006) proposed expressing the wave field by superposition of a number of parameterized wave spectra. He compared the proposed method and the Bayesian method and discussed their applicability. Chen et al. (2018) proposed a new parametric method using the longitudinal hull girder bending stresses as the measured inputs and showed that the wave spectrum could be predicted more accurately by this method. Chen et al.

(2021b) further extended this method so that the multi-peaked directional wave spectrum could be estimated; however, the authors noted further development of the methodology is required remove limitations in predicting directional spectrums with different wave systems from the same direction.

The deformation of a whole ship can be calculated using linear summation of the natural vibration modes. The time history of modal displacement can then be estimated using measured responses, including strains, and applying the standard KF procedure. Then, the time history of the structural responses at any location can be obtained by a linear summation of the modal vectors (Fig. 6b). Miyake et al. (2023) demonstrated the validity of the KF method for DTSS by analysing the onboard hull monitoring data of an 8,600 TEU container ship. Komoriyama et al. (2023) applied the KF method to the prediction of real-time wave profiles using measured responses. This makes it possible to predict not only structural responses, but also the performance any ship in waves, including cargo responses, added resistance, and so forth.

The inverse finite element method (iFEM) uses a least-square variational principle to reconstruct the full-field structural displacement field from a limited number of stress sensors, given solely the meshed geometry and the boundary conditions applied to the structure (Kefal et al., 2018). Using the measured strains, the element equations are assembled into a global system of equations that can then be solved to obtain real-time deformed structural shapes, including the strains at unmeasured locations (Fig. 6c).

5.5.4 Prediction for Decision-Making Support

One of the promising benefits of a DT is its ability to predict a vessel's condition to support decisions. Erikstad (2019) outlines how digital services can be designed to aid design and operational decisions. However, Li and Brennan (2024) note that despite examples of data transfer from physical to virtual objects, the literature lacks guidance on how to make decisions informed by DTs. In the studies where data transfer methods to physical objects were identified, the approaches required human intervention.

Van Der Horn et al. (2022) provide an example of DT predictions that require a human involvement to make decisions. Combining ship track data, metocean hindcast data, and spectral analysis, they presented a way to virtually monitor fatigue damage accumulation at local structures of interest. Those results can be used by maintainers to guide inspections or justify life extensions. This virtual monitoring approach aligns with previous efforts (Thompson, 2018; Aarsnes et al., 2019; Hulkkonen et al., 2019), but Van Der Horn et al. (2022) extend it to address various DT issues.

Fujikubo et al. (2024) calculated structural responses based on wave spectra derived from measured hull responses, and also with responses based on strain sensor outputs using either Kalman filters or the inverse finite element method. Their visual presentation of calculated responses could inform operators' routing decisions or support staff's maintenance and fleet management decisions. Additionally, they examined a variety of predictions such a system could provide to inform decision support.

5.5.5 Review of Existing DT (and on-Going Research Projects)

Giering and Dyck (2021) outline a Maritime Digital Twin Architecture (MDTA) and its application in the maritime industry (Fig. 7). The importance of digital tools and collaborative support is emphasized thorough the maritime product lifecycle, from conceptual design to basic and detailed design and operation. The proposed MDTA aims to supply open-source and modular software framework for industrial application, focusing on improving digital solutions for the maritime market. The need for standardized data management, reuse of information from past projects, and the potential for automated evaluation of system design choices is addressed. DTs are deemed pivotal to transform traditional maritime businesses towards a highly connected and digitalized future, despite challenges and scepticism within the industry.

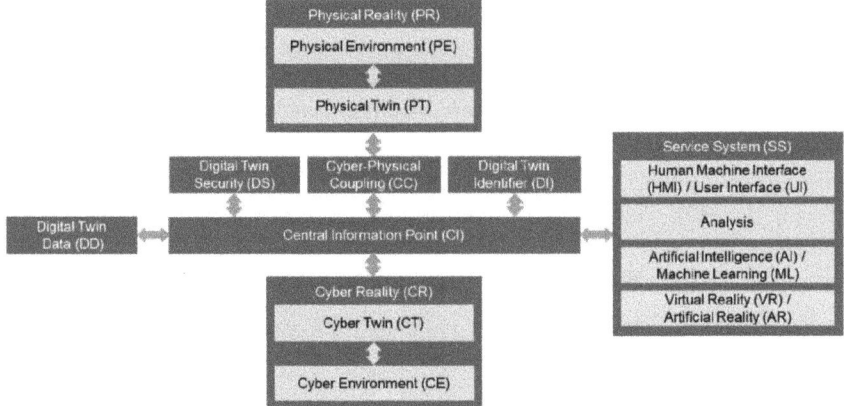

Fig. 7. Maritime Digital Twin Architecture (Giering & Dyck, 2021).

A review by Papanikolaou et al. (2024) identifies a gap in the research and application of DT in ships, particularly in the design and decommissioning phases. Currently, DTs in shipping are most mature in the operational phase, in tasks including predictive maintenance, fault detection, and system performance monitoring. The development of tailored multi-physics model libraries for maritime applications is deemed crucial to unlocking the potential of DTs. The importance of developing robust DT-based design methodology is emphasized.

Explicit examples of ship DT aggregates are sparse and most public examples pertain to research vessels.

Zhang et al. (2023b) present a pioneering DT of the RV Gunnerus posed to enhance maritime operations through the integration of subsystems and operational data into a unified digital platform (Fig. 8). The primary functionality of the DT optimizes ship performance across her lifecycle, including design to operation and maintenance. The DT boasts seamless data exchange between the ship subsystems and applications, providing a comprehensive view to support advanced control, predictive maintenance, and decision-making. The system is equipped with a co-simulation mechanism which allows for distributed modelling of the ship subsystems, ensuring that they work together in a coordinated manner. The DT collects real-time data from the ship and transmits it to a remote control centre, where it is analysed to improve operational planning and decision-making. The DT supports predictive maintenance by using historical data and machine learning to anticipate behaviour and potential issues.

Bekker (2018) outlines a blue skies concept in which a DT of the SA Agulhas II PSRV could comprise digital representations of the vessel, her environment and personnel on board. To date, published work on this effort presents progress on DTI for propulsion system services (Purcell et al., 2024 and Bekker & De Koker, 2023) and personnel onboard (Taylor et al., 2024) which have not progressed to a vessel-wide DTA. The DLR Institute for Maritime Energy Systems present a Virtual Ship Department to develops innovative methods and solutions for the modelling, simulation, validation, optimization and visualization of complex ship systems (DLR Institute for Maritime Energy Systems, 2024). The British Antarctic Survey (BAS) website relates a project entitled *AI and Digital Twinning for Decarbonization* which entails a DT for ship navigation towards the fastest and most energy efficient routes for the PSRV Sir David Attenborough.

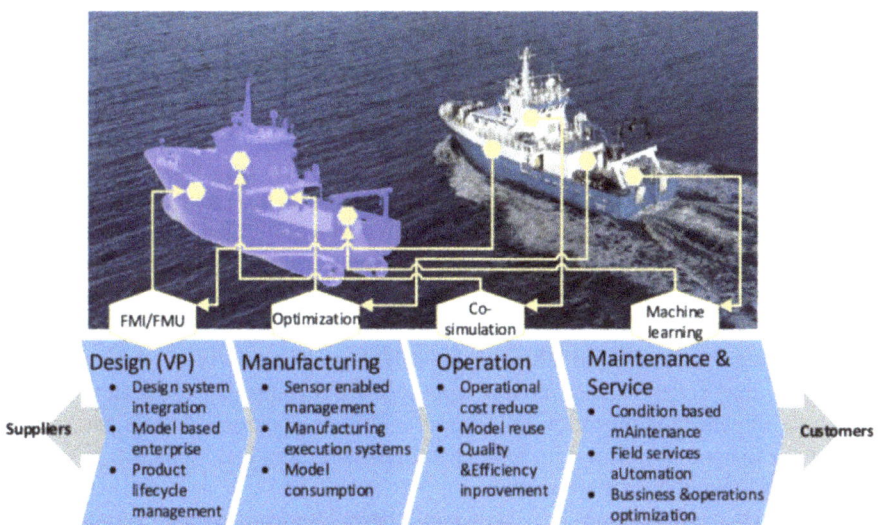

Fig. 8. DT of the RV Gunnerus (Zhang et al., 2023c).

5.6 Concluding Remarks

Developments are presented across Classification Society rules, with introduction of hull notations where Structural Health Monitoring Systems are implemented, and categorisations across levels of autonomy within associated decision making of such systems.

Example applications across commercial, defence and research vessels are presented, demonstrating a variety of objectives to the systems, including quantifying bending moments for operator guidance, assisting operators in ice navigation situational awareness, monitoring torsional moments, measuring accumulated fatigue damage in service, monitoring vibrations, and assessment of fatigue from wave impacts.

DT applications are further discussed, considering methods for data assimilation and reflection of the data in the associated structural model, approaches to decision making from accumulated data. Considerations are presented for standardising data formats for ease of transfer or re-use between software tools and studies to maximise the benefit of the collated data.

6 Conclusions and Recommendations

6.1 Conclusions

The Committee broadly reviewed literature presenting methodologies, physical systems and sensors, and processing approaches to aid with structural assessment and decision making in relation to the operation and maintenance of ship and offshore structures.

No specific definition was stipulated for the duration of an operation that was relevant to the review. As a result, the review ranges from short-term operational benefits (e.g., informing offshore wind turbine decisions to feather blades and reduce loading, or guidance to high-speed craft operators to avoid exceedance of hull loading at sea) to long-term corrosion and fatigue damage monitoring.

By not defining the duration of an operation, some content of the report may be considered to have moved towards assessments akin to structural longevity assessment, with longer term processing that would be away from the asset. This has been difficult to avoid in the process of reviewing applicable data, and a critical review of the report may conclude that some methodologies or subject areas are too slow in their development to actively influence the operation of the asset vs. influencing the maintenance of the asset. However, Structural Health Monitoring (SHM) and the variety of systems that this terminology may cover, enables moving from reactive maintenance to predictive maintenance to influence the operation of the asset within its operation. This is particularly relevant to offshore platforms with long operational lifetimes and high costs to remove from service for survey or maintenance.

Digital twin methodologies have shown rapid development with application moving from the conceptual research phases to commercial application. The term digital twin has been discussed, and its broad use within literature. In reality, there are levels of DTs that need to be recognised and clearly presented in their descriptions. The ultimate goal for the DT is to enable two-way transfer between the asset and computational representation and decision-making system; however, in reality this is rarely implemented due to the complexity of achieving this technically as well as ethically (are the wider population or assuring bodies able to accept transfer of decision making from humans?).

DTs with one way data transfer from the asset to computational representation (referred to as a Digital Shadow in some literature), are more common in commercial application, utilising appropriate structural models and algorithms to process data and present this for human consideration and action to influence the operation of the asset. An interim DT exists between this and the two-way DT, where a level of decision making is transferred to the digital system, with human oversight provided to ensure swift action should intervention be required, or to influence or confirm proposed actions within critical phases of operation.

A broad range of sensor technologies have been reviewed and presented, covering the more widely known strain sensors, accelerometers and Global Positioning System (GPS) sensors to facilitate processing of bending moments and stresses, ship motions, speed, course, etc. More specialist sensors such as those for acoustic emissions monitoring were also discussed.

With such a broad range of sensors available for application, the combination of sensors and the data collected is shown to be able to facilitate a variety of structural assessments, including bending stresses, torsional loading, corrosion degradation and fatigue damage. Processing of what can be a significant quantity of broad data requires use of appropriate models to process, as well as to continue to train the system. Surrogate models, Artificial Intelligence and Machine Learning techniques have been presented as implemented within various systems.

In the continued drive towards increased automation, discussion is presented on the future need to process data onboard the asset and avoid the requirement to transfer large quantities of data for remote processing. This will naturally require increased decision making to be transferred to the asset, processing the data and only relaying essential data, or anomalies, back to the monitoring station rather than status-quo data being continuously streamed.

It is finally concluded that the subject of structural assessment in relation to influencing the operation of a ship or offshore asset during the timescales of its operation is very relevant to the maritime industry, and will continue to become more important as technologies develop, increasing reliability and in their acceptance by the wider population.

6.2 Recommendations

In forming the Committee and developing the strategy for development of this report, the Committee referred to the structure of the report of the previous Committee V.7 Structural Longevity (ISSC V.7 2021), progressing to consider shorter timescale operations and applicable aspects. It may be that a future Committee looks to consider a different approach.

It has been noted that the timescales bounding the operational lifetime were not defined in the context of this Committee's approach, and therefore there may be consideration by a future committee to define a timescale and therefore to focus the report more specifically to shorter- or longer-term operations and the aspects that are more specifically relevant to those timescales.

An aspect highlighted as having not been presented in detail in this report is computational techniques for DTs, for example reduced order modelling, computational techniques to speed up processing, edge processing techniques, use of surrogate models, etc. These may be aspects that a future committee may wish to consider focus on.

Further to this the uncertainties within numerical models and the influence this might have on decision making, coupled with "experimental error" from real world sensors and methods for processing sensor data has not been considered specifically within this report. Uncertainties, variations, known simplifications, etc., will all impact the correlation between the physical asset and digital representation upon which decisions (via human or automated system) are made.

In compiling the literature associated with this report, the Committee has perceived the extent of application of Structural Health Monitoring systems to be greater than the quantity of literature might reflect. Classification Societies have implemented notations associated with SHM, which may be assumed to be adopted for assets implementing such systems. A future Committee might therefore consider whether they should undertake a study to discuss application of SHM systems with Classification Societies, operators of ship and offshore structures and companies developing SHM systems, to gather views on the types of systems being implemented, the practicalities of implementing SHM systems to new or existing assets, the reliability of the systems and collated data, and the decision-making processes implemented following processing of the collected data.

Within the report it was highlighted that recent rule updates for uncrewed or remotely operated vessels require reporting and display of stress information within the control station. It is not clear what form such reporting should take. A future committee may wish to consider a study to consider what data might be presented, how the data might be presented, and how an operator might be able to use such data within the operation of a vessel.

References

4C Offshore: Global Offshore Wind Farms Database – 4C Offshore. http://www.4coffshore.com/windfarms/ (2023). Accessed 4 Oct 2023

Aarsnes, L.H., Storhaug, G., Radon, M.: Utilization of structural design models in operation to monitor fatigue strength performance. In: Practical Design of Ships and Other Floating Structures, pp. 622–636. Springer, Singapore (2019)

Abdelmaksoud, A.M., Oudah, F., Nikolaidis, I., Adeeb, S., Attia, S.: A fuzzy-random fields approach for assessing the natural frequencies of corroded pipes conveying fluids. Eng. Struct. **306**, 117823 (2024)

ABS: Advisory on Structural Health Monitoring: The Application of Sensor-Based Approaches. American Bureau of Shipping (ABS), Houston (2020a)

ABS: Guide for Hull Condition Monitoring System. American Bureau of Shipping (ABS), Houston (2020b)

ABS: Guide for Ice Loads Monitoring Systems. American Bureau of Shipping (ABS), Houston (2021)

ABS: Guide for Smart Functions for Marine Vessels and Offshore Units. American Bureau of Shipping (ABS), Houston (2022)

Adams, M., Huijer, A., Kassapoglou, C., Vaders, J.A.A., Pahlavan, L.: In situ non-destructive stiffness assessment of fiber reinforced composite plates using ultrasonic guided waves. Sensors. **24**(9), 2747 (2024). https://doi.org/10.3390/s24092747

Akyuz, E., Cicek, K., Celik, M.: A comparative research of machine learning impact to future of maritime transportation. Procedia Comput. Sci. **158**, 275–280 (2019). https://doi.org/10.1016/j.procs.2019.09.052

Albuquerque, R.Q., Sarhadi, A., Demleitner, M., Ruckdäschel, H., Eder, M.A.: Fatigue damage reconstruction in glass/epoxy composites via thermal analysis and machine learning: a theoretical study. Compos. Struct. **331**, 117855 (2024). https://doi.org/10.1016/j.compstruct.2023.117855

Al-Karawi, H.: Corrosion effect on the efficiency of high-frequency mechanical impact treatment in enhancing fatigue strength of welded steel structures. J. Mater. Eng. Perform. **31**(11), 9151–9158 (2022)

Alokita, S., et al.: Recent advances and trends in structural health monitoring. In: Structural Health Monitoring of Biocomposites, Fibre-Reinforced Composites and Hybrid Composites, pp. 53–73. Woodhead Publishing, Kidlington (2018). https://doi.org/10.1016/B978-0-08-102291-7.00004-6

Alqurashi, F.S., Trichili, A., Saeed, N., Ooi, B.S., Alouini, M.-S.: Maritime communications: a survey on enabling technologies, opportunities, and challenges. IEEE Internet Things J. **10**(4), 3525–3547 (2022). https://ieeexplore.ieee.org/abstract/document/9939173/

Alsulieman, A., Ge, X., Zeng, Z., Butenko, S., Khan, F., El-Halwagi, M.: Dynamic risk analysis of evolving scenarios in oil and gas separator. Reliab. Eng. Syst. Saf. **243**, 109834 (2024)

Altabey, W.A., Noori, M.: Artificial-intelligence-based methods for structural health monitoring. Appl. Sci. **12**(24), 12726 (2022). https://doi.org/10.3390/app122412726

Al-Zareer, M., Dincer, I., Rosen, M.A.: Performance improvement study of an integrated photovoltaic system for offshore power production. Int. J. Energy Res. **45**(1), 772–785 (2021). https://doi.org/10.1002/er.5900

Ambarita, E.E., Karlsen, A., Scibilia, F.: Industry 4.0 digital twins in offshore wind farms. In: Wind Energy Science Discussions [preprint] (2023). https://doi.org/10.5194/wes-2023-108

American Bureau of Shipping: Guide for hull condition monitoring system. July 2020.

Andresen-Paulsen, G., et al.: The integrity of corrosion protection systems of welded maritime structures under cyclic loading. In: Advances in the Analysis and Design of Marine Structures. CRC Press, Boca Raton (2023)

Animah, I.: Application of Bayesian network in the maritime industry: comprehensive literature review. Ocean Eng. **302**, 117610 (2024)

Antony Jacob, A., Ravichandran, S., Upadhyay, V., Rajagopal, P., Balasubramaniam, K.: Thickness estimation of marine structures using an ROV-based pulsed eddy current technique. In: Advances in Non-destructive Evaluation: Proceedings of NDE 2019, pp. 133–143. Springer, Singapore (2021)

Anyfantis, K.N.: An abstract approach toward the structural digital twin of ship hulls: a numerical study applied to a box girder geometry. In: Proceedings of the Institution of Mechanical Engineers, Part M: Journal of Engineering for the Maritime Environment, vol. 235(3), pp. 718–736. Sage (2021). https://doi.org/10.1177/1475090221989188

API: API-RP-2SIM. Structural Integrity Management of Fixed Offshore Structures, 1st edn. American Petroleum Institute (API) Publishing Services, Washington, DC (2014)

Attallah, O., Ibrahim, R.A., Zakzouk, N.E.: CAD system for inter-turn fault diagnosis of offshore wind turbines via multi-CNNs & feature selection. Renew. Energy. **203**, 870–880 (2023). https://doi.org/10.1016/j.renene.2022.12.064

Ukato, A., Sofoluwe, O.O., Jambol, D.D., Ochulor, O.J.: Optimizing maintenance logistics on offshore platforms with AI: current strategies and future innovations. World J. Adv. Res. Rev. **22**(1), 1920–1929 (2024). https://doi.org/10.30574/wjarr.2024.22.1.1315

Bagherabadi, K.M., Skjong, S., Bruinsma, J., Pedersen, E.: Investigation of hybrid power plant configurations for an offshore vessel with co-simulation approach. Appl. Energy. **343**, 121211 (2023). https://doi.org/10.1016/j.apenergy.2023.121211

Bashir, M., Xu, Z., Wang, J., Guedes Soares, C.: Data-driven damage quantification of floating offshore wind turbine platforms based on multi-scale encoder-decoder with self-attention mechanism. J. Mar. Sci. Eng. **10**(12), 2022 (2022)

Batalden, B.M., Wide, P., Røds, J.F., Haugseggen, Ø.: Enhanced human-machine interaction by fuzzy logic in semi-autonomous maritime operations. In: Chen, J. (ed.) Advances in Human Factors in Robots and Unmanned Systems. AHFE 2018. Advances in Intelligent Systems and Computing, vol. 784. Springer, Cham (2019). https://doi.org/10.1007/978-3-319-94346-6_5

Bekker, A.: Exploring the blue skies potential of digital twin technology for a polar supply and research vessel. Marine Design. **XIII**(1), 135–146 (2018)

Bekker, A., Erikstad, S.O.: Digital Twin Service Patterns for Maintenance Management of Operational Rail Assets. International Heavy Haul Association Conference, Rio de Janeiro Brazil (2023). https://www.researchgate.net/publication/373735113

Bekker, A., De Koker, N.: Towards a digital twin to inform propulsion safety margins in ice. In: Bertram, V. (ed.) Computer Applications and Information Technology Conference, pp. 354–366 (2023)

Bhange, P., et al.: Real-time fatigue crack growth rate estimation methodology for structural health monitoring of ships. IEEE Sensors J. **22**(20), 19729–19738 (2022). https://doi.org/10.1109/JSEN.2022.3204146

Bhardwaj, U., Teixeira, A.P., Guedes Soares, C.: Casualty analysis methodology and taxonomy for FPSO accident analysis. Reliab. Eng. Syst. Saf. **218**, 108169 (2022)

Bhardwaj, U., Teixeira, A.P., Guedes Soares, C., Ariffin, A.K., Singh, S.S.: Evidence-based risk analysis of fire and explosion accident scenarios in FPSOs. Reliab. Eng. Syst. Saf. **215**, 107904 (2021)

Bhattacharya, S., et al.: Physical modelling of offshore wind turbine foundations for TRL (technology readiness level) studies. J. Mar. Sci. Eng. **9**(6), 589 (2021)

Black, I.M., Yeter, B., Häckell, M.W., Kolios, A.: Assessing structural homogeneity and heterogeneity in offshore wind farms: a population-based structural health monitoring approach. Ocean Eng. **311**, 118842 (2024)

Borgelt, J., Possekel, J., Schaumann, P., Dreger, D., Ghafoori, E.: Innovative test concept for evaluating fatigue degradation process of submerged grouted connections. Eng. Struct. **302**, 117335 (2024)

Bossau, J.C., Bekker, A.: Line detection techniques to pinpoint slamming impulses in time-frequency images of hull acceleration measurements. Ocean Eng. **249** (2022). https://doi.org/10.1016/j.oceaneng.2022.110841

Brandtstaedter, H., Ludwig, C., Hubner, L., Tsouchnika, E., Jungiewicz, A., Wever, U.: Digital twins for large electric drive trains. In: 2018 Petroleum and Chemical Industry Conference Europe (PCIC Europe), pp. 1–5. IEEE (2018)

Branlard, E., Jonkman, J., Brown, C., Zhang, J.: A digital twin solution for floating offshore wind turbines validated using a full-scale prototype. Wind Energy Sci. **9**(1), 1–24 (2024). https://doi.org/10.5194/wes-9-1-2024

Branner, K., Di Lorenzo, E., Vettori, S., Berring, P., Haselbach, P.U., Peeters, B.: Executable digital twin demonstrator of wind turbine blade. In: Proceedings of Isma 2022 – International Conference on Noise and Vibration Engineering and Usd 2022 – International Conference on Uncertainty in Structural Dynamics, pp. 1799–1808 (2022)

Branner, K., Eder, M.A., Danielsen, H.K., Chen, X., McGugan, M.: Towards more smart, efficient and reliable wind-turbine structures. In: DTU International Energy Report 2021: Perspectives on Wind Energy, pp. 115–124. DTU Wind Energy (2021). https://doi.org/10.11581/DTU.00000212

Brijder, R., et al.: Review of corrosion monitoring and prognostics in offshore wind turbine structures: current status and feasible approaches. Front. Energy Res. **10** (2022). https://doi.org/10.3389/fenrg.2022.991343

Brijder, R., Helsen, S., Ompusunggu, A.P.: Switching Kalman filtering-based corrosion detection and prognostics for offshore wind-turbine structures. Wind. **3**(1), 1–13 (2023)

British Antarctic Survey: Seeing double: digital twins and net zero. Retrieved August 23, 2024, from https://www.bas.ac.uk/blogpost/seeing-double-digital-twins-and-net-zero/ (2024)

Bronson, J.A., Fonseca, Í.A., Gaspar, H.M.: Challenges towards an integrated digital twin platform for maritime systems: tackling shifts in data ownership. In: ASME 43rd International Conference on Ocean, Offshore and Arctic Engineering, pp. 1–13 (2024b)

Bronson, J.A., Fonseca, Í.A., Gaspar, H.M., Luz, F.H.P.: Data models in ship design and construction – insights from 4D BIM. In: International Marine Design Conference (2024a). https://doi.org/10.59490/imdc.2024.842
BS7910:2019: Guide to Methods for Assessing the Acceptability of Flaws in Metallic Structures. British Standards Institution (2019)
Bureau Veritas: Risk-Based Structural Integrity Management for Topside Structures, Guidance Note NI 653 DT R00. Bureau Veritas Marine and Offshore (2018)
BV: Ice Load Monitoring System (MON-ICE). Bureau Veritas, France (2015)
BV: Hull Stress and Motion Monitoring MON-HULL; Part F, Chapter 5, Section 1. Bureau Veritas, France (2022)
CCS: Rules for Classification of Sea-Going Steel Ships. China Classification Society (2023)
Chen, X., Okada, T., Kawamura, Y.: Sea state estimation using measured hull responses on 14,000 TEU large container ships. In: Proceedings of the 7th International Maritime Conference on Design for Safety, pp. 160–168 (2018)
Chen, X., Okada, T., Kawamura, Y., Mitsuyuki, T.: Estimation of directional wave spectra and hull structural responses based on measured hull data on 14,000 TEU large container ships. Mar. Struct. **80**, 103087 (2021b). https://doi.org/10.1016/j.marstruc.2021.103087
Chen, B.Q., Videiro, P.M., Guedes, S.C.: Opportunities and challenges to develop digital twins for subsea pipelines. J. Mar. Sci. Eng. **10**(6), 739 (2022). https://doi.org/10.3390/jmse10060739
Chen, S., et al.: Marine structural health monitoring with optical fiber sensors: a review. Sensors. **23**(5), 1877 (2023c). https://doi.org/10.3390/s23041877. MDPI
Chen, W., Milosevic, Z., Rabhi, F.A., Berry, A.: Real-time analytics: concepts, architectures, and ML/AI considerations. IEEE Access. **11**, 71634–71657 (2023b). https://doi.org/10.1109/ACCESS.2023.3295694
Chen, X., Eder, M.A.: A critical review of damage and failure of composite wind turbine blade structures. IOP Conf. Ser.: Mater. Sci. Eng. **942**(1), 012001 (2020). https://doi.org/10.1088/1757-899X/942/1/012001
Chen, X., Eder, M.A., Shihavuddin, A., Zheng, D.: A human-cyber-physical system toward intelligent wind turbine operation and maintenance. Sustainability. **13**(2), 1–10 (2021a). https://doi.org/10.3390/su13020561
Chen, X., Sheiati, S., Shihavuddin, A.S.M.: AQUADA PLUS: automated damage inspection of cyclic-loaded large-scale composite structures using thermal imagery and computer vision. Compos. Struct. **318**, 117085 (2023a). https://doi.org/10.1016/j.compstruct.2023.117085
Chen, Y., Cui, H.: Research on the identification of modal parameters of ship structure based on Fiber Bragg grating sensor. In: Proceedings of SPIE – The International Society for Optical Engineering, p. 12790 (2023). https://doi.org/10.1117/12.2690015
Cheng, J., Liu, Y., Li, W., Li, T.: Deep reinforcement learning for cost-optimal condition-based maintenance policy of offshore wind turbine components. Ocean Eng. **283**, 115062 (2023). https://doi.org/10.1016/j.oceaneng.2023.115062
Cheng, J., Liu, Y., Cheng, M., Li, W., Li, T.: Optimum condition-based maintenance policy with dynamic inspections based on reinforcement learning. Ocean Eng. **261**(2022), 112058 (2022)
Chueh, C.-J., Chien, C.-H., Lin, C., Lin, T.-Y., Chiang, M.-H.: Dynamic co-simulation analysis and control of an IEA 15 MW offshore floating semi-submersible wind turbine under Taiwan offshore-wind-farm conditions of wind and wave. J. Mar. Sci. Eng. **11**(1), 173 (2023). https://doi.org/10.3390/jmse11010173
Collette, M., Bielski, R., Rohrer, P., Magistro, A., Sulkowski, B., Van Houten, J.: Needs exploration for long-term autonomous marine systems: working report. In: MSDL Report Number 2021-001. Marine Structures Design Lab Department of Naval Architecture and Marine Engineering University of Michigan (2022). http://deepblue.lib.umich.edu/handle/2027.42/173163
Collu, M., Kolios, A., Chahardehi, A., Brennan, F.P.: Developments in wind, wave, tidal and current technology. Mar. Offshore Renew. Energy. (2010)

Davies, P., Dumergue, N., Arhant, M., Nicolas, E., Paboeuf, S., Mayorga, P.: Material and structural testing to improve composite tidal turbine blade reliability. Int. Mar. Energy J. **5**, 57–65 (2022)

DeepOcean: ROVs. https://www.deepoceangroup.com/what-we-do/assets-rovs (2023) Accessed 4 Oct 2023

Demleitner, M., Albuquerque, R.Q., Sarhadi, A., Ruckdäschel, H., Eder, M.A.: Bayesian optimization-based prediction of the thermal properties from fatigue test IR imaging of composite coupons. Compos. Sci. Technol. **248**, 110439 (2024). https://doi.org/10.1016/j.compscitech.2024.110439

Det Norske Veritas: DNV-SE-04369, Certification of Condition Monitoring. (2021–10)

Diao, Y., Yan, L., Gao, K.: Improvement of the machine learning-based corrosion rate prediction model through the optimization of input features. Mater. Des. **198**, 109326 (2021). https://www.sciencedirect.com/science/article/pii/S0264127520308625

Dihan, M.S., et al.: Digital twin: data exploration, architecture, implementation and future. Heliyon. **10**(5), e26503 (2024). https://doi.org/10.1016/j.heliyon.2024.e26503

Ding, G., Yan, X., Gao, X., Zhang, Y., Jiang, S.: Reconstruction of propeller deformation based on FBG sensor network. Ocean Eng. **249** (2022). https://doi.org/10.1016/j.oceaneng.2022.110884

Disabato, S.: Deep and wide tiny machine learning. In: Riva, C.G. (ed.) Special Topics in Information Technology Springer Briefs in Applied Sciences and Technology. Springer, Cham (2023). https://doi.org/10.1007/978-3-031-15374-7_7

DLR Institute for Maritime Energy Systems: Department – Virtual Ship. Retrieved August 23, 2024, from https://www.dlr.de/en/ms/about-us/departments/department-vis (2024)

DNV: Class Guideline DNV-CG-0264: Autonomous and Remotely Operated Ships, September 2021 (2021a)

DNV: DNV-RP-0416: Corrosion Protection for Wind Turbines. Norway (2021b)

DNV: Rules for Classification. Part 6 Additional Class Notations Chapter 9 Survey Arrangements. Det Norske Veritas, Norway (2022)

DNV: Structural Health Monitoring by Use of Sensor Data. DNV. report no. 2023-1241, Rev 1, 21/02/2024 (2024)

DNV-RP-A204: Assurance of Digital Twins, Det Norske Veritas. https://www.dnv.com/oilgas/download/dnv-rp-a204-assurance-of-digital-twins/ (2023). Last accessed 31 July 2024

Dyer, A.S., et al.: Applied machine learning model comparison: predicting offshore platform integrity with gradient boosting algorithms and neural networks. Mar. Struct. **83**(2022), 103152 (2022)

Eder, M.A., Chen, X.: FASTIGUE: a computationally efficient approach for simulating discrete fatigue crack growth in large-scale structures. Eng. Fract. Mech. **233**, 107075 (2020). https://doi.org/10.1016/j.engfracmech.2020.107075

Eder, M.A., Semenov, S., Sala, M.: Multiaxial stress based high cycle fatigue model for adhesive joint interfaces. Mech. Mach. Sci. **75**, 621–632 (2019). https://doi.org/10.1007/978-3-030-27053-7_53

Ehlers, S., Braun, M. (eds.): Fatigue and Fracture Mechanics of Marine Structures. MDPI – Multidisciplinary Digital Publishing Institute, Basel (2023). https://directory.doabooks.org/handle/20.500.12854/100862

Eichner, L., Gerards-Wünsche, P., Herrmann, R., Schneider, R.: A framework for data and structural integrity management for support structures in offshore wind farms based on building information modelling. In: The 8th International Symposium on Reliability Engineering and Risk Management, Hannover, Germany, September 2022 (2022)

El-Reedy, M.A.: Integrity management versus reliability, Chapter 2. In: El-Reedy, M.A. (ed.) Asset Integrity Management for Offshore and Onshore Structures, pp. 31–93. Gulf Professional Publishing, Cambridge, MA (2022)

Erguido, A., Crespo, M.A., Castellano, E., Gómez Fernández, J.F.: A dynamic opportunistic maintenance model to maximize energy-based availability while reducing the life cycle cost of wind farms. Renew. Energy. **114**, 843–856 (2017). https://doi.org/10.1016/j.renene.2017.07.017

Erikstad, S.O.: Designing ship digital services. In: Proceedings of the Conference on Computer and IT Applications in the Maritime Industries, vol. 18, pp. 459–469 (2019)

Erives Anchondo, R.I.: Delamination of composite blade structures using the cohesive zone approach. In: DTU Wind and Energy Systems (2022). https://doi.org/10.11581/dtu.00000257

Esfidan, M.R., Ranji, A.R.: Corrosion models for steel plates in ship structure based on statistical data. Proc. Inst. Mech. Eng., Part M: J. Eng. Marit. Environ., 14750902241253319 (2024). https://journals.sagepub.com/doi/10.1177/14750902241253319

Fahim, M., Sharma, V., Cao, T.V., Canberk, B., Duong, T.Q.: Machine learning-based digital twin for predictive modeling in wind turbines. IEEE Access. **10**, 14184–14194 (2022). https://doi.org/10.1109/ACCESS.2022.3147602

Fajuyigbe, A., Brennan, F.: Fitness-for-purpose assessment of cracked offshore wind turbine monopile. Mar. Struct. **77**, 102965 (2021). https://doi.org/10.1016/j.marstruc.2021.102965

Falkenberg, R.: Modelling of environmentally assisted material degradation in the crack phase-field framework. Proc. Inst. Mech. Eng. L J. Mater. Des. Appl. **233**(1), 5–12 (2019)

Fang, X., Wang, H., Li, W., Liu, G., Cai, B.: Fatigue crack growth prediction method for offshore platform based on digital twin. Ocean Eng. **244**, 110320 (2022). https://www.sciencedirect.com/science/article/pii/S0029801821016218

Farhan, M., Schneider, R., Thöns, S.: Predictive information and maintenance optimization based on decision theory: a case study considering a welded joint in an offshore wind turbine support structure. Struct. Health Monit. **21**(1), 185–207 (2021). https://doi.org/10.1177/1475921720981833

Farid, M.: Data-driven method for real-time prediction and uncertainty quantification of fatigue failure under stochastic loading using artificial neural networks and Gaussian process regression. Int. J. Fatigue. **155**, 106415 (2022). https://doi.org/10.1016/j.ijfatigue.2021.106415

Finnegan, W., et al.: Numerical modelling, manufacture and structural testing of a full-scale 1 MW tidal turbine blade. Ocean Eng. **266**(1), 112717 (2022)

Firdaus, M.I., Adiputra, R.: Deterioration and imperfection of the ship structural components and its effects on the structural integrity: a review. Curved Layered Struct. **11**(1), 20240008 (2024). https://www.degruyter.com/document/doi/10.1515/cls-2024-0008/html

Fonseca, Í.A., Gaspar, H.M., de Mello, P.C., Sasaki, H.A.U.: A standards-based digital twin of an experiment with a scale model ship. Comput. Aided Des. **145**, 103191 (2022). https://doi.org/10.1016/J.CAD.2021.103191

Fujikubo, M., et al.: A digital twin for ship structures -R&D project in Japan. Data-Centric Eng. **5**, e7 (2024). https://doi.org/10.1017/dce.2024.3

Garcia-Teruel, A., Rinaldi, G., Thies, P.R., Johanning, L., Jeffrey, H.: Life cycle assessment of floating offshore wind farms: an evaluation of operation and maintenance. Appl. Energy. **307**, 118067 (2022)

Gausdal, A.H., Czachorowski, K.V., Solesvik, M.Z.: Applying blockchain technology: evidence from Norwegian companies. Sustainability. **10**(6), 1985 (2018). https://doi.org/10.3390/su10061985

George, B., Loo, J., Jie, W.: Recent advances and future trends on maintenance strategies and optimisation solution techniques for offshore sector. Ocean Eng. **250**, 110986 (2022). https://doi.org/10.1016/j.oceaneng.2022.110986

Ghasemzadeh, M., Mokhtari, M., Bilgin, M.H., Kefal, A.: Pitting corrosion identification approach based on inverse finite element method for marine structure applications. Ocean Eng. **273**, 113953 (2023)

Giering, J.E., Dyck, A.: Maritime digital twin architecture: a concept for holistic digital twin application for shipbuilding and shipping. At-Automatisierungstechnik. **69**(12), 1081–1095 (2021). https://doi.org/10.1515/auto-2021-0082

Giordano, P.F., Quqa, S., Limongelli, M.P.: The value of monitoring a structural health monitoring system. Struct. Saf. **100**(2023), 102280 (2023)

Golahmar, A., Pañeda, E.M., Niordson, C.F.: Phase field modelling of environmentally assisted fatigue. In: International Congress of Theoretical and Applied Mechanics (ICTAM) (2021)

Gomes, C., Thule, C., Broman, D., Larsen, P.G., Vangheluwe, H.: Co-Simulation: State of the Art. Cornell University USA (2017). https://arxiv.org/pdf/1702.00686.pdf

Gong, C., Frangopol, D.M.: System reliability of corroded ship hull girders. Struct. Infrastruct. Eng. **16**(9), 1302–1310 (2020). https://www.tandfonline.com/doi/full/10.1080/15732479.2019.1703761

Gregori, A., Di Giampaolo, E., Di Carlofelice, A., Castoro, C.: Presenting a new wireless strain method for structural monitoring: experimental validation, 5370838. J. Sens. (2019). https://doi.org/10.1155/2019/5370838. Hindawi

Gres, S., Döhler, M., Dertimanis, V.K., Chatzi, E.N.: Subspace-based noise covariance estimation for Kalman filter in virtual sensing applications. Mech. Syst. Signal Process. **222**, 111772 (2024). https://doi.org/10.1016/j.ymssp.2024.111772

Guédé, F.: Risk-based structural integrity management for offshore jacket platforms. Mar. Struct. **63**(2019), 444–461 (2019)

Güemes, Alfredo. Fiber Optic Strain Sensors: NATO- STO Lecture Series, 2014.

Hageman, R.B., Thompson, I.: Virtual hull monitoring using hindcast and motion data to assess frigate-size vessel stress response. Ocean Eng. **245**, 110338 (2022)

Haghshenas, A., Hasan, A., Osen, O., Mikalsen, E.T.: Predictive digital twin for offshore wind farms. Energy Inform. **6**(1), 1 (2023). https://doi.org/10.1186/s42162-023-00257-4

Haider, A.: Efficiency enhancement techniques in finite element analysis: navigating complexity for agile design exploration. Aircr. Eng. Aerosp. Technol. **96**(5), 662–668 (2024). https://doi.org/10.1108/AEAT-02-2024-0053

Hamada, R., et al.: Structural health monitoring of CFRP propellers by piezoelectric line sensors. In: Farhangdoust, S., Guemes, A., Chang, F.-K. (eds.) Structural Health Monitoring, pp. 408–416. DEStech Publications, Stanford (2023)

Han, H., Siebert, J.: TinyML: a systematic review and synthesis of existing research. In: International Conference on Artificial Intelligence in Information and Communication (ICAIIC), pp. 269–274, Jeju Island, Korea, Republic of (2022). https://doi.org/10.1109/ICAIIC54071.2022.9722636

Han, C., Liu, K., Ma, Y., Qin, P., Zou, T.: Multiaxial fatigue assessment of jacket-supported offshore wind turbines considering multiple random correlated loads. Renew. Energy. **169**, 1252–1264 (2021). https://doi.org/10.1016/j.renene.2021.01.093

Han, C., Mo, C., Tao, L., Ma, Y., Bai, X.: An efficient fatigue assessment model of offshore wind turbine using a half coupling analysis. Ocean Eng. **263**, 112318 (2022). https://doi.org/10.1016/j.oceaneng.2022.112318

Han, Z., Liu, Y., Wu, X., Oh, D., Jang, J.: Ultrasonic non-destructive inspection method for glass fibre reinforcement polymer (GFRP) hull plates considering design and construction characteristics. Proc. Inst. Mech. Eng., Part M: J. Eng. Marit. Environ. (2024). https://doi.org/10.1177/14750902241245539. Sage

Hasan, A., Hu, Z., Haghshenas, A., Karlsen, A., Alaliyat, S., Cali, U.: An interactive digital twin platform for offshore wind farms' development. In: Digital Twin Driven Intelligent Systems and Emerging Metaverse, pp. 269–281. Springer, Singapore (2023)

Hatledal, L.I., Chu, Y., Styve, A., Zhang, H.: Vico: an entity-component-system based co-simulation framework. Simul. Model. Pract. Theory. **108**, 102243 (2021). https://doi.org/10.1016/j.simpat.2020.102243

Heo, T., Liu, D.P., Manuel, L.: Weather window analysis in operations and maintenance policies for offshore floating multi-purpose platforms. J. Offshore Mech. Arct. Eng. **145**(4), 2023 (2023)

Hlaing, N., Pablo, G., Nielsen, J.S., Amirafshari, P., Kolios, A., Rigo, P.: Inspection and maintenance planning for offshore wind structural components: integrating fatigue failure criteria with Bayesian networks and Markov decision processes. Struct. Infrastruct. Eng. **18**(7), 983–1001 (2022). https://doi.org/10.1080/15732479.2022.2037667

Hlaing, N., Morato, P., Rigo, P.: Farm-wide virtual load monitoring for offshore wind structures via Bayesian neural networks. Struct. Health Monit. (2023). https://doi.org/10.1177/14759217231186048

HSE: Structural Integrity Management, vol. 2023. Health and Safety Executive (HSE) (2023)

Hu, X., Ng, C.-T., Kotousov, A.: Early damage detection of metallic plates with one side exposed to water using the second harmonic generation of ultrasonic guided waves. In: Thin-Walled Structures, vol. 176, p. 109284. Elsevier BV (2022). https://doi.org/10.1016/j.tws.2022.109284

Huijer, A.J., Kassapoglou, C., Pahlavan, L.: Acoustic emission monitoring of composite marine propellers in submerged conditions using embedded piezoelectric sensors. Smart Mater. Struct. **33**(9), 095018 (2024). https://doi.org/10.1088/1361-665X/ad6739

Hulkkonen, T., Manderbacka, T., Sugimoto, K.: Digital twin for monitoring remaining fatigue life of critical hull structures. In: 18th Conference on Computer and IT Applications in the Maritime Industries, pp. 415–427 (2019)

Huynh, T.-C., Dang, N.-L., Kim, J.-T.: Advances and challenges in impedance-based structural health monitoring. Struct. Monit. Maint. **4**(4), 301–329 (2017). https://doi.org/10.12989/SMM.2017.4.4.301

IACS: Common Structural Rules for Bulk Carriers and Oil Tankers. IACS (2024)

Ilardi, D., Kalikatzarakis, M., Oneto, L., Collu, M., Coraddu, A.: Computationally aware surrogate models for the hydrodynamic response characterization of floating spar-type offshore wind turbine. IEEE Access. **12**, 6494–6517 (2024)

Im, I., Shin, D., Jeong, J.: Components for smart autonomous ship architecture based on intelligent information technology. In: 15th International Conference on Mobile Systems and Pervasive Computing (MobiSPC 2018), Computer Science, vol. 134, pp. 91–98 (2018)

IMO: Recommendations for the Fitting of Hull Stress Monitoring Systems. MSC/Circ. 646. International Maritime Organization (IMO), London (1994)

Improving Blade Reliability through Application of Digital Twins over Entire Life Cycle – ReliaBlade project website: https://www.reliablade.com/ (2024). Last accessed July 2024

Imran, M.M.H., et al.: Application of artificial intelligence in marine corrosion prediction and detection. J. Mar. Sci. Eng. **11**(2), 256 (2023). https://www.mdpi.com/2077-1312/11/2/256

ISO 10303-2: Industrial Automation Systems and Integration – Product Data Representation and Exchange, Part 2: Vocabulary. International Organization for Standardization, Geneva (2024)

ISO 10303-239: Industrial Automation Systems and Integration – Product Data Representation and Exchange, Part 239: Application protocol: Product Life Cycle Support. International Organization for Standardization, Geneva (2024)

ISO 15926: Industrial Automation Systems and Integration – Integration of Life-Cycle Data for Process Plants Including Oil and Gas Production Facilities, Part 13: Integrated Asset Planning Life-Cycle. International Organization for Standardization, Geneva (2018)

ISO 19847: Ships and Marine Technology – Shipboard Data Servers for Sharing Field Data at Sea. International Organization for Standardization, Geneva (2024)

ISO 19848: Ships and Marine Technology – Standard Data for Shipboard Machinery and Equipment. International Organization for Standardization, Geneva (2024)

ISO 19900: Petroleum and Natural Gas Industries – General Requirements for Offshore Structures. International Organization for Standardization, Geneva (2019)

ISO 19901-1: Petroleum and Natural Gas Industries – Specific Requirements for Offshore Structures, Part 1: Metocean Design and Operating Considerations. International Organization for Standardization, Geneva (2015)

ISO 19901-3: Petroleum and Natural Gas Industries – Specific Requirements for Offshore Structures, Part 3: Topsides Structure. International Organization for Standardization, Geneva (2024)

ISO 19901-9: Petroleum and Natural Gas Industries – Specific Requirements for Offshore Structures, Part 9: Structural Integrity Management. International Organization for Standardization, Geneva (2019)

ISO 19902: Petroleum and Natural Gas Industries – Fixed Steel Offshore Structures. International Organization for Standardization, Geneva (2020)

ISO 19904-1: Petroleum and Natural Gas Industries – Floating Offshore Structures, Part 1: Ship-Shaped, Semi-Submersible, Spar and Shallow-Draught Cylindrical Structures. International Organization for Standardization, Geneva (2019)

ISO 19905-1: Oil and Gas Industries Including Lower Carbon Energy – Site-Specific Assessment of Mobile Offshore Units, Part 1: Jack-Ups: Elevated at a Site. International Organization for Standardization, Geneva (2023)

ISO 23247-4: Automation Systems and Integration – Digital Twin Framework for Manufacturing – Part 4: Information Exchange. International Organization for Standardization, Geneva (2021)

ISO 81346-1: Industrial Systems, Installations and Equipment and Industrial Products – Structuring Principles and Reference Designations, Part 1: Basic Rules. International Organization for Standardization, Geneva (2022)

ISO/TR 24464: Automation Systems and Integration – Industrial Data – Visualization Elements of Digital Twins. International Organization for Standardization, Geneva (2020)

ISSC: International Ships and Offshore Structures Congress. Committee V.7: structural longevity. In: Soares, C.G., Garbatov, Y. (eds.) 19th International Ship and Offshore Structures Congress, 7–10 September 2015, pp. 817–864, Cascais, Portugal (2015)

ISSC: International Ships and Offshore Structures Congress. Committee V.7: structural longevity. In: Kaminski, M., Rigo, P. (eds.) 20th International Ship and Offshore Structures Congress, 9–13 September 2018, vol. 2, pp. 391–460, Liege, Belgium/Amsterdam, The Netherlands (2018)

ISSC: International Ships and Offshore Structures Congress. Committee V.7: structural longevity. In: Wang, X., Pegg, N. (eds.) 21st International Ship and Offshore Structures Congress, 11–15 September 2022, pp. 445–502, Vancouver, Canada (2022)

Jacob, A., Mehmanparast, A.: Crack growth direction effects on corrosion-fatigue behaviour of offshore wind turbine steel weldments. Marine Structures. **75**, 102881 (2021). https://doi.org/10.1016/j.marstruc.2020.102881

Jang, J.-K., Abbas, S.H., Lee, J.-R.: Investigation of underwater environmental effects in rotating propeller blade tracking laser vibrometric measurement. Opt. Laser Technol. **132**, 106460 (2020). https://doi.org/10.1016/j.optlastec.2020.106460

Jiao, P., Ye, X., Zhang, C., Li, W., Wang, H.: Vision-based real-time marine and offshore structural health monitoring system using underwater robots. Comput. Aided Civ. Inf. Eng. **39**, 281–299 (2024). https://doi.org/10.1111/mice.12993

Jurdana, I., Lopac, N., Wakabayashi, N., Liu, H.: Shipboard data compression method for sustainable real-time maritime communication in remote voyage monitoring of autonomous ships. Sustainability. **13**(15), 8264 (2021). https://www.mdpi.com/2071-1050/13/15/8264

Kalaitzakis, M., Vitzilaios, N., Rizos, D.C., Sutton, M.A.: Drone-based StereoDIC: system development, experimental validation and infrastructure application. Exp. Mech. **61**, 981–996 (2021). https://doi.org/10.1007/s11340-021-00710-z

Kargar, S.M., Hao, G.: A drifter-based self-powered piezoelectric sensor for ocean wave measurements. Sensors. **22**(1), 5050 (2022). https://doi.org/10.3390/s22015050. MDPI

Karyotakis, A., Bucknall, R.: Planned intervention as a maintenance and repair strategy for offshore wind turbines. J. Mar. Eng. Technol. **9**(1), 27–35 (2010). https://doi.org/10.1080/20464177.2010.11020229

Katsoulakos, T., Tsiochantari, G., O'Donncha, F., Kaklamanis, E., Maccari, A., Mucharski, M.: Shipping digital twin landscape. In: State-of-the-Art Digital Twin Applications for Shipping Sector Decarbonization, pp. 1–25. IGI Global, Hershey (2024). https://doi.org/10.4018/978-1-6684-9848-4.ch001

Kefal, A., Mayang, J.B., Oterkus, E., Yidis, M.: Three dimensional shape and stress monitoring of bulk carriers based on iFEM methodology. Ocean Eng. **147**, 256–267 (2018). https://doi.org/10.1016/j.oceaneng.2017.10.040

Kennedy, K., Walsh, P., Mastaglio, T.W., Scully, T.: Genetic optimisation for a stochastic model for opportunistic maintenance planning of offshore wind farms. In: 2016 4th International Symposium on Environmental Friendly Energies and Applications (EFEA), 2016 4th International Symposium on Environmental Friendly Energies and Applications (EFEA), 9/14/2016 – 9/16/2016, pp. 1–6. IEEE, Belgrade, Serbia (2016)

Khana, A.A., Zafarb, S., Khanc, N.S., Mehmoodd, Z.: History, current status and challenges to structural health monitoring system aviation field. J. Space Technol. **4**, 1 (2014)

Khayatazad, M., De Pue, L., De Waele, W.: Detection of corrosion on steel structures using automated image processing. Dev. Built Environ. **3**, 100022 (2020). https://www.sciencedirect.com/science/article/pii/S2666165920300181

Kim, B.C., Kim, H.C., Han, S., Park, D.K.: Inspection of underwater hull surface condition using the soft voting ensemble of the transfer-learned models. Sensors. **22**, 4392 (2022a)

Kim, C., Oterkus, S., Oterkus, E., Kim, Y.: Probabilistic ship corrosion wastage model with Bayesian inference. Ocean Eng. **246**, 110571 (2022b). https://www.sciencedirect.com/science/article/pii/S0029801822000452

Kim, D.K., Lim, H.L., Cho, N.-K.: An advanced technique to predict time-dependent corrosion damage of onshore, offshore, nearshore and ship structures: part II= application to the ship's ballast tank. Int. J. Nav. Archit. Ocean Eng. **12**(1), 645–656 (2020b). https://koreascience.kr/article/JAKO202003440878753.page

Kim, D.K., Wong, E.W.C., Cho, N.-K.: An advanced technique to predict time-dependent corrosion damage of onshore, offshore, nearshore and ship structures: part I – generalisation. Int. J. Nav. Archit. Ocean Eng. **12**, 657–666 (2020a). https://www.sciencedirect.com/science/article/pii/S2092678220300236

Komoriyama, Y., Iijima, K., Tatsumi, A., Fujikubo, M.: Identification of wave profiles encountered by a ship with no forward speed using Kalman filter technique and validation by tank tests-long-crested irregular wave case. Ocean Eng. **271**, 113627 (2023). https://doi.org/10.1016/j.oceaneng.2023.113627

KR: Guidance for Smart Systems. Korean Register (KR), Busan (2023)

Kritzinger, W., Karner, M., Traar, G., Henjes, J., Sihn, W.: Digital twin in manufacturing: a categorical literature review and classification. IFAC-PapersOnLine. **51**(11), 1016–1022 (2018). https://doi.org/10.1016/j.ifacol.2018.08.474

Kurniawan, H., Pitana, T., Siswantoro, N.: Risk-based integrity management system of oil tanker hull structure due to corrosion. Key Eng. Mater. **940**(2023), 121–129 (2023)

Kutzke, D.T., Carter, J.B., Hartman, B.T.: Subsystem selection for digital twin development: a case study on an unmanned underwater vehicle. Ocean Eng. **223**, 108629 (2021). https://doi.org/10.1016/J.OCEANENG.2021.108629

Kwon, K., Frangopol, D.M.: System reliability of ship hull structures under corrosion and fatigue. J. Ship Res. **56**(4), 2012 (2012)

Lang, X., Wu, D., Tian, W., Zhang, C., Ringsberg, J.W., Mao, W.: Fatigue assessment comparison between a ship motion-based data-driven model and a direct fatigue calculation method. J. Mar. Sci. Eng. **11**, 2269 (2023)

Le, T.-C., Luu, T.-H.-T., Nguyen, H.-P., Nguyen, T.-H., Ho, D.-D., Huynh, T.-C.: Piezoelectric impedance-based structural health monitoring of wind turbine structures: current status and future perspectives. Energies. **15**, 5459 (2022). https://doi.org/10.3390/en15155459

Lee, S., Park, M., Oh, M.-H., Lee, P.-S.: Virtual sensing for real-time strain field estimation and its verification on a laboratory-scale jacket structure under water waves. Comput. Struct. **298**, 107344 (2024). https://doi.org/10.1016/j.compstruc.2024.107344

Li, M., Jiang, X., Carroll, J., Negenborn, R.R.: A closed-loop maintenance strategy for offshore wind farms: incorporating dynamic wind farm states and uncertainty-awareness in decision-making. Renew. Sust. Energ. Rev. **184**, 113535 (2023). https://doi.org/10.1016/j.rser.2023.113535

Li, M., Wang, M., Kang, J., Sun, L., Jin, P.: An opportunistic maintenance strategy for offshore wind turbine system considering optimal maintenance intervals of subsystems. Ocean Eng. **216**, 108067 (2020b). https://doi.org/10.1016/j.oceaneng.2020.108067

Li, H., Diaz, H., Guedes Soares, C.: A developed failure mode and effect analysis for floating offshore wind turbine support structures. Renew. Energy. **164**(2021), 133–145 (2021)

Li, H., Huang, C., Guedes Soares, C.: A real-time inspection and opportunistic maintenance strategies for floating offshore wind turbines. Ocean Eng. **256**, 111433 (2022)

Li, H., Teixeira, A.P., Guedes Soares, C.: A two-stage failure mode and effect analysis of offshore wind turbines. Renew. Energy. **162**, 1438–1461 (2020a)

Li, L., Zou, G.: A novel computational approach for assessing system reliability and damage detection delay: application to fatigue deterioration in offshore structures. Ocean Eng. **297**(2024), 117023 (2024)

Li, S., Brennan, F.: Digital twin enabled structural integrity management: critical review and framework development. Proc. Inst. Mech. Eng., Part M: J. Eng. Marit. Environ., 1–21 (2024a). https://doi.org/10.1177/14750902241227254

Li, S., Brennan, F.: Implementation of digital twin-enabled virtually monitored data in inspection planning. Appl. Ocean Res. **144**, 103903 (2024b)

Liao, W., Wu, J., Wang, Z., Yan, K., Ouyang, F.: Experimental investigation of monopile in over-consolidated marine clay subjected to multiple cyclic lateral loading events. Mar. Georesour. Geotechnol. **40**(8), 953–966 (2022)

Lin, J., et al: Characterization of model uncertainty of the py method for reliability-based design of offshore wind turbines under the serviceability limit state. Ocean Eng. **293**, 116662 (2024)

Lin, B., Dong, X.: Ship hull inspection: a survey. Ocean Eng. **289** (2023). https://doi.org/10.1016/j.oceaneng.2023.116281

Liu, Q., et al.: Digital twin-based designing of the configuration, motion, control, and optimization model of a flow-type smart manufacturing system. J. Manuf. Syst. **58**(Part B), 52–64 (2021a). https://doi.org/10.1016/j.jmsy.2020.04.012

Liu, Z., Chu, Y., Li, G., Zhang, H.: A co-simulation-based system using vico for marine operation. In: Masci, P., Bernardeschi, C., Graziani, P., et al. (eds.) Software Engineering and Formal Methods. SEFM 2022 Collocated Workshops. AI4EA, F-IDE, CoSim-CPS, CIFMA, Berlin, Germany, September 26–30, 2022, Revised Selected Papers, pp. 228–241. Springer International Publishing; Imprint Springer, Cham (2023d)

Liu, B., Doan, V.T., Garbatov, Y., Wu, W., Guedes Soares, C.: Study on ultimate compressive strength of aluminium-alloy plates and stiffened panels. J. Mar. Sci. Appl. **19**(4), 534–552 (2020). https://doi.org/10.1007/s11804-020-00170-2

Liu, G., et al.: Review of wireless RFID strain sensing technology in structural health monitoring. Sensors. **23**(15), 6925 (2023b). https://doi.org/10.3390/s23156925. MDPI

Liu, J., et al.: Prediction of human-machine interface (HMI) operational errors for maritime autonomous surface ships (MASS). J. Mar. Sci. Technol. **27**, 293–306 (2022). https://doi.org/10.1007/s00773-021-00834-w

Liu, M., Fang, S., Dong, H., Xu, C.: Review of digital twin about concepts, technologies, and industrial applications. J. Manuf. Syst. **58**, 346–361 (2021b). https://doi.org/10.1016/j.jmsy.2020.06.017

Liu, M., Li, H., Zhou, H., Zhang, H., Huang, G.: Development of machine learning methods for mechanical problems associated with fibre composite materials: a review. Compos. Commun. **49**, 101988 (2024). https://doi.org/10.1016/j.coco.2024.101988

Liu, S., Ren, S., Jiang, H.: Predictive maintenance of wind turbines based on digital twin technology. Energy Rep. **9**, 1344–1352 (2023a). https://doi.org/10.1016/j.egyr.2023.05.052

Liu, Y., Bao, Y.: Automatic interpretation of strain distributions measured from distributed fiber optic sensors for crack monitoring. Measurement. **224**, 112629 (2023). https://doi.org/10.1016/j.measurement.2023.112629. Elsevier

Liu, Y., Ren, H.: Acquisition method of evaluation stress for the digital twin model of ship monitoring structure. Appl. Ocean Res. **129**, 103368 (2022). https://doi.org/10.1016/j.apor.2022.103368

Liu, Y., Ren, H.: Rapid acquisition method for structural strength evaluation stresses of the ship digital twin model. Ocean Eng. **285**(P1), 115323 (2023). https://doi.org/10.1016/j.oceaneng.2023.115323

Liu, Y., Zhang, J., Min, Y., Yu, Y., Lin, C., Hu, Z.: A digital twin-based framework for simulation and monitoring analysis of floating wind turbine structures. Ocean Eng. **283**, 115009 (2023c)

Lloyd's Register: Ship Event Analysis. Lloyd's Register, London (2012)

Lloyd's Register: ShipRight Design and Construction Additional Design Procedures: LR Code for Unmanned Marine Systems. Lloyd's Register, London (2017)

Lv, Z., Fersman, E.: Digital Twins: Basics and Applications. Springer International Publishing, Cham (2022)

Ma, Y., Liu, X., Guo, Z., Wang, L., Naiwei, L.: Predicting corrosion fatigue crack propagation behavior of HRB400 steel bars in simulated corrosive environments. J. Mater. Civ. Eng. **33**(6), 04021127 (2021)

Maetz, T., et al.: Microwave structural health monitoring of the grouted connection of a monopile-based offshore wind turbine: fatigue testing using a scaled laboratory demonstrator. Struct. Control. Health Monit. **2023**, 1981892 (2023)., 18 pages

Maharana, K., Mondal, S., Nemade, B.: A review: data pre-processing and data augmentation techniques. Global Transitions Proc. **3**(1), 91–99 (2022). https://doi.org/10.1016/j.gltp.2022.04.020

Mahmoud, M., Semeraro, C., Abdelkareem, M.A., Olabi, A.G.: Designing and prototyping the architecture of a digital twin for wind turbine. Int. J. Thermofluids. **22**, 100622 (2024). https://doi.org/10.1016/j.ijft.2024.100622

Makris, P., Silionis, N.E., Anyfantis, K.N.: Spectral fatigue analysis of ship structures based on a stochastic crack growth state model. Int. J. Fatigue. **176**, 107878 (2023). https://www.sciencedirect.com/science/article/pii/S0142112323003791

Marques, V., Sattar, T.P.: Robotic deployment of stabilized shearography unit for wind turbine blade inspection. Lect. Notes Netw. Syst. **530**, 367–379 (2023). https://doi.org/10.1007/978-3-031-15226-9_35

Martinez-Luengo, M., Shafiee, M.: Guidelines and cost-benefit analysis of the structural health monitoring implementation in offshore wind turbine support structures. Energies. **12**(6), 1–26 (2019)

Martínez-Pañeda, E.: Progress and opportunities in modelling environmentally assisted cracking. Rilem Tech. Lett. **6**, 70–77 (2021)

MCA: The Workboat Code Edition 3, 3rd edn. Maritime & Coastguard Agency (2024)

McMorland, J., Collu, M., McMillan, D., Carroll, J., Coraddu, A.: Opportunistic maintenance for offshore wind: a review and proposal of future framework. Renew. Sust. Energ. Rev. **184**, 113571 (2023). https://doi.org/10.1016/j.rser.2023.113571

McMorland, J., Collu, M., McMillan, D., Carroll, J.: Operation and maintenance for floating wind turbines: a review. Renew. Sust. Energ. Rev. **163**, 112499 (2022)

Mehrjoo, A., Song, M., Moaveni, B., Papadimitriou, C., Hines, E.: Optimal sensor placement for parameter estimation and virtual sensing of strains on an offshore wind turbine considering sensor installation cost. Mech. Syst. Signal Process. **169**, 108787 (2022). https://doi.org/10.1016/j.ymssp.2021.108787

Melia, N., Haines, K., Hawkins, E.: Sea ice decline and 21st century trans-Arctic shipping routes. Geophys. Res. Lett. **43**, 9720–9728 (2016). https://doi.org/10.1002/2016GL069315

Mendes, P., et al.: Fatigue assessments of a jacket-type offshore structure based on static and dynamic analyses. Pract. Period. Struct. Des. Constr. **26**(1), 04020054 (2021). https://doi.org/10.1061/(ASCE)SC.1943-5576.0000533

Meucci, A., Young, I.R., Aarnes, O.J., Breivik, Ø.: Comparison of wind speed and wave height trends from twentieth-century models and satellite altimeters. J. Clim. **33**, 611–624 (2020). https://doi.org/10.1175/JCLI-D-19-0540.1

Mgbemena, C.E., Onuoha, D.O., Godwin, H.C.: Development of a novel virtual reality-enabled remote monitoring device for maintenance of cathodic protection systems on oil and gas pipelines. Sci. Rep. **13**(1), 15874 (2023). https://doi.org/10.1038/s41598-023-43159-x

Miao, X.-Y., Chen, X., Lu, R., Eder, M.A.: Multi-site crack initiation in local details of composite adhesive joints. Compos. Part B. **242**, 110055 (2022). https://doi.org/10.1016/j.compositesb.2022.110055

Mieloszyk, M., Majewska, K., Ostachowicz, W.: Application of embedded fibre Bragg grating sensors for structural health monitoring of complex composite structures for marine applications. In: Marine Structures, vol. 76, p. 102903. Elsevier BV (2021). https://doi.org/10.1016/j.marstruc.2020.102903

Min, R., Liu, Z., Pereira, L., Yang, C., Sui, Q., Marques, C.: Optical fiber sensing for marine environment and marine structural health monitoring: a review. Opt. Laser Technol. **138**, 106848 (2021). https://doi.org/10.1016/j.optlastec.2021.107082. Elsevier

Minerva, R., Lee, G.M., Crespi, N.: Digital twin in the IoT context: a survey on technical features, scenarios, and architectural models. Proc. IEEE. **108**(10), 1785–1824 (2020). https://doi.org/10.1109/JPROC.2020.2998530

Mishael, J., Morato, P., Rigo, P.: Numerical fatigue modeling and simulation of interacting surface cracks in offshore wind structural connections. Mar. Struct. **92**, 103472 (2023). https://doi.org/10.1016/j.marstruc.2023.103472

Miyake, Y., Iijima, K., Tatsumi, A., Fujikubo, M.: Estimation of ship hull girder deformation and load by using sensors and numerical model. J. Jpn. Soc. Nav. Archit. Ocean Eng. **17**, 47–56 (2023). https://doi.org/10.2534/jjasnaoe.37.47. (in Japanese)

Mogeke, M., Magoga, T.: Towards improved understanding of naval ship structural performance via virtual hull monitoring. Procedia Struct. Integrity. **45**, 36–43 (2023)

Moghadam, F.K., Nejad, A.R.: Online condition monitoring of floating wind turbines drivetrain by means of digital twin. Mech. Syst. Signal Process. **162**, 108087 (2022)

Moheimani, S.R., Savkin, A.V., Petersen, I.R.: Robust filtering, prediction, smoothing, and observability of uncertain systems. IEEE Trans. Circuits Systems I Fund. Theory Appl. **45**(4), 446–457 (1998)

Moi, T., Cibicik, A., Rølvåg, T.: Digital twin-based condition monitoring of a knuckle boom crane: an experimental study. Eng. Fail. Anal. **112**, 104517 (2020). https://doi.org/10.1016/j.engfailanal.2020.104517

Momber, A.W., Wilms, M., Brün, D.: The use of meteorological and oceanographic sensor data in the German offshore territory for the corrosion monitoring of marine structures. Ocean Eng. **257**, 110994 (2022)

Momber, A.W.: Corrosion and Corrosion Protection of Wind Power Structures in Marine Environments. Academic, Amsterdam (2021)

Morato, P.G., Papakonstantinou, K.G., Andriotis, C.P., Hlaing, N., Kolios, A.: Interpretation and analysis of deep reinforcement learning driven inspection and maintenance policies for engineering systems. In: 14th International Conference on Applications of Statistics and Probability in Civil Engineering, ICASP14, Dublin, Ireland, 9–13 July 2023 (2023)

Morim, J., et al.: Global-scale changes to extreme ocean wave events due to anthropogenic warming. Environ. Res. Lett. **16**, 074056 (2021). https://doi.org/10.1088/1748-9326/ac1013

Moynihan, B., Tronci, E.M., Hughes, M.C., Moaveni, B., Hines, E.: Virtual sensing via Gaussian Process for bending moment response prediction of an offshore wind turbine using SCADA data. Renew. Energy. **227**, 120466 (2024). https://doi.org/10.1016/j.renene.2024.120466

Murawski, L., Opoka, S., Majewska, K., Mieloszyk, M., Ostachowicz, W.M., Weintrit, A.: Investigations of Marine Safety Improvements by Structural Health Monitoring Systems, p. 6. TransNav: International Journal on Marine Navigation and Safety of Sea Transportation (2012)

Nelli, F., et al.: Reconstructing sea-states in the Southern Ocean using ship motion data. In: Proceedings of the International Conference on Offshore Mechanics and Arctic Engineering – OMAE, 6, art. no. V006t06a027 (2021). https://doi.org/10.1115/OMAE2021-62757

Ng, S.M., Khan, R., Riffin, A., Chen, K., Bucknell, J., Azam, B.: Structural Integrity Management (SIM) for steel gravity-based structures. In: 33rd International Ocean and Polar Engineering Conference, ISOPE-I-23-134, Ottawa, Canada (2023)

Nguyen, T.B., Nguyen, M.H.: New practical approach to assess the robustness and redundancy of offshore structures: application in life extension of fixed steel offshore platforms. In: Proceedings of the 2nd Vietnam Symposium on Advances in Offshore Engineering, pp. 325–332 (2022). https://doi.org/10.1007/978-981-16-7735-9_36

Nguyen, T.A.T., Chou, S.-Y.: Maintenance strategy selection for improving cost-effectiveness of offshore wind systems. Energy Convers. Manag. **157**, 86–95 (2018). https://doi.org/10.1016/j.enconman.2017.11.090

Nicard, C., Farin, M., Moulin, E., Balloy, D., Serre, I.P.: Monitoring of generalised corrosion: ultrasonic coda wave interferometry technique applied to steel corrosion in aqueous NaCl solutions. Mater. Chem. Phys. **305**, 127908 (2023)

Nichols, N.W., Khan, R., Ng, S.M., Lee, L.A.: A strengthening, modification & repair (SMR) decisionmaking toolkit for structural integrity management (SIM) of ageing offshore structures. In: SPE/IATMI Asia Pacific Oil & Gas Conference and Exhibition, Bali, Indonesia (2015)

Nickerson, B.M., Bekker, A.: Inverse model for the estimation of ice-induced propeller moments using modal superposition. Appl. Math. Model. **102**, 640–660 (2022). https://doi.org/10.1016/j.apm.2021.10.005

Niederer, S.A., Sacks, M.S., Girolami, M., Willcox, K.: Scaling digital twins from the artisanal to the industrial. Nat. Comput. Sci. **1**(5), 313–320 (2021)

Nielsen, U.D.: Estimations of on-site directional wave spectra from measured ship responses, onboard monitoring of fatigue damage rates in the hull girder. Mar. Struct. **19**, 33–69 (2006). https://doi.org/10.1016/j.marstruc.2006.06.001

NK: Guidelines for Hull Monitoring. Nippon Kaiji Kyokai (Class NK) (2021)

NK: Rules for Hull Monitoring Systems. Nippon Kaiji Kyokai (Class NK) (2023)

Nomikos, N., Gkonis, P.K., Bithas, P.S., Trakadas, P.: A survey on UAV-aided maritime communications: deployment considerations, applications, and future challenges. IEEE Open J. Commun. Soc. **4**, 56–78 (2023). https://ieeexplore.ieee.org/document/9966921/

Nugroho, F.A., Braun, M., Ehlers, S.: Probability analysis of pit distribution on corroded ballast tank. Ocean Eng. **228**, 2021 (2021)

OCEANEERING: Remotely Operated Vehicles (ROVs). https://www.oceaneering.com/rov-services/ (2023). Accessed 4 Oct 2023

OCIUS Bluebottle Specifications: OCIUS. https://ocius.com.au/usv/ (2024)

Okpokparoro, S., Sriramula, S.: Uncertainty modeling in reliability analysis of floating wind turbine support structures. Renew. Energy. **165**, 88–108 (2021)

Ólafsson, Ó.M.: Improved design bases of welded joints in seawater. PhD thesis, Technical University of Denmark 2016)
Ozguc, O.: A new risk-based inspection methodology for offshore floating structures. J. Mar. Eng. Technol. **19**(1), 40–55 (2020). https://doi.org/10.1080/20464177.2018.1508804
Page, A.M., Klinkvort, R.T., Bayton, S., Zhang, Y., Petter Jostad, H.: A procedure for predicting the permanent rotation of monopiles in sand supporting offshore wind turbines. Mar. Struct. **75**, 102813 (2021)
Papanikolaou, A., Boulougouris, E., Erikstad, S.-O., Harries, S., Kana, A.A.: Ship Design in the era of digital transition: a state-of-the-art report. Int. Mar. Des. Conf. (2024). https://doi.org/10.59490/imdc.2024.784
Partovi-Mehr, N., et al.: Fatigue analysis of a jacket-supported offshore wind turbine at Block Island Wind Farm. Sensors. **24**(10), 3009 (2024). https://doi.org/10.3390/s24103009
Patra, S.K., Haldar, S., Bhattacharya, S.: Predicting tilting of monopile supported wind turbines during seismic liquefaction. Ocean Eng. **2022**(252), 111145 (2022)
Peeters, B., Orlowitz, E., Di Lorenzo, E., Mastrodicasa, D.: Operational modal testing of large wind turbine blades. Lect. Notes Civ. Eng. **514**, 550–558 (2024). https://doi.org/10.1007/978-3-031-61421-7_53
Peruń, G.: Advances in non-destructive testing methods. Materials. **17**(3), 554 (2024). https://doi.org/10.3390/ma17030554
Petrosjan, M.: Automated bridge inspection with drone and artificial intelligence. Bauingenieur. **96**(7–8), A27–A30 (2021)
Pezeshki, H., Adeli, H., Pavlou, D., Siriwardane, S.C.: State of the art in structural health monitoring of offshore and marine structures. Proc. Inst. Civ. Eng. Marit. Eng. **176**(2), 89–108 (2023). https://doi.org/10.1680/jmaen.2022.027
Pferdekämper, K.H., Bekker, A.: Investigation of vessel slamming and fatigue using a full-scale test sequence. Appl. Ocean Res. **144**, 103883 (2024). https://doi.org/10.1016/j.apor.2024.103883
Phelps, B., Morris, B (2013). Review of Hull Structural Monitoring Systems for Navy Ships. DSTO-TR-2818.
Preethichandra, D.M.G., Suntharavadivel, T.G., Pushpitha, K., Piyathilaka, L., Izhar, U.: Influence of smart sensors on structural health monitoring systems and future asset management practices. Sensors. (2023). https://doi.org/10.3390/s23198279
Priyanka, E.B., Thangavel, S., Gao, X.-Z., Sivakumar, N.S.: Digital twin for oil pipeline risk estimation using prognostic and machine learning techniques. J. Ind. Inf. Integr. **26**, 100272 (2022). https://doi.org/10.1016/j.jii.2021.100272
PSA: The Use of Digital Solutions and Structural Health Monitoring for Integrity Management of Offshore Structures. REN2021N00. Petroleum Safety Authority Norway (2022)
Purcell, E., Nejad, A.R., Bekker, A.: Methodology for real-time torque estimation in a ship propulsion digital twin. J. Offshore Mech. Arct. Eng. **147**, 1–29 (2024)
Qin, J.: Preposterior analysis considering uncertainties and dependencies of information relevant to structural performance. ASCE-ASME J. Risk Uncertainty Eng. Syst. Part A: Civ. Eng. **8**, 4021085 (2022)
Qureshi, A., Alaloul, W., Musarat, M.A., Rian, A.: Risk-based inspection in offshore structures: a systematic approach. In: Syawitri, T.P. (ed.) 6th Mechanical Engineering, Science and Technology International Conference, pp. 561–569 (2023)
Qvale, P., Zarandi, E.P., Ås, A.K., Skallerud, B.H.: Digital image correlation for continuous mapping of fatigue crack initiation sites on corroded surface from offshore mooring chain. Int. J. Fatigue. **151**, 106350 (2021). https://doi.org/10.1016/j.ijfatigue.2021.106350
Raed, K., Teixeira, A.P., Guedes Soares, C.: Uncertainty assessment for the extreme hydrodynamic responses of a wind turbine semi-submersible platform using different environmental contour approaches. Ocean Eng. **195**, 106719 (2020). https://doi.org/10.1016/j.oceaneng.2019.106719

Reilly, G., Jorgensen, J.: Classification considerations for cyber safety and security in the smart ship era. In: RINA, Royal Institution of Naval Architects – Smart Ship Technology 2016, Papers, January, pp. 33–39 (2016). https://doi.org/10.3940/rina.sst.2016.03

Rhakasywi, D., Marasabessy, A., Hatuwe, M.R., Kotahatuhaha, S.: Safety factor of pump vibrations on ships based on the natural frequency of pump vibrations according to ISO 10816-3. J. Mech. Eng. Res. Dev. **43**(7), 180–192 (2020)

Riccioli, F., Pahlavan, L.: Non-contact acoustic emission monitoring of corrosion-fatigue damage in submerged steel structures. Struct. Health Monit. (2023). https://doi.org/10.12783/shm2023/36856

Riccioli, F., Huijer, A., Grasso, N., Rizzo, C.M., Pahlavan, L.: Development of a retrofit layer with an embedded array of piezoelectric sensors for transient pressure measurement in maritime applications. Mar. Struct. **89**, 103395 (2023). https://doi.org/10.1016/j.marstruc.2023.103395

Rigby, J., Christmas, J.: Changing definitions of digital twins in ship design. In: Conference Proceedings of Engine as a Weapon (EAAW 2023), November 2023 (2023). https://doi.org/10.24868/11075

Ringsberg, J.W., et al.: The ISSC 2022 committee III.1-Ultimate strength benchmark study on the ultimate limit state analysis of a stiffened plate structure subjected to uniaxial compressive loads. Mar. Struct. **79**, 103026 (2021). https://www.sciencedirect.com/science/article/pii/S0951833921000836

Ritto, T.G., Rochinha, F.A.: Digital twin, physics-based model, and machine learning applied to damage detection in structures. Mech. Syst. Signal Process. **155**(2021), 107614 (2021)

Roberts, G.W., Hancock, C.M., Lienhart, W., Klug, F., Zuzek, N., de Ligt, H.: Displacement and frequency response measurements of a ship using GPS and fibre optic-based sensors. Appl. Geomat. **13**, 51–61 (2021). https://doi.org/10.1007/s12518-020-00338-z

Roch, E., Zima, B., Woloszyk, K., Garbatov, Y.: Guided waves in ship structural health monitoring – a feasibility study. Pol. Marit. Res. **30**(2), 76–84 (2023). https://doi.org/10.2478/pomr-2023-0023

Rovelli, D., Brondi, C., Andreotti, M., Abbate, E., Zanforlin, M., Ballarino, A.: A modular tool to support data management for life cycle assessment (LCA) in industry: methodology, application and potentialities. Sustainability. **14**(7), 3746 (2022). https://www.mdpi.com/2071-1050/14/7/3746

Saccone, C., Pahlavan, L.: Influence of stiffeners on acoustic emission monitoring of ship structures. In: Proceedings of the 10th European Workshop on Structural Health Monitoring (EWSHM 2024), June 10–13, 2024 in Potsdam, Germany. e-Journal of Nondestructive Testing (2024). https://doi.org/10.58286/29587

Samareh-Mousavi, S.S., et al.: Monitoring fatigue delamination growth in a wind turbine blade using passive thermography and acoustic emission. Struct. Health Monit. (2024). https://doi.org/10.1177/14759217231217179

Sarhadi, A., Albuquerque, R.Q., Demleitner, M., Ruckdäschel, H., Eder, M.A.: Machine learning based thermal imaging damage detection in glass-epoxy composite materials. Compos. Struct. **295**, 115786 (2022). https://doi.org/10.1016/j.compstruct.2022.115786

Scheeren, B., Kaminski, M.L., Pahlavan, L.: Acoustic emission monitoring of naturally developed damage in large-scale low-speed roller bearings. Struct. Health Monit. **23**(1), 360–382 (2024). https://doi.org/10.1177/14759217231164912

Scheu, M., Matha, D., Hofmann, M., Muskulus, M.: Maintenance strategies for large offshore wind farms. Energy Procedia. **24**, 281–288 (2012). https://doi.org/10.1016/j.egypro.2012.06.110

Schirmann, M., Collette, M., Gose, J.: Ship motion and fatigue damage estimation via a digital twin. In: Life Cycle Analysis and Assessment in Civil Engineering: Towards an Integrated Vision, pp. 2075–2082. CRC Press, London (2019)

Schoone, S., Laumen, F., Leaman, F.: Experimental study on attenuation of acoustic emission signals in metallic plates for deep-sea mining. Min. Metall. Explor. **40**(3), 555–562 (2023). https://doi.org/10.1007/s42461-023-00741-1

Shafto, M., et al.: Draft modelling, simulation, information technology & processing roadmap. Technol. Area. **11**, 1–32 (2010). NASA

Shamsuddoha, M., Prusty, G.B., Maung, P.T., Phillips, A.W., St John, N.A.: Smart monitoring of a full-scale composite hydrofoil manufactured using automated fibre placement under high cycle fatigue. Smart Mater. Struct. **31** (2022). https://doi.org/10.1088/1361-665X/ac3b21

Sheikh, M.F., Kamal, K., Rafique, F., Sabir, S., Zaheer, H., Khan, K.: Corrosion detection and severity level prediction using acoustic emission and machine learning-based approach. Ain Shams Eng. J. **12**(4), 3891–3903 (2021). https://doi.org/10.1016/j.asej.2021.03.024

Shen, W., Pang, Q., Fan, L., Li, P., Zhao, X.: Monitoring and quantification of non-uniform corrosion induced mass loss of steel piles with distributed optical fiber sensors. Autom. Constr. **148**, 104769 (2023)

Shi, S., Wu, G., Chen, H., Zhang, S.: Acoustic emission monitoring of fatigue crack growth in Hadfield steel. Sensors. **23**(14), 6561 (2023)

Shibu, M., Kumar, K.P., Pillai, V.J., Murthy, H., Chandra, S.: Structural health monitoring using AI and ML based multimodal sensors data. Meas.: Sens. **27**, 100762 (2023). https://doi.org/10.1016/j.measen.2023.100762

Shojai, S., Schaumann, P., Brömer, T.: Probabilistic modelling of pitting corrosion and its impact on stress concentrations in steel structures in the offshore wind energy. Mar. Struct. **84**, 2022 (2022a). https://doi.org/10.1016/j.marstruc.2022.103232

Shojai, S., et al.: Assessment of corrosion fatigue in welded joints using 3D surface scans, digital image correlation, hardness measurements, and residual stress analysis. Int. J. Fatigue. **176**, 2023 (2023)

Shojai, S., Kabha, K., Woitzik, C., Braun, M., Ghafoori, E.: Fatigue Analysis of Corroded Welded Offshore Steel Joints. International Institute of Welding IIW-Doc. XIII-3080-2024 (2024)

Shojai, S., Schaumann, P., Braun, M., Ehlers, S.: Influence of pitting corrosion on the fatigue strength of offshore steel structures based on 3D surface scans. Int. J. Fatigue. **164**, 107128 (2022b). https://doi.org/10.1016/j.ijfatigue.2022.107128

Silionis, N.E., Liangou, T., Anyfantis, K.N.: Deep learning-based surrogate models for spatial field solution reconstruction and uncertainty quantification in structural health monitoring applications. Comput. Struct. **301**, 107462 (2024). https://doi.org/10.1016/j.compstruc.2024.107462

Silionis, N.E., Anyfantis, K.: On the detection of thickness loss in ship hull structures through strain sensing. Lect. Notes Civ. Eng. **254 LNCE**, 207–216 (2023)

Silva-Campillo, A., Pérez-Arribas, F., Suárez-Bermejo, J.C.: Health-monitoring systems for marine structures: a review. Sensors. **23**(4), 1–19 (2023). https://doi.org/10.3390/s23042099

Singh, S., Singh, R.: Re-assessment of existing offshore platform for life extension. In: Proceedings of International Conference on Structural Engineering and Construction Management (SECON'23), pp. 723–735 (2023). https://doi.org/10.1007/978-3-031-39663-2_61

Singh, V., Osen, O.L., Bye, R.T.: Explainable artificial intelligence for autonomous surface vessels by fuzzy-based collision avoidance system. In: Senjyu, T., So-In, C., Joshi, A. (eds.) Smart Trends in Computing and Communications. SmartCom 2023 Lecture Notes in Networks and Systems, vol. 650. Springer, Singapore (2023). https://doi.org/10.1007/978-981-99-0838-7_12

Sireta, F.-X., Storhaug, G.: A modal approach for holistic hull structure monitoring from strain gauges measurements and structural analysis. In: Paper Presented at the Offshore Technology Conference, Houston, Texas, USA, May 2022 (2022). https://doi.org/10.4043/31789-MS

Slatnick, S., et al.: Bow-ties use for high-consequence marine risks of offshore structures. Process Saf. Environ. Prot. **165**, 396–407 (2022)

Soal, K., Govers, Y., Bienert, J., Bekker, A.: System identification and tracking using a statistical model and a Kalman filter. Mech. Syst. Signal Process. **133**, 106127 (2019). https://doi.org/10.1016/j.ymssp.2019.05.011

Sousa, I., et al.: Sensing system based on FBG for corrosion monitoring in metallic structures. Sensors. **22**(16), 5947 (2022)

Sözen, L., Yurdakul, M., İç, Y.T.: Risk-based inspection planning for internal surface-defected oil pipelines exposed to fatigue. Int. J. Press. Vessel. Pip. **200**, 104804 (2022)

Spahić, R., Poolla, K., Hepsø, V., Lundteigen, M.A.: Image-based and risk-informed detection of subsea pipeline damage. Discov. Artif. Intell. **3**(1) (2023). https://doi.org/10.1007/s44163-023-00069-1

Stadtmann, F., et al.: Digital twins in wind energy: emerging technologies and industry-informed future directions. IEEE Access. **11**, 110762–110795 (2023). https://doi.org/10.1109/ACCESS.2023.3321320

Steele, B., Chandler, J., Reddy, S.: Real-time analytics. In: Algorithms for Data Science. Springer, Cham (2016). https://doi.org/10.1007/978-3-319-45797-0_12

Sujawat, R.S., Kumar, R.: Stochastic numerical analyses to investigate the effects of the spatial nonuniformity of offshore ground on the serviceability of monopile foundations. Georisk: Assess. Manage. Risk Eng. Syst. Geohazards. **2023**, 1–18 (2023)

Sun, Y., Li, H., Sun, L., Guedes Soares, C.: Failure analysis of floating offshore wind turbines with correlated failures. Reliab. Eng. Syst. Saf. **238**, 109485 (2023)

Suresh, S., Zamiski, G.F., Ritchie, R.O.: Oxide-induced crack closure: an explanation for near-threshold corrosion fatigue crack growth behaviour. Metall. Trans. A. **12A**(8), 1435–1443 (1981)

Svendsen, B.T., Tygesen, U.T., Kelly-Rosenville, J., Azam, N.: The use of digital solutions and structural health monitoring for integrity and management of offshore structures, 14 (2022)

Swartz, R.A., et al.: Hybrid wireless hull monitoring system for naval combat vessels. J. Struct. Infrastruct. Eng. **8**(7) (2010). https://doi.org/10.1080/15732479.2010.495398

Szelezinski, A., Muc, A., Murawski, L., Kluczyk, M., Muchowski, T.: Application of laser vibrometry to assess defects in ship hull's welded joints' technical condition. Sensors (Switzerland). **21**(3), 895, 1–18 (2021). https://doi.org/10.3390/s21030895

Tao, F., Xiao, B., Qi, Q., Cheng, J., Ji, P.: Digital twin modelling. J. Manuf. Syst. **64**, 372–389 (2022)

Tarpø, M., Amador, S., Katsanos, E., Skog, M., Gjødvad, J., Brincker, R.: Data-driven virtual sensing and dynamic strain estimation for fatigue analysis of offshore wind turbine using principal component analysis. Wind Energy. **25**(3), 505–516 (2022). https://doi.org/10.1002/we.2683

Taylor, N.C., Bekker, A., Kruger, K.: The operational development of diagnostic seasickness criteria through a human cyber-physical system. Appl. Ergon. **119** (2024). https://doi.org/10.1016/j.apergo.2024.104316

TechnipFMC: ROV systems. https://www.technipfmc.com/en/what-we-do/subsea/robotics/rov-systems/ (2023). Accessed 4 Oct 2023

Tekgoz, M., Garbatov, Y., and Soares, C. G. (2020). Review of ultimate strength assessment of ageing and damaged ship structures. J. Mar. Sci. Appl., 19(4):512–533. doi: 10.1007/s11804-020-00179-7.

Thibbotuwa, U.C., Cortés, A., Irizar, A.: Ultrasound-based smart corrosion monitoring system for offshore wind turbines. Appl. Sci. **12**(2), 808 (2022). https://doi.org/10.3390/app12020808

Thompson, I.: Fatigue damage variation within a class of naval ships. Ocean Eng. **165**, 123–130 (2018a). https://doi.org/10.1016/j.oceaneng.2018.07.036

Thompson, I.: Virtual Hull monitoring: continuous fatigue assessment without additional instrumentation. Int. J. Marit. Eng. **160**(Part A3), A-293–A-297 (2018b)

Thompson, I.: Virtual hull monitoring of a naval vessel using hindcast data and reconstructed 2-D wave spectra. Mar. Struct. **71**, 102730 (2020). https://doi.org/10.1016/j.marstruc.2020.102730

Thompson, N.C., Greenewald, K., Lee, K., Manso, G.F.: The computational limits of deep learning. Mit Initiative on the Digital Economy Research Brief. **33**(4) (2022)

Tibaduiza Burgos, D.A., Gomez Vargas, R.C., Pedraza, C., Agis, D., Pozo, F.: Damage identification in structural health monitoring: a brief review from its implementation to the use of data-driven applications. Sensors (Switzerland). **20**(3), 733 (2020). https://doi.org/10.3390/s20030733

Tonelli, D., Cappello, C., Zonta, D.: Performance-based design of structural health monitoring systems. Lect. Notes Civ. Eng. **128**, 238–247 (2021). https://doi.org/10.1007/978-3-030-64908-1_22

Tong, W.J., Kar, S., Bin Taib, H., Bin Abdul Rahman, A.: Floating structure integrity management program (FSIMP) towards digitalization. In: International Petroleum Technology Conference, IPTC-22857-MS, 2023, Bangkok, Thailand (2023)

Torkildsen, H.E., et al.: Development and applications of full-scale ship hull health monitoring systems for the Royal Norwegian Navy. NATO Science & Technology Organisation (2005)

Turnbull, A., Thomas, J.G.N.: A model of crack electrochemistry for steels in the active state based on mass-transport by diffusion and ion migration. J. Electrochem. Soc. **129**(7), 1412–1422 (1982)

Turnbull, A., Ferriss, D.H.: Mathematical modelling of the electrochemistry in corrosion fatigue cracks in structural steel cathodically protected in sea water. Corros. Sci. **26**(8), 601–628 (1986)

Uhlenkamp, J.-F., Hribernik, K., Wellsandt, S., Thoben, K.-D.: Digital twin applications: a first systemization of their dimensions. In: 2019 IEEE International Conference on Engineering, Technology and Innovation (ICE/ITMC), pp. 1–8 (2019). https://doi.org/10.1109/ICE.2019.8792579

Valet, A., Altenmüller, T., Waschneck, B., May, M.C., Kuhnle, A., Lanza, G.: Opportunistic maintenance scheduling with deep reinforcement learning. J. Manuf. Syst. **64**, 518–534 (2022). https://doi.org/10.1016/j.jmsy.2022.07.016

Van Der Horn, E., Valluri, S.: Guidance on the verification and validation of digital twins. In: SNAME 29th Offshore Symposium, Texas Texas Section of the Society of Naval Architects and Marine Engineers, 2024 (2024). https://doi.org/10.5957/tos-2024-004

Van Der Horn, E., Wang, Z., Mahadevan, S.: Towards a digital twin approach for vessel-specific fatigue damage monitoring and prognosis. Reliab. Eng. Syst. Saf. **219**, 108222 (2022). https://www.sciencedirect.com/science/article/pii/S0951832021007006

Van Zijl, C., Soal, K., Volkmar, R., Govers, Y., Böswald, M., Bekker, A.: The use of operational modal analysis and mode tracking for insight into polar vessel operations. Mar. Struct. **79**, 103043 (2021). https://doi.org/10.1016/j.marstruc.2021.103043

Verhelst, J., Coudron, I., Ompusunggu, A.P.: SCADA-compatible and scaleable visualization tool for corrosion monitoring of offshore wind turbine structures. Appl. Sci. **12**(3), 1762 (2022). https://doi.org/10.3390/app12031762

Viswanathan, S., Holden, C., Egeland, O., Sten, R.: Cosimulation of a direct-acting riser-tensioner system – validation with field measurements and sample simulations. Ocean Eng. **276**, 114241 (2023). https://doi.org/10.1016/j.oceaneng.2023.114241

von Houwald, B., Sarhadi, A., Eitzinger, C., Eder, M.A.: Layer-by-layer reconstruction of fatigue damages in composites from thermal images by a residual U-Net. Compos. Sci. Technol. **255**, 110712 (2024). https://doi.org/10.1016/j.compscitech.2024.110712

Vukelic, G., Vizentin, G., Brnic, J., Brcic, M., Sedmak, F.: Long-term marine environment exposure effect on butt-welded shipbuilding steel. J. Mar. Sci. Eng. **9**(5), 491 (2021). https://www.mdpi.com/2077-1312/9/5/491

Wahlstrom, M., Hakulinen, J., Karvonen, H., Lindborg, I.: Human factors challenges in unmanned ship operations – insights from other domains. In: 6th International Conference on Applied Human Factors and Ergonomics (AHFE 2015) and the Affiliated Conferences, 2015, pp. 1038–1045 (2015)

Wang, S., Moan, T.: Serviceability limit state assessment of semi-submersible floating wind turbines. J. Offshore Mech. Arct. Eng. **146**(2) (2024)

Wang, S., Moan, T., Gao, Z.: Methodology for global structural load effect analysis of the semsubmersible hull of floating wind turbine under still water, wind, and wave loads. Mar. Struct. **91**(2023), 103463 (2023)

Wang, H., Mao, W., Zhang, D.: Voyage optimization for mitigating ship structural failure due to crack propagation. Proc. Inst. Mech. Eng., Part O: J. Risk Reliab. **233**(1), 5–17 (2019)

Wang, H., et al.: A comprehensive survey on training acceleration for large machine learning models in IoT. IEEE Internet Things J. **9**(2), 939–963 (2022a). https://doi.org/10.1109/JIOT.2021.3111624

Wang, L., Kolios, A., Liu, X., Venetsanos, D., Rui, C.: Reliability of offshore wind turbine support structures: a state-of-the-art review. Renew. Sust. Energ. Rev. **161**, 112250 (2022b)

Wang, M., et al.: Structural health monitoring on offshore jacket platforms using a novel ensemble deep learning model. Ocean Eng. **301**, 117510 (2024). https://doi.org/10.1016/j.oceaneng.2024.117510

Wang, M., Wang, C., Hnydiuk-Stefan, A., Feng, S., Atilla, I., Li, Z.: Recent progress on reliability analysis of offshore wind turbine support structures considering digital twin solutions. Ocean Eng. **232**, 109168 (2021b)

Wang, P.: Civil Aircraft Electrical Power System Safety Assessment. https://www.sciencedirect.com/topics/engineering/system-safety-assessment (2017)

Wang, S., Moan, T., Nejad, A.R.: A comparative study of fully coupled and de-coupled methods on dynamic behaviour of floating wind turbine drivetrains. Renew. Energy. **179**, 1618–1635 (2021a). https://doi.org/10.1016/j.renene.2021.07.136

Waszak, M., Cardaillac, A., Elvesæter, B., Rødølen, G., Ludvigsen, M.: Semantic segmentation in underwater ship inspections: benchmark and data set. IEEE J. Ocean. Eng. **48**(2), 462–473 (2022a). https://doi.org/10.1109/JOE.2022.3219129

Waszak, M., Lam, A.N., Hoffmann, V., Elvesater, B., Mogos, M.F., Roman, D.: Let the asset decide: digital twins with knowledge graphs. In: IEEE 19th International Conference on Software Architecture Companion, ICSA-C 2022, pp. 35–39 (2022b). https://doi.org/10.1109/ICSA-C54293.2022.00014

Wilson, C.L., Lonkar, K., Roy, S., Kopsaftopoulos, F., Chang, F.K.: Structural health monitoring of composites. Comp. Comp. Mater. II. **7–8**, 382–407 (2017). https://doi.org/10.1016/B978-0-12-803581-8.10039-6

Wines, C., Egil Jenden, A., Sagvolden, G.: Ship Hull Monitoring System Applied as a Real Time Decision Support System on Fast Ferries. NATO Science & Technology Organisation (2023)

Woloszyk, K., Garbatov, Y.: Advances in modelling and analysis of strength of corroded ship structures. J. Mar. Sci. Eng. **10**(6), 807 (2022). https://www.mdpi.com/2077-1312/10/6/807

Woloszyk, K., Garbatov, Y.: A probabilistic-driven framework for enhanced corrosion estimation of ship structural components. Reliability Engineering & System Safety. **242**, 109721 (2024). https://www.sciencedirect.com/science/article/pii/S095183202300635X

Wu, B., et al.: A four-dimensional digital twin framework for fatigue damage assessment of semi-submersible platforms and practical application. Ocean Eng. **301**, 117273 (2024)

Wu, Y., Nong, J., Yuan, S., Guo, Q., Xu, J.: Acoustic emission behaviour during the evolution of a single pit on stainless steels. Corros. Sci. **183**, 109308 (2021)

Xia, J., Zou, G.: Operation and maintenance optimization of offshore wind farms based on digital twin: a review. Ocean Eng. **268**, 113322 (2023)

Xiang, Z., et al.: Vibration-based health monitoring of the offshore wind turbine tower using machine learning with Bayesian optimisation. Ocean Eng. **292**, 116513 (2024)

Xin, S., Qi, Z., Yang, L., Yi, H., Ziguang, J.: Deep-learning approach based on multi-data fusion for damage recognition of marine platforms under complex loads. In: Ocean Engineering, vol. 303, p. 116604. Elsevier BV (2024). https://doi.org/10.1016/j.oceaneng.2023.116604

Xu, L., Hu, P., Li, Y., Qiu, N., Chen, G., Liu, X.: Improved fatigue reliability analysis of deepwater risers based on RSM and DBN. J. Mar. Sci. Eng. **11**(4), 2023 (2023)

Yang, Y., Liang, F., Zhu, Q., Zhang, H.: An overview on structural health monitoring and fault diagnosis of offshore wind turbine support structures. J. Mar. Sci. Eng. **12**(3), 377 (2024). https://doi.org/10.3390/jmse12030377

Yao, X., Vien, B.S., Davies, C., Chiu, W.K.: Acoustic emission source characterisation during fatigue crack growth in Al 2024-T3 specimens. Sensors. **22**(22), 8796 (2022)

Yao, Y., Yang, Y., Wang, Y., Zhao, X.: Artificial intelligencebased hull structural plate corrosion damage detection and recognition using convolutional neural network. Appl. Ocean Res. **90**, 101823 (2019). https://www.sciencedirect.com/science/article/pii/S0141118718302694

Yara: Yara Birkeland Press kit 2024. Yara International. https://www.yara.com/news-and-media/media-library/press-kits/yara-birkeland-press-kit/ (2024)

Yeter, B., Garbatov, Y.: Structural integrity assessment of fixed support structures for offshore wind turbines: a review. Ocean Eng. **244**, 110271 (2022). https://doi.org/10.1016/j.oceaneng.2021.110271

Yosri, A., Leheta, H., Saad-Eldeen, S., Zayed, A.: Accumulated fatigue damage assessment of side structural details in a double hull tanker based on spectral fatigue analysis approach. Ocean Eng. **251**, 111069 (2022). https://www.sciencedirect.com/science/article/pii/S0029801822004826

Young, I.R., Ribal, A.: Multiplatform evaluation of global trends in wind speed and wave height. Science. **364**, 548–552 (2019). https://doi.org/10.1126/science.aav9527

Yüce, C., et al.: Prognostics and health management of wind energy infrastructure systems. ASCE-ASME J. Risk Uncertainty Eng. Syst. Part B: Mech. Eng. **8**(2), 2022 (2022)

Zhang, Z., Wu, X.: Interpreting electrochemical noise signal arising from stress corrosion cracking of 304 stainless steel in simulated PWR primary water environment by coupling acoustic emission. J. Mater. Res. Technol. **20**, 3807–3817 (2022)

Zhang, Z., De Risi, R., Sextos, A.: Multi-hazard fragility assessment of monopile offshore wind turbines under earthquake, wind and wave loads. Earthq. Eng. Struct. Dyn. **52**(9), 2658–2681 (2023a)

Zhang, H., et al.: A digital twin of the research vessel Gunnerus for lifecycle services: outlining key technologies. IEEE Robot. Autom. Mag. **30**(3), 6–19 (2023b). https://doi.org/10.1109/MRA.2022.3217745

Zhang, M., Sun, L., Xie, Y.: A monitoring method of hull structural bending and torsional moment. Ocean Eng. **291**, 116344 (2024b)

Zhang, W.-H., Qin, J., Lu, D.-G., Liu, M., Faber, M.H.: Optimizing utilization strategy of real-time SHM information for structural integrity management based on preposterior decision analysis. Ocean Eng. **297**, 117044 (2024a)

Zhao, T., Liu, Z., Cuiwei, D., Sun, M., Li, X.: Effects of Cathodic polarization on corrosion fatigue life of E690 steel in simulated seawater. Int. J. Fatigue. **110**(105–14), 2018 (2018)

Zhao, Z., Chen, N.: Acoustic emission based damage source localization for structural digital twin of wind turbine blades. Ocean Eng. **265**, 112552 (2022)

Zhiqiang, X., Yujiu, S., Mingjing, T.: Crack closure induced by corrosion products and its effect in corrosion fatigue. Int. J. Fatigue. **13**(1), 69–72 (1991)

Zhou, G., Zhang, G., Xue, B.: A maximum-information-minimum-redundancy-based feature fusion framework for ship classification in moderate-resolution SAR image. Sensors. **21**(2), 519 (2021). https://doi.org/10.3390/s21020519. MDPI

Zhou, W., Duan, M., Chen, R., Wang, S., Li, H.: Test study on vortex-induced vibration of Deep-Sea riser under bidirectional shear flow. J. Mar. Sci. Eng. **10**(11), 1689 (2022). https://doi.org/10.3390/jmse10111689

Zhou, P., Yin, P.T.: An opportunistic condition-based maintenance strategy for offshore wind farm based on predictive analytics. Renew. Sust. Energ. Rev. **109**, 1–9 (2019). https://doi.org/10.1016/j.rser.2019.03.049

Zhu, D., Ding, Z., Huang, X., Li, X.: Probabilistic modelling for long-term fatigue reliability of wind turbines based on Markov model and subset simulation. Int. J. Fatigue. **173**(2023), 107685 (2023)

Zhu, F., Yeter, B., Brennan, F., Collu, M.: Time-domain fatigue reliability analysis for floating offshore wind turbine substructures using coupled nonlinear aero-hydro-servo-elastic simulations. Eng. Struct. **318**, 118759 (2024). https://doi.org/10.1016/j.engstruct.2024.118759

Zima, B., Roch, E., Moll, J.: Non-destructive corrosion degradation assessment based on asymmetry of guided wave propagation field. Ultrasonics. **138**, 107243 (2024)

Open Access This chapter is licensed under the terms of the Creative Commons Attribution-NonCommercial-NoDerivatives 4.0 International License (http://creativecommons.org/licenses/by-nc-nd/4.0/), which permits any noncommercial use, sharing, distribution and reproduction in any medium or format, as long as you give appropriate credit to the original author(s) and the source, provide a link to the Creative Commons license and indicate if you modified the licensed material. You do not have permission under this license to share adapted material derived from this chapter or parts of it.

The images or other third party material in this chapter are included in the chapter's Creative Commons license, unless indicated otherwise in a credit line to the material. If material is not included in the chapter's Creative Commons license and your intended use is not permitted by statutory regulation or exceeds the permitted use, you will need to obtain permission directly from the copyright holder.

Committee V.8: Uncertainty Modelling in Waves and Wave Responses

Carlos Guedes Soares[1(✉)], Elzbieta Bitner-Gregersen[2], Apostolos Papanikolaou[3], Josko Parunov[4], Wei Qiu[5], Tomoki Takami[6], Solomon Yim[7], Xueliang Wang[8], Takuji Waseda[9], and Huidong Zhang[10]

[1] Lisboa, Portugal
c.guedes.soares@centec.tecnico.ulisboa.pt
[2] DNV, Høvik, Norway
[3] National Technical University of Athens (NTUA), Athens, Greece
[4] University of Zagreb, Zagreb, Croatia
[5] Memorial University of Newfoundland, St. John's, Canada
[6] Kobe, Japan
[7] Oregon State University, Corvallis, Oregon, USA
[8] Wuxi, China
[9] Tokyo, Japan
[10] Shanghai, China

Committee Mandate. The Committee is concerned with the quantification of uncertainties in measuring and characterizing the wave environment and measuring and predicting the loads and responses of ships and offshore structures to wave excitation. It should address the determination of the accuracy and uncertainty of existing calculation methods for comparison with experimental results and their associated uncertainty promoting benchmark and comparative studies. The scope of work of the Committee should address aspects such as: physical and experimental methods of modelling and representing the environment and the response of ships and offshore structures to the environmental excitations; sources of sensitivity of results to uncertainties in physical tests, including modelling of the environment, response of floating structures, input data, assumptions and the procedures themselves and should include benchmark studies on wave modelling and on measurement and prediction of linear and nonlinear loads and responses, including whipping and springing.

Keywords: Uncertainty modelling · wave environment · wave-induced motions · wave-induced loads · benchmark study · experimental methods · CFD

1 Introduction

The ISSC-ITTC Joint Committee was created by both parent organisations in the sequence of various cooperative activities. The ISSC committees I.1. Environment and II.1 Loads have traditionally kept in contact with the corresponding committees from ITTC. In addition to these interactions among the Committees, the first joint workshop between ISSC and ITTC was held at ISSC2012 in Rostock, promoted by ISSC Committees I.1 Environment and II.1 Loads. The second workshop was held at ITTC2014 in Copenhagen and was promoted by the ITTC Committee on Seakeeping. Both workshops addressed uncertainty modelling, a topic that interests both organisations, although with a different emphasis. There was the intention that those workshops would help bring light to the treatment differences, eventually leading to a common approach or at least a clear understanding of the differences.

After these initiatives, it became clear that those isolated meetings would not be able to achieve any standardisation of the approaches adopted in ISSC and in ITTC, so the idea of creating a joint committee took force and was adopted by the Standing Committee of ISSC 2015, which initiated conversations with ITTC in that direction. An agreement was reached in 2014, and the ITTC members were designated in March 2015. In April 2015, there was an agreement about the first Chair of the Committee so that the Joint Committee would have conditions to start working. The first Committee was to be chaired by Carlos Guedes Soares, representing ISSC, with the plan that an ITTC representative would chair the second term of this committee. Given the proximity of the ISSC Congress, the ISSC membership of the Committee was only confirmed in September 2015 at the ISSC Congress.

It was agreed that this Committee would have a mandate of five years, thus out of phase with the timings of the Congresses of both organisations, so that it would have a role that would go beyond the regular committees of both organisations.

The ISSC-ITTC Joint Committee consists of members from both organisations. Its aim is to undertake and promote studies that enable the determination of the accuracy and uncertainty of existing calculation methods from comparisons with experimental results (and their associated uncertainty).

The Joint Committee should not be concerned with aspects of work routinely undertaken by the standard committees from both organisations. Still, it should focus on aspects, like benchmark and comparative studies, that may require more time to complete. The outcome of the work of the joint committee should contribute to the understanding of the state of the art and, in this way, contribute to the work of the traditional committees in each of the parent organisations.

In the initial documents justifying the creation of this Joint Committee, the following was indicated as the main aim of the Committee:

"Quantification of uncertainties in defining and measuring the environment and measuring and predicting the loads and responses of ships and offshore structures".

In the initial documents justifying the creation of this Joint Committee, the following was indicated as the scope of work of the Committee:

Physical and experimental methods of modelling and representing the environment and the response of ships and offshore structures to environmental excitations
Sources of sensitivity of results to uncertainties in physical tests, including modelling of the environment, response of floating structures input data, assumptions and procedures.
Benchmark studies relating to measurement and prediction of linear and nonlinear loads and responses

Benchmark studies relating to the prediction of whipping and springing effects on torsional loads

Further discussions about the work plan have converged to a two-step approach. Initially, it should aim at dealing with relatively simple problems of uncertainty quantification, which would be used as a benchmark study in which one would aim at making how both organisations address these problems more uniform, aiming eventually to some common principles and terminology in addressing these questions. It might be appropriate to define two issues, one related to wave models and the other related to ship responses. In the second step, more up-to-date problems should be addressed.

The interaction of this Joint Committee with the two parent organisations has not been pre-defined. Thus, it was established in the early work of the Joint Committee and agreed with the parent organisations. Initial ideas agreed in the first formal committee meeting were the following:

Annual progress reports should be delivered to both parent organisations. The timing of this report will differ between the organisations, so the reporting may need to be adapted.

The Joint Committee should aim to organise workshops in conjunction with ISSC and ITTC Congresses so that members of both organisations can more easily participate and be involved in the discussions about the Joint Committee's various partial results.

The Joint Committee should aim to have yearly coordination meetings in various countries.

As the summer break was approaching and it would not be feasible to organise the first meeting of the Committee before summer, informal ad-hoc meetings were held in association with two conferences, OMAE2016 in Pusan and MARTECH2016 in Lisbon, where several committee members were present.

The first ISSC-ITTC Joint Committee Meeting was held in Lisbon on 14/15 November 2016, and it was attended by 14 committee members and about 10 observers. At the meeting, aspects related to the functioning of the Joint Committee were discussed and agreed upon, plans were made for dissemination workshops, state-of-the-art or review papers were made, and benchmark studies were conducted.

The first open dissemination initiative was the organisation of a workshop at ITTC2017 in Wuxi, China, and the second Workshop took place at ISSC2018 in Egmond aan Zee, Netherlands. Very little work was done during the pandemic as the conditions were not very good. However, towards the end of the period, a third workshop was organised virtually in 2021 in Lisbon. This was a successful initiative attended by about 180 persons and included presentations of Committee joint works, presentations made by some committee members with other co-workers and presentations made by authors outside the committee. This represented a good way of disseminating the work of the Committee and obtaining an external view of the topic. Finally, many of the presentations were published as journal papers, and many made a special issue of the Marine Structures journal, while others were published elsewhere.

At ISSC 2022, the present Committee was created with the objective that the ISSC members of the ISSC-ITTC Joint Committee would have an opportunity to report the ongoing work to the Congresses more formally than through workshops, as had been the practice until then. A short time after that, when the issue of terminating the first mandate of the Joint Committee was being raised, and a second mandate now chaired by an ITTC member would be starting, ITTC made the unilateral decision to terminate their participation in the Joint Committee. This decision affected the termination of some ongoing activities as the ITTC members were instructed not to contribute to the work any longer. After several months of negotiations and no progress in the work, several former ITTC members joined the present ISSC Committee to finish the ongoing work, including its reporting.

Thus, the present report provides an overview of the work done by the Joint Committee since its start in 2016/2017. As this work is described in published papers, this report will draw heavily on those papers to summarise what has been achieved.

As a direct result of the Committee work, the following was achieved:

3 journal papers addressing state-of-the-art reviews on uncertainties in waves and wave-induced responses. (Bitner-Gregersen et al. 2022; Parunov et al. 2022; Hirdaris et al. 2023)
3 journal papers addressing benchmark studies (Parunov et al. 2020, 2022, 2024)
Motivated about 30 journal papers with specific uncertainty modelling studies.

2 Framework for Uncertainty Quantification

The framework for uncertainty quantification has been addressed in the initial sections of several papers published by the Joint Committee. This section closely follows the text provided by Parunov et al. (2022), which is in line with the description by Bitner-Gregersen et al. (2022).

Uncertainty is the lack of certainty, a state of limited knowledge where it is impossible to exactly describe the existing state, the future outcome, or more than one possible outcome. This is a concept that, in the early days of probability theory, was applied to random experiments with an enumerable number of outcomes and to the more generic formulations of frequentist probability applicable to experiments with random output. While these concepts apply to the analysis of experimental data, the concept has been generalised to include subjective probabilities commonly used in risk assessment studies (Aven and Reniers, 2013), where random experiments cannot be conducted. The different types of uncertainty can be modelled by random variables and operated using the framework of probability theory (Benjamin and Cornell, 1970; Ang and Tang, 2017). Sometimes, expressions like fuzzy variables and imprecise probabilities appear in the literature, but all those variables can and should be represented as uncertainty and modelled by a random variable to ensure the consistency of treatment using probability theory.

A breakthrough was initiated in the 1970s with the formulation of structural reliability theory, mainly in civil engineering (Cornell, 1969; Hasofer and Lind, 1974). This addresses the combination of the uncertainty in load variables with the uncertainty in strength variables to assess the considered structure's reliability (Ditlevsen and Madsen, 1996; Melchers, 1999; Der Kiureghian, 2022). The application of these concepts to ships was pioneered by Mansour (1972), and Faulkner and Sadden (1979), based on the ship section modulus and the ultimate strength format for hull girder reliability was formulated by Guedes Soares et al. (1996). Overviews of the developments of reliability theory applied to marine structures were presented by DNV (1992) and Guedes

Soares et al. (2011). These developments prompted a drive to identify and quantify the uncertainty associated with various assessment methods and experiments, resulting in a deeper understanding of the nature of uncertainty. At ISSC, many committees have started including the specification of uncertainties related to various fields of knowledge in their mandates.

Uncertainties employed in structural reliability theory are represented by random variables, which probabilistic models fully describe. These models can be parametric or non-parametric and may include objective information from experimental results and subjective information resulting from interpreting the nature of the process. Subjective information that influences the analyst is given by the physical understanding of the problem and the analyst's judgement and experience. In contrast, statistical information can be visualised in histograms and approximated by probability distribution models using the fitting methods described in many statistical textbooks.

Uncertainties are generally divided into two broad groups, namely (a) aleatory (natural, inherent or physical) and (b) epistemic (knowledge-based) or model uncertainty. The former is also known as inherent uncertainty or variability, representing the natural variability of the physical process. A typical example of aleatory uncertainty is the variability of the wave elevation in time, consequently causing variability of wave-induced loads. Aleatory uncertainties are always present when dealing with wave-induced loads. The latter type of uncertainty (epistemic) represents an uncertainty that can be reduced by collecting more information about a considered quantity and improving the measuring methods.

Epistemic uncertainty has been traditionally classified into (a) measurement, (b) statistical and (c) model uncertainty. Measurement uncertainty results from using instruments to measure a quantity, sometimes one that is recorded indirectly and related to the wanted variable. The measuring equipment operates similarly to a mathematical model that transforms input into output. This type of uncertainty is unavoidable in experimental measurements of wave-induced loads in a towing tank or seakeeping basin. Statistical uncertainty, often also referred to as estimation uncertainty, is due to limited information, such as a limited number of quantity observations and an estimation technique applied to evaluate the distribution parameters. Approximating the shape of the wave spectrum by an analytical expression is often considered to involve statistical uncertainty. Spectral ordinates are determined based on a limited number of observations, and it is usually further assumed that the variability around a mean wave spectrum is constant along the whole frequency range. Model uncertainty is due to imperfections, simplifications and idealisations made in the mathematical model formulations of physical processes. The statistical uncertainty referred to earlier can also be considered model uncertainty as it is reflected in the parameters of a chosen probability distribution model used to represent uncertainties. Therefore, epistemic uncertainties can also be generically designated as model uncertainties in this new interpretation, which differs from natural processes' variability.

The accuracy of measurements or predictions characterises how much a measured or predicted quantity agrees with the true value. To characterise the latter, it is necessary to indicate a systematic error (also known as bias) and a precision (or random) error. The systematic error, or bias, of an estimator for a quantity refers to a systematic deviation from the quantity's actual value. The precision of the estimator considered refers to random variations and is usually represented by the standard deviation. Choosing a distribution type based on a limited amount of data is usually very difficult, so normal distribution is commonly adopted to describe model uncertainty. Uncertainty in the computation of transfer functions of wave-induced loads is a typical example of model uncertainty.

Bitner-Gregersen and Hagen (1990) discussed the uncertainties related to wave models, while Olufsen and Bea (1990) addressed the uncertainty associated with loads. The uncertainties in waves and loads are reflected in the probabilistic or stochastic models describing them, which are then used in ship design. Guedes Soares (1990) presented the stochastic models of loads used for the ship design specifications. In the structural reliability formulation adopted for ship design (Guedes Soares et al. 1996), the models of the strength variables use the same concepts of uncertainty (Guedes Soares 1988).

ISSC is interested in the design of marine structures, and thus, critical uncertainty is associated with the predictions of future events to which the structures will be subjected. These predictions can only be made based on mathematical and probabilistic models; thus, knowledge of model uncertainty is crucial for short-term predictions (Guedes Soares, 1991) and long-term ones (Guedes Soares and Moan, 1991). The structural reliability formulations used by ISSC are in line with the specifications of the ISO Standards 22111 and 13824 on the basis for the design of structures and 2394 on general principles of reliability for structures. As these standards are updated every 5 years, the later versions do not explicitly cover the definitions of uncertainty, which can be found in the structural reliability textbooks referred to earlier.

ITTC is concerned with the precision of laboratory experiments that reproduce important full-scale situations related to the motions and loads of these structures. Thus, uncertainty in measurements is essential for evaluating the accuracy of experiments or measurements, including seakeeping experiments (Kim and Hermansky 2014). The primary purpose of uncertainty analysis for ITTC is to quantify errors and to obtain an objective index that can be used to evaluate the confidence level of the measured data. The ISO Guide 98-1 on the expression of uncertainty in measurements (GUM) establishes the main principles to quantify the uncertainty in measurements and, instead of using the designations of aleatory and epistemic uncertainty, classify them in A and B.

ITTC follows the GUM and recommends grouping the uncertainty components into two categories, namely, type A and type B, based on the method by which they are evaluated. Type A uncertainty is based on repeated measurements and is characterised by the randomness of the process that is expressed as an experimental variance. The type B components are estimated by means other than repeated observations. Typical type B sources of uncertainty include the following: the geometry of used models, measuring devices, the calibration and installation of equipment, data acquisition systems, and data processing. Type B uncertainties usually are constant for the duration of an experiment.

Although the definitions of the type of uncertainties do not use precisely the same type of concepts, it is fair to say that, in general, type A uncertainties will be very similar to inherent or aleatory uncertainty, while type B is, in some sense, model or epistemic uncertainties.

These two emphases on these organisations' interests and the different nomenclature they developed raised the necessity of better understanding their differences and of an effort to standardise their approaches to the maritime sector. This motivated the creation of the Joint Committee and the work developed.

3 Uncertainty Modelling in Experiments

3.1 General

Predicting wave-induced loads on ships involves model tests, full-scale trials and numerical simulations. In general, uncertainties in the predictions with ship-model experiments arise from various sources, including the modelling of environmental conditions (wave theory and associated wave generation), ship model (rigid or elasticity model), measurements and instrumentation, and the method used to extrapolate the model test results to the full scale. Sources of uncertainties in sea trial measurements include environments (waves, current, and wind in uncontrolled conditions), operation conditions, measurements, and instrumentation. This section will focus on ship model tests in a tank or wave basin.

3.1.1 Wave Generation in Model Basins

Linear and second-order Stokes wave theories are typically used to generate deep water and finite-depth waves, and various types of wavemakers are considered in towing tanks and wave basins. Flap-type wavemakers, rotating about a horizontal hinge near the bottom of the tank, are suitable for deep water and relatively short waves; piston-type wavemakers, moving horizontally back and forth, are ideal for shallow water and long waves; paddle wavemakers, combining flap and piston motions, generate a broad range of wave conditions; and multi-segment wavemakers, consisting of multiple paddles that can operate independently, generate directional and short-crested waves. Various wave spectra are incorporated in wavemaker controls to generate irregular waves.

Control algorithms are used to adjust the motion of the wavemaker to produce the desired wave profile and spectrum more accurately and correct unwanted waves. For example, spurious wave cancellation corrects for unwanted free waves generated due to the nonlinear interactions in the wavemaker mechanism since spurious harmonics generation can occur in wave basins when strongly nonlinear waves are generated with linear wavemaker theory (Pierella et al., 2021).

The wave theory and control algorithms for wavemakers contribute to uncertainties in wave generation in a towing tank or a wave basin with given dimensions. Another critical factor in wave generation is wave absorption and reflection control due to the limited dimensions of a testing facility. Two types of wave absorption are typically used: passive absorption, such as beaches, wave absorbers and damping materials at the end of the tank to minimise reflections, and active absorption, which is a wave maker adjusting its motions based on the feedback from wave sensors to absorb reflected waves dynamically. Naito (2006) summarised the theory of wave generation and wave absorption in wave basins.

In summary, wave generation uncertainties depend on wave generation systems' capabilities and limitations. The literature on quantifying uncertainties in wave generation due to wave modelling and wavemaker mechanisms and their control is limited.

3.1.2 Rigid and Hydroelastic Ship Models

Rigid ship models are widely used to measure wave-induced motions, while hydroelastic models are employed to measure ship structural responses in waves. Hydroelastic models are designed according to the purposes of tests. For example, a hydroelastic model with a backbone beam of a large containership can be used to test the springing response, i.e., resonant wave-induced hydroelastic ship responses, and whipping, i.e., transient hydroelastic ship responses due to impulsive loading, such as slamming. Marón and Kapsenberg (2014) described the development of a segmented scaled model of a containership for springing and whipping experiments. The hull was divided into six rigid segments connected with a flexible beam. Due to the open structure of the containership and its shear centre well below the keel line of the vessel, the beam was built into the model as low as possible. Accelerometers and rotation rate gyroscopes were installed on each segment, along with relative wave height meters and pressure gauges in the bow area. The beam was also instrumented with strain gauges to measure the internal loads. Experiments have been carried out in regular waves and long-crested irregular waves.

Grammatikopoulos (2023) presented a comprehensive review of recent flexible ship models used in experiments for various responses, including transient, linear and non-linear ones. Tang et al. (2022) tested a 1:50 hydroelastic model with seven segments connected by a flexible backbone beam for a modern ultra-large container ship. Vertical bending moments were measured at different locations along the ship's length. The study revealed significant nonlinear effects on the vertical bending moments, and whipping was observed to contribute significantly to the total bending moments in extreme wave conditions.

Advanced manufacturing techniques like additive manufacturing are being explored to make ship models with realistic structural configurations. For example, the US Naval Surface Warfare Center is conducting research to evaluate the feasibility of using large-scale additive manufacturing for naval model testing. Grammatikopoulos et al. (2020) investigated the design and production of elastic ship models with realistic, thin-walled structural configurations. In their work, 3D printing was suggested as a plausible manufacturing solution for the scaled models. Houtani et al. (2018) employed a hydroelastic model using urethane foam, eliminating the need for model segmentation and introducing a backbone.

Uncertainties in measurements of hydroelastic model tests are, in general, larger than those of rigid models. Uncertainties are due to additional resources in comparison to tests of rigid ship models, such as the calibration of the backbone beam to determine the relationship between the external load and the measured strain. Si et al. (2022) have recently investigated the uncertainties in measured linear bending moments of a hydroelastic model of a 20,000 TEU containership. They found that the uncertainty in model calibration was the largest, followed by the moment of inertia of the ship model.

3.1.3 Measurements and Instrumentations

A summary of instrumentation commonly used in wave-induced ship motions and loads experiments is given as follows.

1. Motion measurements

   - Inertial measurement units, measuring linear accelerations and angular rates;
   - Optical motion capture systems, tracking reflective markers on the ship model;
   - Gyroscopes, measuring angular velocities and orientations; and
   - Accelerometers, measuring linear accelerations in different directions.

2. Wave measurements

   - Resistance-type wave probes to measure water surface elevation; and
   - Capacitance-type wave probes as an alternative for measuring wave heights

3. Load cells

   - Multi-axis load cells, measuring forces and moments in multiple directions;

4. Strain gauges

   - Electrical resistance strain gauges, measuring local strains and stresses of hydroelastic models; and
   - Fibre Bragg Grating (FBG) strain gauges, measuring local strains and stresses of hydroelastic models by connecting a large series of strain gauges on a single optical fibre.

5. Pressure sensors:

   - Piezoelectric or piezoresistive pressure transducers, measuring hydrodynamic pressures on the hull, for example, local pressure due to slamming; and
   - FBG sensors measure hydrodynamic pressure distributions on the hull by connecting the sensors with a single optical fibre.

6. Data acquisition systems:

   - Data acquisition hardware and associated software to simultaneously record measurements from transducers, such as those mentioned.

These instruments are typically integrated into a measurement system for ship model tests in wave tanks or towing tanks. Precisions of these instrumentation, along with their installation (related to human factors), contribute to uncertainties in measurements.

3.1.4 Extrapolation and Scale Effect

Model test results of rigid ship models can be extrapolated to full scale according to the Froude law since the inertia forces dominate for moderate nonlinear wave-induced motions and loads. The scale effect is minimal, and the uncertainty in extrapolation is relatively small. For extreme sea conditions, highly nonlinear phenomena, such as slamming and green water on deck, will occur, and non-linearity in wave-induced responses will be greater. Due to the multi-phase flow, uncertainties in extrapolation increase with the scale effect.

For hydroelastic model tests, the scale effect is more significant due to the scaling of structural properties and calibration of the backbone beam (stiffness and damping). Therefore, uncertainties in extrapolation are considered to be larger than those for rigid models. Li et al. (2022) presented an experimental investigation on the stern slamming and global responses of a segmented hydroelastic model of a cruise ship. However, uncertainties were not quantified. The assessment of uncertainties in extrapolation remains a challenging topic.

3.2 Quantification of Uncertainties in Model Tests

Guidance on uncertainty assessment of uncertainties in measurements has been provided by several organisations, such as the International Organization for Standardization (ISO), the American Society of Mechanical Engineers (ASME), and the Towing Tank Conference (ITTC). Their guidance is essentially the same.

Typical model tests for wave-induced motions and loads on ships and offshore floating structures follow ITTC procedures and guidelines for model tests and CFD simulations. The ITTC-recommended procedures for model tests cover aspects related to the manufacture of ship hull, propeller and offshore structure models, instrumentation and its calibration, and testing and extrapolation methods for resistance, propulsion, manoeuvring, loads and responses.

In the category of loads and responses, procedures and guidelines include those for environmental modelling (for example, 7.5-02-07-01.1 Laboratory modelling of multidirectional irregular wave spectra; 7.5-02-07-01.2, Laboratory modelling of waves: regular, irregular and extreme events), seakeeping tests, and tests of offshore floating structures.

According to 7.5-02-01-01 Guide to the Expression of Uncertainty in Experimental Hydrodynamics, evaluation and expression of uncertainty in measurements for naval architectural experiments, offshore technology testing and experimental hydrodynamics are based on the ISO Guide to the Expression of Uncertainty in Measurements. JCGM 100: 2008, or ISO-GUM (2008), based on GUM 1995.

In ISO-GUM (2008), measurement uncertainty is categorised into Type A and Type B. The Type A uncertainties are those evaluated by applying statistical methods to the results of repeated measurements. The Type B uncertainties are those evaluated by means other than the statistical methods. The associated estimated variance or the standard uncertainty is evaluated by scientific judgement according to the available information, which may include the previous measurement data, experience or knowledge of instruments or materials, information provided by the manufacturers, data obtained in calibration or from other certificates, and uncertainties assigned to reference data from handbooks or manuals. The calculation of Type B uncertainties depends not only on the experience but also on the analysis methodology.

The ISO-GUM method has been used in various model tests, for example, ship inclining experiments (Woodward et al., 2016), high-speed planning hull tests in calm water (Nikolov and Judge, 2017) and resistance tests (Delen and Bal, 2015). The following literature review will focus on the model tests in waves.

Based on the ISO-GUM methodology, uncertainties due to various sources in wave-induced ship model tests can be evaluated. For example, Kim and Hermansky (2014) quantified uncertainties in seakeeping experiments.

The ITTC Ocean Engineering Committee (Qiu et al., 2014) presented the parameters that may cause uncertainties in ocean engineering model tests, full-scale tests and numerical simulations in terms of physical properties of the fluid, initial conditions, model definition, environment, scaling, instrumentation and human factors, which are all applicable to ship model and full-scale tests. The ISO-GUM method can be used to calculate uncertainty in motions, air gap, and mooring line tension. From the results obtained for Type A and Type B uncertainty analysis, it is concluded that the Type B uncertainty may play a more important role in model test results. The Committee also concluded that quantifying uncertainties may be challenging in model tests and numerical simulations of ocean and offshore structures considering deep water mooring lines, risers and dynamic positioning systems. It is particularly challenging to extrapolate model test results to full scale and utilise complex numerical models, especially if the effects of hydrodynamic nonlinearities are significant.

Recently, a combined experimental and numerical method was developed by Qiu et al. (2019) to quantify the uncertainties in two-body interaction tests. The comprehensive study considered uncertainty sources due to model geometry, model mass properties, models' locations, the mooring system's set-up, wave generation, instrumentation, scaling and human factors related to the set-up of the mooring system, model position and orientation and sensor installation. Considerable efforts were made to quantify Type B uncertainties from pre-tests and calibrations, such as those due to wave probes, load cells and the Qualisys system, as well as from wave calibration and swing tests. Type B instrumentation uncertainties were also quantified from the manufacturer's specifications for the power supply and the DAQ system. A numerical method was used to determine the uncertainties due to model geometries, mass properties, mooring spring stiffness and layout. It is also observed that the Type B uncertainties are generally larger than the Type A ones.

Several studies have been conducted to quantify the uncertainties in added wave resistance measurements. Zhang et al. (2025) investigated the uncertainties in added resistance and pressure distribution of a VLCC2 model in regular short waves. They found that the relative uncertainty in added resistance increases significantly with increased wave frequency. Type A uncertainty increases notably with wave frequency, reflecting greater dispersion in the measured results. Precision limitations in measuring short waves are a major contributor to overall uncertainty, while other sources are negligible. Using higher-precision wave probes and ensuring accurate calibration could effectively reduce uncertainty and enhance the overall accuracy of the experiments. Type B uncertainty was found to be mainly related to the sampling time, and the increase in the sampling frequency of the wave height gauge could reduce this Type B uncertainty.

Sogihara et al. (2020) conducted repeated tests to measure resistance in calm water and added resistance in short regular waves at three NMRI facilities using the same ship model. The uncertainty analysis revealed that the sources of uncertainty in the resistance coefficient for calm water differed between the towing tanks and the basin. In the towing tanks, speed was the dominant source of uncertainty, whereas, in the basin, both speed and resistance had comparable effects. This difference stemmed from greater speed variation in the basin than the towing tank. The uncertainty components of the added resistance coefficient in short regular waves also varied between the towing tanks and the basin. In the towing tanks, wave height was the dominant uncertainty factor, whereas uncertainty in calm water resistance had the greatest influence in the basin. In their study, the Synchronous Measurement System, designed to ensure high reproducibility, was used to control the carriage and wave generator but could not regulate the residual current profile in the facilities. They found that using this system led to lower uncertainty in the added resistance coefficient in short regular waves, although it did not improve uncertainty in calm water resistance measurements.

Robertson et al. (2020) studied the uncertainty in hydrodynamic testing of a semi-submersible wind turbine by considering the numerical propagation of systematic uncertainty. They identified and quantified sources of uncertainty. They found that the uncertainties in low-frequency responses are more sensitive to system properties, such as the mooring line stiffness and the platform's centre of gravity.

Orphin et al. (2021) investigated the development of a comprehensive WEC-specific UA methodology by conducting a 1:30-scale experiment on a case study oscillating water column (OWC) WEC. Results show that Type B uncertainty tended to be slightly larger than Type A uncertainty.

Abdelwahab and Guedes Soares (2023) conducted comprehensive research on experimental uncertainty analysis for a physical model of the Esso Osaka tanker moored at a terminal in a Portuguese port. They quantified uncertainties related to model parameters, metacentric height, roll decay tests, significant wave heights at various locations within the port, the dominant wave direction, mooring line pretensions, a nonlinear spring system, and the natural period of the ship, including mooring systems, as well as wave elevations, ship motions, and mooring loads. Their findings concluded that the physical tests satisfactorily replicated wave measurements, ship responses, and mooring loads under irregular sea conditions inside the port, with quantified uncertainties. Similar to previous studies, Type B uncertainties in wave measurements, model geometry, and mass distribution were found to be more significant than Type A uncertainties. Additionally, it was noted that nonlinear effects and extreme conditions could influence uncertainties in shallow-water wave measurements. Nonlinear effects in the mooring system, external forces, and ship response significantly contributed to uncertainties in mooring load characteristics. Abdelwahab et al. (2024) further investigated the influence of sea severity and mooring line pretension configuration on the operability of a moored vessel at a modified berthing site inside a port.

The final uncertainty shall combine Type A and Type B results according to the ISO recommendations. As discussed above, Type B uncertainties are typically larger than Type A uncertainties in most measurements. The combined values of Type A and Type B results may be unrealistic. More studies in this area are recommended.

While most experimental work is conducted in wave tanks and basins with small models, the technology is evolving to have much larger self-propelled models tested in actual sea conditions, even using segmented models (Jiao et al. 2021). Uncertainty quantification studies have not yet been made for these situations.

4 Uncertainties in Wave Modelling

4.1 General

Wave modelling includes long- and short-term descriptions. The long-term wave data and models describe the variability of wave parameters in a long-time history for the design durations of the order of the lifetimes of marine structures. In contrast, short-term wave data and models address water surface elevation variability and associated individual wave parameters in periods of 20- to 30-min. A design wave is determined based on long-term wave characteristics. The duration of wave records is 20- to 30 min, which allows the assumption of stationarity of sea surface elevation. In design, this assumption is commonly extended to 3- to 6-h periods called sea states (see e.g. DNV RP C-205, 2019a). The stationarity assumption should be considered carefully when modelling extreme waves such as rogue waves (Bitner-Gregersen et al. 2020b) and assessing marine structure performance at sea, e.g. ships sailing with speed over 15 knots (Nielsen and Ikonomakis 2021).

Field measurements and laboratory experiments are used to validate wave models. They also provide an important input to the wave models in terms of wave characteristics. Thus, wave description will be affected by uncertainties associated with wave data and wave models. Knowledge of these uncertainties is crucial for assessing the performance of ships, offshore platforms, and renewable energy structures at sea and their corresponding design. However, the relevance of different uncertainties for design and marine operations will be case-dependent.

4.2 Wave Data

4.2.1 Data Sources and Types

Available data sources have evolved, from visual observations to buoys, radars, lasers and other instrumental measurements, numerical wave models and remotely sensed data. Wave data used in the design and specification of operational criteria include field measurements and laboratory and numerical data. Uncertainties associated with laboratory data are addressed in Sect. 3. Herein, only uncertainties related to field measurements and numerical data are discussed.

Wave data are affected by the inaccuracy of a recording instrument or a simulating model. If wave data are post-processed before engineering applications, the uncertainty of a post-processing procedure will impact their accuracy. A sufficient length of wave records is essential for reducing statistical uncertainty (sampling variability).

Traditionally, temporal wave data were collected and used in design work. Recently, attention has been given to synchronised space-time data, which modern instruments and numerical wave models can collect. The latter data have been shown to significantly affect return values of wave parameters compared to temporal data.

Uncertainties associated with different data sources have been summarised in the various reports of ISSC Committee I.1 Environment and the corresponding ITTC Committee reports, as well as in papers related to those reports and presented in the 1st and the 2nd ISSC-ITTC Workshops on uncertainty modelling (Bitner-Gregersen et al.

2014a, 2014b, 2016). They have also been a topic of the investigations of the ISSC-ITTC Committee, which were completed in 2023.

The present section summarises the ISSC-ITTC Committee investigations documented partly in Bitner-Gregersen et al. (2022) and supplements them with new recent findings.

4.2.2 Visual Observations

Since 1854, Voluntary Observing Ship (VOS) data (wind speed and wave height) have been collected from ships in regular service worldwide. Wave height, period, and direction observations started in 1949 (see BMT, 1986). The VOS data are collected following guidance notes of the World Meteorological Organization (WMO, 2001, 2003). Until the end of the 1990s, they were the only global wave data available. The VOS data recorded from merchant ships along standard shipping routes account implicitly for the worldwide practices of shipmasters operating along these routes. Accordingly, they suggest avoiding geographical regions with the most severe weather conditions.

The VOS data, calibrated against buoy measurements, gathered from 1949 to 1986 were summarised in the Global Wave Statistics (GWS) Atlas by British Maritime Technology (BMT, 1986), which includes 104 ocean zones (see Fig. 1). The GWS Atlas was updated in the 1990s and is available online on the BMT website. However, the last decades are missing, which may impact extreme wave statistics, as discussed in Bitner-Gregersen et al. (2014b).

The North Atlantic wave climate was regarded as the most severe, and GWS data were used in the design of ships until 2023. The GWS average wave climate of four ocean zones (8, 9, 15, 16 in Fig. 1) in the form of a scatter diagram of significant wave height and zero-crossing wave period, with some correction introduced by Bitner-Gregersen et al. (1995a) due to inaccuracy of zero-crossing wave period, was recommended by the International Association of Classification Societies (IACS 2001) for ship design.

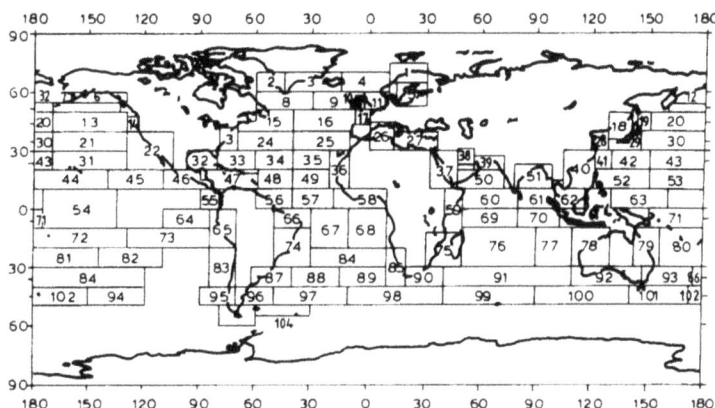

Fig. 1. Global Wave Statistics zone designation, BMT (1986).

Visual observations collected by trained ship officers are affected by measurement uncertainty (see e.g., Gulev and Grigorieva, 2006). The accuracy of the observations has been questioned in the literature for several decades, particularly regarding wave period (e.g., Guedes Soares, 1986a, b; Bitner-Gregersen et al. 1995b; Bitner-Gregersen et al. 2013; Olsen et al. 2006; Vettor and Guedes Soares, 2017). This situation remains and must be considered when using VOS and GWS data.

VOS data has a higher frequency where the ship traffic is more intense (Vettor and Guedes Soares, 2015). This will bias the accuracy of data and statistics derived in those areas.

Another noteworthy point is that the areas used in GWS (Fig. 1) seem to have been decided based on geometric considerations. Recent studies using Regional Frequency Analysis allowed the identification of ocean areas with similar statistical properties of the long-term wave data sets, and they were not entirely in agreement with the areas in GWS (Lucas et al. 2020; Vanem, 2021).

The significant discrepancies between predictions of different global wave databases (see, e.g. Bitner-Gregersen and Guedes Soares, 2007; Bitner-Gregersen et al. 2016; Campos and Guedes Soares 2016) led to delays in updating the IACS scatter diagram for several years. The recent improvement in the accuracy of these databases started investigations on validating the IACS scatter diagram for ship design and its possible revision by Classification Societies (de Hauteclocque et al. 2020). As part of these efforts more accurate, high-resolution wave databases (e.g. ERA5 (Bidlot et al. 2019), IOWAGA (Ardhuin et al. 2011b), and TodaiWW3-NK (Sasmal et al. 2019), see Table 1), buoy and satellite data, together with routes of different ship types, were analysed using the AIS (Automatic Identification System) ship position data. Note that the AIS data provide, in addition to ship positions, ship operational characteristics, e.g., ship speed and heading, allowing estimates of the effect of weather on ship speed (Eisinger et al. 2016; Vitali et al. 2020; Miratsu et al. 2022, Sasmal et al. 2021a).

These investigations resulted in the proposal of a new scatter diagram by IACS (IACS, 2022). The main changes include a part of a new scatter diagram, modification of the wave spectral model and energy spreading function, and a probability level of design load. The Pierson-Moskowitz wave spectrum is replaced by the JONSWAP spectrum, an energy spreading function $\cos^2$ by $\cos^3$, and the minimum ship zero speed by 5 knots speed. For the evaluation of design wave loads for fatigue assessment, 75% of the design speed is recommended. A return period of 25 years is proposed instead of using the VWBM design, which corresponds to probability levels 10^{-8}. A definition of the extended area of the North Atlantic revised IACS scatter diagram is shown in Fig. 2.

A comparative analysis of the long-term extreme VWBM (Vertical Wave Bending Moment) of an oil tanker and a container ship calculated according to IACS Rec. No 34 rev. 1 (2001) and rev. 2 (2022) showed that the obtained discrepancy of VWBM is associated primarily with wave periods, IACS Rec. 34 Rev.1 gives steeper waves (Mikulić et al., 2023). Depending on the vessel and response type, the authors found a reduction of extreme loads from 10% to 30% when rev. 2 of the IACS scatter diagram was applied. Thompson et al. (2024) investigated a naval destroyer's operational history of 22 (years) on fatigue damage estimation using ERA5 hindcast data as a baseline.

Table 1. Characteristics of databases. After Bitner-Gregersen et al. (2022).

	Product Name & Provider	Wave Model & Physics pacakge	Wind Forcing	Horizontal resolution	Duration	Sampling Step	Access
Reanalysis	**ERA-Interim** (ECMWF)	WAM	Coupled IFS	111 km × 111 km	1979–2018	6 h	Open
	ERA5 (ECMWF)	WAM cycle 6	Coupled IFS	30 km × 30 km	1950–2018	1 h	Open
Global Hindcast	**WaveWatch III** (NOAA)	WaveWatch III	NCEP/CFSRR	50 km × 50 km	1979–2009	3 h	Open
	IOWAGA (IFREMER)	WaveWatch III ST4	NCEP/CFSRR ERA5	50 km × 50 km	1990–2016	3 h	Open
	WW3-ST6v (U. Melborne)	WaveWatch III ST6	ERA5	25 km × 25 km	1979–2019	3 h	Open
Regional Hindcast	**NORA10** (Met Norway)	WAM	Coupled	10 km × 10 km	1955–2019	3 h	Not Open
	HIPOCAS (GKSS/IST)	WAM	REMO	110 km × 110 km	1950–2019	3 h	Open
	WorldWaves (Fugro-Oceanor)	WAM/SWAN	ECMWF-wind	50 km × 50 km	1992–2018	6 h	Not Open
	TodaiWW3-NK (U. Tokyo)	WaveWatch III ST4	NCEP/CFSRR	22 km × 28 km	1994–2018	1 h	Not Open

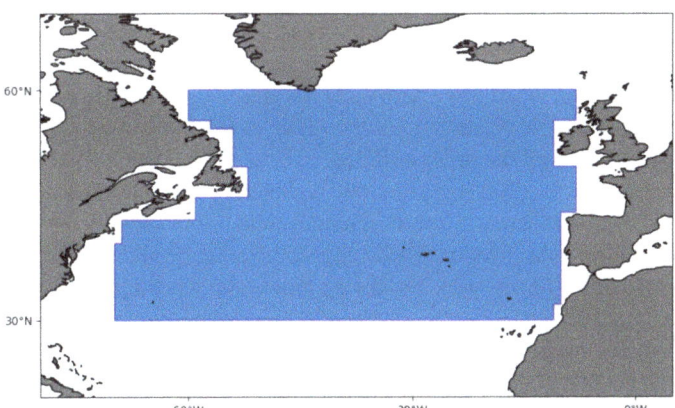

Fig. 2. Definition of the extended area of the North Atlantic IACS scatter diagram Rev. 2.2022. After IACS (2022).

The authors concluded that the IACS and BMT scatter diagrams provide conservative results.

4.2.3 Locally Sensed Instrumental Measurements

To date, wave buoys, wave staff, radars, ADCP, lasers, LASAR and step gauges remain the most widely used devices for in-situ measurements. Lidar and stereo camera systems have grown in use in the last decade. The collection of in-situ wave data is undertaken mainly as a part of on-going national coastal wave monitoring and by oil companies for use in remote areas, at existing oil and gas fields and in support of exploration for new fields. Some data sources in coastal areas are open, e.g. the 39-year (1979–2018) directional wave time series recorded at the fixed location research CNR-ISMAR tower Aqua Alta in the Northern Adriatic Sea (Pomaro et al. 2028). Following the EU (European Union) policy, some met-offices have opened their databases, e.g., the FROST database of the Norwegian Meteorological Institute (MET Norway).

The disadvantage of in-situ wave measurements relates to the lack of global coverage. Nevertheless, a commercial company has substantially developed in-situ wave measurement using lightweight drifting wave buoys (https://weather.sofarocean.com/). Hundreds of buoys now cover the global ocean and are assimilated into an ocean spectral model (Smit et al. 2021).

A recent investigation by Magnusson et al. (2021) shows an inter-comparison between three wave sensors: radar, laser, and a Waverider buoy, based on measurements made at the Ekofisk platform in the North Sea. The authors found that the radar underestimates wave spectral energy in frequency bands higher than 0.125 Hz (8-s waves) up to 4% and that higher underestimation occurs when the sensor is in the lee of the platform. Further, laser measurements show approximately 2% more energy than the Waverider in the most energetic bands, while the Waverider buoy has slightly more energy compared to both altimeters in the lower frequency bands, especially in the higher sea states. The investigation has also shown that the laser (OptechTM laser) is biased 0.1 m higher than the Datawell Waverider in the mean, and the radar, WaveRadar Rex sensor (previously known as a Saab radar), is biased 0.1 m lower than Waverider in terms of significant wave height H_{m01}. The mean wave periods (T_{m01} and T_{m02}) measured from a laser are lower than those from the Waverider (of the order of 0.1 s relative to around 6 s), and radar values are higher (of the order of 0.2 s).

Radhakrishnan et al. (2023) carried out dedicated full-scale measurements using the research vessel Gunnerus in the Breisundet field, to retrieve ocean wave spectra from X-band marine radar images. The image spectra were validated using wave buoy measurements, and the results were promising, however, better agreement was obtained for swell than wind sea.

Apart from the radar, LASAR and stereo camera systems (see e.g., Fedele et al., 2011, 2013; Benetazzo et al. 2015, 2017; Vieira et al. 2020, 2025), in-situ measurements recorded in a single point location giving temporal information about sea surface elevation. They are commonly restricted to 20 or 30 min, allowing for assuming sea state stationarity. Therefore, sampling variability represents an essential source of uncertainty in in-situ temporal data. For the H_s and T_z can be from 3% (low sea states) to 10% (higher sea states), lower for T_z than for H_s (Bitner-Gregersen and Hagen (1990), Bitner-Gregersen and Magnusson, 2014).

Attention to spatial measurements was initiated by Krogstad et al. (2004), who introduced the Piterbarg theorem (1996) to oceanography, showing that single-point measurements may greatly underestimate (especially in short-crested seas) the actual maximum surface wave displacements. The study has brought awareness of the importance of spatial measurements in the marine industry (see e.g., Hagen et al., 2018; Forristall, 2011, 2015; NORSOK, 2017; Bitner-Gregersen et al., 2020a, b). Bitner-Gregersen et al. (2020c) showed using HOSM (Higher Order Spectral Method) for a sea state with H_s

= 3.4 m, T_p = 8.4 s and directional spreading 35.8° that the maximum temporal surface elevation can be up to 2.5 m lower than the corresponding spatial elevation. Besides, the temporal wave profile of the maximum wave can be largely different from that of the space-time one (Zhang et al., 2022a, 2022b).

To reduce the effect of sampling variability, Bitner-Gregersen et al. (2020b) suggested extending the duration of 20–30 min wave records to 1–6 h, provided that the assumption of stationarity can be satisfied, or to apply a simplified approach proposed by Bitner-Gregersen and Magnusson (2014). Numerical simulations and laboratory experiments are of particular importance for investigations of rogue-prone sea states (Bitner-Gregersen et al., 2020b). Having 20–30-min in-situ temporal or spatial wave records may prove challenging to determine the importance of nonlinearity of surface elevation as the sampling variability may dominate over the nonlinear effects.

During the execution of operations, some onboard ship support decision systems rely on wave data collected by a wave radar system (e.g. Izquierdo and Guedes Soares (2005), or using ships as sailing wave buoys (e.g. Nielsen et al., 2006; Pascoal and Guedes Soares, 2008; Iseki and Ohtsu, 2000; Tannuri et al., 2003; Nielsen and Stredulinsky, 2012; Hinostroza and Guedes Soares, 2016; Pascoal et al. 2017; Mak and Düz, 2019; Nielsen and Dietz 2020, Chen et al., 2020; Scholcz and Mak, 2020; Hinostroza and Guedes Soares, 2020). The latter technologies have been shown to compute reasonably accurate estimates of integrated sea state parameters compared to corresponding estimates from real Waverider buoys, providing essential data for modern Machine Learning methods used recently for forecasting ship responses for operational purposes.

4.2.4 Remote Sensed Measurements

The use of satellite observations in engineering applications is growing. Satellite observations offer global coverage but suffer from temporal intermittency, with return periods between 10 days and about one month, and spatial scarcity with interface ranging between about 100 km and 300 km, depending on the mission. Thus, they are affected by sampling variability, which makes the estimation of long-term distributions and extreme analysis challenging. To reduce this uncertainty, multi-satellite missions have started to be utilized in building metocean satellite databases from scatterometers, altimeters and synthetic aperture radars (SAR).

An example of a multiplatform evaluation is a recent comprehensive investigation of Young and Ribal (2019), covering satellite data collected in the period 1985–2019, to specify global trends due to climate change in the average wind speed and significant wave height.

Altimeters record significant wave height, H_s, within one to a few centimetres error (see e.g., ISSC, 2009, 2012, 2015), but not wave period, while scatterometers collect the average wind speed. Synthetic aperture radars, in addition to providing directional wave spectra, record waves with a length longer than about 100 m (i.e. swell). Therefore, to account for wind sea, the SAR data should be combined with wave spectral model data, as de Valk et al. (2004) suggested. This limitation of SAR measurements is currently being investigated by the European Space Agency (ESA).

4.2.5 Numerical Data

Phase-averaged numerical wave spectral models used for forecasting and hindcasting are under continuous development; see, e.g. ISSC (2009, 2012, 2015, 2018, 2022), Bitner-Gregersen et al. (2022). The great advantage of data simulated by these models is their global coverage. That and the continuous improvement of the accuracy of these data has led to increased use of them in design over the last two decades.

The output from phase-averaged wave models is a grid-point ensemble average frequency-directional wave spectrum from which integrated wave parameters for wind sea and swell are calculated (e.g. significant wave height, H_{m0}, spectral peak period, T_p, zero-crossing wave period, T_{m02}, mean wave period, T_{m01}, mean wave direction and spectral peak direction). These parameters have traditionally been provided by met-offices and used in the specification of design and operational criteria for marine structures. Due to information technology's development, wave frequency-directional spectra have been archived by some met-offices, opening new opportunities for environmental modelling and design work.

The WAM and WAVEWATCH-III models are the most used and well-tested wave spectral models for deep and intermediate water depths. Although both WAM and WAVEWATCH-III are 3rd generation (3G) wave models, they differ in several physical and numerical aspects and may give different predictions (Cavaleri et al. 2007). Furthermore, hindcasts (simulated by operational forecasting systems) and reanalysis (provided by assimilation of measurements in the atmospheric model and re-running wave spectral model) wave fields based on WAVEWATCH-III are available in several different versions accounting for various physical phenomena associated with wave generation by different degrees of modelling advances what affects their accuracy. Users should pay special attention to these differences.

In shallow water, the 3G SWAN or Mike code are commonly used. There has been a continuous impetus to push exclusively non-stationary models such as WAM and WAVEWATCH-III closer to shore (e.g. Tolman, 2009, 2013). A nested approach combining deep/intermediate water wave spectral models with shallow water ones has also been applied (e.g. Rusu et al. 2008; Rusu and Guedes Soares, 2013; Guedes Soares et al. 2014; Sasmal et al. 2019; Webb et al. 2020).

The quality of numerical wave hindcasts/reanalysis for offshore and coastal areas depends to large extent on the quality and accuracy of the upper boundary conditions, i.e., particularly, on the quality of the driving wind fields (e.g. Teixeira et al. 1995; Holthuijsen, et al. 1996; Cavaleri and Bertotti 2006; Ponce de León and Guedes Soares, 2008; Campos et al., 2022). It will also be much affected by model resolution in space and time. For example, a coarse resolution of a wave model may result in a few meters lower extreme significant wave height compared to a high-resolution one, e.g., Bitner-Gregersen et al. (2016).

Traditionally, wind fields are used to drive wave models, which result from the output of atmospheric models. However, the relatively recent availability of scatterometer wind fields has allowed wave models to be forced by those wind fields (Silva et al. 2022). Scatterometer wind fields can also be used to be blended with numerically generated wind fields (Salvacao and Guedes Soares, 2018), producing significant differences in the final wind fields (Salvacao, et al. 2022).

Meteorological offices and research organisations are updating their hindcast/reanalysis databases using higher-resolution wave spectral models. For example, MET Norway uses 4 km resolution in the operational WAM wave model MyWave-Wam4 (MWW4), while the European Center for Medium Weather Forecast (ECMWF) 30 km WAM (ERA5). It should be noted that the accuracy of wave spectral model predictions may vary in extreme ocean conditions, e.g., the prediction of extra-tropical storms is more accurate than cyclones.

The wave spectral model data quality is usually validated against in-situ and/or remotely sensed data. Different wave communities adopt different wave measurement strategies (e.g., meteorological agencies, academia, and research organisations). Some assimilate the altimeter data to correct wave model results (e.g. European Center for Medium Weather Forecast, ECMWF). In contrast, others use satellite data (verified by wave buoy data) to calibrate hindcasts (e.g. Ifremer, Ardhuin et al., 2011a). At the same time, some put efforts to modelling the atmospheric forcing without assimilating/calibrating against satellite data (MET Norway, Reistad et al. 2011). The accuracy of data sources used will affect the application of these different strategies, which will consequently impact engineering applications (e.g., return values of wave parameters).

Increased resolution of wave spectral models, as well as the number of wave measurements used for their validation/calibration, has led to significant improvement in their accuracy (Cavaleri and Bertotti, 2006; Shimura and Mori, 2019; Shi et al., 2019). However, for design work, wave model data, particularly extremes, should be compared with measurements at an ocean location/area under consideration to investigate whether correction for a possible bias is necessary.

Although the accuracy of integrated wave parameters derived from hindcast/reanalysis wave spectra is generally recognised, the accuracy of wave spectra is questioned, e.g., Ardag and Resio (2019) and Cavaleri et al. (2020). Some concern exists that the wave spectral models may provide spectra that are too broad compared to the measurements' spectra. The investigations of Aarnes et al. (2018) support this concern. The DIA (Discrete Interaction Approximation) parameterisation of the non-linear interactions remains the dominant source of errors in hindcast predictions. It is currently challenging to eliminate the error caused by the DIA parametrisation because high computational cost prevents using the exact solution.

Further, quasi-resonance wave interactions responsible for the generation of rogue waves were incorporated in spectral models on a research basis (Gramstad and Babanin 2016) but are not implemented in the operational wave spectral models yet, the models do not provide surface elevation. Therefore, to study rogue waves, coupling the wave spectral model with the nonlinear phase-resolving model is necessary (Bitner-Gregersen et al., 2014c; Fujimoto et al. 2019; Bitner-Gregersen et al., 2020c, Bitner-Gregersen et al., 2024).

Several wave databases developed by different organisations are available today. They cover the global ocean or selected ocean areas. Examples are ECMWF ERA-40, ECMWF ERA-Interim (Dee et al., 2011), ECMWF ERA5 (Bidlot et al. 2019), NORA10 (Reistad et al. 2011), HIPOCAS (Weiss et al. 2002; Weiss and Feser, 2003; Guedes Soares et al. 2002; Pilar et al. 2008), WorldWaves (Barstow et al. 2003, 2011), Ifremer IOWAGA (Ardhuin et al. 2011b), NOAA (https://www.ncdc.noaa.gov/cdo-web/ National Oceanic and Atmospheric Administration), Japanese TodaiWW3-NK (Sasmal et al. 2019), WW3-ST6v (Liu et al. 2021) and GWS (BMT, 1986, updated in the 1990's). These databases include different data sources, cover different periods, and differ by sampling steps and resolution of the wave spectral model used to simulate them, see Table 1.

A recent comparison of the ERA5 and IOWAGA database with buoy and satellite data carried out by de Hauteclocque et al. (2020) shows good agreement, indicating that ERA5 slightly underestimates the extremes. This was also confirmed by Sasmal et al. (2019, 2021b) by comparing predictions of the ERA5, IOWAGA and TodaiWW3-NK databases. TodaiWW3-NK provides predictions closer to IOWAGA but gives the highest extremes than IOWAGA because of its finest spatial and spectral resolution, despite the same physics package and wind forcing as used by IOWAGA.

4.3 Long-Term Wave Modelling

4.3.1 General Remarks

The long-term description discussed herein is limited to statistical modelling of variability of sea state characteristics (e.g., H_s and T_p or T_z). A review of long-term distributions of individual wave parameters (e.g., wave crest, wave height) is outside the scope of the present study.

Traditionally, the long-term description of wave climate used in design and operational criteria has been based on time-independent statistics that do not account for climate change. It is affected by the uncertainty associated with applied wave data sources, the period they covered, and the adopted statistical distributions and fitting techniques for the estimation of distribution parameters. These uncertainties impact extreme distributions and return values derived from them, but their relevance for engineering applications is case-dependent.

Particular attention should be paid to selecting the time duration that data covers. A period that is too short would not capture sufficiently natural variability of wave climate and would affect not only the body of distributions of wave parameters but also the distribution tails. A period covering years with severe wave climate only will result in over-design, while unconservative design would occur if the period represents years with mild weather conditions. Bitner-Gregersen and Hagen (1999) showed that the 100-year H_s derived from the Norwegian Sea 0.5-year data set of severe wave conditions was about 3.0 m higher than the one calculated from the extended 2.5-year data set. Including/removing single storm events can also significantly impact return values, see Waseda (2024). Thus, up to a few meters, the discrepancy in return values can be expected if the time duration data cover is not representative of wave climate at a location considered or extreme events are removed from a data set.

To reduce sampling variability, data sets used for design should at least cover a period beyond 10 years for extra-tropical storms and 100 years for hurricanes/typhoons. Accounting for climate change will require more extended periods.

4.3.2 Probability Distributions

Two different analysis strategies are commonly applied in design work: global and event models. The global model (or initial distribution method) utilises all available data from a long series of subsequent observations (e.g. all 3-h data). In the event model observations over some threshold level (peak over threshold, POT, or storm analysis method), annual or seasonal extremes are used.

The use of global models reduces sampling variability but introduces uncertainties due to the correlation of the variables. The impact of the correlation on return values is still controversial, particularly regarding the degree to which it would affect design. A recent study by Vanem (2022) showed that it may affect return values. Furthermore, applying the global models may not adequately discriminate the tail behaviour (DNV, 2019a).

Peak Over Threshold (POT) models (e.g. Ferreira and Guedes Soares, 1998) allow for a better fit of the tails of distributions of wave parameters but are based on a more limited number of observations that result in higher sampling variability than global models. Furthermore, POT models are sensitive to adopted threshold levels and interarrival time, i.e. the minimum time separation of selected events, to ensure that the storms are mutually independent (Katalinić and Parunov, 2020). This sensitivity to separation time should always be investigated as they may significantly impact the H_s return period, as shown by Katalinić and Parunov (2020). A too low threshold is likely to violate the asymptotic basis of the model, while a too high threshold would generate fewer data excesses for the estimation of the model. The authors varied the minimum interarrival time and the threshold to determine their optimal values, see Fig. 3.

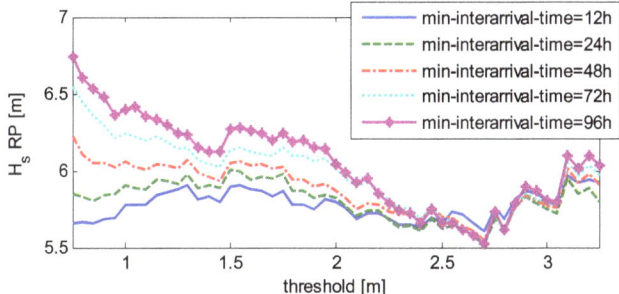

Fig. 3. Hs estimate for the 50-year return period depending on the threshold and the minimum interarrival time (defined as peak-to-peak); after Katalinić and Parunov (2020)

Storm and typhon statistics (see Laface et al., 2019, DNV GL, 2019a, b; Tao et al. 2013), similarly to annual extreme models, can be used if sufficient data exists to avoid biased results. Laface et al. (2015) have also discussed the directionality associated to the storms.

Choosing an approach for long-term description depends on the data set available for an engineering application. In practice, different organisations follow different strategies. The global approach is commonly adopted for establishing environmental contours (Winterstein et al. 1993; Huseby et al., 2013; Ross et al. 2020; DNV, 2019a). Notwithstanding this, the POT method may also be used, see Ross et al. (2020). A comparison of the different approaches to environmental contour estimation and associated uncertainties, as well as design recommendations, are given by Ross et al. (2020). A similar benchmark study using various methods to construct environmental contours was carried out later by Haselsteiner et al. (2021) using different data sets. Such benchmark studies can be used to describe model uncertainty, which is challenging to specify.

There is no theoretical method for selecting distribution functions for metocean parameters. Thus, their choice will always be subjective (hence variable), see Bitner-Gregersen et al. (2022). For the case of the initial distribution method, unless data indicate otherwise, a 3-parameter Weibull distribution $F_{Hs}(h_s)$ can be assumed for the marginal distribution of significant wave height H_s, with ocean location-dependent parameters.

The Gumbel distribution is commonly used to fit annual observations, while the asymptotic Generalized Pareto Distribution (GPD) has become widely used for calculating extremes when the POT method is applied. A challenge with the latter method is that a set threshold needs to be high enough to justify the application of GPD but low enough to reduce sampling variability (see e.g., MacDonald et al. 2011). Because of this limitation, Næss (2011) recommended using the up-crossing rate methodology instead of POT one in design.

Feld (2014) showed statistically significant differences between return values for different seasons and directions. Jonathan et al. (2013, 2014) proposed incorporating the effect of covariates in the conditional extremes model of Heffernan and Tawn (2004). Challenges associated with accounting for seasonality and directionality were demonstrated recently by Feld et al. (2019). Different approaches of accounting for wave directionality in design work are recommended by, e.g. DNV RP-205 (2019a). Full utilisation of seasonality and directionality in engineering applications still requires further research.

A regional frequency analysis (RFA) introduced by Van Gelder et al. (2000) can be used to create more extensive data sets with the data of homogenous neighbouring locations. This technique was successfully applied to different ocean areas by, e.g., Lucas et al. (2017, 2020), Vanem (2017, 2021), and Campos et al. (2019) for significant wave height and wind speed.

In the specification of design and operational criteria of marine structures, multivariate distributions are applied and recommended to be used by industrial standards, e.g. DNV GL (2019a), if reliable simultaneous data exist. Both Hs and other metocean parameters, e.g., Tp, Tz (or Tm02), wave direction, and wind speed, are necessary for predicting loads and responses.

Different approaches exist for establishing joint probabilities (Ochi 1978; Athanassoulis et al. 1994; Ferreira and Guedes Soares, 2002; Dong et al. 2013). Other approaches in current use are the Conditional Modelling Approach (CMA), the conditional extreme model (Jonathan et al. 2010), and copula models. For a review see, e.g. Bitner-Gregersen (2012), Ross et al. (2020).

In CMA a joint density function is defined by a marginal distribution and series of conditional density functions, each modelled using parametric functions. The CMA (hierarchical model), which was proposed by Haver (1985), has been used by academia and the marine industry worldwide, such as Guedes Soares et al. (1988) and Bitner-Gregersen and Haver (1989, 1991), showing satisfactory results, was extended by Bitner-Gregersen (2015). The model which used the 3-parameter Weibull distribution for H_s and log-normal distribution for T_p/T_z is included in the DNV RP-C205 (2019a). Recently, it was used to fit data from the Adriatic Sea (Katalinić and Parunov, 2021). The limitation of the CMA model is that it may give an unsatisfactory fit if a single distribution is used for wave period (commonly log-normal) in ocean regions where two or more wave systems (wind sea and swell/swells) with significantly different wave spectral periods are present. A dual distribution is recommended in ocean areas (Bitner-Gregersen, 2015). A model modification by Horn et al. (2018), where wind speed, and not Hs, is the main parameter on which other metocean parameters are conditioned, is also currently applied by academia and the engineering community.

Copula models are based on the marginal distributions of the considered variables and their dependency, e.g., the Nataf model, which was applied to waves by Bitner-Gregersen and Hagen (1999). Thus, marginal distributions and copulas (correlation) are estimated separately. There are various ways to model a correlation between the variables. Vanem (2016) showed that straightforward application of any of the standard parametric families of copulas fails to give good joint models, confirming the results of Bitner-Gregersen and Hagen (1999) and Silva-González et al. (2013). Vanem (2016) also demonstrated that a CMA model can be expressed within the copula framework and that the asymmetric copulas may provide an adequate model (see also Zhang et al. 2018). It should be noted that copulas can be applied to joint distributions of more than two variables (Dong et al. 2015; Heredia-Zavoni and Montes-Iturrizaga, 2019).

Alternatively, the conditional extreme model is used, motivated by the existence of an asymptotic form for the limiting conditional distribution of one or more conditioned random variables. Such a model applies to a large class of distributions (e.g., Heffernan and Tawn, 2004; Heffernan and Resnick, 2007); see Jonathan et al. (2010) for an outline of the approach. A sufficiently large sample is required to estimate the marginal and conditional tails.

Wada et al. (2016) suggested a Likelihood Weighted Method for describing the uncertainty of extreme value analysis. This method provides epistemic uncertainty for a small sample size (N10) with large data uncertainty. Wada and Waseda (2020) used this method to show how different wave climates in the Gulf of Mexico, Northwest Pacific, Adriatic Sea, and North Sea affect extreme wave estimation. The authors demonstrated that the epistemic uncertainty inherited in the extreme sample data can dominate other uncertainty sources.

Mackay et al. (2025) present the application of a new multivariate extreme value model SPAR (Semi-Parametric Angular-Radial) for the estimation of joint metocean variables and argue that it is superior compared to the commonly used models, global hierarchical models and copula models. The study demonstrates that the SPAR model can reproduce response distributions satisfactorily. Simple response functions for floating structures and an offshore wind turbine with a monopile foundation illustrate the model's application. The model still needs to be further validated before using it in design work.

Marine industry standards, e.g. ISO Standards and DNV GL (2019a), provide recommendations for a choice of distributions of metocean variables. However, different organisations have different practices in deriving the return values. Having a data set, alternative methods for extreme value estimates should be used, and results should be compared. This identifies biases (systematic errors) between different methods and provides a mean value for all methods, which is useful information for a decision process.

4.3.3 Estimations of Distribution Parameters

Three recognised techniques are used to derive parameters of wave distributions from data: the Least Squares (LS), the Method of Moments (MM) and the Maximum Likelihood Method (MLM). Recently, weighted fitting techniques, such as the Weighted LS (WLS), the Likelihood Weighted Method (LWM) and STM-E (Space Time Maxima and Exposure) have received attention.

It is a common agreement that, even though the MLM method is most theoretically appropriate, it should be used carefully as it may yield a biased fit to data if the chosen underlying distribution is incorrect. An assessment of the uncertainty associated with different methods of fitting the data was given by Guedes Soares and Scotto (2001). Vanem et al. (2019) showed that the 100-year H_s derived using MLM were a few meters lower than those obtained using the MM or LS fitting technique. Generally, the MM method is a good choice in most cases. For the 3-parameter Weibull distribution, the LS method often yields a better tail fit than the MM method (DNV, 2019a). See also a benchmark study, e.g. by Vanem (2024), for comparing different fitting techniques. Estimated distribution parameters will be affected by an adopted fitting technique and sampling variability. Parametric and non-parametric bootstrap methods can be used to evaluate the latter's effect.

Vanem et al. (2019) showed that uncertainties associated with estimating the distribution parameters depend on whether we assume a "known" parametric distribution and simulate from this statistical model or simulate from actual data. The authors also demonstrated that sampling variability affects the LS method more than the MLM one. Katalinić and Parunov (2020) demonstrated that for relatively large data sets, the Annual Maximum and POT approaches, in combination with the LS fitting technique, proved to be the most consistent.

No general recommendations exist for selecting a fitting technique. Various organisations use different strategies and software tools, often in-house. For comparison of results, standardisation of fitting techniques is called for, e.g., by recommending modern software packages that provide such techniques.

4.3.4 Climate Change Projection

Global warming of Earth's surface leads to climate change. The observed climate change includes natural variability of climate and anthropogenic climate change, which interact with each other (IPCC, 2007, 2013, 2021). Anthropogenic climate change brings trends in the mean values of metocean parameters, which is a more limited time period (not thousands or millions of years) that can be neglected when the natural variability of wave climate is concerned.

Climate changes may result in wave climate changes in some ocean regions, which are expected to have the largest impact on marine structure design compared with other environmental phenomena, e.g., current, since wave-induced loads are dominant for most marine structures. It should be noted that the shapes of distributions of wave parameters may change because of climate change, which will impact fatigue and extremes (Bitner-Gregersen et al. 2018).

Several studies were carried out in the last two decades addressing wave climate changes in different ocean locations. The investigations of Aarnes et al. (2017) and Bitner-Gregersen et al. (2018) based on CMIP5 winds showed large variability between the different Global Climate Models (GCM) and across different ensemble members of the same model as well as among the analysed emission scenarios. The results were strongly geographical location dependent. Large changes were observed in the Arctic regions because of retreating sea ice and changes in wind tracks. These results were confirmed later by Gonçalves et al. (2021) and Bernardino et al. (2016). Recent investigations of the global wave climate by Cabral et al. (2022). showed that more significant changes can be expected in the Southern Ocean. The investigations of Ewans and Jonathan (2023) of the variability of the 100-year return value of H_s due to climate change for the ocean locations east of Madagascar and south of Australia showed significant variations between different GCM supporting findings reported by others. The Authors used both CMIP5 GCM and CMIP6 GCM.

It should be noticed that physical wave models are also affected by various uncertainties associated with assumptions and resolutions of different models that will impact results. The topic is still under investigation; therefore, firm conclusions regarding climate change are not reached today. However, the results developed so far indicate that the range of possible changes in the operational environment of ships and marine structures, in general, is sufficiently large to influence the safety of marine operations and the design of ships (Bitner-Gregersen et al., 2013, 2018) and marine structures (Hagen et al. 2013) in some ocean regions.

The Norwegian Offshore Standard NORSOK (2017), to be on the safe side in design, proposed increasing extreme significant wave height and wind speed by 4% on q-probability values because of climate change.

Climate change will require adopting time-dependent statistics, where distributions of wave parameters change in time, see, e.g. Bitner-Gregersen et al. (2011) and Vanem and Bitner-Gregersen (2012), Laface et al. (2024). Today, such statistics are not included in current design practice. The topic requires still further research.

4.4 Short-Term Wave Modelling

Short-term wave description will be affected by uncertainties associated with the inaccuracy of wave data and wave models (analytical and statistical) and statistical uncertainty (sampling variability) due to the limited duration of wave records. The work on uncertainties associated with short-term wave modelling is still ongoing in the V.8 Committee and will be documented in a journal paper. Some findings are reported herein.

4.4.1 Stationarity Assumption and Sampling Variability

In contrast to long-term wave description addressing variability of wave characteristics in a period 1–1000 years, short-term wave data and models deal with the variability of water surface elevation and associated wave parameters in periods of 20- to 30-min. The short-term wave models provide design waves in extreme sea states derived from the long-term wave distributions.

Currently, most short-term wave models assume the stationarity of a wave field. Twenty (20) min duration of wave records, for which stationarity can be assumed to be valid, has commonly been adopted by the wave community and used in the modelling of ocean waves. Spectral methods often reduce the 20 to about 17 min because a sampling frequency of wave time series of 2 Hz (together with fast Fourier transform 2^N sampling points requirement) is widely used. Note that a record that includes samples twice (1,024 (2,048) at 2 Hz is 17.07 min long. As mentioned in Sect. 4.1, a wave field within the period when the stationarity assumption can be applied is commonly referred to as a sea state. Another convenient and widely used basic assumption in short-term ocean wave modelling is an assumption of homogeneity of the wave field, i.e. the statistical properties of waves at one location are the same as at another location of a wave field. For steep sea states, stationarity can often be applied to a shorter period than 20 min only (e.g., Bitner-Gregersen et al. (2020b).

All sea state wave characteristics derived from 20- to 30-min wave records will be affected by sampling variability. An example of this uncertainty for H_{M0} and T_{M02}, in terms of standard deviations, for the Northern North Sea scatter diagram was derived by Bitner-Gregersen and Hagen (1990), assuming Gaussian sea surface and the JONSWAP spectrum with $\gamma = 3.3$. Those theoretical values of sampling variability were later compared with North Sea field data, showing good agreement (Bitner-Gregersen and Magnusson, 2014).

The effect of sampling variability on sea state characteristics will depend on wave steepness $k_p H_s/2$ (k_p denotes wavenumber associated with the spectral wave period), the bandwidth of the wave frequency spectrum and wave directional energy spreading (Bitner-Gregersen et al. 2020b). The relation of wave steepness with other sea state parameters has been studied by Antao and Guedes Soares (2014, 2015, 2016). For example, the standard deviation of the maximum spatial wave crest in a unidirectional wave field can reach 10% of the crest height for the JONSWAP spectrum with the spectral parameter $\gamma = 3.3$ for a wave field beyond the second order and 8% for the second order waves (see Bitner-Gregersen and Gramstad, 2019). Sampling variability will affect temporal data more than spatial data (Bitner-Gregersen et al., 2020a, b).

Numerical models and laboratory experiments represent useful tools for deriving sampling variability. Sampling variability associated with sea state characteristics should be considered in the design process to secure safety at sea.

4.4.2 Wave Models

Wave models may be categorized as follows: linear, second order and higher-order wave models (Nonlinear Schrödinger Equations, NLS, (Peregrine, 1983), Modified NLS equation, MNLS, (also known as the Dysthe equation (Dysthe, 1979), Boundary Modified Nonlinear Schrödinger equations, BNLS (Trulsen and Dysthe, 1996), Higher Order Spectral Method, HOSM, (Dommermuth and Yue, 1987; West et al., 1987) for deep and intermediate water depth, and the Boussinesq (Whitham, 1974), Kadomtsev–Petviashvili, KP, (Kadomtsev–Petviashvili, 1970) and the Korteweg de Vries, KdV, (Korteweg de Vries, 1895) models, or their higher order extension, for shallow water, and models based on CFD (Computational Fluid Mechanics) methodology.

Short-term wave models will be affected by uncertainties associated with long-term sea state characteristics providing input to the short-term wave models and by the short-term wave data and model uncertainty (due to an adopted wave model and its assumptions). Furthermore, instrumental, laboratory, and numerically simulated wave records will be affected by sampling variability due to their limited duration.

The linear wave model, being the simplest representation of ocean waves, can approximate satisfactory ocean waves in a sea state with low wave steepness only. With increasing wave steepness, higher-order solutions are required. Short-term wave models and associated uncertainties are discussed partly in the papers of the 1st and 2nd ISSC-ITTC workshops on uncertainty modelling (Bitner-Gregersen et al., 2014a, 2014b, 2016). Herein, more attention is given to extreme and rogue waves, which have recently become of concern to the marine industry.

Numerical tools based on the potential theory for simulation of second-order short- and long-crested irregular seas are available in academia and in the marine industry. They are applied widely by the offshore industry, while traditionally, the shipping industry has used irregular linear waves as input to numerical codes for calculations of structural loads and responses. Neither linear nor second-order wave models can realistically describe very steep waves, and both models fail to correctly predict abnormal events, such as rogue waves, which are larger and steeper than the surrounding waves in a wave record.

The simplified criteria due to Haver (2000), $C_{max}/H_s > 1.25$ or/and $H_{max}/H_s > 2$ (C_{max} and H_{max} denote the maximum wave crest and height, respectively, in a 20-min wave record) have been commonly adopted to define rogue waves. However, several authors alternatively use the coefficient of kurtosis as an indicator of rogue wave occurrence (e.g. Bitner-Gregersen et al. (2020b), see Fig. 4. When a rogue wave is present in a wave record, the value of kurtosis is beyond 3.0, while it is equal 3 for the Gaussian process (Guedes Soares, et al. 2003, 2004).

Figure 4 demonstrates that kurtosis for the linear waves (the order of nonlinearity $M = 1$) is ≈ 3 and >3 when nonlinearities beyond the second order are present ($M = 3$). Note that kurtosis is primarily the third-order effect. Therefore, the second-order nonlinearities do not make a significant contribution to it, while skewness is mainly a second-order effect (Petrova et al. 2007, 2013).

Several expressions for parametrised kurtosis have been proposed by different authors since the end of the 1980s and are reviewed in Gramstad and Lian (2024), where their accuracy is evaluated using model-test results and phase-resolved numerical simulation. Gramstad and Lian (2024) proposed an efficient numerical method for calculating the sea surface skewness and kurtosis for arbitrary wave spectra when second- and third-order non-resonant nonlinear interactions are accounted for. Using this method, the authors developed new parametrisations for skewness and kurtosis as functions of wave steepness, water depth, spectral bandwidth and directional spreading. These new parametrisations, derived using HOSM and the JONSWAP spectrum, significantly improved over existing alternatives. They allow to avoid computationally demanding phase-resolved time domain Monte Carlo simulations. The authors argue that these new parametrisations can be conveniently included in higher-order distributions for crest heights, wave heights and surface elevation.

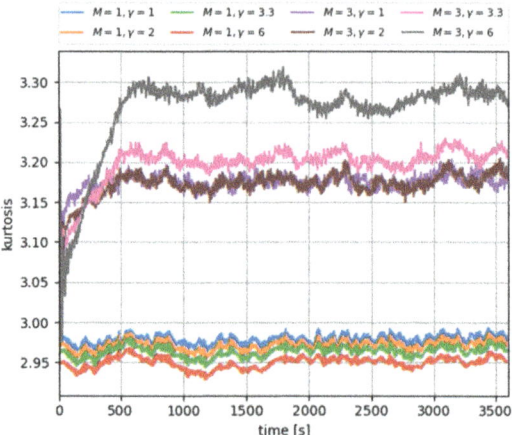

Fig. 4. Spatial kurtosis as a function of simulation time calculated as an average over 1000 repetitions of the 60-min simulation. HOSM and linear simulations; Andrea rogue wave sea state H_S = 9.2 m and T_p = 13.2 s, $k_p H_S/2 = 0.11$. M denotes the order of simulations, $M = 1$ refers to the linear simulations, $M = 3$ to the nonlinear third-order HOSM simulations. After Bitner-Gregersen et al. (2020b).

Today, NLS, MNLS, BMNLS, and HOSM are well-established nonlinear wave models that are used to study extreme and rogue waves. Chabchoub et al. (2017) compared the prediction of NLS, MNLS, and BEM (Boundary Element Method) and the experiment, showing slightly different results. These nonlinear wave models have advantages and limitations which have been discussed by several authors, e.g. Trulsen and Dysthe (1996), Gramstad et al. (2018), Bitner-Gregersen et al. (2024), Ducrozet (2024). Recently published books by Mori et al. (2023) and Zhang and Guedes Soares (2025) also address this topic. Particularly, the 2D cubic NLS equation is an integrable system and thus has many analytical solutions, such as the family of breather solutions, which can be considered as the prototypes of extreme waves. Moreover, the NLS and MNLS equations have another special characteristic that can separate the dynamics of free waves from bound wave effects and explicitly demonstrate the function of each fourth-order nonlinear term in the energy-focusing process. Furthermore, even though disturbed by an irregular background wave, the Peregrine breather solution of the NLS model still has a more significant probability of evolving into a rogue wave (Zhang et al., 2019), casting some light on the generation mechanism of real extreme waves. NLS models have the advantage that they are much less computationally demanding than HOSM models.

The advantage of HOSM compared to other fully nonlinear numerical methods is that it allows the simulation of many random realisations of the surface elevation within a reasonable computational time. Further, compared to NLS models, it does not have a constraint on wave spectral model bandwidth. However, HOSM does not include wave breaking or wind input. Despite these limitations, HOSM has been shown to approximate sea surfaces with acceptable accuracy. Figure 5 presents the reproduction of the triple rogue waves "Justine Three Sisters" (Bitner-Gregersen et al. 2020c), which were recorded by WaveRadar REX on 30 November 2018 between 18:20, and 18:40

UTC (the wave group occurred at 18:34) in the central North Sea (Magnusson et al., 2019). Note that a weak energy dissipation due to wave breaking was included to the HOSM model, using the dissipation model suggested in Xiao et al. (2013). The breaking option was used only for runs where high wave steepness brought difficulties in the code convergence.

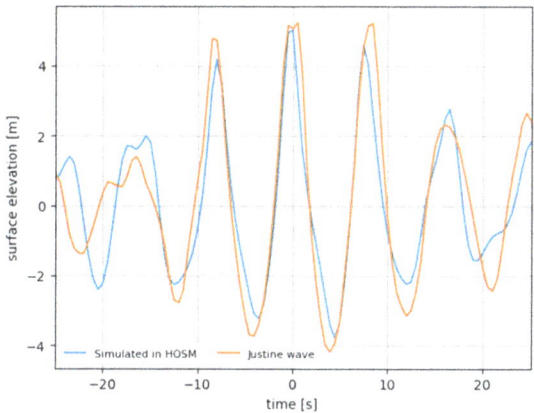

Fig. 5. Justine Three Sisters observed in nature (the orange line) and reproduced in the HOSM simulation (blue line). After Bitner-Gregersen et al. (2020c).

Kirezci et al. (2021) compared HOSM and the 2-D fully nonlinear Chalikov-Sheinin (CS) model (Chalikov and Sheinin, 1996, 2005). The authors claim that HOSM, with its truncation of nonlinearity, produces a larger maximum wave height because it cannot produce wave breaking (they did not include artificial damping). However, Houtani et al. (2022) showed that without breaking, the experiment shows even larger wave heights than HOSM, which contradicts the findings of Kirezci et al. (2021). Further experimental investigation plays an important role in resolving these contradictions.

Zhang et al. (2022) have identified the natural, knowledge, and mixed uncertainties in the initial conditions of the Chalikov-Sheinin model and analysed their effects on the profile of the largest wave in the time series.

According to Ducrozet et al. (2016), HOSM codes have been developed by academia and the marine industry, and open-source packages exist. However, Gramstad et al. (2018) noted that a different implementation of the HOSM model (grid resolution, numerical domain considered) may provide different results.

Bai et al. (2028) present the first attempt to carry out the uncertainty analysis of the numerical simulation of a focused wave by solving the Reynolds averaged Navier Stokes (RANS) equations. The numerical simulation errors caused by the grid and time-step size in the convergence studies were found to have the same order of magnitude.

4.4.3 Wave Crest

Wave crest height represents an important parameter in the design and operations of marine structures. As mentioned in Sect. 4.2.3, a difference between the maximum temporal and spatial wave crest height C_{max} simulated by HOSM can reach beyond 2 m. Zhang et al. (2016) used the MNSL equations to investigate weakly nonlinear random waves' temporal and spatial evolutions. The maximum crest height derived from the temporal evolution (sampling in space) tends to be larger than that obtained from the spatial evolution (sampling in time) in the case of strong nonlinearity. Compared with the measured values in the corresponding laboratory experiments (Fig. 6), neither numerical model can accurately capture the maximum crest height for directional waves, which is mainly attributed to the sampling variability.

Bitner-Gregersen and Gramstad (2019) documented using HOSM simulation, a significant difference between the temporal and spatial C_{max} (η_{max} in Fig. 7) in the unidirectional rogue-prone sea states, see Fig. 7. The authors proposed parametric expressions for C_{max} being a function of wave steepness and spectral peak parameter γ.

Field measurements have shown that in steep sea states, surface elevation deviates significantly from the normal distribution and wave crest height does not follow the Rayleigh distribution (Petrova et al. 2006). Several empirical and analytical distributions for wave crests have been proposed in the past few years. The second-order numerically derived Forristall distributions of wave crest for short- and long-crested waves (Forristall, 2000) are commonly used in engineering applications. These distributions have shown to fit satisfactory field data in deep and intermediate water depths up to some probability level and are recommended by marine industry standards, e.g. DNV (2021).

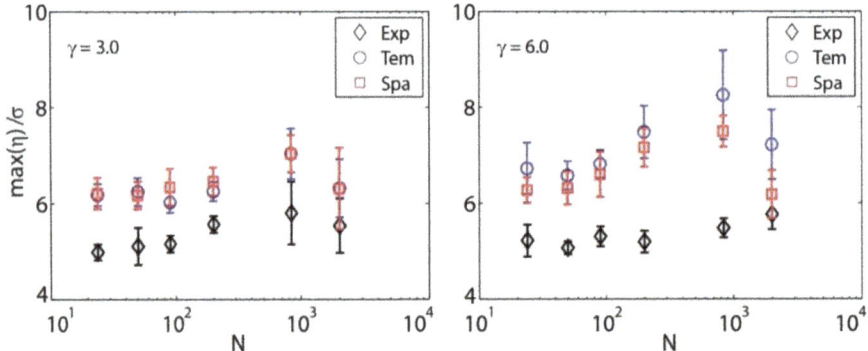

Fig. 6. Maximum crest heights as a function of initial directional spreading N.

However, significant deviations from second-order statistics have been observed in field data. Recently, Kim et al. (2023) experimentally investigated the probability of exceedance (POE) of wave crest and demonstrated that the Forristall distribution was always steeper than the numerically or experimentally estimated ensemble POE, indicating that accounting for effects beyond the second-order is important.

Vanem et al. (2024) reviewed the distribution of crest heights and showed that current state-of-the-art models have notable shortcomings and do not describe the distribution of crest heights well in all sea states. The authors focused mainly on the third-order crest distribution of Tayfun and Fedele (2007). They utilised datasets from high-frequency laser altimeter measurements of the sea surface collected at an offshore platform in the

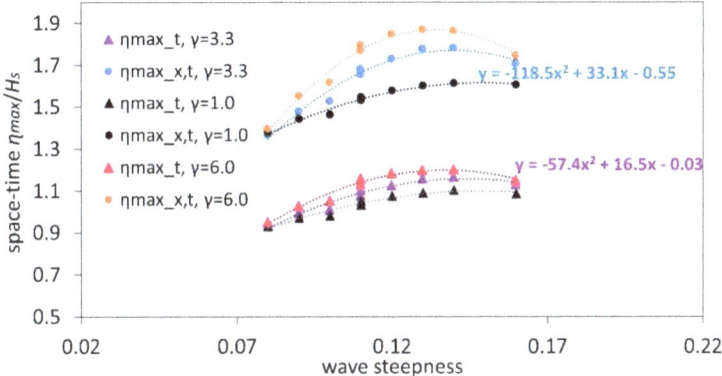

Fig. 7. Spatial and temporal estimators of the maximum surface elevation kurtosis as a function of wave steepness $k_p H_s/2)^2$. Unidirectional HOSM simulations for the JONSWAP spectrum with $\gamma = 1.0, 3.3, 6.0$. After Bitner-Gregersen and Gramstad (2019), see also Bitner-Gregersen et al. (2024).

North Sea and by the Norwegian Meteorological Institute. The data cover a period of 18 years, 2003–2020, and include a range of sea states with the highest significant wave height of 12.7 m.

Vanem et al. (2024) found that the third-order Tayfun–Fedele crest distribution performed well in about 93% of the cases and represented a slight improvement compared to the Forristall distribution. The study demonstrated that the Forristall fails to accurately describe the distribution of extreme crest heights in wave records with significant empirical skewness and kurtosis. It showed that the same was true for the Tayfun–Fedele distribution when it is parameterised in terms of skewness and kurtosis calculated from spectral parameters of Gramstad and Lian (2024). However, the Tayfun–Fedele distribution performed well if it was parameterised with the empirical skewness and kurtosis estimates derived from field wave records. The authors concluded that this is partly because of sampling variability. The distributions of crests in the particular cases of bimodal seas were studied by Petrova et al. (2013).

Development of a wave crest distribution where distribution parameters are not derived from a considered 20-min wave record will be challenging because of sampling variability. Therefore, it can be expected that any general distribution parameters will not be able to capture all sea states and provide good estimates of the wave crest distribution in all situations; sampling variability will need to be accounted for.

4.4.4 Rogue Waves and Design of Marine Structures

Several investigations in the last two decades have shown that accounting for rogue waves in loads and response calculations will impact design. Bitner-Gregersen et al. (2014) indicated that 10% increase in the extreme wave bending moment of a tanker due to wave nonlinearities may require a corresponding 10% increase of a ship deck weight (steel weight of the deck in the midship region) to maintain the reliability level in design.

Despite the mechanisms generating rogue waves and their dynamical properties being better understood, their occurrence in the ocean is still debated in the literature. Rogue waves are not systematically accounted for in the marine industry standards, as the consensus about their probability of occurrence in the ocean has not been achieved yet. However, some progress has been made on the inclusion impact of rogue wave effects in the design process, as seen by Bitner-Gregersen (2024).

The marine industry has software for assessing marine structures' performance in a nonlinear wave field up to the third order. For example, the DNV HOSM solver has become a part of the DNV SESAM Software System.

Following Equinor's lead, the revised version of the Norwegian Standard introduced a simple requirement that accounts for rogue waves when designing the height of a platform deck (NORSOK, 2017). NORSOK (2017) recommends a 10% increase in estimated extreme crest height compared to second-order point statistics to account for nonlinear waves and spatial effects.

It should also be mentioned that practical forecasting of warning criteria for extreme and rogue waves is still very challenging due to the inherent randomness of sea surface elevation and associated with it sampling variability, as demonstrated by Bitner-Gregersen et al. (2024).

5 Uncertainties in Ship Performance Prediction

Accurate prediction of ship performance is essential for optimising vessel design, enhancing operational efficiency, and ensuring safety. However, various uncertainties affect these predictions, including environmental conditions, computational models, and physical phenomena. This section delves into the uncertainties associated with ship performance predictions, mainly focusing on ship loads and responses. Recent advancements in predictive methodologies will be discussed to provide a comprehensive overview.

5.1 Uncertainties in the Prediction of Loads and Responses of Ships and Offshore Structures

Predicting ship performance involves intricate interactions between hydrodynamic forces and structural responses. The primary sources of uncertainty include limitations of numerical models, simulation methods, and environmental variability. This subsection explores these uncertainties, drawing on recent advancements to address them.

5.1.1 Numerical Modelling and Computational Techniques

Huang et al. (2022) performed a thorough uncertainty analysis on CFD-FEA co-simulations for predicting ship wave loads and whipping responses. This study utilised a coupled CFD-FEA approach where the CFD tool simulates fluid dynamics and FEA models structural responses. The iterative coupling of these models aimed to capture the interaction between hydrodynamic forces and structural dynamics. As a result, Huang et al. (2022) identified key sources of uncertainty, including mesh resolution, boundary condition representation, and the fidelity of fluid-structure interaction models. Discrepancies often arose from approximations in wave-induced loads and structural responses. The study suggested enhancing mesh resolution, refining boundary condition treatments, and validating models with experimental data to reduce uncertainties. These steps are crucial for improving prediction accuracy. Gu et al. (2023) introduced an integrated CFD-FEA method for floating structures' hydroelastic analysis. The approach enhanced accuracy in predicting hydroelastic responses. CFD-FEA integration offers more reliable predictions for floating structures.

Duan et al. (2022) explored the Time-Domain Equivalent Boundary Element Method (TEBEM) for predicting springing responses. TEBEM accounted for nonlinear effects and time-domain interactions in the simulation of container ship springing responses. The study demonstrated that TEBEM significantly improves accuracy compared to linear models by incorporating nonlinearities and dynamic effects. The enhanced accuracy provided by TEBEM is essential for evaluating the structural integrity of ships under dynamic loading conditions, leading to better design practices.

Vijith and Rajendran (2023) investigated hydroelastic effects on vertical bending moments. A coupled hydroelastic model simulated vertical bending moments, considering both hydrodynamic and structural interactions. The study highlighted significant hydroelastic effects on bending moments, especially in the head and oblique seas. Incorporating hydroelastic effects into predictions is crucial for accurately assessing bending moments and structural integrity.

5.1.2 Advanced Methods and Predictive Techniques

Shao et al. (2022) developed a consistent second-order hydrodynamic model. The study introduced a second-order model incorporating nonlinear effects beyond traditional linear models. The model demonstrated improved accuracy in simulating floating structures' dynamic responses. Incorporating second-order effects enhances the reliability of performance predictions for floating structures.

Zhai et al. (2022) investigated the hydrodynamic responses of barge-type turbines under the impact of aquaculture cages. Variations in performance due to additional structures were observed. Incorporating additional structures into predictions ensures accurate assessments.

Mukhlas et al. (2023) proposed an efficient time simulation regression procedure for extreme response prediction. Advanced regression techniques were used to estimate extreme responses based on time-domain simulations. The procedures improved efficiency in predicting extreme events such as severe storms. Efficient prediction of extreme responses is crucial for offshore structures' safety and performance.

Sim et al. (2023) analysed multi-linked floating offshore structures. Fluid-structure coupled analysis was employed to assess system performance. Complex interactions between components and environmental loads were identified. Accurate prediction requires accounting for interactions and environmental factors in multi-linked systems.

Wang et al. (2023) studied the response of floating wind turbines to collision loads. A coupled analysis approach considered wind-wave-mooring load interactions. The resilience of floating structures to collision events was highlighted. Comprehensive load coupling analyses are necessary for enhancing safety and performance.

Liu and Papanikolaou (2023) improved the semi-empirical SHOPERA-NTUA-NTU-MARIC method for the prediction of the added resistance of ships in waves (Liu & Papanikolaou, 2020; ITTC, 2021) for the consideration of ships with extreme dimensional ratios. Numerical experiments were designed to investigate the joint effect of CB, L/B, and B/T variation using well-established numerical tools for added wave resistance. The patterns observed in the results are then correlated to the parameters through explicit, readily usable mathematical expressions. The improved formula was benchmarked with an experimental database of 11 ships with extreme dimensional ratios of various types. The validation study showed that the improved formula achieved a higher correlation coefficient and a smaller mean percentage error. Thus, the introduced expressions can significantly improve the prediction of the added resistance of various types of ships in waves, including those with extreme dimensional ratios. The study demonstrates how to combine the merits of both numerical and physical experiments to gain insight into complex physical problems.

Moreira et al. (2021) used the output of a weather routing model to train a neural network capable of consistently predicting fuel consumption and emissions with little computing power. The weather routing system performance depends on a detailed hydrodynamic model (Prpic-Oršic, et al. 2016) and on the input wind and wave fields. Vettor and Guedes Soares, (2022) assessed the uncertainties of ensemble weather forecasts on the predictions of ship fuel consumption.

Mao et al. (2024) introduced a spectral fatigue assessment framework. The framework combined spectral analysis with fatigue assessment for floating offshore wind turbines. The framework demonstrated improved accuracy in predicting fatigue life. Spectral fatigue assessment is vital for evaluating the structural integrity and longevity of floating wind turbines.

5.1.3 Comparative Studies and Model Validation

Amaechi et al. (2022) investigated fluid-structure interactions from wave diffraction using numerical and analytical solutions on CALM buoys. An improved understanding of interactions informs better design and operational practices.

Gledić et al. (2022) compared full-scale seakeeping measurements for research vessels with numerical predictions. Discrepancies between measurements and calculations highlighted areas for improvement. Validation against full-scale measurements is essential for enhancing prediction accuracy.

Jiang et al. (2023) developed a boundary element method for Ship-Ice-Wave interactions. The method improved accuracy in simulating complex interactions. Enhanced ship–ice wave interaction methods provide better predictions for operational scenarios.

Petranović et al. (2023) compared seakeeping methods. Closed-form expressions and other methods were analysed for seakeeping predictions. Variations in prediction accuracy were identified, emphasising the need for method refinement. Comparing methods provides insights into improving seakeeping predictions.

5.1.4 Model Uncertainty and Benchmark Studies

Parunov et al. (2020) assessed numerical predictions for damaged ships. The study simulated different damage scenarios to evaluate their impact on wave load predictions. Challenges in modelling damaged ships were identified, highlighting the need for incorporating damage-related uncertainties. Addressing damage-related uncertainties is crucial for accurate performance predictions and ensuring safety.

Parunov et al. (2022) benchmarked global linear wave load predictions for container ships. Different prediction methods for global wave loads were compared under various operational conditions. Variations in prediction accuracy were observed depending on the method used and the operational conditions. Refinement of prediction methods is necessary to account for diverse ship configurations and operational scenarios.

Parunov et al. (2024) conducted a comprehensive benchmark study on whipping response predictions. Various numerical methods were benchmarked against experimental data to assess their accuracy in predicting whipping responses. The study identified discrepancies between different methods and suggested improvements for model accuracy. Ongoing benchmarking and model refinement are necessary to enhance prediction accuracy and reliability. Integrating experimental data into simulations reduces discrepancies and improves model reliability.

Mittendorf et al. (2022) examined a semi-empirical framework for estimating added resistance in arbitrary wave heading under uncertainty quantification. In this respect, the formula's parameter vector is calibrated based on particle swarm optimisation and a database of model test results comprising 25 different ships and around 1100 samples. In the first iteration, the minimisation of reducible systematic uncertainty is of interest, and the effect of four objective functions on prediction accuracy is evaluated. Moreover, two different parameter combinations were obtained for blunt and slender-type ships. Conversely, the irreducible statistical uncertainty, i.e. the inherent noise of the experimental data, is considered by a quantile regression procedure. Applying this approach, a 90% prediction interval for the formula's estimates is implemented using the skewed version of the superior loss function in the previous iteration. The practical relevance of an uncertainty estimate for predicting the added resistance was emphasised in the final part, in which the proposed approach was validated in regular waves against model test data and other well-established prediction methods. The validation studies generally suggest satisfactory performance and reliability of the adapted semi-empirical formulation.

5.2 Conclusions and Recommendations for Future Research Directions

Addressing uncertainties in ship performance prediction requires a multifaceted approach. Recent advancements in computational techniques, benchmark studies, and predictive methods have significantly improved accuracy and reliability. By integrating these advancements, researchers can better understand and mitigate uncertainties, leading to more robust and efficient ship designs. Ongoing research and development will continue to refine our understanding of ship performance and enhance safety and efficiency in maritime operations.

Advancements in numerical modelling, benchmarking, and predictive techniques have significantly enhanced the accuracy of ship performance predictions. However, further research is required to address the remaining uncertainties. Future directions include:

- Enhanced Computational Models: Continued development of advanced models to capture complex interactions and environmental factors.
- Benchmarking and Validation: Regular benchmarking against experimental data to validate and refine models.
- Integration of New Technologies: Adoption of emerging technologies, such as AI, digital twins and data-driven predictive methods, to improve accuracy and efficiency.

6 Uncertainties in Wave Load Assessment

Different engineering models are used to predict wave loads on marine structures. Although hydrodynamic tools for predicting wave loads nowadays tend to be very sophisticated, they simplify the complex physical reality. Because of these reasons, the accurate definition of structural wave loads is among the major challenges in designing and analysing marine structures (Parunov et al. 2022b).

The role of uncertainties in a broad context is placed in risk-based design and analysis, which includes probabilities and costs of adverse events. If these probabilities cannot be estimated from the frequencies, e.g., if the adverse events are rare, then risk analysis relies on the calculated failure probabilities. This is where uncertainties play a crucial role, as they are unavoidable in limiting state functions for the probability of failure calculation. Based on the reliability analysis, partial safety factors are calibrated in structural design codes to quantify how characteristic values of loads and strength can be related to the uncertainty of the parameters to have acceptable risk involved in structural design (Guedes Soares and Moan, 1985).

This work aims to present a literature review on uncertainties in wave load assessment of ship structures. The review concerns the methods based on potential flow theory, either strip theory or 3D methods. Computational Fluid Dynamics has also been increasingly used to assess wave-induced loads of ships, but these methods are covered in Sect. 7. The review can be useful for loads on large-volume floating offshore structures, such as FPSOs and semi-submersible platforms, while the load assessment for fixed structures and loads on small-volume offshore structures, governed by the Morison equation is outside the scope of this review.

The present review follows the traditional approach to categorising wave load uncertainties, i.e. the uncertainty of wave loads calculated under linear assumptions, uncertainty of non-linear effects and uncertainty of hydroelastic responses. Other uncertainties important in the computation of wave loads, such as the choice of the theoretical wave spectra, the choice of the wave scatter diagram, methods for prediction of long-term extreme values and uncertainty associated with the influence of human actions on ship operation, are covered in Sect. 4 dealing with the environmental uncertainties.

It should be clarified that wave load uncertainties may also be divided in other ways, e.g. according to the limit state of ship structure to which wave loads refer, i.e. ultimate limit state (ULS), fatigue limit state (FLS) and accidental limit state (ALS). Generally, each of these limit states requires separate uncertainty modelling of corresponding loads. For example, probability levels are of interest for ULS and FLS. While exceeding the probability of 10^{-8} is considered for the definition of design wave loads for ULS, a probability level of 10^{-2} is relevant for FLS.

6.1 Uncertainty in Transfer Functions of Wave Loads

The transfer functions represent the complex amplitudes of the motion or the load effects induced by waves of unit amplitude on a ship. Transfer functions are inherently linear and obtained by linear assumptions, meaning that the wetted body surface is defined by the mean position of the hull under the corresponding position of the free surface where the free surface boundary conditions are applied. Transfer functions of ship motion and load effects are usually calculated in the frequency domain using linear numerical hydrodynamics methods or in the time domain using waves of small steepness. 3D panel methods are the industry standard for fixed offshore structures with large displacements. Two two-dimensional strip theory and 3D panel methods are employed for ship structures with forward speed. The transfer function may be found experimentally in the towing tanks using small amplitude waves of different frequencies (Parunov et al. 2022b).

Even though linear seakeeping codes are generally considered reasonably accurate for small wave amplitudes, uncertainties are found between calculated transfer functions and those measured on ship models. The reasons for the uncertainties are difficult to explain as they depend on the seakeeping method used, mesh size used in the modelling, inaccuracies in the mass distribution, and human errors (Kim and Hermansky, 2014). As the uncertainties in the transfer function result in inaccuracies in the predicted extreme values, there is a clear interest in the methods to quantify uncertainties. Several uncertainty quantification methods have been used, as reviewed and compared by Parunov et al. (2022b). These are the Frequency Independent Model Error (FIME), the Total Factored Error (TFE) and the Total Difference Method (TDM). The latest method was improved by Abdelwahab et al. (2023) by comparing transfer functions to the experimental results instead of to the mean values of computation, as was done in the original TDM (Kim and Kim, 2016). The effect of uncertainty on transfer functions during ship lifetime can be calculated by combining the short-term transfer function uncertainty with long-term distributions, as proposed by Schellin et al. (1996).

Among the methods for uncertainty quantification, FIME appears to be the most convenient, as it does not depend on the number of measured frequencies, it directly corrects the calculated transfer functions, and it can be easily implemented in the long-term distribution of wave loads (Parunov et al. 2022b). The FIME was initially proposed by Guedes Soares (1991), having quite simple physical interpretation as the slope of the regression line, forced to pass through the origin if the transfer functions are plotted such that the abscissae and ordinates represent calculated and measured values, respectively. Frequency-Dependent Model Error (FDME) is more detailed than the FIME and more complex to implement, as Nielsen et al. (2016) presented. It was shown, however, that both FIME and FDME decrease the error between experimental and theoretical transfer functions.

A practical application of FIME is presented in Fig. 8. The measured transfer functions and transfer functions calculated by two theories (Theory 1 and Theory 2) are shown in Fig. 8 (left). Visual observation indicates that Theory 1 agrees well with the measurements, while Theory 2 overestimates measured values. A plot of calculated vs.

measured transfer functions for both theories is presented in Fig. 8 (right). Regression lines, forced to pass through the origin, are presented in Fig. 8 (right) for both theories and their equations and coefficient of determination (R^2). It may be seen that FIME reads 1.06 and 0.65 for Theories 1 and 2, respectively. A large coefficient of determination, above 0.9, indicates high confidence in the approximation of points by the regression line.

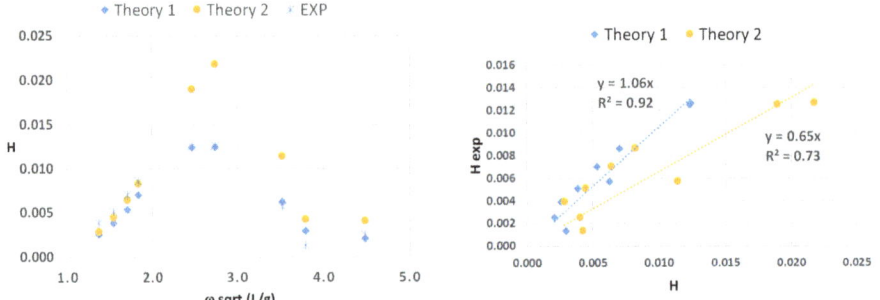

Fig. 8. Transfer functions obtained by measurements, Theory 1 and Theory 2 (left) and FIME (right) (from Parunov et al. (2020))

The review of relevant benchmark studies to assess the accuracy of seakeeping codes is presented by Parunov et al. (2022b). Recently, different model uncertainty measures have been compared within the scope of the benchmark study on the numerical prediction of global wave loads on a damaged ship (Parunov et al., 2020). FIME appeared in that study as the most convenient uncertainty measure, and it was shown how it could be used to compare the accuracy of predictions by different codes and how to compare motions and global wave loads or different global wave load components. That was the first benchmark study of that kind, i.e., aiming to compare different uncertainty measures extensively and to provide conclusions of practical relevance. Much better agreement was achieved with ship motion transfer functions than with global wave-induced load effects. Horizontal wave load components showed larger scatter than vertical global wave loads, while the prediction of torsional moments was the worst compared to all other load components. On average, predicting motions and loads for a damaged ship was worse than for an intact ship. The study setup and results have been useful in planning future benchmarks.

Another benchmark study was on global linear wave loads on a container ship with forward speed (Parunov et al. 2022a). One of the study's aims was to compare different seakeeping methods with measurements. The average transfer functions of VWBM at midship for different heading angles, obtained by strip theory (ST) codes, 3D frequency-domain (3D-F) codes, 3D time-domain (3D-T) codes and measurements is presented in Fig. 9. It is interesting to notice from Fig. 9 that the ST provides the best estimate of VWBM at midship.

Another aim of the benchmark was to compare the results with a similar benchmark in 1996 (Schellin et al. 1996). Fig. 10 compares method average transfer functions and measurements for VWBM at midship for the present and previous benchmarks.

The conclusion from Fig. 10 is that the improvement of the uncertainties of linear seakeeping methods compared to the experimental results cannot be confirmed since the 1996 benchmark.

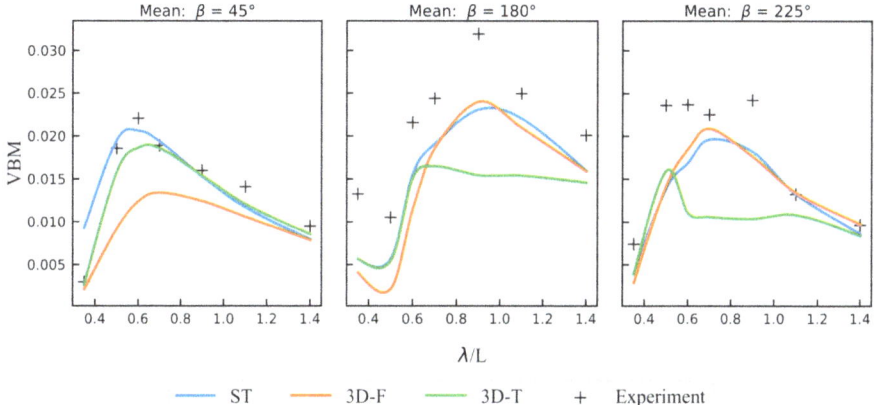

Fig. 9 Comparison of transfer functions of VWBM at midship for different method averages and experiments (ST Strip theory, 3D-F Frequency-domain panel codes, 3D-T Time-domain panel codes) (from Parunov et al. (2022))

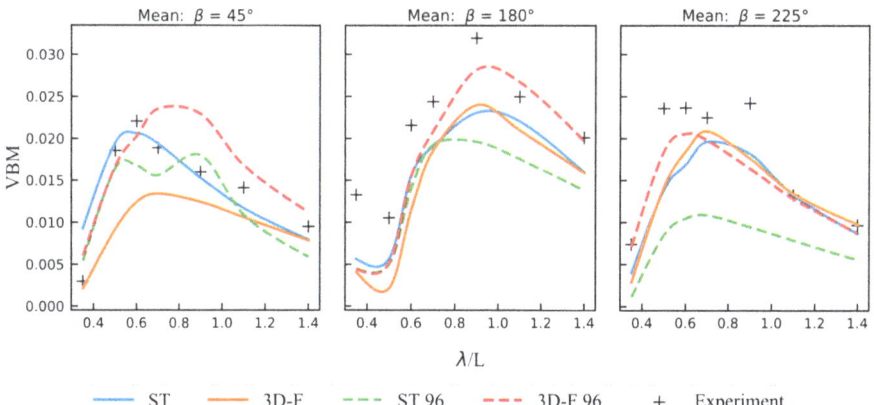

Fig. 10 Comparison of method averages and experiments of transfer functions of VWBM at midship for the present benchmark and 1996 benchmark (ST Strip theory, 3D-F Frequency-domain panel codes) (from Parunov et al. (2022))

Based on the experience from the earlier benchmark, it was decided to focus on FIME as the uncertainty measure. Furthermore, FIME was split into two parts: one part of FIME was the error of the individual codes with respect to the average prediction of the theory to which that code belongs, while the other part was the error of the average of each theory compared to the experimental results. It was shown that FIME could improve the accuracy of the long-term predictions of vertical wave bending moments (VWBM) at midship by correcting the predictions with the average error.

The coefficient of variation of the predictions by 12 codes is reduced from 21% to 10% if FIME is employed. Also, the average long-term extreme value of all predictions by seakeeping codes becomes almost identical to the long-term prediction based on the experimentally obtained transfer functions. Without FIME, the average long-term predictions by seakeeping codes are about 32% lower compared to the prediction based on the experimental transfer functions. This improvement in the uncertainty of the long-term predictions is shown in Fig. 11. The results in Fig. 9 (left) are obtained without FIME application, while results in Fig. 11 (right) are based on the transfer functions modified by FIME.

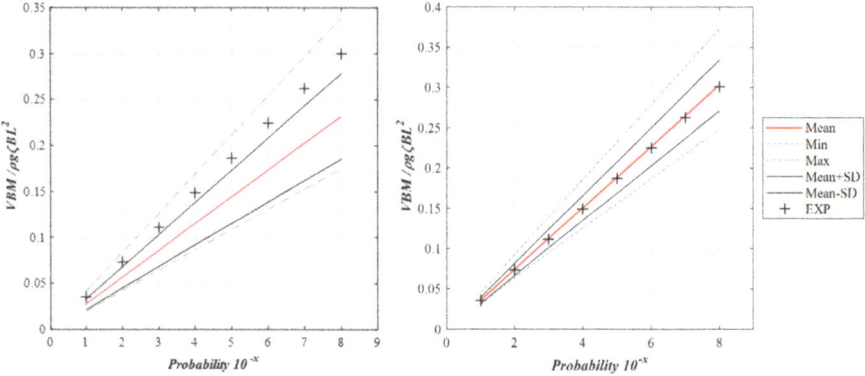

Fig. 11 Long-term distributions based on calculated and experimentally obtained transfer functions. (left – without FIME, right – calculated transfer functions modified by FIME) (from Parunov et al. 2022b)

The ship used in the benchmark was an old design of the container ship, tested in 1974 (Flokstra, 1974), so this benchmark was criticised in the sense that the modern ship hull should be used in the benchmark and modern test facilities. Also, only one benchmark ship was used, but more ships should be studied to have confident results.

Jafaryeganeh et al. (2016) applied the same uncertainty models to study the wave loads in a frigate and compared the uncertainty associated with a trip theory code and a 3D panel code.

The same ship was used to test a new uncertainty quantifier for linear wave-induced ship motions and loads by Abdelwahab et al. (2023). They proposed improving the Total Difference method, which was first used by Kim and Kim (2016). The improvement consisted of using measured transfer functions instead of the mean of the computed transfer functions. Five different 2D and 3D codes were applied, and the uncertainty analysis showed that none were superior to others for all motion and load components and heading angles. Despite that newly proposed uncertainty measure, the Modified Total Difference Method has some advantages in revealing discrepancies between numerical methods and experiments, and the method has the general drawback that the result is dependent on the number of measurement frequencies because of the cumulative nature of the method. FIME is not directly sensitive to the number of calculation points, as it represents a regression line passing through the origin regardless of the number of measurement frequencies.

Rodrigues et al. (2024) used various uncertainty measures to compare seakeeping codes for two ships, an LNG Carrier and a Chemical Tanker. They obtained useful findings by comparing results for different ships and seakeeping codes. Concerning uncertainty measures, the present review focuses on the Modified Total Difference Method, which is preferred over the Total Difference Method, while the FIME offers a more comprehensive evaluation.

Closed-form expressions (CFE) are semi-analytical transfer functions for ship wave-induced responses based only on main ship particulars, formulated by Jensen et al. (2004). Transfer functions of heave, pitch, and VWBM at midship bending moment (VWBM) calculated by CFE are systematically compared to the transfer functions obtained from model-scale experiments by Petranović et al. (2023), enabling the definition of the model uncertainty of the method. Furthermore, all the model-scale experiments chosen from the literature included comparisons with standard methods for seakeeping analysis, such as Strip Theory and 3D panel methods, which allowed for uncertainty quantification with respect to these methods. FIME was used as the uncertainty measure, and it was found that the mean value and standard deviation of FIME were expectedly larger for CFE compared to the commonly used seakeeping methods. Thirteen ships of varied sizes and types, with various speeds and heading angles, were considered. The multivariate linear regression analysis was employed to estimate FIME best based on these parameters, as given by

$$\text{FIME} = A + B\beta + CF_n + DC_b \tag{1}$$

where β is a wave heading angle in °, F_n is a Froude number, and C_b is a block coefficient. Regression coefficients A-B are given in Table 2 for each method, while associated standard deviations are provided in Table 2.

The regression equation (1) with coefficients given in Table 2 may be used to determine the mean value of FIME for application in SRA. The standard deviation of FIME is given in Table 3. It is a widespread practice to assume Gaussian distribution to represent modelling uncertainty in SRA. The approach to implementing FIME in SRA was initially presented by Guedes Soares and Moan (1985).

Table 2 Regression coefficients of FIME used in Eq. 1 for different methods (CFE – Closed Form Expressions, ST – Strip Theory, 3D – 3D panel methods) (Petranović et al. 2023)

	CFE				ST				3D			
	A	B	C	D	A	B	C	D	A	B	C	D
Heave	0.623	−0.000447	0.076	0.990	1.383	−0.000366	−0.107	−0.549	1.098	0	−0.177	0
Pitch	0.540	−0.000527	0.135	0.950	1.033	0.000086	0.540	−0.159	0.686	−0.000313	−0.184	0.589
VWBM	1.149	0.000040	0.640	−0.003	0.497	−0.000446	0.637	0.859	1.100	−0.000418	1.602	−0.151

Onboard ship response measurements could be utilised to mitigate the uncertainty in the transfer functions. Nielsen et al. (2021) presented an online tuning method for the transfer functions of ship motions. The method combines the response-based sea state estimation method, known as the Wave Buoy Analogy (WBA, Nielsen (2006)), with the ERA5 database. Parametrised transfer functions based on the CFE were employed to attain rapid computation. Mounet et al. (2024) demonstrated the effectiveness of the online transfer function estimation method for unmanned surface vehicle's heave and pitch motions (USV). Effective as the method was, issues pertaining to the computational burden associated with the identification of optimal parameters were highlighted. Furthermore, the inaccuracy of the transfer function of roll motion by CFE were pointed out. Using novel CFE, such as Matsui et al. (2023), could solve the inaccuracy in roll motion. Uncertainty in the transfer functions also leads to uncertainty in wave spectra estimated via the WBA (Chen et al. 2023).

Table 3: Standard deviation of FIME (Petranović et al. 2023)

	σ_{CFE}	σ_{ST}	σ_{3D}
Heave	0.27	0.09	0.13
Pitch	0.19	0.13	0.08
VWBM	0.31	0.16	0.20

6.2 Uncertainty of Non-linear Wave Loads

A range of non-linear methods of different complexity for ships with forward speed are in use nowadays. One may distinguish between methods based on potential theory and those solving the Reynolds – Averaged – Navier – Stokes (RANS) equations. Hirdaris et al. (2014) classified non-linear methods into six levels, namely:

- Level 1 (linear)
- Level 2 (Froude–Krylov non-linear)
- Level 3 (Body non-linear)
- Level 4 (Body exact – weak scattered)
- Level 5 (Fully non-linear – Smooth waves) and
- Level 6 (Fully non-linear).

Modelling uncertainty quantification of non-linear wave loads is challenging, as each class has a different uncertainty level than the physical reality. Furthermore, each code will have its own uncertainty with respect to the average of all codes belonging to the same class.

The most important, and hence the most often studied non-linear wave load uncertainty, are differences between hogging and sagging VWBM. The most important cause of these differences is non-vertical ship sides, which is more pronounced in fine-form ships like container ships than full-form ships such as tankers and bulk carriers. Pioneering experimental uncertainty assessment of sag/hog differences for tanker and warship is published by SSC-156 (1964). Due to the variation of the ship's cross-section, the nonlinearity of the restoring force at the ship's fore and stern parts can explain most differences between sag and hog VWBM. Therefore, even pure hydrostatic calculations could provide a reasonable estimate of this effect. Level 2 methods are very popular, as they are based on integrating non-linear Froud-Krylov forces over the instantaneous wetted hull surface and, therefore, can capture sag/hog differences (Guedes Soares 1991). Several approximate procedures to estimate sag/hog VWBM ratio are reviewed by Parunov et al. (2022b).

The sag/hog ratio of VWBM is reflected in the Rules of Classification societies, namely in IACS UR S11 (IACS 2010) for all ship types and in IACS UR S11A (IACS 2015) for container ships only. Although the latter rule formula was based on the comparative analysis of state-of-the-art non-linear methods used by different class societies, the uncertainty of computations compared to the rule formula was only slightly decreased from UR S11 (Parunov et al. 2022b).

An earlier study on using nonlinear codes was presented by (Watanabe and Guedes Soares, 1999). More recently, thirteen non-linear codes were compared within the scope of the benchmark study on the non-linear wave responses of a 6750 TEU container ship (Kim and Kim 2016). Non-linear VWBM at midship and its distribution along the ship, were compared. The uncertainty measure used was the Amplitude Difference, similar to the Total Difference, except that it is not cumulative for specific locations along the ship (Parunov et al. 2022b).

The uncertainty assessment of non-linear wave loads can be done for short-term responses by comparing statistics of sagging and hogging peaks. These peaks are found from the time history of the response process as minima and maxima between consecutive zero up-crossings. This approach was used by Ley and El Moctar (2021), where the short-term sea state was selected to capture long-term extreme response by the Coefficient of Contribution method. A large dispersion of statistical distributions of VWBM was reported, leading to the conclusion that the methods based on the potential theory were suitable for small and moderate ship responses, while CFD analysis was necessary for strong non-linearities.

6.3 Uncertainties Associated with Hull Girder Hydroelastic Response

Hydroelasticity concerns the interaction between hydrodynamic and inertia forces and their effects on flexible hull dynamics. The two most important hydroelastic phenomena of marine structures are *springing* and *whipping*. Springing is a steady-state resonant in and out of-plane elastic response with encounter wave frequencies. It occurs at slender ships with low bending stiffness, such as ultra-large container ships and Great Lakes bulk carriers. Whipping is a transient elastic response caused by nonlinear impulsive wave action, known as *slamming*. The phenomenon is clear on a broad range of ship types, including passenger ships, container ships, LNG carriers and warships that can experience whipping caused by bow flare-, bottom- or stern-slamming (Hirdaris et al. 2023). The fact that more ship types may experience whipping compared to springing is probably why much more attention was paid to the uncertainties of the former type of the hydroelastic response.

Uncertainty in hydroelastic responses comprises three analyses: dry analysis, wet analysis and slamming computations. Dry analysis represents the structural dynamic response by 1D or 3D Finite Element Modelling or by 1D Finite Difference scheme. The wet analysis is a hydrodynamic analysis of the different levels of complexity, as given in Sect. 1.2. Hydrodynamic models may be formulated in the frequency domain for springing analysis or the time domain for whipping analysis. However, it should be mentioned that distinguishing between springing and whipping responses is uncertain in many cases. Models of different complexity have been used to compute the slamming impact, an important source of uncertainty in predicting whipping responses. Experimental methods prescribe multiplying pressure coefficients, given as a function of the deadrise angle, by the square of relative velocity to calculate maximum slamming pressure.

A very popular approach to computing slamming forces is the Momentum theory, based on von Karman added mass variation concepts (Wang and Guedes Soares, 2017). The spatial added mass derivatives are multiplied by the relative velocity squared to obtain the impulsive force. The major drawback of the method is the inability to account for the important pile-up effect. This can be resolved by the Boundary Element Method, which is formulated in 2D and employed in a way that a 3D ship bow is divided into multiple stripes with various locations and orientations. In recent years, local slamming problems have been investigated by numerical methods, namely Smoothed Particle Hydrodynamics (SPH), Arbitrary Lagrangian Eulerian algorithm (ALE) and RANS CFD (Hirdaris et al. 2023). The damping coefficient is another important source of uncertainty, influencing whipping and springing responses. Damping for fundamental global vibration modes of typical ship types is studied by Storhaug et al. (2017).

The benchmark studies dealing with slamming and whipping are scarce. The study by Kim and Kim (2016), although it included some hydroelastic responses, was focused on the rigid-body seakeeping methods. The ISSC 2012 Dynamic Response Committee performed a benchmark study on the uncertainty of the predicted structural responses caused by impulsive loads by Drummen and Holtmann (2014). Even if the prescribed analytical force impulses were applied to the model, six benchmark participants obtained significant differences in structural responses, influencing fatigue life by a factor of five. Large uncertainties were also found in the computations considering an impulse induced by a regular head wave. Uncertainty analysis of springing responses for different structural modelling and potential methods was presented by Yang et al. (2021). They found that a Timoshenko beam would be preferred for the hydroelastic analysis compared to the Euler-Bernoulli beam, as the latter overestimates natural frequencies. They also found that a Frequency-Domain method with speed correction and Rankine Source time-domain method performed better than the time-domain Green function method.

The results of a benchmark study on motion and global wave loads on a warship model in regular waves were presented to quantify the uncertainty in numerical whipping predictions (Parunov et al. 2024a). Compared seakeeping methods include non-linear strip theory, 3D BEM formulated in frequency and time domain, and CFD. At the same time, Euler and Timoshenko's beams are used for modelling the hull girder stiffness. Experimentally based methods, such as CFD and momentum theories, are employed to calculate slamming loads. Coupled CFD and FEM provided results consistent with measurements, but such simulations are prohibitively computationally expensive, and the interpretation of results can be challenging. Therefore, combining the potential theory seakeeping method with correction based on coupled CFD-FEM simulation for limiting the number of cases is recommended as a promising alternative.

The methods for uncertainty quantification similar to rigid body responses are used for hydroelastic responses, namely FIME and TDM (Sect. 1.2). Hirdaris et al. (2023) presented the application of the methods on the example of the flexible vertical response of an ultra-large containership in high seas using a body nonlinear time-domain method (Rajendran et al. 2016). FIME for whipping response calculated in the benchmark of the Canadian Patrol Frigate was compared to the FIME of rigid-body responses by Parunov et al. (2024a). The uncertainty of rigid body responses was found to be similar to earlier benchmark studies without important discrepancies among different seakeeping codes. However, calculated whipping responses in regular waves are considerably different, mostly because of the various models used for slamming simulation. Quantification of uncertainty showed that the average standard deviation of FIME of different seakeeping codes was 0.14 for rigid-body vertical wave bending moment, while it is 0.36 and 0.64 for whipping bending moment for lower and higher speeds, respectively. As an illustration, a comparison of transfer functions of vertical wave bending moments at midships and whipping bending moments at midships, measured and predicted by different seakeeping and hydroelastic methods, is presented in Fig. 12.

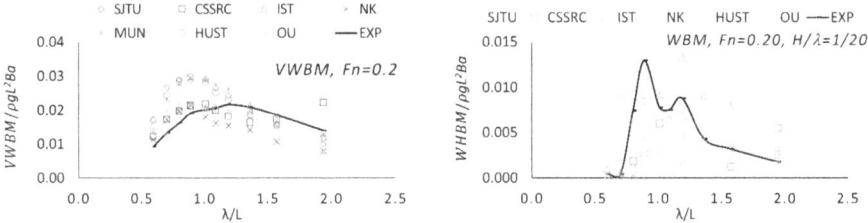

Fig. 12. Comparison of measured and predicted transfer functions for vertical wave bending moment at midships (left) and measured and predicted whipping bending moments in regular waves at midships (right) (from Parunov et al. 2024a)

6.4 Uncertainties of Wave Loads in Offshore Platforms

Although the work of the Committee has been more focused on wave loads in ships, the same methods can be applied to offshore structures, and some relevant work can be reported here.

One specific feature of fixed or floating offshore structures is the presence of cylindrical elements such as legs or bracings of semisubmersible platforms. Typically, the wave loading on bracings and small-diameter elements is determined by the Morison equation, and a relevant issue is the value of the coefficients to adopt in the Morison equation in different situations. Raed et al. (2016) and Raed and Guedes Soares (2018) have analysed the uncertainty of estimating the Morison equation drag and inertia coefficients for wave loading in cylinders. The uncertainty in the wave loads of a DeepCwind semi-submersible platform was assessed by applying the Morison equation to the different members and combining their contributions to the overall uncertainty with first-order second-moment methods (Raed et al. 2025).

For platforms with larger diameter cylinders, the diffraction forces need to be considered, and several codes are available for those calculations. The National Renewable Energy Laboratory (NREL) in USA, has been promoting several very useful studies within the program Offshore Code Comparison, Collaboration, Continued, with Correlation project., in which many research groups have compared the results of the wave loads in various structures, which has been used to validate the Codes. A review of all their studies will not be provided here, but a reference will be made to one of the studies as an example.

The DeepCwind floating semisubmersible wind system was validated using measurement data from a $1/50^{th}$-scale campaign performed at the Maritime Research Institute Netherlands offshore wave basin (Robertson et al. 2017). The models were validated by comparing the calculated ultimate and fatigue loads for eight wave-only and combined wind/wave test cases against the measured data after calibration was performed using free-decay, wind-only, and wave-only tests. Some codes used the Morison-only hydrodynamic modelling approach, while others included the diffraction effects for the larger diameter cylinders. The results of all calculations have been used to study the model uncertainty associated with the codes (Uzunoglu and Guedes Soares, 2018). The motions (i.e., surge, heave, and pitch) were compared based on the hydrodynamics model used to obtain them. The groups that use similar primary approaches were compared within themselves. The inclusion of potential flow theory seemed to consolidate results. However, by removing resonant areas resulted in the Morison equation performing closer to the potential flow theory.

The semi-submersible platforms are prone to wave impact on the deck, which requires providing enough air gap. Various models are available for their assessment, and Hallak et al. (2022) have studied the associated epistemic uncertainties. Wave converters and floating wind platforms have also been used as cases for assessing the uncertainty in wave loads (Kamarlouei and Guedes Soares, 2022; Raed et al., 2025).

6.5 Practical Applications of Wave Load Uncertainties

The main practical application of wave load uncertainties is Structural Reliability Analysis (SRA), including the evaluation of a probability of failure, whose knowledge is often part of the more general risk-based framework that is used for different purposes (e.g. risk-based ship design, inspection planning, route optimisation, manoeuvring in heavy weather). Thus, accurately calculating the probability of failure is vital to many risk-based procedures in the shipping and shipbuilding industry. Sensitivity analyses performed within many SRA showed that the failure probability is the most sensitive to the modelling of wave load uncertainty, indicating the necessity of the research in that field (Parunov et al. 2022b).

The improvement by IMO and IACS is related to the probabilistically defined ultimate longitudinal strength requirement for oil tankers as given in IMO MSC 81/INF.6 (Hørte, T. et al. 2007). The document, for the first time, provided SRA of ultimate bending moment capacity agreed upon among classification societies, presenting the basis for the risk-based calibration of the partial safety factors (PSF) for the ultimate longitudinal strength of oil tankers adopted in the Common Structural Rules (CSR) (IACS, 2023). Rather, a large PSF for VWBM of 1.3 is introduced to cover uncertainties in wave load calculation, serving as another proof of the importance of that uncertainty.

PSFs in CSR were defined based on the standardised IACS North Atlantic (NA) wave scatter diagram, being the most important input parameter for the long-term prediction of wave loads, given in IACS Rec. No, 34 Rev .1 (IACS, 2001). That scatter diagram was developed for the North Atlantic Sea environment and was based on visual observations published in the wave atlas Global Wave Statistics (GWS) (Hogben et al., 1986). As such, a dataset has been collected in the past century using relatively inaccurate visual observations from ships, and it has been criticised by the engineering and research community for decades (Parunov et al. 2022b). As a response, IACS developed the revised document Recommendation No. 34 Rev. 2 (IACS 2022), including the new IACS NA scatter diagram, based on the hindcast wave database, filtered for actual shipping routes using the Automatic Identification System (AIS), a satellite ship positioning system available for all merchant ships. The new IACS NA scatter diagram is milder than the old one, with lower extreme significant wave height and wave steepness (de Hauteclocque et al., 2023).

The change in the notional hull girder failure probability of oil tankers with respect to their ultimate bending moment capacity when IACS Rec. No.34 Rev. 2 is used for the long-term computation of VWBM instead of Rev. 1, which was studied by Parunov et al. (2024b). The modelling uncertainty of VWBM is defined in a way that gets comparable results, as IMO for Rev 1. The structural reliability of oil tankers is increased if the extreme wave loads are calculated using the revised IACS recommendation. The partial safety factor for vertical wave bending moment may be reduced by 15%, and so, the required cross-sectional area of the main deck may be decreased by 10% to get the same notional structural reliability through using the "new" recommendations as opposed to the "old" procedure for wave loads computation.

Another example of the practical application of wave load uncertainties is the SRA of container ships under combined wave and whipping loads (Ćorak et al. 2018). The commonly used limit state function for structural reliability analysis of oil tankers and bulk carriers was extended to consider the whipping response due to the bow flare slamming. Compared to oil tankers, the SRA of container ships needs to consider modelling uncertainty of whipping loads and probabilistic load combination factors between low- and high-frequency loads. Careful uncertainty modelling is required to compare results with those of IMO for oil tankers. SRA may be useful for classification societies in developing probabilistic PSFa for the ultimate strength of container ships.

Besides probabilistic PSF and rule improvement, wave load uncertainties may be useful in developing risk-based decision support systems (DSS) for voyage optimisation or those to support manoeuvring in harsh weather. The necessity of embedding an uncertainty analysis into operational guidance has been found and studied (Vettor et al. 2021; Prpić-Oršić et al. 2020). In risk-based DSS, wave-induced loads and seakeeping limiting parameters are to be defined probabilistically, considering their uncertainties (Parunov et al. 2022b).

Wave load uncertainties are also applied to the structural health monitoring (SHM) systems, which have been intensively developing. The SHM sensors must be placed optimally to detect any structural damage with high reliability before turning critical. The method that Guratzsch and Mahadevan (2006) proposed achieved this goal by combining probabilistic finite element analysis, structural damage detection algorithms, and reliability-based optimisation concepts. Relevant sources of uncertainty and a method to quantify the overall uncertainty in a ship performance indicator are provided by Aldous et al. (2015). Quantifying and accurately representing these uncertainties within the complex mathematical framework of SHM was essential for system identification and damage detection purposes (Dana et al. 2018).

As digital twin (DT) technology is defined as dynamic digital models of physical assets and systems, the transformation and deduction between them will undoubtedly result in the corresponding uncertainty. Application of the DT approach for the prognosis of remaining fatigue life of ship structural details is among the most frequent applications of the DT concept, where wave load uncertainties play a significant role (Van Der Horn et al. 2022). Different types of uncertainties appear in calculating ship responses to random seas for the need of DT, such as assumed wave spectrum formulation, uncertainty in the transfer function calculation, uncertainty in the actual ship loading condition and uncertainty in the still-water loading (Parunov et al. 2022b). Fujikubo et al. (2024) presented utilisation scenarios for the DT in the maritime industry, including navigation support, maintenance support, rule improvement, and product value improvement, together with future research needs for implementation in the maritime industry.

7 Uncertainty in Numerical Field Models

7.1 Uncertainty Analysis

The application of numerical methods in hydrodynamics has continued to develop considerably as faster computing power and more extensive storage capabilities become available. Computational fluid dynamics (CFD) is becoming more widely accessible and a tool used in industry to support the design of ships and offshore structures. One can see a phenomenon similar to what happened with finite elements (FE) becoming an industrial tool for conducting structural analysis several years ago. However, one crucial difference is that the number of elements or degrees of freedom used in CFD is significantly larger than the ones typically used in FE. Consequently, the results of CFD are very sensitive to the mesh adopted, which has given an essential role in quantifying the uncertainty of CFD predictions. The description of the main aspects of uncertainty quantification presented here follows Islam and Guedes Soares (2019).

Uncertainty analysis is a systematic study performed to assess the consistency and accuracy of a CFD solver in solving the intended problem. Uncertainty analysis is subdivided into two processes: verification and validation. Verification assesses the consistency of the solver, whereas validation evaluates its accuracy. According to Wikipedia, Verification evaluates whether or not a product, service, or system complies with a regulation, requirement, specification, or imposed condition. It is often an internal process. Validation is the assurance that a product, service, or system meets the needs of the customer and other identified stakeholders. It often involves acceptance and suitability with external customers.

The American Institute of Aeronautics and Astronautics (AIAA, 1998) gave a more elaborate and generalised definition for verification, validation, and calibration.

Verification is the process of determining that a model implementation accurately represents the developer's conceptual description of the model and the solution to the model.

Validation is the process of determining the degree to which a model accurately represents the real world from the perspective of its intended uses.

Calibration is the process of adjusting numerical or physical modelling parameters in the computational model to improve agreement with real-world data.

Roache (1998) gave a very simplified definition of verification and validation, saying that verification is solving the equations right and validation is solving the right. According to Oberkampf and Roy (2010), verification provides evidence that the computational model is solved correctly and accurately, and validation provides evidence that the mathematical model accurately relates to experimental measurements. They emphasised that for verification, the standard is the "conceptual model" and for validation, the standard is the "real world", and calibration is a response to the "degree of representation of the real world" directed toward improvement of the agreement.

7.2 Methodology for Verification and Validation

Following the ITTC (2008) guidelines for Uncertainty Analysis in CFD, simulation error δ_S may be defined as the difference between actual value T and simulated value S; which may be further elaborated as the sum of modelling error δ_{SM} and numerical error δ_{SN} of CFD ($\delta_s = S - T = \delta_{SM} + \delta_{SN}$). For certain conditions, both the sign and magnitude of the numerical error can be assumed as $\delta_{SN} = \delta_{SN}^* + \varepsilon_{SN}$, where, δ_{SN}^* is an estimation of the sign and magnitude of simulation numerical error and ε_{SN} is the error in that estimate.

The simulation value is corrected to provide a numerical benchmark S_C, which is defined as, $S_C = S - \delta_{SN}^*$. Verification of numerical data involves assessing simulation numerical uncertainty and determining the error sign and magnitude when conditions permit. It also involves assessing the uncertainty in the error estimation, U_{S_cN}. For the uncorrected simulation approach, numerical errors are decomposed into iterative errors δ_I, grid size errors δ_G, time step errors δ_T, and other parameters related errors δ_P. In this case, the simulation numerical uncertainty is provided as,

$$U_{SN}^2 = U_I^2 + U_G^2 + U_T^2 + U_P^2. \qquad (2)$$

For the corrected simulation approach, the solution is corrected to produce a numerical benchmark S_C and the estimated simulation numerical error δ_{SN}^* and corrected uncertainty U_{S_cN} are given by,

$$\delta_{SN}^* = \delta_I^* + \delta_G^* + \delta_T^* + \delta_P^* \qquad (3)$$

$$U_{S_cN}^2 = U_{I_C}^2 + U_{G_C}^2 + U_{T_C}^2 + U_{P_C}^2 \qquad (4)$$

Validation is defined as a process for assessing simulation modeling uncertainty U_{SM} by using benchmark experimental data and, when conditions permit, estimating the sign and magnitude of the modelling error δ_{SM} itself. The comparison error E is given by the difference in the data D and simulation S values, $E = D - S = \delta_D - (\delta_{SM} + \delta_{SN})$.

Modelling errors δ_{SM} can be decomposed into modelling assumptions and the use of previous data. To determine if validation has been achieved, E is compared to the validation uncertainty U_V given by, $U_V^2 = U_D^2 + U_{SN}^2$.

If $|E| < U_V$, the combination of all the errors in D and S is smaller than U_V and validation is achieved at the U_V level. If $U_V << |E|$, the sign and magnitude of $E \approx \delta_{SM}$ can be used to make modelling improvements. For the corrected simulations, the equations can be represented as:

$$E = D - S_C = \delta_D - (\delta_{SM} + \delta_{SN}) \tag{5}$$

$$U_{Vc}^2 = U_{Ec}^2 - U_{SM}^2 = U_D^2 + U_{ScN}^2 \tag{6}$$

However, in practical applications, quantifying modelling errors are very difficult. Thus, attention is provided just to the simulation of numerical errors. As for numerical errors, errors related to other parameters are mostly ignored unless something specific is mentioned. As for errors related to iteration, grid and time steps they are investigated by varying each one, assuming that the other parameters remain unaffected, according to the ITTC guidelines. However, in practice, iterative uncertainties are negligible for steady-state simulations. Thus, mostly grid size and time step uncertainties are analysed

Convergence studies for parameters are performed following a systematic refinement process to create multiple solutions (at least 3) for the parameter under investigation, holding all other parameters constant. Say, for grid refinement, if r_i is the refinement ratio, $r_i = \Delta x_{i,2}/\Delta x_{i,1} = \Delta x_{i,3}/x_{i,2} = \Delta x_{i,m}/x_{i,m-1}$. Here, in $\Delta x_{i,m}$, x is the unit cell size in the mesh, i is the refinement ratio and m is the number of solutions. The process is the same for time or other parameter study. As for the refinement ratio, r_i, the recommended value is $\sqrt{2}$, since the value is large enough to be sensitive to parameter changes and small enough to be used to generate at least three successive solutions to maintain the economy. A larger refinement ratio may be used, however, m has to be at least three. Two solutions may be able to check parameter sensitivity but not convergence.

If $\hat{S}_{i,1}, \hat{S}_{i,2}$ and $\hat{S}_{i,3}$ represent three different solutions with fine, medium and coarse input parameters, respectively, then changes between medium-fine and coarse-medium input are represented by, $\varepsilon_{i,21} = \hat{S}_{i,2} - \hat{S}_{i,1}$ and $\varepsilon_{i,32} = \hat{S}_{i,3} - \hat{S}_{i,2}$. So, the convergence ratio for the input parameter is defined as,

$$R_i = \varepsilon_{i,21}/\varepsilon_{i,32} \tag{7}$$

Depending on the sign and magnitude of R_i, three convergence conditions are possible:

1. Monotonic convergence: $0 < R_i < 1$
2. Oscillatory convergence: $R_i < 0$
3. Divergence: $R_i > 1$

If the convergence is monotonic, Richardson extrapolation is used to estimate U_i or δ_i^* and U_{ic}. For oscillatory convergence, uncertainties are estimated by bounding the error within the average of oscillating maxima (S_U) and minima (S_L), $U_i = \frac{1}{2}(S_U - S_L)$. When there is divergence the errors and uncertainties cannot be estimated.

After the convergence study, for monotonous convergence, the order of convergence rate or accuracy (p_i) is estimated, next, Richardson extrapolation is used to estimate normalised discretisation error ($\delta_{RE\ i,1}^*$) in the solution.

$$p_i = \frac{\ln(\varepsilon_{i,32}/\varepsilon_{i,21})}{\ln(r_i)} \tag{8}$$

$$\delta_{RE\ i,1}^* = \frac{\varepsilon_{i,21}}{r_i^{p_i} - 1} \tag{9}$$

Only one term error and accuracy order can be estimated with three solutions. With five solutions, two terms can be estimated, which provides two-term estimates for error and orders of accuracy. Equations for two-term estimates can be found in ITTC guidelines (2008).

Finally, the Grid Convergence Index (GCI) and corrected simulation uncertainty are calculated by,

$$GCI_{fine}^{21} = Fs \left|\delta_{RE\ i,1}^*\right| \tag{10}$$

$$U_{ic} = (Fs - 1) \left|\delta_{RE\ i,1}^*\right| \tag{11}$$

Thus, GCI is defined as a multiple of Richardson normalised discretisation error by Fs (factor of safety). The value of Fs is suggested to be 1.25 for systematic parameter refinement study with at least three inputs or 3 for simple convergence studies with two input parameter values. The multiplication of the factor of safety makes the error band wider, thus making the prediction less conservative. However, some researchers believe that the value of Fs is ambiguous, and the approach is over-conservative when solutions are close to the asymptotic range and under-conservative when solutions are far from the asymptotic range.

After predicting the simulation numerical uncertainty, validation uncertainty is calculated by adding the uncertainty involved in the data. So, with validation uncertainty, U_v, error $|E|$ and the required level of validation uncertainty U_{reqd}, six possible scenarios can be observed.

1. $|E| < U_v < U_{reqd}$
2. $|E| < U_{reqd} < U_v$
3. $U_{reqd} < |E| < U_v$
4. $U_v < |E| < U_{reqd}$
5. $U_v < U_{reqd} < |E|$
6. $U_{reqd} < U_v < |E|$

For cases, 1, 2, and 3, $|E| < U_v$; so validation is achieved at U_v level. In the case of 4, 5, and 6, $U_v < |E|$, so the comparison error is above the noise level, and validation cannot be achieved. For 4, though, validation is achieved from a programmatic perspective.

Two of the most popular methods to analyse the grid and time step uncertainties are the Correction Factor (Stern et al. 2001, 2006 and Wilson et al. 2001) and the Factor of Safety (Celik et al. 2008) based method. Both methods are based on Richardson extrapolation and are recommended procedures in the ITTC (2017) guidelines for uncertainty quantification in CFD studies.

Celik et al. (2008) used a slightly different approach for predicting the apparent order and also proposed further calculations for predicting extrapolated values, approximate relative error, and extrapolated relative error. As an alternative Wilson et al. (2004) proposed a correction factor (Cf) based approach

ITTC guidelines recommend estimating grid and time step uncertainty separately for numerical uncertainty prediction. However, grid size and time step are closely related in transient CFD simulations. For solution convergence, cells in mesh must conform to the Courant-Friedrichs-Lewy (CFL) condition. Say, if a cell has a length of Δx in flow direction and the time step is Δt, for a domain with a flow velocity of u, the CFL value would be CFL = $u\Delta t/\Delta x$. For solution convergence, the CFL value should be less than 1. Therefore, changing the grid and time step separately, as recommended by ITTC, may lead to situations that violate the CFL condition. To avoid this, Wang et al. (2021a), following Oberhagemann (2016), have proposed that a constant CFL number-based approach should be followed where the uncertainty related to the grid and time step is evaluated together, ensuring simulation stability.

7.3 Application Studies

After a brief literature review, Islam and Guedes Soares (2019) conducted a systematic verification and validation study for four ship models using the open-source CFD toolkit OpenFOAM. A systematic verification study was performed, and the total corrected uncertainty was estimated. Three different mesh resolutions and three-time steps were used for each case. Uncertainty for the cases was estimated using the safety factor (Fs) and correction factor (Ci) based approach. The results show that, despite having geometric similarities, mesh dependency for each hull form is different from the other. Overall, it may be concluded from the study that the safety factor and correction factor-based approach differ significantly from each other in uncertainty prediction. Furthermore, it is not just a case-specific validation study; the verification study is also case-specific.

Wang et al. (2021a) studied the numerical uncertainty due to discretisation on the Arbitrary Lagrangian-Eulerian (ALE) Finite Element method. They quantified uncertainty using two ITTC-recommended methods and applied a constant Courant-Friedrichs-Lewy (CFL) number-based discretisation approach instead of performing the independent grid and time-based discretisation recommended by ITTC. As a case study, water entry of a flat bottom rigid and flexible plate was simulated. Results indicate that numerical errors due to discretisation differ in the various parameters and from case to case. The uncertainty quantification procedures with a constant CFL number-based refinement were recommended to investigate the uncertainty compared to the individual grid and time step study, in particular for the ALE solution, where the time step was adjusted automatically as the grid changed.

Their results showed that the uncertainties based on the Cf method are generally higher than the calculations using the Fs method. The calculations of discretisation errors based on the Cf method are very different from case to case. Generally, the based method predicts a higher uncertainty for the ALE simulations than Fs-based methods. The low level of uncertainty found following the Fs-based method indicates a good verification of the ALE results. However, the low margin of uncertainty also limits the margin of deviation expected from the experimental data while doing validation. Thus, a relatively higher level of uncertainty, as found from the Cf-based method, can be more favourable while performing a validation study. As such, concluding which method may be more suitable for estimating uncertainties is difficult due to discretisation in ALE.

The uncertainty study concludes that the uncertainty due to discretisation in ALE is not just case-specific but also parameter-specific. In this case, the pressure uncertainty is much higher than for forces.

Islam and Guedes Soares (2021) compared the constant Courant–Friedrichs–Lewy (CFL) number-based approach for uncertainty estimation proposed by Wang et al. (2021a) to the ITTC (2017) recommended grid and time-independent procedures. They produced CFD results for wave-induced loads on a fixed vertical cylinder and performed detailed uncertainty analysis to understand simulation-related uncertainties in wave load prediction on these structures. They concluded that a constant CFL number-based uncertainty study provides more stable results and is better suited for uncertainty estimation in CFD than the ITTC recommended individual grid and time-step uncertainty study.

Wang et al. (2021b) studied the free-falling water entry of 2D rigid sections by using the dynamic mesh method and VOF method in the open-source library OpenFOAM. Validation and verification studies were performed for the case of a symmetric wedge section with a deadrise angle of 30°. Grid uncertainty analysis for the CFD results is performed with a constant CFL number. The Correction Factor (Cf) based approach, Factor of Safety (Fs) based approach, and Least-square Fit (Lf) based approach are applied for the uncertainty quantification due to mesh resolution, showing that each of the methods has their advantages and limitations, while all of them can provide reliable uncertainty estimation for CFD simulations, as long as the correct mesh configuration has been used.

Wang et al. (2022) compared the uncertainty quantification methods and the accuracy of mesh-free and mesh-based simulations due to discretisation. Two open-source tools, DualSPHysics (mesh-free) and OpenFOAM (mesh-based), were used to investigate problems related to wave energy converters, such as Point Absorbers and Oscillating Water Columns. Two cases were studied: a sphere free-falling into the water and a heaving box subjected to regular waves. Numerical uncertainties were quantified using the Factor of Safety, Correction Factor, and least square fit methods. The validation and verification analysis showed that all three uncertainty estimation methods are reliable when a constant CFL number is used. Numerical results from the higher spatial resolution were compared with published experiments, showing that both numerical methods converge as the spatial resolution increases. The accuracy of the results obtained with DualSPHysics and OpenFOAM using high spatial resolutions were similar.

Huang et al. (2022) systematically analysed the influence of fluid grid density, time step, fluid domain size, fluid viscosity, and hull structural element count on ship response results using the CFD–FEA coupling method. The numerical uncertainties of the CFD and FEA solvers were assessed separately, and the numerical predictions were validated against experimental data and benchmark results for the S175 container ship.

Although the ITTC recommends that the downstream region in a numerical wave tank should be at least three times the ship's length (3L), this study found that a shorter length (1.5L) does not significantly impact wave calculations or ship responses when using the EOM wave-making technique. The uncertainty analysis revealed that the 2-node whipping loads exhibit greater uncertainty than the 1st and 2nd harmonic responses. Additionally, significant deviations were observed in ship wave load predictions across different methods, highlighting the need for greater attention to uncertainty quantification in numerical simulations.

Other authors have been working on this subject area, but for the review of that literature reference is made to the ISSC Committee II.1 Loads, while the present report deals with the work developed in relation to Committee V8.

8 Conclusions

This report provides an overview of the work performed by the ISSC-ITTC Joint Committee since its start in 2017 up to the present. As initially planned, the committee has performed some review papers and benchmark studies. Committee members have conducted several papers, often in collaboration with other authors, demonstrating different applications of uncertainty modelling.

The various earlier ISSC committees included the assessment of uncertainty in their respective subject areas in their mandates. The committee reviews and the benchmark studies' results are published in the refereed papers. We hope this report has provided a balanced overview of the results achieved.

References

Aarnes, O.J., et al.: Projected changes in significant wave height towards the end of the 21st century – Northeast Atlantic. J. Geophys. Res. Oceans. **122**(4), 3394–3403 (2017). https://doi.org/10.1002/2016JC012521

Abdelwahab, H.S., Guedes Soares, C.: Experimental uncertainty of a physical model of a tanker moored to a terminal in a port. Mar. Struct. **87**, 103331 (2023)

Abdelwahab, H.S., Pinheiro, L.V., Santos, J.A., Fortes, C.J.E.M., Guedes Soares, C.: Experimental investigation of wave severity and mooring pretension on the operability of a moored tanker in a port terminal. Ocean Eng. **291**, 116243 (2024)

Abdelwahab, H.S., Wang, S., Parunov, J., Guedes Soares, C.: A new model uncertainty measure of wave-induced motions and loads on a container ship with forward speed. J. Mar. Sci. Eng. **11**(5), 1042 (2023). https://doi.org/10.3390/jmse11051042

Abels, W.: Integration of software packages with different programming paradigms, exemplified by a ship model generating tool. In: 6th International Conference on Computer and IT Applications in the Maritime Industries (Compit 2007), 23–25 April, Cortona, Italy (2007)

AIAA: Guide for the Verification and Validation of Computational Fluid Dynamics Simulations (AIAA G-077-1998(2002)) (1998). https://doi.org/10.2514/4.472855.001

Aldous, L., Smith, T., Bucknall, R., Thompson, P.: Uncertainty analysis in ship performance monitoring. Ocean Eng. **110**, 29–38 (2015)

Alonso, F., González, C., Pastor, L.: Collaborative design – a practical approach based on the use of replicated databases for concurrent design. In: 6th International Conference on Computer and IT Applications in the Maritime Industries (Compit 2007), 23–25 April, Cortona, Italy (2007)

Amaechi, C.V., Wang, F., Ye, J.: Understanding the fluid-structure interaction from wave diffraction forces on CALM buoys: numerical and analytical solutions. Ships Offshore Struct. **17**(11), 2545–2573 (2022)

Ang, A., Tang, W.H.: Probability Concepts in Engineering, 2nd edn. Wiley (2017)

Antao, E., Guedes Soares, C.: Approximation of bivariate probability density of individual wave steepness and height with copulas. Coast. Eng. **89**, 45–52 (2014). https://doi.org/10.1016/j.coastaleng.2014.03.009

Antao, E., Guedes Soares, C.: Joint distributions of wave steepness in narrow band sea states. Ocean Eng. **101**, 201–210 (2015). https://doi.org/10.1016/j.oceaneng.2015.04.007

Antao, E., Guedes Soares, C.: Approximation of the joint probability density of steepness and height of individual waves with a bivariate gamma distribution. Ocean Eng. **126**, 402–410 (2016). https://doi.org/10.1016/j.oceaneng.2016.09.015

Ardag, D., Resio, D.T.: Inconsistent spectral evolution in operational wave models due to inaccurate specification of nonlinear interactions. J. Phys. Oceaonogr. **49**, 705–722 (2019)

Ardhuin, F., et al.: Semi-empirical dissipation source functions for ocean waves. Part 1: definition, calibration and validation. J. Phys. Oceanogr. **40**, 1917–1941 (2011a)

Ardhuin, F., et al.: Calibration of the IOWAGA global wave hindcast (1991–2011) using ECMWF and CFSR winds. In: Proceedings of 2011 International Workshop on Wave Hindcasting and Forecasting and 3rd Coastal Hazard Symposium, 30 October–4 November, Kona, HI, USA (2011b)

Athanassoulis, G.A., Skarsoulis, E.K., Belibassakis, K.A.: Bivariate distributions with given marginals with an application to wave climate description. Appl. Ocean Res. **16**, 1–17 (1994)

Aven, T., Reniers, G.: How to define and interpret a probability in a risk and safety setting. Saf. Sci. **51**(1), 223–231 (2013). https://doi.org/10.1016/j.ssci.2012.06.005

Bai, J.L., Ma, N., Gu, X.C.: Numerical simulation of focused wave and its uncertainty analysis. J. Shanghai Jiao Tong Univ. (Sci.). **23**(4), 475–481 (2028)

Barstow, S., et al.: WORLDWAVES: high quality coastal and offshore wave data within minutes for any global side. In: Proceedings of 22nd Int Conf on Ocean, Offshore & Arctic Engineering (OMAE 2003), June 8–13, Cancun, Mexico, paper OMAE2003-37297, vol. 3 (2003). https://doi.org/10.1115/OMAE2003-37297

Barstow, S., Mørk, G., Lønseth, L., Mathisen, J.P.: WorldWaves wave energy resource assessments from the deep ocean to the coast. J. Energy Power Eng. **5**(8), 730–742 (2011)

Benetazzo, A., Barrariol, F., Bergamasco, F., Torsello, A., Carniel, S., Sclavo, M.: Observation of extreme sea waves in a space–time ensemble. Am. Meteorol. Soc. **45**, 2261–2275 (2015). https://doi.org/10.1175/JPO-D-15-0017.1

Benetazzo, A., Barrariol, F., Bergamasco, F., Torsello, A., Carniel, S., Sclavo, M.: On the shape and likelihood of oceanic rogue waves. Sci. Rep. **7**, 8276 (2017). https://doi.org/10.1038/s41598-017-07704-9

Benjamin, J.R., Cornell, C.A.: Probability, Statistics and Decision for Civil Engineers. McGraw-Hill, New York (1970)

Bernardino, M., Gonçalves, M., Guedes Soares, C.: Assessing climate change in the North Atlantic wave regimes. J. Offshore Mech. Arct. Eng. **143**(6), 061201 (2016). https://doi.org/10.1115/1.4050698

Bidlot, J.-R., Lemos, G., Semedo, A.: ERA5 reanalysis and ERA5 based ocean wave hindcast. In: 2nd International Workshop on Waves, Storm Surges and Coastal Hazards, 10–15 November, Melbourne, Australia (2019)

Bitner-Gregersen, E.M.: Joint long-term models of met-ocean parameters. In: Guedes Soares, C., Garbatov, Y., Fonseca, N., Texeira, A.P. (eds.) Marine Technology and Engineering, pp. 19–34. Taylor & Francis Group, London (2012)

Bitner-Gregersen, E.M.: Joint met-ocean description for design and operations of marine structures. Appl. Ocean Res. **51**, 279–292 (2015). https://doi.org/10.1016/j.apor.2015.01.007

Bitner-Gregersen, E.M., et al.: Recent developments of ocean environmental description with focus on uncertainties. Ocean Eng. **86**, 26–46 (2014a)

Bitner-Gregersen, E.M., Cramer, E.H., Korbijn, F.: Environmental description for long-term load response of ship structures. In: Proceedings of the Fifth International Offshore and Polar Engineering Conference (ISOPE 1995), 11–16 June, The Hague, the Netherlands (1995a)

Bitner-Gregersen, E.M., Cramer, E.H., Løseth, R.: Uncertainty of load characteristics and fatigue damage of ship structures. Mar. Struct. **8**, 97–117 (1995b)

Bitner-Gregersen, E.M., et al.: Sea state conditions for marine structures' analysis and model tests. Ocean Eng. **119**, 309–322 (2016)

Bitner-Gregersen, E.M., Eide, L.I., Hørte, T., Vanem, E.: Impact of climate change and extreme waves on tanker design. In: SNAME Maritime Convention, October 22–24, Houston, Texas, USA SNAME Transactions SNAME-SMC-2014-T43 (2014d)

Bitner-Gregersen, E.M., Ewans, K.C., Johnson, M.C.: Some uncertainties associated with wind and waves description and their importance for engineering applications. Ocean Eng. **86**, 11–25 (2014b)

Bitner-Gregersen, E., Fernandez, L., Lefevre, J.-M., Toffoli, A.: The North Sea Andrea storm and numerical simulations. Nat. Hazards Earth Syst. Sci. **14**, 1407–1415 (2014c). https://doi.org/10.5194/nhess-14-1407-2014

Bitner-Gregersen, E.M., Gramstad, O.: Comparison of temporal and spatial statistics of nonlinear waves. In: Proceedings of the 38th International Conference on Ocean, Offshore and Arctic Engineering (OMAE 2019), 9–14 June, Glasgow, UK, paper: OMAE2019-95357, V003T02A001 (2019)

Bitner-Gregersen, E.M., Gramstad, O., Magnusson, A.K., Malila, M.: Challenges in description of nonlinear waves due to sampling variability. J. Mar. Sci. Eng. **8**, 279 (2020a). https://doi.org/10.3390/jmse8040279

Bitner-Gregersen, E.M., Gramstad, O., Magnusson, A.K., Malila, M.: Extreme wave events and sampling variability. Ocean Dyn. **71**, 81–95 (2020b). https://doi.org/10.1007/s10236-020-01422-z

Bitner-Gregersen, E.M., Gramstad, O., Magnusson, A.K., Sames, P.: Occurrence frequency of the triple rogue waves in the ocean. In: Proceedings of 39th International Conference on Ocean, Offshore & Arctic Engineering (OMAE 2020), 28 June-3 July, Fort Lauderdale, FL, USA, paper OMAE2020-19314, V02AT02A009 (2020c). https://doi.org/10.1115/OMAE2020-19314

Bitner-Gregersen, E.M., Guedes Soares, C.: Uncertainty of average wave steepness prediction from global wave databases. In: Guedes Soares, C., Das, P.K. (eds.) Advancements in Marine Structures, pp. 3–10. Taylor & Francis Group, London (2007)

Bitner-Gregersen, E.M., Hagen, Ø.: Uncertainties in data for the offshore environment. Struct. Saf. **7**, 11–34 (1990). https://doi.org/10.1016/0167-4730(90)90010-M

Bitner-Gregersen, E.M., Hagen, Ø.: Extreme value analysis of wave steepness and crest using a joint environmental description. In: Proceedings of the 18th International Conference on Ocean, Offshore & Arctic Engineering (OMAE 1999), 11–16 July, St. John's, Newfoundland, Canada (1999)

Bitner-Gregersen, E.M., Haver, S.: Joint long term description of environmental parameters for structural response calculations. In: Proceedings of 2nd International Workshop on Wave Hindcasting and Forecasting, 24–29 September, Vancouver, B.C., Canada (1989)

Bitner-Gregersen, E.M., Haver, S.: Joint environmental model for reliability calculations. In: Proceedings of the First International Offshore and Polar Engineering Conference (ISOPE 1991) 11–16 August, Edingburgh, UK, pp. 246–253 (1991)

Bitner-Gregersen, E.M., Lars, I.E., Hørte, T., Skjong, R.: Ship and Offshore Structure Design in Climate Change Perspective. Monograph, Springer Brief in Climate Studies (2013)

Bitner-Gregersen, E.M., Magnusson, A.K.: Effect of intrinsic and sampling variability on wave parameters and wave statistics. Ocean Dyn. **64**, 1643–1655 (2014) http://link.springer.com/book/10.1007/978-3-642-34138-0/page/1#page-1

Bitner-Gregersen, E.M., et al.: Climate change and safe design of ship structures. Ocean Eng. **149**, 226–237 (2018). https://doi.org/10.1016/j.oceaneng.2017.12.023

Bitner-Gregersen, E.M., et al.: Uncertainties in long-term wave modelling. Mar. Struct. **84**, 103217 (2022)

Bitner-Gregersen, E.M.: Application 2: shipping and offshore industry. In: Mori, N., Waseda, T., Chabchoub, A. (eds.) Science and Engineering of Freak Waves. Elsevier (2023)

Bitner-Gregersen, E.M., et al.: Rogue waves: results of the ExWaMar project. Ocean Eng. **292**, 116543 (2024). https://doi.org/10.1016/j.oceaneng.2023.116543

BMT (British Maritime Technology), Primary Contributors: Hogben, N., Da Cunha, L.F., Oliver, H.N.: Global Wave Statistics. Unwin Brothers Limited, London (1986)

Cabral, I.S., Young, I.R., Toffoli, A.: Long-term and seasonal variability of wind and wave extremes in the Arctic ocean. Front. Mar. Sci. **9**, 802022 (2022). https://doi.org/10.3389/fmars.2022.802022

Campos, R.M., Alves, J.H.G.M., Guedes Soares, C., Guimarães, L.G., Parente, C.E.: Extreme wind-wave modeling and analysis in the South Atlantic Ocean. Ocean Model. **124**, 75–79 (2018)

Campos, R.M., D'Agostini, A., Franca, B.R.L., Damiao, A.L.A., Guedes Soares, C.: Implementation of a multi-grid operational wave forecast in the South Atlantic Ocean. Ocean Eng. **243**, 110173 (2022)

Campos, R.M., Guedes Soares, C.: Comparison and assessment of three wave hindcasts in the North Atlantic Ocean. J. Oper. Oceanogr. **9**(1), 26–44 (2016)

Campos, R.M., Guedes Soares, C.: Assessment of three wind reanalysis in the North Atlantic Ocean. J. Oper. Oceanogr. **10**(1), 30–44 (2017)

Campos, R.M., Guedes Soares, C., Alves, J.H.G.M., Parente, L.G., Guimarães, L.G.: Regional long-term extreme wave analysis using hindcast data from the South Atlantic Ocean. Ocean Eng. **179**, 202–212 (2019)

Cavaleri, L., Bertotti, L.: The improvement of modelled wind and wave fields with increasing resolution. Ocean Eng. **33**, 553–565 (2006)

Cavaleri, L., et al.: Wave modelling – the state of the art. Prog. Oceanogr. **75**(4), 603–674 (2007)

Cavaleri, L., Barbariol, F., Benatazzo, A.: Wind-wave modelling: where we are, where we go. J. Mar. Sci. Eng. **8**, 260 (2020). https://doi.org/10.3390/jmse8040260

Celik, I.B., Ghia, U., Roache, P.J., Coleman, H., Raad, P.E.: Procedure for estimation and reporting of uncertainty due to discretization in CFD application. ASME J. Fluid Eng. **130**(7), 078001–078004 (2008)

Chabchoub, A., Waseda, T., Kibler, B., Akhmediev, N.: Experiment on higher-order and degenerate Akhmediev breather-type rogue waves. J. Ocean Eng. Mar. Energy. **3**, 385–394 (2017)

Chalikov, D.V., Sheinin, D.: Numerical modelling of surface waves based on principal equations of potential wave dynamics. Tech. Note NOAA/NCEP/OMB. **54**(139) (1996)

Chalikov, D., Sheinin, D.: Modeling extreme waves based on equations of potential flow with a free surface. J. Comput. Phys. **210**(1), 247–273 (2005)

Chen, X., Okada, T., Kawamura, Y., Mitsuyuki, T.: Estimation of on-site directional wave spectra using measured hull stresses on 14,000 TEU large container ships. J. Mar. Sci. Technol. **25**, 690–706 (2020)

Chen, X., Li, Q., Zhang, H.: A novel approach to predicting ship motion responses in irregular wave fields. J. Offshore Mech. Arct. Eng. **145**(4) (2023a)

Chen, X., Takami, T., Oka, M., Kawamura, Y., Okada, T.: Stochastic wave spectra estimation (SWSE) based on response surface methodology considering uncertainty in transfer functions of a ship. Mar. Struct. **90**, 103423 (2023b)

Ćorak, M., Parunov, J., Guedes Soares, C.: Long-term prediction of combined wave and whipping bending moments of container ships. Ships Offshore Struct. **10**(1), 4–19 (2015)

Ćorak, M., Parunov, J., Guedes Soares, C.: Structural reliability analysis of container ships under combined wave and whipping loads. J. Ship Res. **62**, 115–133 (2018)

Cornell, C.A.: Structural safety specifications based on second moment reliability analysis. In: Symposium on Concepts of Safety of Structures and Methods of Design (IAB-SE), London, pp. 235–246 (1969)

Dee, D.P., et al.: The Era-Interim reanalysis: Configuration and performance of the data assimilation system. Q. J. R. Meteorol. Soc. **137**(656), 553–597 (2011)

Delen, C., Bal, S.: Uncertainty analysis of resistance tests in Ata Nutku Ship Model Testing Laboratory of Istanbul Technical University. J. Marit. Mar. Sci. **1**(2), 8–27 (2015)

Der Kiureghian, A.: Structural and System Reliability. Cambridge University Press (2022)

Ditlevsen, O., Madsen, H.O.: Structural Reliability Methods. Wiley (1996)

Det Norske Veritas (DNV): Classification Notes No. 30.6., Structural Reliability Analysis of Marine Structures (1992)

DNV: RP-C205 Environmental Conditions and Environmental Loads, DNV GL, September (2019a)

DNV: Planning and Execution of Marine Operations – DNVGL-ST-N001 A Worldwide Standard for Marine Operations (2019b)

Dommermuth, D.G., Yue, D.K.: A high–order spectral method for the study of nonlinear gravity waves. J. Fluid Mech. **184**, 267–288 (1987)

Dong, S., Wang, N., Liu, W., Guedes Soares, C.: Bivariate maximum entropy distribution of significant wave height and peak period. Ocean Eng. **59**, 86–99 (2013)

Dong, S., Tao, S., Li, X., Guedes Soares, C.: Trivariate maximum entropy distribution of significant wave height, wind speed and relative direction. Renew. Energy. **78**, 538–549 (2015)

Drummen, I., Holtmann, M.: Benchmark study of slamming and whipping. Ocean Eng. **86**, 3–10 (2014)

Duan, W., Wang, H., Chen, J., Wang, H.: TEBEM for springing responses of a container ship with forward speed and nonlinear effects in time domain. Eng. Anal. Bound. Elem. **140**, 406–420 (2022)

Ducrozet, G., Bonnefoy, F., Le Touzé, D., Ferrant, P.: Hos-ocean: open-source solver for nonlinear waves in open ocean based on high-order spectral method. Comput. Phys. Commun. **203**, 245–254 (2016)

Ducrozet, G.: Nonlinear wave modelling with higher order spectral method and applications to ocean engineering. In: OMAE 2024 Short-Course, 43rd International Conference on Ocean, Offshore and Arctic Engineering (OMAE 2024), 9–14 June, Singapore (2024)

Dysthe, K.B.: Note on a modification to the nonlinear Schrödinger equation for application to deep water waves. Proc. R. Soc. Lond. A. **369**, 105–114 (1979)

Eisinger, E., Bloch Helmers, J., Storhaug, G.: A method for describing ocean environments for ship assessment. In: Proceedings of 6th International Maritime Conference on Design for Safety, 24–30 November, Hamburg, Germany (2016)

Ewans, K.C., Bitner-Gregersen, E., Guedes Soares, C.: Estimation of wind-sea and swell components in a bimodal sea state. J. Offshore Mech. Arct. Eng. **128**(4), 265–270 (2006)

Ewans, K., Jonathan, P.: Uncertainties in estimating the effect of climate change on 100-year return value for significant wave height. Ocean Eng. **272**, 113840 (2023). https://doi.org/10.1016/j.oceaneng.2023.113840

Faulkner, D., Sadden, J.A.: Toward a unified approach to ship structural safety. Trans. Soc. Nav. Archit. Mar. Eng. **4**, 1–38 (1979)

Fedele, F., Benetazzo, A., Forristall, G.: Space-time waves and spectra in the Northern Adriatic Sea via a Wave Acquisition Stereo System. In: Proceedings of 30th International Conference on Ocean, Offshore & Arctic Engineering (OMAE 2011), 19–24 June, Rotterdam, The Netherlands, paper OMAE2011-49924 (2011)

Fedele, F., et al.: Space–time measurements of oceanic sea states. Ocean Model. **70**, 103–115 (2013)

Feld, G., Randell, D., Wu, Y., Ewans, K.C., Jonathan, P.: Estimation of storm peak and intrastorm directional-seasonal design conditions in the North Sea. In: Proceedings of the 33rd International Conference on Ocean, Offshore & Arctic Engineering (OMAE 2014), 7–14 June, San Francisco, USA, paper OMAE2014-23157, V04AT02A014 (2014). https://doi.org/10.1115/OMAE2014-23157

Feld, G., Jonathan, P., Randell, D.: On the estimation and application of directional design criteria. In: Proceedings of the 38th International Conference on Ocean, Offshore & Arctic Engineering (OMAE 2019), 9–14 June, Glasgow, Scotland, UK, paper OMAE2019-96586, V07BT06A011 (2019)

Ferreira, J.A., Guedes Soares, C.: An application of the peaks over threshold method to predict extremes of significant wave height. J. Offshore Mech. Arct. Eng. **120**(3), 165–176 (1998)

Ferreira, J.A., Guedes Soares, C.: Modelling bivariate distributions of significant wave height and mean wave period. Appl. Ocean Res. **24**(1), 31–45 (2002)

Flokstra, C.: Comparison of ship motion theories with experiments for a container ship. Int. Shipbuild. Prog. **21**, 168–189 (1974). https://doi.org/10.3233/ISP-1974-2123802

Forristall, G.: Maximum crest heights over an area: laboratory measurements compared to theory. In: Proceedings of the 34th International Conference on Ocean, Offshore and Artic Engineering (OMAE 2015), 31 May–5 June, St. John's, NL, Canada, paper OMAE2015-41061, V003T02A044 (2015)

Forristall, G.Z.: Wave crest distributions: observations and second order theory. J. Phys. Oceanogr. **30**, 1931–1943 (2000)

Fujikubo, M., et al.: A digital twin for ship structures – R&D project in Japan. Data-Centric Eng. **5**, e7 (2024)

Fujimoto, W., Waseda, T., Webb, A.: Impact of the four-wave quasi-resonance on freak wave shapes in the ocean. Ocean Dyn. **69**(1), 101–121 (2019)

Gonçalves, M., Bernardino, M., Guedes Soares, C.: Assessing climate change effects on the wave energy in the Canary Islands. In: Guedes Soares, C. (ed.) Developments in Renewable Energies Offshore, pp. 19–25. Taylor and Francis, London (2021)

Gramcianinov, C.B., Campos, R.M., Guedes Soares, C., Camargo, R.: Extreme waves generated by cyclonic winds in the western portion of the South Atlantic Ocean. Ocean Eng. **213**, 107745 (2020a)

Gramcianinov, C.B., Campos, R.M., Camargo, R., Hodges, K.I., Guedes Soares, C., Silva Dias, P.L.: Analysis of Atlantic extratropical storm tracks characteristics in 41 years of ERA5 and CFSR/CFSv2 Databases. Ocean Eng. **216**, 108111 (2020b)

Grammatikopoulos, A.: A review of physical flexible ship models used for hydroelastic experiments, marine structures. Mar. Struct. **90**, 103436 (2023)

Grammatikopoulos, A., Banks, J., Temarel, P.: Prediction of the vibratory properties of ship models with realistic structural configurations produced using additive manufacturing. Mar. Struct. **73**, 102801 (2020)

Gramstad, O., Babanin, A.: The generalized kinetic equation as a model for the nonlinear transfer in third-generation wave models. Ocean Dyn. **66**(4), 509–526 (2016)

Gramstad, O., Bitner-Gregersen, E.M., Trulsen, K., Nieto Borge, J.C.: Modulational instability and rogue waves in crossing sea states. J. Phys. Oceanogr. **48**, 1317–1331 (2018). https://doi.org/10.1175/JPO-D-18-0006.1

Gramstad, O., Lian, G.: Parametrization of sea surface skewness and kurtosis with application to crest distributions. J. Fluid Mech. **979**, A4 (2024). https://doi.org/10.1017/jfm.2023.1047

Gu, N., Liang, D., Zhou, X., Ren, H.: A CFD-FEA method for hydroelastic analysis of floating structures. J. Mar. Sci. Eng. **11**(4), 737 (2023)

Guedes Soares, C.: Assessment of the uncertainty in visual observations of wave height. Ocean Eng. **3**(1), 37–56 (1986a)

Guedes Soares, C.: Calibration of visual observations of wave period. Ocean Eng. **13**(6), 539–547 (1986b)

Guedes Soares, C.: Uncertainty modelling in plate buckling. Struct. Saf. **5**(1), 17–34 (1988)

Guedes Soares, C.: Stochastic models of load effects for the primary ship structure. Struct. Saf. **8**(1–4), 353–368 (1990)

Guedes Soares, C.: Effect of transfer function uncertainty on short-term ship responses. Ocean Eng. **18**(4), 329–362 (1991). https://doi.org/10.1016/0029-8018(91)90018-L

Guedes Soares, C., Bento, A.R., Gonçalves, M., Silva, D., Martinho, P.: Numerical evaluation of the wave energy resource along the Atlantic European coast. Comput. Geosci. **71**, 37–49 (2014)

Guedes Soares, C., Cherneva, Z., Antão, E.: Characteristics of abnormal waves in north sea storm sea states. Appl. Ocean Res. **25**(6), 337–344 (2003)

Guedes Soares, C., Cherneva, Z., Antão, E.: Steepness and asymmetry of the largest waves in storm sea states. Ocean Eng. **31**(8–9), 1147–1167 (2004)

Guedes Soares, C., Dogliani, M., Ostergaard, C., Parmentier, G., Pedersen, P.T.: Reliability-based ship structural design. Trans. Soc. Nav. Archit. Mar. Eng. **104**, 357–389 (1996)

Guedes Soares, C., Lopes, L.C., Costa, M.: Wave climate modelling for engineering purposes. In: Schreffler, B.A., Zienkiewicz, O.C. (eds.) Computer Modelling in Ocean Engineering, pp. 169–175. A.A. Balkema, Rotterdam (1988)

Guedes Soares, C., Moan, T.: Uncertainty analysis and code calibration of the primary load effects in ship structures. In: Konishi, K., Ang, A.H.-S., Shinozuka, M. (eds.) Structural Safety and Reliability, vol. III, pp. 501–512 (1985)

Guedes Soares, C., Moan, T.: Model uncertainty in the long term distribution of wave induced bending moments for fatigue design of ship structures. Mar. Struct. **4**, 295–315 (1991)

Guedes Soares, C., Scotto, M.G.: Modelling uncertainty in long-term predictions of significant wave height. Ocean Eng. **28**(3), 329–342 (2001)

Guedes Soares, C., Weiss, R., Carretero, J.C., Alvarez, E.: A 40 years hindcast of wind, sea level and waves in European Waters. In: Proceedings of the 21st International Conference on Offshore Mechanics and Arctic Engineering, (OMAE 2002), ASME, New York, paper OMAE2002-28604 (2002). https://doi.org/10.1115/OMAE2002-28604

Gulev, K., Grigorieva, V.: Extreme wind waves worldwide from the VOS data over the last 50 years. In: 9th International Workshop on Wave Hindcasting and Forecasting, 25–29 September, Victoria, B.C., Canada (2006)

Guratzsch, R.F., Mahadevan, S.: Health monitoring sensor placement optimization under uncertainty. In: 11th AIAA/ISSMO Multidisciplinary Analysis and Optimization Conference, 6–8 September, Portsmouth, Virginia, USA (2006)

Hagen, Ø., Garrè, L., Friis-Hansen, P.: DNV ADAPT framework for risk-based adaptation: a test case for the offshore industry. In: Proceedings of 11th International Conference on Structural Safety and Reliability (ICOSSAR 2013), 16–20 June, New York, USA (2013)

Hagen, Ø., Birknes-Berg, J., Grue, I.H., Lian, G., Bruserud, K., Vestbøstad, T.: Long-term area statistics for maximum crest height under fixed platform deck. In: Proceedings of the 37th

International Conference on Ocean, Offshore & Arctic Engineering (OMAE 2018), 17–22 June, Madrid, Spain, OMAE2018-77263, V003T02A037 (2018)

Hallak, T.S., Teixeira, A.P., Guedes Soares, C.: Epistemic uncertainties on estimating minimum air gap for semi-submersible platforms. Mar. Struct. **85**, 103244 (2022)

Haselsteiner, A.F., et al.: Benchmarking exercise for environmental contours. Ocean Eng. **236**, 109504 (2021)

Hasofer, A.M., Lind, N.C.: An exact and invariant first-order reliability format. J. Eng. Mech. Div. ASCE. **100**, 111–121 (1974)

de Hauteclocque, G., Zhu, T., Johnson, M., Austefjord, H., Bitner-Gregersen, E.: Assessment of global wave dataset for long term response of ships. In: Proceedings of OMAE 2020, 28 June–3 July, Fort Lauderdale, Florida, USA, paper OMAE 2020-18874, V02AT02A026 (2020)

de Hauteclocque, G., Maretic, N.V., Derbanne, Q.: Hindcast based global wave statistics. Appl. Ocean Res. **130**, 103438 (2023)

Haver, S.: Wave climate off northern Norway. Appl. Ocean Res. **7**, 85–92 (1985)

Heffernan, J.E., Tawn, J.A.: A conditional approach for multivariate extreme values. J. R. Stat. Soc. B. **66**, 497–546 (2004)

Heffernan, J.E., Resnick, S.I.: Limit laws for random vectors with an extreme component. Ann. Appl. Probab. **17**, 537–571 (2007)

Heredia-Zavoni, E., Montes-Iturrizaga, R.: Modeling directional environmental contours using three dimensional vine copulas. Ocean Eng. **187**, 106102 (2019)

Hinostroza, M.A., Guedes Soares, C.: Parametric estimation of the directional wave spectrum from ship motions. Int. J. Marit. Eng. **158**(Part A2), A121–A130 (2016)

Hinostroza, M.A., Guedes Soares, C.: Uncertainty analysis of parametric wave spectrum estimation from ship motions. In: Georgiev, P., Guedes Soares, C. (eds.) Sustainable Development and Innovations in Marine Technologies, pp. 70–78. Taylor & Francis (2020)

Hirdaris, S.E., et al.: Loads for use in the design of ships and offshore structures. Ocean Eng. **78**, 131–174 (2014)

Hirdaris, S., et al.: Review of the uncertainties associated to hull girder hydroelastic response and wave load predictions. Mar. Struct. **89**, 103383 (2023). https://doi.org/10.1016/j.marstruc.2023.103383

Hogben, N., Dacunha, N.M.C., Olliver, G.F.: Global Wave Statistics. Unwin Brothers, London (1986)

Horn, J.T., Bitner-Gregersen, E., Krokstad, J.R., Leira, B.J., Amdahl, J.: A new combination of conditional environmental distributions. Appl. Ocean Res. **73**, 17–26 (2018)

Hørte, T., Wang, G., White, N.: Calibration of the hull girder ultimate capacity criterion for double hull tankers. In: 10th International Symposium on Practical Design of Ships and Other Floating Structures (PRADS 2007), 30 September–5 October, American Bureau of Shipping, Houston, TX, USA, vol. 1, pp. 553–564 (2007)

Houtani, H., et al.: Designing a hydro-structural model ship to experimentally measure its vertical bending and torsional vibrations. J. Adv. Res. Ocean Eng. **4**(4), 174–184 (2018)

Houtani, H., Sawada, H., Waseda, T.: Phase convergence and crest enhancement of modulated wave trains. Fluids. **7**(8), 275 (2022). https://doi.org/10.3390/fluids7080275

Huang, S., Jiao, J., Guedes Soares, C.: Uncertainty analyses on the CFD–FEA co-simulations of ship wave loads and whipping responses. Mar. Struct. **82**, 103129 (2022). https://doi.org/10.1016/j.marstruc.2022.103129

Huang, Y., Wu, J., Zhang, Q.: Efficient simulation of wave-induced loads on floating structures with high-resolution CFD methods. Mar. Struct. **87**, 103338 (2023)

Huseby, A.B., Vanem, E., Natvig, B.: A new approach to environmental contours for ocean engineering applications based on direct Monte Carlo simulations. Ocean Eng. **60**, 124–135 (2013)

International Association of Classification Societies (IACS): Recommendation No. 34: Standard Wabe Data. Rev. 1. IACS, London (2001)
International Association of Classification Societies (IACS): UR S11 Longitudinal Strength Standard. Rev 6. IACS, London (2010)
International Association of Classification Societies (IACS): UR S11A Longitudinal Strength Standard for Container Ships. London (2015)
IACS: Standard Wave Data. (2001) IACS Rec. 34. 2000. Rev.1
IACS: Standard Wave Data. (2022) IACS Rec. 34. 2022. Rev.2
International Association of Classification Societies (IACS): Recommendation No. 34: Standard Wave Data. Rev. 2. IACS, London (2022)
International Association of Classification Societies (IACS): Common Structural Rules for Bulk Carriers and Oil Tankers. London (2023)
IPCC: The Fourth Assessment Report: Climate Change (AR4). The AR4 Synthesis Report, the Working Group I Report: The Physical Science Basis (ISBN 978 0521 88009-1 Hardback; 978 0521 70596-7 Paperback), the Working Group II Report Impacts: Adaptation and Vulnerability, the Working Group III Report: Mitigation of Climate Change (2007)
IPCC: Climate Change 2013: The Physical Science. Contribution of Working Group 1 to the Fifth Assessment Report of the Intergovernmental Panel on Climate Change (2013)
IPCC: AR6 Climate Change 2021: The Physical Science Basis. July 2021 (2021)
Iseki, T., Ohtsu, K.: Bayesian estimation of directional wave spectra based on ship motions. Control. Eng. Pract. **8**(2), 215–219 (2000). https://doi.org/10.1016/S0967-0661(99)00156-2
Islam, H., Guedes Soares, C.: Uncertainty analysis in ship resistance prediction using OpenFOAM. Ocean Eng. **191**, 105805 (2019)
Islam, H., Guedes Soares, C.: Assessment of uncertainty in the CFD simulation of the wave-induced loads on a vertical cylinder. Mar. Struct. **80**, 103088 (2021)
ISO/IEC Guide 98-3:2008 Uncertainty of Measurement – Part 3: Guide to the Expression of Uncertainty in Measurement (GUM:1995)
ISSC: ISSC 2009 Committee I.1 Environment Report. Proc. ISSC 2009, 1, 2009;1–126, Seoul, Korea, 2009
ISSC: ISSC 2012 Committee I.1 Environment Report. Proc. ISSC 2012, 2012;1:1–78, Rostock, Germany, 2012
ISSC: ISSC 2015 Committee I.1 Environment Report. Proc. ISSC 2015. 2015;1, Lisbon, Portugal, 2015
ISSC: ISSC 2018 Committee I.1 Environment Report. Proc. ISSC 2018, 2018;1, Amsterdam, The Netherlands, 2018
ISSC: ISSC 2022 Committee I.1 Environment Report. Proc. ISSC 2022, 2022;1, Vancouver, Canada, 2022
ITTC: Recommended Procedures and Guidelines. Uncertainty Analysis in CFD Verification and Validation Methodology and Procedures. ITTC – 7.5-03-01-01 (2017)
Izquierdo, P., Guedes Soares, C.: Analysis of sea waves and wind from X-band radar. Ocean Eng. **32**(11–12), 1404–1409 (2005)
Jafaryeganeh, H., Teixeira, A.P., Guedes Soares, C.: Uncertainty on the bending moment transfer functions derived by a three-dimensional linear panel method. In: Guedes Soares, C., Santos, T.A. (eds.) Maritime Technology and Engineering 3, pp. 295–302. Taylor & Francis Group, London (2016)
Jensen, J.J., Mansour, A.E., Smærup Olsen, A.: Estimation of ship motions using closed-form expressions. Ocean Eng. **31**, 61–85 (2004)
Jiang, Z., Li, F., Mikkola, T., Kujala, P., Hirdaris, S.: A boundary element method for the prediction of hydrodynamic ship–ice–wave interactions in regular waves. J. Offshore Mech. Arct. Eng. **145**(6), 061601 (2023)

Jiao, J.L., Ren, H.L., Guedes Soares, C.: A review of large-scale model at-sea measurements for ship hydrodynamics and structural loads. Ocean Eng. **227**, 108863 (2021)

Jonathan, P., Ewans, K.C., Randell, D.: Joint modelling of environmental parameters for extreme sea states incorporating covariate effects. Coast. Eng. **79**, 22–31 (2013)

Jonathan, P., Flynn, J., Ewans, K.C.: Joint modelling of wave spectral parameters for extreme sea states. Ocean Eng. **37**, 1070–1080 (2010)

Jonathan, P., Randell, D., Wu, Y., Ewans, K.: Return level estimation from non-stationary spatial data exhibiting multidimensional covariate effects. Ocean Eng. **88**, 520–532 (2014)

Kadomtsev, B.B., Petviashvili, V.I.: On the stability of solitary waves in weakly dispersive media. Sov. Phys. Dokl. **15**, 539–541 (1970)

Kamarlouei, M., Guedes Soares, C.: Uncertainty analysis in the frequency domain simulation of a hinged wave energy converter. In: Guedes Soares, C., Santos, T.A. (eds.) Trends in Maritime Technology and Engineering, pp. 435–444. Taylor and Francis, London (2022)

Katalinić, M., Parunov, J.: Uncertainties of estimating extreme significant wave height for engineering applications depending on the approach and fitting technique – Adriatic Sea case study. J. Mar. Sci. Eng. **8**, 259 (2020)

Kim, Y., Hermansky, G.: Uncertainties in seakeeping analysis and related loads and response procedures. Ocean Eng. **86**, 68–81 (2014)

Kim, Y., Kim, J.-H.: Benchmark study on motions and loads of a 6750-TEU containership. Ocean Eng. **119**, 262–273 (2016)

Kirezci, C., Babanin, A.V., Chalikov, D.V.: Modelling rogue waves in 1D wave trains with the JONSWAP spectrum, by means of the high order spectral method and a fully nonlinear numerical model. Ocean Eng. **231**, 108715 (2021)

Kodaira, T., Sasmal, K., Miratsu, R., Fukui, T., Zhu, T., Waseda, T.: Uncertainty in wave hindcasts in the North Atlantic Ocean. Mar. Struct. **89**, 103370 (2023). https://doi.org/10.1016/j.marstruc.2023.103370

Korteweg, D.J., de Vries, G.: XLI. On the change of form of long waves advancing in a rectangular canal, and on a new type of long stationary waves. Philos. Mag. J. Sci. **39**(240), 422–443 (1895)

Krogstad, H.E., Liu, J., Socquet-Juglard, H., Dysthe, K.B., Trulsen, K.: Spatial extreme value analysis of nonlinear simulations of random surface waves. In: Proceedings of the 23rd International Conference on Offshore Mechanics and Arctic Engineering (OMAE 2004), 20–25 June, Vancouver, Canada, paper OMAE2004-51336, vol. 2, pp. 285–295 (2004)

Laface, V., Arena, F., Bitner-Gregersen, E.: Non-stationary equivalent storm model for long-term statistics of ocean storms. In: Proceedings of the 43rd International Conference on Ocean, Offshore and Arctic Engineering (OMAE 2024), 9–14 June 2024, Singapore, paper OMAE2024-127907, V002T02A004 (2024). https://doi.org/10.1115/OMAE2024-127907

Laface, V., Arena, F., Guedes Soares, C.: Directional analysis of sea storms. Ocean Eng. **107**, 45–53 (2015). https://doi.org/10.1016/j.oceaneng.2015.07.027

Laface, V., Bitner-Gregersen, E.M., Arena, F., Romolo, A.: A parametrization of DNVGL storm profile for long-term analysis of ocean storms: equivalent trapezoidal storm model. In: Proceedings of the 38th International Conference on Ocean, Offshore and Arctic Engineering (OMAE 2019), 9–14 June, Glasgow, Scotland, UK, OMAE2019-95880, V003T02A048 (2019)

Ley, J., El Moctar, O.: A comparative study of computational methods for wave-induced motions and loads. J. Mar. Sci. Eng. **9**, 83 (2021). https://doi.org/10.3390/jmse9010083

Li, H., Zou, J., Deng, B., Liu, R., Sun, S.: Experimental study of stern slamming and global response of a large cruise ship in regular waves. Mar. Struct. **86**, 103294 (2022)

Li, J., Zhang, Y., Wu, X.: Investigation of extreme wave loads on large floating structures using a new hybrid numerical approach. Mar. Struct. **93**, 103444 (2024)

Liu, J., Yang, S., Chen, Y.: A hybrid model for predicting the effects of fluid-structure interactions on offshore wind turbine performance. J. Offshore Mech. Arct. Eng. **145**(2) (2023)

Liu, Q., et al.: Global wave hindcasts using the observation-based source terms: description and validation. J. Adv. Model. Earth Syst. **13**(8) (2021). https://doi.org/10.1029/2021MS002493

Liu, S., Papanikolaou, A.: Regression analysis of experimental data for added resistance in waves of arbitrary heading and development of a semi-empirical formula. Ocean Eng. **206**, 107357 (2020)

Liu, S., Papanikolaou, A.: Improving the empirical prediction of the added resistance in regular waves of ships with extreme main dimension ratios through dimensional analysis and parametric study. Ocean Eng. **273**, 113963 (2023)

Liu, Y., Xu, Y., Wang, X.: Time-domain simulations of ship motions in extreme sea states: A comparative study. Ocean Eng. **261**, 113989 (2022)

Lucas, C., Guedes Soares, C.: Bivariate distributions of significant wave height and mean wave period of combined sea states. Ocean Eng. **106**, 341–353 (2015)

Lucas, C., Muraleedharan, G., Guedes Soares, C.: Regional frequency analysis of extreme waves in a coastal area. Coast. Eng. **126**, 81–95 (2017)

Lucas, C., Muraleedharan, G., Guedes Soares, C.: Assessment of the uncertainty of estimated extreme quantiles by regional frequency analysis. Ocean Eng. **190**, 106347 (2019)

Lucas, C., Muraleedharan, G., Guedes Soares, C.: Assessment of extreme waves in the North Atlantic Ocean by regional frequency analysis. Appl. Ocean Res. **100**, 102165 (2020)

MacDonald, A., Scarrott, C.J., Lee, D., Darlow, B., Reale, M., Russell, G.: A flexible extreme value mixture model. Comput. Stat. Data Anal. **55**, 2137–2157 (2011)

Mackay, E.B.L., Murphy-Barltrop, C.J.R., Jonathan, P.: The SPAR model: a new paradigm for multivariate extremes application to joint distributions of metocean variables. J. Offshore Mech. Arct. Eng. **47**(1), 011205 (2025)., paper OMAE-24-1018

Magnusson, A.K., Jensen, R.E., Swail, V.R.: Spectral shapes and parameters from three different wave sensors. Ocean Dyn. **71**, 893–909 (2021). https://doi.org/10.1007/s10236-021-01468-7

Magnusson, A.K., Trulsen, K., Aarnes, O.J., Bitne-Gregersen, E.M., Malila, M.: Three sisters measured as a triple wave group. In: Proceedings of the of the 38th International Conference on Ocean, Offshore and Arctic Engineering (OMAE 2019), 9–14 June, Glasgow, Scotland, paper: OMAE2019-96837, V003T02A008 (2019)

Mak, B., Düz, B.: Ship as a wave buoy: Estimating relative wave direction from in-service ship motion measurements using machine learning. In: Proceedings of the 38th International Conference on Ocean, Offshore & Arctic Engineering (OMAE 2019), 9–14 June, Glasgow, Scotland, UK, Paper No: OMAE2019-96201, V009T13A043 (2019)

Mansour, A.E.: Probabilistic Design concepts in ship structural safety and reliability. Trans. Soc. Nav. Archit. Mar. Eng. **80**, 64–97 (1972)

Mao, Y., Xu, Y., Xu, J., Liu, Y.: Spectral fatigue assessment of floating offshore wind turbines. J. Mar. Sci. Eng. **12**(2), 262 (2024)

Marón, A., Kapsenberg, G.: Design of a ship model for hydro-elastic experiments in waves. Int. J. Nav. Archit. Ocean Eng. **6**(4), 1130–1147 (2014)

Matsui, S., Sugimoto, K., Shinomoto, K.: Simplified estimation formula for frequency response function of roll motion of ship in waves. Ocean Eng. **276**, 114187 (2023)

Melchers, R.E.: Structural Reliability and Analysis Prediction, 2nd edn. Wiley (1999)

Mikulić, A., Ćorak, M., Parunov, J.: Comparative analysis of the long-term extreme VWBM calculated according to IACS Rec. No 34 rev 1 and rev 2. IOP Conf. Ser. Mater. Sci. Eng. **1288**, 012030 (2023). https://doi.org/10.1088/1757-899X/1288/1/012030

Miratsu, R., Sasmal, K., Kodaira, T., Fukui, T., Zhu, T., Waseda, T.: Evaluation of ship operational effect based on long-term encountered sea states using wave hindcast combined with storm avoidance model. Mar. Struct. **86**(1), 103293 (2022). https://doi.org/10.1016/j.marstruc.2022.103293

Mittendorf, M., Nielsen, U.D., Bingham, H.B., Liu, S.: Towards the uncertainty quantification of semi-empirical formulas applied to the Added Resistance of Ships in Waves. Ocean Eng. **251**, 111040 (2022)

Moreira, L., Vettor, R., Guedes Soares, C.: Neural network approach for predicting ship speed and fuel consumption. J. Mar. Sci. Eng. **9**, 119 (2021)

Mori, N., Waseda, T., Chabchoub, A.: Science and Engineering of Freak Waves. Published by Elsevier (2023)

Mounet, R.E.G., Nielsen, U.D., Brodtkorb, A.H., Øveraas, H., Dallolio, A., Johansen, T.A.: Data-driven method for hydrodynamic model estimation applied to an unmanned surface vehicle. Measurement. **234**, 114724 (2024)

Mukhlas, M., Kaefer, F., Amini, M.: Efficient time simulation regression procedures for predicting extreme responses of fixed offshore structures. J. Offshore Mech. Arct. Eng. **145**(5) (2023)

Naess, A.: An Introduction to Extreme Value Prediction for Engineering Applications, Centre for Ships and Ocean Structures, NTNU, No.2011-749 Trondheim, Norway (2011)

Naito, S.: Wave generation and absorption in wave basins: theory and application. Int. J. Offshore Polar Eng. **16**(2), 81–89 (2006)

Nasr, D.E., Slika, W.G., Saad, G.A.: Uncertainty quantification for structural health monitoring applications. Smart Struct. Syst. **22**(4), 399–411 (2018). https://doi.org/10.12989/sss.2018.22.4.399

Nielsen, U.D.: Estimations of on-site directional wave spectra from measured ship responses. Mar. Struct. **19**(1), 33–69 (2006)

Nielsen, U.D., Dietz, J.: Estimation of sea state parameters by the wave buoy analogy with comparisons to third generation spectral wave models. Ocean Eng. **216**, 107781 (2020)

Nielsen, U.D., Fønss Bach, K., Iseki, T.: Improved wave-vessel transfer functions by uncertainty modelling. Nihon Kokai Gakkai Ronbunshu. **134**, 134–137 (2016)

Nielsen, U.D., Ikonomakis, A.: Wave conditions encountered by ships—a report from a larger shipping company based on ERA5. Ocean Eng. **237**, 109584 (2021)

Nielsen, U.D., Mounet, R.E.G., Brodtkorb, A.H.: Tuning of transfer functions for analysis of wave–ship interactions. Mar. Struct. **79**, 103029 (2021)

Nielsen, U.D., Stredulinsky, D.C.: Sea state estimation from an advancing ship – a comparative study using sea trial data. Appl. Ocean Res. **34**, 33–44 (2012)

Parunov, J., et al.: Benchmark study and uncertainty assessment of numerical predictions of global wave loads on damaged ships. Ocean Eng. **197**, 106876 (2020). https://doi.org/10.1016/j.oceaneng.2019.106876

Nikolov, M.C., Judge, C.Q.: Uncertainty analysis for calm water tow tank measurements. J. Ship Res. **61**, 177–197 (2017)

NORSOK: Standard N-003: Action and Action Effects. Rev. Jan. 2017 (2017)

Oberhagemann, J.: On prediction of wave-induced loads and vibration of ship structures with finite volume fluid dynamic methods. Doctoral dissertation, University of Duisburg-Essen (2016)

Oberkampf, W.L., Roy, C.J.: Verification and Validation in Scientific Computing. Cambridge University Press (2010). https://doi.org/10.1017/CBO9780511760396

Ochi, M.K.: Wave statistics for the design of ships and ocean structures. Trans. Soc. Nav. Archit. Mar. Eng. **60**, 47–76 (1978)

Olsen, A.S., Schrøter, C., Jensen, J.J.: Wave height distribution observed by ships in the North Atlantic. Ships Offshore Struct. **1**, 1–12 (2006)

Olufsen, A., Bea, R.G.: Loading uncertainties in extreme waves. Mar. Struct. **3**(3), 237–260 (1990)

Orphin, J., Nader, J.R., Penesis, I.: Uncertainty analysis of a WEC model test experiment. Renew. Energy. **168**, 216–233 (2021)

Parunov, J., et al.: Benchmark on the prediction of whipping response of a warship model in regular waves. Mar. Struct. **94**, 103549 (2024a). https://doi.org/10.1016/j.marstruc.2023.103549

Parunov, J., et al.: Benchmark study of global linear wave loads on a container ship with forward speed. Mar. Struct. **84**, 103162 (2022a)

Parunov, J., Guedes Soares, C., Hirdaris, S., Wang, X.L.: Uncertainties in modelling the low-frequency wave-induced global loads in ships. Mar. Struct. **86**, 103307 (2022b)

Parunov, J., Mikulić, A., Ćorak, M.: Consequences of the improved wave statistics on a hull girder reliability of double hull oil tankers. J. Mar. Sci. Eng. **12**(4), 642 (2024b). https://doi.org/10.3390/jmse12040642

Pascoal, R., Guedes Soares, C.: Non-parametric wave spectral estimation using vessel motions. Appl. Ocean Res. **30**(1), 46–53 (2008)

Pascoal, R., Perera, L.P., Guedes Soares, C.: Estimation of directional sea spectra from ship motions in sea trials. Ocean Eng. **132**, 126–137 (2017)

Peregrine, D.H.: Water waves, nonlinear schrödinger equations and their solutions. J. Aust. Math. Soc. Ser. B Appl. Math. **25**, 16–43 (1983). https://doi.org/10.1017/S0334270000003891

Petrova, P.G., Cherneva, Z., Guedes Soares, C.: Distribution of crest heights in sea states with abnormal waves. Appl. Ocean Res. **28**(4), 235–245 (2006)

Petrova, P.G., Cherneva, Z., Guedes Soares, C.: On the adequacy of second-order models to predict abnormal waves. Ocean Eng. **34**(7), 956–961 (2007)

Petrova, P.G., Tayfun, M.A., Guedes Soares, C.: The effect of third order nonlinearities on the statistics distributions of wave heights, crests and troughs in bimodal crossing seas. J. Offshore Mech. Arct. Eng. **135**(2), 021801-1 (2013)

Petranović, T., Mikulić, A., Gledić, I., Parunov, J.: Comparative analysis of closed-form expressions and other commonly used seakeeping methods. Ocean Eng. **281**, 114977 (2023)

Pierella, F., Bredmose, H., Dixen, M.: Generation of highly nonlinear irregular waves in a wave flume experiment: spurious harmonics and their effect on the wave spectrum. Coast. Eng. **164**, 103816 (2021)

Pilar, P., Guedes Soares, C., Carretero, J.C.: 44-year wave hindcast for the North East Atlantic European coast. Coast Eng. **55**, 861–871 (2008)

Piterbarg, V.I.: Asymptotic Methods in the Theory of Gaussian Processes and Fields, vol. 148. American Mathematical Society, Providence (1996)

Pomaro, A., Cavaleri, L., Papa, A., Lionello, P.: 39 years of directional wave recorded data and relative problems, climatological implications and use. Sci. Data. **5**, 180139 (2028). https://doi.org/10.1038/sdata.2018.139

Ponce de León, S., Guedes Soares, C.: Sensitivity of wave model predictions to wind fields in the Western Mediterranean Sea. Coast. Eng. **55**, 920–929 (2008)

Prpić-Oršic, J., Sasa, K., Valčić, M., Faltinsen, O.M.: Uncertainties of ship speed loss evaluation under real weather conditions. ASME J. Offshore Mech. Arct. Eng. **142**(3), 031106 (2020). https://doi.org/10.1115/1.4045790

Qiu, W., et al.: Uncertainties related to predictions of loads and responses for ocean and offshore structures. Ocean Eng. **86**, 58–67 (2014)

Prpic-Oršic, J., Vettor, R., Faltinsen, O.M., Guedes Soares, C.: The influence of route choice and operating conditions on fuel consumption and CO_2 emission of ships. J. Mar. Sci. Technol. **21**(3), 434–457 (2016)

Qiu, W., Meng, W., Peng, H., Li, J., Rousset, J.M., Rodríguez, C.A.: Benchmark data and comprehensive uncertainty analysis of two-body interaction model tests in a towing tank. Ocean Eng. **171**, 663–676 (2019)

Radhakrishnan, G., Leira, B.J., Gao, Z., Sævik, S., Christakos, K.: Retrieval of ocean spectra from X-Band marine radar images using inversion schemes based on auto-spectral analysis. In: Proceedings of the 42nd International Conference on Ocean, Offshore & Arctic Eng (OMAE 2023), 11–16 June, Melbourne, Australia, Paper No. OMAE2023-104877, V005T06A060 (2023)

Raed, K., Karmakar, D., Guedes Soares, C.: Effect of the wind turbine floater geometry on the uncertainty associated with the hydrodynamic loading. Ocean Eng. (2025). https://doi.org/10.1016/j.oceaneng.2025.121134

Rajendran, S., Fonseca, N., Guedes Soares, C.: A numerical investigation of the flexible vertical response of an Ultra Large Containership in high seas compared with experiments. Ocean Eng. **122**, 293–310 (2016)

Rascle, N., Ardhuin, F.: A global wave parameter database for geophysical applications. Part 2: model validation with improved source term parameterization. Ocean Model. **70**, 174–188 (2013)

Reistad, M., Breivik, Ø., Haakenstad, H., Aarnes, O.J., Furevik, B.R., Bidlot, J.-R.: A high-resolution hindcast of wind and waves for the North Sea, the Norwegian Sea and the Barents Sea. J. Geophys. Res. **116**, C05019 (2011)

Roache, P.J.: Verification and Validation in Computational Science and Engineering. Hermosa Publishers, Albuquerque (1998)

Robertson, A., Bachynski, E., Gueydon, S., Wendt, F., Schünemann, P.: Total experimental uncertainty in hydrodynamic testing of a semisubmersible wind turbine, considering numerical propagation of systematic uncertainty. Ocean Eng. **195**, 106605 (2020)

Robertson, A.N., et al.: OC5 project Phase II validation of global loads of the deepcwind floating semisubmersible wind turbine. Energy Procedia. **137**, 38–57 (2017)

Rodrigues, M.I.P., Wang, S., Guedes Soares, C.: Uncertainty assessment for linear transfer functions from different numerical methods. In: Guedes Soares, C., Santos, T.A. (eds.) Advances in Marine Technology and Engineering, vol. 2, pp. 227–236. CRC Press, London (2024)

Roland, A., Ardhuin, F.: On the developments of spectral wave models: numeric and parameterization for the coastal ocean. Ocean Dyn. **64**, 833–846 (2014). https://doi.org/10.1007/s10236-014-0711-z

Ross, E., et al.: On environmental contours for marine and coastal design. Ocean Eng. **195**, 106194 (2020). https://doi.org/10.1016/j.oceaneng.2019.106194

Rusu, L., Guedes Soares, C.: Evaluation of a high-resolution wave forecasting system for the approaches to ports. Ocean Eng. **58**, 224–238 (2013)

Rusu, L., Pilar, P., Guedes Soares, C.: Hindcast of the wave conditions along the West Iberian coast. Coast. Eng. **55**(11), 906–919 (2008)

Salvacao, N., Guedes Soares, C.: Wind resource assessment offshore the Atlantic Iberian coast with the WRF model. Energy. **145**, 276–287 (2018)

Salvacao, N., Bentamy, A., Guedes Soares, C.: Developing a new wind dataset by blending satellite data and WRF model wind predictions. Renew. Energy. **198**, 283–295 (2022)

Sasmal, K., et al.: Wave climate in the North Atlantic Ocean and extreme value analysis. In: 2nd International Workshop on Waves, Storm Surges and Coastal Hazards, 10–15 November, Melbourne, Australia (2019)

Sasmal, K., Miratsu, R., Kodaira, T., Fukui, T., Zhu, T., Waseda, T.: Statistical model representing storm avoidance by merchant ships in the North Atlantic Ocean. Ocean Eng. **235**, 109163 (2021a)

Sasmal, K., et al.: Modeled and satellite-derived extreme wave height statistics in the North Atlantic Ocean reaching 20 m. Earth Space Sci. (2021b). Under review

Schellin, T.E., Östergaard, C., Guedes Soares, C.: Uncertainty assessment of low-frequency load effects for containerships. Mar. Struct. **9**, 313–332 (1996)

Scholcz, T.P., Mak, B.: Ship as a wave buoy – estimating full directional wave spectra from in-service ship motion measurements using Deep Learning. In: Proceedings of the 39th International Conference on Ocean, Offshore & Arctic (OMAE 2020). Virtual (2020)

Shao, X., Wu, W., Zhang, X., Zhang, H., Liu, Y.: A consistent second-order hydrodynamic model for floating structures with large horizontal motions. Ocean Eng. **251**, 111809 (2022)

Shi, J., et al.: A 39-year high resolution wave hindcast for the Chinese coast: model validation and wave climate analysis. Ocean Eng. **183**, 224–235 (2019)

Shimura, T., Mori, N.: High-resolution wave climate hindcast around Japan and its spectral representation. Coast. Eng. **151**, 1–9 (2019)

Si, H., et al.: Uncertainty analysis of linear vertical bending moment in model tests and numerical prediction. Mech. Syst. Signal Process. **178**, 109331 (2022)

Silva, D., Gonçalves, M., Bentamy, A., Guedes Soares, C.: Assessment of the use of scatterometer wind data to force wave models in the North Atlantic Ocean. Ocean Eng. **266**, 112803 (2022)

Silva-González, F., Heredia-Zavoni, E., Montes-Iturrizaga, R.: Development of environmental contours using Nataf distribution model. Ocean Eng. **58**, 27–34 (2013)

Sim, K., Lee, K.: A comparative study on the structural response of multi-linked floating offshore structures between digital model and physical model test for digital twin implementation. J. Mar. Sci. Eng. **12**(2), 262 (2023)

Smit, P.B., et al.: Assimilation of significant wave height from distributed ocean wave sensors. Ocean Model. **159**, 101738 (2021)

Sogihara, N., Tsujimoto, M., Fukasawa, R., Hamada, T.: Uncertainty analysis for measurement of added resistance in short regular waves: its application and evaluation. Ocean Eng. **216**, 107823 (2020)

SSC-156: Ship Structure Committee: an investigation of midship bending moments experienced in extreme regular waves by models of a tanker and a destroyer (1964)

Stamatelopoulos, S., Sapsis, T.P.: Can diffusion models capture extreme event statistics? Comput. Methods Appl. Mech. Eng. **435**, 117589 (2025)

Stern, F., Wilson, R.V., Coleman, H.W., Paterson, E.G.: Comprehensive approach to verification and validation of CFD simulations—part 1: methodology and procedures. J. Fluids Eng. **123**(4), 793–802 (2001)

Stern, F., Wilson, R., Shao, J.: Quantitative V&V of CFD simulations and certification of CFD codes. Int. J. Numer. Methods Fluids. **50**, 1335–1355 (2006)

Storhaug, G., Laanemets, K., Edin, I., Ringsberg, J.W.: Estimation of damping from wave induced vibrations in ships. In: Guedes Soares, C., Garbatov, Y. (eds.) Progress in the Analysis and Design of Marine Structures, pp. 121–130. Taylor & Francis Group, London (2017)

Tang, Y., Sun, S.L., Yang, R.S., Ren, H.L., Zhao, X., Jiao, J.L.: Nonlinear bending moments of an ultra large container ship in extreme waves based on a segmented model test. Ocean Eng. **243**, 110335 (2022)

Tannuri, E.A., Sparano, J.V., Simos, A.N., da Cruz, J.J.: Estimating directional wave spectrum based on stationary ship motion measurements. Appl. Ocean Res. **25**, 243–261 (2003)

Tao, S., Dong, S., Wang, N., Guedes Soares, C.: Estimating storm surge intensity with poisson bivariate maximum entropy distributions based on copulas. Nat. Hazards. **68**(2), 791–807 (2013)

Tayfun, M.A., Fedele, F.: Wave-height distributions and nonlinear effects. Ocean Eng. **34**(11–12), 1631–1649 (2007)

Teixeira, A.P., Parunov, J., Guedes Soares, C.: Assessment of ship structural safety. In: Guedes Soares, C., Garbatov, Y., Fonseca, N., Teixeira, A.P. (eds.) Marine Technology and Engineering, pp. 1377–1394. Taylor & Francis Group, London (2011)

Teixeira, J.C., Abreu, M.P., Guedes Soares, C.: Uncertainty of ocean wave hindcasts due to wind modelling. J. Offshore Mech. Arct. Eng. **117**, 294–297 (1995)

Thompson, I., Chiritoiu, R., Magoga, T., Mondoro, A., Smith, M.: Effects of sampling from a naval destroyer's operational history on fatigue damage estimation. Int. J. Nav. Archit. Ocean Eng., PII: S2092-6782(24)00029-3 (2024). https://doi.org/10.1016/j.ijnaoe.2024.100610. Reference: IJNAOE 100610

Tolman, H.L.: User manual and system documentation of WAVEWATCH III version 3.14. NOAA/NWS/NCEP/MMAB Technical Note 276 (2009). http://polar.ncep.noaa.gov/mmab/papers/tn276/MMAB_276.pdf

Tolman, H.L.: A generalized multiple discrete interaction approximation for resonant four-wave interactions in wind wave models. Ocean Model. **70**, 11–24 (2013)

Trulsen, K., Dysthe, K.: A modified nonlinear Schrödinger equation for broader bandwidth gravity waves on deep water. Wave Motion. **24**(3), 281–289 (1996)

Uzunoglu, E., Guedes Soares, C.: On the model uncertainty of offshore code comparison collaboration continuation within IEA wind task 30 Phase II results regarding a floating semisubmersible wind system. In: Guedes Soares, C. (ed.) Progress in Renewable Energies Offshore, pp. 785–794. Taylor & Francis Group, London (2016)

Uzunoglu, E., Guedes Soares, C.: On the model uncertainty of wave induced platform motions and mooring loads of a semisubmersible based wind turbine. Ocean Eng. **148**, 277–285 (2018)

de Valk, C., Groenewoud, P., Hulst, S., Klopman, G.: Building a global resource for rapid assessment of the wave climate. In: Proceedings of the 23rd International Conference on Offshore Mechanics and Arctic Engineering (OMAE 2004), 20–25 June, Vancouver, Canada, paper OMAE2004-51308 (2004)

Van Der Horn, E., Wang, Z.H., Mahadevan, S.: Towards a digital twin approach for vessel-specific fatigue damage monitoring and prognosis. Reliab. Eng. Syst. Saf. **219**, 108222 (2022)

Van Gelder, P., De Ronde, J., Neykov, N.M., Neytchev, P.: Regional frequency analysis of extreme wave heights: trading space for time. In: Edge, B.L. (ed.) Coastal Engineering 2000 Proceedings 27th International Conference on Coastal Engineering, 16–21 July, Sydney, Australia. 2000, pp. 1099–1112. American Society of Civil Engineers (2000)

Vanem, E., Bitner-Gregersen, E.M.: Stochastic modelling of long-term trends in the wave climate and its potential impact on ship structural loads. Appl. Ocean Res. **37**, 235–248 (2012)

Vanem, E.: Copula-based bivariate modelling of significant wave height and wave period and the effects of climate change on the joint distribution. In: Proceedings of the 35th International Conference on Offshore Mechanics and Arctic Engineering (OMAE 2016), 19–24 June, Busan, Korea, paper OMAE2016-54314, V003T02A033 (2016)

Vanem, E.: A regional extreme value analysis of ocean waves in a changing climate. Ocean Eng. **144**, 277–295 (2017)

Vanem, E.: A simple approach to account for seasonality in the description of extreme ocean environments. Mar. Syst. Ocean Technol. **13**, 63–73 (2018). https://doi.org/10.1007/s40868-018-0046-6

Vanem, E., Gramstad, O., Bitner-Gregersen, E.M.: A simulation study on the uncertainty of environmental contours due to sampling variability for different estimation methods. Appl. Ocean Res. **91**, 101870 (2019)

Vanem, E.: Bivariate Regional frequency analysis of sea state conditions. In: Proceedings of the 40th International Conference on Offshore Mechanics and Arctic Engineering (OMAE 2021), 21–30 June, Virtual, Online, paper OMAE2021-61988, V002T02A013 (2021)

Vanem, E.: Analyzing extreme sea state conditions by time-series simulation. In: Proceedings of the 40th International Conference on Offshore Mechanics and Arctic Engineering (OMAE 2022), 5–10 June, Hamburg, Germany, paper No. OMAE2022-78795, V002T02A041 (2022)

Vanem, E., Fekhari, E., Dimitrov, N., Kelly, M., Cousin, A., Guiton, M.: A joint probability distribution for multivariate wind-wave conditions and discussions on uncertainties. J. Offshore Mech. Arct. Eng. **146**(6), 061701 (2024a)., paper OMAE-23-1131

Vanem, E., Gramstad, O., Babanin, A., De Bin, R., Trulsen, K.: On the distribution of ocean wave crest heights in varying wave conditions. J. Ocean Eng. Mar. Energy. **10**, 797–815 (2024b). https://doi.org/10.1007/s40722-024-00350-0

Vettor, R., Guedes Soares, C.: Detection and analysis of the main routes of voluntary observing ships in the North Atlantic. J. Navig. **68**(2), 397–410 (2015)

Vettor, R., Guedes Soares, C.: Characterisation of the expected wave conditions in the main European coastal traffic routes. Ocean Eng. **140**, 244–257 (2017)

Vettor, R., Guedes Soares, C.: On the accuracy of voluntary observing ship's records. J. Offshore Mech. Arct. Eng. **143**(5), 054501 (2021)

Vettor, R., Guedes Soares, C.: Reflecting the uncertainties of ensemble weather forecasts on the predictions of ship fuel consumption. Ocean Eng. **250**, 111009 (2022)

Vettor, R., Bergamini, G., Guedes Soares, C.: A comprehensive approach to account for weather uncertainties in ship route optimization. J. Mar. Sci. Eng. **9**, 1434 (2021). https://doi.org/10.3390/jmse9121434

Vieira, M., Guedes Soares, C., Guimaraes, P.V., Bergamasco, F., Campos, R.M.: Nearshore space-time ocean wave observation using low-cost video-cameras. Coast. Eng. **197**, 104694 (2025)

Vieira, M., Guimaraes, P.V., Violante-Carvalho, N., Benetazzo, A., Bergamasco, F., Pereira, H.A.: Low-Cost stereo video system for measuring directional wind waves. J. Mar. Sci. Eng. **8**, 831 (2020)

Vijith, P.P., Rajendran, S.: Hydroelastic effects on the vertical bending moment of a container ship in head and oblique seas. Ocean Eng. **285**, 115385 (2023)

Vitali, N., Prpić-Oršić, J., Guedes Soares, C.: Coupling voyage and weather data to estimate speed loss of container ships in realistic conditions. Ocean Eng. **210**, 106758 (2020)

Wada, R., Waseda, T., Jonathan, P.: Extreme value estimation using the likelihood-weighted method. Ocean Eng. **124**, 241–251 (2016)

Wada, R., Waseda, T.: Assessment of data-inherited uncertainty in extreme wave analysis. J. Offshore Mech. Arct. Eng. **142**(2), 021204 (2020)

Wang, H., Zhang, L., Zhang, X.: Advanced numerical techniques for predicting hydrodynamic responses of large offshore structures. J. Mar. Sci. Eng. **11**(10), 2190 (2023a)

Wang, S., Gadelho, J.F.M., Islam, H., Guedes Soares, C.: CFD modelling and grid uncertainty analysis of the free-falling water entry of 2D rigid bodies. Appl. Ocean Res. **115**, 102813 (2021b)

Wang, S., Guedes Soares, C.: Review of ship slamming loads and responses. J. Mar. Sci. Appl. **16**(4), 427–445 (2017)

Wang, S., Gonzalez-Cao, J., Islam, H., Gómez-Gesteira, M., Guedes Soares, C.: Uncertainty estimation of mesh-free and mesh-based simulations of the dynamics of floaters. Ocean Eng. **256**, 111386 (2022)

Wang, S., Islam, H., Guedes Soares, C.: Uncertainty due to discretisation on the ALE algorithm for predicting water slamming loads. Mar. Struct. **80**, 103086 (2021a)

Wang, X., Wu, X., Zhang, L.: Collision load response of floating wind turbines considering wind-wave-mooring load coupling. Ocean Eng. **292**, 110078 (2023b)

Waseda, T., et al.: Uncertainty in the extreme waves in the North Atlantic Ocean and ship responses. In: Proceedings of the 43rd International Conference on Offshore Mechanics and Arctic Engineering (OMAE 2024), 9–14 June, Singapore, paper OMAE2024–128240, V002T02A001 (2024)

Watanabe, I., Guedes Soares, C.: Comparative study on time domain analysis of non-linear ship motions and loads. Mar. Struct. **12**(3), 153–170 (1999)

Webb, A., Waseda, T., Kiyomatsu, K.: A high-resolution, long-term wave resource assessment of Japan with wave-current effects. Renew. Energy. **161**, 1341–1358 (2020)

Weiss, R., Feser, F., Günther, H.: A 40-year high-resolution wind and wave hindcast for the Southern North Sea. In: Proceedings of the 7th International Workshop on Wave Hindcasting and Forecasting, Banff, Alberta, Canada, pp. 97–104 (2002)

Weiss, R., Feser, F.: Evaluation of a method to reduce uncertainty in wind hindcasts performed with regional atmosphere model. Coast Eng. **48**(4), 211–225 (2003)

West, B.J., Brueckner, K.A., Jand, R.S., Milder, D.M., Milton, R.L.: A new method for surface hydrodynamics. J. Geophys. Res. **92**, 11803–11824 (1987)

Whitham, G.B.: Linear and Nonlinear Waves. Wiley Collection, New York (1974)
Wilson, R.V., Stern, F., Coleman, H.W., Paterson, E.G.: Comprehensive approach to verification and validation of CFD simulations – part 2: application for RANS simulation of a cargo/container ship. ASME J. Fluids Eng. **123**(4), 803–810 (2001)
Winterstein, S.R., Ude, T.C., Cornell, C.A., Bjerager, P., Haver, S.: Environmental parameters for extreme response: inverse FORM with omission factors. In: Proceedings of the 6th International Conference on Structural Safety and Reliability (ICOSSAR'93), 9–13 August, Innsbruck, Austria (1993)
WMO: Guide to Marine Meteorological Services. World Meteorological Organization, Geneva (2001)
WMO: Manual on the Global Observing System. World Meteorological Organization, Geneva (2003)
Woodward, M.D., Rijsbergen, M., Hutchinson, K.W., Scott, A.: Uncertainty analysis procedure for the ship inclining experiment. Ocean Eng. **114**, 79–86 (2016)
Xiao, W., Liu, Y., Wu, G., Yue, D.K.P.: Rogue wave occurrence and dynamics by direct simulations of nonlinear wave-field evolution. J. Fluid Mech. **720**, 357–392 (2013)
Yang, P., et al.: Uncertainty analysis of hydroelastic responses and wave loads for different structural modeling and potential methods. Ocean Eng. **222**, 108529 (2021)
Yang, X., Liu, Q., Chen, Z.: Numerical analysis of wave impact forces on floating structures using advanced CFD techniques. Ocean Eng. **280**, 115475 (2023)
Young, I.R., Ribal, A.: Multiplatform evaluation of global trends in wind speed and wave height. Science. **364**(6440), 548–552 (2019)
Zhai, Y., Xu, Y., Zhang, J.: Hydrodynamic responses of barge-type floating offshore wind turbines integrated with aquaculture cages. Mar. Struct. **85**, 103257 (2022)
Zhang, J., Li, W., Zhang, L.: A hybrid approach for estimating wave-induced responses of floating offshore platforms. J. Offshore Mech. Arct. Eng. **144**(4) (2022a)
Zhang, H., Cui, J., Liao, X., Shi, H., Guedes Soares, C.: Numerical study on the vertical response of LNG carrier in abnormal waves generated with different mechanisms. Ocean Eng. **262**, 112090 (2022d)
Zhang, H., Cui, J., Shi, H., Guedes Soares, C.: Analysis of the peaks of ship motions in linear and nonlinear focused waves. Ocean Eng. **266**, 113028 (2022e)
Zhang, H., Guedes Soares, C., Chalikov, D., Toffoli, A.: Modelling the spatial evolutions of nonlinear unidirectional surface gravity waves with fully nonlinear numerical method. Ocean Eng. **125**, 60–69 (2016a)
Zhang, H., Guedes Soares, C., Onorato, M., Toffoli, A.: Modelling of the temporal and spatial evolutions of weakly nonlinear random directional waves with the modified nonlinear Schrödinger equations. Appl. Ocean Res. **55**, 130–140 (2016b)
Zhang, H., Liao, X.M., Shi, H.D., Babanin, A., Guedes Soares, C.: Effect of initial condition uncertainty on the profile of maximum wave. Mar. Struct. **82**, 103127 (2022b)
Zhang, H., Liu, Y., Zhang, X.: Efficient simulation techniques for predicting ship slamming loads in extreme sea conditions. Mar. Struct. **90**, 103589 (2023)
Zhang, H.D., Shi, H.D., Guedes Soares, C.: Evolutionary properties of mechanically generated deepwater extreme waves induced by nonlinear wave focusing. Ocean Eng. **186**, 106077 (2019)
Zhang, H.D., Guedes Soares, C.: Numerical Modelling of Extreme Waves. Published by Springer Nature, Cham (2025)
Zheng, J., Wang, L., Zhang, L.: A comprehensive review of numerical methods for predicting wave-induced loads on offshore structures. Mar. Struct. **83**, 103151 (2022)
Zhang, J.C., Zhao, X.W., Jin, S., Greaves, D.: Phase-resolved real-time ocean wave prediction with quantified uncertainty based on variational Bayesian machine learning. Appl. Energy. **324**, 119711 (2022c)

Zhang, L., Duan, W.Y., Wu, K., Cui, X.M., Guedes Soares, C., Huang, L.: Optimized WAVEWATCH III for significant wave height computation using machine learning. Ocean Eng. **312**, 119004 (2024a)

Zhang, X., Yang, L., Wang, Y.: An advanced method for predicting the hydroelastic response of floating platforms under complex wave conditions. J. Mar. Sci. Eng. **12**(6), 915 (2024b)

Zhang, Y., Kim, C.-W., Beer, M., Dai, H.L., Guedes Soares, C.: Modeling multivariate ocean data using asymmetric copulas. Coast. Eng. **135**, 91–111 (2018)

Zhang, Z., Ma, N., Shi, Q., Zhang, Y., Wen, Y.: Experimental study and uncertainty analysis on added resistance and pressure distribution of KVLCC2 in regular short waves. Ocean Eng. **317**, 120093 (2025)

Open Access This chapter is licensed under the terms of the Creative Commons Attribution-NonCommercial-NoDerivatives 4.0 International License (http://creativecommons.org/licenses/by-nc-nd/4.0/), which permits any noncommercial use, sharing, distribution and reproduction in any medium or format, as long as you give appropriate credit to the original author(s) and the source, provide a link to the Creative Commons license and indicate if you modified the licensed material. You do not have permission under this license to share adapted material derived from this chapter or parts of it.

The images or other third party material in this chapter are included in the chapter's Creative Commons license, unless indicated otherwise in a credit line to the material. If material is not included in the chapter's Creative Commons license and your intended use is not permitted by statutory regulation or exceeds the permitted use, you will need to obtain permission directly from the copyright holder.

Author Index

A
Abdussamie, Nagi   254
Abrahamsen, Asger Bech   254
Ali-Lavroff, J.   359
An, Chen   170
Andersen, Michael Rye   1
Ås, Sigmund K.   92

B
Barbato, A.   533
Bayindir, Cihan   170
Begovic, E.   359
Bekker, A.   533
Bingham, Harry   442
Bitner-Gregersen, Elzbieta   626
Boote, D.   359
Braun, M.   533
Busiello, Ciro   442

C
Catipovic, Ivan   254
Choisnet, Thomas   254
Chujo, Toshiki   254

D
de Carvalho Pinheiro, Bianca   1
De Luca, Gaetano   1
Drummen, Ingo   442

E
Eder, M. A.   533
Egorov, A.   359
Ehlers, Sören   92

F
Ferguson, Tom Mitchell   1
Fonseca, Nuno   442

G
Gaspar, Jose   254
Grammatikopoulos, Apostolos   92
Greco, Luca   254

H
Hermundstad, Ole A.   254
Hess, P.   533
Hu, Zhiqiang   442

J
Jeong, Han-Koo   254
Jochum, C.   533

K
Karmakar, Debabrata   442
Kim, Do Kyun   1
Kim, Ekaterina   442
Konno, Yoshihiro   170
Kurt, Rafet Emek   1
Kwan, Seo Jung   92

L
Lara, Paul   92
Le Sourne, Herve   1
Li, Shen   254
Lindemann, Thomas   1
Liu, Bin   92

M
Mao, Wengang   254
Marquez, Lucas   1
Mohapatra, Sarat   442
Montewka, Jakub   1
Montoya, Carlos   170
Morato, D.   533
Morooka, Celso Kazuyuki   170

Murai, Motohiko 442
Murayama, Hideaki 92

N
Nicholls-Lee, Rachel 254
Nilva, V. 533
Notaro, Gabriele 254

O
Ogus, Elif 254
Osawa, N. 533

P
Pahlavan, P. 533
Papanikolaou, Apostolos 626
Parsa, Kourosh 170
Parunov, Josko 626
Pasqualino, Ilson Paranhos 92
Prebeg, Pero 92

Q
Qiu, Wei 626
Quinton, Bruce W. T. 1

R
Rahman, Tauhid 170
Ralph, Freeman 254
Rawson, Charles 254
Rizzo, Cesare Mario 92
Rodriguez, M. 359
Rudan, Smiljko 1

S
Sævik, Svein 170
Salcedo, A. 359
Sarsoza Burgos, D. 533
Sawamura, J. 359
Seyffert, H. 359
Shim, Chunsik 170

Sielski, Robert 442
Soares, Carlos Guedes 626
Souppez, J. B. 359
Sun, Shali 254
Suominen, Mikko 92

T
Tabri, Kristjan 1
Takami, Tomoki 626
Teixeira, Â. P. 533
Thompson, I. 533
Tian, Chao 442
Toshkov, L. 359

U
Underwood, J. M. 533

V
Vladimir, N. 359

W
Walters, Carey 1
Wang, F. 359
Wang, Xueliang 626
Wang, Y. 533
Waseda, Takuji 626

X
Xia, Yi 170

Y
Yamada, Yasuhira 1
Yim, Solomon 626
Yu, Zhaolong 1

Z
Zhang, Huidong 626
Zhu, Ling 1

GPSR Compliance

The European Union's (EU) General Product Safety Regulation (GPSR) is a set of rules that requires consumer products to be safe and our obligations to ensure this.

If you have any concerns about our products, you can contact us on

ProductSafety@springernature.com

In case Publisher is established outside the EU, the EU authorized representative is:

Springer Nature Customer Service Center GmbH
Europaplatz 3
69115 Heidelberg, Germany

www.ingramcontent.com/pod-product-compliance
Ingram Content Group UK Ltd.
Pitfield, Milton Keynes, MK11 3LW, UK
UKHW022202230426
470311UK00001BA/2